Aron Simis, Zaqueu Ramos

Graded Algebras in Algebraic Geometry

De Gruyter Expositions in Mathematics

Volume 70

Aron Simis, Zaqueu Ramos

Graded Algebras in Algebraic Geometry

DE GRUYTER

Mathematics Subject Classification 2010
Primary: 13A02, 13A30, 13B25, 13C14, 13C15, 13C40, 13D02, 13D05, 13E05, 13E15, 13H10, 13H15, 13N05, 13N15, 13P10; Secondary: 14A05, 14E05, 14E07, 14M05, 14M10, 14M12

Authors
Prof. Dr. Aron Simis
Departamento de Matemática, CCEN
Universidade Federal de Pernambuco
50740-640 Recife
PE Brazil
aron@dmat.ufpe.br

Prof. Dr. Zaqueu Ramos
Departamento de Matemática, CCET
Universidade Federal de Sergipe
49100-000 São Cristovão
SE Brazil
zaqueu@mat.ufs.br

ISBN 978-3-11-063754-0
e-ISBN (PDF) 978-3-11-064069-4
e-ISBN (EPUB) 978-3-11-063775-5
ISSN 0938-6572

Library of Congress Control Number: 2021950450

Bibliographic information published by the Deutsche Nationalbibliothek
The Deutsche Nationalbibliothek lists this publication in the Deutsche Nationalbibliografie; detailed bibliographic data are available on the Internet at http://dnb.dnb.de.

Typesetting: VTeX UAB, Lithuania
Printing and binding: CPI books GmbH, Leck

www.degruyter.com

In memoriam
This book is dedicated to Wolmer V. Vasconcelos,
who taught us that a beautiful mind
is the socle of it all

Acknowledgments

The senior author thanks numerous collaborators whose ideas and style, if not the actual results, will be quite visible in the text. By and large, he is indebted to Paulo Brumatti, Winfried Bruns, Ricardo Burity, Marc Chardin, Yairon Cid-Ruiz, Barbara Costa, André Doria, Philippe Gimenez, Hamid Hassanzadeh, Jürgen Herzog, Gary Kennedy, Marcel Morales, Maral Mostafazadehfard, Abbas Nasrollah Nejad, Francesco Russo, Karen Smith, Stefan Tohăneanu, Ngô-Viet Trung, Bernd Ulrich, Wolmer Vasconcelos and Rafael Villarreal, among his close coauthors, for allowing to share many parts of the collaboration. An effort has been made to honor them by pointing out these parts right on the spot, but there are so many of them throughout that often became impossible to keep. Both authors are indebted to the personnel in De Gruyter for their patience and courtesy in listening to our needs and allowing for a flexible delivery schedule in the face of the pandemic. This is especially true for Nadja Schedensack.

https://doi.org/10.1515/9783110640694-201

Foreword

This book grew out of a series of earlier shorter accounts by the senior author, motivating the use of graded algebras as a tool for certain geometric problems. Some of these accounts appeared sporadically in informal proceedings of meetings, in Brazil, Chile, Italy, Mexico, Peru, Portugal and Spain. One of them was even published [178].

The original format underwent several modifications, some motivated by criticism regarding its too overly compact nature to be of regular use by students and prospective researchers. The idea of turning it into a longer, more detailed, account was cherished for a few years. Of course, as time went by, many other results on the subject vicinity grew very fast in the hands of many other people, including the senior author's associates. It became totally clear that the whole idea would not take off without a partnership, and happily turned into a concrete joint venture. As a natural consequence of this cooperation, the material underwent many changes in order to accommodate as well the work of the junior author during the first 10 years since his PhD under my guidance. Thus, many new topics were included that did not appear before, driving the overall look of the contents a lot more closer to commutative algebra, though always keeping an adamant eye to the geometric side.

The book is about certain graded algebras that appear frequently in algebraic geometry whose role goes beyond the mere language usefulness. Graded algebras have an importance that can hardly be exaggerated. They cover a large spectrum ranging from the classical theory of projective varieties (as homogeneous coordinate rings) to the modern systematization of intersection theory (as associated graded rings and related algebras) through the refined parts of blowup techniques (as Rees algebras and alike). They play a distinguished role in other theories, such as toric varieties (geometric collage of semigroup algebras) and stretch as far as the eye can see through the combinatorics of simplicial complexes and matroids and Hodge algebras (algebras with a straightening law). Last, but surely not least, they are the natural examples of classical invariant theory. At the other end, graded algebras, whose grading is not necessarily over $\mathbb{N}$ (or $\mathbb{Z}$) became increasingly germane in recent years. Among those, bigraded and finely graded algebras play a special role.

The book trails much beyond standard algebras over a field, though the latter come around quite often. On one hand, nurturing more general ground rings make the algebras become more and more useful. On the other hand, a book is supposed to explain well all of its contents, even at the cost of becoming long. Therefore, the original inception has been endowed with many other sections for the appropriate handling of these algebras. One is here more inclined toward those algebras related to blowups (so-called Rees algebras) and their close associates obtained thereof by various kinds of algebraic manipulations. It often happens that notable standard algebras will show up that may have been overlooked before. Thus, for instance, tangential varieties, secant varieties and embedded joins are some of the topics where putting systematically together a few algebraic methods may prove useful. By and large, the background is

https://doi.org/10.1515/9783110640694-202

somewhat pervaded by the structure of correspondences and, as far as various algebraic aspects are inherent to the theory, also is the structure of incidence varieties. Since there exists quite an extensive literature on the subject, one has modestly engaged in adjusting the terminology if only to acquire an insight into some underlying ideas not yet fully disclosed.

It is not easy to keep the right proportion between the algebraic and the geometric sides. When the first notes by the senior author appeared more than 25 years ago, the computational curiosity was then at its peak. Therefore, there was then an effort to emphasize that facet of the objects and their effectiveness. The present authors decided to keep some of this air throughout as it may still be of interest even in these days of sophisticated computer assistance. To compromise, most of these aspects are deferred to the exercises, which play an important role in the book. The senior author once heard from the late George Kempf: "I don't care if one cannot actually compute it, I only care to know whether its computation is effective." For some of us, theoretic mathematicians, this is all there is.

As to guiding exercises, we made an effort to avoid filling items that ask for extensions of the theory as we believe the intended goal is that of having the reader become able to solve concrete data problems, which will help him (her) to get a grip of the exposed theory. Since the book is not essentially a classroom item, it would seem that a count of ten exercises per chapter will fulfill our intent. However, even if parts of the book are used in classroom then the listed items seem to be fit for the purpose. As usual, some of the exercises will require some restriction on the base field, but it will be rewarding to the reader self to take care of this point.

A word about notation. By and large, we chose to denote an algebra by the letter S as a loose reminder of the forerunner symmetric algebra. An unspecified ambient ring is usually denoted by the letter R, honoring both the English and the German spellings. Clearly, often such general rules may be trespassed on a given context.

January 2022

Contents

1 Geometric motivation

1.1 Bits from intersection theory

Since this book flirts with the intertwining of graded algebras and parts of algebraic geometry, it sounds appropriate to give some motivation from a classical topic such as intersection theory. We chose two aspects thereof: Segre classes and intersection multiplicities. Some of these preliminaries may eventually help to point an eye to the discussed topics. A small related piece is taken up in Chapter 12 based on P. Aluffi's work on Segre classes. It is our hope that such a motivation will serve to prevent the book from becoming an orphan of readers from the geometric spectrum.

First, let us discuss Segre classes. For a smooth projective variety X, one defines the ith *Segre class* of X as the ith *Segre class* of the sheaf of Kähler differentials $\Omega_{X/k}$. Note that the latter is locally-free in this case.

However, if X is only assumed to be a pure dimensional variety, one ought to look for a different approach as $\Omega_{X/k}$ is no longer locally-free. Now, in general one has $\Omega_{X/k} = \mathbb{D}/\mathbb{D}^2$, where $\mathbb{D}$ is the ideal sheaf of the image of the diagonal map $X \to X \times X$. To circumvent the problem, one replaces the conormal sheaf of the diagonal by its asymptotic version, the normal cone $\oplus_{j\geq 0}\mathbb{D}^j/\mathbb{D}^{j+1}$.

A main point about the usefulness of this construction is perhaps the following result.

Proposition ([119]). *Let $X \subset \mathbb{C}^n$ be a complex affine variety, let $\Delta : X \to X \times X$ denote the diagonal map and $\mathbb{D}$, the ideal sheaf of the subvariety $\Delta(X) \subset X \times X$. Let further $\mathbb{G}$ denote the Grassmannian of lines in $\mathbb{P}^n$ and consider the map*

$$\mathfrak{L}: (X \times X) \setminus \Delta(X) \longrightarrow \mathbb{G},$$

which takes a point $(x, y), x \neq y$ to the line $\overline{xy} \in \mathbb{G}$. Then:
(i) *The closure of the graph of $\mathfrak{L}$ in $(X \times X) \times \mathbb{G}$ is isomorphic to $\mathbb{P}(\oplus_{j\geq 0}\mathbb{D}^j)$.*
(ii) *For any closed point $x \in X$, the fiber $\mathbb{P}(\oplus_{j\geq 0}\mathbb{D}^j/\mathbb{D}^{j+1}) \times_X \mathbf{k}(x)$ can be identified set theoretically with the union of lines $l \subset \mathbb{P}^n$ for which there exist sequences $\{y_i\}, \{y_i'\}$ of points in X converging to x such that the sequence of lines $\overline{y_i, y_i'}$ converges to l.*

Thus, roughly, the fibers of the exceptional locus of the blowup along the diagonal embedding are limits of secant lines to the variety. That was the original viewpoint of H. Whitney, who in [221] and [222] exhibited various alternatives for the notion of a cone over a point, assuming an analytic setup. He was mainly interested in this kind of object for results of stratification nature.

First, systematically employed by K. Johnson [119]—who suggested the name *tangent star*—this sort of cone has been considered from a different perspective by G. Kennedy [122] and L. van Gastel [70]. The subject ties naturally with the geometry of secants. The fiber of the *star cone* over the vertex of the affine cone over X is the affine

https://doi.org/10.1515/9783110640694-001

cone over the secant variety. Quite a bit of this geometric environment will be translated back to algebra in terms of the Rees algebra and the associated graded ring of the ideal of the biprojective cone over the ideal sheaf $\mathbb{D}$ (Chapter 10, Section 8.2).

So much for Segre classes as motivation. Let us turn to intersection multiplicities. Among the most accepted approaches to a general theory of intersection products is the one developed by W. Fulton, as one gathers from [64, Chapters 5 and 6]. We quote from his summary: "The existence of such a deformation (to the normal cone), together with the 'principle of continuity' that intersection products should vary nicely in families, explains the prominent role to be played by the normal cone in constructing intersection products."

The normal cone is a geometric patching of associated graded rings of ideals. In the same vein of ideas, there is a deformation of a ring A to the associated graded ring $\mathrm{gr}_I(A) = \oplus_{n\geq 0} I^n/I^{n+1}$ with deformation family the so-called (extended) Rees algebra $A[It, t^{-1}]$—this idea goes back to Rees in the mid-1950s.

What Fulton calls "the basic construction" in the subsequent parts has the following algebraic translation (we are retaining but the main core of the idea).

The basic element is the following "fiber square"

$$\begin{array}{ccc} R/(I,J) & \longleftarrow & R/J \\ \uparrow & & \uparrow \\ R/I & \longleftarrow & R \end{array}$$

where R is an "ambient" ring, $I, J \subset R$ are ideals, and the arrows are the natural surjections. One should think of $(I, J) \subset R$ as the ideal of the intersection of two subvarieties of some ambient variety, whereby one is looking for an appropriate notion of intersection multiplicities along each component of the intersection. Now, algebraically, the notion that yields such a component is a minimal associated prime of the ring $R/(I, J)$. However, as it turns this is insufficient, a point that has been made by van der Waerden in the late 1920s. To detect the missing contribution off the minimal primes themselves, Fulton proposed to use the associated graded ring in the following way.

Let $P_1, \dots, P_s \subset \mathrm{gr}_{(I,J)/J}(R/J)$ be the minimal primes of the associated graded ring $\mathrm{gr}_{(I,J)/J}(R/J)$. The respective contractions $P_1 \cap (R/(I,J)), \dots, P_s \cap (R/(I,J)) \subset R/(I,J)$ are the *distinguished primes* of $R/(I,J)$. Note that any minimal prime $\wp$ of $R/(I,J)$ is a distinguished prime: let $Q \in \mathrm{Spec}\,(\mathrm{gr}_{(I,J)/J}(R/J))$ such that $Q \cap R/(I,J) = \wp$ (such a prime exists since $R/(I,J)$ is the zeroth piece in the ordinary grading of the associated graded ring). If $P \subset Q$ is a minimal prime of $\mathrm{gr}_{(I,J)/J}(R/J)$, then the contraction of P contains $\wp$; hence, they must be equal. However, there are in general distinguished primes, which are not minimal (in fact, maximal ideals of $R/(I,J)$ can often be distinguished).

Now, in the ordinary setup of intersection multiplicities one typically has that R/I is a complete intersection and that R/J is *equidimensional* (i. e., $\dim R/p = \dim R/J$ for every minimal prime p of R/J). The first hypothesis implies that $\mathrm{gr}_I(R)$ is a polynomial ring over R/I while the second hypothesis implies that the associated graded ring

$\mathrm{gr}_{I,J/J}(R/J)$ is also equidimensional. By base change, $\mathrm{gr}_I(R)\otimes_{R/I} R/(I,J)$ is a polynomial ring over $R/(I,J)$.

The following estimate is then basic to the rest of the theory.

Proposition 1.1.1. *Under the above hypotheses, for every distinguished prime $\wp$ of $R/(I,J)$ one has*

$$\dim(R/J) - \mathrm{ht}(I) \le \dim(R/\wp) \le \dim(R/J).$$

Proof. The rightmost inequality is clear as $\dim(R/\wp) \le \dim(R/(I,J)) \le \dim(R/J)$. For the other inequality, note first that by flatness and base change, one has

$$\dim(\mathrm{gr}_I(R)\otimes_{R/I} R/\wp) = \dim(R/\wp) + \mathrm{rank}(I/I^2) = \dim(R/\wp) + \mathrm{ht}(I).$$

However, from the natural surjection $\mathrm{gr}_I(R)\otimes_{R/I} R/(I,J) \twoheadrightarrow \mathrm{gr}_{(I,J)/J}(R/J)$, we have similarly a surjection $\mathrm{gr}_I(R)\otimes_{R/I} R/\wp \twoheadrightarrow \mathrm{gr}_{(I,J)/J}(R/J)/P$, where P is a minimal prime of $\mathrm{gr}_{(I,J)/J}(R/J)$ contracting to $\wp$. Since the minimal primes of $\mathrm{gr}_{(I,J)/J}(R/J)$ have the same dimension ($= \dim(R/J)$), the above equality yields the desired inequality. □

The ideal

$$\bigoplus_{r>0}(J\cap I^r, I^{r+1})/(JI^r, I^{r+1}) = \ker(\mathrm{gr}_I(R)\otimes_{R/I} R/(I,J) \twoheadrightarrow \mathrm{gr}_{(I,J)/J}(R/J))$$

has been considered by many authors since Hironaka's work and it ties with many other questions.

However, Fulton seems to be the first to realize that it could be used in such a form for intersection multiplicities. He has applied the above construction in a special case in order to introduce what he called *refined intersections* (cf. [64, Chapter 8]). The game is to cast the previous idea on the sheaf of the diagonal map in algebraic terms. Namely, let $S = k[x_1,\dots,x_n]$ be an affine k-algebra (k a field) and $R = S\otimes_k S$. Let $\mathbb{D}\subset R$ denote the ideal generated by the differences $x_i\otimes 1 - 1\otimes x_i$, for $i=1,\dots,n$—the *diagonal ideal* of R. Let $\mathfrak{a},\mathfrak{b}\subset S$ be ideals. The fiber square for the data $R := S\otimes_k S$, $I := \mathbb{D}$, $J := (\mathfrak{a}\otimes 1, 1\otimes\mathfrak{b})$ is

$$\begin{array}{ccccccc} S/(\mathfrak{a},\mathfrak{b}) & \simeq & R/(\mathfrak{a}\otimes 1, 1\otimes\mathfrak{b}, \mathbb{D}) & \longleftarrow & R/(\mathfrak{a}\otimes 1, 1\otimes\mathfrak{b}) & = & S/\mathfrak{a}\otimes_k S/\mathfrak{b} \\ & & \uparrow & & \uparrow & & \\ S & \simeq & R/\mathbb{D} & \longleftarrow & R & & \end{array}$$

By setting $\mathbb{D} := (\mathfrak{a}\otimes 1, 1\otimes\mathfrak{b}, \mathbb{D})/(\mathfrak{a}\otimes 1, 1\otimes\mathfrak{b})$, one sees in particular that the zeroth piece of the associated graded ring $\mathrm{gr}_{\mathbb{D}}(R)$ is again $S/(\mathfrak{a},\mathfrak{b})$. Therefore, the latter has distinguished varieties as contractions of minimal primes of $\mathrm{gr}_{\mathbb{D}}(R)$. This allows to define the refined intersection cycle (by assuming that S is regular, so that $R/\mathbb{D}$ is locally a complete intersection).

On itself, Spec $(\mathrm{gr}_{\mathbb{D}}(R))$ is fairly central, as a convenient tool in Segre classes and in intersection theory. The book will expand quite a bit on the algebraic side of these constructions (Chapter 10, Section 8.2).

1.2 Correspondences and incidence varieties

Loosely speaking, a *correspondence* between two varieties V and W over a field k is a cycle (or a class of cycles) on the product $V \times W = V \times_{\mathrm{Spec}(k)} W$. There is a classical algebra of correspondences using various natural maps that arise in intersection theory (cf., e. g., [64, Chapter 16]). Our interest in this book lies on the structure of an individual structure and its close associates rather than on the formalism developed in intersection theory.

Of course, a great many known correspondences arise as the closure of the graph of a rational map $V \dashrightarrow W$. Others are obtained by means of some incidence relation between subvarieties of V and of W—loosely called *incidence correspondences* or *incidence varieties*. In this book, it is mainly the two latter kinds that will trap our interest, and for all purposes, they can be though of as a (biprojective) subvariety of $V \times W$, where V and W are projective varieties.

1.2.1 Examples

In this portion, we will contend ourselves in briefly surveying a few examples of such correspondences. Many of these are to be found in [87].

Example 1.2.1. Let X, Y be subvarieties of the same ambient space. Let V (resp., W) be quasi-projective varieties parametrizing a certain subset of subvarieties of X (resp., of Y) (one assumes that they do so in a bijective way). Then $\Sigma := \{(P,Q) \in V \times W \mid P \subset Q\}$ comprises many of the well-known incidence correspondences. Here, of course, we identify the point $P \in V$ (resp., $Q \in W$) with the corresponding (uniquely defined) subvariety of X (resp., of Y).

The most notable of these examples is obtained when $X = Y = \mathbb{P}^n$, while $V = \mathbb{G}(m,n)$ and $W = \mathbb{G}(d,n)$ are Grassmann varieties, parameterizing the linear subspaces of $\mathbb{P}^n$ of dimensions m and d, respectively. For example, with $m = 0$, hence $\mathbb{G}(m,n) \simeq \mathbb{P}^n$, one has

$$\Sigma = \{(p, q_0 \wedge \cdots \wedge q_d) \mid p \wedge q_0 \wedge \cdots \wedge q_d = 0\}.$$

We may find the defining equations of Σ, as a subvariety of the product $\mathbb{P}^n \times \mathbb{P}^N$, after embedding $\mathbb{G}(d,n)$ in $\mathbb{P}^N$ via the Plücker embedding. Say, $\mathbf{X}$ and $\mathbf{T}$ are coordinates for $\mathbb{P}^n$ and for $\mathbb{P}^N$, respectively. Then Σ is defined by the Plücker relations in $\mathbf{T}$ (of bidegree

$(0,2))$ and by the quadrics of bidegree $(1,1)$ coming from expanding every maximal minor of the augmented matrix $((\mathbf{X})^t|\mathbf{Z})$, where $\mathbf{Z}$ is a generic $(n+1)\times(d+1)$ matrix, and replacing a maximal minor of $\mathbf{Z}$ by the corresponding Plücker coordinate.

All of these incidence correspondences have the following common feature, provided V is a projective variety: for any algebraic subset $W' \subset W$, the union over the subvarieties of V corresponding to the points of W' is again an algebraic subset of V. This seemingly innocuous observation allows in fact a good deal of important constructions in intersection theory and in enumerative geometry.

Example 1.2.2. As above, we consider a correspondence between Grassmann varieties of the same space, namely, for a fixed number $r \in \mathbb{N}$, we consider $\{(L,L') \in \mathbb{G}(d,n)\times\mathbb{G}(m,n) \mid \dim(L\cap L') \geq r\}$.

When $r = d < m$, one obtains a so-called *flag manifold.*

Example 1.2.3. Think about $\mathbb{P}^{mn-1}$ as the projective space of nonzero $m\times n$ matrices over a field k (modulo scalars). For any $0 \leq r \leq \min\{m,n\}$, consider the incidence $\Sigma = \{(M,L) \in \mathbb{P}^{mn-1}\times\mathbb{G}(n-r,n) \mid L \subset \ker M\}$. Here, M is looked upon as a k-linear map $k^n \to k^m$.

Note that the projection of Σ to the first factor maps onto the subvariety of $\mathbb{P}^{mn-1}$ whose points correspond to matrices of rank $\leq r$ – one of the *determinantal varieties.*

There are of course examples of incidence varieties in products with more than two factors—such is the convenient setup, e. g., for Chow varieties and general flag varieties.

Example 1.2.4 (Jacobian duality). Consider a correspondence given by a closed subvariety of a biprojective space $\mathbb{P}^{d-1}\times\mathbb{P}^{n-1}$. As such, its defining ideal is generated by bihomogeneous polynomials (or biforms) in the bigraded polynomial ring $k[\mathbf{X},\mathbf{T}] = k[X_1,\dots,X_d;T_1,\dots,T_n]$, in which, X_i (resp., T_j) has bidegree $(1,0)$ (resp., $(0,1)$). Among these, biforms of bidegree $(1,1)$ play an outstanding role. On one hand, they define subvarieties whose singular locus conveys a rather moderate complexity, at least empirically; on the other hand, their constituent terms are square-free monomials, which from both combinatorial and computational viewpoints are a lot more tractable.

Algebraically, there is yet another way of setting up these data so as to bring in well-known constructions from the ring and module theory. Thus, let $\mathcal{J}$ stand for an ideal of the bigraded polynomial ring $k[\mathbf{X},\mathbf{T}]$ generated by bihomogeneous polynomials $G_1,\dots,G_m$ of bidegree $(1,1)$. Then one can consider two modules, one over the ring $R = k[\mathbf{X}]$ and the other over the ring $S = k[\mathbf{T}]$, both defined in terms of a free representation as follows:

$$R^m \xrightarrow{\varphi} R^n \longrightarrow E \to 0, \quad S^m \xrightarrow{\psi} S^d \longrightarrow F \longrightarrow 0,$$

where φ (resp., ψ) is the Jacobian (resp., transposed Jacobian) matrix of $G_1, \dots, G_m$ with respect to the variables $\mathbf{T}$ (resp., $\mathbf{X}$).

One might say that the above Jacobian matrices (or the modules they present) are *exchangeable*. The prevailing terminology these days is that of *Jacobian dual*, a slightly unfortunate choice as the construction refers to shifting from one fibration to the other, so *Jacobian dual fibration* would sound better.

It follows that the bigraded residue ring $k[\mathbf{X}, \mathbf{T}]/\mathcal{J}$ is isomorphic as k-algebra to both $\mathcal{S}_R(E)$ and $\mathcal{S}_S(F)$, the respective symmetric algebras of E and F (for the bigraded structure of the symmetric algebra, see Section 2.2).

Conversely, given a finitely generated R-module E defined as the cokernel of a matrix φ whose entries are linear forms, one can present the symmetric algebra $\mathcal{S}_R(E) \simeq k[\mathbf{X}, \mathbf{T}]/\mathcal{J}(\varphi)$, where $\mathcal{J}(\varphi)$ is generated by the entries of the product $(\mathbf{T})\varphi$. This gives a way of moving back and forth between such modules and certain bigraded ideals of $k[\mathbf{X}, \mathbf{T}]$.

As it turns out, there is quite a number of cases where one can say more about the two exchangeable Jacobian matrices.

1.2.2 Algebraic counterpart

Let $S = k[\mathbf{X}, \mathbf{Y}]$ be a polynomial ring over a field k in two sets $\mathbf{X}$ and $\mathbf{Y}$ of mutually independent indeterminates, with $|\mathbf{X}| = |\mathbf{Y}| = n+1$. In this part, we index the elements of $\mathbf{X}$ as $X_0, \dots, X_n$ (similarly for $\mathbf{Y}$) to conform with the prevailing notation in algebraic geometry while thinking of S as the bigraded coordinate ring of the product $\mathbb{P}^n \times \mathbb{P}^n$.

S has a (standard) bigrading $S = \oplus_{(u,v)\in\mathbb{N}} S_{(u,v)}$, where $S_{(u,v)}$ is the k-vector space spanned by the monomials $X_0^{a_0} \cdots X_n^{a_n} Y_0^{b_0} \cdots Y_n^{b_n}$ of bidegree $(\sum_i a_i, \sum_i b_i) = (u, v)$.

Let $\mathcal{J} \subset S$ stand for an ideal. A major interest rests on the case where $\mathcal{J}$ is *bihomogeneous*, i. e., where $\mathcal{J} = \oplus_{(u,v)\in\mathbb{N}\times\mathbb{N}} \mathcal{J} \cap S_{(u,v)}$ is a bigraded submodule of S, or alternatively, where $\mathcal{J}$ has a set of ideal generators each homogeneous separately in the variables $\mathbf{X}$ and $\mathbf{Y}$. We remark that in the latter case, the ring $A := S/\mathcal{J}$ can be thought of as the bigraded coordinate ring of a correspondence $\mathcal{T} \subset \mathbb{P}^n \times \mathbb{P}^n$; hence, we will sometimes call $\mathcal{J}$ a *correspondence ideal*.

On the other hand, $A = S/\mathcal{J}$ (with $\mathcal{J}$ bihomogeneous) is an $\mathbb{N}$-graded ring by the total degree, i. e., $A_d = \oplus_{u+v=d} A_{(u,v)}$. To distinguish between the two gradings, we denote S^*, $\mathcal{J}^*$, A^* the $\mathbb{N}$-graded versions of S, $\mathcal{J}$, A, respectively. Though no distinction exists at the level of the strict ring-theoretic properties, the corresponding geometry will differ from one setting to the other, as will the associated Hilbert functions: the bigraded version is a rational polynomial in two variables of total degree $\dim A - 2$ (actually this is true under some additional restrictions on A, cf. [184]), while the ordinary graded version is a polynomial in one variable of degree $\dim A - 1$.

Since in any case one expects the degree of the Hilbert polynomial to coincide with the dimension of the corresponding algebraic variety, the above seems to agree with the geometric situation. In fact, given a subvariety $\mathcal{V} \subset \mathbb{P}^n \times \mathbb{P}^n$, the homogeneous coordinate ring A_Δ of its Segre embedding is the so-called diagonal subalgebra of the bigraded coordinate ring A of $\mathcal{V}$ under the given embedding $\mathcal{V} \subset \mathbb{P}^n \times \mathbb{P}^n$. (A more appropriate designation for this algebra would be the $(1,1)$-*directional subalgebra* of A; cf. [177]. This terminology not only sounds closer to the nature of the objects involved, it also avoids a potential confusion with other notion of diagonal which will be among our main themes).

In [184], the relation between the (bigraded) Hilbert polynomial $P_A(s,t)$ of A and the (graded) Hilbert polynomial $P_{A_\Delta}(t,t)$ of A_Δ is developed. In particular, one obtains $\dim \mathcal{V} = \dim \operatorname{Proj} A_\Delta = \dim A - 2$ while $\dim \operatorname{Proj} A^* = \dim A - 1$. Both $\operatorname{Proj} A^*$ and the affine cone over it are important in the Vogel theory of intersections [70]. For the intersection theory of irreducible (or, more generally, pure-dimensional) correspondences, both in the classical formulation of Pieri [64] and in that of Vogel, what seems to be important is the relation between $P_A(s,t)$ and the (graded) Hilbert polynomial $P_{A^*}(s+t)$.

1.3 Ruled joins

The construction below is based on F. Gaeta's work [67], largely adapted to the present landscape. Some of it has also been considered later by van der Waerden in [214], in connection with his earlier work on multiprojective geometry.

Let k be an algebraically closed field. The natural biregular morphism

$$\tau : \mathbb{A}^{2n+2} \to \mathbb{A}^{n+1} \times \mathbb{A}^{n+1}$$
$$(a_0, \dots, a_n, b_0, \dots, b_n) \mapsto ((a_0, \dots, a_n), (b_0, \dots, b_n))$$

induces a rational map

$$\gamma = \gamma_{1,n} : \mathbb{P}^{2n+1} \dashrightarrow \mathbb{P}^n \times \mathbb{P}^n,$$

called the *Gaeta map*, also known as the *Gaeta join construction*.

Let $L_{\mathbf{X}} = V(X_0, \dots, X_n) \subset \mathbb{P}^{2n+1}$ and $L_{\mathbf{Y}} = V(Y_0, \dots, Y_n) \subset \mathbb{P}^{2n+1}$. Then $\gamma = \gamma_{n,n}$ is regular on the open set $\mathbb{P}^{2n+1} \setminus (L_{\mathbf{X}} \cup L_{\mathbf{Y}})$.

Given a biprojective subvariety $\mathcal{V} \subset \mathbb{P}^n \times \mathbb{P}^n$, take the set-theoretic inverse image of $\mathcal{V}$ by the regular map $\mathbb{P}^{2n+1} \setminus (L_{\mathbf{X}} \cup L_{\mathbf{Y}}) \to \mathbb{P}^n \times \mathbb{P}^n$ and close it in $\mathbb{P}^{2n+1}$. The resulting projective subvariety will be denoted by ${}^{\gamma}\mathcal{V}$.

Recall that $k[\mathbf{X}] \otimes_k k[\mathbf{X}]$ is a bigraded k-algebra, where a typical tensor $f \otimes g$ of homogeneous elements $f, g \in k[\mathbf{X}]$, of respective degrees u, v, has bidegree (u,v). It can be endowed with a usual $\mathbb{N}$-grading by taking total degrees. When a bihomogeneous ideal $J \subset k[\mathbf{X}] \otimes_k k[\mathbf{X}]$ is given total degree, the notation J^* will be sometimes used to denote the corresponding $\mathbb{N}$-graded ideal (see Subsection 1.2.2). This process is shown in the following.

Lemma 1.3.1. *Let $\mathcal{J} \subset k[\mathbf{X}] \otimes_k k[\mathbf{X}]$ be the bihomogeneous defining ideal of $\mathcal{V} \subset \mathbb{P}^n \times \mathbb{P}^n$. Then the homogeneous defining ideal of ${}^{\gamma}\mathcal{V}$ is $\mathcal{J}^*$.*

Proof. Let τ^* be the k-algebra isomorphism corresponding to the biregular map $\tau : \mathbb{A}^{2n+2} \to \mathbb{A}^{n+1} \times \mathbb{A}^{n+1}$. That is,

$$\tau^* : k[\mathbf{X}] \otimes_k k[\mathbf{X}] \longrightarrow k[\mathbf{X}, \mathbf{Y}]$$

is the uniquely defined k-isomorphism induced by the multiplication in $k[\mathbf{X}, \mathbf{Y}]$ (i. e., by the bilinear map $k[\mathbf{X}] \times k[\mathbf{X}] \to k[\mathbf{X}, \mathbf{Y}]$ such that $(f(\mathbf{X}), g(\mathbf{X})) \mapsto f(\mathbf{X})g(\mathbf{Y})$, with inverse given by $X_i \mapsto X_i \otimes 1$, $Y_i \mapsto 1 \otimes X_i$. Then the homogeneous defining ideal of ${}^{\gamma}\mathcal{V}$ is $\mathcal{J}^* = \tau^*(\mathcal{J})$. □

Remark 1.3.2. In the particular case where $\mathcal{V}$ is the product $V \times W$, with $V, W \subset \mathbb{P}^n$ projective subvarieties, then ${}^{\gamma}(V \times W) \subset \mathbb{P}^{2n+1}$ is called the *ruled join* of V and W (see Chapter 11 for the relation to the notion of embedded join). The ruled join is an important tool in intersection theory, mainly in regard to Bézout theorem (cf. [64, Chapter 8]).

We collect the main features of this map in the next proposition.

Proposition 1.3.3. *The Gaeta map satisfies the following properties:*

(a) *The fibers of the restriction map*

$$\mathbb{P}^{2n+1} \setminus (L_{\mathbf{X}} \cup L_{\mathbf{Y}}) \to \mathbb{P}^n \times \mathbb{P}^n$$

are copies of $\mathbb{A}^1 \setminus \{0\}$.

(b) *Let $\delta : \mathbb{P}^n \to \mathbb{P}^n \times \mathbb{P}^n$ denote the diagonal map*

$$(a_0 : \cdots : a_n) \mapsto ((a_0 : \cdots : a_n), (a_0 : \cdots : a_n))$$

and let $\sigma = \sigma_{1,n} : \mathbb{P}^1 \times \mathbb{P}^n \to \mathbb{P}^{2n+1}$ denote the Segre map

$$((u_0 : u_1), (b_0 : \cdots : b_n)) \mapsto (u_0 b_0 : \cdots : u_0 b_n : u_1 b_0 : \cdots : u_1 b_n),$$

with image the Segre variety $\mathfrak{S} = \mathfrak{S}_{1,n}$. Then:

(i) *The image $\delta(\mathbb{P}^n)$ supports a closed subvariety $\mathcal{D} \subset \mathbb{P}^n \times \mathbb{P}^n$ whose inverse image in $\mathbb{P}^{2n+1}$ by the Gaeta map is $\mathfrak{S}$. Moreover, $\mathfrak{S} \setminus (L_{\mathbf{X}} \cup L_{\mathbf{Y}})$ is an $\mathbb{A}^1 \setminus \{0\}$-bundle over $\mathcal{D}$.*

(ii) *Setting $\Delta = \sigma(\{(1 : 1)\} \times \mathbb{P}^n)$, the restriction of the Gaeta map y to Δ is a biregular isomorphism onto $\mathcal{D}$.*

Proof. (a) For mnemonic, we denote a point of $\mathbb{P}^{2n+1}$ by $(a_0 : \cdots : a_n : b_0 : \cdots : b_n) = (\mathbf{a} : \mathbf{b})$.

Since the Gaeta map fits into the natural diagram

$$\begin{array}{ccccc} \mathbb{A}^{2n+2}\setminus \mathbf{0} & \subset \mathbb{A}^{2n+2} & \xrightarrow{\tau} & \mathbb{A}^{n+1}\times\mathbb{A}^{n+1} & \supset (\mathbb{A}^{n+1}\setminus \mathbf{0})\times(\mathbb{A}^{n+1}\setminus \mathbf{0}) \\ \downarrow & & & & \downarrow \\ \mathbb{P}^{2n+1} & & \xrightarrow{\gamma} & & \mathbb{P}^n\times\mathbb{P}^n \end{array},$$

it clearly suffices to check the condition for two points

$$(\mathbf{a}:\mathbf{b}) = (a_0:\cdots:a_n:b_0:\cdots:b_n), \quad (\mathbf{a}':\mathbf{b}') = (a_0':\cdots:a_n':b_0':\cdots:b_n')$$

in $\mathbb{P}^{2n+1}\setminus(L_{\mathbf{X}}\cup L_{\mathbf{Y}})$ to have the same image under γ, namely, that there exist nonzero $\lambda,\mu\in k$ such that $\mathbf{a}'=\lambda\mathbf{a}$ and $\mathbf{b}'=\mu\mathbf{b}$. Saying it in other words, fixing a bipoint $(\mathbf{a},\mathbf{b})\in\mathbb{P}^n\times\mathbb{P}^n$, then $\gamma^{-1}(\mathbf{a},\mathbf{b})=\{(\lambda\mathbf{a}:\mu\mathbf{b})\}$, with both $\lambda,\mu\in k$ nonzero. This shows that the fiber is a copy of $\mathbb{A}^1\setminus\{0\}$, as intended.

(b) (i) To show that a subset of $\mathbb{P}^n\times\mathbb{P}^n$ supports a closed subvariety, one must check that it is the zero set of a bihomogeneous ideal $\mathcal{J}$ in the bigraded ring $k[\mathbf{X}]\otimes_k k[\mathbf{X}]$ (here, $k[\mathbf{X}]$ has the standard grading and the piece of $k[\mathbf{X}]\otimes_k k[\mathbf{X}]$ of bidegree (u,v) is the k-vector space generated by the elements $f\otimes g$ with $\deg f=u$, $\deg g=v$).

For that, one takes for $\mathcal{J}$ the ideal of 2×2 minors of the matrix

$$\begin{pmatrix} X_0 & \dots & X_n \\ Y_0 & \dots & Y_n \end{pmatrix}.$$

This ideal is obviously bihomogeneous and vanishes on the points of $\delta(\mathbb{P}^n)$. On the other hand, it is well known that $\mathcal{J}$ is prime and of codimension $n=n+1-2+1$. Since $\dim\delta(\mathbb{P}^n)=n$, we are through.

Applying Lemma 1.3.1 to the case where $\mathcal{V}=\mathcal{D}$, one gets that the homogeneous defining ideal of the Segre model $\mathfrak{S}\subset\mathbb{P}^{2n+1}$ is $\mathcal{J}$ above considered as homogeneous ideal by total degree. The claim on the bundle identification is clear from the Segre map σ.

(ii) Since $\Delta\subset\mathbb{P}^{2n+1}\setminus(L_{\mathbf{X}}\cup L_{\mathbf{Y}})$, it will be enough to show that for a point $p\in\Delta$, the fiber $\gamma^{-1}(\gamma(p))$ is the reduced point p. But, if $\{b_0,\dots,b_n\}$ are coordinates of a current point on the right factor of the product $\mathbb{P}^1\times\mathbb{P}^n$, then $p=(b_0:\cdots:b_n:b_0:\cdots:b_n)$ with some $b_i\neq 0$. Therefore,

$$\begin{aligned} \gamma^{-1}(\gamma(p)) &= \gamma^{-1}(b_0:\cdots:b_n,b_0:\cdots:b_n) \\ &= (\{(\lambda b_0:\cdots:\lambda b_n:\mu b_0:\cdots:\mu b_n)\mid\lambda\neq 0,\mu\neq 0,\})\cap\Delta \\ &= (b_0:\cdots:b_n:b_0:\cdots:b_n), \end{aligned}$$

which is clearly a scheme-theoretic equality of reduced structures. □

The projective variety $\Delta \subset \mathbb{P}^{2n+1}$ is the natural *projective* diagonal one is looking for, as opposed to $\mathcal{D}$ which is the *biprojective* diagonal. This distinction would avoid a great deal of potential confusion.

An obvious advantage of Gaeta's model is to turn a problem about maps $\mathbb{P}^n \times \mathbb{P}^n \to \mathbb{P}^m$ into a problem of ordinary maps between projective spaces.

Remark 1.3.4. Proposition 1.3.3 is readily extendable to the multigraded case. We feature the highlights of the analogous construction: first, there is an analogous Gaeta rational map

$$\gamma_{m-1,n} : \mathbb{P}^{m(n+1)-1} \dashrightarrow \mathbb{P}^n \times \mathbb{P}^n \times \cdots \times \mathbb{P}^n \ (m \text{ times}).$$

As before, the image of the regular map $\delta : \mathbb{P}^n \to \mathbb{P}^n \times \mathbb{P}^n \times \cdots \times \mathbb{P}^n$ supports a subvariety $\mathcal{D}$ (called sometimes *small diagonal*) and pulls back to the Segre model $\mathfrak{S}_{m-1,n} \subset \mathbb{P}^{m(n+1)-1}$. Again, the projective diagonal

$$\Delta = \sigma_{m-1,n}(\{(1:1:\ldots:1)\} \times \mathbb{P}^n)$$

maps isomorphically onto $\mathcal{D}$ by the Gaeta map. Ring-theoretically, if we take a multicoordinate set of variables $\mathbf{X}_0, \ldots, \mathbf{X}_{m-1}$ for $\mathbb{P}^n \times \mathbb{P}^n \times \cdots \times \mathbb{P}^n$, then the defining homogeneous ideal of $\mathfrak{S}_{m-1,n}$ is the ideal generated by the 2×2 minors of the $m \times (n+1)$ matrix whose rows are $\mathbf{X}_0, \ldots, \mathbf{X}_{m-1}$, while the defining ideal of Δ is generated by the linear forms $X_{0j} - X_{ij}$ with $1 \le i \le m-1$, $0 \le j \le n$. These ideals have codimension mn and $m(n+1)$, respectively.

The details of the verification are left to the exercises.

1.4 Exercises

Exercise 1.4.1 (van der Waerden). Let $I \subset R = k[x_0, x_1, x_2, x_3]$ be the defining ideal of the nonnormal rational quartic and let $J = (x_0, x_3) \subset R$.

(1) Prove that $R/(I, J)$ has (Hilbert) multiplicity 5.

(2) Let $Q \subset R[t, u]$ denote the presentation ideal of the associated graded ring of J over R/I. Show that Q has multiplicity 4.

(3) Which of the two is the correct intersection multiplicity of the respective cones in $\mathbb{P}^4$ of the corresponding varieties at the point $(0:0:0:0:1)$?

Exercise 1.4.2 (Hartshorne). Let $I = (x_0, x_1) \cap (x_2, x_3) \subset R = k[x_0, x_1, x_2, x_3]$ be the defining ideal of two nonconcurrent lines in $\mathbb{P}^3$ and let $J = (x_0 - x_2, x_1 - x_3) \subset R$.

(1) Prove that $R/(I, J)$ has (Hilbert) multiplicity 3.

(2) Let $Q \subset R[t, u]$ denote the presentation ideal of the associated graded ring of J over R/I. Show that Q has multiplicity 2.

(3) Which of the two is the correct intersection multiplicity of the respective cones in $\mathbb{P}^4$ of the corresponding varieties at the point $(0:0:0:0:1)$?

Exercise 1.4.3. Find the defining equations of the incidence variety of points and straight lines in $\mathbb{P}^3$ as embedded by the Plücker coordinates.
(*Hint*: use Example 1.2.1).

Exercise 1.4.4. Find the defining equations of the incidence variety of lines and planes in $\mathbb{P}^4$ as embedded by the Plücker coordinates.

Exercise 1.4.5. Let $(xu : yu : yv)$ define a bigraded rational map $\mathfrak{F} : \mathbb{P}^1 \times \mathbb{P}^1 \dashrightarrow \mathbb{P}^2$, where x, y (resp., u, v) are the coordinates of the first factor (resp., of the second factor).

(1) Prove that $\mathfrak{F}$ is birational.
(*Hint*: given a general point $(a : b : c)$ in the target, the existence of an inverse implies the vanishing of the 2-minors of the matrix

$$\begin{pmatrix} a & b & c \\ xu & yu & yv \end{pmatrix}.$$

Based on this, determine the quotients x/y and u/v in terms of a, b, c).
Letting $\gamma : \mathbb{P}^3 \dashrightarrow \mathbb{P}^1 \times \mathbb{P}^1$ denote the Gaeta map, consider the two hyperplanes H_v, H_u of $\mathbb{P}^3$ of equations $x = v$ and $x = u$, respectively. Let γ_v (resp., γ_u) denote the restriction of γ to H_v (resp., H_u).
(2) Prove that the composite of both γ_v and γ_u with $\mathfrak{F}$ is birational.
(3) By identifying both γ_v and γ_u with $\mathbb{P}^2$, identify which Cremona maps are obtained in the previous item.

Exercise 1.4.6. Verify the details of Remark 1.3.4.

Exercise 1.4.7. Let $\mathfrak{F} : \mathbb{P}^n \dashrightarrow \mathbb{P}^n$ be a rational map and let $I \subset R = k[\mathbf{X}] = k[X_0, \dots, X_n]$ denote the ideal generated by the forms defining $\mathfrak{F}$. Consider the closure $\mathcal{G}$ of the graph of $\mathfrak{F}$ in $\mathbb{P}^n \times \mathbb{P}^n$, the biprojective variety defined by the blowup of $\mathbb{P}^n$ along Spec (R/I). Elaborate on the properties of the Gaeta inverse image of $\mathcal{G}$ and compare the various gradings.

2 Graded algebras

2.1 Arbitrary gradings

We start with the following general definition.

Definition 2.1.1. Let A be a commutative ring and let $\mathbb{M}$ be a commutative monoid (additive notation). An $\mathbb{M}$-*graded algebra* is a commutative A-algebra S with a decomposition into a direct sum of additive subgroups

$$S = \bigoplus_{u \in \mathbb{M}} S_u,$$

such that $S_0 = A$ and $S_u S_v \subset S_{u+v}$ for all $u, v \in \mathbb{M}$. In particular, each S_u is an A-module called the *component* or the *homogeneous part of degree u* of S. The above decomposition is called an $\mathbb{M}$-*grading*.

Throughout, we tacitly assume that the A-modules S_u ($u \in \mathbb{M}$) are finitely generated. We note this assumption does not imply that S be finitely generated as an A-algebra in the sense that there exists a finite subset $M \subset \mathbb{M}$ with $S = A[\oplus_{u \in M} S_u]$. Typically, however, we will deal with finitely generated graded algebras.

An $\mathbb{M}$-*graded S-module E* is defined similarly as an S-module possessing a decomposition into a direct sum of subgroups

$$E = \bigoplus_{u \in \mathbb{M}} E_u,$$

such that $S_u E_v \subset E_{u+v}$ for all $u, v \in \mathbb{M}$. It follows, in particular, that each E_u is an A-module. Again, we will mostly assume that all graded modules are finitely generated.

There is a fairly general and useful mechanism to create new graded algebras from old ones. Namely, let $\mathbb{M}' \subset \mathbb{M}$ be a submonoid and let $\varphi : \mathbb{M} \to \mathbb{M}''$ be a homomorphism of monoids. Given an $\mathbb{M}$-graded algebra S, we attach to it an $\mathbb{M}'$-graded algebra S' and an $\mathbb{M}''$-graded algebra S^{φ} defined as

$$S' = \bigoplus_{u' \in \mathbb{M}'} S_{u'}, \quad S^{\varphi} = \bigoplus_{u'' \in \mathbb{M}''} S^{\varphi}_{u''},$$

where $S^{\varphi}_{u''} = \oplus_{\varphi(u)=u''} S_u$, with the understanding that $S^{\varphi}_{u''} = (0)$ if $\varphi^{-1}(u'') = \emptyset$. It is an exercise to check that S' is naturally a subring of S and S^{φ} gets identified with S as rings by the "principle of summation" (identifying a homogeneous element $x_u \in S_u$ with the element $(y_v) \in \oplus_{\varphi(v)=\varphi(u)} S_v \in S^{\varphi}_{\varphi(u)}$ such that $y_v = x_u$ if $v = u$, $y_v = 0$ otherwise).

A similar construction works starting with an $\mathbb{M}$-graded S-module.

We remark that this construction is general enough so as to encompass some strange gradings.

https://doi.org/10.1515/9783110640694-002

Example 2.1.2. Let $S = k[X]$ (k a field) be graded by setting $S_u = kX^u$, as usual. Then S is $\mathbb{Z}$-graded with $S_u = (0)$ for $u \notin \mathbb{N}$. Consider the canonical residue map $\varphi : \mathbb{Z} \to \mathbb{Z}/2\mathbb{Z}$. Then the ring S^φ is $(\mathbb{Z}/2\mathbb{Z})$-graded with $S_{\bar{0}}$ the subring generated by the even powers of X and $S_{\bar{1}}$ the additive subgroup generated by the odd powers of X. This is a rather unusual grading, which finds no use in these notes.

Often dealing with general versions of $\mathbb{M}$ takes a dramatic path (see, e. g., [37, Lemma 2.4]). However, in this book the most important situation is when $\mathbb{M} = \mathbb{Z}^m = \mathbb{Z}\times\cdots\times\mathbb{Z}$ (m times) (or any of its notable submonoids, such as $\mathbb{N}^m = \mathbb{N}\times\cdots\times\mathbb{N}$). In this case, we will adhere to the traditional terminology of calling $\mathbb{Z}^m$-graded algebras and modules *multigraded*, if the rank of the group is sufficiently clear. By the same token, the degree $(u_1, \dots, u_m)$ of a homogeneous element of S will be often called *multidegree*.

Typically, in most geometric applications, $S_{(0,\dots,0)} = k$ will turn to be a field and the algebra generators of S will usually be homogeneous of multidegree of the form $(0, \dots, 0, u_i, 0, \dots, 0)$, for various choices of $u_i \in \mathbb{N}$. If further the u_i's can all be chosen to be 1, then S is said to be *multistandard*.

Example 2.1.3. The prototype is the ring of polynomials $S = k[\mathbf{X}_1, \dots, \mathbf{X}_m]$ in m sets of mutually independent variables (possibly of different sizes), which is multigraded by setting $S_{(u_1,\dots,u_m)}$ to be the k-vector space with basis the set of monomials $\mathbf{X}_1^{\alpha_1} \cdots \mathbf{X}_m^{\alpha_m}$ where $|\alpha_j| = u_j$, for $j = 1, \dots, m$. The $\mathbb{N}$-graded k-algebra S^φ obtained by taking the homomorphism

$$\varphi : \mathbb{N}^m \to \mathbb{N}, \quad (u_1, \dots, u_m) \mapsto u_1 + \cdots + u_m$$

is the prototype of a standard algebra; it is said to be graded by *total degree* from the multidegree of S.

Had one in the above example chosen a homomorphism mapping $(0, \dots, 0, 1, 0, \dots, 0)$ to $d_i \in \mathbb{N}$, for given d_i's, then one would obtain the notion of a quasi-homogeneous grading with weights d_i's (roughly, one would be redefining new weights for the variables). Alternatively, one could have considered the multipolynomial ring as a $(d_1)\mathbb{N} \times \cdots \times (d_m)\mathbb{N}$-graded ring by decreeing a monomial $\mathbf{X}_1^{\alpha_1} \cdots \mathbf{X}_m^{\alpha_m}$ to be of multidegree $(a_1\alpha_1 d_1, \dots, a_m\alpha_m d_m)$, where $a_i = |\alpha_i|$, and then have applied the homomorphism $(0, \dots, 0, d_i, 0, \dots, 0) \mapsto d_i \in \mathbb{N}$.

A homogeneous ideal in a multigraded algebra will often be called multihomogeneous. The residue ring of a multigraded algebra by a multihomogeneous ideal is again multigraded. Such mechanism as applied to the multipolynomial ring above yields in particular the multicoordinate ring of a correspondence in products $\mathbb{P}^{n_1} \times \cdots \times \mathbb{P}^{n_m}$ of projective spaces, the case $m = 1$ yielding the homogeneous coordinate ring of a projective variety.

We will next exemplify the submonoid mechanism by giving two constructions that reflect some quite important objects in geometry. The first is as follows.

Definition 2.1.4. Let S be $\mathbb{N}$-multistandard and let $p_1,\dots,p_m \geq 0$ be given integers. The $(p_1,\dots,p_m)$-*directional* subalgebra of S is the algebra

$$\bigoplus_{u\in\mathbb{N}} S_{(p_1u,\dots,p_mu)}.$$

The vector $\Delta := (p_1,\dots,p_m)$ is called the *direction* of the directional algebra, which will be denoted S_Δ. In the case where $\Delta = (1,\dots,1)$, the directional algebra is called the *diagonal subalgebra* of S (cf. [184]).

In terms of generators, if $S = k[\mathbf{x}_1,\dots,\mathbf{x}_m]$, with $\mathbf{x}_1,\dots,\mathbf{x}_m$ multistandard generators, then

$$S_\Delta = k[\mathbf{x}_1^{\alpha_1}\cdots\mathbf{x}_m^{\alpha_m} \mid |\alpha_j| = p_j, 1 \leq j \leq m].$$

In particular, if S is a toric algebra then S_Δ is a toric algebra (in the wide sense, but not necessarily normal).

Example 2.1.5. Let $S = k[\mathbf{X}_1,\dots,\mathbf{X}_m]$ be a multistandard polynomial ring as above. Note that $S \simeq k[\mathbf{X}_1] \otimes_k \cdots \otimes_k k[\mathbf{X}_m]$ as multigraded k-algebras, where

$$(k[\mathbf{X}_1] \otimes_k \cdots \otimes_k k[\mathbf{X}_m])_{(u_1,\dots,u_m)} := (k[\mathbf{X}_1])_{u_1} \otimes_k \cdots \otimes_k (k[\mathbf{X}_m])_{u_m}.$$

Under this identification, the diagonal subalgebra of S is the Segre multiproduct of the standard polynomial rings $k[\mathbf{X}_1],\dots,k[\mathbf{X}_m]$. Explicitly, $S_\Delta = k[X_{i_1}\cdots X_{i_m} \mid 0 \leq i_j \leq n_j, 1 \leq j \leq m]$. Applying the monoid isomorphism $\mathbb{N}^r \simeq (m\mathbb{N})^r$, with $r = \prod_{1\leq j\leq m}(n_j+1) - 1$, changes S_Δ into a standard k-algebra. As such, S_Δ is the homogeneous coordinate ring of the Segre embedding of the multiprojective space $\mathbb{P}^{n_1} \times \cdots \times \mathbb{P}^{n_m}$ into $\mathbb{P}^r$.

The second example of the submonoid machine is based on starting from an $\mathbb{N}$-graded (resp., $\mathbb{N}^m$-graded) algebra S and taking the subalgebra S' corresponding to a submonoid $d\mathbb{N}$ (resp., $d_1\mathbb{N} \times \cdots \times d_m\mathbb{N}$). To simplify, we will only explain the polynomial case. Thus, let $S = k[X_0,\dots,X_n]$.

Definition 2.1.6. Let S be $\mathbb{N}$-graded (say, by the standard grading or by a quasi-homogeneous grading). Let $d \in \mathbb{N}$. The *dth Veronesean subalgebra* of S is the subalgebra $S^{(d)} := \oplus_{u\in d\mathbb{N}} S_u = \oplus_{v\in\mathbb{N}} S_{dv}$.

In case S is standard, it follows immediately from the definition that $S^{(d)}$ is generated by the monomials of degree d. Since $d\mathbb{N} \simeq \mathbb{N}$, one can regrade $S^{(d)}$ so that it becomes standard. As such, it is the homogeneous coordinate ring of the d-uple embedding of $\mathbb{P}^n$ into $\mathbb{P}^{\binom{n+d}{d}}$.

Definition 2.1.7. Consider the multistandard k-algebra $S = k[X_0,\dots,X_n]$, where each variable constitutes a block (this grading is also called the *fine grading* of S as the homogeneous parts are one-dimensional vector spaces). Let $d_0,\dots,d_n \in \mathbb{N}$. The

$(d_0, \ldots, d_n)$-*multiweighted subalgebra* of S is the subalgebra

$$S^{(d_0,\ldots,d_n)} := \bigoplus_{(u_0,\ldots,u_n)\in d_0\mathbb{N}\times\cdots\times d_n\mathbb{N}} S_{(u_0,\ldots,u_n)} = \bigoplus_{(v_0,\ldots,v_n)\in\mathbb{N}\times\cdots\times\mathbb{N}} S_{(d_0v_0,\ldots,d_nv_n)}.$$

Since $S_{(d_0v_0,\ldots,d_nv_n)} = kX_0^{d_0v_0}\cdots X_n^{d_nv_n}$, one readily sees that $S^{(d_0,\ldots,d_n)}$ is generated as a k-algebra by the pure powers $X_0^{d_0}, \ldots, X_n^{d_n}$. From the algebraic viewpoint, this looks a rather trivial ring as it is isomorphic to $k[X_0, \ldots, X_n]$ itself! Applying further the monoid homomorphism $\varphi : d_0\mathbb{N}\times\cdots\times d_n\mathbb{N} \to \mathbb{N}$ that maps $(0, \ldots, 0, d_i, 0, \ldots, 0)$ to d_i, the resulting ring $(S^{(d_0,\ldots,d_n)})^\varphi$ is a finitely generated ($\mathbb{N}$-graded) quasi-homogeneous k-algebra. As a counterpart to the multiweighted original, this latter algebra is called the $(d_0, \ldots, d_n)$-*weighted* (or $(d_0, \ldots, d_n)$-*quasi-homogeneous*) subalgebra of S. Let $s \in \mathbb{N}$ be such that $(S^{(d_0,\ldots,d_n)})^\varphi = k[(S^{(d_0,\ldots,d_n)})^\varphi_s]$. Then the sth Veronesean subalgebra of $(S^{(d_0,\ldots,d_n)})^\varphi$ can be regraded to become a standard k-algebra, which is generated by $(S^{(d_0,\ldots,d_n)})^\varphi_s$ (spanned by a certain subset of monomials of degree s).

2.2 Notable bigraded algebras

2.2.1 The symmetric algebra

Among multigraded algebras, a few bigraded ones have a special role. One of the most basic examples is the symmetric algebra $\mathcal{S}(E)$ of an S-module E generated by elements of fixed degree $r \geq 1$.

At the outset, the symmetric algebra is defined over any commutative ring R as a standard graded R-algebra; namely, we have the following.

Definition 2.2.1. Let $\mathcal{T}_R(E) = \bigoplus M^{\otimes t}$ denote the graded tensor algebra of E—also known as the free algebra generated by E. Then the *symmetric algebra* of E is $\mathcal{S}_R(E) := \mathcal{T}_R(E)/\mathfrak{C}$, where $\mathfrak{C}$ is the two-sided ideal generated by the elements of the form $e\otimes e' - e'\otimes e$, for all $e, e' \in M$.

Clearly, any iterated "symmetrization"

$$\cdots\otimes e\otimes\cdots\otimes e'\otimes\cdots - \cdots\otimes e'\otimes\cdots\otimes e\otimes\cdots$$

lies in $\mathfrak{C}$, and hence the latter can be given as the two-sided ideal generated by all such iterations.

Note that $\mathcal{S}_R(E)$ acquires a grading from the tensor algebra. Let $\mathcal{S}_R^t(E) \subset \mathcal{S}_R(E)$ denote its tth graded piece, called the tth *symmetric power* of E. In particular, $\mathcal{S}_R^0(E) = R$, $\mathcal{S}_R^1(E) = E$. As such, $\mathcal{S}_R(E)$ is a standard graded R-algebra since it is generated in degree 1 over R.

To get a hold of the defining equations of $\mathcal{S}_R(E)$, i. e., of the nature of the ideal $\ker(R[\mathbf{X}_\alpha] \twoheadrightarrow \mathcal{S}_R(E) = \mathcal{S}_R(\sum R\mathfrak{g}_\alpha))$, $\mathbf{X}_\alpha \mapsto \mathfrak{g}_\alpha$, where $\{\mathfrak{g}_\alpha\}$ is a set of generators of E, one proceeds in the following way.

Lemma 2.2.2. *Let $F_1 \xrightarrow{\Phi} F_0 \longrightarrow E \to 0$ stand for a free presentation of E as an R-module, associated to a given set $\{\mathfrak{g}_\alpha\}$ of generators of E. Then*

$$\mathcal{S}_R(E) \simeq R[\mathbf{X}_\alpha]/\mathfrak{J},$$

where one sets $\mathcal{S}_R(F_0) = R[\mathbf{X}_\alpha]$ and $\mathfrak{J}$ is the ideal generated by the image of the R-algebra map $\mathcal{S}(\Phi) : \mathcal{S}_R(F_1) \longrightarrow R[\mathbf{X}_\alpha]$ in degree 1.

A more detailed description can be given in the case $R = S$ is a graded algebra. Precisely, given an $\mathbb{M}$-graded ring S and an $\mathbb{M}$-graded S-module E, where $\mathbb{M}$ is an arbitrary abelian group, one defines for $d \in \mathbb{M}$ the $\mathbb{M}$-graded S-module $E(d)$ by setting $E(d)_u = E_{u+d}$, for $u \in \mathbb{M}$ – $E(d)$ is often called the *d-shifted* module of E. Also, if $E_1, \dots, E_m$ are $\mathbb{M}$-graded S-modules, the direct sum $\oplus_{j=1}^m E_j = E_1 \oplus \cdots \oplus E_m$ has an induced grading by setting $(\oplus_{j=1}^m E_j)_u = (E_1)_u \oplus \cdots \oplus (E_m)_u$ (where the symbol $\oplus$ on the right-hand side indicates direct sum of abelian groups or S_0-modules). This means that a homogeneous element of degree u of the direct sum is a “vector” whose “coordinates” are all homogeneous of degree u in the respective summands. These operations commute in the sense that $(E_1 \oplus \cdots \oplus E_m)(d) = E_1(d) \oplus \cdots \oplus E_m(d)$.

In particular, $S^m = S \oplus \cdots \oplus S$ is graded, generated by elements in $(S^m)_0$ (the vectors of the canonical basis) and $(S^m)(d) = \oplus_{j=1}^m S(d)$ is generated in degree d.

Now, let $E = \sum_{j=1}^m Sx_j$ be a graded S-module, where x_j is homogeneous of degree $d_j \in \mathbb{M}$. For each j, the homomorphism of S-modules $S(-d_j) \to Sx_j$ sending 1 to x_j is degree preserving (i. e., it is a homogeneous homomorphism). It induces a surjective homogeneous homomorphism $\oplus_{j=1}^m S(-d_j) \to E$. The kernel of the latter homomorphism is a graded submodule of $\oplus_{j=1}^m S(-d_j)$, say, generated by homogeneous elements of degrees $r_1, \dots, r_s$. Repeating the procedure, one obtains a short exact sequence of graded modules and homogeneous homomorphisms

$$S(-r_1) \oplus \cdots \oplus S(-r_s) \xrightarrow{\varphi} S(-d_1) \oplus \cdots \oplus S(-d_m) \longrightarrow E \longrightarrow 0,$$

called a *homogeneous presentation* of E.

By the obvious universal property of the symmetric algebra, one readily sees that $\mathcal{S}_S(E)$ is the homomorphic image of $\mathcal{S}_S(S(-d_1) \oplus \cdots \oplus S(-d_m))$ and the latter is isomorphic as an S-algebra to $\mathcal{S}_S(S(-d_1)) \otimes_S \cdots \otimes_S \mathcal{S}_S(S(-d_m)) \simeq S[T_1, \dots, T_m] = S[\mathbf{T}]$, where $\mathbf{T} = \{T_1, \dots, T_m\}$ are variables over S. Now, with this identification, the kernel of that algebra homomorphism is the ideal $\mathcal{J} \subset S[\mathbf{T}]$ generated by the polynomials of the form $\sum_{j=1}^m g_j T_j$, where $(g_1, \dots, g_m)$ is a homogeneous generator of the image of φ in $S(-d_1) \oplus \cdots \oplus S(-d_m)$ as a graded submodule.

A typical setup is then when $S = S_0[x_1, \dots, x_n]$ is $\mathbb{N}$-graded and standard. Then $S[\mathbf{T}]$ is a standard bigraded S_0-algebra with $\operatorname{bideg}(x_i) = (1,0)$ and $\operatorname{bideg}(T_j) = (0,1)$. If, moreover, $d_1 = \cdots = d_m$, then $\mathcal{J}$ is bihomogeneous in the bigrading of $S[\mathbf{T}]$ and $\mathcal{S}_S(E)$ is up to a degree normalization a standard bigraded S_0-algebra.

Needless to say, for many geometric applications, S_0 will be a field.

Example 2.2.3. If E is an $S = k[\mathbf{X}]$-module admitting a linear presentation (i. e., if the image of φ above is generated by vectors of degree $d+1$ in the grading of $k[\mathbf{X}]^m(-d)$), then the algebra $\mathcal{S}_S(E)_\Delta$ is the homogeneous coordinate ring of a linear section of the Segre embedding of $\mathbb{P}^{n-1} \times \mathbb{P}^{m-1}$, where $n = \dim S$. Graded modules over the polynomial ring admitting linear presentation and locally-free off the maximal irrelevant ideal very often give rise to interesting vector bundles over $\mathbb{P}^{n-1}$ (cf. [187], where some considerations along this line have been developed).

2.2.2 Other bigraded algebras

Let E be a graded module over the standard polynomial ring $S = k[\mathbf{X}]$, generated in fixed degree. We have seen that $\mathcal{S}_S(E)$ is a standard bigraded k-algebra. The S-torsion submodule of $\mathcal{S}_S(E)$ is a bihomogeneous (prime) ideal. The corresponding factor ring is then again a standard bigraded k-algebra, called the *Rees algebra* of E. These algebras will be considered at length in Section 2.3.2.

The case in which $E = I \subset S$ deserves special mention. Here, $\mathcal{R}(I) \simeq \sum_{k\geq 0} I^k t^k \subset k[\mathbf{X}][t]$ is the ring of a suitable blowup. The factor ring $\operatorname{gr}_I(R) = \sum_{k\geq 0} I^k/I^{k+1}$ is an additionally important graded algebra, called the *associated graded ring* of I. However, this algebra can have torsion as an R/I-algebra. We refer to Subsection 2.3.1 for the main properties of these algebras.

2.3 Standard graded algebras

2.3.1 Rees algebras of ideals

Let R denote a ring and $I \subset R$ an ideal. The Rees algebra of I, denoted $\mathcal{R}_R(I)$, is a well-known construct (see [179, Section 7.2.2]). It can be taken to be, equivalently, either of the following standard $\mathbb{N}$-graded R-algebras:

- (External-like) $\bigoplus_{l\geq 0} I^l$.
- (Embedded-like) $R[It] \subset R[t]$, t an indeterminate over R.

They are indeed isomorphic as R-algebras—actually, as standard graded R-algebras. At first sight, the first of these forms looks independent of embeddings, but actually it depends on having $I \subset R$ as one needs to consider the iterated inclusions $I^{l+1} \subset I^l$ in R.

2.3.1.1 Basic knowledge

A very basic condition on an ideal is as follows.

Definition 2.3.1. An ideal I in a ring R is said to be *of linear type* if the natural R-algebra homomorphism $\mathcal{S}_R(I) \twoheadrightarrow \mathcal{R}_R(I)$ is injective.

Here, $\mathcal{S}_R(I)$ and $\mathcal{R}_R(I)$ denote the symmetric and the Rees algebra of I, respectively. The literature on this is quite extensive. Reasonably comprehensive treatment can be found in [189], [215] and [179].

A case where the behavior of this property is fairly understood is when I is an almost complete intersection, i. e., when I is generated by $\operatorname{ht} I+1$ elements. The designation derives from the fact that, when R is Cohen–Macaulay, then height and grade coincide, so typically $I = (f_1, f_2, f_3, f)$, with f_1, f_2, f_3 a regular sequence. For convenience, we file the following general result, which collects in a more detailed version the several known facts about an almost complete intersection in this context (see [192, Proposition 3.7], also [96, Proposition 8.4, Proposition 10.4, Remark 10.5]).

Proposition 2.3.2. *Let R denote a Cohen–Macaulay local ring and let $I \subset R$ denote a proper ideal. Assume that:*

(1) *I is a strict almost complete intersection (i. e., minimally generated by* $\operatorname{ht} I + 1$ *elements).*
(2) *R/I is equidimensional (i. e., all minimal primes of R/I have height* $\operatorname{ht} I$*).*
(3) $\operatorname{depth} R/I \geq \dim R/I - 1$.

Let $R^m \xrightarrow{\varphi} R^{\operatorname{ht} I+1} \longrightarrow I \longrightarrow 0$ stand for a minimally free presentation of I as an R-module. The following conditions are equivalent:

(a) $\operatorname{ht} I_1(\varphi) \geq \operatorname{ht} I + 1$.
(b) *I is generically a complete intersection.*
(c) *I is of linear type.*

Proof. (a) $\Rightarrow$ (b) Localize at a minimal prime $P \subset R$ of R/I, which has height $h := \operatorname{ht} I$ by assumption. Since R is local, some entry of φ is invertible, so up to an elementary transformation on φ_P the local presentation has the form

$$R_P^{m-1} \oplus R_P \xrightarrow{\varphi_P} R_P^h \oplus R_P \longrightarrow I_P \longrightarrow 0,$$

with

$$\varphi_P = \left(\begin{array}{c|c} 1 & 0 \\ \hline 0 & \psi \end{array} \right)$$

Therefore, we get a free presentation $R_P^{m-1} \xrightarrow{\psi} R_P^h \longrightarrow I_P \longrightarrow 0$, thus showing that I_P is generated by (at most) h elements.

(b) $\Rightarrow$ (c) By [96, Proposition 10.4] the symmetric algebra of I is a Cohen–Macaulay ring. Therefore, by [96, Proposition 8.4] it suffices to show that $\mu(I_Q) \leq \operatorname{ht} Q$ for every prime $Q \subset R$ containing I. Let P be a minimal prime of R/I contained in Q. If $P = Q$, the hypothesis guarantees the inequality. Otherwise, $\operatorname{ht}(Q) \geq h+1$. But then $\mu(I_Q) \leq h+1 \leq \operatorname{ht} Q$.

(c) $\Rightarrow$ (a) By [96, Lemma 8.2 and Proposition 8.4], one has $\operatorname{ht} I_t(\varphi) \geq \operatorname{rank}(\varphi) - t + 2$ for every $1 \leq t \leq \operatorname{rank}(\varphi) = h$. In particular, $\operatorname{ht} I_1(\varphi) \geq h - 1 + 2 = \operatorname{ht} I + 1$. □

We digress a bit more in the graded case. Thus, let R denote a standard graded k-algebra over a field k. Let $\mathbf{f} = \{f_0, \ldots, f_m\} \subset R$ be forms of a fixed degree $d \geq 1$. We will henceforth denote by $\varphi = \varphi_{\mathbf{f}}$ a graded matrix whose columns generate the first syzygy module of $\mathbf{f}$. Thus, one has a free graded presentation of the homogeneous ideal $I := (\mathbf{f}) \subset R$.

$$\oplus_s R(-d_s) \xrightarrow{\varphi} R^{m+1}(-d) \to R \tag{1}$$

We denote by $\varphi_1 = \varphi_1(\mathbf{f})$ the matrix whose columns form a set of generators of the submodule $\operatorname{Im}(\varphi)_{d+1}$ (i. e., the coordinates of such vectors as elements of the standard graded ring R are linear forms). Clearly, φ_1 can always be taken as a submatrix of φ, so we informally say that φ_1 is the *linear part* of the whole syzygy matrix φ. Not infrequent is the case where $\varphi_1 = 0$. At the other end of the spectrum, an important special case is when $\varphi_1 = \varphi$, in which case I is said to be *linearly presented*.

A weaker condition is a certain rank requirement. Recall that φ has a well-defined rank r if and only if $I_s(\varphi) = \{0\}$ for $s > r$ and $I_r(\varphi)$ has a regular element. This is equivalent to requiring that the R-submodule $Z := \operatorname{Im}(\varphi) \subset R^m$ have rank r, i. e., that Z_p be free of rank r for every associated prime of R. One also has that Z (hence, φ, too) has a rank if and only if I has a regular element.

We will say that φ_1 has maximal (possible) rank if φ has a rank and $\operatorname{rank}(\varphi_1) = \operatorname{rank}(\varphi)$. We could also say in this situation that φ has *maximal linear rank*.

2.3.1.2 Filtrations of ideals

From the strict algebraic side, the Rees algebra of an ideal $I \subset R$ is a way of gluing together the information coming from its individual powers. As such it admits a version for a (multiplicative) *filtration* of ideals in a ring R, namely, $\mathcal{F} = \{I_r\}_{r \in \mathbb{N}}$, where I_r are ideals of R such that $I_r \subset I_{r-1}$ and $I_r I_s \subset I_{r+s}$, for any r, s. The Rees algebra of such a filtration is the $\mathbb{N}$-graded R-subalgebra,

$$\mathcal{R}(\mathcal{F}) := R + I_1 t + \cdots + I_r t^r + \cdots \subset R + Rt + \cdots + Rt^r + \cdots = R[t].$$

We say that an element $f \in I_r t^r$ is an *essential element of order r* if it does not belong to the graded piece of degree r of the R-subalgebra $R[I_1 t, \ldots, I_{r-1} t^{r-1}] \subset \mathcal{R}(\mathcal{F})$. It amounts to saying that f does not belong to the "preceding" part

$$\sum_{\substack{s_1+2s_2+\cdots+(r-1)s_{r-1}=r\\ 0\le s_j<r}} RI_1^{s_1}I_2^{s_2}\cdots I_{r-1}^{s_{r-1}} = \sum_{1\le s\le r-1} RI_sI_{r-s}.$$

In other words, an essential element of order r is a generator of $\mathcal{R}(\mathcal{F})$ in degree r not expressible as an R-combination of elements of degree $\le r-1$.

Besides the filtration of ordinary powers of an ideal, which is central in this book, another important filtration is the *symbolic filtration* associated to a given ideal $I \subset R$, defined by $I_\ell = I^{(\ell)} := \iota^{-1}(\mathfrak{S}^{-1}I^\ell)$, where $\mathfrak{S} \subset R$ denotes the set of elements of R that are regular on R/I and $\iota : R \to \mathfrak{S}^{-1}R$ denotes the structural homomorphism of fractions. If I is a pure (unmixed) ideal, $I^{(r)}$ is precisely the intersection of the primary components of the ordinary power I^r relative to the associated primes of R/I, i. e., the unmixed part of I^r.

A slightly different way to envisage symbolic powers is by noting that the module $(I^{(r)} \cap I^{r-1})/I^r$ is the R/I-torsion of the conormal module I^{r-1}/I^r of order r. Taking the direct sum over all $r \ge 0$ yields the R/I-torsion of the associated graded ring of I; hence, the nontriviality of symbolic powers gives a measure of the torsion of the latter. In particular, there is no nonzero torsion if and only if $I^{(r)} = I^r$ for every $r \ge 0$, in which case one says that the ideal I is *normally torsion-free*.

The associated Rees algebra is the *symbolic Rees algebra* of I, possibly denoted $\mathcal{R}_R^{(I)}$ (notation of Marcel Morales). However, unlike the ordinary Rees algebra, this algebra may not be finitely generated over R and there are no definite effective ways to check when $\mathcal{R}_R^{(I)}$ is Noetherian. The necessary and sufficient conditions of Huneke [115, Theorems 3.1 and 3.25] obtained in dimension 3 are not effective and neither is the necessary condition of Cowsik–Vasconcelos ([48], [187, Proposition 3.5]). Nevertheless, the latter becomes quite effective provided one has a good guess about what finitely generated subalgebra looks like a strong candidate.

2.3.1.3 The fiber cone

Our general reference for this part is [179, Section 7.3]. Besides the notion of the Rees algebra and the associated graded ring brought up earlier, one takes for granted here the knowledge of the fiber cone (or special fiber) and the analytic spread. Thus, if $(R, \mathfrak{m})$ is a local ring and $I \subset R$ is an ideal, the fiber cone of I is the algebra $\mathcal{F}_R(I) := \mathcal{R}_R(I)/\mathfrak{m}\mathcal{R}_R(I)$. Note that this is an algebra over the residue field of R. The analytic spread of I, denoted $\ell(I)$, is the dimension of its fiber cone.

Another important notion is that of a reduction of an ideal. We briefly recall it here for the reader's convenience. For further details, see [179, Subsection 7.3.3].

An ideal $J \subset I$ is called a reduction of I if $JI^n = I^{n+1}$ for some integer n. The reduction J is said to be a *minimal reduction* if it does not contain properly another reduction of I. The *reduction number* of I with respect to the minimal reduction J, denoted by $r_J(I)$, is the least integer $n \geq 0$ such that $JI^n = I^{n+1}$. The (absolute) reduction number of I is $\min\{r_J(I) : J \text{ is a minimal reduction of } I\}$. If A is a local ring with infinite residue field, then the analytic spread coincides with the minimum number of the generators of a minimal reduction of I.

The idea of a fiber cone when the base ring is local can be transported to a graded setup. Thus, let A be a Noetherian ring of finite Krull dimension and let $(R, \mathfrak{m})$ denote a standard graded algebra over a $A = [R]_0$ and its graded irrelevant ideal $\mathfrak{m} = ([R]_1)$. If $I \subset R$ is a homogeneous ideal, the (relative) fiber cone of I is the A-algebra $\mathcal{R}_R(I)/\mathfrak{m}\mathcal{R}_R(I)$.

The following special case is important throughout: let $I \subset R$ be an ideal generated by $s+1$ forms $\{f_0, \dots, f_s\} \subset R$ of the same degree $d > 0$ – in particular, $I = ([I]_d)$. One often says that I is *equigenerated* in degree d or *d-equigenerated.*

Consider the bigraded A-algebra

$$\mathfrak{A} := R \otimes_A S = R[y_0, \dots, y_s],$$

where $\operatorname{bideg}([R]_1) = (1, 0)$ and $\operatorname{bideg}(y_i) = (0, 1)$. By setting $\operatorname{bideg}(t) = (-d, 1)$, then $\mathcal{R}_R(I) = R[It]$ inherits a bigraded structure over A. One has a bihomogeneous R-homomorphism (of degree zero)

$$\mathfrak{A} \longrightarrow \mathcal{R}_R(I) \subset R[t], \quad y_i \mapsto f_i t.$$

Thus, the bigraded structure of $\mathcal{R}_R(I)$ is given by

$$\mathcal{R}_R(I) = \bigoplus_{c,n \in \mathbb{Z}} [\mathcal{R}_R(I)]_{c,n} \quad \text{and} \quad [\mathcal{R}_R(I)]_{c,n} = [I^n]_{c+nd} t^n.$$

One is often interested in the R-grading $[\mathcal{R}_R(I)]_c = \bigoplus_{n=0}^{\infty} [\mathcal{R}_R(I)]_{c,n}$, and of particular interest is

$$[\mathcal{R}_R(I)]_0 = \bigoplus_{n=0}^{\infty} [I^n]_{nd} t^n = A[[I]_d t] \simeq A[[I]_d] = \bigoplus_{n=0}^{\infty} [I^n]_{nd} \subset R.$$

Clearly, $\mathcal{R}_R(I) = [\mathcal{R}_R(I)]_0 \oplus (\bigoplus_{c \geq 1} [\mathcal{R}_R(I)]_c) = [\mathcal{R}_R(I)]_0 \oplus \mathfrak{m}\mathcal{R}_R(I)$. Therefore, one gets

$$A[[I]_d] \simeq [\mathcal{R}_R(I)]_0 \simeq \mathcal{R}_R(I)/\mathfrak{m}\mathcal{R}_R(I) \tag{2}$$

as graded A-algebras.

2.3.1.4 Theorems on the structural map $\mathcal{S}(J) \twoheadrightarrow \mathcal{R}(J)$

Let R be a Noetherian ring. Most of the time R will be the residue of a locally regular (or at least locally a complete intersection) ring S by an ideal $I \subset S$. We will thus reserve the notation I for such an ideal, while a typical ideal of R will be denoted by J. By and large, it will be assumed that J has grade ≥ 1, although often one can go through with a weaker assumption or no assumption at all.

We have already considered in the previous section a few preliminaries of the connection between the Rees algebra and some of its related algebras. In this part, we wish to focus on its relation to its degree one part, the previously introduced symmetric algebra.

Thus, consider the canonical surjective map

$$\mathcal{S}_R(J) \twoheadrightarrow \mathcal{R}_R(J), \tag{3}$$

where the source denotes the symmetric algebra of $J = (f_0, \ldots, f_m)$. This map is induced by the universal property of the symmetric algebra. Fix algebra presentations $\mathcal{R}_R(J) = T/\mathcal{J}$ and $\mathcal{S}_R(J) = T/\mathcal{L}$, where $T := R[\mathbf{y}] = R[y_0, \ldots, y_m]$ is a standard graded polynomial ring over $T_0 = R$ and $\mathcal{L} \subset \mathcal{J}$ are homogeneous ideals of T—so-named *presentation ideals*. It is well known, or easy to see, that $\mathcal{L}$ is generated by the forms $g_0y_0 + \cdots + g_my_m$ of degree one such that $g_0f_0 + \cdots + g_mf_m = 0$. In other words, $\mathcal{L} = I_1(\mathbf{y} \cdot \varphi)$ where φ is the defining matrix of a free R-presentation of J:

$$R^s \xrightarrow{\varphi} R^{m+1} \longrightarrow J \to 0. \tag{4}$$

It follows that $\mathcal{L}$ is identified with the homogeneous subideal of $\mathcal{J}$ generated in degree one.

A property of sufficient interest when J is equigenerated is the *fiber-type* property, which says that $\mathcal{J}$ is fully generated by $\mathcal{L}$ and the defining generators of the fiber cone of J.

Quite generally, if J has a regular element a then $\mathcal{J} = \mathcal{L} : (a)^{\infty}$. Supposing, for simplicity, that R is a domain then $\mathcal{J}$ is an associated prime of $\mathcal{L}$. But, since $\mathcal{S}_{R_{\wp}}(J_{\wp}) = \mathcal{R}_{R_{\wp}}(J_{\wp})$ for every $\wp \in \operatorname{Spec} R \setminus V(J)$, a power of J kills the kernel of (3); hence, any prime containing L contains either J or $\mathcal{J}$. It follows that $\mathcal{J}$ is actually a minimal prime of $\mathcal{L}$.

The fundamental theorems regarding the nature of the map $\mathcal{S}(J) \twoheadrightarrow \mathcal{R}(J)$ can be arranged in the following diagram, where the equality signs mean that the natural surjective maps are injective:

$$\begin{array}{ccc}
\mathcal{S}_R(J) = \mathcal{R}_R(J) & \overset{\text{(Valla)}}{\Longleftrightarrow} & \mathcal{S}_{R/J}(J/J^2) = \operatorname{gr}_J(R) \\
\Big\Downarrow & & \Big\Downarrow \\
\mathcal{S}_R(J)_{\text{red}} = \mathcal{R}_R(J)_{\text{red}} & \overset{\text{(Herzog-Simis-Vasconcelos)}}{\Longleftrightarrow} & \mathcal{S}_{R/J}(J/J^2)_{\text{red}} = \operatorname{gr}_J(R)_{\text{red}} \\
\Big\Updownarrow \text{(Huneke)} & & \\
J \text{ is 'locanin'} & &
\end{array} \tag{5}$$

where “locanin” is short for *locally everywhere generated by analytically independent elements*. For the notion of analytic independence, see [179, Definition 7.3.15]. While the bottom most horizontal equivalence is valid in general, the bottom most vertical one requires a priori that R be reduced.

Though none of the above equivalences is obvious, the respective proofs are not extremely difficult and there are a few different ways to go about them (see [213, 96, 116]). At the other end, their impact in blowup theory is quite pervasive.

Provided R has local infinite residue fields, the bottom most condition means that J has no proper reductions locally everywhere (see [179, Proposition 7.3.17]). In particular, due to the bounds of the analytic spread, the condition implies that $\mu(J_{\wp}) \leq \operatorname{ht} \wp$, for every prime ideal $\wp \supset J$ – a property that has variously been called (G_{∞}) after [7] or (F_1) after [96]. Its role here comes around in the following converse.

Proposition 2.3.3. *Let R denote a Noetherian ring with infinite residue fields everywhere, and let $J \subset R$ be an ideal such that:*
(i) *J satisfies property (F_1).*
(ii) *For every prime ideal $\wp$ containing J and every ideal $L \subsetneq J_{\wp}$, one has* $\operatorname{ht}(L{:}J_{\wp}) \leq \mu(L)$.

Then J is locally everywhere generated by analytically independent elements.

Proof. Assume to the contrary. Let $\wp$ be a prime ideal of smallest height for which $J_{\wp}$ has a proper reduction L. Rename $R_{\wp}$ as R and $J_{\wp}$ as J. The minimality of choosing $\wp$ now implies that $L_p = J_p$ for any prime $p \in \operatorname{Spec}(R)$ except the maximal ideal $\mathfrak{m}$ of R. Therefore, $L : J$ is $\mathfrak{m}$-primary. On the other hand, $\mu(J) \leq \operatorname{ht} \mathfrak{m}$ by (i). Moreover, a minimal set of generators of L can be extended to one of J and, since the reduction $L \subset J$ is proper, one has $\mu(L) < \mu(J)$. Therefore, $\operatorname{ht} L : J \leq \mu(L) < \mu(J) \leq \operatorname{ht} \mathfrak{m}$, which is a contradiction. □

Example 2.3.4. Perhaps the simplest example where (F_1) alone does not imply analytic independence above is the defining ideal

$$J = (x_1, x_2) \cap (x_3, x_4) \subset A = k[x_1, x_2, x_3, x_4] \tag{6}$$

of the union of two skew lines in $\mathbb{P}^3$. One can readily verify that J satisfies (F_1), but clearly has an algebraic relation of degree 2 (actually, the analytic spread of J is 3).

Looking closely at (6), one sees how condition (b) of Proposition 2.3.3 fails. Since the analytic spread of J is 3, J has a minimal reduction L generated by 3 elements—one can actually take L to be generated by 3 general k-linear combinations of the generators of J. Then $L : J$ is the irrelevant ideal of A. That is, one has $\operatorname{ht} L : J = 4 > \mu(L)$.

Question 2.3.5. Assuming sufficient hypotheses on the ring R, is property (ii) in the proposition implied by local analytic independence?

An affirmative answer would have some impact on the specialization of local analytic independence (see Exercise 2.4.7).

More basic than the content of (5) is the question about the dimension. Actually, (F_1) is a dimension-like property. Its role in this regard is as follows.

Proposition 2.3.6. *Let R be a Noetherian ring of finite Krull dimension and let $J \subset R$ be an ideal. Consider the following two conditions:*
(i) *J satisfies (F_1).*
(ii) $\dim S_{R/J}(J/J^2) = \dim \mathrm{gr}_J(R)$.

Then (i) $\Rightarrow$ (ii). *The converse holds if R satisfies the following additional conditions:* (a) $\dim R = \operatorname{ht} P + \dim R/P$ *for every* $P \in \operatorname{Spec}(R)$; (b) *$R$ is equicodimensional (or J is contained in the Jacobson radical of R).*

Proof. (i) $\Rightarrow$ (ii). Since there is a surjective homomorphism

$$\dim S_{R/J}(J/J^2) \to \dim \mathrm{gr}_J(R),$$

it suffices to show that $\dim S_{R/J}(J/J^2) \leq \dim \mathrm{gr}_J(R)$. By the Huneke–Rossi formula [114], one has

$$\begin{aligned}\dim S_{R/J}(J/J^2) &= \sup_{P\supseteq J} \{\mu(J_P/J_P^2) + \dim(R/J)/(P/J)\}\\ &= \sup_{P\supseteq J} \{\mu(J_P) + \dim R/P\}\\ &= \leq \sup_{P\supseteq J}\{\operatorname{ht} P + \dim R/P\}. \end{aligned} \tag{7}$$

Now, for each $P \supseteq J$, choose a maximal ideal M of R containing P and such that $\dim R/P = \dim R_M/P_M$. We get

$$\operatorname{ht} P + \dim R/P = \operatorname{ht} P_M + \dim R_M/P_M \leq \dim R_M. \tag{8}$$

Replacing the estimate (8) in (7) yields

$$\begin{aligned}\dim S_{R/J}(J/J^2) &\leq \sup_{M\supseteq J}\{\dim R_M\} \quad M \text{ maximal}\\ &= \dim \mathrm{gr}_J(R),\end{aligned}$$

by the well-known formula for the dimension of the associated graded ring.

(ii) $\Rightarrow$ (i). Since R is equicodimensional (or J is contained in the Jacobson radical of R), using again the Huneke–Rossi formula, the assumption reads

$$\sup_{P\supseteq J}\{\mu(J_P) + \dim R/P\} = \dim R.$$

Thus, using hypothesis (b), for every prime $P \supseteq J$, one has $\mu(J_P) \leq \dim R - \dim R/P = \operatorname{ht} P$. □

In addition to the property (F_1), a weaker property (F_0) has been introduced which states that, for an ideal $J \subset R$, the bounds $\mu(J_{\wp}) \leq \operatorname{ht}\wp + 1$ hold for every prime ideal $\wp \supset J$. Its role is as follows.

Proposition 2.3.7. *Let R be a Noetherian ring of finite Krull dimension and let $J \subset R$ be an ideal. Consider the following two conditions:*
(i) *J satisfies* (F_0).
(ii) $\dim \mathcal{S}_R(J) = \dim \mathcal{R}_R(J)$.

Then (i) $\Rightarrow$ (ii). *The converse holds if R satisfies the property to the effect that* $\dim R = \operatorname{ht} P + \dim R/P$ *for every* $P \in \operatorname{Spec}(R)$.

Proof. The argument is similar to the one in Proposition 2.3.6 (see [179, Proposition 7.2.21]). □

Both (F_1) and (F_0) can be restated in terms of Fitting ideals (see [96]) if J has grade ≥ 1, as follows. Let φ denote a presentation matrix of J over R. Then:

$$J \text{ satisfies } (F_0) \text{ if and only if } \operatorname{ht} I_t(\varphi) \geq \operatorname{rank}\ \varphi - t + 1, \quad \text{for all } 1 \leq t \leq \operatorname{rank}\ \varphi,$$

and

$$J \text{ satisfies } (F_1) \text{ if and only if } \operatorname{ht} I_t(\varphi) \geq \operatorname{rank}\ \varphi - t + 2, \quad \text{for all } 1 \leq t \leq \operatorname{rank}\ \varphi.$$

If we extend the above notions to a general (F_i) condition, with $i \geq 1$, the following descent relationship is quite useful.

Lemma 2.3.8 ([179, The proof of Proposition 7.2.29]). *Let R be a finitely generated domain over a field and let $J \subset R$ be a nonzero ideal. Then J satisfies the property (F_{i+1}) if and only if the R/J-module J/J^2 satisfies the property (F_i).*

Proof. Let $R^m \xrightarrow{\varphi} R^n \longrightarrow J \to 0$ stand for a free presentation of J. Suppose that $\operatorname{ht} I_t(\varphi) \geq \operatorname{rank} \varphi - t + (i+1)$, for $1 \leq t \leq \operatorname{rank}\varphi$. Tensoring with R/J over R yields a free presentation $\overline{R}^m \xrightarrow{\overline{\varphi}} \overline{R}^n \longrightarrow J/J^2 \to 0$ of J/J^2 over $\overline{R} = R/J$. Then $\operatorname{rank}\overline{\varphi} = n - \operatorname{rank} J/J^2 = \operatorname{rank}\varphi + 1 - \operatorname{ht} J$. On the other hand, $\operatorname{ht} I_t(\overline{\varphi}) = \operatorname{ht} I_t(\varphi) - \operatorname{ht} J$, for $1 \leq t \leq \operatorname{rank}\overline{\varphi}$, because R is a domain of finite type over a field. Putting together yields

$$\operatorname{ht} I_t(\overline{\varphi}) \geq \operatorname{rank}\varphi - t + (i+1) - \operatorname{ht} J = \operatorname{rank}\varphi + 1 - \operatorname{ht} J - t + i = \operatorname{rank}\overline{\varphi} - t + i,$$

for $1 \leq t \leq \operatorname{rank}\overline{\varphi}$. Therefore, J/J^2 satisfies condition (F_i).

The converse is similar. □

Remark 2.3.9. As usual, the hypothesis on R can be replaced by assuming that R is Cohen–Macaulay or, for that matter, any ring satisfying the dimension/codimension complementary condition, as was assumed in Proposition 2.3.6.

2.3.1.5 Complements on the dimension of the symmetric algebra

The following notion goes back to Artin and Nagata.

Definition 2.3.10. Let R be a Noetherian ring and let $I \subset R$ be an ideal. Given an integer $s \geq 1$, one says that I satisfies condition G_s if $\mu(I_p) \leq \operatorname{ht} p$ for every prime $p \in V(I)$ with $\operatorname{ht} p \leq s - 1$.

If $s \geq \dim R + 1$, the property is denoted (G_∞).

This sort of property and its analogues have an impact on certain dimension theoretic aspects. It is more effective under additional hypotheses, such as when R is Cohen–Macaulay and/or when I has a rank, in which case its effectiveness is expressed in terms of the ideals of minors of a matrix φ of a presentation of I as in the case of (F_i).

Here, we intend to consider lower cases of s to see how they impact in a few situations.

The symmetric algebra is very sensitive to it (see, e. g., [179, Section 7.2.3 and ff]). The next result is an encore of [114, Corollary 2.13] in the present terminology. The conclusion of the statement was originally conceived by G. Valla, which spurred the subsequent work by C. Huneke and M. E. Rossi.

For the present purpose, recall that a Noetherian ring of finite Krull dimension is said to be *equicodimensional* if all its maximal ideals have the same height ($= \dim R$).

Proposition 2.3.11. *Let R be a Noetherian ring of finite Krull dimension $\mathfrak{d}$ and let $I \subset R$ be an ideal generated by r elements. Suppose that the following conditions hold:*

(i) *R is equicodimensional and* $\operatorname{grade} I \geq 1$.
(ii) *The number of generators of I locally in height $\mathfrak{d} - 1$ is at most $r - 1$.*
(iii) *I satisfies condition $G_{\mathfrak{d}-1}$.*

Then $\dim \mathcal{S}_R(I) = \max\{\mathfrak{d} + 1, r\}$.

Proof. The proof, based on the Huneke–Rossi formula, follows the same pattern with the more or less obvious adaptation. Thus, first since $\operatorname{grade} I \geq 1$ then for a minimal prime $P \in \operatorname{Spec} R$ of maximal dimension one has $I \not\subset P$, and hence $\dim R/P + \mu(I_P) = \dim R + 1 = \mathfrak{d} + 1$, so $\dim \mathcal{S}_R(I) \geq \mathfrak{d} + 1$. On the other hand, $\dim \mathcal{S}_R(I) \geq r$ [114, Corollary 2.8].

Therefore, it suffices to show the upper bound $\dim \mathcal{S}_R(I) \leq \max\{\mathfrak{d}+1, r\}$. For $I \not\subset P$, there is nothing to prove, so assume that $I \subset P$. If P is a maximal ideal, equicodimensionality takes care since $\dim R/P = 0$ and $\mu(I_P) \leq r$. If P has height $\mathfrak{d} - 1$ assumption

(ii) takes care again since $\dim R/P \leq 1$. Finally, for P of height at most $\eth - 2$ the $G_{\eth-1}$ assumption takes over. □

Remark 2.3.12. It is curious that a small piece of the proof in [114, Corollary 2.13] depends on knowing that the Noetherian base ring R is equicodimensional. This hypothesis is understated there possibly because the authors assume throughout that R is a ring with sufficient conditions for equicodimensionality to hold. Since it has been shown afterwards that these hypotheses are superfluous for the main dimension formula, so it seemed that all the corollaries would dispense as well with any extra hypotheses on R (unless otherwise explicit in the statements).

Particular cases of interest are as follows.

Corollary 2.3.13. *Let $I \subset R$ denote an ideal of grade ≥ 2 in a Noetherian equicodimensional ring R of dimension 3. Let I be generated by $r \geq 4$ elements. Then the following are equivalent:*
(i) $\mu(I_P) \leq r - 1$ *for every height 2 prime $P \in \operatorname{Spec} R$.*
(ii) $\dim \mathcal{S}_R(I) = r$.

Proof. (i) ⇒) (ii). This implication follows from Proposition 2.3.11.

(ii) ⇒) (i). Let $\operatorname{ht} P = 2$. Then $\dim R/P = 1$ since R is equicodimensional. Therefore,

$$1 + \mu(I_P) = \dim R/P + \mu(I_P) \leq \dim \mathcal{S}_R(I) = r,$$

hence $\mu(I_P) \leq r - 1$. □

Example 2.3.14. The following simple example illustrates the above corollary. Let $R = k[x,y,z]$ (k a field) and let $I = (x,y)^3 \cap (x,z)^3$, a 4-generated ideal. Then the number of generators of I locally at the minimal primes (x,y) and (x,z) does not drop, and in fact, as it turns out, $\dim \mathcal{S}_R(I) = 5 > 4$.

Corollary 2.3.15. *Let R be a Noetherian equicodimensional, equidimensional catenary ring and let φ denote an $n \times (n-1)$ matrix with entries in R. Suppose that $\dim R := \eth < n$ and let $\mathcal{L} = I_1(\mathbf{y}.\varphi)$ denote a presentation ideal of $\mathcal{S}_R(I)$. If φ satisfies condition $G_{\eth-1}$ and $I_1(\varphi)$ has height $\eth$, then $\operatorname{ht} \mathcal{L} = \eth$ (maximal possible).*

Proof. Since $n > \eth$, by Proposition 2.3.11 one has $\dim \mathcal{S}_R(I) = n$. By the assumptions on R, one has $\operatorname{ht} \mathcal{L} = \eth + n - n = \eth$. □

2.3.2 Rees algebras of modules

A notion of Rees algebra of an abstract R-module E is not as pliable as the one of the symmetric algebra of E, at least if one wishes to mimic the ideal case. The obstruction is double at the outset: one lacks a notion of powers of a module and no embedding of E in a free R-module may be available. Consequently, the notion for a module is not completely stable. If one assumes that E has some additional properties, then varying definitions have been proposed, while if no such properties are required, then a general definition turns out to be slightly embarrassing (see [59], also [140] as a predecessor for these ideas).

Let us look closer at the ideal case, focusing on the second of the above definitions. By the universal property of the symmetric algebra, there is a natural surjective R-algebra homomorphism $\rho : \mathcal{S}_R(I) \longrightarrow \mathcal{R}_R(I)$ such that $\rho(a) = at$, for $a \in I$. Since the target is R-torsion-free, the kernel $\mathcal{A}$ of ρ contains the R-torsion of $\mathcal{S}_R(I)$. Conversely, if I has positive grade, the R-torsion is contained in the kernel—to see this, note that locally at any prime ideal of R not containing I, the map ρ is the identity; hence, its kernel is killed by a power of I; next pick a regular element in I.

Recall that the module E is said to *have a rank*, say, r, if E_p is R_p-free of rank r for every prime $p \in \operatorname{Ass} R/I$. Now, I contains a regular element if and only if it has a rank (= 1). Thus, one learns that two relevant ingredients for the definition of the Rees algebra of an R-module E are the existence of a rank and the R-torsion of $\mathcal{S}_R(E)$. Since the second of these is mandatory even in the ideal case, one takes it at the outset. The *Rees algebra* $\mathcal{R}_R(E)$ of E is the residue algebra of the symmetric algebra $\mathcal{S}_R(E)$ by its R-torsion $\mathcal{A}$. This notion only works fully if one assumes moreover that E has a rank, which we will tacitly take for granted.

One reason to look at such objects—even before the traditional ideal case shows up— is that they give a handy algebraic way of considering certain incidences in biprojective space.

2.3.2.1 Dimension

This is the first place where the notion of a rank has impact. If the module E has a rank, we denote it by $\operatorname{rank} E$.

Proposition 2.3.16. *Let R be a noetherian ring of finite Krull dimension. Let E be a finitely generated module R-module having a rank. Then*

$$\dim \mathcal{R}_R(E) = \dim R + \operatorname{rank} E. \tag{9}$$

Proof. Write $\dim \mathcal{R}_R(E) = \mathcal{R}(E)$ for short.

(1) $\dim \mathcal{R}(E) \geq \dim R + \operatorname{rank} E$.

First, observe that $\mathcal{R}(E)$ is a graded R-algebra with a grading naturally inherited from $\mathcal{S}(E)$. Now, $\mathcal{R}(E)/\mathcal{R}(E)_+ \simeq R$ implies the inequality $\dim \mathcal{R}(E) \geq \dim R + \operatorname{ht} \mathcal{R}(E)_+$, so it suffices to show that $\operatorname{ht} \mathcal{R}(E)_+ = \operatorname{rank} E$. Thus, let $\mathcal{P} \supset \mathcal{R}(E)_+$ be a prime in $\mathcal{R}(E)$ such that $\operatorname{ht} \mathcal{P} = \operatorname{ht} \mathcal{R}(E)_+$. We have $\mathcal{P} = (\wp, \mathcal{R}(E)_+)$ where $\wp = \mathcal{P} \cap R \in \operatorname{Min} R$. Therefore, by the rank hypothesis,

$$\mathcal{R}(E)_{\mathcal{P}} = (\mathcal{R}_{R_\wp}(E_\wp))_{\mathcal{P}} \simeq R_\wp[\mathbf{T}]_{(\wp_\wp, \mathbf{T})},$$

where $\#\mathbf{T} = \operatorname{rank} E$. This shows that

$$\operatorname{ht} \mathcal{R}(E)_+ \; (= \operatorname{ht} \mathcal{P} = \operatorname{ht} \mathcal{P}_{\mathcal{P}} = \dim \mathcal{R}(E)_{\mathcal{P}}) = \dim R_{\wp} + \operatorname{rank} E = \operatorname{rank} E,$$

as required.

(2) $\dim \mathcal{R}(E) \le \dim R + \operatorname{rank} E$.
Let $\mathcal{P} \subset \mathcal{R}(E)$ be a minimal prime with $\dim \mathcal{R}(E) = \dim \mathcal{R}(E)/\mathcal{P}$. Set $\mathcal{P} \cap R = \wp$. Applying the well-known consequence of Cohen's dimension inequality [179, Corollary 2.5.38] one obtains:

$$\begin{aligned}
\dim \mathcal{R}(E) &= \dim \mathcal{R}(E)/\mathcal{P} \le \dim R/\wp + \operatorname{trdeg}_{R/\wp} \mathcal{R}(E)/\mathcal{P} \\
&\le \dim R/\wp + \dim(\mathcal{R}(E)/\mathcal{P}) \otimes_R R_{\wp}/\wp_{\wp} \le \dim R/\wp + \dim \mathcal{R}(E) \otimes_R R_{\wp}/\wp_{\wp} \\
&\le \dim R/\wp + \dim \mathcal{R}(E) \otimes_R R_{\wp} = \dim R/\wp + \dim \mathcal{R}_{R_{\wp}}(E_{\wp}) && (10)\\
&= \dim R/\wp + \dim \mathcal{S}_{R_{\wp}}(E_{\wp}) = \dim R/\wp + \dim R_{\wp}[\mathbf{T}] && (11)\\
&= \dim R/\wp + \operatorname{ht} \wp + \operatorname{rank} E \le \dim R + \operatorname{rank} E.
\end{aligned}$$

Note that to pass from (10) to (11) one uses that $\wp \subset \mathfrak{q}$ for some $\mathfrak{q} \in \operatorname{Ass} R$ since $\mathcal{R}_R(E)$ is R-torsion-free, and hence $E_{\wp} = (E_{\mathfrak{q}})_{\wp_{\mathfrak{q}}}$ is $R_{\wp}$-free. □

2.3.2.2 Universal property

$\mathcal{R}_R(E)$ has a kind of universal mapping property parallel to that of $\mathcal{S}(E)$, to wit, given a torsion-free R-algebra T and an R-module homomorphism $E \to T$ there exists a unique R-algebra map $\mathcal{R}_R(E) \to T$ making the obvious diagram commutative. Alas, torsion behaves quite badly with respect to base change $R \to R'$. As a result, one of the drawbacks of the Rees algebra, in contrast to the symmetric algebra, is that it is not quite functorial (i. e., it does not commute) with respect to base change.

Definition 2.3.17. An R-module E is said to be of *linear type* if the canonical map $\mathcal{S}_R(E) \twoheadrightarrow \mathcal{R}_R(E)$ is an isomorphism.

Proposition 2.3.18. *Let R be a noetherian ring, let E be a finitely generated R-module and let $I \subset R$ be an ideal such that E_p is of linear type for every $p \in \operatorname{Ass} R/I$. Then the canonical surjection $\mathcal{S}_R(E) \twoheadrightarrow \mathcal{S}_{R/I}(E/IE)$ descends to a map $\mathcal{R}_R(E)/I\mathcal{R}_R(E) \twoheadrightarrow \mathcal{R}_{R/I}(E/IE)$.*

Proof. It clearly suffices to show that the map of symmetric algebras descend to a map $\mathcal{R}_R(E) \twoheadrightarrow \mathcal{R}_{R/I}(E/IE)$. Let $\mathcal{A} \subset \mathcal{S}_R(E)$ denote the R-torsion of $\mathcal{S}_R(E)$. We show that the image of $\mathcal{A}$ under the map $\mathcal{S}_R(E) \twoheadrightarrow \mathcal{S}_{R/I}(E/IE)$ is contained in the R/I-torsion submodule of $\mathcal{S}_{R/I}(E/IE)$. Let then $f \in \mathcal{A}$. It suffices to show that $(0 : f) \cap R \not\subset \wp$ for every $\wp \in \operatorname{Ass} R/I$, where the annihilator is taken in $\mathcal{S}_R(E)$. By assumption, $\mathcal{A}_{\wp} = (0)$ for every $\wp \in \operatorname{Ass} R/I$, i. e., $\wp \notin \operatorname{Supp}_R(\mathcal{A}) = \operatorname{Supp}(R/(0 : \mathcal{A}) \cap R)$ for every $\wp \in \operatorname{Ass} R/I$. Therefore, $(0 : \mathcal{A}) \cap R \not\subset \wp$, hence also $(0 : f) \cap R \not\subset \wp$ for every $\wp \in \operatorname{Ass} R/I$. □

A typical situation is when both E and E/IE have ranks over R and R/I, respectively (e. g., when $E = I$ and I is an equidimensional ideal which is generically a complete intersection), in which case one has $\mathrm{rk}_{R/I}E/IE \geq \mathrm{rk}_R E$, since if $\wp \in \mathrm{Ass}\, R/I$ then

$$\mathrm{rk}_{R/I}E/IE = \nu_{R_\wp/I_\wp}(E_\wp/I_\wp E_\wp) = \nu_{R_\wp}(E_\wp) \geq \mathrm{rank}\, E.$$

Now focus on the conormal module I/I^2 of an ideal I.

Corollary 2.3.19. *Let R be a Noetherian ring and let $I \subset R$ be an ideal of linear type locally at the associated primes of R/I. Then there is an isomorphism of R/I-algebras*

$$\mathcal{R}_{R/I}(I/I^2) \simeq \frac{\mathrm{gr}_I(R)}{(R/I)\text{-torsion}}.$$

Proof. Applying Proposition 2.3.18, we get a surjective map $\mathrm{gr}_I(R) \twoheadrightarrow \mathcal{R}_{R/I}(I/I^2)$. Since $\mathcal{R}_{R/I}(I/I^2)$ is torsion-free over R/I, this map induces yet another surjective map

$$\mathrm{gr}_I(R)/((R/I)\text{-torsion}) \twoheadrightarrow \mathcal{R}_{R/I}(I/I^2). \tag{12}$$

Now the canonical map $\mathcal{S}_{R/I}(I/I^2) \twoheadrightarrow \mathrm{gr}_I(R)$ induces naturally a map

$$\mathcal{S}_{R/I}(I/I^2)/((R/I)\text{-torsion}) \twoheadrightarrow \mathrm{gr}_I(R)/((R/I)\text{-torsion}).$$

Therefore, we get a map going the opposite direction of (12), clearly an inverse. □

2.3.2.3 Analytic spread

We specify the following settings throughout this part: $(R, \mathfrak{m})$ is a local ring and its maximal ideal or a standard graded ring $R = R_0[R_1]$ over a Noetherian ring R_0 and its irrelevant ideal $\mathfrak{m} = (R_1) = (R_+)$. All R-modules will be finitely generated (and R-graded if R is graded).

Definition 2.3.20. The (relative) *fiber cone* of E is $\mathcal{F}_R(E) := \mathcal{R}_R(E)/\mathfrak{m}\mathcal{R}_R(E)$.

Germs of this notion in the local case or when R_0 is a field appear in [176, Section 2.4] and [125]. A more thorough account appears in [189]. In such cases, the *analytic spread* of E is defined to be $\ell(E) := \dim \mathcal{F}_R(E)$, thus extending the usual notion in the case of an ideal.

Since $\dim \mathcal{F}_R(E)$ is a residue ring of $\mathcal{S}(E)/\mathfrak{m}\mathcal{S}(E)$, one gets $\dim \mathcal{F}_R(E) \leq \nu(E)$. In particular, if R is local or else R_0 is a field, one obtains the upper bound $\ell(E) \leq \nu(E)$, as an analogue of the ideal case. In the tradition of ideals, one searches for yet another upper bound.

Corollary 2.3.21. *Suppose that either R is local or else R_0 is a field and that E has a rank. Then $\ell(E) \leq \dim R + \mathrm{rank}\, E - 1$ (resp., $\ell(E) = \mathrm{rank}\, E$) if $\dim R \geq 1$ (resp., if $\dim R = 0$).*

Proof. The case where $\dim R = 0$ is clear since E is then automatically free. Let $\dim R \geq 1$. The result is immediate if E is free since then $\mathcal{R}_R(E)$ is a polynomial ring over R in $\operatorname{rank} E$ variables and the extension of the maximal ideal has height at least one. Thus, assume that E is not free. We still claim that $\mathfrak{m}\mathcal{R}_R(E)$ has height at least one. Indeed, a minimal prime of $\mathcal{R}_R(E)$ containing $\mathfrak{m}$ would have to contract to $\mathfrak{m}$. But since $\mathcal{R}_R(E)$ is torsion-free, $\mathfrak{m}$ would necessarily be an associated prime of R; hence, $E = E_{\mathfrak{m}}$ would be free. In any case, the result follows immediately from Proposition 2.3.16. □

If R is local or else R_0 is a field and E has a rank, mimicking the ideal case one says that E has *maximal analytic spread* provided

$$\ell(E) = \min\{\dim R + \operatorname{rank} E - 1, \nu(E)\}. \tag{13}$$

Here is such a case.

Proposition 2.3.22. *Let $\dim R \geq 1$, where R is either local or R_0 is a field, and let E be a finitely generated module R-module with a rank having the following properties:*
(i) *E is of linear type on the punctured spectrum of R.*
(ii) $\nu(E) = \dim R + \operatorname{rank} E$.
(iii) $\operatorname{Spec}(\mathcal{S}_R(E))$ *is connected in codimension one (e.g., $\mathcal{S}_R(E)$ is a Cohen–Macaulay ring).*

Then $\ell(E) = \dim R + \operatorname{rank} E - 1$.

Proof. By (i), $\mathcal{A}$ is annihilated by a power of $\mathfrak{m}$. Since $\dim R \geq 1$, $\mathfrak{m}\mathcal{S}(E)$ is not nilpotent. On the other hand, by (ii) and Corollary 2.3.21, $\ell(E) \leq \nu(E) - 1$. Therefore, $\mathcal{A}$ is surely nonnilpotent. Since $\operatorname{Spec}(\mathcal{S}(E))$ is connected in codimension one, it must be the case that $\operatorname{ht}(\mathfrak{m}\mathcal{S}, \mathcal{A}) \leq 1$ by the Abhyankar–Hartshorne result (cf. Definition 10.1.1). But then

$$\mathcal{A}/\mathfrak{m} \cap \mathcal{A}(= (\mathfrak{m}\mathcal{S}, \mathcal{A})/\mathfrak{m}\mathcal{S}) = \ker(\mathcal{S}(E)/\mathfrak{m}\mathcal{S}(E) \to \mathcal{R}(E)/\mathfrak{m}\mathcal{R}(E))$$

also has height at most 1. Thus, we get inequality in the other direction, namely, $\ell(E) \geq \nu(E) - 1 = \dim R + \operatorname{rank} E - 1$. □

Even back to the ideal case, one gets something.

Proposition 2.3.23. *Let $(R, \mathfrak{m})$ as above be either local or else standard graded over a field and $\mathfrak{m} = (R_1)$, and let $I \subset R$ be an ideal (homogeneous in the graded case) of linear type locally at the associated primes of R/I. Then $\ell(I) \geq \ell(I/I^2)$.*

Proof. Note that $\ell(I)$ is the analytic spread of an ideal over R while $\ell(I/I^2)$ is the analytic spread of an R/I-module (graded by the induced grading of R/I when I is homogeneous).

Setting $G := \mathrm{gr}_I(R)$, by Corollary 2.3.19 one has an isomorphism:

$$\frac{\mathcal{R}_{R/I}(I/I^2)}{(\mathfrak{m}/I)\mathcal{R}_{R/I}(I/I^2)} = \frac{\mathcal{R}_{R/I}(I/I^2)}{\mathfrak{m}\mathcal{R}_{R/I}(I/I^2)} \simeq \frac{G}{(\mathfrak{m}G, (R/I)\text{-torsion})}.$$

Since $\ell(I) = \dim G/\mathfrak{m}G$, the assertion is clear. □

2.3.2.4 Embedding into a polynomial ring

Let E be a torsion-free R-module with a rank along with a given embedding $E \subset R^r$ – recall that any finitely generated torsion-free module with a rank admits such an embedding [179, Proposition 3.4.3].

Let $E = \sum_{j=1}^{m} Rv_j \subset R^r$. Let $\{e_1, \ldots, e_r\}$ stand for the canonical basis of R^r and let $\mathcal{S}_R(R^r) \simeq R[\mathbf{t}] = R[t_1, \ldots, t_r]$ be the symmetric algebra of R^r, via the homomorphism $t_i \mapsto e_i$. Write $v_j = \sum_i a_{i,j}e_i \in R^r$ for suitable $a_{i,j} \in R$. By the universal property of the symmetric functor, there follows a surjective R-algebra map:

$$\mathcal{S}_R(E) \twoheadrightarrow R\Big[\sum_i a_{i,1}t_i, \ldots, \sum_i a_{i,m}t_i\Big] \subset R[\mathbf{t}].$$

By assumption, this map is an isomorphism generically on R. Since the R-torsion is the kernel of the canonical map $\mathcal{S}(E) \to \mathcal{S}(EU^{-1})$, where

$$U = R \setminus \bigcup_{\wp \in \mathrm{Ass}\, R} \wp,$$

the kernel is contained in the R-torsion of $\mathcal{S}(E)$. This shows that there is an R-algebra isomorphism $\mathcal{R}_R(E) \simeq R[f_1, \ldots, f_m]$, where $f_j = \sum_i a_{i,j}t_i$.

2.3.2.5 Equigenerated modules

Aspects of the results of this part have been stated for ideals in Subsection 2.3.1.3.

Suppose now that $R = R_0[R_+]$ is a standard graded ring over a Noetherian ring R_0 and let E be a graded R-submodule of a free R-module R^r equipped with the natural grading inherited from R. One says that E is *equigenerated* if it is generated by homogeneous vectors of a fixed degree.

Proposition 2.3.24. *If E as above is, in addition, equigenerated then $\mathcal{F}_R(E)$ has a structure of graded R_0-algebra and is graded-isomorphic to a graded R_0-subalgebra of $\mathcal{R}_R(E)$ making $\mathcal{F}_R(E)$ into a direct summand of $\mathcal{R}_R(E)$ as $\mathcal{F}_R(E)$-module.*

Proof. As above, we have identified $\mathcal{R}_R(E)$ with $R[f_1, \ldots, f_m] \subset R[\mathbf{t}]$, where the f_is are forms of the same degree. $\mathcal{R}_R(E)$ has a natural bigraded structure as R_0-algebra; hence, also a structure of graded R_0-algebra (by taking "total" degrees). Since

$(R_+)\mathcal{R}_R(E)$ is a homogeneous ideal for this grading, $\mathcal{R}_R(E)/(R_+)\mathcal{R}_R(E)$ is a graded R_0-algebra as well.

Clearly,

$$R[f_1,\dots,f_m]/(R_+)R[f_1,\dots,f_m] \simeq R_0[f_1,\dots,f_m]$$

as graded R_0-algebras. Moreover, $(R_+)\mathcal{R}_R(E)$ is obviously closed under multipliers from $R_0[f_1,\dots,f_m]$. Therefore, the graded decomposition

$$\mathcal{R}_R(E) = R_0[f_1,\dots,f_m] \bigoplus (R_+)\mathcal{R}_R(E)$$

is a direct sum of $R_0[f_1,\dots,f_m]$-modules, with $\mathcal{F}_R(E) \simeq R_0[f_1,\dots,f_m]$ as graded R_0-algebras. □

The above representation of $\mathcal{R}_R(E)$ as an R-subalgebra of $R[\mathbf{t}]$ generated by finitely many biforms of bidegree $(d,1)$, for some fixed $d \geq 1$, was a consequence of the given embedding of $\mathcal{R}_R(E)$ and the additional equigeneration hypothesis. Once this is accomplished, we can forget about the original nature of $\mathcal{R}_R(E)$ and of $\mathcal{F}_R(E)$ and stick to the algebra $R[f_1,\dots,f_m]$ and $R_0[f_1,\dots,f_m]$, respectively. Let $K \subset Q$ denote, respectively, the field of fractions of $R_0[f_1,\dots,f_m]$ and of $R[f_1,\dots,f_m]$.

Corollary 2.3.25. *Under the hypothesis of* Proposition 2.3.24, *if R is a domain then $\mathcal{F}_R(E)$ is a domain. Moreover, if $\mathcal{R}_R(E)$ is a normal domain, then so is $\mathcal{F}_R(E)$.*

Proof. By Proposition 2.3.24, one has a direct sum decomposition

$$\mathcal{R}_R(E) = R_0[f_1,\dots,f_m] \bigoplus (R_+)R[f_1,\dots,f_m],$$

where $\mathcal{F}_R(E) \simeq R_0[f_1,\dots,f_m]$ and $\mathcal{R}_R(E) \simeq R[f_1,\dots,f_m]$. The first assertion is then clear. Now, assume that $R[f_1,\dots,f_m]$ is a domain (i. e., R is a domain).

The argument will be split into two claims.

Claim 1. $R[f_1,\dots,f_m] \cap K = R_0[f_1,\dots,f_m]$.

One inclusion being obvious, let $a_0 + a_+ = f/g$, where $a_0 \in R_0[f_1,\dots,f_m]$, $a_+ \in (R_+)R[f_1,\dots,f_m]$, $f, g \in R_0[f_1,\dots,f_m]$, with $g \neq 0$. Then $f = ga_0 + ga_+$, with f, ga_0 both in $R_0[f_1,\dots,f_m]$, hence $ga_+ \in R_0[f_1,\dots,f_m]$. But $ga_+ \in (R_+)R[f_1,\dots,f_m]$ as well. Therefore, we must have $ga_+ = 0$, i. e., $a_+ = 0$.

For the second claim, let a bar over a domain denote its integral closure in the field of fractions.

Claim 2. $\overline{R_0[f_1,\dots,f_m]} = \overline{R[f_1,\dots,f_m]} \cap K$ as subrings of Q.

The inclusion $\overline{R_0[f_1,\dots,f_m]} \subset \overline{R[f_1,\dots,f_m]} \cap K$ is obvious as an element integral over $R_0[f_1,\dots,f_m]$ is certainly so over the larger ring $R[f_1,\dots,f_m]$.

Conversely, let $h := f/g \in K$ be integral over $R[f_1,\dots,f_m]$. Say, $h^m + a_1h^{m-1} + \cdots + a_m = 0$, with $a_i \in R[f_1,\dots,f_m]$. Write $a_i = a_{i,0} + a_{i,+}$, with $a_{i,0} \in R_0[f_1,\dots,f_m]$ and

$a_{i,+} \in (R_+)R[f_1,\dots,f_m]$. Substituting and clearing denominator yields

$$(f^m + a_{1,0}gf^{m-1} + \cdots + a_{m,0}g^m) + L = 0,$$

where $L \in (R_+)R[f_1,\dots,f_m]$. Since $f^m + a_{1,0}gf^{m-1} + \cdots + a_{m,0}g^m \in R_0[f_1,\dots,f_m]$, we must conclude that $f^m + a_{1,0}gf^{m-1} + \cdots + a_{m,0}g^m = 0$. Therefore, $h^m + a_{1,0}h^{m-1} + \cdots + a_{m,0} = 0$, with $a_{i,0} \in R_0[f_1,\dots,f_m]$ by construction. This means that $h \in \overline{R_0[f_1,\dots,f_m]}$, as claimed.

To conclude, suppose that $R[f_1,\dots,f_m] = \overline{R[f_1,\dots,f_m]}$. Then the two claims imply that $\overline{R_0[f_1,\dots,f_m]} = R_0[f_1,\dots,f_m]$. □

For the next result, let $I \subset R$ be an equigenerated homogeneous ideal, where R is a standard graded domain over an infinite field k. We will shorten $\mathcal{R}_R(I)$ and $\mathcal{F}_R(I)$ to $\mathcal{R}(I)$ and $\mathcal{F}(I)$, respectively.

Recall that, given a filtration $\mathfrak{F} := \{I_s \mid s \geq 0, I_0 = R\}$ of ideals, one can consider the Rees algebra $\mathcal{R}(\mathfrak{F}) := \bigoplus_s I_s t^s$ of this filtration. If the ideals I_s are homogeneous, we say that $\mathfrak{F}$ is *balanced* provided there exists an integer $d \geq 1$ such that I_s is generated by forms of degree sd, for every s (one allows some of the ideals I_s to vanish).

A characteristic property of the Rees algebra of a balanced filtration is the same as that of the usual Rees algebra of an ideal generated by forms of the same degree as given in Proposition 2.3.24; namely, one has

$$\mathcal{R}(\mathfrak{F}) = k\big[[I_s]_{sd}t^s\big] \oplus \mathfrak{m}\mathcal{R}(\mathfrak{F}), \tag{14}$$

so that $\mathcal{R}(\mathfrak{F})/\mathfrak{m}\mathcal{R}(\mathfrak{F}) \simeq k[[I_s]_{sd}t^s]$.

Theorem 2.3.26. *Suppose that the following conditions hold:*
(i) $\mathcal{R}(I)_{\mathfrak{m}\mathcal{R}(I)}$ *is regular.*
(ii) $\overline{\mathcal{R}(I)} \subset Q$ *is the Rees algebra of a balanced filtration.*

If $\mathcal{F}(I)$ is normal then $\mathcal{R}(I)$ is normal.

Proof. Start with the exact sequence of $\mathcal{R}(I)$-modules

$$0 \to \mathcal{R}(I) \longrightarrow \overline{\mathcal{R}(I)} \longrightarrow C \to 0. \tag{15}$$

Tensoring with $R/\mathfrak{m}$ over R yields the exact sequence

$$0 \to \mathcal{R}(I)/\mathfrak{m}\mathcal{R}(I) \longrightarrow \overline{\mathcal{R}(I)}/\mathfrak{m}\overline{\mathcal{R}(I)} \longrightarrow C/\mathfrak{m}C \to 0, \tag{16}$$

where the leftmost map is injective since $\mathcal{R}(I) \cap \mathfrak{m}\overline{\mathcal{R}(I)} = \mathfrak{m}\mathcal{R}(I)$. Indeed, let $a = a_0 + a_+ \in \mathcal{R}(I) \cap \mathfrak{m}\overline{\mathcal{R}(I)}$, with $a_0 \in k[I_d t]$ and $a_+ \in \mathfrak{m}\mathcal{R}(I)$. Obviously, $a_+ \in \mathfrak{m}\overline{\mathcal{R}(I)}$, hence $a_0 \in \mathfrak{m}\overline{\mathcal{R}(I)}$. By (ii), one has the splitting in (14) for $\overline{\mathcal{R}(I)}$. But $k[I_d t]$ is clearly contained in the k-algebra summand of (14). It follows that $a_0 = 0$.

Next, localizing the sequence (15 at the prime ideal $\mathfrak{m}\mathcal{R}(I)$, by (i) it follows that $C_{\mathfrak{m}\mathcal{R}(I)} = 0$. Therefore, localizing (16) at $\mathfrak{m}\mathcal{R}(I)$ yields

$$K = (\mathcal{R}(I)/\mathfrak{m}\mathcal{R}(I))_{\mathfrak{m}\mathcal{R}(I)} \simeq (\overline{\mathcal{R}(I)}/\mathfrak{m}\overline{\mathcal{R}(I)})_{\mathfrak{m}\mathcal{R}(I)}.$$

This means that the inclusion $\mathcal{R}(I)/\mathfrak{m}\mathcal{R}(I) \subset \overline{\mathcal{R}(I)}/\mathfrak{m}\overline{\mathcal{R}(I)}$ has the same field of fractions. Since it is integral and $\mathcal{F}(I) = \mathcal{R}(I)/\mathfrak{m}\mathcal{R}(I)$ is normal by assumption, then we have an equality. This implies that $C/\mathfrak{m}C = 0$; hence, $C = 0$ by Nakayama. □

Corollary 2.3.27. *Suppose that the following conditions hold:*
(i) *The rational map Φ defined by I is birational onto the image.*
(ii) *$\overline{\mathcal{R}(I)} \subset Q$ is the Rees algebra of a balanced filtration.*

If $\mathcal{F}(I)$ is normal then $\mathcal{R}(I)$ is normal.

Proof. Since Φ is birational, in particular $\dim \mathcal{F}(I) = \dim R$. Thus, the prime ideal $\mathfrak{m}\mathcal{R}(I)$ has height one, i. e., $\dim \mathcal{R}(I)_{\mathfrak{m}\mathcal{R}(I)} = 1$.

On the other hand, $\deg(\Phi) = e(\mathcal{R}(I)_{\mathfrak{m}\mathcal{R}(I)})$ (see [131, Proof of Theorem 5.3]), where $e(\_)$ stands for the graded Hilbert multiplicity. Since Φ is birational, this multiplicity is one. Since $\mathcal{R}(I)_{\mathfrak{m}\mathcal{R}(I)}$ is a domain, it must be regular. Thus, the result follows from the previous theorem. □

Remark 2.3.28. Clearly, the strong assumption in the last two results is the balancedness of the integral closure. Typically, for a Cremona transformation Φ, the fiber is trivially a polynomial ring, while the Rees algebra is seldom normal, thereby failing to be the Rees algebra of a balanced filtration. For a simplest example, consider the Jonquières transformation defined by the monomials x^4, x^3y, y^3z in $R = k[x,y,z]$. A computation with Normaliz (we thank R. Villarreal) gives $\overline{\mathcal{R}(I)} = R[It, x^2y^2zt]$, thus $\overline{\mathcal{R}(I)}$ lives in degree one, but has minimal generators of different degrees. And, in fact, one knew a priori that $\mathcal{R}(I)$ could not be normal since it is not Cohen–Macaulay in this degree.

2.3.3 Complements on the fiber cone

In this part, we establish some notable relations between the fiber cone and the associated graded ring in the graded setup.

Let $I \subset R$ be a homogeneous ideal in a standard graded ring R over a field and set $\mathfrak{m} := (R_+) \subset R$. The image of an element $f \in I$ in $I/I^2 \subset \mathrm{gr}_I(R)$ will be denoted by f^o. In addition, set $\mathrm{gr}_I(R)_+ := (I/I^2)\mathrm{gr}_I(R)$ for the ideal generated in positive degrees. If I is moreover generated by forms of the same degree, we say that it is *equigenerated*; in addition, if the common degree of these forms becomes relevant, we say d-equigenerated.

Similarly, the image of an element $f \in I$ in $I/\mathfrak{m}I \subset \mathcal{F}(I)$ will be denoted by f^*. However, in this case, the notation will become obsolete throughout due to the identification of graded k-algebras $\mathcal{F}(I) \simeq k[I_d] = \bigoplus_{n\geq 0}[I^n]_{nd} \subset R$ established in (2), whereby f_i^* gets identified with $f_i \in [I]_d \subset k[I_d]$.

Lemma 2.3.29. *Let R denote a standard graded ring over an infinite field k and let $I \subset R$ be a d-equigenerated homogeneous ideal. Let $J \subset I$ be a minimal reduction. One has:*
(i) *$\mathcal{F}(I) \simeq k[I_d]$ is Cohen–Macaulay if and only if $k[I_d]$ is a free $k[J_d]$-module via the natural extension $k[J_d] \subset k[I_d]$.*
(ii) *If $\mathcal{F}(I) \simeq k[I_d]$ is Cohen–Macaulay, then $\{1\} \cup B_1 \cup \cdots \cup B_{\mathfrak{r}-1}$ is a free basis of $k[I_d]$ over $k[J_d]$, where B_n is the lifting of a homogeneous basis of the k-vector space $[I^n/JI^{n-1}]_{nd}$ and $\mathfrak{r}$ denotes the reduction number of I.*

Proof. (i) This is a well-known argument, as applied to the graded Noether normalization $k[J_d] \subset k[I_d]$.

(ii) Since $JI^{\mathfrak{r}} = I^{\mathfrak{r}+1}$, it is clear that the stated set generates $k[I_d]$ as $k[J_d]$-module. By the Krull–Nakayama lemma, this set is a minimal set of generators; hence, it must be a free basis. □

Lemma 2.3.30. *With $\mathrm{gr}_I(R)_+$ as above, one has:*
(1) *If R is Cohen–Macaulay or a domain (equidimensional catenary suffices) then*

$$\mathrm{ht}\,\mathrm{gr}_I(R)_+ \leq \mathrm{ht}\,I.$$

(2) *If $\mathrm{gr}_I(R)$ is Cohen–Macaulay, then $\mathrm{ht}\,\mathrm{gr}_I(R)_+ \geq \mathrm{ht}\,I$.*
(3) *If both R and $\mathrm{gr}_I(R)$ are Cohen–Macaulay, then $\mathrm{ht}\,\mathrm{gr}_I(R)_+ = \mathrm{ht}\,I$.*

Proof. (1) and (2) come out from the isomorphism $\mathrm{gr}_I(R)/\mathrm{gr}_I(R)_+ \simeq R/I$, while (3) is the conjunction of (1) and (2). □

Lemma 2.3.31. *Let R denote a standard graded ring over an infinite field and let $I \subset R$ be a d-equigenerated homogeneous ideal of height g. Let $f_1,\ldots,f_g \in I_d$ be a maximal regular sequence in I such that*

$$(f_1,\ldots,f_g) \cap I^n = (f_1,\ldots,f_g)I^{n-1}, \tag{17}$$

for every $n \geq 1$. Then:
(i) *$f_1^*,\ldots,f_g^*$ is a regular sequence in $\mathcal{F}(I)$; in particular, $\mathrm{depth}\,\mathcal{F}(I) \geq g$.*
(ii) *$(f_1,\ldots,f_g)_P$ is a minimal reduction of I_P for every prime $P \supset I$ with $\mathrm{ht}(P) = \mathrm{ht}(I)$.*
(iii) *The sequence $f_1,\ldots,f_g$ can be extended to a sequence generating a minimal reduction of I.*

Proof. (i) Having identified $\mathcal{F}(I) \simeq k[I_d] = \bigoplus_{n\geq 0}[I^n]_{nd} \subset R$, the assertion is merely a property of the graded inclusion $k[I_d] \subset R$, whereby one has to check that, for $0 \leq i \leq$

$g-1$, f_{i+1} is a regular element on $\bigoplus_{n\geq 0}[I^n/(f_1,\ldots,f_i)I^{n-1}]_{nd}$. Thus, let there be given n and $h \in [I^n]_{nd}$ such that $hf_{i+1} \in (f_1,\ldots,f_i)[I^n]_{nd} \subset (f_1,\ldots,f_i)R$.

Since $\{f_1,\ldots,f_{i+1}\}$ is a regular sequence in R, then $h \in (f_1,\ldots,f_i)R$; hence, $h \in [(f_1,\ldots,f_i)R \cap I^n]_{nd}$. Then, by [212, Corollary 2.7], $h \in [(f_1,\ldots,f_i)I^{n-1}]_{nd}$, as was to be shown.

(ii) Localizing (17) at a prime $P \supset I$ gives

$$(f_1,\ldots,f_g)_P \cap I_P^n = (f_1,\ldots,f_g)_P I_P^{n-1}.$$

If, moreover, $\operatorname{ht} P = \operatorname{ht} I$ then, since $\operatorname{ht}(f_1,\ldots,f_g) = \operatorname{ht} I$, it follows that I_P is contained in the radical of $(f_1,\ldots,f_g)_P$. Therefore, $I_P^n \subset (f_1,\ldots,f_g)_P$ for some $n \gg 0$, and hence, $I_P^n = (f_1,\ldots,f_g)_P I_P^{n-1}$. This shows that $(f_1,\ldots,f_g)_P$ is a reduction of I_P; hence, a minimal one since any minimal reduction is minimally generated by $\ell(I_P) \geq \operatorname{ht} I_P = g$ elements.

(iii) Since k is infinite, and, by (i), $f_1,\ldots,f_g$ is a homogeneous regular sequence in $k[I_d]$, then it is a subset of a homogeneous system of parameters of $k[I_d]$. The latter generates a minimal reduction of the irrelevant maximal ideal of $k[I_d]$. By a well-known fact, this reduction lifts to a minimal reduction of I extending the R-regular sequence $f_1,\ldots,f_g$. □

By [212], one knows that the hypotheses of the previous lemma imply (actually, are equivalent to the assertion) that the residues $f_1^o,\ldots,f_g^o$ constitute a regular sequence in $\operatorname{gr}_I(R)$. The next lemma gives a condition on $\operatorname{gr}_I(R)$ for as to when such sequences exist at all.

Lemma 2.3.32. *Let R denote a standard graded ring over an infinite field and let $I \subset R$ be a d-equigenerated homogeneous ideal of height g. If the associated graded ring $\operatorname{gr}_I(R)$ is Cohen–Macaulay, then there exist forms $f_1,\ldots,f_g \in I_d$ such that $f_1^o,\ldots,f_g^o$ is a regular sequence in $\operatorname{gr}_I(R)$; if, in addition, R is Cohen–Macaulay, then $\operatorname{gr}_I(R)_+$ has grade g.*

Proof. It is basically a prime avoidance argument. To wit, letting $I = (\mathfrak{f}_1,\ldots,\mathfrak{f}_r)$, one has

$$\begin{aligned}\operatorname{ht}(\mathfrak{f}_1^o,\ldots,\mathfrak{f}_r^o) &= \dim \operatorname{gr}_I(R) - \dim \frac{\operatorname{gr}_I(R)}{(\mathfrak{f}_1^o,\ldots,\mathfrak{f}_r^o)}, \quad \text{since } \operatorname{gr}_I(R) \text{ is Cohen–Macaulay}\\ &= \dim R - \dim R/I \geq g.\end{aligned}$$

Now, since $\operatorname{gr}_I(R)$ is Cohen–Macaulay then $\operatorname{ht}\mathfrak{P} = 0$ for every associated prime $\mathfrak{P}$ of $\operatorname{gr}_I(R)$. Since $\operatorname{ht}(\mathfrak{f}_1^o,\ldots,\mathfrak{f}_r^o) \geq g > 0$ then $(\mathfrak{f}_1^o,\ldots,\mathfrak{f}_r^o)$ is not contained in any of these primes. By prime avoidance, a sufficiently general k-linear combination f_1^o of $\mathfrak{f}_1^o,\ldots,\mathfrak{f}_r^o$ is such that f_1^o is not contained in any $\mathfrak{P}$, hence is $\operatorname{gr}_I(R)$-regular. Let $f_1 \in I_d$ be any lifting of f_1^o.

The general inductive is similar, whereby one assumes the existence of $f_1,\ldots,f_i \in I_d$, for $1 \leq i \leq g-1$, such that $f_1^o,\ldots,f_i^o$ is a regular sequence in $\operatorname{gr}_I(R)$. Since

$\mathrm{gr}_I(R)/(f_1^o,\dots,f_i^o)$ is Cohen–Macaulay then again,

$$\mathrm{ht}(\mathfrak{f}_1^o,\dots,\mathfrak{f}_r^o)\mathrm{gr}_I(R)/(f_1^o,\dots,f_i^o) = g-1 \geq 1,$$

hence, by prime avoidance, there exists a form $f_{i+1} \in I_d$ such that f_{i+1}^o is regular on $\mathrm{gr}_I(R)/(f_1^o,\dots,f_i^o)$.

The supplementary assertion is a consequence of Lemma 2.3.30 (3). □

Although Lemma 2.3.32 claims the existence of *some* appropriate d-forms filling the request, it is clear from the proof that by choosing general forms in I_d the result is equally obtainable. However, the following example shows that even general forms may not work if $\mathrm{gr}_I(R)$ is not Cohen–Macaulay.

Example 2.3.33. Let $I = (z^6, yz^5, y^2z^4, xy^2z^3, x^2y^2z^2, x^3y^3) \subset R = k[x,y,z]$, a height two perfect ideal of analytic deviation one. A calculation with [79] gives that both $\mathrm{gr}_I(R)$ and $\mathcal{F}(I)$ are almost Cohen–Macaulay (i. e., both have depth 2). If $f_1, f_2 \in I_6$ are general forms, an additional computation outputs $(f_1,f_2)\cap I^2 \not\subset (f_1,f_2)I$ (alternatively, the grade of $\mathrm{gr}_I(R)_+$ is 1).

Corollary 2.3.34. *Let R denote a standard graded ring over an infinite field and let $I \subset R$ be a d-equigenerated homogeneous ideal. Suppose that I is equimultiple (i. e., $\mathrm{ht}\,I = \ell(I)$). If $\mathrm{gr}_I(R)$ is Cohen–Macaulay then so is the special fiber $\mathcal{F}(I)$.*

Remark 2.3.35. In the last corollary, if R is Cohen–Macaulay, the assumption that $\mathrm{gr}_I(R)$ is Cohen–Macaulay can be weakened by just assuming that there exist $\mathrm{ht}\,I$ forms in I_d whose images in $\mathrm{gr}_I(R)$ constitute a regular sequence.

2.3.4 Cohomological invariants

Let $(A,\mathfrak{m})$ denote a Noetherian local ring of dimension d and let I stand for a proper ideal of A. As before, $\mathcal{R}_A(I) := A[It] \subset A[t]$ is the Rees algebra of I over A and $\mathrm{gr}_I(A) := \mathcal{R}_A(I)/I\mathcal{R}_A(I)$ is the associated graded algebra of I. Let $S = A[y_1,\dots,y_r]$ denote a standard graded polynomial ring over $A = S_0$ and let $\mathcal{J}$ stand for a presentation ideal of $\mathcal{R}_A(I)$ over S. Clearly, then $\mathcal{R}_A(I) \simeq S/\mathcal{J}$ is a standard graded algebra over A, with graded maximal ideal $\mathfrak{N} := (\mathfrak{m}, (\mathcal{R}_A(I))_+)$, while $\mathcal{J}$ is homogeneous ideal in this grading.

The maximum degree of the elements of a minimal generating set of $\mathcal{J}$ is called the *relation type* of I. The relation type depends only on the ideal I and is independent of its generators. Similarly, the residue algebra $\mathrm{gr}_I(A)$ is a standard graded algebra over A/I, with graded maximal ideal $\mathfrak{M} := (\mathfrak{m}/I, \mathrm{gr}_I(A)_+)$.

We will be often interested in the case where $\mathcal{R}_A(I)$ is a Cohen–Macaulay ring. For this, recall more generally that if S is a finitely generated $\mathbb{N}$-graded algebra over a local ring $(A,\mathfrak{m})$ and its maximal ideal or a standard graded algebra over a field and its irrelevant ideal, with $S_0 = A$, then S is a Cohen–Macaulay ring if and only if the local ring $T := S_{(\mathfrak{m},S_+)}$ is Cohen–Macaulay ([106] or [21, Exercise 2.1.17]).

Clearly, then this happens if and only if $\dim T = \operatorname{depth}(T)$. Since $\dim S = \dim T$ and $\operatorname{depth}_{(\mathfrak{m},S_+)}(S) = \operatorname{depth}(T)$ if no confusion arises we will be talking about the dimension and the depth of S instead.

Now, quite more generally, let $R = \bigoplus_{n\geq 0} R_n$ be standard graded Noetherian ring over $(A,\mathfrak{m}) = R_0$, with graded maximal ideal $\mathfrak{N} := (\mathfrak{m}, R_+)$. For a graded R-module M, $\operatorname{indeg}(M) := \inf\{i : M_i \neq 0\}$ and $\operatorname{end}(M) := \sup\{i : M_i \neq 0\}$.

Let M be a finitely generated graded R-module.

Definition 2.3.36. The (strict) *Castelnuovo–Mumford regularity* (or, simply, the *regularity*) of M is the integer

$$\operatorname{reg}(M) := \max\{\operatorname{end}(H^i_{R_+}(M)) + i\}. \tag{18}$$

The ith *a-invariant* of M is the integer $a_i(M) := \operatorname{end}(H^i_{R_+}(M))$; for $i = \dim M$, one gets the more usual *a-invariant* of M:

$$a(M) := a_{\dim(M)}(M). \tag{19}$$

There are nonstrict versions with respect to the graded maximal ideal $\mathfrak{N}$ instead. To make a distinction, we append an asterisk to the respective designations. Thus, for example:

$$a_i^*(M) := \operatorname{end}(H^i_{\mathfrak{N}}(M)). \tag{20}$$

A major interest is when R is either the Rees algebra of an ideal $I \subset A$ or the corresponding associated graded ring. Moreover, if $(A,\mathfrak{m})$ happens to be itself standard graded over an infinite field and I is equigenerated, one can take, e. g., the regularity of $\mathcal{R}_A(I)$ with respect to the extended ideal $(\mathfrak{m}) = \mathfrak{m}\mathcal{R}_A(I)$. That is, $\operatorname{reg}_{\mathfrak{m}}(\mathcal{R}_R(I) := \max\{\operatorname{end}(H^i_{(\mathfrak{m})}(\mathcal{R}_R(I))) + i\}$, where the grading has been reset so as to have the degree one for the generators of $\mathfrak{m}$ and zero for the presentation generators of the Rees algebra. This slightly convoluted notion will be strongly used in Section 7.2.2.

The role of the Castelnuovo–Mumford regularity of a standard graded ring A over a field, denoted $\operatorname{reg}(A)$, can be hardly exaggerated—it is of course impossible to lift a reasonable list of references, but see [58] and [210] as "classical" ones, and [130], of a more recent vintage.

The following result seems to have gone unnoticed in this exact form.

Lemma 2.3.37. *Let A denote a standard graded domain over an algebraically closed field k such that* $\operatorname{reg}(A) \leq 1$. *Then either A is Cohen–Macaulay with minimal multiplicity or else* $\operatorname{depth} A \leq \dim A - 2$.

Proof. We may assume that $\operatorname{reg}(A) = 1$ as otherwise A is a polynomial ring over k. Then, writing $A = S/I$, where S is a polynomial ring over k, A has a 2-linear resolution over S. Suppose as if it were, that $\operatorname{depth} A = \dim A - 1$. Then, by [98, Theorem 1.5] one has

$$e(A) = \frac{p!}{(p-1)!} - \frac{b_p}{(p-1)!} < p,$$

where $e(A)$ denotes the multiplicity of A, p the homological dimension of A over S and b_p is the last Betti number of the resolution. On the other hand, $p = \operatorname{edim} A - \operatorname{depth} A$. Therefore, $e(A) < \operatorname{edim} A - \dim A + 1 = \operatorname{ecod} A + 1$. Since A is a domain over an algebraically closed field, this contradicts a well-known result [58]. It follows that A is Cohen–Macaulay and, necessarily, has minimal multiplicity. □

Example 2.3.38. The following well-known example shows that the above result does not hold in general if A is not a domain: $A = R/I$, where $I = (x_1, x_2) \cap (x_3, x_4) \subset R = k[x_1, x_2, x_3, x_4]$. Here, $\operatorname{depth} A = 1 = \dim A - 1$ and $\operatorname{reg} A = 1$.

Next is an application in our context. Recall that the *analytic deviation* of an ideal I (in a local or standard graded ring) is the difference $\ell(I) - \operatorname{ht} I$.

Corollary 2.3.39. *Let R denote a standard graded domain over an algebraically closed field and let $I \subset R$ be a d-equigenerated homogeneous ideal of analytic deviation one. If $\operatorname{gr}_I(R)$ is Cohen–Macaulay and has regularity at most one, then $\mathcal{F}(I)$ is Cohen–Macaulay with minimal multiplicity.*

Proof. By [152] (see also [210, Theorem 1.1]), $\operatorname{reg} \operatorname{gr}_I(R) = \operatorname{reg} \mathcal{R}(I)$, and by [165, Theorem 1.3 and Theorem 3.2] or [60, Section 1], $\operatorname{reg} \mathcal{F}(I) \le \operatorname{reg} \mathcal{R}(I)$. We may assume that $\operatorname{reg}\mathcal{F}(I) = 1$ as otherwise there is nothing to prove. Since $\dim \mathcal{F}(I) = \ell(I) = \operatorname{ht} I + 1$ by the analytic deviation assumption, and $\operatorname{depth} \mathcal{F}(I) \ge \operatorname{ht} I$, it follows that $\operatorname{depth} \mathcal{F}(I) \ge \dim \mathcal{F}(I) - 1$. On the other hand, as R is a domain and $\mathcal{F}(I) \simeq k[I_d t] \subset R[It] \subset R[t]$, then $\mathcal{F}(I)$ is a domain. Therefore, the result is a consequence of the previous lemma. □

2.4 Exercises

Exercise 2.4.1. Let $\mathcal{V} \subset \mathbb{P}^{n_1} \times \cdots \times \mathbb{P}^{n_m}$ be a correspondence, with the multihomogeneous coordinate ring $S = k[\mathbf{X}_1, \ldots, \mathbf{X}_m]/I(\mathcal{V})$. Show that the diagonal subalgebra S_Δ can be identified with the homogeneous coordinate ring of the image of $\mathcal{V}$ under the Segre embedding of $\mathbb{P}^{n_1} \times \cdots \times \mathbb{P}^{n_m}$. Can you generalize this result to the relative case (i. e., replacing the field k by any ring A)?

Exercise 2.4.2. Let $\operatorname{Aut}(S)$ stand for the group of k-linear automorphisms of the polynomial ring $S = k[\mathbf{X}]$ and let there be given $d_0, \ldots, d_n \in \mathbb{N}$. For each d_i, fix a primitive (d_i)th root of unity ζ_i. Consider the group homomorphism $\mathbb{Z}^n \to \operatorname{Aut}(S)$ that maps

$(0,\dots,0,\underbrace{1}_{i},0,\dots,0)$ to the following linear automorphism:

$$(X_0:\dots:X_i:\dots:X_n)\mapsto(X_0:\dots:\zeta_iX_i:\dots:X_n).$$

- Show that this homomorphism descends to a homomorphism

$$G:=\mathbb{Z}/d_0\mathbb{Z}\times\dots\times\mathbb{Z}/d_n\mathbb{Z}\to\mathrm{Aut}(S).$$

- Let $S^G\subset S$ denote the corresponding ring of polynomial invariants. Show that S^G and $(S^{(d_0,\dots,d_n)})^{\varphi}$ are isomorphic as $\mathbb{N}$-graded k-algebras (the corresponding geometric object is the quotient of $\mathbb{P}^n$ by the associated action of G—called *weighted projective space* and denoted $\mathbb{P}(d_0,\dots,d_n)$).
- Show that taking the sth Veronesean subalgebra as above corresponds to taking the s-uple embedding of $\mathbb{P}(d_0,\dots,d_n)$.

Exercise 2.4.3. Let $I=(x^2,xy,xz,yz)\subset R=k[x,y,z]$.
(1) Give the minimal free graded resolution of I.
(2) Determine the defining ideal of the fiber cone of I and show that the Rees algebra $\mathcal{R}_R(I)$ is of fiber type.
(3) Prove that $\mathcal{R}_R(I)$ is a Gorenstein ring of codimension 3.
(*Hint*: write an explicit alternate matrix whose maximal Pfaffians generate the presentation ideal of $\mathcal{R}_R(I)$ – note that if computer assistance is used, one may have to edit the result.)
(4) What is the (absolute) reduction number of I?

Exercise 2.4.4. Let $I=(x^2-y^2,xy,xz,yz)\subset R=k[x,y,z]$.
(1) Give the Betti numbers and the shifts of the minimal free graded resolution of I.
(2) Prove that $\mathcal{R}_R(I)$ is of fiber type, but not Cohen–Macaulay.
(3) What is the (absolute) reduction number of I?

Exercise 2.4.5. Consider the monomial ideal $I=(x,y)^3\cap(x,z)^3\subset R=k[x,y,z]$.
(1) Show that I is minimally generated by four monomials of degrees 3, 4, 5, 6, respectively.
(2) Argue that I does not satisfy property (F_0).
(3) Show directly that the toric k-subalgebra $A\subset R$ generated by the monomials in item (1) admits a presentation $A\simeq k[t,u,v,w]/\mathfrak{J}$, for a suitable grading of the polynomial ring $k[t,u,v,w]$ and where $\mathfrak{J}$ is the ideal of 2-minors of a 2×3 Hankel matrix.
(4) Give a presentation ideal of the (weighted) Rees algebra $\mathcal{R}_R(I)$.
(*Hint*: this should come out by an easy manipulation of the syzygies of I.)
(5) Deduce that the fiber cone of I is isomorphic to A as weighted k-algebras.
(6) Is the symmetric algebra $\mathcal{S}_{R/I}(I/I^2)$ torsion-free over R/I? What would be a reasonable version of the Rees algebra of the conormal module I/I^2 over R/I?

(*Hint*: note that I/I^2 has no rank over R/I.)

Exercise 2.4.6. Let $I \subset R = k[x_0, x_1, x_2, x_3]$ denote the homogeneous defining ideal of the nonnormal rational quartic curve.

(1) Show that I is generated by the minimal generators among the 2-minors of the matrix

$$\begin{pmatrix} x_0 & x_2 & x_1^2 & x_1x_3 \\ x_1 & x_3 & x_0x_2 & x_2^2 \end{pmatrix}.$$

(2) Prove that the above generators of I are not analytically independent. (*Hint*: consider the Plücker equation associated to the matrix in (1).)
(3) Deduce that the analytic spread of I is 3.
(4) ($\mathrm{char}(k) = 0$) Show that the dimension of the k-subalgebra of R generated by the generators of I is 4. (*Hint*: compute the determinant of the Jacobian matrix of the generators.)
(5) Show, by via of the hint in the previous item, that the Jacobian determinant is divisible by the form $x_0x_3 - x_1x_2 \in I$. What is the geometric meaning of this phenomenon?

Exercise 2.4.7. Let $\widetilde{R}$ be an equidimensional Noetherian local ring and let $\widetilde{J} \subset \widetilde{R}$ be an ideal for which question 2.3.5 is affirmative. Let there be given elements $a_1, \ldots, a_n$ of $\widetilde{R}$ such that:

- $a_1, \ldots, a_i$ belong to $\widetilde{J}$.
- $\{a_{i+1}, \ldots, a_n\}$ is a regular sequence on $\widetilde{R}/\widetilde{J}$.
- $\mathrm{ht}(a_1, \ldots, a_n) = n$ (i. e., $(a_1, \ldots, a_n)$ is an ideal of the principal class).

Write $R = \widetilde{R}/(a_1, \ldots, a_n)$ and $J = \widetilde{J}R$. Prove: if $\widetilde{J}$ is locally generated by analytically independent elements then J is locally generated by analytically independent elements if and only if it satisfies property (F_1).

(*Hint*: By inducting on n, assume that $n = 1$ and $a_1 = a$, an element either in $\widetilde{J}$ or else regular modulo $\widetilde{J}$. Localize at a prime ideal $\wp$ and rename ring and ideal accordingly. Let $L \subsetneq J$ be an ideal. Let $\widetilde{L} \subset \widetilde{R}$ denote the ideal generated by any liftings of a set of minimal generators of L. Without loss of generality, assume that $\mu(\widetilde{L}) = \mu(L)$ and that $\widetilde{L}$ is a proper ideal of $\widetilde{R}$. Case 1: $a \in \widetilde{J}$. Apply the conjectured result of Question 2.3.5 to the "tilde" environment yielding $\mathrm{ht}((\widetilde{L}, a){:}\widetilde{J}) \leq \mu(L) + 1$, and obviously $L{:}J = ((\widetilde{L}, a){:}\widetilde{J})/(a)$; hence, $\mathrm{ht}(L{:}J) \leq \mu(L)$. Case 2: a is regular on $\widetilde{R}/\widetilde{J}$. By a similar token, $\mathrm{ht}(\widetilde{L}{:}\widetilde{J}) \leq \mu(L)$, and $\sqrt{L{:}J} = \sqrt{(\widetilde{L}{:}\widetilde{J}, a)/(a)}$ by [117, Lemma 4.1]. Hence again, $\mathrm{ht}(L{:}J) \leq \mu(L)$. Now apply Proposition 2.3.3.)

Exercise 2.4.8. Given $f := x^2y + y^2z + xz^2 \in k[x,y,z]$ and $\mathfrak{m} := (x,y,z)/(f) \subset R := k[x,y,z]/(f)$, let $\mathcal{R}_R(\mathfrak{m})$ be the Rees algebra of $\mathfrak{m}$, with presentation ideal $\mathcal{J}$ over the polynomial ring $R[t,u,v]$.

(1) Determine a set of minimal generators of $\mathcal{J}$.
(2) Prove that $\mathcal{R}_R(\mathfrak{m})$ is normal, but not Cohen–Macaulay.
(3) Provide a generalization of the above for any form $f \in k[x,y,z]$ defining a smooth projective plane curve.

Exercise 2.4.9 (Algebraic side of the Segre rational map and its graph). Let $\{x_1,\dots,x_n\}$ and $\{y_1,\dots,y_m\}$ two sets of mutually independent indeterminates over a field k and let $I = (x_1,\dots,x_n)(y_1,\dots,y_m) \subset R := k[x_1,\dots,x_n;y_1,\dots,y_m]$.

(1) Show that depth $R/I = 1$.
(2) Show that the fiber cone of I is defined over k by the 2-minors of an $n \times m$ generic matrix.
(3) Is I an ideal of fiber type?

Exercise 2.4.10. Let $I \subset R = k[x_0,x_1,x_2,x_3]$ denote the homogeneous defining ideal of the nonnormal rational quartic curve, as in the previous exercise. Let $\{y_0,y_1,y_2,y_3\}$ be new variables and consider the grading of the polynomial ring $R[y_0,y_1,y_2,y_3]$ by giving weights 1, 1, 1, 1, 1, 2, 2, 2 to x_0, x_1, x_2, x_3, y_0, y_1, y_2, y_3, respectively.

(1) Prove that a presentation ideal $\mathcal{I} \subset R[y_0,y_1,y_2,y_3]$ of the symmetric algebra $\mathcal{S}_R(I)$ is generated by four forms of weighted degree 3.
(2) Show that the four generators in the previous item can be taken to be four of the five Pfaffians of the skew-symmetric matrix

$$\begin{pmatrix} 0 & -x_1 & -x_3 & x_2 & -x_0 \\ x_1 & 0 & -x_2y_0 & -y_3 & -y_2 \\ x_3 & x_2y_0 & 0 & -y_1 & y_3 \\ -x_2 & y_3 & y_1 & 0 & x_1y_0 \\ x_0 & y_2 & -y_3 & -x_1y_0 & 0 \end{pmatrix}.$$

(3) Show that the fifth Pfaffian of the matrix in the previous item is a relation of the generators of I having standard $\mathbf{y}$-degree 2.
(*Hint*: get a handle to item (2) of the previous exercise.)
(4) Prove that the ideal $\mathcal{J} \subset R[y_0,y_1,y_2,y_3]$ generated by the five Pfaffians is a presentation ideal of the Rees algebra $\mathcal{R}_R(I)$.
(*Hint*: show that $\mathcal{J}$ is normal (e. g., by looking at its Jacobian matrix), hence prime.)

Exercise 2.4.11. (This is the sketch of a project, intended as a partial generalization of the two previous exercises. For the full structural results, see [143, Theorem 4.2].) Consider the rational curve of $\mathbb{P}^3 = \mathrm{Proj}(k[x_0,x_1,x_2,x_3])$ parameterized by the binary monomials $u^{b+c} : u^b v^c : u^c v^b : v^{b+c}$, where b, c are integers such that $b > c \geq 1$ and $\gcd(b,c) = 1$. Let $P \subset R := k[x_0, x_1, x_2, x_3]$ stand for its homogeneous defining ideal. Consider the following matrix:

$$\mathcal{M} = \begin{pmatrix} x_0 & x_2 & x_1^{b-1} & x_3x_1^{b-2} & \cdots & x_3^{b-c-2}x_1^{c+1} & x_3^{b-c-1}x_1^{c} \\ x_1 & x_3 & x_0^{b-c-1}x_2^{c} & x_0^{b-c-2}x_2^{c+1} & \cdots & x_0x_2^{b-2} & x_2^{b-1} \end{pmatrix},$$

and let $Q : I_2(\mathcal{M}) \subset k[x_0, x_1, x_2, x_3]$.

(1) Prove that $Q \subset P$ and that Q is minimally generated by the 2-minors fixing the first columns and the minor of second and last columns.

(2) Prove that $Q = P$ locally at the powers of every x_i.
(*Hint*: argue that the following are syzygies of the minimal generators of Q:

$$\begin{matrix} x_1^{b-1} & x_1^{b-2}x_3 & \cdots & x_1^{c}x_3^{b-c-1} & x_0^{b-c-1}x_2^{c} & \cdots & x_0x_2^{b-2} & x_2^{b-1} \\ -x_2 & 0 & \cdots & 0 & -x_3 & \cdots & 0 & 0 \\ x_0 & -x_2 & \cdots & 0 & x_1 & \cdots & 0 & 0 \\ 0 & x_0 & \cdots & 0 & 0 & \cdots & 0 & 0 \\ \vdots & \vdots & \ddots & \vdots & \vdots & \ddots & \vdots & \vdots \\ 0 & 0 & \cdots & -x_2 & 0 & \cdots & x_1 & -x_3 \\ 0 & 0 & \cdots & x_0 & 0 & \cdots & 0 & x_1 \end{matrix}$$

.)

(3) Deduce that P is the only associated prime of Q other than $\mathfrak{m} = (x_0, x_1, x_2, x_3)$.

(4) Prove that depth $R/Q = 1$.
(*Hint*: mount a free resolution of R/Q drawing upon the syzygies in the hint for of the previous item.)

(5) Deduce that $Q = P$.

3 Rational maps, I

3.1 Basics of rational maps

The goal of this part is to underpin the algebraic side of the theory of rational maps. The reader interested in a more geometric, i. e., scheme theoretic approach is referred to [51, Lecture 1, Section 1.1]. For rational maps with source a projective space over the field of complexes, we refer to the beautiful differential-like preliminaries of [80, Appendix A (b)].

3.1.1 Terminology

Let k be an arbitrary field. For the purpose of the full geometric picture, we may have to assume k to be algebraically closed. However, k will have arbitrary characteristic except for a few clearly marked passages. We denote by $\mathbb{P}^n = \mathbb{P}^n_k$ the nth projective space with homogeneous coordinates $(a_0 : \cdots : a_n)$ over k. Recall that $\mathbb{P}^n$ is the quotient space of $\mathbb{A}^{n+1} \setminus \{\mathbf{0}\}$ by the homothetical action.

Classical rational maps have to do with projective varieties over a fixed ground field. The "relative" problem where the ground coefficients lie on a Noetherian domain will be considered in Section 4.4.

Let $X \subset \mathbb{P}^n$ be a projective subvariety, assumed throughout this section to be integral (i. e., reduced and irreducible or, equivalently, its largest defining ideal $I(X) \subset k[\mathbf{x}] = k[x_0, \dots, x_n]$ is prime). Thus, X admits a *function field* $k(X)$ and one has the main dimension theorem of these preliminaries, namely, $\dim X = \operatorname{tr.deg}_{\cdot k} k(X) = \dim k[\mathbf{x}]/I(X) - 1$, where $\dim X$ is a combinatorial-like dimension defined in terms of lengths of chains of subvarieties.

Let $\mathbb{P}^m$ be yet another projective space. We wish to consider rational maps with "source" X and target $\mathbb{P}^m$. The quotation signs indicate that there will be some difficulties as to what will be exactly the realm of definition of these maps.

Of course, the terminology comes from having these maps defined in terms of rational functions on the domain, i. e., fractions of polynomial functions thereof. As is known, there are two different fields around in the theory of projective varieties: the "big" field of fractions of the homogeneous coordinate ring $k[X] = k[\mathbf{x}]/I(X)$ of X and the "small" field of fractions of a suitable affine open set of X. While the first gives perfectly defined regular functions on the structural affine cone over X, it is the latter that is called the function field of X because it makes sense to say for its elements that they induce locally well-defined functions on X—while the elements of $k[X]$ do not. The notation $k(X)$ for the function field is highly incoherent with the previous notation $k[X]$ but we will leave it at that since the go-between the two fields is well known from elementary courses.

We first put somewhat naively the following.

https://doi.org/10.1515/9783110640694-003

Definition 3.1.1. A *rational datum* on X is a collection of elements $f_1/f_0, \ldots, f_m/f_0 \in k(X)$, where $f_0, f_1, \ldots, f_m \in k[X]$ are forms of the same degree, with $f_0 \neq 0$.

In a slightly imprecise way, the above datum induces a map with target $\mathbb{A}^m \subset \mathbb{P}^m$ a certain subset of closed points of X as source, namely, given $p = (a_0 : \cdots : a_n) \in X$, with $a_i \in k$, such that $f_0(p) = f_0(a_0, \ldots, a_n) \neq 0$, the image of p is the point with affine coordinates $(f_1(p)/f_0(p), \ldots, f_m(p)/f_0(p)) \in \mathbb{A}_k^m$.

Remark 3.1.2. If one thinks of points in the language of Weil's school, the generic point of X is $P := (\bar{x}_0 : \cdots : \bar{x}_n)$, where $\bar{x}_i$ denotes the residue of the variable x_i in $k[X]$. In this version, the image of the generic point is the "affine point"

$$(f_1(P)/f_0(P), \ldots, f_m(P)/f_0(P)) = (f_1/f_0, \ldots, f_m/f_0).$$

Thus, the coordinates of the image of the generic point of X generate a k-subfield of $k(X)$ and can be thought of as the generic point of a subvariety whose affine coordinate ring is $k[f_1/f_0, \ldots, f_m/f_0]$.

The map thus obtained is called a *rational map*, denoted $F: X \dashrightarrow \mathbb{P}^m$, where the dotted tail of the arrow has the purpose to remind us of the partial domain of the map.

3.1.2 Basic properties

The following facts highlight the main features of the concept:

- The finitely generated k-domain $k[f_1/f_0, \ldots, f_m/f_0] \subset k(f_1/f_0, \ldots, f_m/f_0) \subset k(X)$ is the affine coordinate ring of a uniquely defined subvariety of $\mathbb{A}^m \subset \mathbb{P}^m$. The closure of the latter $Y \subset \mathbb{P}^m$ is called the *image* of the rational map F. By abuse, we also say that the rational map F is defined by the homogeneous datum $f_0, f_1, \ldots, f_m$ and frequently use the notation $F = (f_0 : f_1 : \ldots : f_m)$, which as we will indicate below conveys the lack of uniqueness of choosing these forms.
- We emphasize en passant that the image of a rational map on an integral projective subvariety is an integral projective subvariety of the target. As such, its homogeneous coordinate ring is isomorphic to the subring $k[f_0, \ldots, f_m] \subset k[X]$ as graded k-algebras–we say that one obtains the first from the second by a degree renormalization. Thus, although the dealing with rational maps is rather subtle, the image of such maps is a rather readily understood object in algebraic terms.
- One says that F is *birational onto its image* if there exists a rational map $G: Y \dashrightarrow \mathbb{P}^n$ whose image is X such that F and G are inverses to each other as field maps, i. e., if they define a k-isomorphism of fields $k(X) \simeq k(Y)$ and its inverse.
- One can also express the latter behavior at the level of the homogeneous data, namely, if $F = (f_0 : f_1 : \ldots : f_m)$ and $G = (g_0 : g_1 : \ldots : g_n)$ then F is birational

onto its image with inverse map G if and only if the following relations are satisfied:

$$(f_0(g_0,g_1,\dots,g_n):\cdots:f_m(g_0,g_1,\dots,g_n)) = (\overline{y_0}:\cdots:\overline{y_m})$$

and

$$(g_0(f_0,f_1,\dots,f_m):\cdots:g_n(f_0,f_1,\dots,f_m)) \equiv (\overline{x_0}:\cdots:\overline{x_n})$$

as tuples of homogeneous coordinates in $\mathbb{P}^m_{K(Y)}$ (resp., $\mathbb{P}^n_{K(X)}$) where $K(Y)$ is the field of fractions of $k[Y]$ (resp., of $k[X]$).

- If d is the common degree of $f_0,f_1,\dots,f_m$, it is not difficult to see that this is the case if and only if the natural inclusion $k[f_0,f_1,\dots,f_m] \subset k[k[X]_d] = k[(\mathbf{x})_d]/I(X)_d$ is an equality at the level of the respective fields of fractions.
- There is nothing unique about the choice of $f_0,f_1,\dots,f_m$, as one learns quite early in a first course. Nevertheless, any other choice of such a representative of the map F, say, $f'_0,f'_1,\dots,f'_m$, necessarily satisfies the condition that the 2×2 minors of the matrix

$$\begin{pmatrix} f_0 & f_1 & \cdots & f_m \\ f'_0 & f'_1 & \cdots & f'_m \end{pmatrix}$$

vanish (as elements of the homogeneous coordinate ring $k[X]$ of X). A trivial remark as it is it represents a notable clarification in the theory. We believe it first explicitly appeared in a paper of Zariski in an issue of the American J. Math from the early 1940s, the precise reference of which we failed to confirm. The condition also means that the above matrix has rank 1 over the field of fractions $K(X)$ of $k[X]$, i. e., the two row vectors are proportional over this field. This justifies the previous notation $F = (f_0 : \cdots : f_m)$ (rigorously, an element of $\mathbb{P}^m_{K(X)}$).
- When considering the *base locus* of a rational map there is a potential confusion as to whether one is looking at the projective scheme defined by the forms $\mathbf{f}$ or more accurately at the homogeneous ideal generated by these forms—of course, for the set theoretic version of this locus one may as well go all the way to the radical of the ideal. However, since saturation is a truly ideal theoretically operation one may loose track of the nature of the subalgebra $k[f_0,\dots,f_m]$, which should be left intact as it defines the image of the given map. Also, we will be using the subalgebra and the defining ring of the blowup on the same foot; hence, one had better keep the Rees algebra intact in terms of a minimal set of generators of a presentation ideal thereof.

Quite a bit of the nature of the statements and their proofs appeal to the syzygies of an ideal representative of F; however, the very format of the conditions does not directly involve them. This is because the condition for birationality rather requires dwelling on the "hidden side" of the Rees algebra, as will be made more precise.

It is all more remarkable, as we will see, that the theory is in terms of a certain Jacobian matrix with respect to variables that only appear linearly in the pertinent equations, thereby saving us from any a priori restriction on the field characteristic, certainly a non-negligible point in birational geometry as is well known.

3.2 A characteristic-free criterion of birationality

3.2.1 Prolegomena: the Koszul–Hilbert lemma

The following elementary piece of linear algebra seems to have gone unnoticed before its appearance in [53].

Proposition 3.2.1. *Let A be a commutative ring and n a positive integer. Let ρ stand for an $(n+1)\times n$ matrix with arbitrary entries in A. Then the Koszul relations of the (ordered, signed) n-minors of ρ belong to the submodule*

$$\mathrm{Im}(\rho) \subset A^{n+1}.$$

Proof. Let $B = A[\mathbf{X}]$ be a polynomial ring in $(n+1)\times n$ indeterminates and let $\mathcal{X}$ denote the corresponding $(n+1)\times n$ matrix whose entries are $\mathbf{X}$. Let D denote the prime subring of A, i. e., the image of the natural map $\mathbb{Z} \to A$. Consider the matrix $\mathcal{X}$ over $D[\mathbf{X}]$.

Claim. The ideal $I_n(\mathcal{X})$ of n-minors in $D[\mathbf{X}]$ has grade ≥ 2.

First, any n-minor Δ thereof is a regular element since its coefficients are ± 1. By a similar token, any other n-minor is a regular element modulo Δ.

As a consequence, the well-known Hilbert–Burch principle says that the sequence

$$0 \to D[\mathbf{X}]^n \xrightarrow{\mathcal{X}} D[\mathbf{X}]^{n+1} \xrightarrow{\Delta} D[\mathbf{X}]$$

is exact, where one denotes the minors over $D[\mathbf{X}]$ by the same letters as those over $A[\mathbf{X}]$. Clearly, then the Koszul relations of $\boldsymbol{\Delta}$ belong to the image of the map $\mathcal{X}$ as this is exactly the matrix of first syzygies of $\boldsymbol{\Delta}$.

Now specialize the variables $\mathbf{X}$ to the entries of ρ by means of the obvious homomorphism $D[\mathbf{X}] \longrightarrow A$ extending the inclusion $D \subset A$. Since the formation of determinants and the Koszul complex specialize, it is clear that the specialization preserves the above relation. □

Remark 3.2.2. Note that one is not asserting that ρ coincides with the matrix of syzygies of its n-minors, which is only true if these minors generate an ideal of grade at least 2 as well (as follows from the Hilbert–Burch theorem [179, Theorem 6.2.32]).

3.2.2 Preliminaries on rational maps on reduced varieties

We first fix our standing setup and notation.

Let k denote an arbitrary ground field. As usual, having in mind a meaningful geometric background, in order that a projective set of zeros have enough points one allows coordinates to lie on an algebraically closed field containing k.

Let $R = k[\mathbf{x}] = k[x_0,\dots,x_n] \simeq k[\mathbf{X}]/\mathfrak{a} = k[X_0,\dots,X_n]/\mathfrak{a}$ denote a standard graded k-algebra over k, with $\dim R \geq 1$. In the sequel, we always write $\mathbf{X}$ and $\mathbf{x}$ for $X_0,\dots,X_n$ and $x_0,\dots,x_n$, respectively. The same notation principle will apply to other sets of variables as needed.

3.2.2.1 Rational data and rational maps

Let $K(R)$ denote the homogeneous total ring of quotients of R, i. e., the ring of fractions $\mathfrak{M}^{-1}R$, where $\mathfrak{M}$ is the set of homogeneous regular elements of R not belonging to any of the minimal primes of R.

We will be dealing with ordered $(m+1)$-sets $\mathbf{f} = \{f_0,\dots,f_m\}$ of forms of the same degree in R. Let us say that two such ordered $(m+1)$-sets $\mathbf{f} = \{f_0,\dots,f_m\}$ and $\mathbf{f}' = \{f'_0,\dots,f'_m\}$, are *equivalent* if the corresponding tuples $(f_0,\dots,f_m)$ and $(f'_0,\dots,f'_m)$ are proportional by means of a homogeneous invertible element of $K(R)$. This means that there exist regular homogeneous elements $F, F' \in R$ such that $Ff'_j = F'f_j$, for $j = 0,\dots,m$. This is readily seen to be an equivalence relation.

Note that equivalent ordered homogeneous $(m+1)$-sets $\mathbf{f}, \mathbf{f}' \subset R$ induce a graded k-isomorphism of the respective representative algebras $k[\mathbf{f}]$ and $k[\mathbf{f}']$, such that $f_j \longleftrightarrow f'_j$.

Definition 3.2.3. A *rational* $(m+1)$*-datum of degree* d on R is an ordered set of $m+1$ forms $\mathbf{f} = \{f_0,\dots,f_m\} \subset R$ of the same degree ≥ 1 satisfying the following conditions:
(i) R is torsion-free over $k[\mathbf{f}]$ via the ring extension $k[\mathbf{f}] \subset R$.
(ii) The ideal $(\mathbf{f})R \subset R$ is not contained in any minimal prime of R.

The first condition reads that every minimal prime of R contracts to a minimal prime of $k[\mathbf{f}]$, while the second says that the ideal $(\mathbf{f}) \subset R$ has a regular element. Obviously, both conditions are trivially met if R is a domain.

Definition 3.2.4. An *m-targeted rational map* with source $\mathrm{Proj}(R) \subset \mathbb{P}^n_k$ is a collection of equivalent rational $(m+1)$-data.

Geometrically, choosing a representative yields a rational map

$$\mathrm{Proj}(R) \dashrightarrow \mathbb{P}^m,$$

where the arrow indicates that the map is not defined everywhere.

Lemma 3.2.5. *If k is an infinite field, two homogeneous $(m+1)$-sets $\mathbf{f}, \mathbf{f}' \subset R$ satisfying condition* (ii) *of the above definition are equivalent if and only if the matrix*

$$\begin{pmatrix} f_0 & f_1 & \dots & f_m \\ f'_0 & f'_1 & \dots & f'_m \end{pmatrix}$$

has rank 1.

Proof. Indeed, the rank condition means that the 2×2 minors vanish (which is readily implied by equivalence as defined) and that the ideal generated by the entries has a regular element (which is obviously the case if one of the ideals generated by the rows has a regular element). To see the converse, since k is infinite one can choose suitable $\lambda_j \in k$ such that both $\lambda_0 f_0 + \dots + \lambda_m f_m$ and $\lambda_0 f'_0 + \dots + \lambda_m f'_m$ are regular elements. By hypothesis, $f_j f'_i = f_i f'_j$ for all pairs i, j. Therefore, for arbitrary i, one has $F' f_i = F f'_i$, where $F = \lambda_0 f_0 + \dots + \lambda_m f_m$ and $F' = \lambda_0 f'_0 + \dots + \lambda_m f'_m$. □

Any individual rational $(m+1)$-datum is said to be a *representative* of the map, and in an informal speech, we say that it defines the rational map. Here, and subsequently, we will omit the piece "m-targeted" if the dimension of the target is clear from the context. To cut on notation, we often trade the notation $\mathfrak{F} : \mathrm{Proj}(R) \dashrightarrow \mathbb{P}^m$, for $\mathfrak{F} : \mathbf{f} \subset R$, so as to enhance a chosen representative. This abuse perpetration of identifying the map with one representative will cause no confusion.

Remark 3.2.6. As remarked above, equivalent rational $(m+1)$-data $\mathbf{f}, \mathbf{f}' \subset R$ induce a k-isomorphism of the respective representative algebras $k[\mathbf{f}]$ and $k[\mathbf{f}']$, such that $f_j \longleftrightarrow f'_j$. Although this isomorphism is graded, it is not homogeneous (of degree 0). Nevertheless, one can renormalize the gradings of all these algebras so they become standard k-algebras. What this actually means is that one can take a common polynomial representation $k[\mathbf{Y}]/\mathfrak{b} \simeq k[\mathbf{f}]$ for all the representatives of the same rational map, with $k[\mathbf{Y}]/\mathfrak{b}$ standard.

3.2.2.2 The base ideal

The ideal $(\mathbf{f}) \subset R$ generated by a representative $\mathbf{f} \subset R$ is called a *base ideal* of the map. Clearly, the rational datum conveys more information than the base ideal in the sense that the ideal can be generated by lesser forms. However, as will be seen, properties of the ideal give a bona fide picture of the corresponding rational map. At the level of base ideals, two representatives $\mathbf{f}$ and $\mathbf{f}'$ of the same map give an isomorphism of $(\mathbf{f})$ and $(\mathbf{f}')$ as R-submodules of $K(R)$, i. e., as fractional ideals.

The notion of a representative or of a base ideal is pretty stable in the following sense.

Lemma 3.2.7. *Let $\mathbf{f} = \{f_0, \dots, f_m\}$ and $\mathbf{f}' = \{f'_0, \dots, f'_m\}$ be equivalent ordered sets in R. Then $\mathbf{f}$ is a rational datum if and only if $\mathbf{f}'$ is a rational datum.*

Proof. By the obvious symmetry in an argument, it suffices to show one implication. Thus, assume that $\mathbf{f}$ satisfies the conditions stated in Definition 3.2.3. As noted above, $k[\mathbf{f}] \simeq k[\mathbf{f}']$, so there is a bijection between the respective sets of minimal primes. This shows that condition (i) of Definition 3.2.3 holds for $\mathbf{f}'$. Since the k-isomorphism induces an isomorphism of the ideals $(\mathbf{f}), (\mathbf{f}') \subset R$ as R-modules (also remarked before), condition (ii) follows as well. □

3.2.2.3 Total rings of quotients

Recall that the *homogeneous total rings of quotients* $K(R)$ of a graded rings such as R is the ring of fractions of R by the multiplicatively closed subset of homogeneous regular elements of R. Note that $K(R)$ is naturally graded as well; when R is a domain, its degree zero part $K(R)_0$ gives the usual function field of the projective variety $\mathrm{Proj}(R)$.

Let $S = k[\mathbf{f}]$. Since $S \hookrightarrow R$ is torsion-free, it induces the following maps:

(1) An injective graded homomorphism $K(S) \hookrightarrow K(R)$ of degree $d-1$ of the respective homogeneous total rings of quotients.
(2) An injective *homogeneous* homomorphism (i. e., a graded homomorphism of degree zero) $K(S) \hookrightarrow K(R^{(d)})$, where $R^{(d)} \subset R$ is the dth Veronese subring of R.
(3) An injective (nongraded) homomorphism $K(S)_0 \hookrightarrow K(R)_0$ in degree zero.

Note that, by definition, $K(R^{(d)})_0 = K(R)_0$, so passing to the Veronese in degree zero is tantamount to doing it "on the nose" over R.

3.2.2.4 Composition of maps

Two such objects can be "composed" in the following sense.

Lemma 3.2.8. *Let $R = k[\mathbf{X}]/\mathfrak{a}$ be a reduced standard graded k-algebra and let $\mathfrak{F} : \mathbf{f} \subset R$ be a rational $(m+1)$-datum of degree d on R. Take a presentation $S := k[\mathbf{Y}]/\mathfrak{b} \simeq k[\mathbf{f}]$, with $\mathbf{Y} = \{y_0, \dots, y_m\}$ variables over k. Let $\mathbf{g} := \{g_0, \dots, g_r\} \subset S$ denote a rational $(r+1)$-datum of degree e on S. Then:*

(i) *$\mathbf{g}(\mathbf{f}) := \{g_0(\mathbf{f}), \dots, g_r(\mathbf{f})\}$ is a rational $(r+1)$-datum of degree de on R.*
(ii) *If $\mathbf{f}' \subset R$ and $\mathbf{g}' \subset S$ are representatives of $\mathbf{f}$ and $\mathbf{g}$, respectively, then $\mathbf{g}'(\mathbf{f}') := \{g_0'(\mathbf{f}'), \dots, g_r'(\mathbf{f}')\}$ is a representative of the rational map defined by $\mathbf{g}(\mathbf{f})$.*

Proof. (i) By assumption, R is torsion-free over $k[\mathbf{f}] \simeq S$ and S is torsion-free over $k[\mathbf{g}]$. Therefore, R is torsion-free over $k[\mathbf{g}]$. Likewise, by assumption, $(\mathbf{g})S$ is not contained in any minimal prime of S, i. e., its image $\mathbf{g}(\mathbf{f})k[\mathbf{f}]$ via the isomorphism $S \simeq k[\mathbf{f}]$ is not contained in any minimal prime of $k[\mathbf{f}]$. Therefore, by torsion-freeness of R over $k[\mathbf{f}]$, the ideal $(\mathbf{g}(\mathbf{f}))R$ is not contained in any minimal prime of R.

(ii) Let $S = k[\mathbf{Y}]/\mathfrak{b}$ be a common presentation of the algebras $k[\mathbf{f}]$ and $k[\mathbf{f}']$. By assumption, there exist $F, F' \in R$ homogeneous such that $Ff_j' = F'f_j$, for $j = 0, \dots, m$ and, likewise, $G = G(\mathbf{y})$, $G' = G'(\mathbf{y}) \subset S$ homogeneous such that $G(\mathbf{y})g_\ell'(\mathbf{y}) = G'(\mathbf{y})g_\ell(\mathbf{y})$, for

$\ell = 0,\dots,r$. We have appended $\mathbf{y}$ to the latter to emphasize that they are elements of $S = k[\mathbf{Y}]/\mathfrak{b}$. Using the isomorphism $k[\mathbf{Y}]/\mathfrak{b} \simeq k[\mathbf{f}']$, for instance, one has

$$G(\mathbf{f}')g'_\ell(\mathbf{f}') = G'(\mathbf{f}')g_\ell(\mathbf{f}'). \tag{21}$$

Substituting for $f'_j = (F'/F)f_j$, $j = 0,\dots,m$ in $g_\ell(\mathbf{f}')$ and pulling out the obvious factor yields $F^e \cdot g_\ell(\mathbf{f}') = F'^e \cdot g_\ell(\mathbf{f})$, where e stands for the common degrees of the g_ℓ's. Multiplying by F^e both members of (21) and bringing into it the last equality gives

$$(F^e\, G(\mathbf{f}'))g'_\ell(\mathbf{f}') = G'(\mathbf{f}')(F^e\, g_\ell(\mathbf{f}')) = (G'(\mathbf{f}')\, F'^e)g_\ell(\mathbf{f}),$$

for $\ell = 0,\dots,r$. Now, $G(\mathbf{f}')$ and $G'(\mathbf{f}')$ are regular (homogeneous) elements of $k[\mathbf{f}]$ and $k[\mathbf{f}']$, respectively, as they are both images of regular (homogeneous) elements of $S = k[\mathbf{Y}]/\mathfrak{b}$ under a graded k-isomorphism of algebras. Since R is torsion-free over $k[\mathbf{f}']$ they are still regular (homogeneous) elements of R. Therefore, $F^e\, G(\mathbf{f}')$ and $G'(\mathbf{f}')\, F'^e$ are regular homogeneous elements of R and we are done.

Caveat. As a slight control, if d (resp., d') and e (resp., e') are the respective degrees of f_j (resp., f'_j) and of g_ℓ (resp., g'_ℓ) and, likewise D (resp., D') and E (resp., E') are the respective degrees of F (resp., F') and of G (resp., G'), then

$$\begin{aligned}\deg\,(G'(\mathbf{f}')\, F'^e) + \deg(g_\ell(\mathbf{f})) &= d'E' + eD' + de = d'(E + e' - e) + e(D + d' - d) + de\\ &= eD + d'E + d'e' = \deg\,(F^e\, G(\mathbf{f}')) + \deg(g'_\ell(\mathbf{f}')).\end{aligned}$$ □

3.2.2.5 Image of a rational map

We have seen that all representatives of a rational map have a common presentation $S = k[\mathbf{Y}]/\mathfrak{b}$, a standard graded reduced k-algebra. In the geometric terminology, $\mathrm{Proj}(S) \subset \mathbb{P}^m_k$ the *image* of this map.

3.2.2.6 Restriction of a map

One can introduce an additional notion, that of a restriction of a rational map $\mathfrak{F} : \mathbf{f} \subset R$. Given a radical ideal $I \subset R$ such that $(\mathbf{f})$ is not contained in any minimal prime of R/I on R and such that R/I is torsion-free over $k[\mathbf{f}]/I \cap k[\mathbf{f}]$, then we say that the map *restricts* to $\mathrm{Proj}(R/I)$. Thus, being able to restrict a rational map requires quite a bit—note that, if I is a prime ideal, then torsion-freeness of $k[\mathbf{f}]/I \cap k[\mathbf{f}] \subset R/I$ is trivially satisfied, so we just have to worry about the condition on the relative minimal primes, which is unavoidable in the classical geometric setup.

There is one easy situation where restriction works as expected.

Lemma 3.2.9. *Let $\mathfrak{F} : \mathbf{f} \subset R$ be a rational map with source* $\mathrm{Proj}(R)$.

(1) *If $P \subset R$ is a minimal prime of R, then the map restricts to* $\mathrm{Proj}(R/P)$ *and the image of the restriction is* $\mathrm{Proj}(k[\mathbf{f}]/P \cap k[\mathbf{f}])$.

(2) *Conversely, if* $\wp \subset k[\mathbf{f}]$ *is a minimal prime, then* $\mathrm{Proj}(k[\mathbf{f}]/\wp)$ *is the image of the restriction of* $\mathbf{f}$ *to* R/P *for some minimal prime* $P \subset R$.

Proof. (1) This follows immediately from the properties of the rational map.

(2) It suffices to show that $\wp$ is contracted from some prime $\mathcal{P} \subset R_+$, as in this case a minimal prime of R contained in $\mathcal{P}$ will contract to $\wp$ as well. For that, it suffices to show that $(\wp)R$ is not an irrelevant ideal in R. But if that were the case, one would have $\wp = (\mathbf{f})k[\mathbf{f}]$, which is impossible since $k[\mathbf{f}]$ is a reduced ring of positive dimension (e. g., because $(\mathbf{f})$ has a regular element contained in $k[\mathbf{f}]$, see Remark 3.2.5). □

The lemma allows many arguments to be reduced to the case where R is a integral domain. Of course, this has to be used judiciously.

Remark 3.2.10 (Geometric transcription). Suppose that k is algebraically closed. By a version of the projective Nullstellensatz, (2) implies that for every $P \in \mathrm{Min}(R)$ there exists some homogeneous *submaximal* ideal (i. e., generated by $n-1$ linearly independent 1-forms in $\mathbf{x}$) $\mathfrak{n} \subset R$ containing P but not $(\mathbf{f})$. For such an $\mathfrak{n}$, the contraction $\mathfrak{n} \cap k[\mathbf{f}] \subset k[\mathbf{f}]$ is an ideal generated by $m-1$ linearly independent 1-forms in $\mathbf{f}$. That is to say, if we write $k[\mathbf{Y}]/\mathfrak{b} \simeq k[\mathbf{f}]$, then $\mathfrak{n} \cap k[\mathbf{f}]$ is the residual image of a homogeneous submaximal ideal in $k[\mathbf{Y}]$. Geometrically, this implies that $f_0(\mathbf{a}), \dots, f_m(\mathbf{a})$ makes sense as a point in $\mathbb{P}^m$ for at least one point $\mathbf{a}$ on any given irreducible component of the projective subset $V \subset \mathbb{P}^n$ whose homogeneous coordinate ring is R. Therefore, on every such component there is a nonempty open subset where $\mathbf{f}$ affords a well-defined map.

In particular, the so obtained map $\mathfrak{F}$ is well defined on any open set $\mathcal{U} \subset V$ that intercepts nonemptily every irreducible component. The image of $\mathfrak{F}$ is the closure $W \subset \mathbb{P}^m$ of the set of images of points from any such $\mathcal{U}$. This is the projective set of $\mathbb{P}^m$ whose homogeneous coordinate ring is $k[\mathbf{Y}]/\mathfrak{b}$, i. e., after the obvious degree renormalization, $k[\mathbf{f}]$ itself.

3.2.2.7 Birational maps and "field" extension criteria

The simplest instance of a rational map is $\mathfrak{F} : \mathbf{x} \subset R$, represented by the residues in R of the variables $\{X_0, \dots, X_n\}$—called *the identity* rational map of $\mathrm{Proj}(R)$. The closest rational maps to the identity are the so-called birational maps. We give a formal definition.

Definition 3.2.11. An m-targeted rational map $\mathbf{f} \subset R = k[X_0, \dots, X_n]/\mathfrak{a}$ of degree d is said to be *birational* (*onto the image*) if there exists an n-targeted rational map $\mathfrak{G} : \mathbf{g} \subset S = k[Y_0, \dots, Y_m]/\mathfrak{b} \simeq k[\mathbf{f}]$ such that the composite $\mathbf{g}(\mathbf{f}) \subset R$ is equivalent to the identity on $\mathrm{Proj}(R)$.

The rational map $\mathfrak{G}$ is called an *inverse* to $\mathfrak{F}$.

The classical field theoretic criterion can be extended via the following formulation.

Proposition 3.2.12. *Let $\mathfrak{F} : \mathbf{f} \subset R = k[X_0, \ldots, X_n]/\mathfrak{a}$ denote an m-targeted rational map of degree d. The following conditions are equivalent:*

(a) *$\mathfrak{F} : \mathbf{f} \subset R$ is birational onto the image.*

(b) *For every minimal prime P of R, the induced m-targeted rational map*

$$\mathfrak{F}|P : \bar{\mathbf{f}} \subset \bar{R} = R/P$$

is birational onto its image, and, moreover, R/P is the unique minimal component of R mapping to the latter by $\mathfrak{F}$.

(c) *The induced homomorphism in degree zero $K(S)_0 \hookrightarrow K(R)_0$ is an isomorphism.*

(d) *The induced homogeneous homomorphism $K(S) \hookrightarrow K(R^{(d)})$ is an isomorphism.*

Proof. The equivalence of (c) and (d) is fairly known; see, e. g., [190, Proof of Theorem 6.6] or [195, Proof of Proposition 2.1] for the argument in a more special setup. Here, we trade the nonhomogeneous total ring of quotients for the homogeneous one, but the reasoning is the same, namely, $K(S) = K(S)_0(f)$ and $K(R^{(d)}) = K(R^{(d)})_0(f) = K(R)_0(f)$ for some regular 1-form $f \in S$ or, up to identification $S = k[\mathbf{f}]$, a suitable k-linear combination of $f_0, \ldots, f_m$. Since the maps in the statement are natural, the equivalence is clear.

Thus, we are left with the mutual equivalence of (a), (b) and (c).

We note that, quite generally, if R is any reduced Noetherian ring, its ordinary total ring of quotients decomposes as the direct product of the respective fields of fractions of the domains R/P, where P runs through the minimal primes of R. If, moreover, R is standard graded over a field k (possibly even more generally) then, by restriction, this decomposition implies a similar decomposition of the *homogeneous* total ring of fractions $K(R) = \prod_{P\in \mathrm{Min}(R)} K(R/P)$. Clearly, this decomposition induces one in degree zero as well:

$$K(R)_0 = \prod_{P\in \mathrm{Min}(R)} K(R/P)_0, \tag{22}$$

where the factors are now fields.

(a) ⇒ (b) This piece follows from the first statement of Lemma 3.2.9.

(b) ⇒ (c) By the classical field criterion in the domain case, one knows that he induced homomorphism in degree zero $K(S/P \cap S)_0 \hookrightarrow K(R/P)_0$ is an isomorphism for every minimal prime P of R. This is well known, but for the sake of completion, we repeat the argument. In fact, it is now easy to work directly with coordinates. Updating the notation, assume that $R = k[\mathbf{x}]$ is a domain. Then one has

$$\begin{aligned} K(S)_0 = k(\mathbf{y})_0 &= k\left(\frac{y_1}{y_0}, \ldots, \frac{y_m}{y_0}\right) \simeq k(\mathbf{f})_0 \\ &= k\left(\frac{f_1}{f_0}, \ldots, \frac{f_m}{f_0}\right) \subset k(\mathbf{x})_0 = k\left(\frac{x_1}{x_0}, \ldots, \frac{x_n}{x_0}\right), \end{aligned}$$

where we assume $y_0 \neq 0$, $x_0 \neq 0$ without lost of generality. By assumption, there are nonzero (homogeneous) $H, H' \in R$ such that $H \cdot g_\ell(\mathbf{f}) = H' \cdot x_\ell$, for $\ell = 0, \ldots, n$. Then

$$\frac{x_\ell}{x_0} = \frac{g_\ell(\mathbf{f})}{g_0(\mathbf{f})} = \frac{f_0^e(g_\ell(f_1/f_0, \ldots, f_m/f_0))}{f_0^e(g_0(f_1/f_0, \ldots, f_m/f_0))} = \frac{g_\ell(f_1/f_0, \ldots, f_m/f_0)}{g_0(f_1/f_0, \ldots, f_m/f_0)}, \quad \ell = 0, \ldots, n,$$

thus showing the required equality $k(f_1/f_0, \ldots, f_m/f_0) = k(x_1/x_0, \ldots, x_n/x_0)$.

Next, note that the map $K(S/P \cap S)_0 \hookrightarrow K(R/P)_0$ is naturally induced by the map $K(S)_0 \hookrightarrow K(R)_0$.

Then, up to identification, we only need to show that if $K(S/S \cap P)_0 = K(R/P)_0$ for every $P \in \operatorname{Min}(R)$, then $K(S)_0 = K(R)_0$. To see this, note that the hypothesis implies that, for every $\wp \in \operatorname{Min}(S)$, exactly one prime $P \in \operatorname{Min}(R)$ contracts to $\wp$, i. e., the map $\operatorname{Min}(R) \to \operatorname{Min}(S)$ is surjective (hence, bijective by the second statement of Lemma 3.2.9). This is due to the fact that in the injection $K(S)_0 \hookrightarrow K(R)_0$ the decomposition (22) forces the restriction to an individual factor $K(S/\wp)$ to map diagonally to the product of the factors of $K(R/P)_0$, one for each $P \in \operatorname{Min}(R)$ contracting to $\wp$. By the same token, this implies the sought global isomorphism $K(S)_0 \simeq K(R)_0$.

(c) $\Rightarrow$ (a) (Assuming k infinite) Fix a regular homogenous element $x \in (\mathbf{x}) = R_+$ of the form $x_0 + \sum_{\ell=1}^n \lambda_\ell x_\ell$, for suitable $\lambda_\ell \in k$. For $\ell = 1, \ldots, n$, let $g_\ell(\mathbf{y})/g(\mathbf{y}) \in K(S)_0$ denote the (uniquely defined) inverse image of x_ℓ/x under the isomorphism $K(S)_0 \simeq K(R)_0$, for suitable homogeneous elements $g(\mathbf{y}), g_\ell(\mathbf{y}) \in (\mathbf{y}) = S_+$, with $g(\mathbf{y})$ regular. Since the isomorphism $K(S)_0 \simeq K(R)_0$ is induced by $y \mapsto f_j$, one has $g_\ell(\mathbf{f})/g(\mathbf{f}) = x_\ell/x$, for $\ell = 1, \ldots, n$. Thus, $x \cdot g_\ell(\mathbf{f}) = g(\mathbf{f}) \cdot x_\ell$, for $\ell = 1, \ldots, n$, with x, $g(\mathbf{f})$ regular homogeneous elements in R. Now set $g_0(\mathbf{y}) := g(\mathbf{y}) - \sum_{\ell=1}^n \lambda_\ell g_\ell(\mathbf{y})$. Then

$$\begin{aligned} g(\mathbf{f})x_0 &= g(\mathbf{f})\left(x - \sum_{\ell=1}^n \lambda_\ell x_\ell\right) = xg(\mathbf{f}) - \sum_{\ell=1}^n \lambda_\ell (g(\mathbf{f})x_\ell) = xg(\mathbf{f}) - \sum_{\ell=1}^n \lambda_\ell (xg_\ell(\mathbf{f})) \\ &= x\left(g(\mathbf{f}) - \sum_{\ell=1}^n \lambda_\ell g_\ell(\mathbf{f})\right) = x \cdot g_0(\mathbf{f}). \end{aligned}$$

Finally, note that $g_0(\mathbf{y}), \ldots, g_n(\mathbf{y}) \subset S = k[\mathbf{y}]$ is a rational $(n+1)$-datum by Lemma 3.2.7, hence defines an n-targeted rational map. Thus, the rational $(n+1)$-data $g_0(\mathbf{f}), \ldots, g_n(\mathbf{f})$ and $x_0, \ldots, x_n$ are equivalent, as was to be shown. □

Corollary 3.2.13. *Let $\mathfrak{F} : \mathbf{f} \subset R = k[X_0, \ldots, X_n]/\mathfrak{a}$ denote an m-targeted rational map which is birational onto its image, with inverse $\mathfrak{G} : \mathbf{g} \subset S = k[Y_0, \ldots, Y_m]/\mathfrak{b} \simeq k[\mathbf{f}]$. Then the composite $\mathbf{f}(\mathbf{g}) \subset S$ is equivalent to the identity on* Proj(S)*. Consequently, the inverse is uniquely defined.*

Proof. The first statement follows from the equivalence of (a) and (c) in the previous theorem. The second statement follows from the first in the usual way, namely writing

$\equiv$ for equivalence, if $\mathbf{g}'$ is another inverse, we have

$$\mathbf{x} \equiv \mathbf{g}(\mathbf{f}) \Rightarrow \mathbf{g}' \equiv \mathbf{x}(\mathbf{g}') \equiv \mathbf{g}(\mathbf{f})(\mathbf{g}') \equiv \mathbf{g}(\mathbf{f}(\mathbf{g}')) \equiv \mathbf{g}(\mathbf{y}) \equiv \mathbf{g}. \qquad \square$$

The corollary means, as it should, that a birational map on Proj(R) onto its image Proj(S) has a unique "two-sided" inverse and the inverse is a birational map on Proj(S) onto Proj(R).

3.2.2.8 Relation to the graph

Recall the notion of the Rees algebra of an ideal $I \subset R$ from Section 2.3.1, denotes $\mathcal{R}_R(I)$. The next result relates the Rees algebras of respective base ideals of a birational map and its inverse.

Proposition 3.2.14. *Let $\mathfrak{F}$ denote a rational map with source* Proj(R), *where* $R = k[X_0, \ldots, X_n]/\mathfrak{a}$. *Suppose that $\mathfrak{F}$ is birational onto* Proj(S), *where* $S = k[Y_0, \ldots, Y_m]/\mathfrak{b}$, *and let $\mathfrak{G}$ denote the inverse map. Then the identity map of $k[\mathbf{X}, \mathbf{Y}]/(\mathfrak{a}, \mathfrak{b})$ induces a bigraded isomorphism of k-algebras $\mathcal{R}_R(I) \simeq \mathcal{R}_S(J)$, where $I \subset R$ and $J \subset S$ are any given base ideals of $\mathfrak{F}$ and $\mathfrak{G}$, respectively.*

Proof. Take equigenerated sets of forms $\mathfrak{f} \subset k[\mathbf{X}]$ and $\mathfrak{g} \subset k[\mathbf{Y}]$ whose residues $\mathbf{f} \in R$ and $\mathbf{g} \in S$ are respective representatives of $\mathfrak{F}$ and $\mathfrak{G}$.

As a preliminary, we note that $\mathcal{R}_R((\mathbf{f}, \mathfrak{a})/\mathfrak{a})$ (resp. $\mathcal{R}_S((\mathbf{g}, \mathfrak{b})/\mathfrak{b})$) contains a copy of $k[\mathbf{y}]/\mathfrak{b}$ (respectively, of $k[\mathbf{x}]/\mathfrak{a}$) as a k-subalgebra generated in bidegree $(0,1)$ (resp., $(1,0)$). In particular, both algebras are residue algebras of the standard bigraded k-algebra $k[\mathbf{X}, \mathbf{Y}]/(\mathfrak{a}, \mathfrak{b})$.

As such, we have

$$\mathcal{R}_R((\mathbf{f})) = \frac{k[\mathbf{X}, \mathbf{Y}]/(\mathfrak{a}, \mathfrak{b})}{\mathcal{J}_{\mathbf{f}}} \simeq \frac{k[\mathbf{X}]}{\mathfrak{a}}[\mathbf{f}t], \quad \text{by } x_i \mapsto x_i (\mathrm{mod}\ \mathfrak{a}),\ y_j \mapsto f_j t$$

and, similarly,

$$\mathcal{R}_R((\mathbf{g})) = \frac{k[\mathbf{X}, \mathbf{Y}]/(\mathfrak{a}, \mathfrak{b})}{\mathcal{J}_{\mathbf{g}}} \simeq \frac{k[\mathbf{Y}]}{\mathfrak{b}}[\mathbf{g}u], \quad \text{by } y_i \mapsto y_i (\mathrm{mod}\ \mathfrak{b}),\ x_i \mapsto g_i u$$

Claim. Under the identity map of $k[\mathbf{X}, \mathbf{Y}]/(\mathfrak{a}, \mathfrak{b})$, the ideal $\mathcal{J}_{\mathbf{f}}$ maps to $\mathcal{J}_{\mathbf{g}}$ and, conversely.

To see this, let $F(\mathbf{X}, \mathbf{Y}) \in k[\mathbf{X}, \mathbf{Y}]$ be bihomogeneous such that $F \in \mathcal{J}_{\mathbf{f}}\ (\mathrm{mod}\ (\mathfrak{a}, \mathfrak{b}))$, i. e., $F(\mathbf{X}, \mathbf{f}t) \equiv 0 \pmod{\mathfrak{a}} k[\mathbf{X}, t]$.

Say, $F(\mathbf{X}, \mathbf{f}t) = \sum_l h_l(\mathbf{X}) b_l(\mathbf{X}, t)$, with $h_l(\mathbf{X}) \in \mathfrak{a}$ homogeneous and $b_l(\mathbf{X}, t)$ bihomogeneous in $\mathbf{X}$, t. If we evaluate $F(\mathbf{X}, \mathbf{f}t)$ under the homomorphism $k[\mathbf{X}, t] \to k[\mathbf{Y}, u, t]$ that maps $x_i \mapsto g_i u$ and $t \mapsto t$, we find that $F(\mathbf{g}u, \mathbf{f}(\mathbf{g}u)t) = \sum_l h_l(\mathbf{g}u) b_l(\mathbf{g}u, t)$. But since $k[\mathbf{X}]/\mathfrak{a} \simeq k[\mathbf{g}] \subset k[\mathbf{Y}]/\mathfrak{b}$, it follows that $h_l(\mathbf{g}u) = u^{\deg(h_l)} h_l(\mathbf{g}) \in \mathfrak{b} k[\mathbf{Y}, u]$. Therefore, $F(\mathbf{g}u, \mathbf{f}(\mathbf{g}u)t) \in \mathfrak{b} k[\mathbf{Y}, u, t]$. Pulling out t again by $\mathbf{Y}$-homogeneity, we find $F(\mathbf{g}u, \mathbf{f}(\mathbf{g}u)) \in$

$\mathfrak{b}k[\mathbf{Y},u]$. Clearly, then $F(\mathbf{g}u, g\,\mathbf{f}(\mathbf{g}u)) \in \mathfrak{b}k[\mathbf{Y},u]$ for any form $g \in k[\mathbf{Y}]$ On the other hand, by the third fact above, there are forms $g, g' \in k[\mathbf{Y}]$ of the same degree such that $g'\,\mathbf{Y} = g\,\mathbf{f}(\mathbf{g})$ as ordered tuples. Substituting, we find $F(\mathbf{g}u, g'\,\mathbf{Y}) \in \mathfrak{b}k[\mathbf{Y},u]$ and, once more by $\mathbf{Y}$-homogeneity, $F(\mathbf{g}u, \mathbf{Y}) \in \mathfrak{b}k[\mathbf{Y},u]$. Therefore, $F(\mathbf{g}u, \mathbf{Y}) \in \mathcal{J}_{\mathbf{g}}$, as was to be shown.

The proof of the reverse inclusion is entirely analogous. These induce k-algebra homomorphisms between the respective algebras $(k[\mathbf{X}]/\mathfrak{a})[\mathbf{f}t]$ and $(k[\mathbf{Y}]/\mathfrak{b})[\mathbf{g}t]$ that are inverses of each other by construction. Clearly, by construction, the maps preserve the bigrading in either direction. □

3.2.3 The Jacobian dual rank versus birationality

Let us review the present algebraic tool for birationality that goes in parallel to the geometric notion of the linear fibers of a rational map. The concept has evolved continuously from its early inception, of which the cradle is the Jacobian dual fibration of [187], and the subsequent [166] and [174].

3.2.3.1 Definition and background

Let $\mathfrak{F} : \mathbf{f} \subset R = k[\mathbf{X}]/\mathfrak{a}$ be an m-targeted rational map with source Proj(R), where $k[\mathbf{X}]$ is standard graded over k, with $\dim R \geq 1$. Let $R[\mathbf{Y}] = R[Y_0, \ldots, Y_m]$ be a polynomial ring over R with the standard bigrading where $\deg(x_i) = (1,0)$ and $\deg(Y_j) = (0,1)$. Since $\mathbf{f}$ is generated in a fixed degree, the Rees algebra $\mathcal{R}_R((\mathbf{f})) \simeq R[It] \subset R[t]$ is a standard bigraded k-algebra. Mapping $Y_j \mapsto f_j t$ yields a presentation $R[\mathbf{Y}]/\mathcal{J} \simeq \mathcal{R}_R((\mathbf{f}))$, with $\mathcal{J}$ a bihomogeneous presentation ideal.

Just as the image of a rational map, one realizes that $\mathcal{J}$ depends only on the rational map and not on any of its representatives. Write

$$\mathcal{J} = \bigoplus_{(p,q)\in\mathbb{N}^2} \mathcal{J}_{p,q},$$

where $\mathcal{J}_{p,q}$ denotes the k-vector subspace of forms of bidegree (p,q).

First, focus on $\mathcal{J}_{0,*}$, which is spanned by the forms of bidegree $(0,q)$, for all $q \geq 1$. Since these forms have coefficients only in k, one can view them as elements of $k[\mathbf{Y}]$. As such they generate the ideal of all polynomial relations (over k) of $\mathbf{f}$. Thus,

$$k[\mathbf{Y}]/(\mathcal{J}_{0,*})k[\mathbf{Y}] \simeq k[\mathbf{f}],$$

as standard graded k-algebras, after a degree renormalization in $k[\mathbf{f}]$. Thus, the ideal $(\mathcal{J}_{0,*})k[\mathbf{Y}]$ recovers the ideal $\mathfrak{b}$ of the image brought up in the previous section. We write $S = k[\mathbf{Y}]/(\mathcal{J}_{0,*})k[\mathbf{Y}] \simeq k[\mathbf{f}]$ throughout, thinking interchangeably of S as one or the other.

For birationality, the following bihomogeneous piece is no less important:

$$\mathcal{J}_{1,*} := \bigoplus_{r\in\mathbb{N}} \mathcal{J}_{1,q}$$

with $\mathcal{J}_{1,q}$ denoting the bigraded piece of $\mathcal{J}$ spanned by the forms of bidegree $(1,q)$ for all $q \geq 0$. Note that $\mathcal{J}_{1,*} = \{0\}$ may very well be the case in general.

Now, a form of bidegree $(1,*)$ can be written as $\sum_{i=0}^{n} Q_i(\mathbf{Y})\, x_i$, for suitable homogeneous $Q_i(\mathbf{Y}) \in k[\mathbf{Y}] \subset R[\mathbf{Y}]$ of the same degree. Since $\mathbf{Y}$ are indeterminates over R, two such representations of the same form imply a syzygy of $\{x_0,\dots,x_n\}$ with coefficients in k. Thus, the representation is unique up to k-linear dependency of $\{x_0,\dots,x_n\}$, i. e., up to elements of $\mathfrak{a}_1$.

Recall that the *embedding dimension* of R is the k-vector space dimension $\operatorname{edim}(R) := \dim_k(R_1) = n+1-\dim_k \mathfrak{a}_1$. If $\operatorname{edim} R = n+1$ then every such form as above has a unique expression.

Pick a minimal set of generators of the ideal $(\mathcal{J}_{1,*})$ consisting of a finite number of forms of bidegree $(1,q)$, for various q's. Let $\{P_1,\dots,P_s\} \subset k[\mathbf{X},\mathbf{Y}]$ denote liftings of these biforms and let $\{\ell_1,\dots,\ell_r\} \subset k[\mathbf{X}]$ be a k-vector space basis of $\mathfrak{a}_1$ and consider the Jacobian matrix Ψ of the polynomials $\{\ell_1,\dots,\ell_r,P_1,\dots,P_s\}$ with respect to $\mathbf{X}$, a matrix with entries in $k[\mathbf{Y}]$.

Definition 3.2.15. The *weak Jacobian matrix* associated to the given set of generators of $(\mathcal{J}_{1,*})$ is the matrix Ψ considered as a matrix over the residual ring $S = k[\mathbf{Y}]/\mathfrak{b}$.

To avoid confusion, we will denote the latter matrix by small ψ.

Remark 3.2.16. (1) We will be concerned with the rank of ψ. Note that ψ has a distinguished $(n+1-\operatorname{edim}(R)) \times (n+1)$ submatrix ρ with entries in $k[\mathbf{Y}]_0 = k$, which is nonsingular. Its number of rows $r = n+1-\operatorname{edim}(R)$ can be dubbed the *degeneration index* $\operatorname{dgi}(\mathfrak{a})$ of the presentation $R \simeq k[\mathbf{X}]/\mathfrak{a}$. Clearly, this index is a lower bound for the rank of ψ over S whenever this rank is well-defined.

(2) In addition, by definition one clearly has $\mathbf{X} \cdot \psi^t = 0$ over $\mathcal{R}_R(\mathbf{f})$, in particular the vector $(\mathbf{X})^t$ belongs to the null space of ψ over $\mathcal{R}_R(\mathbf{f})$. If R is a domain, then the rank of ψ over $\mathcal{R}_R(\mathbf{f})$ is at most n; hence, for even more reason also over the residue ring $\mathcal{R}_R((\mathbf{f})) \twoheadrightarrow S$. A consequence of Proposition 3.2.19 below will show that this upper bound holds true even if R is not integral, provided ψ still has a rank.

One seemingly drawback of a weak Jacobian matrix ψ is that it is not uniquely defined by a given representative of the rational map, due to the lack of uniqueness in the expression of an individual form and to the choice of bihomogeneous generators. Actually, it can be fairly "slack," in the sense that some of its columns may be zero—actually the whole matrix itself can turn out to be the zero matrix for that matter.

On the bright side however, one has the following.

Lemma 3.2.17. *If the weak Jacobian matrix associated to a set of bihomogeneous minimal generators of* $(\mathcal{J}_{1,*})$ *has a rank over S, then the weak Jacobian matrix associated to any other set of bihomogeneous minimal generators of* $(\mathcal{J}_{1,*})$ *has a rank over S and the two ranks coincide.*

Proof. One shows, namely, that two such matrices have the same column submodule over S.

Let $\{P_1,\dots,P_s\},\{Q_1,\dots,Q_s\}\subset k[\mathbf{X},\mathbf{Y}]$ denote liftings of two sets of minimal biforms of various bidegrees $(1,q)$, for suitable $q\geq 1$, generating $(\mathcal{J}_{1,*})$. Note that though the sets may be different, the set of corresponding bidegrees is uniquely defined up to reordering of generators. Say, $\deg(P_t)=\deg(Q_t)=(1,d_t)$, with $1\leq t\leq s$ and $1\leq d_1\leq d_2\leq\cdots\leq d_s$. Therefore, for $t=1,\dots,s$ we may write

$$Q_t=\sum_{j=1}^{s}\lambda_{tj}p_{tj}(\mathbf{Y})P_j+\sum_{|\alpha_t|=d_t}a_{\alpha_t}(\mathbf{X})\,\mathbf{Y}^{\alpha_t},$$

for suitable $\lambda_{tj}\in k$, $p_{tj}(\mathbf{Y})\in k[\mathbf{Y}]$ of degree $d_t-\deg_{\mathbf{Y}}(P_j)$ and $a_{\alpha_t}(\mathbf{X})\in\mathfrak{a}_1$. Setting $a_{\alpha_t}(\mathbf{X})=\sum_{k=1}^{r}\mu_{tk}\ell_k(\mathbf{X})$, we obtain

$$\frac{\partial Q_t}{\partial X_i}=\sum_{j=1}^{s}\lambda_{tj}p_{tj}(\mathbf{Y})\frac{\partial P_j}{\partial X_i}+\sum_{|\alpha_t|=d_t}\left(\sum_{k=1}^{r}\mu_{tk}\,\mathbf{Y}^{\alpha_t}\frac{\partial\ell_k(\mathbf{X})}{\partial X_i}\right)$$

This shows that the tth column of the Jacobian matrix of $\{\ell_1,\dots,\ell_r,Q_1,\dots,Q_s\}$ is contained in the column $k[\mathbf{Y}]$-submodule of the Jacobian matrix of $\{\ell_1,\dots,\ell_r,P_1,\dots,P_s\}$. By symmetry, the two column $k[\mathbf{Y}]$-submodules coincide, hence the S-submodule $\operatorname{Im}(\psi)$ through any one of the two sets of generators is the same, as was to be shown. □

Because of the previous lemma, we say that any weak Jacobian matrix associated to a set of minimal generators of $(\mathcal{J}_{1,*})$ is a *Jacobian dual matrix* of $\mathbf{f}$ (or of the corresponding m-targeted rational map $\mathfrak{F}$). The lemma unveils a numerical character of such a construction.

Definition 3.2.18. If some (hence, every) Jacobian dual matrix of an m-targeted rational map $\mathfrak{F}:\mathbf{f}\subset R\simeq k[\mathbf{X}]/\mathfrak{a}$ has a rank over S, we will say that $\mathfrak{F}$ has a *Jacobian dual rank* and write $\operatorname{jdrank}(\mathfrak{F})$ to denote it. In this situation, the *nondegenerate Jacobian dual rank* of $\mathfrak{F}$ is $\operatorname{jdrank}_+(\mathfrak{F}):=\operatorname{jdrank}(\mathfrak{F})-\operatorname{dgi}(\mathfrak{a})\geq 0$.

Note that the nondegenerate Jacobian dual rank is well-defined, provided the Jacobian dual rank is.

The next result shows that the rank just introduced is sensitive to the dimension difference between source and target.

Proposition 3.2.19. *Let $\mathfrak{F} : \mathbf{f} \subset R$ stand for an m-targeted rational map. Set $S = k[\mathbf{Y}]/\mathfrak{b} \simeq k[\mathbf{f}] \subset R$. If $\mathfrak{F}$ has a Jacobian dual rank then*

$$\dim R - \dim S \leq n - \operatorname{jdrank}(\mathfrak{F}) = \operatorname{edim}(R) - 1 - \operatorname{jdrank}_+(\mathfrak{F}).$$

Proof. Up to a nonsingular k-linear map on R, one may assume that the forms in $\mathbf{f}$ involve effectively at most the first $r+1 = \operatorname{edim}(R)$ elements $x_0, \ldots, x_r$. More precisely, let

$$A = \left(\begin{array}{c|c} \mathbf{1}_{r+1} & \mathbf{a} = (a_{h,i}) \\ \hline & \mathbf{1}_{n-r} \end{array}\right)$$

be an invertible k-matrix such that $(X_0, \ldots, X_n) \cdot A = (X_0, \ldots, X_r, \ell_{r+1}, \ldots, \ell_n)$, with $\ell_h = X_h - \sum_{i=0}^{r} a_{h,i} X_i$, $r+1 \leq h \leq n$ and $\{\ell_{r+1}, \ldots, \ell_n\}$ forming a basis of the k-vector space $\mathfrak{a}_1$. This defines a homogeneous k-algebra isomorphism

$$R = k[X_0, \ldots, X_n]/\mathfrak{a} \simeq \tilde{R} = k[X_0, \ldots, X_r]/\tilde{\mathfrak{a}}$$

which in turn restricts to homogeneous k-algebra isomorphism of the respective subalgebras $k[\mathbf{f}]$ and $k[\tilde{\mathbf{f}}]$, where $f_j \mapsto \tilde{f}_j$ under this map; however, note that the last isomorphism is not induced by equivalence of representatives but rather changing the rational map by "source and target" transformations.

Thus, this operation does not affect either $\dim R$ or $\dim S$ and, likewise, will not affect the numbers in the rightmost side of the stated inequality.

Updating the notation, consider the S-module $E := \operatorname{coker}_S(\psi^t)$ defined as the cokernel of the dual map associated to a weak Jacobian matrix ψ of $\mathfrak{F}$ with respect to the presentation ideal of $\mathcal{R}_R((\mathbf{f}))$. Since ψ^t has a rank over S, so does E. Then

$$\dim \mathcal{R}_S(E) = \dim S + \operatorname{rank}_S(E) = \dim S + \operatorname{edim}(R) - \operatorname{rank}_S(\psi),$$

where $\mathcal{R}_S(E)$ stands for the Rees algebra of E as discussed in Section 2.3.2. On the other hand, assumption (ii) of Definition 3.2.3 implies that $\dim \mathcal{R}_R((\mathbf{f})) = \dim R + 1$— it would suffice to know that $(\mathbf{f})$ is not contained in at least one minimal prime of R of maximal dimension.

Then the asserted inequality follows from the inequality

$$\dim \mathcal{R}_R((\mathbf{f})) \leq \dim \mathcal{R}_S(E),$$

coming from the existence of an S-algebra surjection $\mathcal{R}_S(E) \twoheadrightarrow \mathcal{R}_R((\mathbf{f}))$ as follows. First, $\mathcal{R}_R((\mathbf{f}))$ is naturally an S-algebra by identifying $S = k[\mathbf{Y}]/\mathfrak{b} \simeq k[\mathbf{Y}]/(\mathcal{J}_{0,*})$ with a direct summand of $R[\mathbf{Y}]/\mathcal{J} \simeq \mathcal{R}_R((\mathbf{f}))$, where $\mathcal{J}$ is as earlier.

Second, one shows that there is an S-surjection $\mathcal{S}_S(E) \twoheadrightarrow \mathcal{R}_R((\mathbf{f}))$ with source the symmetric algebra of E, and show that its S-torsion maps to zero. Indeed, one has

$$\begin{aligned}\mathcal{S}_S(E) &\simeq S[\mathbf{X}]/I_1(\mathbf{X}\cdot\psi^t) = k[\mathbf{Y}][\mathbf{X}]/(\mathfrak{b}, I_1(\mathbf{X}\cdot\psi^t)) \twoheadrightarrow R[\mathbf{Y}](\mathfrak{b}, I_1(\mathbf{X}\cdot\psi^t))\\ &\twoheadrightarrow R[\mathbf{Y}]/\mathcal{J}.\end{aligned}$$

Now, since $S \subset R$ is a torsionfree inclusion and $\mathcal{R}_R((\mathbf{f}))$ is torsion-free over R, then $\mathcal{R}_R((\mathbf{f}))$ is torsion-free over S. Therefore, the S-torsion maps to zero. □

Corollary 3.2.20. *With the notation of the previous theorem, one has*

$$\operatorname{jdrank}_+(\mathfrak{F}) \le \operatorname{edim}(R) - 1.$$

Moreover, if equality holds then $\dim S = \dim R$.

Thus, if the rank exists, one has the bounds:

$$\begin{aligned}\operatorname{dgi}(\mathfrak{a}) &\le \operatorname{jdrank}(\mathfrak{F}) \le n\\ 0 &\le \operatorname{jdrank}_+(\mathfrak{F}) \le \operatorname{edim}(R) - 1.\end{aligned}$$

Remark 3.2.21. We note that, in the classical integral case in characteristic zero, the corollary says: $\operatorname{jdrank}_+(\mathfrak{F}) = \operatorname{edim}(R)-1$ implies that the general fiber of the map is a finite set of points. Thus, the preliminary estimate in Proposition 3.2.19 puts in evidence that the Jacobian dual rank plays a sort of complementary role to a transcendence degree in the integral case. The subsequent results will actually show that it takes over a complementary role to the degree of the rational map in the classical setup.

3.2.4 Statement and proof of the criterion

We now state the main theorem of this section, which gives a criterion of birationality. After having worked out a purely algebraic setup all the way, it is natural to first convey an algebraic formulation of the criterion.

We keep the essential notation of the previous subsections.

Theorem 3.2.22. *Let $\mathfrak{F} : \mathbf{f} \subset R = k[\mathbf{X}]/\mathfrak{a} = k[X_0, \dots, X_n]/\mathfrak{a}$ stand for an m-targeted rational map and, for every minimal prime P of R, let $\mathfrak{F}|P$ denote the induced m-targeted rational map with source R/P. The following conditions are equivalent:*

(a) *$\mathfrak{F}$ is birational onto the image.*

(b) *$\mathfrak{F}$ has nondegenerate Jacobian dual rank satisfying*

$$\operatorname{jdrank}_+(\mathfrak{F}) = \operatorname{edim}(R) - 1.$$

(c) *For every minimal prime P of R, $\mathfrak{F}\lceil P$ has nondegenerate Jacobian dual rank satisfying*

$$\operatorname{jdrank}_+(\mathfrak{F}\lceil P) = \operatorname{edim}(R/P) - 1.$$

Supplement. *Suppose, moreover, that the above equivalent conditions holds. Let ψ denote a weak Jacobian dual matrix of $\mathfrak{F}$ over $S = k[\mathbf{Y}]/\mathfrak{b} \simeq k[\mathbf{f}]$.*

(i) *A rational $(n+1)$-datum in S representing the inverse map is given by the coordinates of any homogeneous vector of positive degree in the (rank one) null space of ψ over S for which these coordinates generate an ideal containing a regular element.*

(ii) *If further R is a domain, a rational $(n+1)$-datum as in (i) can be taken to be the set of the (ordered, signed) $(\operatorname{edim}(R)-1)$-minors of an arbitrary $(\operatorname{edim}(R)-1)\times\operatorname{edim}(R)$ submatrix of ψ of rank $\operatorname{edim}(R)-1$.*

Proof. As in the proof of Proposition 3.2.19, up to a nonsingular k-linear map on R, one may assume that the forms in $\mathbf{f}$ involve only the first $r+1 = \operatorname{edim}(R)$ elements $x_0, \dots, x_r$. In addition, the resulting isomorphisms at the respective levels of R and S induce homogeneous k-algebra isomorphisms $K(R) \simeq K(\tilde{R})$ and $K(S) \simeq K(\tilde{S})$, where as before $S = k[\mathbf{Y}]/\mathfrak{b} \simeq k[\mathbf{f}] \simeq k[\tilde{\mathbf{f}}]$.

Finally, passing to degree zero and using the field criterion of Proposition 3.2.12, it follows that $\mathfrak{F} : \mathbf{f} \subset R$ is birational with source $\operatorname{Proj}(R)$ if and only if $\tilde{\mathbf{f}} \subset \tilde{R}$ gives a birational map with source $\operatorname{Proj}(\tilde{R})$. This shows that (a) is stable under this change. According to Proposition 3.2.19 and the discussion of its preliminaries, the nondegenerate Jacobian dual rank in both (a) and (b) is also stable under this process.

Thus, we may restart now assuming that $n+1 = \operatorname{edim}(R)$—in other words, $R = k[X_0, \dots, X_n]/\mathfrak{a}$, with $\mathfrak{a} \subset (R_+)^2$ – and $\operatorname{jdrank}(\mathfrak{F}) = \operatorname{jdrank}_+(\mathfrak{F})$.

(a) $\Rightarrow$ (b) By assumption, there is a representative $\mathbf{g}(\mathbf{y}) = \{g_0(\mathbf{y}), \dots, g_n(\mathbf{y})\} \subset S$ of the inverse such that $\mathbf{g}(\mathbf{f})$ is equivalent to the identity on R. Now, there is an isomorphism of k-algebras $\Phi : \mathcal{R}_R((\mathbf{f})) \simeq \mathcal{R}_S((\mathbf{g}))$ induced by the identity map of $k[\mathbf{X}, \mathbf{Y}]/(\mathfrak{a}, \mathfrak{b})$ (Proposition 3.2.14).

Restricting to the bihomogeneous pieces, one has in particular $\Phi((\mathcal{J}_{\mathbf{f}})_{(1,s)}) = (\mathcal{J}_{\mathbf{g}})_{(s,1)}$, for every $s \geq 1$. Note that for varying $s \geq 1$, $(\mathcal{J}_{\mathbf{g}})_{(s,1)}$ generate the presentation ideal of the symmetric algebra of $\mathbf{g}$ over S. Up to a permutation of the variables $\mathbf{x}$, one has, say, $(\mathcal{J}_{\mathbf{g}})_{(s,1)} = I_1((x_0 \cdots x_{\mu-1}) \cdot \mathcal{M})$, where $\mathcal{M}$ denotes the matrix of a minimal graded presentation of $\mathbf{g}$ over S and $\mu = \mu(\mathbf{g})$ is the minimal number of generators of the ideal $(\mathbf{g})$. Further, if ψ denotes a weak Jacobian dual matrix of $\mathbf{f}$ over S then $\Phi((\mathcal{J}_{\mathbf{f}})_{(1,s)}) = I_1((x_0 \cdots x_n) \cdot \psi^t)$ by the definition of ψ. The ensued equality $I_1((x_0 \cdots x_n) \cdot \psi^t) = I_1((x_0 \cdots x_{\mu-1}) \cdot \mathcal{M})$ shows that the last $n-\mu$ rows of ψ^t are null. Hence, if ψ' is the matrix obtained by removing the zero columns of ψ then any column of ψ'^t lives in $\operatorname{Im}(\mathcal{M})$ and vice versa. Therefore, $\operatorname{Im}(\psi') = \operatorname{Im}(\mathcal{M})$.

Now $\mathcal{M}$ has a rank over S. This is because $\operatorname{grade}_S(\mathbf{g}) \geq 1$ by definition of a rational datum, hence a presentation map of $(\mathbf{g})$ over S has a rank as well. Clearly, then ψ' has a rank and $\operatorname{rank}(\psi) = \operatorname{rank}(\psi') = \operatorname{rank}(\mathcal{M}) = \mu - 1 = \operatorname{edim}(R) - 1$, the last equality following from the nondegeneracy of R.

(b) $\Rightarrow$ (c) Note that $n = \operatorname{edim}(R) - 1$ as we are assuming that R is nondegenerate. By the definition of rank, the ideal of minors $I_n(\psi)$ of a weak Jacobian dual matrix ψ has a regular element on S. Let P be minimal prime of R and $Q = P \cap S$, using the isomorphism $S \simeq k[\mathbf{f}]$. Clearly, then $I_n(\psi\lceil Q) \neq \{0\}$, where $\psi\lceil Q$ is the matrix on S/Q obtained by taking the entries of ψ modulo Q. It follows that $\psi\lceil Q$ has rank n over S/Q.

Now consider a weak Jacobian dual matrix $\tilde{\psi}$ of $\mathfrak{F}\lceil P$, as obtained from the Rees algebra $\mathcal{R}_{R/P}((\mathbf{f}, P)/P)$. By definition, $\operatorname{Im}(\psi\lceil Q)$ can be taken to be a submodule of $\operatorname{Im}(\tilde{\psi})$. Therefore, the rank of $\tilde{\psi}$ is at least n, hence it obtains

$$\operatorname{edim}(R/P) - 1 - \operatorname{jdrank}_+(\mathfrak{F}\lceil P) = n - \operatorname{jdrank}(\mathfrak{F}\lceil P) \leq 0,$$

from which ensues $\operatorname{jdrank}_+(\mathfrak{F}\lceil P) \geq \operatorname{edim}(R/P) - 1$ and then equality follows by Proposition 3.2.19, as required.

(c) $\Rightarrow$ (a) The proof will simultaneously argue for item (ii).

The first step is to use Proposition 3.2.12, by which the condition in (a) is equivalent to the condition that every restriction $\mathfrak{F}\lceil P$ of $\mathfrak{F}$ is birational onto its image and, moreover, R/P is the unique minimal component of R mapping to the latter by $\mathfrak{F}$.

To show that one can reduce to the case where R is a domain, we argue as follows. Given a minimal prime P of R, write $R/P \simeq (k[\mathbf{x}]/\mathfrak{a})/(\mathfrak{P}/\mathfrak{a}) \simeq k[\mathbf{x}]/\mathfrak{P}$ and $S/P \cap S \simeq k[\mathbf{y}]/\mathfrak{Q}$, where $\mathfrak{Q}$ is the inverse image of $\mathfrak{P}$ in $k[\mathbf{y}]$. Further, set $\mathfrak{F}_{\mathfrak{P}} = \mathfrak{F}\lceil P$ to emphasize the original polynomial ambient. The assumption then reads as

$$\operatorname{jdrank}_+(\mathfrak{F}_{\mathfrak{P}}) = \operatorname{edim}(R/P) - 1 = n - \operatorname{dgi}(\mathfrak{P}),$$

where $\operatorname{dgi}(\mathfrak{P})$ is the degeneration index of $\mathfrak{P}$ as discussed in the beginning of the subsection. Equivalently, $\operatorname{jdrank}(\mathfrak{F}_{\mathfrak{P}}) = n$. Thus, if we show that this implies that $\mathfrak{F}_{\mathfrak{P}}$ is birational onto its image we will be done.

Therefore, we may assume at the outset that R is a domain and a weak Jacobian dual matrix ψ has rank n. But then one can select an $n \times (n+1)$ submatrix σ of ψ of rank n. Let $\mathbf{\Delta} := \{\Delta_0, \ldots, \Delta_n\}$ stand for the ordered signed n-minors of σ. This set is a rational $(n+1)$-datum. It remains to prove that $\mathbf{\Delta}(\mathbf{f})$ is equivalent to the identity on $\operatorname{Proj}(R)$.

Apply Proposition 3.2.1 with $A := k[\mathbf{y}]$ and with ρ standing for the transpose of the selected $n \times (n+1)$ submatrix σ of ψ of rank n to get that the Koszul forms $\Delta_i x_j - \Delta_j x_i$, one for each pair i, j, belong to the $k[\mathbf{y}]$-submodule of $A^{n+1} = \sum_{i=0}^{n} A x_i$ generated by the forms induced by the rows of σ, i. e., obtained as inner product of these by the transpose $(x_0, \ldots, x_n)^t$. But the latter, by definition, belong to the defining ideal of the Rees algebra of $(\mathbf{f})$ over $R[\mathbf{Y}]$, hence, vanish on the tuple whose coordinates are the constituents of $\mathbf{f}$.

Therefore,

$$\Delta_i(\mathbf{f})\, x_j = \Delta_j(\mathbf{f})\, x_i, \quad \text{for each pair } i, j. \tag{23}$$

Recall that some $\Delta_i(\mathbf{f}) \neq 0$; say, this is the case for $i = n$. On the other hand, $x_n \neq 0$ since otherwise (23) implies that $x_i = 0$ for every i, which is nonsense. In particular, one has the equalities $\Delta_n(\mathbf{f})\, x_i = x_n\, \Delta_i(\mathbf{f})$, with $\Delta_n(\mathbf{f})$, x_n nonzero. Therefore, $\mathbf{\Delta}(\mathbf{f})$ is equivalent to the identity on $\mathrm{Proj}(R)$.

It remains to prove (i).

Let $\mathbf{h}^t \in \ker(\psi^t)$ be given as in the hypothesis of (i). Let $\mathbf{g}$ be a representative of the inverse map such that the vector $\mathbf{g}^t \in \ker(\psi^t)$ as in the proof and notation of the implication (a) $\Rightarrow$ (b). The argument there showed that $\ker(\psi^t)$ has rank 1. Since the ideal $(\mathbf{h})$ has a regular element by assumption, Lemma 3.2.5 implies that $\mathbf{h}$ and $\mathbf{g}$ are equivalent $(n+1)$-sets. Then, by Lemma 3.2.7, $\mathbf{h}$ is rational datum, hence defines the inverse map as well. □

Remark 3.2.23. The argument in the proof of (ii) above goes through regardless of the integrality assumption, provided a weak Jacobian dual matrix of $\mathbf{f}$ admits a submatrix of size $(\mathrm{edim}(R) - 1) \times \mathrm{edim}(R)$ and of rank $\mathrm{edim}(R) - 1$.

We now transcribe the above result into a geometric version. We assume that the ground field k is infinite and that varieties have points in an algebraic closure of k.

Theorem 3.2.24. *Let $V \subset \mathbb{P}^n$ $(n \geq 1)$ denote a reduced, but not necessarily irreducible, projective subvariety. Let $\mathfrak{F} : V \dashrightarrow \mathbb{P}^m_k$ denote a rational map.*

The following conditions are equivalent:

(a) *$\mathfrak{F}$ is a birational map of V onto its image.*

(b) *$\mathfrak{F}$ has nondegenerate Jacobian dual rank and the latter equals the embedding dimension of the affine cone over V.*

(c) *For every irreducible component W of V, the restriction of $\mathfrak{F}$ to W has non-degenerate Jacobian dual rank and the latter equals the embedding dimension of the affine cone over W.*

Moreover, when (a) *holds then*

(i) *The inverse map to $\mathfrak{F}$ is given by the coordinates of any sufficiently general homogeneous vector of positive degree in the null space of a weak Jacobian dual matrix of $\mathfrak{F}$.*

(ii) *If besides, V is irreducible, the coordinates of the inverse to $\mathfrak{F}$ are the* (ordered, signed) *n-minors of an arbitrary $n \times (n+1)$ submatrix rank n of a weak Jacobian dual matrix of $\mathfrak{F}$.*

("Sufficiently general" in the above statement means that the general k-linear combination of the coordinates of the vector ought to be a nonzero-divisor on the homogeneous coordinate ring of the image.)

Remark 3.2.25. (1) A tall order in the computational side of the theory is to get hold of the spanning biforms of $\mathcal{J}_{1,*}$ and $\mathcal{J}_{0,*}$ without actually computing all the generators of $\mathcal{J}_{\mathbf{f}}$. For the second bihomogeneous piece, this is beckoningly the main problem of implicitization in elimination theory. One of the common techniques is to try and derive the implicit equations by means of the so-called Sylvester method starting out from the defining equations of the symmetric algebra of $I = (\mathbf{f})$ (i. e., the bihomogeneous piece of bidegrees $(*, 1)$). In special situations, this method often throws some light on obtaining $\mathcal{J}_{1,*}$; see, e. g., [108, Example 5.7].

(2) Note that the theorem gives no hint on the minimal possible degree of a representative of the inverse map. Even in the case of a Cremona map, the degree will be elusive as the representative given by the minors will typically carry a nontrivial factor. Thus, the question arises as to how one goes about computing the degree of the gcd of the minors in the latter case. Often this can be computed by a careful scrutiny of the Hilbert series (see, e. g., [166, Proposition 3.1]) or by local structure (see [166, Examples 2.4, 2.5] and also [197, Theorem 3.1]).

(3) It would be interesting to understand the meaning of a weak Jacobian dual matrix—or at least of its rank—in purely geometric terms. Some results in this direction will be stated later.

3.2.5 A role of syzygies

In this part, we draw some corollaries to Theorem 3.2.22 in the case the source is a projective space. We emphasize that in this context much of the theory, including the main criterion, gets simplified and more visible. Having $\mathbb{P}^n$ map birationally onto its image $Y \subset \mathbb{P}^m$ is a situation in which the classical literature referred to the rational subvariety Y as a representation of $\mathbb{P}^n$.

The following is a characteristic-free generalized version of [39, Proposition 4.5(iii)].

Theorem 3.2.26. *Let $\mathfrak{F}\colon \mathbb{P}^n \dashrightarrow \mathbb{P}^m$ be a rational map, given by $m+1$ forms $\mathbf{f} = \{f_0, \dots, f_m\}$ of a fixed degree. If the image of $\mathfrak{F}$ has dimension n and if the ideal $(\mathbf{f}) \subset R$ has maximal linear rank ($= m$), then $\mathfrak{F}$ is birational onto its image.*

Proof. Recall the notion of linear rank from Section 2.3.1. Let φ_1 denote the linear part of a presentation matrix of I. Following the *leitmotif* of [187, Proposition 1.1], write $\mathbf{Y} \cdot \varphi_1 = \mathbf{X} \cdot \rho$, where ρ is a uniquely defined matrix whose entries are linear forms in $k[\mathbf{Y}] = k[Y_0, \dots, Y_m]$. Note that the left (resp., right) side generates the presentation ideal of the symmetric algebra $\mathcal{S}_{k[\mathbf{X}]}(\operatorname{coker}(\varphi_1))$ (resp., $\mathcal{S}_{k[\mathbf{Y}]}(\operatorname{coker}(\rho))$). Therefore, the identity map of $k[\mathbf{X}, \mathbf{Y}]$ induces a k-isomorphism between these algebras.

Here, $R = k[\mathbf{X}]$. Let $\mathcal{R}_{k[\mathbf{X}]}((\mathbf{f})) \simeq k[\mathbf{X}, \mathbf{Y}]/\mathcal{J}$ as before. Note the surjection $\operatorname{coker}(\varphi_1) \twoheadrightarrow (\mathbf{f})$, whose kernel is torsion since φ_1 has rank m by hypothesis. Therefore,

$$\mathcal{R}_{k[\mathbf{X}]}(\operatorname{coker}(\varphi_1)) \simeq k[\mathbf{X}, \mathbf{Y}]/\mathcal{J}. \tag{24}$$

Writing $S := k[\mathbf{Y}]/\mathfrak{b}$ for the homogeneous coordinate ring of the image of $\mathfrak{F}$ on $\mathbb{P}^m$, one has $\mathfrak{b} \subset \mathcal{J}$.

Write $E := \operatorname{coker}(\rho)$. Since the symmetric algebra is functorial, one has

$$\mathcal{S}_S(E \otimes_{k[\mathbf{Y}]} S) \simeq \mathcal{S}_{k[\mathbf{Y}]}(E)/\mathfrak{b}\,\mathcal{S}_{k[\mathbf{Y}]}(E) \simeq k[\mathbf{X},\mathbf{Y}]/(\mathfrak{b}, I_1(\mathbf{X}\cdot\rho)) = k[\mathbf{X},\mathbf{Y}]/(\mathfrak{b}, I_1(\mathbf{Y}\cdot\varphi_1)).$$

Thus, there is a surjection $\mathcal{S}_S(E \otimes_{k[\mathbf{Y}]} S) \twoheadrightarrow k[\mathbf{X},\mathbf{Y}]/\mathcal{J}$.

Claim. The surjection $\mathcal{S}_S(E \otimes_{k[\mathbf{Y}]} S) \twoheadrightarrow k[\mathbf{X},\mathbf{Y}]/\mathcal{J}$ induces an isomorphism

$$\mathcal{R}_S(E \otimes_{k[\mathbf{Y}]} S) \simeq k[\mathbf{X},\mathbf{Y}]/\mathcal{J}. \tag{25}$$

To see this, recall that the Rees algebra $\mathcal{R}_S(E \otimes_{k[\mathbf{Y}]} S)$ comes from $\mathcal{S}_S(E \otimes_{k[\mathbf{Y}]} S)$ by killing the S-torsion. Let $\mathcal{T}$ denote the lifting of the latter to $k[\mathbf{X},\mathbf{Y}]$. Thus,

$$\mathcal{R}_S(E \otimes_{k[\mathbf{Y}]} S) \simeq k[\mathbf{X},\mathbf{Y}]/(\mathfrak{b}, I_1(\mathbf{Y}\cdot\varphi_1), \mathcal{T}).$$

First, let $G \in \mathcal{T}$. Then $F(\mathbf{Y})\,G \in (\mathfrak{b}, I_1(\mathbf{Y}\cdot\varphi_1)) \subset \mathcal{J}$ for some $F(\mathbf{Y}) \in k[\mathbf{Y}] \setminus \mathfrak{b}$. Since $\mathcal{J}$ is a prime ideal and $F(\mathbf{Y}) \notin \mathcal{J}$ because $k[\mathbf{Y}] \cap \mathcal{J} = \mathfrak{b}$, necessarily $G \in \mathcal{J}$. This proves the inclusion $(\mathfrak{b}, I_1(\mathbf{Y}\cdot\varphi_1), \mathcal{T}) \subset \mathcal{J}$ as $(\mathfrak{b}, I_1(\mathbf{Y}\cdot\varphi_1)) \subset \mathcal{J}$.

Conversely, let $G = G(\mathbf{X},\mathbf{Y}) \in \mathcal{J}$. By (24), $\mathcal{J}$ is the $k[\mathbf{X}]$-torsion of $k[\mathbf{X},\mathbf{Y}]/I_1(\mathbf{Y}\cdot\varphi_1)$ lifted to $k[\mathbf{X},\mathbf{Y}]$. Therefore, there exists $F(\mathbf{X}) \in k[\mathbf{X}] \setminus I_1(\mathbf{Y}\cdot\varphi_1)$, i. e., $F(\mathbf{X}) \neq 0$, such that $F(\mathbf{X})G(\mathbf{X},\mathbf{Y}) \subset I_1(\mathbf{Y}\cdot\varphi_1)$. Suppose that $G \notin (\mathfrak{b}, I_1(\mathbf{Y}\cdot\varphi_1), \mathcal{T})$. The latter is a prime ideal, hence $F(\mathbf{X}) \in (\mathfrak{b}, I_1(\mathbf{Y}\cdot\varphi_1), \mathcal{T})$. By the definition of $\mathcal{T}$, there exists $H(\mathbf{Y}) \in k[\mathbf{Y}] \setminus \mathfrak{b}$ such that $H(\mathbf{Y})F(\mathbf{X}) \in (\mathfrak{b}, I_1(\mathbf{Y}\cdot\varphi_1))$. Evaluating $\mathbf{Y} \mapsto \mathbf{f}$ would give $H(\mathbf{f})F(\mathbf{X}) = 0$ whence $F(\mathbf{X}) = 0$ since $H(\mathbf{f}) \neq 0$; this is a contradiction.

As a consequence of (25) and Proposition 2.3.16, one gets

$$\dim S + \operatorname{rank}_S(E \otimes_{k[\mathbf{Y}]} S) = \dim \mathcal{R}_S(E \otimes_{k[\mathbf{Y}]} S) = \dim \mathcal{R}_{k[\mathbf{X}]}((\mathbf{f})) = (n+1)+1.$$

Since $\dim S = n+1$ by assumption, we get $\operatorname{rank}_S(E \otimes_{k[\mathbf{Y}]} S) = 1$; hence, $\operatorname{rank}_S(\rho \otimes_{k[\mathbf{Y}]} S) = n$. Since $\operatorname{Im}(\rho \otimes_{k[\mathbf{Y}]} S) \subset \operatorname{Im}(\psi)$ for a weak Jacobian dual matrix ψ, one can apply Theorem 3.2.22. □

A birational map of $\mathbb{P}^n$ onto itself is known as a *Cremona map* or *transformation*. Cremona transformations have a uniquely defined minimal representative (both the map and its inverse) and the corresponding degrees are important invariants in the classical theory. In the remainder of the section, following the classical terminology, we say that a Cremona map has degree r if it is defined by forms of degree r generating an ideal of codimension ≥ 2. A more thorough study of Cremona transformations is deferred to Chapter 6 and Chapter 7.

The following result has undergone several versions.

Corollary 3.2.27. *Let $\mathfrak{F}: \mathbb{P}^n \dashrightarrow \mathbb{P}^n$ be a rational map whose base ideal I is of linear type. The following conditions are equivalent:*
(a) *$\mathfrak{F}$ is a Cremona transformation.*
(b) *The linear part of the syzygy matrix of I has rank n.*

Moreover, if these conditions take place, both $\mathfrak{F}$ and the inverse map $\mathfrak{F}^{-1}$ have degree $\leq n$.

Proof. The implication (a) $\Rightarrow$ (b) follows from Theorem 3.2.22 because the Jacobian dual matrix comes totally from the linear syzygies.

The reverse implication (b) $\Rightarrow$ (a) comes from Theorem 3.2.26 since the linear type assumption for an ideal generated by forms of the same degree implies at least the algebraic independence of these forms.

For the degree bounds, since I has codimension ≥ 2, by R-dualizing a presentation $R^m \xrightarrow{\varphi} R^{n+1} \xrightarrow{\pi} I \to 0$ gives a left exact sequence

$$0 \to I^* \simeq R \xrightarrow{\pi^t} R^{n+1*} \xrightarrow{\varphi^t} R^{m*},$$

where the coordinates of π^t can be taken to be the n-minors of an $n \times (n+1)$ linear submatrix of φ^t of rank n after dividing by their gcd. It follows that I is generated in degree $\leq n$, as was to be shown.

The bound for the inverse map follows from part (ii) of the supplement to Theorem 3.2.22. Indeed, the transpose of an $n \times (n+1)$ matrix of rank n formed by linear syzygies of I can be taken as a submatrix of a weak Jacobian dual matrix of $\mathfrak{F}$. Since the signed n-minors of this matrix define the inverse to $\mathfrak{F}$, the degree bound is clear. □

Remark 3.2.28. The assumption that I is of linear type in the result above is quite strong, implying as we saw that the degree of the Cremona map is quite low. The question as to how or when one can have other bounds on this degree is quite involved. Coming from the graph side of the map, the Cohen–Macaulay assumption on the latter triggers other bounds (Section 7.2.2).

A reading of the above in a particular landscape gives the following.

Corollary 3.2.29. *Let $\mathfrak{F}: \mathbb{P}^n \dashrightarrow \mathbb{P}^n$ be a rational map with a base ideal $I \subset R = k[X_0, \dots, X_n]$. Suppose that $\dim R/I = 1$ and that I is locally a complete intersection at its minimal primes. Then $\mathfrak{F}$ the following are equivalent:*
(a) *$\mathfrak{F}$ is a Cremona transformation.*
(b) *The linear part of the syzygy matrix of I has rank n.*

Moreover, if these conditions take place, both $\mathfrak{F}$ and the inverse map $\mathfrak{F}^{-1}$ have degree $\leq n$.

Proof. It follows from the above via Proposition 2.3.2. □

Applying the last corollary for $n = 2$ gives the following.

Corollary 3.2.30. *Let $\mathfrak{F}: \mathbb{P}^2 \dashrightarrow \mathbb{P}^2$ be a rational map whose base ideal is a codimension 2 local complete intersection. Then $\mathfrak{F}$ is a Cremona map if and only if is defined by the 2×2 minors of a 3×2 matrix of linear forms.*

Proof. Suppose that $\mathfrak{F}$ is a Cremona map. By the above, $\mathfrak{F}$ has degree ≤ 2. Therefore, as the base ideal is of codimension 2, the map is one of the three classical types of plane quadratic Cremona maps, in which case the result is directly verified.

The converse also follows from the previous corollary. □

A homogeneous polynomial $f \in R = k[X_0, \ldots, X_n]$ whose partial derivatives define a Cremona map is called *homaloidal*. The ideal of R generated by the partial derivatives of a polynomial $f \in R$ is called the *gradient ideal* of f. A more detailed study of these ideals is deferred to Chapter 5, particularly, Section 5.1.1.

Here is a bold guess.

Conjecture 3.2.31. *Let $f \in R = k[X_0, \ldots, X_n]$ be a reduced homaloidal homogeneous polynomial and let $J_f \subset R$ denote its gradient ideal. If $\dim R/J_f = 1$, then J_f is locally a complete intersection at its minimal primes.*

The conjecture holds true in characteristic zero for $n = 2$, due to the theorem of Dolgachev [52]. However, even in null characteristic an independent proof would be of interest as it would imply Dolgachev's result by way of Corollary 3.2.30.

In the plane case ($n = 2$), the above local condition on J_f is also equivalent to requiring that the ideal generated by the entries of a syzygy matrix of J_f have codimension three, so the condition is readily verified by computation. Perhaps the simplest example of a non-homaloidal Cremona transformation where the analogue of the above conjecture fails is the plane de Jonquières transformation of degree 3 defined by the monomials x^3, x^2y, y^2z.

We refer to Section 5.1.1 for the relationship to the Dimca–Papadima conjecture (now a theorem proved in [112]).

3.3 Birational combinatorics

This section is devoted to giving a short account of characteristic-free rational maps $\mathbb{P}^{n-1} \dashrightarrow \mathbb{P}^{m-1}$ defined by monomials, along with natural criteria for such maps to be birational onto their image varieties. By and large, we refer to [19] and [219] for the combinatorial background. The material of this section is basically lifted from [195] and [196] and is adapted here for the reader's convenience. It has also partially appeared in [219, Section 10.8].

3.3.1 The underlying arithmetic

Let $R = k[\mathbf{x}] = k[x_1, \ldots, x_n]$ be a polynomial ring over a field k. As usual we set $x^\alpha := x_1^{a_1} \cdots x_n^{a_n}$ if $\alpha = (a_1, \ldots, a_n) \in \mathbb{N}^n$. In the sequel, we consider a finite set of distinct monomials $F = \{x^{v_1}, \ldots, x^{v_q}\} \subset R$ of the same degree $d \geq 2$ and having no nontrivial common factor. We also take for granted throughout that every x_i divides at least one element of F, a condition harmless to be assumed.

The integer matrix $A_F = (v_1, \ldots, v_q)$ where each v_i is regarded as a column vector will be called the *log-matrix* of F.

If C is an integer matrix with r rows, we denote by $\mathbb{Z}C$ (resp., $\mathbb{Q}C$) the subgroup of $\mathbb{Z}^r$ (resp., subspace of $\mathbb{Q}^r$) generated by the columns of C. $\Delta_r(C)$ will denote the greatest common divisor of all the nonzero $r \times r$ minors of C.

An extension $D' \subset D$ of integral domains is said to be birational if it is an equality at the level of the respective fields of fractions.

Lemma 3.3.1. *Let F and G be finite sets of monomials of R such that $F \subset G$. Then $k[F] \subset k[G]$ is a birational extension if and only if $\mathbb{Z}A_F = \mathbb{Z}A_G$. In particular, if the monomials of F have the same degree $d \geq 1$, then they define a birational map onto the image if and only if $\Delta_n(A_F) = d$.*

Proof. In this situation, the ring extension is birational if and only every monomial of G can be written as a fraction whose terms are suitable power products of the monomials of F. Clearing denominators of such a fraction and taking log of both members establishes the required equivalence.

To see that the supplementary assertion follows from this, recall from previous parts that the map in question is birational onto its image if and only if the homogeneous inclusion $k[F] \subset k[\mathbf{x}_d]$ is a birational extension. □

We need the following arithmetical result.

Lemma 3.3.2. *Consider the natural inclusion $\mathbb{Z}^n \subset \mathbb{Z}^{n+1}$ into the first n coordinates. Then, for any $\alpha_1, \ldots, \alpha_m \in \mathbb{Z}^n$ there is an induced injective homomorphism of $\mathbb{Z}$-modules*

$$\mathbb{Z}^n/\mathbb{Z}(\alpha_2 - \alpha_1, \ldots, \alpha_m - \alpha_1) \hookrightarrow \mathbb{Z}^{n+1}/\mathbb{Z}((\alpha_1, 1), \ldots, (\alpha_m, 1)),$$

which is an isomorphism at the level of the respective torsion submodules.

Proof. Clearly, there is an induced map as argued—because $\alpha_j - \alpha_1$ maps to $(\alpha_j - \alpha_1, 0) = (\alpha_j, 1) - (\alpha_1, 1)$—and the induced map is injective because the two equations $a_1 + \cdots + a_m = 0$ and $\alpha = a_1\alpha_1 + \cdots + a_m\alpha_m$ readily imply that $\alpha \in \mathbb{Z}(\alpha_2 - \alpha_1, \ldots, \alpha_m - \alpha_1)$.

Next, clearly any homomorphism maps torsion to torsion, so it remains to check surjectivity at the torsion level. Let then (α, b) be a torsion element of $\mathbb{Z}^{n+1}/\mathbb{Z}((\alpha_1, 1), \ldots, (\alpha_m, 1))$. This implies a relation

$$s(\alpha, b) = \lambda_1(\alpha_1, 1) + \cdots + \lambda_m(\alpha_m, 1) \quad (\lambda_i \in \mathbb{Z}),$$

where $0 \neq s \in \mathbb{N}$, $\alpha \in \mathbb{Z}^n$ and $b \in \mathbb{Z}$. Then

$$
\begin{aligned}
s\alpha &= \lambda_1\alpha_1 + \cdots + \lambda_m\alpha_m, \\
sb &= \lambda_1 + \cdots + \lambda_m, \\
s(\alpha - b\alpha_1) &= \lambda_2(\alpha_2 - \alpha_1) + \cdots + \lambda_m(\alpha_m - \alpha_1).
\end{aligned}
$$

Hence, it follows that the class $\overline{\alpha - b\alpha_1} \in \mathbb{Z}^n/\mathbb{Z}(\alpha_2-\alpha_1,\ldots,\alpha_m-\alpha_1)$ is a torsion element and maps to $\overline{(\alpha, b)}$, as required. □

It affords the following arithmetic criterion of birationality.

Theorem 3.3.3. *Let $F = \{v_1,\ldots,v_q\}$ be a finite set of monomials of the same degree $d \geq 2$. The following conditions are equivalent:*
(i) *The rational map $\mathbb{P}^{n-1} \dashrightarrow \mathbb{P}^{q-1}$ defined by $v_1,\ldots,v_q$ is birational onto the image.*
(ii) *The $\mathbb{Z}$-module $\mathbb{Z}^n/\mathbb{Z}(\{v_1 - v_j \mid 2 \leq j \leq q\})$ is torsion-free of rank 1.*

Proof. Introduce the extended log-matrix of F,

$$
A' = \begin{pmatrix} v_1 & \cdots & v_q \\ 1 & \cdots & 1 \end{pmatrix}.
$$

Letting T denote $\mathbb{Z}$-torsion, there is an exact sequence

$$
0 \to T(\mathbb{Z}^{n+1}/\mathbb{Z}A') \xrightarrow{\varphi} T(\mathbb{Z}^n/\mathbb{Z}A) \xrightarrow{\psi} \mathbb{Z}/d\mathbb{Z} \to 0, \tag{26}
$$

where $\varphi(\overline{(\alpha,b)}) = \bar{\alpha}$ and $\psi(\bar{\alpha}) = \bar{\boldsymbol{\alpha}}$, with $\alpha \in \mathbb{Z}^n$, $b \in \mathbb{Z}$ and $\boldsymbol{\alpha}$ denoting the sum of the entries of α.

If, moreover, A has full rank, then $\mathbb{Z}^n/\mathbb{Z}A$ is torsion; hence, $\mathbb{Z}^n/\mathbb{Z}A \simeq \mathbb{Z}_d$ if and only if $\mathbb{Z}^{n+1}/\mathbb{Z}A'$ is torsion-free, and in this case the 0*th* fitting ideal $\Delta_n(A)$ of $\mathbb{Z}^n/\mathbb{Z}A$ is the same as that of $\mathbb{Z}/d\mathbb{Z}$, i. e., $\Delta_n(A) = (d)$.

On the other hand, we have an exact sequence of $\mathbb{Z}$-modules

$$
0 \to \mathbb{Z}A/\mathbb{Z}(\{v_1 - v_j \mid 2 \leq j \leq q\}) \to \mathbb{Z}^n/\mathbb{Z}(\{v_1 - v_j \mid 2 \leq j \leq q\}) \to \mathbb{Z}^n/\mathbb{Z}A \to 0 \tag{27}
$$

Again, if A has full rank then the leftmost module has rank 1 and, since the rightmost module is torsion, the mid module has rank 1. Now apply Lemma 3.3.2 with $m = q$ and $\alpha_j = v_j$ to get

$$
T(\mathbb{Z}^n/\mathbb{Z}(\{v_1 - v_j \mid 2 \leq j \leq q\})) \simeq T(\mathbb{Z}^{n+1}/\mathbb{Z}A').
$$

Therefore, the equivalence follows from Lemma 3.3.1. □

Corollary 3.3.4. *Let $F = \{v_1,\ldots,v_q\} \subset R := k[x_1,\ldots,x_n]$ be a finite set of monomials of the same degree $d \geq 2$ and let $I \subset R$ denote the ideal generated by F. Suppose that:*

(i) *The log-matrix A of F has rank n.*
(ii) *The Rees algebra $\mathcal{R}_R(I)$ is normal.*

Then the rational map defined by F is birational onto the image.

Proof. By (i), the $\mathbb{Z}$-module $\mathbb{Z}^n/\mathbb{Z}A$ is a finite abelian group, in particular, it is torsion. On the other hand, the order of the torsion is $\Delta_n(A)$. Therefore, by (26) and Lemma 3.3.1, it suffices to prove that $T(\mathbb{Z}^{n+1}/\mathbb{Z}A') = \{0\}$, i. e., that $\Delta_n(A') = 1$. The latter is [61, Theorems 3.9 an 3.15]. □

The next result explains how the inverse matrix over $\mathbb{Q}$ acts upon differences of canonical basis vectors.

Proposition 3.3.5. *Let $v_1,\dots,v_n$ be a set of vectors in $\mathbb{N}^n$ such that $|v_i| = d \geq 1$ for all i and* $\det(A) = \pm d$, *where A is the $n \times n$ matrix with column vectors $v_1,\dots,v_n$. Then $A^{-1}(e_i - e_j) \in \mathbb{Z}^n$ for all i, j.*

Proof. Fixing indices i, j, there are $\lambda_1,\dots,\lambda_n$ in $\mathbb{Q}$ such that $A^{-1}(e_i - e_j) = \sum_{k=1}^n \lambda_k e_k$. Notice that $A^{-1}(e_i)$ is the ith column of A^{-1}. Set $\mathbf{1} = (1,\dots,1)$. Since $\mathbf{1}A = d\mathbf{1}$, we get $\mathbf{1}/d = \mathbf{1}A^{-1}$. Therefore, $|A^{-1}(e_i)| = |A^{-1}(e_j)| = 1/d$ and $\sum_k \lambda_k = 0$. Then we can write

$$A^{-1}(e_i - e_j) = \sum_{k=2}^{n} \lambda_k(e_i - e_1) \implies e_i - e_j = \sum_{k=2}^{n} \lambda_k(v_k - v_1).$$

Thus, there is $0 \neq s \in \mathbb{N}$ such that $s(e_i - e_j)$ belong to $\mathbb{Z}\{v_1 - v_k\}_{k=2}^n$, the subgroup of $\mathbb{Z}^n$ generated by $\{v_1 - v_k\}_{k=2}^n$. By Lemma 3.3.1 and Theorem 3.3.3, the quotient group $\mathbb{Z}^n/\mathbb{Z}\{v_1 - v_k\}_{k=2}^n$ is free, in particular has no nonzero torsion elements. Then we can write

$$e_i - e_j = \eta_2(v_2 - v_1) + \cdots + \eta_n(v_n - v_1),$$

for some η_i's in $\mathbb{Z}$. Since $\mathbb{Z}\{v_1-v_k\}_{k=2}^n$ is also free (of rank $n-1$), the vectors $v_2-v_1,\dots,v_n-v_1$ are linearly independent. Thus, $\lambda_k = \eta_k \in \mathbb{Z}$ for all $k \geq 2$, hence ultimately $A^{-1}(e_i - e_j) \in \mathbb{Z}^n$. □

3.3.2 Monomial birational maps in degree two

The birational theory of monomials of degree two can be completely established using elementary graph theory as we show in the sequel.

For this, it is convenient to interpret a set of monomials $\{x^{v_i}, 1 \leq i \leq m\}$ of degree two in terms of graphs, possibly with loops. Thus, consider the graph $\tilde{\mathcal{G}}$ on the vertex set $X = \{x_1,\dots,x_n\}$ whose set of edges and loops correspond bijectively to the pairs $\{x_i, x_j\}$ such that $x_ix_j \in F$ (possibly $i = j$). Denote by $\mathcal{G}$ the underlying simple graph obtained by omitting all loops. Notice that, in our situation, the log-matrix A of F is the incidence matrix of $\tilde{\mathcal{G}}$ and the monomial subring $k[F]$ is the edge subring $k[\tilde{\mathcal{G}}]$ of the graph $\tilde{\mathcal{G}}$.

We introduce the following general condition for forms.

Definition 3.3.6. A set $F = \{f_1, \dots, f_q\} \subset R = k[\mathbf{x}] = k[x_1, \dots, x_n]$ of forms of fixed degree ≥ 2 will said to be *cohesive* if the forms have no nontrivial common factor and admits no partition $\mathbf{x} = \mathbf{y} \cup \mathbf{z}$ of the variables such that $F = G \cup H$, where the forms in the set G (resp., H) involve only the $\mathbf{y}$-variables (resp., $\mathbf{z}$-variables), if neither G nor H is empty.

Cohesiveness of monomials of degree 2 corresponds to connectedness of the corresponding graph. The following result explains its role for monomials of degree 2. Note that in degree 2 the entries of the linear syzygy matrix $\mathcal{LS}(F)$ are reduced Koszul relations.

Proposition 3.3.7. *Let $F \subset k[x_1, \dots, x_n]$ be a set of monomials of degree 2 with no nontrivial common factor. Then $\mathcal{LS}(F)$ has maximal rank if and only if F is cohesive.*

Proof. One implication is pretty obvious: if F is noncohesive, then $\mathcal{LS}(F)$ is a block-diagonal $(r+s) \times m$ matrix

$$\begin{pmatrix} \mathcal{LS}(G) & 0 \\ 0 & \mathcal{LS}(H) \end{pmatrix},$$

where $F = G \cup H$. Since $\operatorname{rank}(\mathcal{LS}(G)) \leq r-1$ and $\operatorname{rank}(\mathcal{LS}(H)) \leq s-1$, then $\operatorname{rank}(\mathcal{LS}(F)) \leq r+s-2 \leq q-2$.

For the reverse implication, assume that F is cohesive. Then the corresponding graph $\tilde{\mathcal{G}}$ as above is connected, hence the underlying simple graph $\mathcal{G}$ has a spanning tree $\mathcal{T}$. Being a tree, $\mathcal{T}$ has $n-1$ edges. The required result is readily verified in this case by induction on the number n of vertices: consider the subtree $\mathcal{T} \setminus x_n$ obtained by removing a vertex of degree one and the corresponding edge, say, $x_i x_n$. By the inductive assumption, the corresponding linear syzygy matrix has rank $n-3$, so let L denote an $(n-2) \times (n-3)$ submatrix thereof of rank $n-3$. If $x_i x_j$ is any edge of $\mathcal{T} \setminus x_n$ then, by restoring the removed vertex and edge, yields a matrix of linear syzygies of $\mathcal{T}$ involving edges $x_i x_j$ and $x_i x_n$ and a submatrix of the linear syzygy matrix of $\mathcal{T}$ formed by bordering L with the corresponding column syzygy and a last row of zeros. It is clear that this $(n-1) \times (n-2)$ matrix has rank $n-2$.

This takes care of the spanning tree $\mathcal{T}$. Next, one successively restores edges and loops on to $\mathcal{H}$ in order to recover the whole $\tilde{\mathcal{G}}$, this time with no new vertices. By a similar token, adding one such edge or loop at a time to the connected subgraph $\mathcal{H}$, will increase by one the rank of the new submatrix of the linear syzygy matrix of F formed by bordering as before the previous one with the column corresponding to the added edge or loop. □

The next result is an overture towards a monomial Cremona theory, to be expanded in Section 7.3.

Proposition 3.3.8. *Let $I \subset k[x_1,\dots,x_n]$ be an ideal generated by a cohesive set of n squarefree monomials of degree 2 and let $\mathcal{G}$ denote the corresponding edge-graph. The following conditions are equivalent:*
(i) *I is of linear type.*
(ii) *$\mathcal{G}$ has one odd cycle and no other cycles.*
(iii) *The rational map defined by the generators of I is a Cremona transformation of $\mathbb{P}^{n-1}$.*

Proof. (i) $\Leftrightarrow$ (ii) This is the result of [218].

(ii) $\Rightarrow$ (iii) We apply the criterion of Lemma 3.3.1. For this, we induct on the number of vertices of the graph not belonging to the unique odd cycle of length, say, $r < n$. Now, x_n belongs to a unique edge, say, x_ix_n. Consider the log-matrix A_n of the monomials, where, say, the last column comes from x_ix_n. Computing $\det A_n$ by Laplace along this last column yields a sum $x_n \det A_{n-1} + x_iB$ (up to signs), where A_{n-1} is the log-matrix of the subset omitting the term x_ix_n. Since $n > r$, this subset contains the cycle. By the inductive hypothesis, it gives a birational transformation of $\mathbb{P}^{n-2}$. By Lemma 3.3.1, $\det A_{n-1} = \pm 2$. On the other hand, by the obvious construction, $\det B = 0$ since the last row of B has only zeros.

It remains to deal with case where $r = n$. In this case, the corresponding log-matrix is the $r \times r$ matrix,

$$\begin{pmatrix} 1 & 0 & 0 & 0 & \cdots & 0 & 0 & 1 \\ 1 & 1 & 0 & 0 & \cdots & 0 & 0 & 0 \\ 0 & 1 & 1 & 0 & \cdots & 0 & 0 & 0 \\ \vdots & \vdots & \vdots & \vdots & \cdots & 0 & 0 & 0 \\ 0 & 0 & 0 & 0 & \cdots & 1 & 1 & 0 \\ 0 & 0 & 0 & 0 & \cdots & 0 & 1 & 1 \end{pmatrix},$$

with r odd. Laplace expansion along the first row readily shows that the determinant of this matrix is 2, as required.

(iii) $\Rightarrow$ (ii) Since the graph is connected and has exactly n edges, then it must contain a unique cycle. An even cycle would produce a nonzero polynomial relation among the monomials, which is forbidden by the assumption that the map is dominant. □

The next result gives graph theoretic conditions for monomial birationality in degree two.

Theorem 3.3.9. *Let $F \subset R$ be a finite set of monomials of degree two having no nontrivial common factor and let $\mathcal{G} \subset \tilde{\mathcal{G}}$ denote the corresponding graphs as above. Let A denote the incidence matrix of $\tilde{\mathcal{G}}$. The following conditions are equivalent:*

(i) *F is cohesive and A has maximal rank.*
(ii) *The extension $k[F] \subset k[\mathbf{x}_2]$ is birational.*
(iii) *$\mathcal{G}$ is connected and, moreover, either it is nonbipartite or else it is bipartite and $\tilde{\mathcal{G}} \setminus \mathcal{G} \neq \emptyset$.*

Proof. (i) $\Rightarrow$ (ii) By Proposition 3.3.7, $\mathcal{LS}(F)$ has maximal rank. The hypothesis that A has maximal rank means that the rational map defined by the forms of F is generically finite, i. e., $\dim k[F] = \operatorname{rank} A = n$. Therefore, the result follows from Theorem 3.2.26.

The reverse implication is similar.

(iii) $\Rightarrow$ (ii). Since $\mathcal{G}$ is connected, there is a spanning tree T of $\mathcal{G}$ containing all the vertices of $\mathcal{G}$.

Suppose that $\mathcal{G}$ is a bipartite graph and x_n is a loop of $\tilde{\mathcal{G}}$. We may then regard T as a tree with a loop at x_n. Notice that T has exactly $n-1$ simple edges plus a loop. The incidence matrix B of T has order n, is nonsingular, and we may assume that the last column of B is the transpose of $(0,0,\dots,0,2)$. Consider the matrix B' obtained from B by removing the last column. The matrix B' is totally unimodular because it is the incidence matrix of a simple bipartite graph [171, p. 273]. Therefore, $\det(B) = \pm 2$ and $\operatorname{rank}(A) = n$. From Lemma 3.3.1, we obtain that $k[T] \subset k[\mathbf{x}_2]$ is birational; hence, $k[G] \subset k[\mathbf{x}_2]$ is birational as well.

Now, let $\mathcal{G}$ be a nonbipartite graph. Then $\operatorname{rank}(A) = n$. Since $\mathcal{G}$ has a spanning tree and G has at least one odd cycle [82, pp. 37–39 and p. 42], then $\mathcal{G}$ admits a connected simple subgraph $\mathcal{H}$ with n vertices and n edges with a cycle of odd length and no other cycles. By Proposition 3.3.8, the extension $k[\mathcal{H}] \subset k[\mathbf{x}_2]$ is birational; hence, so is $k[\mathcal{G}] \subset k[\mathbf{x}_2]$.

(ii) $\Rightarrow$ (iii). By Proposition 3.3.7, F must be cohesive, i. e., $\mathcal{G}$ is connected. We have already seen that $\operatorname{rank} A = \dim k[F] = n$. Suppose that $\mathcal{G}$ is bipartite. Then the log-matrix of $\mathcal{G}$ has rank $n-1$; hence, $\tilde{\mathcal{G}}$ has at least one loop. □

Example 3.3.10. The geometric version of Theorem 3.3.9 is that any cohesive coordinate projection of the 2-Veronesean that preserves dimension is birational onto the image. This is clearly false if the projection is a noncoordinate cohesive map, e. g., if F is a set of 2-forms forming a cohesive regular sequence (the simplest example with $n = 2$ would be $F = \{x_1x_2, x_1^2 - x_2^2\}$. At the other end, for $d > 2$, a cohesive coordinate projection of the d-Veronesean preserving dimension may fail to be birational for the simple reason that it may be composed with a noncohesive set. The simplest example of this phenomenon is $F = \{x_1^4, x_1^2x_2^2, x_2^4\}$. Here, $k[F] \subset k[(x_1,x_2)_4]$ is not birational, but its “reparametrization” $F' = \{y_1^2, y_1y_2, y_2^2\}$ is the 2-Veronesean. For $n > 2$, one of the simplest examples is $F = \{x_1^3, x_1^2x_2, x_2x_3^2\}$, which is cohesive of maximal rank, non-reparametrizable and non-birational: the ideal $(F) \subset k[x_1,x_2,x_3]$ is of linear type, but the linear syzygy matrix is of rank 1; hence, falls below the needed value 2 (of course, this apparatus in such a simple example is worthless since one immediately sees that x_3^3 does not belong to the field of fractions of $k[x_1^3, x_1^2x_2, x_2x_3^2]$).

Corollary 3.3.11. *Let $\mathcal{G}$ be a connected simple bipartite graph. Assume, without loss of generality, that $x_{n-1}x_n$ is an edge of an even cycle of $\mathcal{G}$. Then $k[\widetilde{G \setminus x_n}] \subset k[(\mathbf{x} \setminus x_n)_2]$ is a birational extension, where $\widetilde{G \setminus x_n}$ is the graph on the vertices $X \setminus x_n = \{x_1, \ldots, x_{n-1}\}$ obtained by contracting the edge $x_{n-1}x_n$ to a loop around the vertex x_{n-1}.*

Proof. By the contracting-looping transformation, the resulting graph $\widetilde{G \setminus x_n}$ acquires an odd cycle. Therefore, the simple subgraph induced by $\widetilde{G \setminus x_n}$ is nonbipartite and the assertion follows from Theorem 3.3.9. □

3.4 On the degree of a rational map

In this section, one deals with the (field theoretic or topological) degree of a rational map, not to be confused with the degree of the forms of a representative.

Let k denote a field of arbitrary characteristic and let $X \subseteq \mathbb{P}_k^n = \operatorname{Proj}(k[x_0, \ldots, x_n])$ be a closed projective variety. One is given a rational map

$$\Phi : X \dashrightarrow \mathbb{P}_k^m = \operatorname{Proj}(k[y_0, \ldots, y_m]),$$

with (closed) image Y. We will often write $k[\mathbf{x}] = k[x_0, \ldots, x_n]$ and $k[\mathbf{y}] = k[y_0, \ldots, y_m]$.

We may also need to consider rational maps $\mathfrak{X} \dashrightarrow \mathbb{P}_k^m$, with source a biprojective variety $\mathfrak{X} \subset \mathbb{P}_k^{n_1} \times_k \mathbb{P}_k^{n_2}$. Given a closed projective (resp., biprojective) variety $X \subset \mathbb{P}_k^n$ (resp., $\mathfrak{X} \subset \mathbb{P}_k^n \times_k \mathbb{P}_k^m$), we denote by $K(X)$ (resp., $K(\mathfrak{X})$) its field of functions. Recall that $K(X)$ consists of the homogeneous fractions of degree zero in the total ring of fractions of the standard graded domain R over k such that $X = \operatorname{Proj}(R)$. The same goes for Y and for $\mathfrak{X}$ one takes the fractions of bihomogeneous elements of the same bidegree in the total ring of fractions of the standard bigraded domain R over k such that $X = \operatorname{BiProj}(R)$.

The contents of this part are based on [36]. The exposé here has a more algebraic flavor.

3.4.1 Basic ingredients

Definition 3.4.1. The *degree* of the rational map $\Phi : X \dashrightarrow Y$ (resp., $\mathfrak{X} \dashrightarrow \mathbb{P}_k^m$) with source a projective (resp., biprojective) variety is the degree of the field extension $K(X)|K(Y)$ (resp., $K(\mathfrak{X})|K(Y)$); Φ is said to be *generically finite* if $[K(X) : K(Y)] < \infty$ (resp., $[K(\mathfrak{X}) : K(Y)] < \infty$).

The following result is the well-known geometric version of the case of a finite degree.

Lemma 3.4.2. *Let $\Phi : X \dashrightarrow Y$ (resp., $\mathfrak{X} \dashrightarrow \mathbb{P}_k^m$) be a rational map with source a projective (resp., biprojective) variety. The following are equivalent:*

(i) Φ *is generically finite.*
(ii) $\dim(X) = \dim(Y)$ (*resp.*, $\dim(\mathfrak{X}) = \dim(Y)$).
(iii) *There exists a Zariski dense open subset* $U \subset Y$ *such that* $\Phi^{-1}(U) \to U$ *is a finite morphism.*

We refer to Section 3.1 for further properties of these maps and the related terminology.

The degree $\deg(X)$ of the projective variety $X = \operatorname{Proj}(R)$ in its embedding is the Hilbert multiplicity $e_0(R)$ of the standard graded k-algebra R.

An essential role will be played in this part by the Rees algebra $\mathcal{R}_R(I) = R[Iu]$ of I and by its special fiber $\mathcal{R}_R(I)/\mathfrak{m}\mathcal{R}_R(I)$, where $\mathfrak{m} = (R_+)$. Here, I is an ideal generated by forms of the same degree. Recall from the preliminaries (Section 2.3.1.3) that $\mathcal{R}_R(I)/\mathfrak{m}\mathcal{R}_R(I) \simeq k[I_d]$ as graded k-algebras, which are further identified up to degree renormalization with the homogeneous defining ring of the closed image Y of Φ.

Remark 3.4.3. It is important to note heretofore that we will fix a base ideal of Φ, but any two base ideals induce graded isomorphisms between the respective Rees algebras and fiber cones, as have been noted before in this chapter. This will give certain slight instability in the main result, but this is a small price to pay as far as applications are concerned.

The first auxiliary result is probably known (see [190, (ii), p. 251] for similar considerations). We give an elementary proof.

Lemma 3.4.4. *Let* $A = \bigoplus_{i=0}^{\infty} A_i$ *denote a standard graded Noetherian domain of dimension 1 with* $k := A_0$ *a field. Let* $K(A)$ *be the subfield of homogeneous fractions of degree zero in the field of fractions of* A. *Then* $[K(A) : k] = e_0(A)$, *where* $e_0(A)$ *is the graded multiplicity of* A.

Proof. Set $e := e_0(A)$. Fix an integer $d \geq 1$ such that $\dim_k(A_i) = e$ for $i \geq d$ and let $\{a_1, \ldots, a_e\}$ be a k-vector basis for A_d. Letting $Q(A)$ denote the fraction field of A, one knows that $Q(A) = K(A)(H)$ for some transcendental linear form. Since $\dim(A) = 1$, then $\operatorname{trdeg}_k Q(A) = 1$, hence $\operatorname{trdeg}_k K(A) = 0$.

Clearly, $\{a_1/H^d, \ldots, a_e/H^d\} \subset K(A)$ and they are linearly independent over k. Thus, it suffices to prove that $K(A)$ is spanned by $\{a_1/H^d, \ldots, a_e/H^d\}$ as a k-vector space.

First, any fraction of the form β/H^{ld}, with $\beta \in A_{ld}$, $l \geq 1$, is an element of the k-vector space V spanned by $\{a_1/H^d, \ldots, a_e/H^d\}$. Indeed, by construction, $\dim_k(A_{ld}) = e$ for any such, hence there exist $\gamma_1, \ldots, \gamma_e \in k$ such that $\beta = \sum \gamma_i H^{(l-1)d} a_i$. Clearly, then $\beta/H^{ld} = \sum \gamma_i a_i/H^d$.

The inverse H^{ld}/β also belongs to V. To see this, note that $K(A)$ is algebraic over k, hence there exist $\theta_i \in k$ such that

$$\frac{\beta}{H^{ld}}\left(\sum \theta_i \left(\frac{\beta}{H^{ld}}\right)^i\right) = 1.$$

Since the terms inside the parenthesis belong to V, so does H^{ld}/β.

The argument for an arbitrary homogeneous fraction β/γ is now clear, by writing $\beta/\gamma = (H^{ld}/\gamma)(\beta/H^{ld})$, where $ld = \deg\beta = \deg\gamma$. □

Next, consider a homogeneous ideal $\mathfrak{a} \subset S := k[x_0, \dots, x_n]$. The degree $e_0(S/\mathfrak{a})$ is classically known to be the sum of the degrees of its primary components of maximal possible dimension. In other words, the classical degree only looks at the minimal primes of $S/\mathfrak{a}$ of same dimension as $S/\mathfrak{a}$. Here, we wish to consider the simplest of the generalized degrees considered in [216, Chapter 9], and further only care about the reduced structure. That is, we consider the prime decomposition of the radical

$$\sqrt{\mathfrak{a}} = \bigcap_{\mathfrak{p}\in \mathrm{Min}(S/\mathfrak{a})} \mathfrak{p}. \tag{28}$$

Definition 3.4.5. The *radical degree* of $S/\mathfrak{a}$ is

$$\delta(S/\mathfrak{a}) := \sum_{\mathfrak{p}\in \mathrm{Min}(R/\mathfrak{a})} e_0(S/\mathfrak{p}).$$

Clearly, then $\delta(S/\mathfrak{a}) = e_0(S/\mathfrak{a})$ if and only if $S/\mathfrak{a}$ is equidimensional and reduced.

The following result explains the behavior of the radical degree, under hypersurface sections.

Proposition 3.4.6. *With the above notation, let $f \in S$ be any nonzero form. Then $\delta(S/(\mathfrak{a},f)) \le \delta(S/\mathfrak{a})\deg(\sqrt{f})$.*

Proof. Drawing upon (28), we have that $(\mathfrak{a},f)$ and $\bigcap_{\mathfrak{p}}(\mathfrak{p},f)$ have the same radical. Thus, the minimal primes of the first are among the minimal primes of the individual intersecting terms of the second, which yields an inequality $\delta(S/(\mathfrak{a},f)) \le \sum_{\mathfrak{p}} \delta(S/(\mathfrak{p},f))$, allowing for possible repetitions.

Repeating the argument this time around for $\sqrt{f} = \prod_p p$ decomposed into its prime factors, one has $\delta(S/(\mathfrak{p},f)) \le \sum_p \delta(S/(\mathfrak{p},p))$, for any minimal prime $\mathfrak{p}$ of $S/\mathfrak{a}$ and any prime factor p of f.

Now, one has

$$\delta(S/(\mathfrak{p},p)) = \begin{cases} \delta(\mathfrak{p}) = e_0(S/\mathfrak{p}), & \text{if } p \in \mathfrak{p} \\ \sum_q e_0(S/q) \le \sum_q i_q(\mathfrak{p},p)e_0(S/q) = e_0(S/\mathfrak{p})\deg(p), & \text{else,} \end{cases}$$

by the classical Bézout theorem [179, Theorem 2.7.32], where q runs trough the minimal primes of $S/(\mathfrak{p},p)$ and $i_q(\mathfrak{p},p)$ is the intersection multiplicity of $S/\mathfrak{p}$ and $S/(p)$ along q.

Putting together all inequalities yields the required result. □

The next result is more involved: it essentially explains how the partial sums of the degrees of the irreducible components of a scheme can be bounded by partial products of the degrees of its defining equations. In its proof, we use the following prime avoidance preliminary.

Lemma 3.4.7. *Let R stand for a standard graded ring over an infinite field and let $\mathfrak{a} \subset R$ be a homogeneous ideal generated by forms of degrees $d_1 \geq \cdots \geq d_r \geq 1$. Then there exists a subset $\{\mathfrak{f}_1, \ldots, \mathfrak{f}_r\} \subset \mathfrak{a}$ with $\deg(\mathfrak{f}_i) = d_i$ such that $\operatorname{ht}(\mathfrak{f}_1, \ldots, \mathfrak{f}_{i-1}, \mathfrak{f}_i)_{\mathfrak{p}} = i$ for all $\mathfrak{p} \in \operatorname{Min}(R/\mathfrak{a})$ and every $i = 1, \ldots, \operatorname{ht} \mathfrak{p}$.*

Proof. We first show that the assertion is valid for a given $\mathfrak{p} \in \mathfrak{P} := \{\mathfrak{p} \in \operatorname{Min}(R/\mathfrak{a})\}$, for which we induct on $\operatorname{ht} \mathfrak{p}$.

For $\operatorname{ht} \mathfrak{p} = 0$, the result is trivially, or vacuously, verified by selecting the set of the given generators of $\mathfrak{a}$.

Thus, assume that $\operatorname{ht} \mathfrak{p} \geq 1$. If $\mathfrak{q}$ is a prime of R with $\dim R = \dim R/\mathfrak{q}$, then $\operatorname{ht} \mathfrak{q} = 0$. If $\mathfrak{q} \subset \mathfrak{p}$, then $\mathfrak{a} \not\subset \mathfrak{q}$ since $\mathfrak{p}$ is minimal over $\mathfrak{a}$.

Letting $\{f_1, \ldots, f_r\}$ be the given generators of $\mathfrak{a}$, by one of the forms of homogeneous prime avoidance (see [121, Theorem 124] for a nonhomogeneous version) there exist forms $h_2, \ldots, h_r \in \mathfrak{a}$, of respective degrees $d_1 - d_i$, such that $\mathfrak{f}_1 := f_1 + \sum_{i\geq 2} h_i f_i \in \mathfrak{a}_{d_1}$ avoids the finitely many minimal prime ideals of R of maximal dimension contained in $\mathfrak{p}$.

Taking residues modulo the principal ideal $(\mathfrak{f}_1)$, one has

$$\operatorname{ht} \mathfrak{p}/(\mathfrak{f}_1) = \dim R_{\mathfrak{p}}/(\mathfrak{f}_1)_{\mathfrak{p}} = \dim R_{\mathfrak{p}} - 1 = \operatorname{ht} \mathfrak{p} - 1$$

since locally at $\mathfrak{p}$, $\mathfrak{f}_1$ avoids the primes of $R_{\mathfrak{p}}$ of maximal dimension. Let a bar over an element of R denote its residue class in $R/(\mathfrak{f}_1)$. Thus, $\bar{\mathfrak{p}}$ is a minimal prime of $\bar{\mathfrak{a}}$ of height $\operatorname{ht} \mathfrak{p} - 1$. Moreover, $\bar{\mathfrak{a}}$ is generated by forms of degrees $d_2 \geq \cdots \geq d_r \geq 1$. By the inductive hypothesis, there exist forms $\bar{\mathfrak{f}}_2, \ldots, \bar{\mathfrak{f}}_r \in \bar{\mathfrak{a}}$ of respective degrees $d_2, \ldots, d_r$ such that $\operatorname{ht}(\bar{\mathfrak{f}}_2, \ldots, \bar{\mathfrak{f}}_i)_{\bar{\mathfrak{p}}} = i - 1$, for $1 \leq i - 1 \leq \operatorname{ht} \mathfrak{p} - 1$. Since $\operatorname{ht}(\mathfrak{f}_1)_{\mathfrak{p}} = 1$, one obtains

$$\operatorname{ht}(\mathfrak{f}_1, \mathfrak{f}_2, \ldots, \mathfrak{f}_i)_{\mathfrak{p}} \geq \operatorname{ht}(\bar{\mathfrak{f}}_2, \ldots, \bar{\mathfrak{f}}_i)_{\bar{\mathfrak{p}}} + 1 = i,$$

and equality ensues by Krull's theorem.

So much for the case of one single minimal prime. Note that at each step of the proof, a general choice has been made by the prime avoidance principle. Therefore, at each such step, identifying the k-vector space $\mathfrak{a}_{d_i}$ with the corresponding affine space $\mathbb{A}^{n_i}$, there is a nonempty open set $\mathcal{O}_{i,\mathfrak{p}} \subset \mathbb{A}^{n_i}$ of these choices. Then, for each $\mathfrak{p} \in \mathfrak{P}$, the set

$$\mathcal{P}_{\mathfrak{p}} := \{(\mathfrak{f}_1 \cdots \mathfrak{f}_r) \in \mathbb{A}^{n_1} \times \cdots \times \mathbb{A}^{n_r} \mid \operatorname{ht}(\mathfrak{f}_1, \mathfrak{f}_2, \ldots, \mathfrak{f}_i)_{\mathfrak{p}} = i, 1 \leq i \leq \operatorname{ht} \mathfrak{p}\}$$

contains a nonempty open set $U_{\mathfrak{p}} := \mathcal{O}_{1,\mathfrak{p}} \times \cdots \times \mathcal{O}_{r,\mathfrak{p}}$. In order to conclude, take the intersection $\bigcap_{\mathfrak{p}\in\mathfrak{P}} U_{\mathfrak{p}} \subset \mathbb{A}^{n_1} \times \cdots \times \mathbb{A}^{n_r}$. Then any set $\{\mathfrak{f}_1 \cdots \mathfrak{f}_r\}$ such that $(\mathfrak{f}_1 \cdots \mathfrak{f}_r) \in \bigcap_{\mathfrak{p}\in\mathfrak{P}} U_{\mathfrak{p}}$ satisfies the request. □

Proposition 3.4.8. *Let D stand for a standard graded domain over a field and let $\mathfrak{a} \subset D$ be a homogeneous ideal generated by forms of degrees $d_1 \geq \cdots \geq d_r \geq 1$. Then, for any* *fixed* $1 \leq t \leq r$, *one has*

$$\sum_{\substack{\mathfrak{p} \in Min(D/\mathfrak{a}) \\ \mathrm{ht}\,\mathfrak{p} \leq t}} e_0(R/\mathfrak{p}) \leq e_0(D)\, d_1 \cdots d_{t-1} d_r.$$

Proof. By passing to a pure algebraic extension, one can assume that k is infinite.

Say, $D \simeq k[x_0, \ldots, x_n]/I$. Let $f_i \in k[x_0, \ldots, x_n]$ denote a homogeneous inverse image of $\mathfrak{f}_i$, so that $\deg f_i = \deg \mathfrak{f}_i$. By an iterated version of Proposition 3.4.6, one has

$$\begin{aligned}\delta(D/(\mathfrak{f}_1, \ldots, \mathfrak{f}_{t-1}, \mathfrak{f}_r)) &= \delta(S/(I, f_1, \ldots, f_{t-1}, f_r)) \\ &\leq \delta(S/I) \deg \sqrt{f_1} \cdots \deg \sqrt{f_{t-1}} \deg \sqrt{f_r} \\ &\leq e_0(D) \deg f_1 \cdots \deg f_{t-1} \deg f_r \\ &\leq e_0(D)\, d_1 \cdots d_{t-1} d_r.\end{aligned}$$

Now use Lemma 3.4.7 by observing that, since D is assumed to be a domain, we can first pass to D modulo a generator of smallest degree d_r, and then apply the contents of the lemma. Clearly, then every $\mathfrak{p} \in \mathrm{Min}(D/\mathfrak{a})$ of height $\leq t$ is a minimal prime of $D/(\mathfrak{f}_1, \ldots, \mathfrak{f}_{t-1}, \mathfrak{f}_r)$ as well; therefore, we are through. □

3.4.2 Degree upper bounds

In this part, we prove the main result on bounding the degree of a rational map. Recall the basic notation: R is a standard graded ring over a field k. For an ideal $I \subset R$ generated by forms of the same degree, the Rees algebra $\mathcal{R}_R(I)$ has a presentation $S/\mathcal{J} \simeq \mathcal{R}_R(I)$, where $S := R[\mathbf{y}] = k[y_0, \ldots, y_m]$ is endowed with the usual standard bigrading, and $\mathcal{J}$ is bihomogeneous in this bigrading. The geometric version is the biprojective variety $\mathrm{BiProj}(\mathcal{R}_R(I)) \subset \mathrm{Proj}(R) \times \mathbb{P}^m_k$.

3.4.2.1 Main result

As seen earlier, one can write $\mathcal{R}_R(I) = [\mathcal{R}_R(I)]_0 \oplus \mathfrak{m}\mathcal{R}_R(I)$, a direct sum decomposition of graded modules over $[\mathcal{R}_R(I)]_0 \simeq k[I_d]$, with $\mathfrak{m} = (R_+)$. By a previous identification, the latter is the special fiber $\mathcal{F}(I)$ of I and also the homogeneous coordinate ring of the closed image Y of Φ. In particular, the field of fractions of $[\mathcal{R}_R(I)]_0$ is identified with the field $Q(Y)$ of fractions of this ring.

Set $B := k[\mathbf{y}] \subset S := R[\mathbf{y}]$ and take respective presentations $[\mathcal{R}_R(I)]_0 \simeq B/\mathfrak{q}$ and $\mathcal{R}_R(I) \simeq S/\mathcal{J}$. In this notation, one has

$$\begin{aligned}\mathcal{R}_R(I) \otimes_{[\mathcal{R}_R(I)]_0} Q(Y) &\simeq \mathcal{R}_R(I) \otimes_{B/\mathfrak{q}} B_q/qB_q \simeq (\mathcal{R}_R(I) \otimes_B B/\mathfrak{q}) \otimes_{B/\mathfrak{q}} B_q/qB_q \\ &\simeq \mathcal{R}_R(I) \otimes_B (B/\mathfrak{q} \otimes_{B/\mathfrak{q}} B_q/qB_q) \simeq \mathcal{R}_R(I) \otimes_B B_q/qB_q \\ &\simeq \mathcal{R}_R(I) \otimes_B Q(Y) \simeq (S/\mathcal{J}) \otimes_B Q(Y). \qquad (29)\end{aligned}$$

Theorem 3.4.9. *Let $X \subseteq \mathbb{P}^n_k = \operatorname{Proj}(k[x_0,\dots,x_n])$ be a reduced and irreducible nondegenerate variety and let $\Phi : X \dashrightarrow Y \subset \mathbb{P}^m_k$ be a generically finite rational map. Fix a base ideal $I \subset R$ of Φ and let $\mathcal{J} \subset S$ as above be a presentation ideal of $\mathcal{R}_R(I)$. Given biforms of $\mathcal{J}$ of $\mathbf{x}$-degrees $d_1 \geq \cdots \geq d_r \geq 1$, suppose that $\mathcal{J}$ is a minimal prime over the ideal $\mathcal{I}$ generated by these biforms. Then*

$$\deg(\Phi) \leq d_1 \cdots d_{t-1} \cdot d_r \cdot e(R),$$

where $t = \dim X$.

Proof. First, note that the statement at least makes sense because $t \leq r$; indeed, since a finitely generated domain over a field is catenary, one has

$$\begin{aligned} t = \dim X = \dim Y \leq m &= \operatorname{ht}(\mathcal{J}) \\ &\leq r, \quad \text{by Krull's prime ideal theorem.} \end{aligned}$$

With the notation of (29), one has the following.

Claim 1. $(S/\mathcal{J}) \otimes_B Q(Y)$ is a one-dimensional domain.

$(S/\mathcal{J}) \otimes_B Q(Y)$ is a domain because it is the ring of fractions of the domain $S/\mathcal{J}$ with respect to the multiplicatively closed subset $\bar{T} \subset S/\mathcal{J}$, image of $T := B \setminus \mathfrak{q} \subset S$ by the natural surjection $S \twoheadrightarrow S/\mathcal{J}$.

As for the dimension, since R is a domain, one has by [194, Lemma 1.1.2]:

$$\dim R + 1 = \dim \mathcal{R}_R(I) = \dim[\mathcal{R}_R(I)]_0 + \operatorname{ht} \mathfrak{m}\mathcal{R}_R(I) = \dim \mathcal{F}(I) + \operatorname{ht} \mathfrak{m}\mathcal{R}_R(I).$$

Since Φ is generically finite, $\dim \mathcal{F}(I) = \dim R$. Then $\operatorname{ht} \mathfrak{m}\mathcal{R}_R(I) = 1$.

But

$$\begin{aligned} \mathcal{R}_R(I) \otimes_{[\mathcal{R}_R(I)]_0} Q(Y) &= ([\mathcal{R}_R(I)]_0 \otimes_{[\mathcal{R}_R(I)]_0} Q(Y)) \bigoplus (\mathfrak{m}\mathcal{R}_R(I) \otimes_{[\mathcal{R}_R(I)]_0} Q(Y)) \\ &= Q(Y) \bigoplus \mathfrak{m}\mathcal{R}_R(I) \otimes_{[\mathcal{R}_R(I)]_0} Q(Y) \\ &= Q(Y)[\mathfrak{m}\mathcal{R}_R(I) \otimes_{[\mathcal{R}_R(I)]_0} Q(Y)] \end{aligned}$$

Since $\mathfrak{m}\mathcal{R}_R(I) \otimes_{[\mathcal{R}_R(I)]_0} Q(Y)$ is the maximal irrelevant ideal of $\mathcal{R}_R(I) \otimes_{[\mathcal{R}_R(I)]_0} Q(Y)$, it suffices to see that it has height one. But this follows from the fact that it is fractions of the prime ideal $\mathfrak{m}\mathcal{R}_R(I)$ and the latter has height one.

So much for the claim.

Claim 2. $\deg \Phi = e_0(S/\mathcal{J} \otimes_B Q(Y))$.

To see this, introduce the blowup diagram

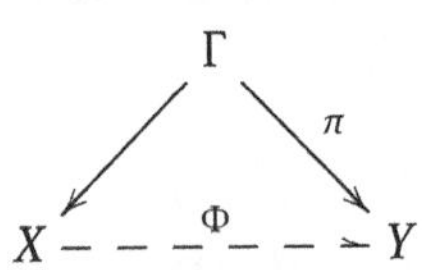

(30)

where $\Gamma = \operatorname{BiProj}(\mathcal{R}_R(I))$.

Since the structural map $\Gamma \to X$ is birational, it suffices to show the equality for $\deg \pi$ instead. By Lemma 3.4.4, it amounts to prove that $\deg \pi = [K(S/\mathcal{J} \otimes_B Q(Y)) : Q(Y)]$.

Note that $Q(Y)|K(Y)$ is pure transcendental of degree one. Similarly, $K(S/\mathcal{J} \otimes_B Q(Y))|K(\Gamma)$ is pure transcendental of degree one because $K(S/\mathcal{J} \otimes_B Q(Y))$ is the subfield of the field of fractions of $S/\mathcal{J}$ whose elements are fractions of forms with same $\mathbf{x}$-degree. Moreover, we can generate both extensions by a single sufficiently general k-linear form $0 \neq H \in B/\mathfrak{q} \subset Q(Y)$.

Clearly, then $\deg(\pi) = [K(\Gamma) : K(Y)] = [K((S/J) \otimes_B Q(Y)) : Q(Y)]$, as was to be shown.

Claim 3. The following properties hold:

- $e_0(S \otimes_B Q(Y)) = e_0(R)$.
  One has $S \otimes_B Q(Y) = R[\mathbf{y}] \otimes_B Q(Y) \simeq (R \otimes_k k[\mathbf{y}]) \otimes_{k[\mathbf{y}]} Q(Y) \simeq R \otimes_k Q(Y)$. Since the Hilbert multiplicity is stable under a ground field extension, $e_0(R) = e_0(R \otimes_k Q(Y))$.
- $S \otimes_B Q(Y) \simeq (S \otimes_B B_\mathfrak{q})/(\mathfrak{q}S \otimes_B B_\mathfrak{q})$ and $\mathcal{J} \otimes_B Q(Y) \simeq (\mathcal{J} \otimes_B B_\mathfrak{q})/(\mathfrak{q}S \otimes_B B_\mathfrak{q})$.
  Since $B_\mathfrak{q}$ is flat over B, one has $(S \otimes_B B_\mathfrak{q})/(\mathfrak{q} \otimes_B B_\mathfrak{q}) \simeq S/\mathfrak{q}S \otimes_B B\mathfrak{q}$. The rest is obvious since $Q(Y)$ is the field of fractions of $B/\mathfrak{q}$. The same argument holds for $(\mathcal{J} \otimes_B B_\mathfrak{q})/(\mathfrak{q}S \otimes_B B_\mathfrak{q})$.
- Let $\mathcal{I} \subset \mathcal{J}$ denote the ideal generated by the biforms in the statement of the theorem. Then $\mathcal{J} \otimes_B Q(Y)$ is a minimal prime over the residue in $S \otimes_B Q(Y)$ of $\mathcal{I} \otimes_B B_\mathfrak{q}$. Clearly, since $\mathcal{J}$ is a minimal prime over $\mathcal{I}$ by assumption,
- $\operatorname{ht} \mathcal{J} \otimes_B Q(Y) \leq \dim X$.
  In fact, one has

$$\begin{aligned}\operatorname{ht} \mathcal{J} \otimes_B Q(Y) &\leq \dim S \otimes_B Q(Y) - \dim S/\mathcal{J} \otimes_B Q(Y)\\ &= \dim R \otimes_k Q(Y) - 1 = \dim R - 1 = \dim X,\end{aligned}$$

  by Claim 1 and a previous remark.

Now apply Proposition 3.4.8 with $D := S \otimes_B Q(Y)$ and $\mathfrak{a}$ the residue of $\mathcal{I} \otimes_B B_\mathfrak{q}$ in $S \otimes_B Q(Y)$. The latter is generated by the residues of the generators of $\mathcal{I}$ and the $\mathbf{x}$-degrees are preserved. Drawing upon Claim 2 and the four items of Claim 3, the stated inequality follows. □

Remark 3.4.10. Note that $\mathcal{R}_R(I) \otimes_{\mathcal{R}_R(I)_0} Q(Y)$ localized at $\mathfrak{m}(\mathcal{R}_R(I) \otimes_{\mathcal{R}_R(I)_0} Q(Y))$ is isomorphic to $\mathcal{R}_R(I)_{\mathfrak{m}\mathcal{R}_R(I)}$. Then birationality of Φ is tantamount to the assumption that the latter local ring have multiplicity one, i. e., that it be regular. In other words, the maximal ideal $\mathfrak{m}\mathcal{R}_R(I)_{\mathfrak{m}\mathcal{R}_R(I)}$ is generated by one single element. Thus, locally at $\mathfrak{m}R[\mathbf{y}]$ modulo $\mathcal{J}$, there is one single generator. This forces $\mathcal{J}$ to admit sufficient biforms of $\mathbf{x}$-degree one. This retro explains the shape of Theorem 3.2.22 in terms of the Jacobian dual rank.

3.4.2.2 Application

Proposition 3.4.11 (Syzygy rank criterion). *Let $\Phi : X \dashrightarrow Y \subset \mathbb{P}^m_k$ be a generically finite rational map with source a nondegenerate reduced and irreducible variety $X = \mathrm{Proj}(R) \subsetneq \mathbb{P}^n_k$. Fix a base ideal $I \subset R$ of Φ and its minimal graded presentation*

$$F_1 = \bigoplus^{r} R(-d-d_i) \xrightarrow{\psi} F_0 = \bigoplus^{m+1} R(-d) \to I \to 0.$$

Let G be a finitely generated graded R-submodule of $\mathrm{Im}(\psi) \subset F_0(+d)$ of rank m, generated in degrees $d_{i_1} \geq \cdots \geq d_{i_s}$. If $\mathrm{grade}_R(I) \geq 2$, then $\deg(\Phi) \leq d_{i_1} \cdots d_{i_{t-1}} \cdot d_{i_s} \cdot e(R)$ where $t = \dim X$.

Proof. Let $\mathcal{N}$ be the $(m+1) \times s$ matrix whose columns are the generators of G. First, note that $r \geq s \geq m \geq \dim Y = \dim X = t$. Since $\mathcal{N}$ has rank m, we can choose an $(m+1) \times m$ submatrix $\mathcal{M}$ of $\mathcal{N}$ of rank m. With a previous notation, let $J \subset S = R[\mathbf{y}]$ denote a presentation ideal of the Rees algebra of I. Then one has inclusions of ideals

$$L_{\mathcal{M}} := I_1(\mathbf{y} \cdot \mathcal{M}) \subset L_{\mathcal{N}} := I_1(\mathbf{y} \cdot \mathcal{N}) \subset L := I_1(\mathbf{y} \cdot \psi) \subset J.$$

Claim. J is a minimal prime over $L_{\mathcal{N}}$.

It suffices to prove that J is a minimal prime over $L_{\mathcal{M}}$. To see this, let $\Delta_0, \ldots, \Delta_m$ be the maximal minors of $\mathcal{M}$. Since $\mathrm{grade}_R(I) \geq 2$, by [179, Proposition 6.2.30], one has $hI = (\Delta_0, \ldots, \Delta_m)$, for some nonzero element $h \in R$. Let $\mathfrak{S} \subset R$ denote the multiplicatively closed set of powers of h. Taking fractions, $\mathfrak{S}^{-1}I = \mathfrak{S}^{-1}(\Delta_0, \ldots, \Delta_m)$ and $\mathrm{grade}_{\mathfrak{S}^{-1}R}\, \mathfrak{S}^{-1}(\Delta_0, \ldots, \Delta_m) \geq 2$ as well, yielding the free presentation (resolution)

$$0 \to \bigoplus^{m} \mathfrak{S}^{-1}R \xrightarrow{\mathcal{M}} \bigoplus^{m+1} \mathfrak{S}^{-1}R \to \mathfrak{S}^{-1}I \to 0.$$

Thus, $\mathfrak{S}^{-1}(L_{\mathcal{M}})$ is the defining ideal of the symmetric algebra of $\mathfrak{S}^{-1}I$. But, $\mathcal{R}_R(hI) = \mathcal{R}_R(I)$ as R is a domain. Therefore, $\mathfrak{S}^{-1}J$ is a presentation ideal of the Rees algebra of $\mathfrak{S}^{-1}I$ on $\mathfrak{S}^{-1}R$, and hence $\mathfrak{S}^{-1}J$ is a minimal prime over $\mathfrak{S}^{-1}q(L_{\mathcal{M}})$. Since $h \notin J$, J is a minimal prime over $L_{\mathcal{M}}$, thence over $L_{\mathcal{N}}$, too.

Now the result follows from Theorem 3.4.9. □

A special case is as follows.

Corollary 3.4.12. *Let $R = k[\mathbf{x}] = k[x_0, \ldots, x_n]$ be a standard polynomial ring over an infinite field and let $I \subset R$ a perfect homogeneous ideal of codimension two, with graded free resolution*

$$0 \to \bigoplus_{i=1}^{n} R(-d-\mu_i) \longrightarrow R(-d)^{n+1} \longrightarrow I \to 0.$$

If $\mathfrak{F} : \mathbb{P}^n_k \dashrightarrow \mathbb{P}^n_k$ denotes the rational map defined by the generators of I, then $\deg(\mathfrak{F}) \leq \mu_1 \cdots \mu_n$. Moreover, if I satisfies the property (F_1) then equality holds.

Proof. The inequality follows immediately from Proposition 3.4.11 by noting that here $t = n$.

Now, since (F_1) holds, then I is an ideal of linear type. Therefore, the Rees algebra $\mathcal{R}_R(I)$ is a complete intersection defined by n forms of bidegree $(\mu_i, 1)$, for $1 \le i \le n$. Letting $\mathfrak{m} = (\mathbf{x})$, as seen above $\deg(\mathfrak{F}) = \deg(\mathcal{R}_R(I)_{\mathfrak{m}\mathcal{R}_R(I)})$. But the latter is a localization of a domain defined by a regular sequence of forms of degrees μ_i, $1 \le i \le n$, over $k(\mathbf{y})[\mathbf{x}]$, where $\mathbf{y}$ denote the presentation variables of $\mathcal{R}_R(I)$. This shows that $\deg(\mathcal{R}_R(I)_{\mathfrak{m}\mathcal{R}_R(I)}) = \mu_1 \cdots \mu_n$. □

3.5 Exercises

Exercise 3.5.1. Consider the map $\mathfrak{F}\colon \mathbb{P}^5 \dashrightarrow \mathbb{P}^5$ defined by the 3-forms

$$X_0X_1X_2,\ X_0X_2X_3,\ X_0X_4X_5,\ X_1X_2X_4,\ X_2X_3X_5,\ X_3X_4X_5.$$

Show:

(1) The matrix φ_1 is of maximal rank (= 5). In fact, φ_1 generates a free submodule.

(3) The three algebras $\mathcal{S}(\mathrm{coker}(\varphi_1))$, $\mathcal{S}(I)$ and $\mathcal{R}(I)$ are all different, of the same dimension.

(3) The base ideal of the inverse of $\mathfrak{F}$ is generated by 4-forms, so this degree is one less than the expected degree were I of linear type (see Corollary 3.2.27 and its supplement).

(*Hint*: computer program access may be needed.)

Exercise 3.5.2. Let $\Phi = (f_0 : f_1 : f_2) : \mathbb{P}^2_k \dashrightarrow \mathbb{P}^2_k$ be a birational map, with $I = (f_0, f_1, f_2) \subset k[x, y, z]$ the corresponding base ideal.

(1) Elaborate on the relation between Φ and the rational map defined by a different set of generators of I.

(2) Give an example to show that Φ can be an involution, whereas changing the generators of I may fail to be so.

Exercise 3.5.3. Consider the map $\Phi : \mathbb{P}^2_k \dashrightarrow \mathbb{P}^2_k$ defined by the monomials $\{x^2, yz, z^2\}$.

(1) Prove that Φ is dominant, but not birational (use minimally the computer). (*Hint*: for birationality use Lemma 3.3.1; for dominance, use that, since the ideal $I := (x^2, yz, z^2)$ is generically a complete intersection, if x^2, yz, z^2 were algebraically dependent then I would be globally a complete intersection (what theorem are you using?).)

(2) Prove that the rank of the Jacobian dual rank is 1.

(3) Argue that $\deg(\Phi) = 2$. (*Hint*: for general $a, b \in k$, the point $(1 : ab : b^2)$ has exactly two preimages.)

(4) Prove that $\deg(\Phi) = 2$ by computing the field extension degree $[k((x,y,z)_2) : k(x^2, yz, z^2)]$.

Exercise 3.5.4. Let $\Phi : \mathbb{P}^2_k \dashrightarrow \mathbb{P}^2_k$ be defined by $\{x^2y, xz^2, y^2z\}$.

(1) Prove that Φ is dominant, but not birational.
(2) Prove that the rank of the Jacobian dual rank is 0.
(*Hint*: show that the ideal I in (1) is of linear type and it has no linear syzygy.)
(3) Prove that $\deg(\Phi) = 3$ (Proposition 3.4.11 only gives $\deg(\Phi) \le 4$).
(*Hint*: Try the following steps: (i) Show that $y/x \notin k([I]_3)$, by showing this would imply that $x^3, z/x \in k([I]_3)$, which implies birationality; (ii) Show that $(y/x)^3 \in k([I]_3)$, but $(y/x)^2 \notin k([I]_3)$.)

Exercise 3.5.5 (Complements Exercise 2.4.3). Let $I = (x^2, xy, xz, yz) \subset k[x,y,z]$.

(1) Prove that the rational map $\mathbb{P}^2 \dashrightarrow \mathbb{P}^3$ with base ideal I is birational onto the image.
(*Hint*: xy, xz, yz define a birational map of $\mathbb{P}^2$.)
(2) Show that the inverse map is the restriction of a coordinate projection and give the corresponding representative.

Exercise 3.5.6 (Complements Exercise 2.4.4). Let $I = (x^2 - y^2, xy, xz, yz) \subset k[x,y,z]$.

(1) Prove that the rational map $\mathbb{P}^2 \dashrightarrow \mathbb{P}^3$ with base ideal I is birational onto the image.
(2) Show that the inverse map admits three independent representatives in degree 2.

Exercise 3.5.7. Let $I \subset k[x_0, \dots, x_n]$ be a codimension 2 perfect ideal with minimal graded free resolution

$$0 \to R(-(n+1))^n \to R(-(n))^{n+1} \to I \to 0$$

Suppose that I is of linear type. Show that the rational map $\mathfrak{F} : \mathbb{P}^n \dashrightarrow \mathbb{P}^n$ is birational and the base ideal of the inverse $\mathfrak{F}^{-1}$ is also of linear type of degree n. Is this conclusion true if I is not of linear type?

Exercise 3.5.8. Consider the following matrices with entries in $R = k[x_0, x_1, x_2, x_3]$:

$$N := \begin{pmatrix} 0 & -x_1 & -x_1 \\ -x_0 & x_0 & x_1 \\ x_0 & 0 & 0 \\ x_2 & 0 & x_3 \end{pmatrix} \quad \text{and} \quad \widetilde{N} := \begin{pmatrix} 2x_0 & 0 & 0 \\ x_1 & 2x_0 & 0 \\ 0 & 3/2x_1 & 2x_0 \\ -x_3 & x_2 & x_1 \end{pmatrix}.$$

(1) Show that the respective ideals of maximal minors of N and $\widetilde{N}$ are the base ideals of birational maps F and $\widetilde{F}$ of $\mathbb{P}^3$.
(2) Show that neither of the base ideals in (1) is of linear type.

(3) Show that the respective inverses to F and $\widetilde{F}$ are defined by forms of degree 2, whose corresponding base ideals are not Cohen–Macaulay.

(4) Show that the base ideal of the inverse map to $\widetilde{F}$ is generated by the partial derivatives of the determinant of the degenerate Hankel matrix

$$\begin{pmatrix} x_0 & x_1 & x_2 \\ x_1 & x_2 & x_3 \\ x_2 & x_3 & 0 \end{pmatrix}.$$

Exercise 3.5.9. Let $I \subset R = k[x,y,z]$ be the ideal generated by the maximal minors of the following matrix:

$$\begin{bmatrix} z^2 & 0 & 0 & 0 \\ x^2 & z^2 & 0 & 0 \\ 0 & y^2 & z & 0 \\ 0 & 0 & x & z \\ 0 & 0 & 0 & y \end{bmatrix}$$

(1) Prove that I is not generically a complete intersection.

(2) (Computer assistance) Prove that the defining ideal of the fiber cone of I is a perfect ideal of height two.

(3) (Computer assistance) Prove that I defines a birational map onto its image.

(4) Is the Rees algebra $\mathcal{R}_R(I)$ Cohen–Macaulay?

(5) What does Proposition 3.4.11 have to say here?

Exercise 3.5.10. Let $I \subset R = k[x,y,z]$ denote a linearly presented codimension 2 perfect ideal minimally generated by n monomials of degree $n-1$. Suppose that R/I has exactly two minimal primes. Show that

(1) Up to conjugation, the syzygy matrix of I has the form

$$\varphi = \begin{pmatrix} z & 0 & 0 & \cdots & 0 & 0 \\ -c_1 & z & 0 & \cdots & 0 & 0 \\ 0 & -c_2 & z & \cdots & 0 & 0 \\ \vdots & \vdots & \vdots & \ddots & \vdots & \vdots \\ 0 & 0 & 0 & \cdots & z & 0 \\ 0 & 0 & 0 & \cdots & -c_{n-2} & z \\ 0 & 0 & 0 & \cdots & 0 & -c_{n-1} \end{pmatrix} \tag{31}$$

where $c_i \in \{x,y\}$ for each $1 \le i \le n-1$.

(2) The rational map $\mathfrak{F} : \mathbb{P}^2 \dashrightarrow \mathbb{P}^{n-1}$ defined by the linear system I_{n-1} is birational onto its image.

(3) Consider $1 \le r < n-1$. Suppose that

$$c_i = \begin{cases} x, & \text{if } 1 \le i \le r \\ y, & \text{if } r+1 \le i \le n-1 \end{cases}$$

Then the image of $\mathfrak{F}$ is an arithmetically Cohen–Macaulay projective variety.

4 Rational maps, II

4.1 Complements on the base ideal

We are given a reduced and irreducible variety $X \subset \mathbb{P}^n$, with homogeneous coordinate ring $R = k[\mathbf{x}]/I(X) = k[x_0, \ldots, x_n]/I(X)$. Let $\mathbf{f} = \{f_0, \ldots, f_m\}$ be forms in R of the same degree d. We will focus on the base ideal $I := (\mathbf{f}) \subset R$ of the rational map $\mathfrak{F}: X \dashrightarrow \mathbb{P}^m$ defined by these forms.

4.1.1 Degree sequences

Consider the minimal graded free presentation of I,

$$\oplus_s R(-d_s) \xrightarrow{\varphi} R^{m+1}(-d) \to I \to 0 \tag{32}$$

As is customary, we will by abuse call φ both the map above or a representative matrix thereof. The next result ties up properties of $\mathfrak{F}$ with algebraic features of I.

Proposition 4.1.1. *Let $\mathfrak{F}$ and $I \subset R$ be as above.*

(i) *The set of representatives of $\mathfrak{F}$ correspond bijectively to the homogeneous vectors in the rank one-graded R-module $\operatorname{Hom}(I, R) = \ker(\varphi^t)$, where ^{t} denotes transpose.*

(ii) *$\mathfrak{F}$ has a unique representative of lowest possible degree (up to elements of $k \setminus \{0\}$) if and only if I has grade at least two—in which case every other representative is a multiple of the latter by a factor in R.*

(iii) *(k algebraically closed) Let $J_\varphi := I_1(\ker(\varphi^t)) \subset R$ stand for the ideal generated by the entries of $\ker(\varphi^t)$. Then the closed points of X at which $\mathfrak{F}$ is well-defined correspond to the homogeneous submaximal ideals of R not containing J_φ. In particular, $\mathfrak{F}$ is a regular (i. e., everywhere defined) map if and only if J_φ is primary to the irrelevant ideal R_+.*

Proof. (i) Given another representative $\mathbf{f}'$ of $\mathfrak{F}$, then by definition

$$I_2\begin{pmatrix} f_0 & f_1 & \cdots & f_n \\ f'_0 & f'_1 & \cdots & f'_n \end{pmatrix} = 0,$$

i. e., the two sets $\mathbf{f}$ and $\mathbf{f}'$ are proportional by a factor, which is a homogeneous element of the field of fractions K of R. This establishes a bijection between the set of representatives of $\mathfrak{F}$ and the set of homogeneous elements of K that drive $I = (\mathbf{f})$ into R. The latter generate the fractional ideal $R:_K I \simeq \operatorname{Hom}(I, R)$.

(ii) It is well known that the natural inclusion $R \subset \operatorname{Hom}(I, R)$ is an equality if and only if I has grade at least two. Therefore, $\mathfrak{F}$ is uniquely represented up to a proportionality factor in $k[X]$ if and only I has grade at least two.

https://doi.org/10.1515/9783110640694-004

(iii) Clearly, $\ker(\varphi^t) \simeq \mathrm{Hom}(I,R)$, so we pick up from there by using part (i). This means that any representative $\mathbf{f} = \{f_0,\ldots,f_m\}$ of $\mathfrak{F}$ corresponds to the unique homogeneous column vector $\ker(\varphi^t)$ whose coordinates are $f_0,\ldots,f_m$ (we view the vectors of $R^{m+1}(-d)$ as column vectors). For any such $\mathbf{f}$, let $Z(\mathbf{f}) \subset X$ denote the set of zeros of $\{f_0,\ldots,f_m\}$. Clearly, the map $\mathfrak{F}$ is exactly defined in the set $\bigcup_{\mathbf{f}}(X \setminus Z(\mathbf{f})) = X \setminus \bigcap_{\mathbf{f}} Z(\mathbf{f}) = X \setminus Z(\sum_{\mathbf{f}}(\mathbf{f}))$, where $\mathbf{f}$ runs through the set of representatives of $\mathfrak{F}$ (equivalently, the set of homogeneous vectors of $\ker(\varphi^t)$) and $(\mathbf{f}) \subset R$ denotes the ideal generated by the entries of the vector. Therefore, the domain of definition of $\mathfrak{F}$ is precisely the complement of the set of zeros of the ideal of R generated by the coordinates of all homogeneous vectors of $\ker(\varphi^t)$. □

Remark 4.1.2. The proposition emphasizes the effective side of determining whether a rational map $\mathfrak{F}: X \dashrightarrow \mathbb{P}^m$ is regular. Thus, the ideal $I_1(\ker(\varphi^t))$ could naturally be taken as definition of the base ideal of $\mathfrak{F}$ though it is rather an ideal with loose ends.

We emphasize that $\mathfrak{F}$ being everywhere defined does not automatically imply the existence of a representative whose coordinates will generate a $k[X]_+$-primary ideal (see Exercise 4.5.3).

Here is a side application that explains why any rational map defined on a smooth projective curve must be a regular map.

Proposition 4.1.3. *Let R be a 2-dimensional standard graded domain over an algebraically closed field k and let $C = \mathrm{Proj}(R) \subset \mathbb{P}^n$ be any embedding of the corresponding curve. If R is regular locally in codimension, one then every rational map with source C is regular.*

Proof. Let $I \subset R$ denote the base ideal of the given rational map with source C. Let $R^m \xrightarrow{\varphi} R^n \to I \to 0$ be a free graded presentation of I as before and consider the dualized sequence $0 \to I^* \to R^{m*} \to R^{n*}$. Now map, say R^r, surjectively onto I^* so as to get the free exact sequence

$$R^r \xrightarrow{\psi} R^{m*} \to R^{n*}$$

of R-modules. Let $P \subset R$ be any height one prime. Since R_P is regular and $\mathrm{rank}(\psi_P) = \mathrm{rank}(\psi) = n - \mathrm{rank}(\varphi) = 1$—hence ψ_P is injective—we get an exact sequence

$$0 \to R_P \xrightarrow{\psi_P} R_P^{m*} \to R_P^{n*}$$

of R_P-modules. But again R_P is regular of dimension 1, hence every R_P-module—in particular, $\mathrm{coker}(R_P^{m*} \to R_P^{n*})$—has projective dimension at most 1. Therefore, the last sequence must split at its tail, i. e., $I_1(\psi)_P = I_1(\psi_P) = R_P$. But P is an arbitrary height one prime, hence $I_1(\psi)$ is R_+-primary. □

Proposition 4.1.1 motivates the following notion.

Definition 4.1.4. The *degree sequence* of the rational map $\mathfrak{F}: X \dashrightarrow \mathbb{P}^m$ is the sequence $d_1, \dots, d_r$ (with $d_1 \leq \cdots \leq d_r$) of degrees of a minimal set of generators of $\operatorname{Hom}((\mathbf{f}), R)$, where $\mathbf{f}$ is a representative of $\mathfrak{F}$—or, equivalently, the ordered sequence of degrees of the column vectors of the syzygy matrix $\ker(\varphi^t)$.

Thus, the knowledge of this sequence alone is not sufficient to detect the complete nature of the true locus where a rational map is defined. However, it is definitely an invariant of the rational map (i. e., does not depend on the chosen representative). In particular, if for one representative $\mathbf{f}$ the corresponding ideal has grade at least two—for example, if $X = \mathbb{P}^n$—then every representative of $\mathfrak{F}$ is a multiple of $\mathbf{f}$ by a factor which is a homogeneous element of $k[X]$. Clearly, this is then the unique (up to a nonzero factor in the base field k) representative with this property.

This is also a condition under which the base ideal and the "true" base ideal coincide. In this case, the degree sequence consists of a unique number, namely, the degree of $\mathbf{f}$. This degree will then be said to be the *coordinate degree* of $\mathfrak{F}$—to distinguish it from the degree of $\mathfrak{F}$ as a map—and the corresponding representative will generate the (uniquely defined) *base ideal* of $\mathfrak{F}$.

Example 4.1.5. The Gauss map associated to a projective embedding $X \subset \mathbb{P}^n$ will be thoroughly studied in Chapter 9. It is defined in terms of the embedded projective tangent spaces to X at its smooth points, by associating to each such point the corresponding tangent space as an element of the Grassmannian $\mathbb{G}(d, n)$, where $d = \dim X$.

This map is not defined everywhere in general. If, moreover, one embeds the Grassmannian in a projective space $\mathbb{P}^m$ by, e. g., the Plücker embedding, then the Gauss map gives place to a rational map $X \dashrightarrow \mathbb{P}^m$. In the case where X is a hypersurface, defined by a form $f \in k[\mathbf{x}]$, one has $\mathbb{G}(n-1, n) = \mathbb{P}^n$ by the well-known point/hyperplane duality; hence, it is not difficult to see that a representative of this rational map in this case is

$$(\partial f/\partial x_0 : \cdots : \partial f/\partial x_n)$$

as viewed modulo (f). Therefore, the image is actually the *dual variety* to the hypersurface. One often identifies the Gauss map with this rational map, which depends on the Plücker embedding.

Corollary 4.1.6 (k perfect). *Let $X \subset \mathbb{P}^n$ be a hypersurface of degree $d \geq 2$. Then the degree sequence of the Gauss map $\mathfrak{F}: X \dashrightarrow \mathbb{P}^n$ is $d-1$ if and only if X is arithmetically normal.*

Proof. In this situation, the base ideal is $J_f = (\partial f/\partial x_0, \dots, \partial f/\partial x_n)k[\mathbf{x}]/(f)$ and is called the Jacobian ideal of $k[\mathbf{x}]/(f)$. By the well-known Jacobian criterion, $k[\mathbf{x}]/(f)$ is a normal ring if and only if J_f has codimension at least two. □

4.1.2 The semilinear relation type

In this part we raise a few questions related to a behavior of a rational map vis-à-vis a certain semilinear relation type of its base ideal.

Proposition 4.1.7. *Let $\mathfrak{F} : \mathbb{P}^n \dashrightarrow \mathbb{P}^m$ be a rational map and let $W \subset \mathbb{P}^m$ denote its image. If the codimension of the ideal $(\mathcal{J}_{1,*})$ over the ring $(k[\mathbf{Y}]/I(W))[\mathbf{X}]$ is at least n then $\mathfrak{F}$ is birational onto W.*

Proof. By Theorem 3.2.24, it suffices to show that the rank of the Jacobian dual matrix ψ over $S := k[\mathbf{Y}]/I(W)$ is at least n (hence $= n$).

For this note that if $C := \operatorname{coker}_S(\psi^t)$, then the symmetric algebra $\mathcal{S}_S(C) \simeq S[\mathbf{X}]/I_1(\mathbf{X}\cdot\psi^t)$. In addition, by the very definition, $(\mathcal{J}_{1,*})S[\mathbf{X}] = I_1(\mathbf{X}\cdot\psi^t)$. It follows that

$$\dim \mathcal{S}_S(C) \le \dim S + n + 1 - \operatorname{ht}(\mathcal{J}_{1,*})S[\mathbf{X}] \le \dim S + n + 1 - n = \dim S + 1.$$

On the other hand, letting $\mathcal{R}_S(C)$ stand for the Rees algebra of C, one has

$$\dim \mathcal{S}_S(C) \ge \dim \mathcal{R}_S(C) = \dim S + \operatorname{rank}_S(C).$$

From this $\operatorname{rank}_S(C) \le 1$, hence $\operatorname{rank}_S(\psi) = \operatorname{rank}_S(\psi^t) \ge n$. □

Regarding the above proof, one sees that since $\operatorname{rank}_S(\psi) = n$ then $\operatorname{rank}_S(C) = 1$. In this notation, the inverse map to $\mathfrak{F}$ is defined by the coordinates of any nonzero homogenous vector in the dual module C^*.

The converse statement to the effect that $\operatorname{rank}_S(\psi) = n$ over $k[\mathbf{Y}]/I(W)$ implies that the codimension of $(\mathcal{J}_{1,*})$ over $(k[\mathbf{Y}]/I(W))[\mathbf{X}]$ is at least n is generally false, even for Cremona maps. The reason is simply given by the fact that there are Cremona maps for which the symmetric algebra of the base ideal has larger dimension than its Rees algebra. No such examples exist for plane Cremona maps as the base ideal is an almost complete intersection. For space Cremona maps, however, even monomial involutions may show this behavior (see Exercise 4.5.5).

In connection to Proposition 4.1.7, the following numerical invariant seems worth looking at.

Definition 4.1.8. The *semilinear relation type* of an ideal $I \subset R = k[\mathbf{X}]$, denoted $\operatorname{srt}(I)$, is the minimum $s \ge 1$ such that $(\mathcal{J}_{1,*}) = (\mathcal{J}_{1,\le s})$, where the ideal on the right side is the subideal generated by all ("semilinear") relations of bidegree $(1, r)$ with $r \le s$.

This numerical invariant expresses the measure of how far off through the successive powers I^m of the ideal I one has to go in order to pick a sufficient number of linear syzygies of I^m to generate the ideal $(\mathcal{J}_{1,*})$. In case I is actually the base ideal of a Cremona map, then this number may have bearing to the degree of the coordinates of the inverse map. Here is a very crude lower bound.

Proposition 4.1.9. *Let $I \subset k[\mathbf{X}]$ denote the base ideal of a Cremona map $\mathfrak{F} : \mathbb{P}^n \dashrightarrow \mathbb{P}^n$. Then* $\mathrm{srt}(I) \geq \deg(\mathfrak{F}^{-1})/n$, *where* $\deg \mathfrak{F}^{-1}$ *denotes the degree of the defining coordinates of the inverse map.*

Proof. Using item (ii) of Theorem 3.2.24, one immediately has $\deg \mathfrak{F}^{-1} \leq \mathrm{srt}(I)\, n$, as required. □

For instance, if I is of linear type then $\deg \mathfrak{F}^{-1} \leq n$, a property already observed in Corollary 3.2.27.

Question 4.1.10. Find upper bounds for $\mathrm{srt}(I)$ in terms of other invariants of I.

A related question was raised by Vasconcelos in connection with elimination theory. Note that the above numerical invariant makes sense even for a general rational map with target some $\mathbb{P}^m$.

Question 4.1.11. Let more generally $\mathfrak{F}: \mathbb{P}^n \dashrightarrow \mathbb{P}^m$ be a rational map, given by $m+1$ k-linearly independent forms $\mathbf{f} = \{f_0, \ldots, f_m\} \subset k[\mathbf{X}]$ of a fixed degree ≥ 2, without proper common factor. Let $I = (f_0, \ldots, f_m)$ denote as above the base ideal of $\mathfrak{F}$ and let d denote the highest degree of a minimal generator of the defining ideal of the image of $\mathfrak{F}$ in $\mathbb{P}^m$. Is $\mathrm{srt}(I) \leq d$?

Note that the answer is trivially affirmative for ideals of fiber type. There is both theoretical and computational evidence for an affirmative answer in the case of an $\mathbf{X}$-primary almost complete intersection (see [108]).

4.2 Inversion factors of birational maps

As a matter of recalling the basic notation of previous sections, let k denote an arbitrary infinite field and let $\mathfrak{F} : \mathbb{P}^{n-1} = \mathbb{P}^{n-1}_k \dashrightarrow \mathbb{P}^{m-1} = \mathbb{P}^{m-1}_k$ be a rational map defined by m forms $\mathbf{f} = \{f_1, \ldots, f_m\} \subset R := k[\mathbf{X}] = k[X_1, \ldots, X_n]$ of the same degree $d \geq 1$, not all null. We naturally assume throughout that $n \geq 2$. We often write $\mathfrak{F} = (f_1 : \cdots : f_m)$ to underscore the projective setup and assume that $\gcd\{f_1, \ldots, f_m\} = 1$ (in the geometric terminology, the linear system defining $\mathfrak{F}$ "has no fixed part"), in which case d is the coordinate degree of $\mathfrak{F}$.

Recall that the image of $\mathfrak{F}$ is the projective subvariety $W \subset \mathbb{P}^{m-1}$ whose homogeneous coordinate ring is the k-subalgebra $k[\mathbf{f}] \subset R$ up to degree renormalization. Write $k[\mathbf{f}] \simeq k[\mathbf{Y}]/I(W)$, where $I(W) \subset k[\mathbf{Y}] = k[Y_1, \ldots, Y_m]$ is the homogeneous defining ideal of the image in the embedding $W \subset \mathbb{P}^{m-1}$.

One of the ways to say that $\mathfrak{F}$ is birational onto the image was by claiming the existence of a rational map $\mathbb{P}^{m-1} \dashrightarrow \mathbb{P}^{n-1}$ with defining coordinate forms $\mathbf{g} = \{g_1, \ldots, g_n\} \subset k[\mathbf{Y}]$ (not simultaneously vanishing modulo $I(W)$) satisfying the relations:

$$(g_1(\mathbf{f}) : \cdots : g_n(\mathbf{f})) = (X_1 : \cdots : X_n), \; (f_1(\mathbf{g}) : \cdots : f_m(\mathbf{g})) \equiv (Y_1 : \cdots : Y_m) \pmod{I(W)}$$

For the purpose of this section, we will stress this point of view of birational maps.

As sufficiently discussed before, the sets of forms $\{g_1,\dots,g_n\}$ defining the inverse map are not unique, anyone of them called a representative thereof. Furthermore, one can choose a finite minimal set $\mathbf{g_1},\dots,\mathbf{g_s}$ of these representatives such that any other representative belongs to the $k[\mathbf{Y}]/I(W)$-submodule generated by $\mathbf{g_1},\dots,\mathbf{g_s}$. More exactly, any such a minimal representative is the transpose of a minimal generator of the syzygy module of the so-named weak Jacobian dual matrix (Definition 3.2.15).

4.2.1 Basic landscape

Having information about the inverse map—e. g., about its coordinate degree—will be quite relevant in the sequel. For instance, the first of the above structural congruences

$$(g_1(f_1,\dots,f_m):\cdots:g_n(f_1,\dots,f_m)) = (X_1:\cdots:X_n) \tag{33}$$

involving the inverse map, in terms of a given representative lifted to $k[\mathbf{y}]$, yields a uniquely defined form $F \in R$ up to a nonzero scalar in k, such that $g_i(f_1,\dots,f_m) = X_iF$, for every $i = 1,\dots,n$. This is merely a consequence of factoriality. Indeed, the congruence means that there are forms F, F' such that $F'g_i(\mathbf{f}) = Fx_i$ for every $1 \le i \le n$. Now, a prime factor of F' that does not divide F would have to divide all $n \ge 2$ variables, which is only possible if $F' \in k$. Otherwise, necessarily F' divides F; in any case we are through.

We call F the *source inversion factor* of $\mathfrak{F}$ associated to the given representative. There is a counterpart *target inversion factor*, defined in a parallel way by exchanging the roles of $\mathbf{f}$ and $\mathbf{g}$ in (33). However, due to the lack of complete symmetry (as the source of the inverse map is a subvariety) one has to be slightly more careful, as is now explained. For convenience, we state the result in the form of a lemma.

Lemma 4.2.1. *Let $\mathfrak{F} : \mathbb{P}^{n-1} \dashrightarrow \mathbb{P}^{m-1}$ be a rational map defined by forms $\mathbf{f} = \{f_1,\dots,f_m\} \subset k[\mathbf{X}] = k[X_1,\dots,X_n]$ of the same degree. Suppose that:*

(i) *$\mathfrak{F}$ is birational onto its image $W \subset \mathbb{P}^{m-1}$.*

(ii) *W is nondegenerate and its homogeneous coordinate ring has a depth ≥ 2.*

Then the inverse map admits a representative by forms $\mathbf{g} = \{g_1,\dots,g_n\} \subset k[\mathbf{Y}]$ of the same degree such that there exists a uniquely defined form $G \subset k[\mathbf{Y}]$ modulo $I(W)$ satisfying the congruences

$$(f_1(\mathbf{g}),\dots,f_m(\mathbf{g})) \equiv G\cdot(Y_1,\dots,Y_m) \pmod{I(W)}.$$

Proof. By Theorem 3.2.22, Supplement (ii), a representative $\mathbf{g} = \{g_1,\dots,g_n\}$ of the inverse map can be taken in which g_i denotes a signed $(n-1)$th order minor of an $(n-1)\times n$

submatrix of the weak Jacobian dual matrix of $\mathfrak{F}$. Next, drawing upon the so-called Koszul–Hilbert lemma (Section 3.2.1), one derives a set of simpler congruences:

$$f_m(\mathbf{g})\, Y_j = f_j(\mathbf{g})\, Y_m \pmod{I(W)}, \quad 1 \le j \le m. \tag{34}$$

Thus, $f_m(\mathbf{g})$ draws the irrelevant maximal ideal of $k[\mathbf{Y}]$ into $(Y_m, I(W))$. Since $W \subset \mathbb{P}^{m-1}$ is nondegenerate and $I(W)$ is prime, then

$$\operatorname{depth}(k[\mathbf{Y}]/(Y_m, I(W))) = \operatorname{depth}(k[\mathbf{Y}]/I(W)) - 1 \ge 1$$

by assumption (ii). It follows that $f_m(\mathbf{g}) \in (Y_m, I(W))$, and hence Y_m can be canceled in (34). To conclude, take $G := f_m(\mathbf{g})/Y_m$. □

We emphasize that both F and G depend on the choice of the representative $\{g_1, \dots, g_n\}$ which as observed is not uniquely determined in general. One case in which both are uniquely determined is when the map is a Cremona transformation. Inversion factors of Cremona maps in characteristic zero have a curious property seemingly first observed in [159].

Proposition 4.2.2 (char(k) = 0). *Let $\mathfrak{F}$ denote a Cremona map of $\mathbb{P}^{n-1}$ defined by forms $\mathbf{f} := \{f_1, \dots, f_n\}$ in R without fixed part and let $\Theta(\mathbf{f})$ denote the Jacobian matrix of $\mathbf{f}$. Then* $\det(\Theta(\mathbf{f}))$ *divides a power of the source inversion factor F of $\mathfrak{F}$. In particular, if* $\det(\Theta(\mathbf{f}))$ *is reduced then it divides F.*

Proof. Let $\mathbf{g} : \{g_1, \dots, g_n\}$ define the inverse map. Applying the chain rule of derivatives to the structural equation $\mathbf{g}(\mathbf{f}) - F \cdot (\mathbf{X})$, it obtains

$$\Theta(\mathbf{g})(\mathbf{f}) \cdot \Theta(\mathbf{f}) = F \cdot \mathcal{I} + (\mathbf{X})^t \cdot \operatorname{Grad}(F) \tag{35}$$

where $\mathcal{I}$ is the identity matrix and $\operatorname{Grad}(F) = (\partial F/\partial X_1 \cdots \partial F/\partial X_n)$. Note that the right side of (35) is the result of evaluating $\lambda \mapsto F$ in the characteristic matrix $\lambda\mathcal{I} - \mathcal{A}$, where $\mathcal{A} = -(\mathbf{X})^t \cdot \operatorname{Grad}(F)$.

Recall that, quite generally the coefficients of the characteristic polynomial $p(\lambda) = \lambda^n + a_1\lambda^{n-1} + \cdots + a_{n-1}\lambda + a_n$ of $\mathcal{A}$ can be recursively computed as

$$\begin{aligned} -a_1 &= s_1 \\ -ra_r &= s_r + \sum_{i=1}^{r-1} s_i a_{r-i} \end{aligned} \tag{36}$$

where s_r is the trace of the matrix $\mathcal{A}^r$, for $1 \le r \le n$.

Now, one has

$$\operatorname{Trace}((\mathbf{X})^t \cdot \operatorname{Grad}(F)) = \sum_{i=1}^{n} X_i \frac{\partial F}{\partial X_i} = dF, \tag{37}$$

where $d = \deg(F)$. On the other hand, a calculation yields the equality $\mathcal{A}^2 = (dF)\mathcal{A}$. By an immediate recursion, it follows that

$$((\mathbf{X})^t \cdot \mathrm{Grad}(F))^r = (d^{r-1}F^{r-1})(\mathbf{X})^t \cdot \mathrm{Grad}(F) \tag{38}$$

From this, applying (36) recursively yields $a_2 = a_3 = \cdots = a_n = 0$, hence $p(t) = t^n + (dF)t^{n-1}$. Evaluating $\lambda \mapsto F$ yields $p(F) = F^n + (dF)F^{n-1} = (d+1)F^n$. But this is $(\det F \cdot \mathcal{I} + (\mathbf{X})^t \cdot \mathrm{Grad}(F))$, hence (35) gives

$$\det(\Theta(\mathbf{g})(\mathbf{f})) \cdot \det(\Theta(\mathbf{f})) = (d+1)F^n. \tag{39}$$

Therefore, $\det(\Theta(\mathbf{f}))$ indeed divides F^n. □

Remark 4.2.3. As applied to the particular case of a polar Cremona map, it says that if the Hessian determinant of a homaloidal homogeneous polynomial $f \in k[\mathbf{X}]$ is reduced then it divides the source inversion factor of the map. For a plane polar Cremona maps, the two coincide (up to a nonzero scalar) as it easily follows from [52]. The question remains as to when this is the case for higher-dimensional polar Cremona maps.

4.2.2 Relation to symbolic powers

For the elements of symbolic powers, see Subsection 2.3.1.

There is a very basic and somewhat surprising result relating inversion factors to symbolic powers, which seems to have gone unnoticed for the base ideal of a birational map.

Proposition 4.2.4. *Let $\mathfrak{F} : \mathbb{P}^n \dashrightarrow \mathbb{P}^m$ be a birational map onto the image, with base ideal $I \subset R = k[\mathbf{X}]$ generated in degree $d \geq 2$. Let $F \in R$ denote the source inversion factor relative to a given representative of the inverse map. If* $\mathrm{depth}(R/I) > 0$, *then:*
(a) *F is an element of the symbolic power $I^{(d')}$, where d' is the degree of the coordinates of the representative. In particular, $I^{(d')} \neq I^{d'}$.*
(b) *If, moreover, $I^{(\ell)} = I^{\ell}$, $\ell \leq d'-1$, then F is an essential symbolic element of order d'.*
(c) *Moreover, if $I^{(d')}$ is generated in standard degree $\geq dd' - 1$, where d is the coordinate degree of $\mathfrak{F}$, then F is a homogeneous minimal generator of the symbolic Rees algebra.*

Proof. (a) The characteristic property of F is that $f_i'(f_0, \ldots, f_m) = x_iF$, for every $i = 0, \ldots, m$. In particular, $F \in I^{d'} : (\mathbf{x})$. We may assume that $\mathfrak{F}$ is not the identity map of $\mathbb{P}^n$. By assumption, every associated prime of R/I has codimension at most $\dim R - 1$; hence, there is a form $h \in (\mathbf{x})\setminus P$, for every associated prime P of R/I, such that $hF \in I^{d'}$. This means that $F \in I^{(d')}$.

The additional assertion follows from a degree counting, since $\deg(F) = dd' - 1$, while $I^{d'}$ is generated in degree dd'.

(b) This follows straightforwardly from (a) because, under the hypothesis that $I^{(\ell)} = I^{\ell}$, $\ell \leq d' - 1$, being essential just means not belonging to the ordinary power of the same order.

(c) This is clear since $\deg(F) = dd' - 1$. □

The above will be later applied to the case of the base ideal of a Cremona map (Subsection 7.2.3).

4.3 Newton complementary dual

This section is based upon the original results of [47] and [54]. Later related contributions to the theme can be found in [8, 14, 23] and [134].

4.3.1 Preliminaries

Let $R := k[\mathbf{x}] = k[x_0, \ldots, x_n]$ denote the standard graded polynomial ring in $n + 1$ indeterminates with coefficients in a field k.

Let f be a form of degree $d \geq 1$ in R. Denote by $N(f)$ (as a reminder of the Newton polygon) the so-called log-matrix of the (finitely many) terms of f having nonzero coefficients in a fixed ordering, i. e., the matrix whose columns are the vectors of the exponents of the nonzero terms of f in a fixed ordering. If f is a nonzero scalar, we set $N(f)$ to be the $(n + 1) \times 1$ zero matrix, thinking of 1 as the trivial monomial with zero exponent vector.

Call $N(f)$ the *Newton log matrix* (or simply the *Newton matrix*) of f.

Given a finite ordered set $\mathbf{f} := \{f_0, \ldots, f_m\} (m \geq 1)$ of such forms of the same degree $d \geq 1$, let $N(\mathbf{f})$ denote the concatenation of the Newton matrices $N(f_0), \ldots, N(f_m)$; accordingly, we call $N(\mathbf{f})$ the *Newton matrix* of the set $\mathbf{f}$. Note that $N(\mathbf{f})$ is an integer stochastic matrix.

The row vector $\mathbf{c}_f$ whose entries are the nonzero coefficients of a form f in a fixed ordering is the *coefficient frame* of f. We write symbolically

$$f = \langle \mathbf{c}_f, \mathbf{x}^{N(f)} \rangle \tag{40}$$

as the inner product of the coefficient frame by the set of the corresponding monomials—it will be called the *Newton representation* of f.

The *Newton complementary dual matrix* (or simply the *Newton dual matrix*) of the Newton matrix $N(\mathbf{f}) = (a_{i,\ell})$ is the matrix

$$\widehat{N(\mathbf{f})} = (\alpha_i - a_{i,\ell}),$$

where $\alpha_i = \max_\ell\{a_{i,\ell}\}$, with $0 \le i \le n$ and ℓ indexes the set of all nonzero terms of all forms in the set $\mathbf{f}$.

In other words, denoting $\boldsymbol{\alpha} := (\alpha_0, \dots, \alpha_n)^t$, one has

$$\widehat{N(\mathbf{f})} = [\,\boldsymbol{\alpha}\,|\cdots|\,\boldsymbol{\alpha}\,]_{(n+1)\times(r_0+\cdots+r_m)} - N(\mathbf{f}),$$

where r_j denotes the number of nonzero terms of f_j, $j = 0, \dots, m$. The vector $\boldsymbol{\alpha}$ is called the *directrix vector* of $N(\mathbf{f})$ (or of $\mathbf{f}$ by abuse). Note that this notion of directrix vector can be defined for any matrix with integer entries and will sometimes be employed as such.

For every $j = 0, \dots, m$, let $\widehat{N(\mathbf{f})}_j$ denote the submatrix of $\widehat{N(\mathbf{f})}$ whose columns come from f_j. Finally, consider the set of forms defined in terms of their Newton representations

$$\widehat{\mathbf{f}} := \{\widehat{f_0} := \langle \mathbf{c}_0, \mathbf{x}^{\widehat{N(\mathbf{f})}_0}\rangle, \dots, \widehat{f_m} := \langle \mathbf{c}_m, \mathbf{x}^{\widehat{N(\mathbf{f})}_m}\rangle\} \tag{41}$$

where $\mathbf{c}_j = \mathbf{c}_{f_j}$ stands for the coefficient frame of f_j.

Definition 4.3.1. $\widehat{\mathbf{f}}$ is the *Newton complementary dual* set of $\mathbf{f}$.

The reason for calling the procedure a *dual* is the following result: If the forms $\mathbf{f}$ have trivial gcd and every variable in $\mathbf{x}$ is in the support of at least one nonzero term of one of the forms, then the Newton dual set $\widehat{\mathbf{f}}$ also satisfies these conditions and, moreover,

$$\widehat{\widehat{\mathbf{f}}} = \mathbf{f}.$$

These two assumptions are called *canonical restrictions*.

Here is a permeating example.

Example 4.3.2. If $\mathbf{x} = \{x_0, \dots, x_n\}$, then its directrix vector is $(1, \dots, 1)^t$ and

$$\widehat{\mathbf{x}} = \{x_1 \cdots x_n, \dots, x_0 \cdots \underline{x_i} \cdots x_n, \dots, x_0 \cdots x_{n-1}\}, \tag{42}$$

where $\underline{x_i}$ means x_i is omitted. In terms of the rational maps defined by the respective forms, this says that the Newton complementary dual of the identity map of $\mathbb{P}^n$ is the reciprocal involution.

This simple example will be quite crucial in expressing basic relations as in the next subsection.

4.3.2 Main results

We start by observing the impact of evaluating forms on the Newton dual $\widehat{\mathbf{x}}$ of the set $\mathbf{x}$ of the variables.

Lemma 4.3.3. *Let $g \in R := k[x_0, \dots, x_n]$ be a form and $\widehat{\mathbf{x}}$ denote the Newton complementary dual set of $\mathbf{x} := \{x_0, \dots, x_n\}$, as described in* Example 4.3.2. *Then $N(g(\widehat{\mathbf{x}})) = N(\widehat{\mathbf{x}}).N(g)$, where the dot indicates matrix multiplication. In particular, the Newton representation of $g(\widehat{\mathbf{x}})$ is*

$$g(\widehat{\mathbf{x}}) = \langle \mathbf{c}_g, \mathbf{x}^{N(\widehat{\mathbf{x}}).N(g)} \rangle.$$

Proof. Let $g_1, g_2 \in R$ be forms with empty common monomial support (i. e., no nonzero term appears in both forms) satisfying the desired equality in the statement. Then, up to suitable ordering of the constituent monomials, one has

$$\begin{aligned} N((g_1 + g_2)(\widehat{\mathbf{x}})) &= N(g_1(\widehat{\mathbf{x}}) + g_2(\widehat{\mathbf{x}})) = N(g_1(\widehat{\mathbf{x}})) \sqcup N(g_2(\widehat{\mathbf{x}})) \\ &= N(\widehat{\mathbf{x}}).N(g_1) \sqcup N(\widehat{\mathbf{x}}).N(g_2) = N(\widehat{\mathbf{x}}).(N(g_1) \sqcup N(g_2)) \\ &= N(\widehat{\mathbf{x}}).N(g_1 + g_2), \end{aligned}$$

where $\sqcup$ denotes matrix concatenation.

The supplementary result about the Newton representation is clear since, in the assumed situation of the forms g_1, g_2, one has the relation $\mathbf{c}_{g_1+g_2} = \mathbf{c}_{g_1} \sqcup \mathbf{c}_{g_2}$ for the respective coefficient frames.

Therefore, we have reduced the problem to the case when $g(\mathbf{x}) = c\, x_0^{a_0} \dots x_n^{a_n}$ is a term. In this situation, the result is obvious because

$$N(g) = \begin{pmatrix} a_0 \\ \vdots \\ a_n \end{pmatrix} \quad \text{and} \quad N(\widehat{\mathbf{x}}) = \begin{pmatrix} 0 & 1 & 1 & \cdots & 1 \\ 1 & 0 & 1 & \cdots & 1 \\ 1 & 1 & 0 & \cdots & 1 \\ \vdots & \vdots & \vdots & \ddots & \vdots \\ 1 & 1 & 1 & \cdots & 0 \end{pmatrix}.$$

□

With the same notation, one has the following.

Proposition 4.3.4. *If $g \in k[x_0, \dots, x_n]$ is a form of degree $d \geq 1$, then*

$$g(\widehat{\mathbf{x}}) = x_0^{d-\beta_0} \cdots x_n^{d-\beta_n}\, \widehat{g},$$

where $\boldsymbol{\beta} := (\beta_0, \dots, \beta_n)^t$ is the directrix vector of $N(g)$.

Proof. By Lemma 4.3.3,

$$N(g(\widehat{\mathbf{x}})) = N(\widehat{\mathbf{x}}).N(g).$$

Therefore, by taking respective Newton representations it suffices to prove the matrix equality

$$N(\widehat{\mathbf{x}}).N(g) = [\boldsymbol{\alpha}|\dots|\boldsymbol{\alpha}]_{(n+1)\times r} + \widehat{N(g)},$$

where r denotes the number of nonzero terms of g and $\boldsymbol{\alpha} = (d-\beta_0,\dots,d-\beta_n)^t$. But this is clear as

$$\begin{aligned}N(\widehat{\mathbf{x}}).N(g) &= (\mathbb{I} - N(\mathbf{x})).N(g) = \mathbb{I}.N(g) - N(\mathbf{x}).N(g) = \mathbb{I}.N(g) - N(g)\\ &= \mathbb{I}.N(g) - [\beta|\dots|\beta] + [\beta|\dots|\beta] - N(g)\\ &= [\boldsymbol{\alpha}|\dots|\boldsymbol{\alpha}] + \widehat{N(g)},\end{aligned}$$

where $\mathbb{I}$ is the $(n+1)\times(n+1)$ matrix whose entries are all 1. □

Thus, the proposition tells us that to get the Newton dual $\widehat{g}$ of a form g one has to evaluate g on the terms of the involution in (42) and multiply out by a certain Laurent monomial. The moral of this statement is that it may be easier to handle the properties of $\widehat{g}$ by evaluation than by applying the original definition. Thus, for example, one can derive the following result.

Corollary 4.3.5. *Let $p, q \in R$ be forms. Then $\widehat{pq} = \widehat{p}\,\widehat{q}$.*

Proof. Set $g := p.q$. Clearly, $g(\widehat{\mathbf{x}}) = p(\widehat{\mathbf{x}}).q(\widehat{\mathbf{x}})$, where $\widehat{\mathbf{x}}$ denote the Newton complementary dual set of $\mathbf{x} = \{x_0,\dots,x_n\}$. By Proposition 4.3.4, one has

$$g(\widehat{\mathbf{x}}) = x_0^{r+s-\gamma_0}\cdots x_n^{r+s-\gamma_n}\,\widehat{g},\quad p(\widehat{\mathbf{x}}) = x_0^{r-\alpha_0}\cdots x_n^{r-\alpha_n}\,\widehat{p},\quad q(\widehat{\mathbf{x}}) = x_0^{s-\beta_0}\cdots x_n^{s-\beta_n}\,\widehat{q},$$

where $r = \deg(p)$, $s = \deg(q)$ and $\boldsymbol{\gamma} := (\gamma_0,\dots,\gamma_n)^t$, $\boldsymbol{\alpha} := (\alpha_0,\dots,\alpha_n)^t$ and $\boldsymbol{\beta} := (\beta_0,\dots,\beta_n)^t$ are the directrix vectors of $N(g)$, $N(p)$ and $N(q)$, respectively. Thus,

$$x_0^{r+s-\gamma_0}\cdots x_n^{r+s-\gamma_n}\,\widehat{g} = x_0^{r+s-\alpha_0-\beta_0}\cdots x_n^{r+s-\alpha_n-\beta_n}\,\widehat{p}.\widehat{q}.$$

On the other hand, one has $\deg_{x_i}(g) = \deg_{x_i}(p) + \deg_{x_i}(q)$, for every $i = 0,\dots,n$. Therefore, $\boldsymbol{\gamma} = \boldsymbol{\alpha} + \boldsymbol{\beta}$, and hence we are through. □

Proposition 4.3.4 extends to a set of forms.

Theorem 4.3.6. *Let $\mathbf{g} := \{g_0,\dots,g_m\} \subset R := k[x_0,\dots,x_n]$ be forms of same degree $d \geq 1$. Then*

$$\mathbf{g}(\widehat{\mathbf{x}}) = x_0^{d-\beta_0}\cdots x_n^{d-\beta_n}\,\widehat{\mathbf{g}},$$

where $\boldsymbol{\beta} := (\beta_0, \ldots, \beta_n)^t$ is the directrix vector of $N(\mathbf{g})$.

Proof. Applying Proposition 4.3.4 with $g := g_i$ gives

$$g_i(\widehat{\mathbf{x}}) = x_0^{d-\beta_{0,i}} \cdots x_n^{d-\beta_{n,i}}\, \widehat{g_i},$$

where $\underline{\beta_i} := (\beta_{0,i}, \ldots, \beta_{n,i})^t$ is the directrix vector of g_i.

By definition, the directrix vector $\boldsymbol{\beta} = (\beta_0, \ldots, \beta_n)^t$ of $\mathbf{g} = \{g_0, \ldots, g_m\}$ satisfies the inequalities $\beta_i - \beta_{i,j} \geq 0$, for all $i = 0, \ldots, n$ and $j = 0, \ldots, m$. Therefore,

$$\begin{aligned} x_0^{d-\beta_{0,i}} \cdots x_n^{d-\beta_{n,i}}\, \widehat{g_i} &= x_0^{d-\beta_{0,i}} \cdots x_n^{d-\beta_{n,i}} \frac{x_0^{\beta_0-\beta_{0,i}} \cdots x_n^{\beta_n-\beta_{n,i}}}{x_0^{\beta_0-\beta_{0,i}} \cdots x_n^{\beta_n-\beta_{n,i}}}\, \widehat{g_i} \\ &= x_0^{d-\beta_0} \cdots x_n^{d-\beta_n}\, x_0^{\beta_0-\beta_{0,i}} \cdots x_n^{\beta_n-\beta_{n,i}}\, \widehat{g_i}. \end{aligned}$$

Note that $N(\widehat{\mathbf{g}})$ is the concatenation of the $m+1$ matrices

$$[\,y_0 | \ldots | y_0\,]_{(n+1)\times r_0} + N(\widehat{g_0}), \ldots, [\,y_m | \ldots | y_m\,]_{(n+1)\times r_m} + N(\widehat{g_m}),$$

where $y_j := (\beta_0 - \beta_{0,j}, \ldots, \beta_n - \beta_{n,j})^t$ and r_j denotes the number of nonzero terms of g_j, $j = 0, \ldots, m$. Therefore, $\mathbf{g}(\widehat{\mathbf{x}}) = x_0^{d-\beta_0} \cdots x_n^{d-\beta_n}\, \widehat{\mathbf{g}}$, as was to be shown. □

Next is a generalization of the previous proposition to the case where one evaluates on the forms of the Newton dual of a set of $n+1 = \dim R$ forms of the same degree (instead of $\widehat{\mathbf{x}}$).

Proposition 4.3.7. *Let $\mathbf{g} := \{g_0, \ldots, g_m\} \subset R$ be arbitrary forms of degree $d \geq 1$ and let $\mathbf{h} := \{h_0, \ldots, h_n\} \subset R$ denote a set of $n+1 = \dim R$ forms of a fixed degree. Then*

$$\mathbf{g}(\widehat{\mathbf{h}}) - x_0^{d\alpha_0-\beta_0} \cdots x_n^{d\alpha_n-\beta_n}\, \widehat{\mathbf{g}(\mathbf{h})},$$

where $\boldsymbol{\alpha} := (\alpha_0, \ldots, \alpha_n)^t$ and $\boldsymbol{\beta} := (\beta_0, \ldots, \beta_n)^t$ are the directrix vectors of $N(\mathbf{h})$ and $N(\mathbf{g}(\mathbf{h}))$, respectively.

Proof. We apply Theorem 4.3.6 twice, first taking $\mathbf{g}$ to be $\mathbf{h}$; it obtains

$$\mathbf{h}(\widehat{\mathbf{x}}) = x_0^{s-\alpha_0} \cdots x_n^{s-\alpha_n}\, \widehat{\mathbf{h}},$$

where s is the common degree of the forms $\mathbf{h}$. Evaluating $\mathbf{g}$ on these n forms yields

$$\begin{aligned} \mathbf{g}(\widehat{\mathbf{h}}) &= (x_0^{ds-d\alpha_0} \cdots x_n^{ds-d\alpha_n})^{-1}\, \mathbf{g}(\mathbf{h}(\widehat{\mathbf{x}})) \\ &= (x_0^{ds-d\alpha_0} \cdots x_n^{ds-d\alpha_n})^{-1}\, (\mathbf{g}(\mathbf{h}))(\widehat{\mathbf{x}}). \end{aligned}$$

Apply Theorem 4.3.6 again, this time around taking $\mathbf{g}$ to be $\mathbf{g}(\mathbf{h})$. It now yields

$$\mathbf{g}(\mathbf{h})(\widehat{\mathbf{x}}) = x_0^{ds-\beta_0} \cdots x_n^{ds-\beta_n}\, \widehat{\mathbf{g}(\mathbf{h})}.$$

Therefore,

$$\mathbf{g}(\widehat{\mathbf{h}}) = (x_0^{ds-d\alpha_0} \cdots x_n^{ds-d\alpha_n})^{-1} x_0^{ds-\beta_0} \cdots x_n^{ds-\beta_n} \widehat{\mathbf{g}(\mathbf{h})} = x_0^{d\alpha_0-\beta_0} \cdots x_n^{d\alpha_n-\beta_n} \widehat{\mathbf{g}(\mathbf{h})}. \qquad \square$$

Corollary 4.3.8. *Let $\mathbf{g} := \{g_0, \ldots, g_m\} \subset R$ be a set of forms of the same degree satisfying the canonical restrictions. Then $k[\mathbf{g}]$ and $k[\widehat{\mathbf{g}}]$ are isomorphic as k-algebras.*

Proof. Because of the duality $\widehat{\widehat{g}} = g$, it suffices to show that any homogeneous polynomial relation of $\mathbf{g}$ is one of $\widehat{\mathbf{g}}$. Let $F(\mathbf{y}) \in k[\mathbf{y}] := k[y_0, \ldots, y_m]$ be a homogeneous polynomial of degree d such that $F(\mathbf{g}) = 0$. Since $\mathbf{g}$ satisfies the canonical restrictions, applying Theorem 4.3.6 to the Newton dual $\widehat{\mathbf{g}}$ yields

$$\widehat{\mathbf{g}}(\widehat{\mathbf{x}}) = M\,\widehat{\widehat{\mathbf{g}}} = M\,\mathbf{g},$$

for suitable monomial $M \in R$. Therefore,

$$F(\widehat{\mathbf{g}}(\widehat{\mathbf{x}})) = F(M\,\mathbf{g}) = M^d\,F(\mathbf{g}) = 0.$$

But then $F(\widehat{\mathbf{g}}(\widehat{\mathbf{x}})) = F(\widehat{\mathbf{g}})(\widehat{\mathbf{x}})$ tells us that the form $F(\widehat{\mathbf{g}})$ further evaluated at the forms of $\widehat{\mathbf{x}}$ vanishes. Since the latter forms are algebraically independent over k it follows that $F(\widehat{\mathbf{g}}) = 0$. $\square$

Corollary 4.3.9. *Let $\mathbf{g} = \{g_0, \ldots, g_m\} \subset R := k[x_0, \ldots, x_n]$ $(m \geq 1, n \geq 1)$ be arbitrary forms of the same degree. Then:*

(a) *$\mathbf{g}(\widehat{\mathbf{x}})$ and $\widehat{\mathbf{g}}$ define the same rational map.*
(b) *If $\mathbf{g}$ defines a birational map onto its image then so does $\widehat{\mathbf{g}}$ and the two maps have the same image.*

Proof. (a) This follows immediately from Theorem 4.3.6 since the two maps are defined by tuples of orderly proportional coordinates stemming from multiplication by a nonzero element of the fraction field $k(x_0, \ldots, x_n)$ of R.

(b) This follows from (a) and Corollary 4.3.8. $\square$

For the next corollary, we note that if $\mathfrak{G} : \mathbb{P}^n \dashrightarrow \mathbb{P}^m$ is a rational map defined by a set of forms $\mathbf{g} = \{g_0, \ldots, g_m\} \subset R$ of the same degree then the rational map obtained by composing the reciprocal involution with $\mathfrak{G}$ is defined by the evaluated forms $\mathbf{g}(\widehat{\mathbf{x}}) = \{g_0(\widehat{\mathbf{x}}), \ldots, g_m(\widehat{\mathbf{x}})\}$. It will be denoted $\widehat{\mathfrak{G}}$.

The next result explains the interaction between Newton duality and the group law of the Cremona group of $\mathbb{P}^n$, here denoted $\circ$.

Proposition 4.3.10. *Let $\mathfrak{G} : \mathbb{P}^n \dashrightarrow \mathbb{P}^n$ denote a Cremona map, let $\widehat{\mathfrak{G}}$ be as above, and let $\mathfrak{M}$ stand for the reciprocal involution. Then:*

(i) $\widehat{\mathfrak{G}} = \mathfrak{G} \circ \mathfrak{M}$ *and* $(\widehat{\mathfrak{G}})^{-1} = \mathfrak{M} \circ \mathfrak{G}^{-1}$.
(ii) $\widehat{\mathfrak{G}^{-1}} = (\widehat{\mathfrak{G}})^{-1}$ *if and only if* $\mathfrak{M} \circ \mathfrak{G} = \mathfrak{G} \circ \mathfrak{M}$.

In particular, if $\mathfrak{G}$ is a monomial Cremona map, then $\widehat{\mathfrak{G}^{-1}} = (\widehat{\mathfrak{G}})^{-1}$.

(iii) *Let* $C, D \in k[\mathbf{x}]$ *denote the source inversion factors of* $\mathfrak{G}$ *and* $\widehat{\mathfrak{G}}$*, respectively, one has*

$$\Gamma_1(\widehat{C})^n = q_1 D \quad \text{and} \quad \Gamma_2(\widehat{D})^n = q_2 C,$$

where $\Gamma_1, \Gamma_2, q_1, q_2 \in k[\mathbf{x}]$*, with* Γ_1 *and* Γ_2 *are monomials.*

Proof. (i) This is a rephrasing of Corollary 4.3.9(a).

(ii) By item (i), $(\widehat{\mathfrak{G}})^{-1} = \mathfrak{M} \circ \mathfrak{G}^{-1}$. By the same token, $\widehat{\mathfrak{G}^{-1}} = \mathfrak{G}^{-1} \circ \mathfrak{M}$. Therefore,

$$\widehat{\mathfrak{G}^{-1}} = (\widehat{\mathfrak{G}})^{-1} \Leftrightarrow \mathfrak{G}^{-1} \circ \mathfrak{M} = \mathfrak{M} \circ \mathfrak{G}^{-1} \Leftrightarrow \mathfrak{M} \circ \mathfrak{G} = \mathfrak{G} \circ \mathfrak{M}.$$

The supplementary assertion follows from the fact that $\mathfrak{G}$ commutes with $\mathfrak{M}$, when $\mathfrak{G}$ is a monomial Cremona map.

(iii) Let $\mathbf{g} = \{g_0, \ldots, g_n\}$ and $\mathbf{g}' = \{g'_0, \ldots, g'_n\}$ denote the representative of $\mathfrak{G}$ and $\mathfrak{G}^{-1}$, respectively, satisfying $\gcd\{g_0, \ldots, g_n\} = \gcd\{g'_0, \ldots, g'_n\} = 1$. Thus, by definition of the source inversion factor, one has

$$g'_i(g_0, \ldots, g_n) = x_i\, C,$$

for every $i = 0, \ldots, n$. By Theorem 4.3.6, $\widehat{\mathbf{g}} = \frac{1}{M}\mathbf{g}(\widehat{\mathbf{x}})$, where M is a suitable monomial.

Let $\mathbf{h} = \{h_0, \ldots, h_n\}$ denote the representative of $(\widehat{\mathfrak{G}})^{-1}$ satisfying $\gcd\{h_0, \ldots, h_n\} = 1$. So

$$h_i(\widehat{\mathbf{g}}) = x_i\, D,$$

for every $i = 0, \ldots, n$. By (i), $p\,\mathbf{h} = \widehat{\mathbf{x}}(\mathbf{g}')$, for a suitable form p.

Therefore, for some nonnegative integer s,

$$\begin{aligned}
x_0\, D = h_0(\widehat{\mathbf{g}}) &= \left(\frac{1}{p}\, g'_1 \cdots g'_n\right) \circ \left(\frac{1}{M}\mathbf{g}(\widehat{\mathbf{x}})\right) = \frac{1}{M^s}\left[\left(\frac{1}{p}\, g'_1 \cdots g'_n\right) \circ (\mathbf{g}(\widehat{\mathbf{x}}))\right] \\
&= \frac{1}{M^s\, p(\mathbf{g}(\widehat{\mathbf{x}}))}[(g'_1 \cdots g'_n) \circ (\mathbf{g}(\widehat{\mathbf{x}}))] = \frac{1}{M^s\, p(\mathbf{g}(\widehat{\mathbf{x}}))}[(g'_1(\mathbf{g}) \cdots g'_n(\mathbf{g})) \circ \widehat{\mathbf{x}}] \\
&= \frac{1}{M^s\, p(\mathbf{g}(\widehat{\mathbf{x}}))}[(x_1 \cdots x_n C^n) \circ \widehat{\mathbf{x}}] = \frac{x_0^n x_1^{n-1} \ldots x_n^{n-1}}{M^s\, p(\mathbf{g}(\widehat{\mathbf{x}}))}[C(\widehat{\mathbf{x}})]^n.
\end{aligned}$$

By Proposition 4.3.4, with $g := C$, $(\beta_0, \ldots, \beta_n)^t$ the directrix vector of C and $d := \deg(C)$,

$$\Gamma_2(\widehat{C})^n = q_2 D,$$

where $\Gamma_2 := x_0^{n-1+nd-n\beta_0} \ldots x_n^{n-1+nd-n\beta_n}$ and $q_2 := M^s\, p(\mathbf{g}(\widehat{\mathbf{x}}))$.

The other equation in (iii) follows by the Newton complementary duality. □

We have seen that taking Newton dual on individual forms is a multiplicative operation. Additivity does not hold, but there is a close relationship. Moreover, one can actually deal with sets in the following sense: given sets of forms $\mathbf{p} = \{p_0, \ldots, p_m\} \subset R$ and $\mathbf{q} = \{q_0, \ldots, q_m\} \subset R$, both in the same degree, we write $\mathbf{p} + \mathbf{q}$ for the set of forms obtained by adding the tuples $(p_0, \ldots, p_m)$ and $(q_0, \ldots, q_m)$ coordinatewise.

Corollary 4.3.11. *Let $\mathbf{p} = \{p_0, \ldots, p_m\} \subset R$ and $\mathbf{q} = \{q_0, \ldots, q_m\}$ be sets of forms of R, both in the same degree d. Then*

$$(x_0^{d-\gamma_0} \cdots x_n^{d-\gamma_n})\,\widehat{\mathbf{p}+\mathbf{q}} = (x_0^{d-\alpha_0} \cdots x_n^{d-\alpha_n})\,\widehat{\mathbf{p}} + (x_0^{d-\beta_0} \cdots x_n^{d-\beta_n})\,\widehat{\mathbf{q}},$$

where $\boldsymbol{\gamma} := (\gamma_0, \ldots, \gamma_n)^t$, $\boldsymbol{\alpha} := (\alpha_0, \ldots, \alpha_n)^t$ and $\boldsymbol{\beta} := (\beta_0, \ldots, \beta_n)^t$ are the directrix vectors of $\mathbf{p} + \mathbf{q}$, $\mathbf{p}$ and $\mathbf{q}$, respectively.

Proof. This is an immediate application of Theorem 4.3.6 to all sets (tuples) of the relation $(\mathbf{p} + \mathbf{q})(\widehat{\mathbf{x}}) = \mathbf{p}(\widehat{\mathbf{x}}) + \mathbf{q}(\widehat{\mathbf{x}})$. □

4.3.3 Relation to the graph of a rational map

Let $\mathbf{g} = \{g_0, \ldots, g_m\} \subset R := k[\mathbf{x}] = k[x_0, \ldots, x_n]$ be arbitrary forms of the same degree. In this section we will see a relationship between the defining ideals of the Rees algebras of $\mathbf{g}$ and $\widehat{\mathbf{g}}$.

Let $R[\mathbf{y}] := R[y_0, \ldots, y_m]$ denote a polynomial ring over R with the standard bigrading where $\deg(x_i) = (1, 0)$ and $\deg(y_j) = (0, 1)$.

Let $p = p(\mathbf{x}, \mathbf{y}) \in R[\mathbf{y}] = k[\mathbf{x}, \mathbf{y}]$ denote a bihomogeneous polynomial of bidegree (d_x, d_y). Write

$$p(\mathbf{x}, \mathbf{y}) = \sum_{\substack{|\mathbf{a}|=d_x \\ |\mathbf{b}|=d_y}} c_{\mathbf{a},\mathbf{b}}\, \mathbf{x}^{\mathbf{a}} \mathbf{y}^{\mathbf{b}},$$

where $\mathbf{a} = (a_0, \ldots, a_n)$, $\mathbf{b} = (b_0, \ldots, b_m)$.

As in the previous part, we let $\mathbf{c}(p)$ denote the row of nonzero coefficients $c_{\mathbf{a},\mathbf{b}}$ of p in a fixed ordering (the coefficient frame of p) and let $N_{\mathbf{x}}(p)$ (resp., $N_{\mathbf{y}}(p)$) denote the (Newton-like) matrix whose columns are the exponent vectors of the corresponding $\mathbf{x}$-terms (resp., $\mathbf{y}$-terms) in the same ordering.

Accordingly, we may write symbolically

$$p(\mathbf{x}, \mathbf{y}) = \langle \mathbf{c}(p), \mathbf{x}^{N_{\mathbf{x}}(p)} \mathbf{y}^{N_{\mathbf{y}}(p)} \rangle \tag{43}$$

and call it the *Newton representation* of p in analogy to the homogeneous case.

The following result is an analogue of Proposition 4.3.4 for bihomogeneous polynomials.

Lemma 4.3.12. *Let $p(\mathbf{x},\mathbf{y}) \in R[\mathbf{y}]$ be a bihomogeneous form such that $d_x \geq 1$. Then*

$$p(\widehat{\mathbf{x}},\mathbf{y}) = x_0^{d_x-\beta_0} \cdots x_n^{d_x-\beta_n} \left\langle \mathbf{c}(p), \mathbf{x}^{\widehat{N_\mathbf{x}(p)}} \mathbf{y}^{N_\mathbf{y}(p)} \right\rangle,$$

where $\boldsymbol{\beta} := (\beta_0,\ldots,\beta_n)^t$ is the directrix vector of $N_\mathbf{x}(p)$ and $\widehat{\mathbf{x}}$ denotes the Newton complementary dual set of $\mathbf{x} = \{x_0,\ldots,x_n\}$.

Proof. For the sake of clarity, denote $P(\mathbf{x}) := p(\mathbf{x},\mathbf{y})$ as an element of $K[\mathbf{x}]$, where $K = k(\mathbf{y})$ stands for the field of fractions of $k[\mathbf{y}]$. Applying Proposition 4.3.4 to $P(\mathbf{x})$ gives

$$P(\widehat{\mathbf{x}}) = (x_0^{d_x-\beta_0} \cdots x_n^{d_x-\beta_n})\widehat{P(\mathbf{x})}, \tag{44}$$

since $N(P(\mathbf{x})) = N_x(p(\mathbf{x},\mathbf{y})) = N_x(p)$ as one readily sees from the definitions, and hence the directrix vector does not change. At the other end, one has the Newton representation of the Newton dual of $P(\mathbf{x})$:

$$\widehat{P(\mathbf{x})} = \left\langle \mathbf{c}(\widehat{P(\mathbf{x})}), \mathbf{x}^{N_x(p)} \right\rangle, \tag{45}$$

where $\mathbf{c}(\widehat{P(\mathbf{x})}) = \mathbf{c}(P(\mathbf{x}))$ denotes the coefficient frame of $P(\mathbf{x})$ over K. The elements of this frame (in the established ordering) are the terms of the form $c_{\mathbf{a},\mathbf{b}}\mathbf{y}^{\mathbf{b}}$. Therefore, taking in account the definition of the Newton matrix $N_y(p)$, the required expression follows from (44) and (45). □

Remark 4.3.13. With the convention that the Newton matrix of a nonzero scalar is the zero matrix, both Proposition 4.3.4 and Lemma 4.3.12 hold trivially for $d = 0$ and $d_x = 0$, respectively. The latter recovers the Newton representation of a form in $k[\mathbf{y}]$.

The basic result of this part is the following.

Proposition 4.3.14. *Let $\mathbf{g} = \{g_0,\ldots,g_m\} \subset R$ stand for a set of forms of the same degree and let $p(\mathbf{x},\mathbf{y}) \in R[\mathbf{y}]$ denote a bihomogeneous form. If $p(\mathbf{x},\mathbf{g}) = 0$ then $p(\widehat{\mathbf{x}},\widehat{\mathbf{g}}) = 0$.*

Proof. By Theorem 4.3.6, there exists a suitable monomial M in $k[\mathbf{x}]$ such that

$$p(\widehat{\mathbf{x}},\widehat{\mathbf{g}}) = p\left(\widehat{\mathbf{x}}, \frac{\mathbf{g}(\widehat{\mathbf{x}})}{M}\right) = \frac{1}{M^{d_y}} p(\widehat{\mathbf{x}},\mathbf{g}(\widehat{\mathbf{x}})) = \frac{1}{M^{d_y}} p(\mathbf{x},\mathbf{g})(\widehat{\mathbf{x}}).$$

Therefore, $p(\widehat{\mathbf{x}},\widehat{\mathbf{g}}) = 0$, as was to be shown. □

With this convention and the notation of Lemma 4.3.12 we introduce the following map of the standard graded polynomial ring $k[\mathbf{x},\mathbf{y}]$.

Definition 4.3.15. Let $p(\mathbf{x},\mathbf{y}) \in R[\mathbf{y}]$ be a bihomogeneous form. Set

$$\psi(p) := \left\langle \mathbf{c}(p), \mathbf{x}^{\widehat{N_\mathbf{x}(p)}} \mathbf{y}^{N_\mathbf{y}(p)} \right\rangle = (x_0^{d_x-\beta_0} \cdots x_n^{d_x-\beta_n})^{-1} p(\widehat{\mathbf{x}},\mathbf{y}).$$

The definition of ψ is naturally extended to the whole of $k[\mathbf{x},\mathbf{y}]$ using the direct sum decomposition $k[\mathbf{x},\mathbf{y}] = \bigoplus_{(a,b)\in\mathbb{N}} k[\mathbf{x},\mathbf{y}]_{a,b}$, where $k[\mathbf{x},\mathbf{y}]_{a,b}$ denotes the k-vector space spanned by the monomials of bidegree (a,b).

Note that, with the above convention, one has $\psi(q) = q$ for any homogeneous form $q \in k[\mathbf{y}]$. In other words, the restriction of ψ to the k-subalgebra $k[\mathbf{y}] \subset k[\mathbf{x},\mathbf{y}]$ is the identity map.

Unfortunately, ψ is not a ring homomorphism, but it is quite close in the following sense.

Lemma 4.3.16. *Let $p, q \in k[\mathbf{x},\mathbf{y}]$ bihomogeneous polynomials. One has:*
(i) *$M\psi(p+q) = M_1\psi(p) + M_2\psi(q)$, where M, M_1 and M_2 are suitable monomials in $k[\mathbf{x}]$.*
(ii) *$M\psi(pq) = M'\psi(p)\psi(q)$, where M and M' are suitable monomials in $k[\mathbf{x}]$.*

Proof. (i) If p and q have different bidegrees, then the result is trivial by the definition of ψ extended to the whole of $k[\mathbf{x},\mathbf{y}]$. Thus, assume they have the same bidegree. Set $g := p + q$. Clearly,

$$g(\widehat{\mathbf{x}},\mathbf{y}) = p(\widehat{\mathbf{x}},\mathbf{y}) + q(\widehat{\mathbf{x}},\mathbf{y}).$$

Then the result follows immediately from Lemma 4.3.12.

(ii) This proof is analogous to the one in item (i). □

Fix presentations

$$\mathcal{R}_R(\mathbf{g}) \simeq R[\mathbf{y}]/\mathcal{J}_{\mathbf{g}} \quad \text{and} \quad \mathcal{R}_R(\widehat{\mathbf{g}}) \simeq R[\mathbf{y}]/\mathcal{J}_{\widehat{\mathbf{g}}} \tag{46}$$

of the respective Rees algebras of $\mathbf{g}$ and $\widehat{\mathbf{g}}$.

The main result of this part is the following.

Theorem 4.3.17. *Let $\mathbf{g} = \{g_0,\dots,g_m\} \subset R$ be a set of forms of the same degree satisfying the canonical restrictions and let $\mathcal{J}_{\mathbf{g}}$ and $\mathcal{J}_{\widehat{\mathbf{g}}}$ be as in (46). Then:*
(a) *ψ maps $\mathcal{J}_{\mathbf{g}}$ to $\mathcal{J}_{\widehat{\mathbf{g}}}$.*
(b) *Given a bihomogeneous $q(\mathbf{x},\mathbf{y}) \in \mathcal{J}_{\widehat{\mathbf{g}}}$ such that no variable of $k[\mathbf{x}]$ is a factor of $q(\mathbf{x},\mathbf{y})$ then $q(\mathbf{x},\mathbf{y}) = \psi(p(\mathbf{x},\mathbf{y}))$ for some bihomogeneous $p(\mathbf{x},\mathbf{y}) \in \mathcal{J}_{\mathbf{g}}$; in particular, a finite subset of $\mathcal{J}_{\mathbf{g}}$ maps onto a set of minimal generators of $\mathcal{J}_{\widehat{\mathbf{g}}}$.*
(c) *If $\widetilde{\mathcal{J}_{\mathbf{g}}} \subset \mathcal{J}_{\widehat{\mathbf{g}}}$ is the subideal generated by the image by ψ of a set of generators of $\mathcal{J}_{\mathbf{g}}$, then $\mathcal{J}_{\widehat{\mathbf{g}}}$ is a minimal prime over $\widetilde{\mathcal{J}_{\mathbf{g}}}$.*

Proof. (a) This follows straightforwardly from Proposition 4.3.14 and the definition of ψ.

(b) Since $\widehat{\widehat{\mathbf{g}}} = \mathbf{g}$, it suffices to prove that $\psi(\psi(q(\mathbf{x},\mathbf{y}))) = q(\mathbf{x},\mathbf{y})$. By hypothesis, no x_i is factor of $q(\mathbf{x},\mathbf{y})$, so every row of the Newton matrix $N_{\mathbf{x}}(q)$ has at least one column such that its ith column entry vanishes. Therefore, $\widehat{\widehat{N_{\mathbf{x}}(q)}} = N_{\mathbf{x}}(q)$, that is, $\psi(\psi(q(\mathbf{x},\mathbf{y}))) = q(\mathbf{x},\mathbf{y})$.

(c) We will show a slightly stronger result, namely, there is a monomial $M \in k[\mathbf{x}]$ such that $\mathcal{J}_{\hat{\mathbf{g}}} \subset \widetilde{\mathcal{J}_{\mathbf{g}}} : M$. Since $M \notin \mathcal{J}_{\hat{\mathbf{g}}}$ because the presentation ideal is prime, then this will force the equality $\mathcal{J}_{\hat{\mathbf{g}}} = \widetilde{\mathcal{J}_{\mathbf{g}}} : M$.

Let $q(\mathbf{x}, \mathbf{y})$ denote a bihomogeneous minimal generator of $\mathcal{J}_{\hat{\mathbf{g}}}$. In particular, by part (b) there exists $p(\mathbf{x}, \mathbf{y}) \in \mathcal{J}_{\mathbf{g}}$ such that $\psi(p(\mathbf{x}, \mathbf{y})) = q(\mathbf{x}, \mathbf{y})$. Now express $p(\mathbf{x}, \mathbf{y})$ in terms of the given (finite) set of generators of $\mathcal{J}_{\mathbf{g}}$. By Lemma 4.3.16, one can write

$$M\psi(p(\mathbf{x}, \mathbf{y})) = \sum_{i=1}^{t} h_i \psi(p_i(\mathbf{x}, \mathbf{y}))$$

for suitable monomial $M \in k[\mathbf{x}]$ and certain biforms $\{h_0, \dots, h_t\} \in k[\mathbf{x}, \mathbf{y}]$. Therefore, $M\, q(\mathbf{x}, \mathbf{y}) \in \widetilde{\mathcal{J}_{\mathbf{g}}}$. Now, letting $q(\mathbf{x}, \mathbf{y})$ run through a finite set of minimal (hence, irreducible) generators of $\mathcal{J}_{\hat{\mathbf{g}}}$, and letting M denote the product of the corresponding finitely many monomials of $k[\mathbf{x}]$, one obtains $\mathcal{J}_{\hat{\mathbf{g}}} \subset \widetilde{\mathcal{J}_{\mathbf{g}}} : M$, as contended. □

Example 4.3.18. Let $\mathbf{g} = \{x_1^2x_3^3, x_0x_1^3x_2, x_1x_2^3x_3, x_2^3x_3^2\} \subset R = k[x_0, x_1, x_2, x_3]$.

A calculation with [79] gives that $\mathcal{J}_{\mathbf{g}}$ is minimally generated by the biforms:

$$x_3y_2 - x_1y_3,\ x_2^3y_0 - x_1^2x_3y_3,\ x_0x_1x_2y_0 - x_3^3y_1,\ x_2^2x_3y_1 - x_0x_1^2y_2,$$
$$x_0x_1y_2^2 - x_2^2y_1y_3,\ x_0x_2y_0y_2 - x_3^2y_1y_3,\ x_0^2y_0y_2^4 - x_2x_3y_1^2y_3^3.$$

Similarly, a computation for $\hat{\mathbf{g}} = \{x_0x_1x_2^3, x_2^2x_3^3, x_0x_1^2x_3^2, x_0x_1^3x_3\}$ gives the following set of minimal generators of $\mathcal{J}_{\hat{\mathbf{g}}}$:

$$x_1y_2 - x_3y_3,\ x_0x_1^2y_1 - x_2^2x_3y_2,\ x_1^2x_3y_0 - x_2^3y_3,\ x_3^3y_0 - x_0x_1x_2y_1,\ x_1x_3^2y_0 - x_2^3y_2,$$
$$x_2^2y_2^2 - x_0x_1y_1y_3,\ x_3^2y_0y_2 - x_0x_2y_1y_3,\ x_2x_3y_0y_2^4 - x_0^2y_1^2y_3^3.$$

Note that the generator $x_1x_3^2y_0 - x_2^3y_2$ of $\mathcal{J}_{\hat{\mathbf{g}}}$ is the image of the form $x_2^3y_0 - x_1x_3^2y_2$ by ψ. The latter is not a element of the above minimal set of generators of $\mathcal{J}_{\mathbf{g}}$. Yet, one has $x_1x_3(x_1x_3^2y_0 - x_2^3y_2) \in \widetilde{\mathcal{J}_{\mathbf{g}}}$. A direct verification gives that the remaining elements of the above set of generators of $\mathcal{J}_{\hat{\mathbf{g}}}$ are images by ψ of elements of the above minimal set of generators of $\mathcal{J}_{\mathbf{g}}$. Therefore, one gets $\widetilde{\mathcal{J}_{\mathbf{g}}} : x_1x_3 = \mathcal{J}_{\hat{\mathbf{g}}}$, hence $\mathcal{J}_{\hat{\mathbf{g}}}$ is minimal prime of $\widetilde{\mathcal{J}_{\mathbf{g}}}$ as a confirmation of item (c) of the theorem.

Remark 4.3.19. The multiplier x_1x_3 in the above example is not accidental. Indeed, the exceptional form $x_1x_3^2y_0 - x_2^3y_2$ can be expressed in terms of the elements of the given set of minimal generators of $\mathcal{J}_{\mathbf{g}}$, namely we have

$$x_1x_3^2y_0 - x_2^3y_2 = x_2^3y_0 - x_1^2x_3y_3 - x_1x_3(x_3y_2 - x_1y_3).$$

Since this expression is an elementary operation on the elements of the given set of minimal generators of $\mathcal{J}_{\mathbf{g}}$, this example answers negatively the question as to whether there is in fact a minimal set of generators of $\mathcal{J}_{\mathbf{g}}$ mapping via ψ *onto* a minimal set of generators of $\mathcal{J}_{\hat{\mathbf{g}}}$.

4.4 Elements of rational maps over a ground ring

Quite a bit of the basic classical theory in Subsection 3.2.2 can be extended to the case where the ground coefficients belong to a Noetherian domain (rather than a field). The purpose of this part is to work out some of the corresponding preliminaries, with no intention of originality except for the resilient use of the algebraic side. The results are basically lifted from [38].

4.4.1 Algebraic data

Fix the following.

Notation 4.4.1. Let A be a Noetherian ring of finite Krull dimension. Let $(R, \mathfrak{m})$ denote a standard graded algebra over a $A = [R]_0$ and its graded irrelevant ideal $\mathfrak{m} = ([R]_1)$.

Proposition 4.4.2. *In the above notation, if A is a domain and* $\mathrm{ht}(\mathfrak{m}) \geq 1$, *then there exists a chain of graded prime ideals*

$$0 = \mathfrak{P}_0 \subsetneq \cdots \subsetneq \mathfrak{P}_{k-1} \subsetneq \mathfrak{P}_d \subset R$$

such that $\mathfrak{P}_d \not\supset \mathfrak{m}$, where $d = \dim A$.

Proof. Proceed by induction on d.

The case $d = 0$ it clear or vacuous. Thus, assume that $d > 0$.

Let $\mathfrak{n}$ be a maximal ideal of A with $\mathrm{ht}(\mathfrak{n}) = d$. Then there is some $0 \neq a \in \mathfrak{n} \subset A$ such that $\mathrm{ht}(\mathfrak{n}/(a)) = \mathrm{ht}(\mathfrak{n}) - 1$—to see this, take a chain of primes $\{0\} \subsetneq \mathfrak{q} \subsetneq \cdots \subsetneq \mathfrak{n}$ of length d and any $a \in \mathfrak{q}$, $a \neq 0$; then $\mathrm{ht}\, \mathfrak{n}/(a) \geq d-1$, while $\mathrm{ht}\, \mathfrak{n}/(a) \leq d-1$ is always true. Note that A/aA is injected as a graded summand of R/aR; pick a minimal prime $\mathfrak{Q}$ of R/aR such that $\mathfrak{q} = \mathfrak{Q} \cap A$—a well-known fact for an inclusion of rings (Zorn lemma).

Clearly, $\mathfrak{m} \not\subseteq \mathfrak{Q}$. Indeed, otherwise $(\mathfrak{q}, \mathfrak{m}) \subseteq \mathfrak{Q}$ and since $\mathfrak{m}$ is a prime ideal of R of height at least 1 then $(\mathfrak{q}, \mathfrak{m})$ has height at least 2; this contradicts Krull's principal ideal theorem since $\mathfrak{Q}$ is a minimal prime of a principal ideal.

Let $R' = R/\mathfrak{Q}$ and $A' = A/\mathfrak{q}$. Then R' is a finitely generated graded algebra over A' with $[R']_0 = A'$ and $\mathfrak{m}' := ([R']_1) = \mathfrak{m}R'$. Since $\mathfrak{Q} \not\supset \mathfrak{m}$, it follows $\mathrm{ht}(\mathfrak{m}R') \geq 1$ and by construction, $\dim(A') = \dim(A) - 1$. So by the inductive hypothesis there is a chain of graded primes $0 = \mathfrak{P}'_0 \subsetneq \cdots \subsetneq \mathfrak{P}'_{d-1}$ in R' such that $\mathfrak{P}'_{d-1} \not\supset \mathfrak{m}R'$. Finally, for $j \geq 1$ define $\mathfrak{P}_j$ as the inverse image of $\mathfrak{P}'_{j-1}$ via the surjection $R \twoheadrightarrow R'$. □

Note that $X := \mathrm{Proj}(R)$ is a closed subscheme of the (relative) rth projective space $\mathbb{P}^r_A$ over A, for some r, satisfying the following properties:

- Its underlying topological space is the set of all homogeneous prime ideals of R not containing $\mathfrak{m}$ and it has a basis given by the open sets of the form $D_+(f) := \{\wp \in X \mid f \notin \wp\}$, where $f \in R_+$ is a homogeneous element of positive degree;
- Its sheaf structure is given by the degree zero part of the homogeneous localizations

$$\Gamma(D_+(f), \mathcal{O}_X|_{D_+(f)}) := R_{(f)} = \left\{\frac{g}{f^l} \mid g, f \in R, \deg(g) = l\deg(f)\right\}.$$

Notation 4.4.3. In addition to 4.4.1, assume that R itself is a domain.

Under this assumption, X admits a field $K(X) := [R_{(0)}]_0$ of rational functions of X, where

$$[R_{(0)}]_0 = \left\{\frac{f}{g} \mid f, g \in R, \deg(f) = \deg(g), g \neq 0\right\},$$

the degree zero part of the homogeneous localization of R at the null ideal (note that $R_{(0)}$ is a graded ring, but not itself a field). The *dimension* $\dim(X)$ of the closed subscheme X is defined to be the supremum of the lengths of chains of irreducible closed subsets (see, e. g., [88, Definition, p. 5 and p. 86]). This topological notion is kind of loose, so we give a working algebraic version based on 4.4.2.

For any domain D, let $\mathrm{Quot}(D)$ denote its field of fractions.

Proposition 4.4.4. *In the notation 4.4.3, one has*

$$\dim(X) = \dim(R) - 1 = \dim(A) + \mathrm{tr.deg}._{\mathrm{Quot}(A)}(K(X)).$$

Proof. For any prime $\mathfrak{P} \in X$, the ideal $(\mathfrak{P}, \mathfrak{m}) \neq R$ is an ideal properly containing $\mathfrak{P}$, hence the latter is not a maximal ideal. Therefore, $\mathrm{ht}(\mathfrak{P}) \leq \dim(R) - 1$ for any $\mathfrak{P} \in X$, which clearly implies that $\dim(X) \leq \dim(R) - 1$.

From [194, Lemma 1.1.2] one gets the equalities

$$\dim(R) = \dim(A) + \mathrm{ht}(\mathfrak{m}) = \dim(A) + \mathrm{tr.deg}._{\mathrm{Quot}(A)}(\mathrm{Quot}(R)).$$

There exists a chain of graded prime ideals $0 = \mathfrak{P}_0 \subsetneq \cdots \subsetneq \mathfrak{P}_{h-1} \subsetneq \mathfrak{P}_h = \mathfrak{m}$ such that $h = \mathrm{ht}(\mathfrak{m})$ (see, e. g., [138, Theorem 13.7], [21, Theorem 1.5.8]). Let $T = R/\mathfrak{P}_{h-1}$. Since $\mathrm{ht}(\mathfrak{m}T) = 1$, 4.4.2 yields the existence of a chain of graded prime ideals $0 = \mathfrak{Q}_0 \subsetneq \cdots \subsetneq \mathfrak{Q}_k$ in T, where $k = \dim(A)$ and $\mathfrak{Q}_k \not\supseteq \mathfrak{m}T$. By taking inverse images along the surjection $R \twoheadrightarrow T$, one obtains a chain of graded prime ideals not containing $\mathfrak{m}$ of length $h-1+k = \dim(R)-1$. Thus, one has the reverse inequality $\dim(X) \geq \dim(R)-1$.

Now, for any $f \in [R]_1$, one has $\mathrm{Quot}(R) = R_{(0)}(f)$ with f transcendental over $K(X) = R_{(0)}$. Therefore,

$$\dim(X) = \dim(A) + \mathrm{tr.deg}._{\mathrm{Quot}(A)}(\mathrm{Quot}(R)) - 1 = \dim(A) + \mathrm{tr.deg}._{\mathrm{Quot}(A)}(K(X)),$$

as was to be shown. □

4.4.2 Rational maps and representatives

In the case of a ground field, we have motivated the idea of a rational map by classical algebraic geometry, but approached a stabilized notion in terms of coordinate forms. In the next few paragraphs, we mimic this procedure by assuming a ground ring instead of a field.

First, we restate the following generally accepted notion of a rational map for (relative) projective varieties. Thus, as in the previous subsection, let X denote a subscheme of projective space $\mathbb{P}^r_A$ over $\operatorname{Spec}(A)$, where A is a ring.

Fixing an integer $s \geq 1$, let $\operatorname{Rat}(X, \mathbb{P}^s_A)$ denote the set of pairs (U, φ) where U is a dense open subscheme of X and where $\varphi : U \to \mathbb{P}^s_A$ is an A-morphism of schemes. Two pairs $(U_1, \varphi_1), (U_2, \varphi_2) \in \operatorname{Rat}(X, \mathbb{P}^s_A)$ are said to be *equivalent* if there exists a dense open subscheme $W \subset U_1 \cap U_2$ such that $\varphi_1|_W = \varphi_2|_W$. This gives an equivalence relation on $\operatorname{Rat}(X, \mathbb{P}^s_A)$. A *rational map* of X to $\mathbb{P}^s_A$, denoted $\mathfrak{F} : X \dashrightarrow \mathbb{P}^s_A$, is defined to be an equivalence class in $\operatorname{Rat}(X, \mathbb{P}^s_A)$ and any element of this equivalence class determines this rational map uniquely.

We proceed to explain how this topological minded notion can be translated in terms of homogeneous coordinate functions, as in the case of a ground field. Letting $X = \operatorname{Proj}(R)$, where R is as in 4.4.1, pick any $s+1$ forms $\mathbf{f} = \{f_0, f_1, \dots, f_s\} \subset R$ contained in $\mathfrak{m} = (R_1)$. Setting $S := A[y_0, y_1, \dots, y_s]$, a polynomial ring over A, let $\mathfrak{h} : S \to R$ stand for the graded homomorphism of A-algebras such that $y_i \mapsto f_i$. Denote by

$$\Phi(\mathbf{f}) = \operatorname{Proj}(\mathfrak{h}) : D_+(\mathbf{f}) \longrightarrow \operatorname{Proj}(S) = \mathbb{P}^s_A$$

the corresponding A-morphism of schemes, where $D_+(\mathbf{f}) = \bigcup_{i=0}^s D_+(f_i)$. Therefore, a set of $s+1$ forms $\mathbf{f} = \{f_0, f_1, \dots, f_s\} \subset R$ of the same positive degree determines a unique rational map, denoted $\mathfrak{F}_{\mathbf{f}}$, given by the equivalence class of $(D_+(\mathbf{f}), \Phi(\mathbf{f}))$ in $\mathfrak{R}(X, \mathbb{P}^s_A)$. But, conversely, we have the following.

Lemma 4.4.5. *Any rational map $\mathfrak{F} : X = \operatorname{Proj}(R) \dashrightarrow \mathbb{P}^s_A$ is of the form $\mathfrak{F}_{\mathbf{f}}$, where $\mathbf{f}$ are forms of the same positive degree.*

Proof. Let U be a dense open subset in X and $\varphi : U \to \mathbb{P}^s_A$ be a morphism determining $\mathfrak{F}$. Set $V := D_+(y_0)$, and $W := \varphi^{-1}(V)$. Pick an affine open set $W' := \operatorname{Spec}(R_{(\ell)}) \subset W$, where $\ell \in \mathfrak{m}$ is a homogeneous element. The restriction $\varphi|_{W'} : W' \to V$ corresponds to a ring homomorphism $\tau : S_{(y_0)} \to R_{(\ell)}$. Say, $\tau(y_i/y_0) = g_i/\ell^{\alpha_i}$, where $\deg(g_i) = \alpha_i \deg(\ell)$. Letting $\alpha := \max_{1 \leq i \leq s}\{\alpha_i\}$, one sets

$$f_0 := \ell^\alpha \quad \text{and} \quad f_i := \ell^\alpha \frac{g_i}{\ell^{\alpha_i}} = \ell^{\alpha - \alpha_i} g_i \quad \text{for } 1 \leq i \leq s.$$

By construction, $\varphi|_{W'} = \Phi(\mathbf{f})|_{W'}$, where $\Phi(\mathbf{f})$ denotes the $\mathbf{f}$-coordinate morphism determined by $\mathbf{f} = \{f_0, \dots, f_s\}$, as above. Therefore, $\mathfrak{F} = \mathfrak{F}_{\mathbf{f}}$, where $\mathbf{f} = \{f_0, \dots, f_s\}$. □

Given a rational map $\mathfrak{F} : X \dashrightarrow \mathbb{P}^s_A$, any ordered $(s+1)$-tuple $\mathbf{f} = \{f_0, f_1, \dots, f_s\}$ of forms of the same positive degree such that $\mathfrak{F} = \mathfrak{F}_{\mathbf{f}}$ is called a *representative* of the rational map $\mathfrak{F}$. We will by abuse say that the representative is $(f_0 : f_1 : \cdots : f_m)$ to stress its role determining the rational map. Note that the notion of a base ideal (or even base scheme) of $\mathfrak{F}$ depends on the choice of a representative **f**, whereby it is defined as the ideal $(\mathbf{f}) \subset R$.

The following results establish some of the relevant properties of the representatives of a rational map. For simplicity, the A-subalgebra of R generated by the individual forms of a representative **f** will be denoted $A[\mathbf{f}]$.

Proposition 4.4.6. *Under the notation of 4.4.3, let $\mathfrak{F} : X \dashrightarrow \mathbb{P}^s_A$ denote a rational map.*

(i) *Any two representatives $(f_0 : \cdots : f_s)$ and $(f'_0 : \cdots : f'_s)$ of $\mathfrak{F}$ are proportional coordinate sets in the sense that there exist homogeneous forms h, h' of positive degree such that $hf'_i = h'f_i$ for $i = 0, \dots, s$.*

(ii) *Fix a representative and let $I \subset R$ denote the corresponding base ideal. Then the set of representatives of $\mathfrak{F}$ correspond bijectively to the nonzero homogeneous vectors in the rank one graded R-module* $\mathrm{Hom}_R(I, R)$. *In particular, if* $\mathrm{grade}(I) \geq 2$, *any representative of $\mathfrak{F}$ is a multiple of* **f** *by a homogeneous element in R (this is the case, e. g., when A is a UFD and R is a polynomial ring over A).*

(iii) *For any two representatives $(f_0 : \cdots : f_s)$ and $(f'_0 : \cdots : f'_s)$ of $\mathfrak{F}$, with respective base ideals $I \subset R$ and $I' \subset R$, one has $A[\mathbf{f}] \simeq A[\mathbf{f}']$ as graded subalgebras of R and $\mathcal{R}_R(I) \simeq \mathcal{R}_R(I')$ as bigraded algebras.*

(iv) *For any representative* **f** *of $\mathfrak{F}$, the closure of the image of $\mathfrak{F}$ is isomorphic to* $\mathrm{Proj}\, A[\mathbf{f}]$ *(normalized degrees).*

Proof. (i) Let $\Phi(\mathbf{f}) : D_+(\mathbf{f}) \to \mathbb{P}^s_A$ and $\Phi(\mathbf{f}') : D_+(\mathbf{f}') \to \mathbb{P}^s_A$ be morphisms as above. Let $V = \mathrm{Spec}(D_+(y_0))$ and choose $W = \mathrm{Spec}(R_{(\ell)})$ such that $W \subset \Phi(\mathbf{f})^{-1}(V) \cap \Phi(\mathbf{f}')^{-1}(V)$ and $\Phi(\mathbf{f})|_W = \Phi(\mathbf{f}')|_W$. The morphisms $\Phi(\mathbf{f})|_W : W \to V$ and $\Phi(\mathbf{f}')|_W : W \to V$ correspond to ring homomorphisms $\tau : S_{(y_0)} \to R_{(\ell)}$ and $\tau' : S_{(y_0)} \to R_{(\ell)}$ such that

$$\tau\left(\frac{y_i}{y_0}\right) = \frac{f_i}{f_0} \quad \text{and} \quad \tau'\left(\frac{y_i}{y_0}\right) = \frac{f'_i}{f'_0},$$

respectively. Since this is now an affine setting, the ring homomorphisms τ and τ' are the same (see, e. g., [88, Proposition II.2.3]). It follows that, for every $i = 0, \dots, s$, $f'_i/f'_0 = f_i/f_0$ as elements of the homogeneous total ring of quotients of R. Therefore, there are homogeneous elements $h, h' \in R$ $(h = f_0, h' = f'_0)$ such that $hf'_i = h'f_i$ for $i = 0, \dots, s$.

(ii) The same argument as in Proposition 4.1.1 works.

(iii) Since $f'_i = (h'/h)f_i$ for every i, an equality taking place in the quotient field of R, it is clear that a polynomial $P(\mathbf{y}) \in S = A[\mathbf{y}]$ vanishes on $\{\mathbf{f}\}$ if and only if it vanishes on $\{\mathbf{f}'\}$. This takes care of the two subalgebras of R. The argument for the Rees algebras is similar.

(iv) The argument is pretty much that over a ground field (alternatively, see [29, Theorem 2.1]), noting that in the present setup the defining ideal of $A[\mathbf{f}]$ is saturated since the latter is a domain (reducedness would suffice). □

The graded algebra $A[\mathbf{f}]$ can be renormalized to become standard graded. As such, it gives as explained above the homogeneous coordinate ring of the image Y of the rational map determined by the representative $\mathbf{f}$. Thus, the function field of the image is given by $K(Y) := [A[\mathbf{f}]_{(0)}]_0$. Note then the naturally induced field extension $K(Y) \subset K(X)$.

The geometric conditions expressing, respectively, the notion of a generically finite rational map and a birational map in terms of the equivalence classes of morphisms from dense open sets can now be converted into their algebraic analogs, pretty much as in the case of a ground field.

Proposition 4.4.7. *Under the notation of* 4.4.3, *let* $\mathfrak{F} : X = \mathrm{Proj}(R) \dashrightarrow \mathbb{P}^s_A$ *be a rational map with image* Y.

(a) *The following conditions are equivalent:*
   (i) $\mathfrak{F}$ *is generically finite.*
   (ii) *The field extension* $K(Y) \subset K(X)$ *is finite.*
   (iii) $\dim(X) = \dim(Y)$.
(b) *The following conditions are equivalent:*
   (i) $\mathfrak{F}$ *is birational onto its image* Y.
   (ii) *The field extension* $K(Y) \subset K(X)$ *is an equality.*
   (iii) *There exists a rational map* $\mathfrak{G} : Y \dashrightarrow X \subset \mathbb{P}^r_A$ *such that, for some (any) representative* $\mathbf{f}$ *of* $\mathfrak{F}$ *and some (any) representative* $\mathbf{g} = (g_0 : \cdots : g_r)$ *of* $\mathfrak{G}$, *one has* $(g_0(\mathbf{f}) : \cdots : g_r(\mathbf{f})) \equiv (x_0 : \cdots : x_r)$, *where* $R = k[x_0, \ldots, x_r]$.

Proof. Here, the respective first items in (a) and (b) follow the geometric definition. Their equivalences to the respective second items of (a) and (b) are left to the reader.

The equivalence (ii) ⇔ (iii) in (a) follows immediately from Proposition 4.4.4.

The implication (iii) ⇒ (ii) in (b) is quite clear by translating the equivalence of representatives to field theoretic equalities. The converse is quite similar. □

Under the equivalent conditions of item (a) above, one defines the *degree* of $\mathfrak{F}$ to be the field extension degree $[K(X) : K(Y)]$. A tall order is to obtain information about the degree—such as upper and lower bounds—from properties of the various algebras related to a representative.

4.4.3 Degree under specialization

Since our study of rational maps is pretty much algebraically minded, the usual notion of specialization of algebras and modules may be transported to the present considerations. As such, we establish the following.

Setup 4.4.8. Let $A = k[z_1, \ldots, z_m]$ be a polynomial ring over a field k. Consider a rational map $\mathfrak{F} : \mathbb{P}^r_A \dashrightarrow \mathbb{P}^s_A$ given by a representative $(f_0 : \cdots : f_s)$, where $\mathbb{P}^r_A = \operatorname{Proj}(R)$ with $R = A[\mathbf{x}] = A[x_0, \ldots, x_r]$. Fix a maximal ideal $\mathfrak{n} = (z_1 - \alpha_1, \ldots, z_m - \alpha_m)$ of A where $\alpha_i \in k$.

Clearly, $k \simeq A/\mathfrak{n}$ and the structure of A-module of k comes from the canonical surjection $A \twoheadrightarrow A/\mathfrak{n} \simeq k$.

Then $R/\mathfrak{n}R \simeq k[x_0, \ldots, x_r]$. Consider the rational map $\mathfrak{f} : \mathbb{P}^r_k \dashrightarrow \mathbb{P}^s_k$ with representative $(\overline{f_0} : \cdots : \overline{f_s})$, where $\overline{f_i}$ is the image of f_i under the canonical map $R \twoheadrightarrow R/\mathfrak{n}R$. Further assume that $\overline{f_i} \neq 0$ for all $0 \le i \le s$.

Finally, let $\mathcal{I} := (f_0, \ldots, f_s) \subset R$ and $I := (\mathcal{I}, \mathfrak{n})/\mathfrak{n} = (\overline{f_0}, \ldots, \overline{f_s}) \subset R/\mathfrak{n}R$.

The problem of the behavior of the degree under specialization in the present sense can be envisaged in the following example.

Example 4.4.9. Let $m \ge 1$ be an integer. Let $A = k[t]$ be a polynomial ring over k. Let $R = A[x,y,z]$ be a standard graded polynomial ring over $A = R_0$ and consider the following matrix:

$$\mathcal{M} = \begin{pmatrix} x & zy^{m-1} \\ -y & zx^{m-1} + y^m \\ tz & zx^{m-1} \end{pmatrix}$$

which is homogeneous in this grading. For any $\beta \in k$, let $\mathfrak{n}_\beta := (t - \beta) \subset A$. Let $\mathcal{I} = (f_0, f_1, f_2) := I_2(\mathcal{M}) \subset R$ and $I_\beta := (\pi_\beta(f_0), \pi_\beta(f_1), \pi_\beta(f_2)) \subset R/\mathfrak{n}_\beta R$ be the specialization of $\mathcal{I}$ via the canonical map $\pi_\beta : R \twoheadrightarrow R/\mathfrak{n}_\beta R$.

Let $\mathfrak{F} : \mathbb{P}^2_A \dashrightarrow \mathbb{P}^2_A$ and $\mathfrak{f}_\beta : \mathbb{P}^2_{A/\mathfrak{n}_\beta} \dashrightarrow \mathbb{P}^2_{A/\mathfrak{n}_\beta}$ be the rational maps with representatives $(f_0 : f_1 : f_2)$ and $(\pi_\beta(f_0) : \pi_\beta(f_1) : \pi_\beta(f_2))$, respectively.

When $\beta = 0$, from [89, Proposition 2.3] one has that $\mathfrak{f}_0 : \mathbb{P}^2_{A/\mathfrak{n}_0} \dashrightarrow \mathbb{P}^2_{A/\mathfrak{n}_0}$ is a de Jonquières map, which is birational. On the other hand, if $\beta \neq 0$, then I_β satisfies the condition G_3 and so one has $\deg(\mathfrak{f}_\beta) = m$.

Therefore, it follows that

$$\deg(\mathfrak{f}_\beta) = \begin{cases} 1 & \text{if } \beta = 0, \\ m & \text{if } \beta \neq 0. \end{cases}$$

Also, note that $\deg(\mathfrak{F}) = m$. By taking $k = \mathbb{Q}$, one can check with [79] that $\dim(\operatorname{gr}_{\mathcal{I}}(R) \otimes_A A/(t)) = 4$ and that, e. g., for $\beta = 1$ one has $\dim(\operatorname{gr}_{\mathcal{I}}(R) \otimes_A A/(t-1)) = 3$.

The above values of the specialized associated graded ring look like they may have an impact in general, and one may wonder whether there is any chance that the degree become stable under specialization.

Here is then a pretty encompassing result trailing the above behavior.

Theorem 4.4.10 (char$(k) = 0$). *In the notation of Setup 4.4.8, suppose that the following conditions hold:*

(a) $\mathfrak{F}$ *and* $\mathfrak{f}$ *are generically finite.*

(b) *The special fiber* $A[f_0, \dots, f_s] \subset R$ *is locally normal at its homogeneous punctured spectrum.*

(c) $\dim \mathrm{gr}_{\mathcal{I}}(R) \otimes_A k \leq \dim R - \dim A + 1 = r + 2$.

Then:

(i) $\deg \mathfrak{f} \leq \deg \mathfrak{F}$.

(ii) *If, moreover,* $\dim \mathrm{gr}_{\mathcal{I}}(R) \otimes_A k \leq r + 1$ *then* $\deg \mathfrak{f} = \deg \mathfrak{F}$.

(iii) *When k is algebraically closed, there is a dense open subset* $\mathcal{W} \subset k^m$ *such that, if* $\mathfrak{n} = (z_1 - \alpha_1, \dots, z_m - \alpha_m)$ *with* $(\alpha_1, \dots, \alpha_m) \in \mathcal{W}$, *one has* $\deg \mathfrak{f} = \deg \mathfrak{F}$.

Supplement. *If* $\ell(\mathcal{I}_\wp) \leq \mathrm{ht}(\wp/\mathfrak{n}R) + 1$ *(resp.,* $\ell(\mathcal{I}_\wp) \leq \mathrm{ht}(\wp/\mathfrak{n}R)$*) for every prime ideal* $\wp \subset R$ *containing* $(\mathcal{I}, \mathfrak{n})$ *then condition* (c) *(resp., the hypothesis of* (ii)*) holds.*

Summing up, provided condition (b) is satisfied (e. g., if $r = s$), then the degree of the specialized rational map is stable if the specialized associated graded ring has the "expected" dimension—this being the case if the specialization is general. If the dimension of the specialized associated graded ring is one worse, then the degree does not decrease under specialization.

Note the dimension values in the above example are the only possible ones.

The proof of this theorem exceeds the scope of the book; for a full proof with a geometric flavor we refer to [38]. We do however state a couple of relevant applications in the usual tradition of perfect height two and Gorenstein height three ideals. Of course, for the applications one is given the special data, so the general data have to be created by introducing those many needed variables $z_1, \dots, z_m$.

4.4.3.1 Perfect ideals of height two

We think of those as being generated by the maximal minors of Hilbert–Burch sort of matrices. Accordingly, the required number of generic variables will be dictated by the sizes of such matrices and the degrees of their entries.

Proposition 4.4.11. *Let* $\mathfrak{f} : \mathbb{P}^r_k \dashrightarrow \mathbb{P}^r_k$ *be a dominant rational map, with base ideal having free minimal resolution*

$$0 \to \bigoplus_{i=1}^{r} B(-d - d_i) \xrightarrow{\varphi} B(-d)^{r+1} \to I \to 0,$$

where $B = k[x_0, \ldots, x_r]$. *If* $\mu(I_\wp) \leq \operatorname{ht}\wp + 1$ *(resp.,* $\mu(I_\wp) \leq \operatorname{ht}\wp$*) for every prime ideal* $\wp \subset B$ *then* $\deg \mathfrak{f} \leq d_1 \cdots d_r$ *(resp.,* $\deg \mathfrak{f} = d_1 \cdots d_r$*).*

The reader may recognize the two assumptions in the statement as properties F_0 and F_1 (or G_∞), respectively.

4.4.3.2 Gorenstein ideals of height three

We think of those as being generated by the maximal Pfaffians of an alternate matrix. Thus, again the number of generic variables is dictated by the details of such matrices.

It can be readily seen that if the Pfaffians have the same degree d, then every nonzero entry of the corresponding alternate matrix has the same degree D as well and they are related by the equation $d = (r/2)D$, where $r + 1$ is the number of Pfaffians (see, e. g., [134, Section 3.1]).

Then a similar result to Proposition 4.4.11 holds by replacing the product $d_1 \cdots d_r$ by the power D^r.

Remark 4.4.12. The degree upper bound in the above applications holds without the F_0 condition, as can be seen in Section 3.4. However, not the exact value under the condition F_1. Since in both applications, this condition forces the base ideal to be of linear type, one wonders if the latter property on itself has any impact whatsoever on the degree of the corresponding rational map.

4.5 Exercises

Exercise 4.5.1. Let k be a field. Show that the rational map $\mathbb{P}^3_k \dashrightarrow \mathbb{P}^3_k$ with representative $(x_0x_3 : x_1x_3 : x_0(x_1 - x_2) : x_1(x_0 - x_1))$ is birational and that the base ideal of inverse map is a codimension 2 perfect ideal.

Exercise 4.5.2. Let k be a field. Consider the rational map $\mathbb{P}^n_k \dashrightarrow \mathbb{P}^n_k$ with representative $(x_0x_1 : x_1x_2 : \cdots : x_{n-1}x_n : x_nx_0)$.

(1) If n is even show that the map is birational and the minimal representative of the inverse map is in degree $(n + 2)/2$. Give an interpretation in terms of an odd cycle graph and its minimal coverings.

(2) (Perazzo hypersurface) If n is odd show that the image of the map is a hypersurface of degree $(n + 1)/2$ and the restriction of the map to a suitable hyperplane in the source is birational onto the hypersurface. Is there a uniquely defined minimal representative of the inverse map?

Exercise 4.5.3. Let $X \subset \mathbb{P}^2$ be the nodal cubic with equation $y^2z - x^2(x + z)$.

Prove:

(1) The rational map $(\bar{x} : \bar{y}) : X \dashrightarrow \mathbb{P}^1$ is defined everywhere except at $(0 : 0 : 1)$.

(*Hint*: the ideal of coordinates J_φ relative to the base ideal $(\bar{x}, \bar{y})$ on the residue algebra $k[x,y,z]/(y^2z - x^2(x+z))$ is (x,y), which vanishes exactly at $(0:0:1)$.)

(2) $(\bar{x} : \bar{y})$ is birational onto $\mathbb{P}^1$. What is the inverse map?
(3) The rational map $(\bar{x}^2 : \bar{y}^2) : X \dashrightarrow \mathbb{P}^1$ is defined everywhere, but is not birational. What is its degree as a map?

Exercise 4.5.4. Let $\mathcal{F} : \mathbb{P}^2 \dashrightarrow \mathbb{P}^3$ be the rational map defined by the (ordered, signed) maximal minors of the 4×3 matrix:

$$\varphi = \begin{pmatrix} x & 0 & z \\ y & x & 0 \\ z & y & x \\ 0 & z & y \end{pmatrix}.$$

Let Y be the image of $\mathfrak{F}$ and let ψ denote the 3×3 submatrix of the weak Jacobian dual matrix of $\mathfrak{F}$ over the homogeneous coordinate ring $k[t_1, t_2, t_3, t_4]/I(Y)$ of Y obtained from the generators of $I_1(\mathbf{t} \cdot \varphi)$. Show that:

(1) $\mathfrak{F}$ is a birational map onto its image. In particular, Y is a hypersurface of $\mathbb{P}^3$.
(2) For each $1 \leq i \leq 3$, denote by ψ_i the 2×3 submatrix of ψ obtained by omitting the ith row of ψ. The (ordered, signed) 2-minors of ψ define a representative of the inverse of $\mathfrak{F}$.
(3) Denote by D_i the source inversion factor of $\mathfrak{F}$ corresponding to representing of the inverse given by the 2-minors of ψ_i. The ideal D_1, D_2, D_3 has codimension 2 and its minimal graded free resolution is

$$0 \to R(-9) \xrightarrow{\partial_1} R(-8)^3 \xrightarrow{\partial_0} R(-5)^3 \to R$$

where ∂_0 is the matrix ψ evaluated on the coordinates of $\mathfrak{F}$ and $\partial_1 = (x\,y\,z)^t$.

Exercise 4.5.5. Consider the rational map

$$\mathfrak{F} = (x_0^4 : x_0^2x_1x_3 : x_0x_1^2x_2 : x_1^3x_2) : \mathbb{P}^3 \dashrightarrow \mathbb{P}^3.$$

Argue or compute:

(1) $\mathfrak{F}$ is birational onto $\mathbb{P}^3$.
(*Hint*: For a quick answer, use Proposition 3.3.1.)
(2) The defining ideal of the symmetric algebra $\mathcal{S}(I)$ has height only 2.
(3) The Jacobian dual rank is 3.
(4) $\mathfrak{F}$ is an involution up to source/target variable permutation.

Exercise 4.5.6. Let $X = \mathbb{P}^1$ and let $F: \mathbb{P}^1 \longrightarrow \mathbb{P}^3$ be the regular parametrization of the nonnormal rational quartic given by $\mathbf{f} = \{x_0^4, x_0^3x_1, x_0x_1^3, x_1^4\}$.

Prove:
(1) **f** has two independent linear syzygies and at least one relation form of bidegree (1, 2) not coming from the syzygies.
(*Hint*: use computer assistance as a last resource.)
(2) The partial Jacobian dual matrix with the three forms in (1) has enough rank for birationality and it affords at least the vectors $(\bar{y}_0, \bar{y}_1)^t$, $(\bar{y}_2, \bar{y}_3)^t$ as syzygies modulo the defining equations of the quartic.
(3) The inverse map to F is regular and F is a biregular map.

Caveat. The following two exercises deal with the Gauss map of a reduced plane curve defined by a form $f \in R = k[x,y,z]$ of degree d. Recall that we identify the actual Gauss map with the rational map obtained after the Plücker embedding (Section 9.1.1). If $\mathrm{char}(k) = 0$, there is a classical formula for the degree of the image of the Gauss map of $X = \mathrm{Proj}(R/(f)) \subset \mathbb{P}^2$ (i. e., the dual curve X^*):

$$\deg(X^*) = d(d-1) - \sum_{p \in \mathrm{Sing}(X)} e_{J_p}(R_p),$$

where $e_{J_p}(R_p)$ is the local Samuel multiplicity of the Jacobian ideal J of f at a singular point $p \in X$. It also known that the Gauss map is birational onto the dual curve.

Exercise 4.5.7 ($\mathrm{char}(k) = 0$). Let $X \subset \mathbb{P}^2$ be the nodal cubic with equation $f = x_0^3 - x_0^2x_2 + x_1^2x_2$ and let $F\colon X \dashrightarrow \mathbb{P}^2$ be the Gauss map.
Prove:
(1) $\deg(X^*) = 4$.
(2) The degree sequence (Definition 4.1.4) of F is 2, 3.
(3) F is birational onto X^* (by means of the criterion in Theorem 3.2.22, with computer assistance).
(4) The degree sequence of F^{-1} is 3, 3, 3, 3.
(5) One of the representatives of F^{-1} is given by the partial derivatives of the form defining the dual curve.
(6) Neither F, nor F^{-1}, is everywhere defined. How general is this phenomenon?

Exercise 4.5.8 ($\mathrm{char}(k) = 0$). Let $X \subset \mathbb{P}^2$ be the elliptic cubic with equation $f = x_0^3 + x_1^3 - x_2^3$ and let $F\colon X \dashrightarrow \mathbb{P}^2$ be the Gauss map.
Prove:
(1) $\deg(X^*) = 6$.
(2) The degree sequence of F is 2.
(3) F is birational onto X^* (with computer assistance).
(4) The degree sequence of F^{-1} is 4, 4, 4, 5. Is a vector in degree 5 the partial derivatives of the dual curve?

Exercise 4.5.9. Let $\{f_0, f_1, f_2\} \subset R = k[x_0, x_1, x_2, x_3]$ be the defining equations of the standard twisted cubic $C \subset \mathbb{P}^3$—i. e., the 2-minors of the generic 2×3 Hankel matrix. Let $X = \{\Delta = 0\} \subset \mathbb{P}^3$ denote a smooth general cubic surface containing C and consider the rational map $F: X \dashrightarrow \mathbb{P}^2$ defined by the residues of f_0, f_1, f_2. Write $I \subset R/(\Delta)$ for the ideal generated by these residues.

Prove:

(1) The defining equation of X can be written as the determinant of a 3×3 matrix whose first two rows are the rows of the Hankel matrix defining C and the third row has entries, which are sufficiently general k-linear combinations of the coordinates of $\mathbb{P}^3$.

(2) F is everywhere defined on X.
(*Hint*: (i) geometrically, one argues that for every point $p \in X$ there exists a twisted cubic not passing through p and whose defining equations still represent F; (ii) algebraically, show that the syzygy matrix φ of I is the transpose of the 3×3 matrix with determinant Δ. Next, show that $\mathrm{Hom}_{R/(\Delta)}(I, R/(\Delta))$ is minimally generated by three vectors of $R/(\Delta)$-degree 2, one of which can be taken to be the transposed of the original representative of F. The transposed of any k-linear combination of the generators yield yet another representative of F, so these infinitely many choices retro explain the geometric argument.)

Exercise 4.5.10. With the notation of the previous exercise, show:

(1) I is an ideal of linear type (on $R/(\Delta)$).
(*Hint*: the symmetric algebra of I is Cohen–Macaulay and the relevant Fitting ideals of φ have the required rank.)

(2) F is birational onto $\mathbb{P}^2$.
(*Hint*: φ is linear.)

(3) The base ideal of the inverse map is the homogeneous ideal of six general points in $\mathbb{P}^2$.
(*Hint*: the inverse map is defined by 4 cubics generating an ideal of degree (multiplicity) 6.)

5 The gradient ideal

The *gradient ideal* (or *Jacobian ideal*) of a polynomial $f \in R = k[x_1, \dots, x_n]$ is the ideal $J_f \subset R$ generated by the partial derivatives of f. It comes up in several different landscapes, both algebraic and geometric, such as polar maps, singularity theory, arrangements, differentials, Milnor numbers and many others.

Throughout the chapter, we deal with the case where the polynomial ring R is standard graded and $f \in R$ is a form of positive degree. In particular, when k has characteristic zero, questions having to do with the relationship between Milnor and Tjurina devices have no purpose here.

For display convenience, we will often shorten $\partial f/\partial x_i$ to f_{x_i}.

5.1 The underlying geometry

The ideal J_f has strong geometric contours as it relates to logarithmic derivations and polar maps. It brings along the corresponding module $\mathrm{Syz}(J_f)$ of syzygies, whose importance can be hardly exaggerated as it is the main summand in the direct sum decomposition

$$\mathrm{Derlog}(f) = \mathrm{Syz}(J_f) \oplus R\theta_E, \tag{47}$$

up to the identification $\mathrm{Der}(R) = R^{n+1}$, with $\theta_E = \sum_{i=1}^n x_i \frac{\partial}{\partial x_i}$ denoting the Euler derivation. Here, $\mathrm{Derlog}(f) = \{\theta \in \mathrm{Der}(R) \mid \theta(f) \in (f)\}$ is the module of logarithmic derivations of f (see Section 5.2 below).

Now, by and large we assume that f is reduced. There is a game changer between the cases where f is irreducible or not. If f has at least two factors, then $\mathrm{ht}\, J_f < 2$, while otherwise J_f can have pretty much any codimension. This basic watershed explains much of the chosen method of approaching the problems. In a more precise way, if $f = p_1 \cdots p_m$ is the factorization of f into prime factors, it is easy to see that

$$J_f \subset (p_1 \cdots \widehat{p_j} \cdots p_m \mid 1 \leq j \leq m)\, I_1(J_{p_1,\dots,p_m}),$$

where $J_{p_1,\dots,p_m}$ denote the Jacobian matrix of $p_1, \dots, p_m$. Since the rightmost ideal has height 2 when $m \geq 2$, so does J_f.

In the case where the polynomial ring R is standard graded and $f \in R$ is a form of positive degree, a basic question asks whether the partial derivatives are algebraically independent over k, in which case it means that the so-called *polar map* is dominant or, equivalently, the special fiber of J_f is a polynomial ring over k.

If f defines a smooth hypersurface then J_f is irrelevant, i. e., is generated by a regular sequence, hence is of linear type. We regard this case as uninteresting and assume that f has singularities. This entails $\mathrm{ht}(J_f/(f)) \leq n-2$. If moreover f is reduced, then

https://doi.org/10.1515/9783110640694-005

$\mathrm{ht}(J_f/(f)) \geq 1$. For $n = 3$, we therefore find $\mathrm{ht}(J_f/(f)) = 1$. Ideals of height 1 in nonregular rings of dimension 2 are a tall order and typically involve a nontrivial primary decomposition. Thus, even for plane projective curves the structure of the gradient ideal seems to be fairly intricate.

5.1.1 Linear type

We regard the linear type case as a limit situation we would like to understand first. Clearly, this property implies that the partial derivatives are algebraically independent over k. In characteristic zero the latter property in turn is tantamount to saying that the Hessian of f does not vanish.

For an almost complete intersection, this property is fairly under control.

Proposition 5.1.1. *Let $f \in R = k[x_1, \dots, x_n]$ stand for a reduced homogeneous polynomial, with k algebraically closed. Assume that the singular locus of $V(f) \subset \mathbb{P}_k^{n-1}$ consists of a nonempty set of points, i. e., that* $\dim R/J_f = 1$. *The following are equivalent:*

(1) *The coordinates of the vector fields of $\mathbb{P}_k^{n-1}$ vanishing on f generate an irrelevant ideal.*
(2) *The ideal J_f is a complete intersection locally at each singular point of $V(f)$.*
(3) *The ideal J_f is of linear type.*

Proof. A vector field $v = \sum_{i=1}^n a_i \, \partial/\partial x_i$ vanishes on f if and only if $\sum_{i=1}^n a_i f_{x_i} = 0$. Therefore, the coordinates of all such vector fields generate the ideal of 1-minors of a syzygy matrix of the gradient ideal. The result then follows from Proposition 2.3.2 once its hypotheses are verified in this setup, as we next proceed to see.

Since f is assumed to be reduced, whose singular locus is a nonempty set of isolated singularities, its gradient ideal is a (homogeneous) ideal of codimension $n-1$; hence, can only have minimal primes of codimension $n-1$ in R. Therefore, it is equidimensional.

Finally, the depth condition is trivially verified for the numbers in question. □

Definition 5.1.2. The rational map defined by the partial derivatives of a form $f \in R = k[x_1, \dots, x_n]$ is called the *polar map* of f. The form f is *homaloidal* provided the associated polar map is homaloidal, i. e., a Cremona transformation of $\mathbb{P}_k^{n-1}$.

We recall the following conjecture, already introduced in Chapter 3.

Conjecture 5.1.3 (k **algebraically closed of characteristic zero).** *Let $f \in R = k[x_1, \dots, x_n]$ denote a reduced homaloidal form such that* $\dim R/J_f = 1$. *Then J_f is a complete intersection locally at each singular point of $V(f)$.*

A consequence of an affirmative answer to this conjecture is a substantial part of the famous Dimca–Papadima conjecture proved by Huh [112].

Corollary 5.1.4 (k algebraically closed, $\mathrm{char}(k) = 0$). *Let $f \in R = k[x_1, \dots, x_n]$ denote a reduced homaloidal form such that* $\dim R/J_f = 1$. *Then;*
(1) $\deg(f) \leq n$.
(2) *If* $\deg(f) = n$ *then* $n = 3$.

Proof. (1) By Proposition 5.1.1, J_f is an ideal of linear type. Since f is homaloidal, Corollary 3.2.27 implies that the linear part of the syzygy matrix of J_f has maximal rank $(= n-1)$.

Claim. J_f is generated by the maximal minors of a linear $(n-1)\times n$ matrix upon dividing by their gcd.

Indeed, since f reduced implies that J_f has codimension ≥ 2, by dualizing a presentation $R^m \xrightarrow{\varphi} R^n \xrightarrow{\pi} J_f \to 0$ into R gives a left exact sequence

$$0 \to J_f^* \simeq R \xrightarrow{\pi^t} R^{n*} \xrightarrow{\varphi^t} R^{m*},$$

where the coordinates of π^t can be taken to be the $(n-1)$-minors of an $(n-1)\times n$ linear submatrix of φ^t of rank $n-1$ after dividing by their gcd. But these coordinates recover a set of generators of J_f. As a consequence, the degree of a partial derivative of f is bounded by $n-1$, as required.

(2) Upon equality, the gcd of the $(n-1)$-minors is trivial; hence, J_f is a perfect ideal of height 2. But since it has height $n-1$ by assumption, necessarily $n = 3$. □

Note that item (1) of the above corollary gives the essential part of Dolgachev's theorem for $n = 3$ [52].

5.1.2 Hyperplane arrangements

Let $\mathcal{A}$ denote a central hyperplane arrangement in affine space $\mathbb{A}^n = k^n$. For our convenience, we perpetrate the abuse of thinking of $\mathcal{A}$ as a set of mutually independent linear forms $l_1, \dots, l_m$ in the polynomial ring R—here m is often called the *size* of $\mathcal{A}$, while $f = l_1 \cdots l_m$ is its *defining form*. We assume throughout that $m \geq 2$. The rank of the arrangement is the vector space dimension of the span of $l_1, \dots, l_m$. As usual, we also assume that the arrangement has rank n.

5.1.2.1 Ideal theoretic properties, I

Lemma 5.1.5. *Let $f = l_1 \cdots l_m$ be the defining polynomial of a central hyperplane arrangement of rank n in affine space $\mathbb{A}^n = k^n$. If* $\mathrm{char}(k)$ *does not divide m, then every minimal prime $\wp$ of R/J_f of height 2 is generated by two of the forms $l_1, \dots, l_m$.*

Proof. Let $\wp \subset R = k[x_1, \dots, x_n]$ denote any such minimal prime. By the Euler relation, $f \in \wp$; hence, some $l_j \in \wp$. Let $a_{i,j} \neq 0$ be a nonzero coefficient of l_j. Since $\partial f/\partial x_i \in \wp$, it follows that $a_{i,j}l_1 \cdots \hat{l}_j \cdots l_m \in \wp$. Therefore, $l_t \in \wp$ for some $t \neq j$. This implies that $\wp = (l_j, l_t)$. □

In particular, if $n = 3$ then every minimal prime thereof is generated by two of the defining linear forms of the arrangement. Of course, if k algebraically closed, as is the usual assumption in geometry, then any homogeneous prime of height $n-1$ of $k[x_1, \ldots, x_n]$ can be generated by $n-1$ linear forms. For $n = 3$, we get in addition the following.

Proposition 5.1.6. *Let $f \in R = k[x,y,z]$ be the defining polynomial of a central hyperplane arrangement of rank 3 and size m. If* char(k) *does not divide m or else k is algebraically closed, then:*
(i) *J_f is generically a complete intersection.*
(ii) *J_f is an ideal of linear type.*
(iii) *f is homaloidal if and only if $m = 3$.*

Proof. (i) Either from Lemma 5.1.5 or else assuming that k is algebraically closed, by a change of variables, any minimal prime of R/J_f can be brought to the form (x,y), corresponding to the plane point $(0 : 0 : 1)$. The result follows from the calculation in [170, Lemma 2.4].

(ii) This follows from (i) and from Proposition 2.3.2 as applied to the almost complete intersection J_f.

(iii) If f is homaloidal, it follows from the previous items, via Corollary 3.2.29, that $m \leq 3$. Since f cannot be homaloidal for $m \leq 2$, then $m = 3$. The converse is easy as by a change of coordinates one may assume that $f = xyz$. □

Remark 5.1.7. The statement of Lemma 5.1.5 breaks down if the characteristic divides m and $m \geq n+1$. Perhaps the simplest example is the arrangement $f = xyz(x+y+z)$ in characteristic 2. Here, the degree of J_f is 7 (instead of $6 = \binom{4}{2}$), with $(x+y, y+z)$ the additional minimal prime. This example has no bearing on item (i) of Proposition 5.1.6 because one can always pass to the algebraic closure of k. Yet, as we will see below, for many questions the hypothesis on the characteristic will prevail.

We next focus on the module $\mathrm{Syz}(J_f)$ of syzygies of J_f for an arrangement f.

A rank n arrangement is said to be *generic* if every subset of $\{l_1, \ldots, l_m\}$ with n elements is k-linearly independent. For such an arrangement, it is known that $\mathrm{Syz}(J_f)$ is equigenerated in degree $m-n+1$ [164, Corollary 4.4.3]. The proof is quite involved as it draws on the detailed analysis of various chain complexes, none of which is trivial.

If one only cares about the initial degree of $\mathrm{Syz}(J_f)$, then a more down to earth argument has been given in [25], which we now reproduce.

Proposition 5.1.8. *Let $f = l_1 \cdots l_m$ be defining polynomial of a generic central hyperplane arrangement of rank n. If* char(k) *does not divide m, then* $\mathrm{indeg}(\mathrm{Syz}(J_f)) = m-n+1$.

Proof. As a preparatory step, a change of variables of $R = k[x_1, \dots, x_n]$ will not affect either the hypotheses or the conclusion. Thus, we may assume that $f := x_1 \cdots x_n l_1 \cdots l_{m-n}$, for suitable linear forms $l_1, \dots, l_{m-n}$.

Induct on m.

For $m = n$, the partial derivatives generate a well-known linearly presented ideal, hence $m - n + 1 = 1$.

Thus, assume that $m \geq n + 1$.

Recall the well-known deletion and restriction constructs associated to an arrangement of forms $l_1, \dots, l_m$ with respect to a chosen form l_i. The first is just the arrangement of the forms $l_1, \dots, l_{i-1}, \widehat{l_i}, l_{i+1}, \dots, l_m$ in the same polynomial ring R. The second is the arrangement of the residual forms of the forms $l_j, j \neq i$ in the residual ring $R/(l_i)$, upon identification of the latter with a polynomial ring over k in $n-1$ variables.

It is easy to verify that if the given arrangement is generic then so are its deletion and its restriction with respect to any chosen form. Let us take both with respect to $i = 1$. Setting $f \setminus \{x_1\} := x_2 \cdots x_n l_1 \cdots l_{m-n}$, by the inductive hypothesis one has $\operatorname{indeg}(\operatorname{Syz}(J_{f\setminus\{x_1\}})) = (m-1) - n + 1 = m - n$.

Likewise, the defining polynomial of the restriction is $f_{|x_1} = x_2 \cdots x_n \widetilde{l_1} \cdots \widetilde{l_{m-n}}$, with $\widetilde{l_i} = l_i|_{x_1=0}$, and $\operatorname{indeg}(\operatorname{Syz}(J_{f_{|x_1}})) = (m-1) - (n-1) + 1 = m - n + 1$ by the inductive hypothesis. Therefore, $\operatorname{indeg}(\operatorname{Syz}(J_{f\setminus\{x_1\}})) < \operatorname{indeg}(\operatorname{Syz}(J_{f_{|x_1}}))$. Then, by [1, Theorem 2.14], one has $\operatorname{indeg}(\operatorname{Syz}(J_f)) = \operatorname{indeg}(\operatorname{Syz}(J_{f\setminus\{x_1\}})) + 1 = m - n + 1$, as required. □

Remark 5.1.9. Though [1] assumes $\operatorname{char}(k) = 0$ throughout, a closer scrutiny of the proof of [1, Theorem 2.14] shows it only requires that $\operatorname{char}(k)$ does not divide $m = \deg(f)$.

An immediate consequence of Proposition 5.1.8 is that, for a generic arrangement of size $m \geq n$ with defining polynomial $f \in k[x_1, \dots, x_n]$, one has $\operatorname{reg}(J_f) \geq 2m - n - 1$. The next result will show quite a bit more.

For it, one introduces yet another ideal in $R = k[x_1, \dots, x_n]$ related to a rank n arrangement $\mathcal{A} = \{l_1, \dots, l_m\}$. Namely, given an integer $r \geq 0$ such that $m \geq r$, let

$$\mathbb{I}_{m-r}(\mathcal{A}) := (l_{i_1} \cdots l_{i_{m-r}} \mid 1 \leq i_1 < \cdots < i_{m-r} \leq m) \subset R \tag{48}$$

denote the ideal of $(m-r)$-fold products of distinct elements of $\mathcal{A}$.

For $r = 1$, the k-algebra generated by the $(m-1)$-fold products of distinct elements of the arrangement has been originally brought up by Orlik and Terao [153], while the ideal version has been thoroughly dealt with in [68]. Clearly, since the ideal is equigenerated, the algebra gets identified with the fiber cone of the latter, thus tying them strongly together. In the sequel, if no confusion arises, we shorten $\mathbb{I}_{m-1}(\mathcal{A})$ to simply $\mathbb{I}$. It is known that $\mathbb{I}$ is a codimension two linearly presented perfect ideal [68, Lemma 3.1].

At the other end, if $r = n$ and $m \geq n$ then one has the following.

Lemma 5.1.10. *Let $f = l_1 \cdots l_m$ denote the defining polynomial of a generic central arrangement $\mathcal{A}$ in $R = k[x_1, \ldots, x_n]$ with $m \geq n$. Then $\mathbb{I}_{m-n}(\mathcal{A}) = \mathfrak{m}^{m-n}$.*

Proof. Induct on m. The statement is trivial or vacuous for $m = n$. Thus, let $m \geq n + 1$ and consider the deletion $\mathcal{A}_i := \mathcal{A} \setminus l_i$, for $1 \leq i \leq m$. By the inductive hypothesis, $\mathbb{I}_{m-1-n}(\mathcal{A}_i) = \mathfrak{m}^{m-1-n}$; hence,

$$l_i\, \mathfrak{m}^{m-1-n} = l_i\, \mathbb{I}_{m-1-n}(\mathcal{A}_i) \subset \mathbb{I}_{m-n}(\mathcal{A}),$$

for all i. Since $\mathcal{A}$ is generic, in particular $\{l_1, \ldots, l_n\}$ generates $\mathfrak{m}$, hence $\mathfrak{m}^{m-n} = (l_1, \ldots, l_n)\mathfrak{m}^{m-1-n}$, so we are through. □

Remark 5.1.11. Note the need of a formal proof of the above lemma. Although one could, by a change of variables, assume that $f = x_1 \cdots x_n l_{n+1} \cdots l_m$, this would not help since one is taking products of distinct linear forms.

For a homogeneous ideal I in the standard graded polynomial ring $k[x_1, \ldots, x_n]$, with $\mathfrak{m} = (x_1, \ldots, x_n)$, we will consider its *saturation* $I^{\mathrm{sat}} := I : \mathfrak{m}^{\infty}$ and its *saturation index* (or *satiety*) $\mathrm{sat}(I)$ defined to be $\mathrm{end}(I^{\mathrm{sat}}/I) + 1$, i. e., the least integer $r \geq 0$ such that $(I^{\mathrm{sat}}/I)_s = 0$ for every $s \geq r$.

We now bring out one of the main statements of [25].

Theorem 5.1.12 (char$(k) \nmid m$). *Let $f = l_1 l_2 \cdots l_m$ as above denote the defining polynomial of a generic central arrangement in $R = k[x_1, \ldots, x_n]$ with $m \geq n \geq 2$. Then:*
(i) $[\mathbb{I}]_{2m-n-1} = [J_f]_{2m-n-1}$.
(ii) *If $n \geq 3$ then* $(J_f)^{\mathrm{sat}} = \mathbb{I}$.
(iii) $\mathrm{reg}(J_f) = 2m - n - 1$; *in addition, if $m \geq n + 1$ then* $\mathrm{sat}(J_f) = \mathrm{reg}(J_f)$.

Proof. (i) It suffices to show the inclusion $[\mathbb{I}]_{2m-n-1} \subset [J_f]_{2m-n-1}$. Thus, let $P \in \mathbb{I}$, with $\deg(P) = 2m - n - 1$. Write

$$P = (l_2 \cdots l_m)P_1 + \cdots + (l_1 \cdots l_{m-1})P_m, \tag{49}$$

for certain forms $P_i \in [\mathfrak{m}^{m-n}]_{m-n}$, where $\mathfrak{m} = (x_1, \ldots, x_n)$.

By Lemma 5.1.10, for each $i = 1, \ldots, m$, we can write

$$P_i = \sum_{1 \leq j_1 < \cdots < j_{m-n} \leq m} c^i_{j_1, \ldots, j_{m-n}}\, l_{j_1} \cdots l_{j_{m-n}},$$

for certain coefficients $c^i_{j_1, \ldots, j_{m-n}} \in k$.

Claim. $(l_1 \cdots \hat{l}_i \cdots l_m)P_i$ belongs to $[J_f]_{2m-n-1}$, for every $i = 1, \ldots, m$.

To see this, due to the obvious symmetry, we can assume that $i = m$. Look at a typical term therein $(l_1 \cdots l_{m-1})l_{j_1} \cdots l_{j_{m-n}}$. If some $j_t = m$, then this term is a multiple of f, hence belongs to J_f by the Euler relation. Otherwise, we can assume, without loss of generality, that the term in question has the shape $\Delta := (l_1 \cdots l_{m-n})^2 l_{m-n+1} \cdots l_{m-1}$. To conclude, we induct on $m \geq n$.

The result is trivial for $m = n$ since, more strongly, $\mathbb{I} = J_f$.

Letting $m \geq n+1$, write

$$\Delta = (l_1)^2 \underbrace{(l_2 \cdots l_{m-n})^2 l_{m-n+1} \cdots l_{m-1}}_{\Delta'}.$$

By the inductive hypothesis, $\Delta' \in J_{f'}$, where $f' = l_2 \cdots l_m$. Clearly, $f_{x_i} = (l_1)_{x_i} f' + l_1 f'_{x_i}$, hence; $(l_1)^2 f'_{x_i} \in J_f$ for $i = 1, \ldots, n$. Consequently, $\Delta = l_1^2 \Delta' \in J_f$, as was to be shown.

(ii) This follows from (i) since any form in $[\mathfrak{m}^{m-n}]_{m-n}$ multiplies any f/l_i into $[\mathbb{I}]_{2m-n-1}$, thus giving $\mathbb{I} \subset (J_f)^{\mathrm{sat}}$. The reverse inclusion is clear since $\mathbb{I}$ itself is saturated, being a codimension two perfect ideal with $n \geq 3$.

(iii) The stated part for $m \geq n$ is again clear when $m = n$ since $\mathbb{I}$ is linearly presented.

Thus, assume that $m \geq n+1$ in order to prove the equality $\operatorname{reg}(J_f) = \operatorname{sat}(J_f)$.

Namely, by Proposition 5.1.8, certainly $\operatorname{reg}(J_f) \geq 2m - n - 1$; hence, $\operatorname{reg}(J_f) > m - 1$ since $m \geq n+1$. If $n \geq 3$, then $\operatorname{reg}(J_f^{\mathrm{sat}}) = \operatorname{reg}(\mathbb{I})$ by (ii). But, since $\mathbb{I}$ is a linearly presented codimension two perfect ideal, $\operatorname{reg}(\mathbb{I}) = m - 1$. This implies the proposed equality by using the well-known relation $\operatorname{reg}(J_f) = \max\{\operatorname{sat}(J_f), \operatorname{reg}(J_f^{\mathrm{sat}})\}$ (see, e. g., [77, Section 1]). If, on the other hand, $n = 2$ then J_f is $\mathfrak{m}$-primary; hence, $J_f^{\mathrm{sat}} = (1)$, whose regularity is zero by convention. Thus, we are home in this case too.

To prove that $\operatorname{reg}(J_f) = 2m-n-1$ it suffices to show the inequality $\operatorname{reg}(J_f) \leq 2m-n-1$ or, by what we already established when $m \geq n+1$, that $\operatorname{sat}(J_f) \leq 2m-n-1$. Note that J_f is not saturated since $J_f \neq \mathbb{I}$. Therefore, if $\operatorname{sat}(J_f) \geq 2m-n$ then, by [13, Proposition 2.1], there exists a minimal generator h of $J_f : \mathfrak{m}$ of degree $d := \operatorname{sat}(J_f) - 1 \geq 2m - n - 1$, not lying on J_f.

But $(J_f : \mathfrak{m}) \subset (J_f)^{\mathrm{sat}} = \mathbb{I}$, so $h = (l_2 \cdots l_m)P_1 + \cdots + (l_1 \cdots l_{m-1})P_m$, for some $P_j \in R$ of degree $d - m + 1 \geq m - n$. Thus, each P_j is a polynomial combination of monomials of degree $m-n$. As in the displayed claim in the proof of item (i), we conclude that $h \in J_f$, contradicting our choice of h. □

The next theorem retrieves part of the result in [164] and [223].

Theorem 5.1.13 (char(k) $\nmid m$). *Let $f = l_1 l_2 \cdots l_m$ as above denote the defining polynomial of a generic central arrangement in $R = \mathbb{K}[x_1, \ldots, x_n]$ with $m \geq n+1 \geq 3$. Then:*

(i) $\operatorname{depth}(R/J_f) = 0$.

(ii) *The minimal graded free resolution of J_f has the following shifts:*

$$0 \to R^{b_n}(-(2m-1)) \to R^{b_{n-1}}(-(2m-2)) \to \cdots \to R^{b_1}(-(2m-n)) \to R^n(-(m-1)).$$

Proof. (i) The assertion is trivial for $n = 2$ since J_f is $\mathfrak{m}$-primary. Thus, assume that $n \geq 3$ and recall that J_f has height two.

Take a reduced primary decomposition $J = J^{\mathrm{un}} \cap \bigcap_i \mathcal{N}_i$, where J^{un} is the unmixed part of J and each $\mathcal{N}_i$ is a primary component of codimension at least 3.

If we assume that $\operatorname{depth}(R/J) > 0$, then $3 \le \operatorname{codim} \mathcal{N}_i < n$, for every i. Therefore, by Theorem 5.1.12(ii), $\mathbb{I} = J : \mathfrak{m}^\infty = (J^{\mathrm{un}} : \mathfrak{m}^\infty) \cap \bigcap_i (\mathcal{N}_i : \mathfrak{m}^\infty) = J^{\mathrm{un}} \cap \bigcap_i \mathcal{N}_i$. Therefore, $J = \mathbb{I}$, which is nonsense since the minimal number of generators of $\mathbb{I}$ is at least $n+1$.

(ii) By definition, $\operatorname{reg}(J_f)$ is the maximum value $\beta_k - k$, where β_k is the maximum shift at step k in the minimal graded free resolution of J_f. By Proposition 5.1.8, the minimum shift at step $k = 1$ is $m - n + 1 + (m-1) = 2m - n$, which equals $\operatorname{reg}(J_f) + 1$ by Theorem 5.1.12(iii). But the maximum shift at each step must increase by at least one and the homological dimension of R/J_f over R is n by item (i). This implies the claimed result by a standard argument (see, e. g., [151, Corollary 9 (2)]). □

Remark 5.1.14. (1) Proposition 5.1.8, Theorem 5.1.12 and Theorem 5.1.13 depend strongly on the hypothesis that the arrangement be generic. For example, consider the following rank 3 nongeneric arrangement in $k[x,y,z]$: $\mathcal{A} = \{x, y, z, x+y\}$, a free arrangement (i. e., J_f is a perfect ideal); here, the initial degree of $\operatorname{Syz}(J_f)$ is $1 < 2 = m - n + 1$.

(2) In [164], the ground field k is assumed to be of characteristic zero throughout, but the authors remark at the end (Section 4.6) that the result is valid provided $\operatorname{char}(k)$ does not divide $m = \deg(f)$. As to our approach, the vast majority of the above results crumbles down without this proviso. A simplest example is as follows: let $R = k[x,y,z]$, where $\operatorname{char}(k) = 2$ and let

$$\mathcal{A} = \{x, y, z, x + y + z\}.$$

Clearly, $\mathcal{A}$ is generic. Calculation with [79] yields:
- $f \notin J_f$.
- J_f is perfect; in particular, $J_f^{\mathrm{sat}} = J_f \neq \mathbb{I}$.
- The free resolution of J_f has the shape $0 \to R(-4) \oplus R(-5) \to R(-3)^3 \to J_f \to 0$; in particular, J_f has a syzygy of degree $1 < 2 = m - n + 1$ – in fact, 2 is the degree of a minimal syzygy of maximal degree, which happens to be given by a reduced Koszul syzygy.
- $\operatorname{Derlog}(\mathcal{A})$ does not split as in (47); in fact, it admits quite many generators of degree 2.
- $\operatorname{Derlog}(\mathcal{A})$ has homological dimension 1, while $\operatorname{Syz}(J_f)$ is a free module.
- J_f is not a reduction of $\mathbb{I}$ since $J_f : \mathbb{I}^\infty \neq (1)$.

Characteristic 2 is not the exception as similar examples are equally obtained in any positive characteristic dividing the size of the arrangement.

5.1.2.2 Ideal theoretic properties, II

The overall theory of the previous section is embraced by the following questions.

Conjecture 5.1.15 (char(k) $\nmid m$). *Let $f \in R = k[x_1,\dots,x_n]$ denote the defining polynomial of a central hyperplane arrangement of size $m \geq n \geq 2$. Then:*

(1) *The Jacobian ideal J_f is of linear type.*
(2) *The Jacobian ideal J_f is a minimal reduction of the ideal $\mathbb{I}$ of $(m-1)$-fold products of f.*

We prove an important case of these conjectures.

Proposition 5.1.16. *Conjecture 5.1.15 holds true for a generic arrangement of rank $n \leq 3$.*

Proof. Let $n = 2$. The linear type property is well known since J_f is generated by a regular sequence. As to the minimal reduction property, we may argue as follows: it suffices to prove that $\mu(\mathbb{I}^2) = \mu(J_f\mathbb{I})$ since $J_f\mathbb{I} \subset \mathbb{I}^2$ and both ideals are equigenerated in the same degree. Now, since J_f is generated by a regular sequence (of two forms), then $\mu(J_f\mathbb{I}) = 2\mu(\mathbb{I}) - \binom{2}{2} = 2m - 1$ (see, e. g., [163, Proposition 2.22]).

On the other hand, one has $\mu(\mathbb{I}^2) = \binom{m+1}{2} - \mu(Q)$, where Q as above is the homogeneous presentation ideal of the Orlik–Terao algebra. It is known that Q is generated by a minimal set of generators among the circuits corresponding to the dependencies [153]—all of size 3 in the rank 2 case; hence, the generators are quadrics. This minimal set has $\binom{m-1}{2}$ elements. Therefore, one gets

$$\mu(\mathbb{I}^2) = \binom{m+1}{2} - \binom{m-1}{2} = 2m - 1,$$

as was to be shown.

Let now $n = 3$. For the linear type property, since J_f is an almost complete intersection it suffices to show that it is generically a complete intersection locally at its minimal primes. Since the arrangement is generic, then J_f and $\mathbb{I}$ admit the same minimal primes by Theorem 5.1.12(ii). But any minimal prime of the latter is generated by two of the defining linear forms of the arrangement. By a change of variables, we may assume that such prime is (x, y), corresponding to the point $(0 : 0 : 1)$, in which case the result follows from [170, Lemma 2.4].

For the reduction property, we claim that $\mathbb{I}^3 \subset J_f\mathbb{I}^2$. The following argument is due to Ş. Tohăneanu.

Let $L_j := l_1 \cdots \hat{l_j} \cdots l_m$, so $\mathbb{I} = (L_1,\dots,L_m)$. By the symmetry of the question, it suffices to deal with the following sort of generators of $\mathbb{I}^3$: $L_1L_2L_3$, $L_1^2L_2$, L_1^3. Moreover, by a change of variables (since the arrangement is assumed to be generic), we may take $l_1 = x$, $l_2 = y$, $l_3 = z$, hence $f = xyzl_4 \cdots l_m$. Then our three types of generators become:

(I) $P := x^2y^2z^2(l_4 \cdots l_m)^3$
(II) $P := xy^2z^3(l_4 \cdots l_m)^3$
(III) $P := y^3z^3(l_4 \cdots l_m)^3$

Type (I) $P := x^2y^2z^2l_4^3\cdots l_m^3$. Then

$$\begin{aligned}
mP &= mf(xyzl_4^2\cdots l_m^2)\\
&= (xf_x + yf_y + zf_z)xyz(l_4\cdots l_m)^2\\
&= f_x x^2yz(l_4\cdots l_m)^2 + f_y xy^2z(l_4\cdots l_m)^2 + f_z xyz^2(l_4\cdots l_m)^2\\
&= f_x(xyl_4\cdots l_m)(xzl_4\cdots l_m) + f_y(xyl_4\cdots l_m)(yzl_4\cdots l_m) + f_z(xzl_4\cdots l_m)(yzl_4\cdots l_m),
\end{aligned}$$

an element of $J_f\mathbb{I}^2$.

Type (II) $P := xy^2z^3l_4^3\cdots l_m^3$. One has

$$f_x = yzl_4\cdots l_m + xyz(a_4l_5\cdots l_m + \cdots + a_ml_4\cdots l_{m-1}),$$

where $a_i := (l_i)_x \in k\setminus\{0\}$, for $i = 4,\dots,m$. Then

$$\begin{aligned}
P &= (xyz^2l_4^2\cdots l_m^2)(f_x - xyz(a_4l_5\cdots l_m + \cdots + a_ml_4\cdots l_{m-1}))\\
&= (xzl_4\cdots l_m)(yzl_4\cdots l_m)f_x - (a_4x^2y^2z^3l_4^2l_5^3\cdots l_m^3 + \cdots + a_mx^2y^2z^3l_4^3\cdots l_{m-1}^3l_m^2).
\end{aligned}$$

Clearly, $(xzl_4\cdots l_m)(yzl_4\cdots l_m) \in \mathbb{I}^2$, while $a_i(x^2y^2z^3l_4^3\cdots l_m^3)/l_i$ is of type (I), for any $i = 4,\dots,m$. Therefore, $P \in J_f\mathbb{I}^2$.

Type (III) $P := y^3z^3l_4^3\cdots l_m^3$. Then

$$\begin{aligned}
P &= (y^2z^2l_4^2\cdots l_m^2)(f_x - xyz(a_4l_5\cdots l_m + \cdots + a_ml_4\cdots l_{m-1}))\\
&= (yzl_4\cdots l_m)^2f_x - (a_4xy^3z^3l_4^2l_5^3\cdots l_m^3 + \cdots + a_mxy^3z^3l_4^3\cdots l_{m-1}^3l_m^2).
\end{aligned}$$

Again, $(yzl_4\cdots l_m)^2 \in \mathbb{I}^2$, while $a_i(xy^3z^3l_4^3\cdots l_m^3)/l_i$ is of type (II) for any $i = 4,\dots,m$; hence $P \in J_f\mathbb{I}^2$. □

For the next corollary, one needs the notion of a free divisor, to be established later in the next section. We draw upon the characterization in (Proposition 5.2.6).

Corollary 5.1.17. *Let $f \in R = k[x,y,z]$ be the defining polynomial of a rank 3 generic central hyperplane arrangement. Then f is a free divisor if and only if the Rees algebra $\mathcal{R}_R(J_f)$ is a complete intersection.*

Proof. Let $\mathcal{I}$ and $\mathcal{J}$ denote the respective presentation ideals of the symmetric algebra $\mathcal{S}(J_f)$ and the Rees algebra $\mathcal{R}(J_f)$. Since J_f is of linear type by Proposition 5.1.16, then $\operatorname{ht}\mathcal{I} = \operatorname{ht}\mathcal{J} = \mu(J_f) - 1 = n - 1$. But f is a free divisor if and only if $\operatorname{Syz}(J_f)$ is a free R-module. This is the case if and only if $\mu(\mathcal{I}) = n-1$, i. e., if and only if $\mu(\mathcal{J}) = n-1$. □

Remark 5.1.18. Conjecture 5.1.15 holds true for $n = 3$ regardless of the generic condition – see [25] for the details.

5.1.3 Behavior along a family

In this part, we focus on families of projective curves having a fixed set of singularities. For such families, the defining form usually has a complicated expression, so one aims to reduce it to a "normal" form. The occurrence of the linear type property for the members of the family may happen to be a difficult question altogether.

For the preliminaries, let $k[\mathbf{u}] = k[u_1, \dots, u_m]$ stand for a polynomial ring over the field k and let $F \in S := k[\mathbf{u}][x,y,z]$ denote a polynomial which is a form on x, y, z with parameter coefficients. Mostly, we endow S with the structure of standard graded ring over $S_0 = k[\mathbf{u}]$, but there may be exceptions. The basic assumption is that the content of F with respect to the $\mathbf{u}$-coefficients is 1. Then F is a nonzero divisor on $k[\mathbf{u}]/I$ for every ideal $I \subset k[\mathbf{u}]$; hence, $\mathrm{Tor}^1_{k[\mathbf{u}]}(k[\mathbf{u}]/I, S/(F)) = \{0\}$ for any such ideal. This gives that the inclusion $k[\mathbf{u}] \subset S/(F)$ is flat; hence, defines a family of curves in $\mathbb{P}^2$ over the parameters $\mathbf{u}$.

We freely abuse the terminology "member" while referring to the specification of a set $\mathbf{a} = \{a_1, \dots, a_m\} \subset k$ and the corresponding polynomial $f_{\mathbf{a}} \in k[x,y,z]$, which is the image of F by the evaluation homomorphism $k[\mathbf{u}][x,y,z] \longrightarrow k[\mathbf{u}]/(\mathbf{u}-\mathbf{a})[x,y,z] = k[x,y,z]$ Often, by abuse, we set $f_{\mathbf{a}} = F(\mathbf{a})$.

Our interest lies on the case where the general member of the family is a reduced singular plane curve and we refer to this family as a family of that singularity. We assume that F has the form

$$F = \varphi_0(x,y,z) + \sum_{j=1}^{m} u_j \varphi_j(x,y,z), \tag{50}$$

where d is the (homogeneous) degree of F in x, y, z and $m \le \binom{d+2}{2} - 1$, with $\{\varphi_j(x,y,z) \mid 0 \le j \le m\}$ is a set of monomials of degree d in x, y, z and $\varphi_0(x,y,z) \neq 0$.

Example 5.1.19. A "normal" form for the family of irreducible singular quartic plane curves whose general member is a simple node looks like

$$F = xyz^2 + x^4 + y^4 + u_1x^3z + u_2y^3z + u_3x^3y + u_4x^2y^2 + u_5xy^3$$

thus, depending on four parameters.

A normal form may have degenerations to other normal forms whose general member has more involved singularities or even acquires new singular points. The following example may illuminate this phenomenon.

Example 5.1.20. Consider the family of irreducible rational plane quartics with exactly three nodes. In [74, Lemma 11.3], a normal form is given of a family whose general member is an irreducible quartic with three double points, namely

$$F = \lambda x^2y^2 + \mu x^2z^2 + \nu y^2z^2 + 2xyz(u_1x + u_2y + u_3z), \quad \lambda\nu\mu \neq 0.$$

To get a normal form whose general member is an irreducible quartic with three nodes, substituting $x = (\nu/\lambda\mu)^{1/4}x$, $y = (\mu/\lambda\nu)^{1/4}y$, $z = (\lambda/\mu\nu)^{1/4}z$ and renaming, one obtains the normal form

$$F = x^2y^2 + x^2z^2 + y^2z^2 + 2xyz(u_1x + u_2y + u_3z).$$

Note that for k-values $u_1 = \pm 1$, one of the nodes degenerates into a cusp and, similarly, for $u_2 = \pm 1$ or $u_3 = \pm 1$. Thus, the general member requires that the k-values of the triple (u_1, u_2, u_3) do not lie on the hypersurface $V((u_1^2 - 1)(u_2^2 - 1)(u_3^2 - 1))$ in order that it have exactly three nodes.

Requiring that the general member will acquire no new singular points besides the three nodes imposes yet another obstruction. Of course, in the present low degree 4, because of genus reason there will be new singular points only if the general member properly factors. This obstruction is precisely given by the hypersurface whose equation is the discriminant $2u_1u_2u_3 + u_1^2 + u_2^2 + u_3^2 - 1$ of a suitable conic (see Section 5.1.3.2).

A difficulty regarding the question as to how the property of being of linear type moves on a family is that this property is neither kept by specialization nor by deformation.

The next result gives a general view of the situation.

Theorem 5.1.21. *Let $F \in S = k[\mathbf{u}][x, y, z]$ denote a family of singular plane curves of degree $d \geq 2$ with general member reduced and irreducible. Let $\mathcal{I} := I_1(\mathrm{Syz}(J_F)) \subset S$ stand for the ideal of 1-minors of the syzygy matrix of the ideal $J_F \in S$ of F generated by the partial derivatives of F with respect to the variables x, y, z. Then:*

(a) *J_F has codimension 2.*
(b) *$\mathcal{I}$ has codimension at most 3.*
(c) *If k is algebraically closed of characteristic zero, the following are equivalent:*
  (i) *$\mathcal{I}$ has codimension 3.*
  (ii) *The contraction of the ideal $\mathcal{I} : (x, y, z)S^{\infty}$ to $k[\mathbf{u}]$ has codimension ≥ 1.*
  (iii) *The plane projective curve $F(\alpha) \in k[x, y, z]$ obtained by evaluating $\mathbf{u} \mapsto \alpha$ off a set of codimension ≥ 1 in $\mathbb{A}_k^m$ has Jacobian ideal of linear type.*
  (iv) *There is some $\alpha \in \mathbb{A}_k^m$ for which the evaluated ideal $\mathcal{I}(\alpha) \in k[x, y, z]$ has codimension 3.*

Proof. (a) Since the general member is singular and reduced its gradient ideal has codimension 2. This forces $\mathrm{codim}(J_F) = 2$.

(b) Since F is a homogeneous polynomial in the grading of S, its partial derivatives are homogeneous of same degree. It suffices to argue that any syzygy of the partial derivatives has coefficients in $(x, y, z)S$. Indeed, since the partial derivatives are homogeneous of the same degree $d \geq 1$ in x, y, z, any syzygy is homogeneous of nonnegative degree in x, y, z. Now, if a syzygy would happen to be of degree 0, i. e., with all its coordinates in the zero degree part $k[\mathbf{u}]$, this would imply that the general member of

F has a nonzero numerical syzygy. But this is impossible since the general member is reduced and irreducible; hence, its partial derivatives are linearly independent.

(c) (i) $\Rightarrow$ (ii) Write $\mathfrak{m} = (x,y,z)S$. Since $\operatorname{codim}(\mathcal{I}) = 3$ and $\mathfrak{m}$ is a minimal prime therein by the proof of (b), the saturation $\mathcal{I} : \mathfrak{m}^\infty$ picks up the primary components of $\mathcal{I}$ not containing $\mathfrak{m}$. This shows that $(\mathcal{I} : \mathfrak{m}^\infty) \cap k[\mathbf{u}] \neq \{0\}$.

(ii) $\Rightarrow$ (iii) Let $g = g(\mathbf{u}) \in (\mathcal{I} : \mathfrak{m}^\infty) \cap k[\mathbf{u}]$ be any nonzero element. By hypothesis, g conducts a power of $(x,y,z)S$ inside $\mathcal{I}$. Giving $\mathbf{u}$ k-values α off $V((\mathcal{I} : \mathfrak{m}^\infty) \cap k[\mathbf{u}])$ yields a power of the maximal ideal $(x,y,z) \subset k[x,y,z]$ inside the image $\mathcal{I}(\alpha)$ of $\mathcal{I}$ by this evaluation. Let $f = F(\alpha) \in k[x,y,z]$ denote the member of the family thus obtained. Then $\mathcal{I}(\alpha) \subset I_1(\varphi)$, where φ denotes the syzygy matrix of the partial derivatives of f. This shows that $I_1(\varphi)$ is (x,y,z)-primary. Therefore, the result follows from Proposition 5.1.1.

(iii) $\Rightarrow$ (iv) The hypothesis is that the gradient ideal of the general member of the family is of linear type. Again by Proposition 5.1.1, this implies that the ideal of 1-minors of such a plane curve has codimension 3. On the other hand, for general value α of $\mathbf{u}$, one has $\mathcal{I}(\alpha) = I_1(\varphi_\alpha)$, where φ_α stands for the syzygy matrix of $F(\alpha)$.

(iv) $\Rightarrow$ (i) By definition, $\mathcal{I}(\alpha) = (\mathcal{I}, \mathbf{u}-\alpha)/(\mathbf{u}-\alpha)$ upon identifying $k[\mathbf{u},x,y,z]/(\mathbf{u}-\alpha) = k[x,y,z]$ under the surjection $k[\mathbf{u}][x,y,z] \twoheadrightarrow k[x,y,z]$ such that $\mathbf{u} \mapsto \alpha$. Since $\mathbf{u}-\alpha$ is a regular sequence in a polynomial ring, we readily get

$$\operatorname{ht}((\mathcal{I}, \mathbf{u}-\alpha)/(\mathbf{u}-\alpha)) = \operatorname{ht}(\mathcal{I}, \mathbf{u}-\alpha) - \operatorname{ht}(\mathbf{u}-\alpha) \leq \operatorname{ht}(\mathcal{I}) + \operatorname{ht}(\mathbf{u}-\alpha) - \operatorname{ht}(\mathbf{u}-\alpha) = \operatorname{ht}(\mathcal{I}),$$

which implies the result. □

Example 5.1.22. Let $F = y^4z + x^5 + ux^3y^2$ be a one-parameter family of plane cusps. A calculation with [79] gives $\mathcal{I} \subset (x,y)S$; hence, the gradient ideal of the general member $F(\alpha)$ of the family is not of linear type. However, the special member obtained by evaluating F at $\mathbf{0}$ is a standard plane cuspid singularity, readily seen to have gradient ideal of linear type (e. g., the linear syzygy $(0,y,-4z)^t$, and one of the Koszul syzygies that has a pure power of x, imply the result). This simple example shows that the property in question does not deform to the generic member. For this reason, one often considers the family over an open set of $\mathbb{A}^m_{\mathbf{u}}$. It can be extended to a two-parameter family $F = y^4z + u_1x^5 + u_2x^3y^2$ with similar behavior.

Under the equivalent assumptions of item (c) of Theorem 5.1.21, one can give the approximate structure of the contracted ideal in item (ii).

Proposition 5.1.23. *Let the assumptions be those of Theorem 5.1.21 and assume that $\mathcal{I}$ has codimension 3. If $\mathcal{I}$ has a minimal prime of codimension 3 other than $\mathfrak{m} = (x,y,z)S$ then $(\mathcal{I} : \mathfrak{m}^\infty) \cap k[\mathbf{u}]$ has codimension 1. If, moreover, $\mathcal{I}$ is pure-dimensional, then this contraction is a principal ideal.*

Proof. Let $\mathfrak{p}$ be a minimal prime of codimension 3 of $\mathcal{I}$ other than $\mathfrak{m}$. Since $I_F \subset \mathcal{I}$ (because of the Koszul relations), then $\mathfrak{p}$ contains a minimal prime of I_F. But the latter are of two sorts: either the extensions of the minimal primes (in $k[x,y,z]$) of the singular points of the general curve of the family, or else minimal primes of codimension 3. In the first case, $\mathfrak{p}$ contains two independent 1-forms in $k[x,y,z]$, which up to a projective change of coordinates, can be assumed to be x, y. Clearly, these forms are part of a minimal set of generators of $\mathfrak{p}$ and $(x,y,\mathcal{I}) \subset \mathfrak{p}$. But $\mathcal{I}$ is homogeneous in the variables $k[x,y,z]$, generated in positive such degrees (see the argument in the proof of (b)); hence, there exist suitable polynomials $g_j(\mathbf{u}) \in k[\mathbf{u}]$ and integers $k_j \geq 1, 1 \leq j \leq s$, such that

$$(x,y,\mathcal{I}) = (x,y, z^{k_j}g_j(\mathbf{u}), 1 \leq j \leq s).$$

Since $z \notin \mathfrak{p}$, necessarily some $g_j(\mathbf{u}) \in \mathfrak{p}$. But then $\mathfrak{p} = (x,y,p(\mathbf{u}))$, for some prime factor of $g_j(\mathbf{u})$.

We now deal with the case where $\mathfrak{p}$ contains a minimal prime of I_F of codimension 3, hence coincides with it. Let $\mathfrak{p}_i, 1 \leq i \leq r$ denote the minimal primes of codimension 3 of $\mathcal{I}$. Each of these, by the previous argument, has a minimal generator $p_i(\mathbf{u}) \in k[\mathbf{u}]$. Then

$$\Big(\prod_i \mathfrak{p}_i\Big) \cap k[\mathbf{u}] = \Big(\prod p_i(\mathbf{u})\Big),$$

a principal ideal. On the other hand,

$$\sqrt{\mathcal{I} : \mathfrak{m}^\infty} \subset \bigcap \sqrt{\mathcal{P}_i : \mathfrak{m}^\infty} \subset \bigcap (\sqrt{\mathcal{P}_i} : \mathfrak{m}^\infty) = \bigcap \sqrt{\mathcal{P}_i} = \bigcap \mathfrak{p}_i = \sqrt{\prod_i \mathfrak{p}_i},$$

where $\mathcal{P}_i$ denotes the $\mathfrak{p}_i$th primary component of $\mathcal{I}$. This proves that the contraction $(\mathcal{I} : \mathfrak{m}^\infty) \cap k[\mathbf{u}]$ has codimension ≤ 1, hence is exactly 1 by Theorem 5.1.21.

The additional assertion at the end of the statement is now clear. □

5.1.3.1 Plane cubics

Assume throughout that $\mathrm{char}(k) = 0$.

Proposition 5.1.24. *Let $f \in k[x,y,z]$ denote a reduced cubic form. Then J_f is of linear type.*

Proof. Since J_f is an almost complete intersection, the strategy is to prove that the ideal of the entries of the syzygy matrix of J_f is (x,y,z)-primary or, alternatively, that J_f is generically a complete intersection (Proposition 2.3.2). We distinguish the cases according to the number of irreducible factors of f.

Three irreducible factors. Then the factors are linear forms. If they are linearly dependent then, up to a change of variables, they depend only on two variables. In this case, one of the derivatives vanishes, hence J_f is a complete intersection of two elements—obviously, of linear type.

Therefore, we may assume that the three linear forms are linearly independent, in which case, up to a change of variables, $f = xyz$. Then J_f is well known or readily seen to be a perfect ideal of linear type.

Two irreducible factors. Say, $f = lq$, where l is a linear form and q a quadric. There are only two possible configurations: either the straight line defined by l is tangent to the conic defined by the quadric or intersect it in two distinct points.

In the first case, by a change of variables, we may assume that $l = x$ and that $q = y^2 + h(x,z)$, for some $a \in k$ and $h(x,z)$ a form of degree 2 in x, z not involving z^2. Say, $h = x(ax + bz)$, where we may assume that $b \neq 0$. Then

$$J_f = \left(y^2 + h + x\frac{\partial h}{\partial x}, 2xy, x\frac{\partial h}{\partial z}\right) = (y^2 + 3ax^2 + 2bxz, xy, x^2).$$

Therefore, J_f admits the Koszul reduced syzygy of the last two generators, generating the ideal (x,y). An additional simple elementary transformation of the generators above yields the following generating shape $J_f = (cy^2 + dxz, xy, x^2)$, where $c \neq 0$, $d \neq 0$. It is now obvious that there is an additional syzygy with last coordinate z. This implies that the ideal of entries has height 3 and, in fact, that J_f is perfect and linearly presented. Here, alternatively, J_f is generated by $\{(c/dz)y^2 + x, xy\}$ locally at its only minimal prime (x,y).

In the transversal case, one can in a similar way focus on the prototype $f = y(y^2 + xz)$. Then the following two syzygies are directly found: $(x, 0, -z)^t$ and $(3y^2, -yz, -z^2)^t$. Alternatively, here again, J_f is readily seen to be a complete intersection at its two minimal primes (x,y) and (y,z). Thus, we are through again.

The irreducible case. We may assume that f has a singular point, as otherwise J_f is a complete intersection. In that case, one knows that, up to change of variables, there are two prototypes: $y^2z - x^3$ (cusp) and $y^2z - x^2(x - z)$ (node).

First, the cusp: since $J_f = (x^2, yz, y^2)$ the reduced Koszul syzygy of $\{yz, y^2\}$ and the Koszul syzygy of $\{x^2, yz\}$ show that the ideal of entries is (x,y,z)-primary. Alternatively, one can see that J_f is locally a complete intersection at its only minimal prime (x,y). For the node case, it is easier to see that J_f is locally a complete intersection at its only minimal prime (x,y). □

5.1.3.2 Plane rational quartics

The explicit classification of rational quartics, showing the invariants one is interested in, is not directly available.

In this part, we show that the singular ideal of the general member of a family of plane *rational* quartics, fixing the singularity type, is of linear type. The proof is lifted from the PhD thesis of A. N. Nejad [148]. It has been conjectured that any irreducible plane quartic has gradient ideal of linear type. The argument would involve a good deal of mixed theory and computation.

For the preliminaries on plane rational quartics, we refer to [220].

An irreducible rational quartic having only double points can be obtained as a rational transform of a nondegenerate conic by means of one of the three basic rational quadratic maps of $\mathbb{P}^2$:

(1) The one with defining coordinates $(yz : zx : xy)$ ("reciprocal involution"), which has three proper base points.
(2) The one with defining coordinates $(xz : yz : y^2)$, which has two proper base points.
(3) The one with defining coordinates $(y^2 - xz : yz : z^2)$, which has one single proper base point.

As in Subsection 5.1.3, let $k[\mathbf{u}] = k[u_1, \ldots, u_m]$ stand for a polynomial ring over the field k (parameters).

Theorem 5.1.25 (k algebraically closed of characteristic zero). *Let $F = F(\mathbf{u}, x, y, z) \in k[\mathbf{u}, x, y, z]$ be a family of rational plane curves of degree 4 with a fixed set of singular points. Then the gradient ideal of the general member $f = F(\alpha) \in k[x, y, z]$ in this family is of linear type.*

Proof. We will actually show a bit more, namely, that any irreducible rational quartic falls within a family whose general member has the required property for its gradient ideal. In this vein, we can and will assume that the members of any family are singular. This is because the gradient ideal of any smooth plane curve is generated by a regular sequence, hence is of linear type.

We will as usual flirt with the abuse of thinking of the u's as both indeterminates or its specialized values in k. To keep the difference, we will below write the specialized form with a tilde above it.

Now, any irreducible rational quartic $f \in k[x, y, z]$ has at least one double singular point and at most a triple point. Let us first consider the situation where f has a double point; hence, has at most 3 such points and no triple point.

In this case, as explained above, f comes from a quadric by means of a rational quadratic map.

Let $Q = u_1x^2 + u_2y^2 + u_3z^2 + 2u_4yz + 2u_5zx + 2u_6xy$ be the equation of the quadric as above, assumed nonsingular, i. e., the corresponding symmetric matrix has nonzero determinant $\Delta = u_1u_2u_3 + 2u_4u_5u_6 - u_1u_4^2 - u_2u_5^2 - u_3u_6^2$ (the *discriminant* of Q).

Distinguishing the three cases of quadratic transforms, we obtain, respectively:

(1) A quartic with exactly three double points at $P_1 = (1 : 0 : 0)$, $P_2 = (0 : 1 : 0)$ and $P_3 = (0 : 0 : 1)$, where P_1 (resp., P_2, P_3) is a node except when the principal minor $u_1u_2 - u_6^2$ vanishes (resp., except when the principal minors $u_1u_3 - u_5^2$, $u_2u_3 - u_4^2$ vanish).
Here, we may harmlessly assume that $u_1 = u_2 = u_3 = 1$ provided they are all nonzero. The following families fit this block:

(a) Three nodes:

$$\tilde{f} = y^2z^2 + x^2z^2 + x^2y^2 + 2xyz(u_4x + u_5y + u_6z), \quad u_4, u_5, u_6 \neq \pm 1, \quad \Delta \neq 0$$

where $\Delta = 2u_4u_5u_6 - u_4^2 - u_5^2 - u_6^2 + 1$.

(b) Two nodes and one cusp:

$$\tilde{f} = y^2z^2 + x^2z^2 + x^2y^2 + 2xyz^2 + 2xyz(u_4x + u_5y), \quad u_4, u_5 \neq \pm 1, \quad \Delta \neq 0$$

where $\Delta = 2u_4u_5 - u_4^2 - u_5^2 = -(u_4 - u_5)^2$.

(c) One node and two cusps:

$$\tilde{f} = y^2z^2 + x^2z^2 + x^2y^2 + 2xy^2z + 2xyz^2 + 2u_4x^2yz, \quad u_4 \neq \pm 1.$$

(d) Three cusps:

$$\tilde{f} = y^2z^2 + x^2z^2 + x^2y^2 - 2xyz(x + y + z).$$

(2) A quartic with a double point at $P_1 = (0:0:1)$ (a node or a cusp according as to whether $u_2 \neq 0$ or $u_2 = 0$) and a double point at $P_2 = (1:0:0)$ (which is either a tacnode or a ramphoid cusp according as to whether the principal minor $u_1u_3 - u_5^2$ is nonzero or vanishes).
Here, we may assume that $u_1 = u_3 = 1$ and $u_6 = 0$.
It comprises the following families of singularities.

(e) One tacnode and one cusp:

$$\tilde{f} = x^2z^2 + y^4 + 2y^3z + 2u_5xy^2z, \quad u_5 \neq \pm 1.$$

(f) One tacnode and one node:

$$\tilde{f} = z^2(x^2 + y^2) + y^4 + 2y^2z(u_4y + 2u_5x), \quad u_5 \neq \pm 1, \quad \Delta = u_4^2 + u_5^2 - 1 \neq 0.$$

(g) One ramphoid cusp and one node:

$$\tilde{f} = x^2z^2 + y^4 + 2zy^3 + 2xy^2z + u_2z^2y^2, \quad u_2 \neq 0.$$

(h) One ramphoid cusp and one cusp:

$$\tilde{f} = x^2z^2 + y^4 + 2zy^3 + 2xy^2z.$$

(3) A quartic with an oscnode at $P_1 = (1:0:0)$ if $u_2 \neq 0$; else, a singularity of type A_6. Here, we may assume that $u_1 = 1$ and $u_5 = u_6 = 0$. Namely, we get the following forms:

(i) Oscnode:

$$\tilde{f} = (y^2 - xz)^2 + y^2z^2 + u_3z^4, \quad u_3 \neq 0.$$

(j) Singularity of type A_6:

$$\tilde{f} = (y^2 - xz)^2 + 2yz^3.$$

We now consider the case where the quartic has a triple point, say, at $P = (0:0:1)$.

Generally, a plane irreducible curve of degree d with a singular point of multiplicity $d-1$ (hence, a rational curve), has the shape $\varphi(x,y)z + \psi(x,y) = 0$, i. e., it is a z-monoid. Here, φ can further be brought up to one of the forms $x(y^2 - x^2)$, xy^2 and y^3, and ψ may be further normalized in such a way that the resulting family has as few parameters as possible.

(4) After these reductions, any irreducible plane quartic having $(0:0:1)$ as a triple point falls within three basic families, according to the nature of the triple point:
(k) An ordinary triple point:

$$\tilde{f} = x(y^2 - x^2)z + y^4 + x^2y(u_1y + u_2x).$$

(l) A triple point with double tangent:

$$\tilde{f} = xy^2z + x^4 + y^4 + u_1x^3y.$$

(m) A higher cusp:

$$\tilde{f} = y^3z + x^4 + u_1x^2y^2.$$

□

Proposition 5.1.26. *For each $\tilde{f}$ in the above list* (a) *through* (m), $J_{\tilde{f}}$ *is of linear type.*

Proof. We argue separately in three blocks.

Block (1)

(a) By Theorem 5.1.21, it suffices to show that $\mathcal{I}(\mathbf{0})$ has codimension 3. From the symmetrical parametric structure of $\tilde{f}$, the inclusion $\mathcal{I}(\mathbf{0}) \subset J_{\tilde{f}(\mathbf{0})}$ is an equality. On the other hand, by direct verification, the latter ideal admits the following syzygies:

$$\begin{pmatrix} xy^2 - xz^2 \\ -y^3 - yz^2 \\ y^2z + z^3 \end{pmatrix}, \quad \begin{pmatrix} -x^3 - xz^2 \\ x^2y - yz^2 \\ x^2z + z^3 \end{pmatrix}$$

Bringing in besides the generators of the gradient ideal, one obtains after a calculation $x^3, y^3, z^3 \in \mathcal{I}(\mathbf{0})$. Thus, $\mathcal{I}(\mathbf{0})$ has codimension 3.

We first note that (b)–(d) are obvious successive strata of the family (a).

(b) In this first stratum, the evaluated ideal $\mathcal{I}(\mathbf{0})$ has codimension 2, so one may try another evaluation. Note that one cannot blindly apply the result of (a) to claim that, here too, there is some general value α for which the assertion holds, since α could lie outside the open set obtained in (a)—it goes without saying that an explicit such open set can be computed by way of Theorem 5.1.21, (ii). Care has to be exercised while using the computer since for that the u's are true indeterminates. Anyhow, resorting to a mixed calculation, one finds the following syzygy of $\mathcal{I}$:

$$\begin{pmatrix} x^2(u_4^2-1)+xy(u_4u_5-1)+2xz(u_4-u_5)+yz(u_4-u_5) \\ y^2(u_5^2-1)+xy(u_4u_5-1)+xz(u_5-u_4)+2yz(u_5-u_4) \\ 3z^2(u_4-u_5)+xy(u_4-u_5)+xz(2u_4^2-u_5u_4-1)+yz(-2u_5^2+u_5u_4+1) \end{pmatrix}$$

Looking at the first summand of each coordinate, one sees that for any "value" $\alpha = (u_4,u_5) \in \mathbb{A}_k^2$ with $u_4 \neq u_5$ and $u_4 \neq \pm 1$, $u_5 \neq \pm 1$, pure powers of x, y, z remain and the ideal generated by the coordinates will have codimension 3, hence, also the corresponding $\mathcal{I}(\alpha)$.

(c) In this stratum by the same token, we look at the following two syzygies:

$$\begin{pmatrix} -xy+xz \\ 3y^2+2xyu_4+xz+3yz \\ -3z2-2xzu_4-2xy-3yz \end{pmatrix}, \quad \begin{pmatrix} x^2(u_4+1)+3/2xy+3/2xz+yz \\ 3/2y^2-xy(u_4+1)+1/2yz \\ 3/2z^2+2/2yz+xz(u_4+1) \end{pmatrix}$$

An identical analysis as above, looking at the pure powers, allow to choose any "valuc" $u_4 \neq -1$.

(d) Here, too, it suffices to look at the following syzygies:

$$\begin{pmatrix} x^2+xy+2/3xz-2/3yz \\ -xy-y^2+2/3xz-2/3yz \\ -1/3xz+1/3yz \end{pmatrix}, \quad \begin{pmatrix} xy+xz-2/3yz \\ -y^2+1/3yz \\ 1/3yz-z^2 \end{pmatrix}$$

Once more, pure powers of the variables are readily located.

Block (2)

(e) As in the proof of (a) here, too, the ideal $\mathcal{I}(\mathbf{0})$ contains the ideal of 1-minors of the syzygy matrix of $I_{\tilde{f}(\mathbf{0})}$. One can check that the latter ideal admits the following syzygies:

$$\begin{pmatrix} 2x^2-3y^2 \\ xz \\ -2xz \end{pmatrix}, \quad \begin{pmatrix} 2xy+3xz \\ yz \\ -2yz-3z2 \end{pmatrix}.$$

Bringing in the generator $\partial\tilde{f}(\mathbf{0})/\partial y = 4y^3+6y^2z$, one readily sees that $\mathcal{I}(\mathbf{0})$ has codimension 3.

(f) This follows the same pattern as (e). The relevant syzygies to look at are

$$\begin{pmatrix} x^2+y^2 \\ 0 \\ -xz \end{pmatrix}, \quad \begin{pmatrix} 2xy^2+xz^2 \\ yz^2 \\ -2y^2z-z3 \end{pmatrix}$$

and the calculation to get suitable powers of x, y, z inside $\mathcal{I}(\mathbf{0})$ is pretty straightforward.

(g) Indulging ourselves with the use of [79], we get $u_2^2 \in (\mathcal{I} : \mathfrak{m}^\infty)$. By Theorem 5.1.21, *every* member in this family has gradient ideal of linear type. Alternatively, one can look for $\mathcal{I}$ evaluated, say, at $u_2 \mapsto 1$. One column turns out to be

$$\begin{pmatrix} 5x^2-y^2+xz-yz \\ y^2+xy+xz+yz \\ -z^2-2y^2-xz-2yz \end{pmatrix}$$

Thus, the ideal generated by the three coordinates above already has codimension 3.

(h) Again, a computation with [79] gives the following syzygies:

$$\begin{pmatrix} 2y^2-3xz \\ -yz \\ 3z^2 \end{pmatrix}, \quad \begin{pmatrix} x^2-27/50xz \\ 1/5xy+1/5y^2+3/25xz-9/50yz \\ -2/5y^2-xz-6/25yz+27/50z^2 \end{pmatrix}.$$

The coordinates $3z^2$ and $x^2-27/50xz = x(x-27/50z)$ are invertible locally at (x, y) and (y, z), respectively. Since the latter are the two singular minimal primes of the quartic, this shows that the gradient ideal is locally a complete intersection at these primes; hence, is of linear type.

Block (3)

(i) The discussion of this case is analogous to the one of (g). A computation with [79] yields $u_3 \in (\mathcal{I} : \mathfrak{m}^\infty)$. Therefore, *every* member in this family has gradient ideal of linear type. As previously enacted, the ideal $\mathcal{I}(1)$ has codimension 3 through a convenient analysis of its terms.

(j) The following vectors are directly seen to be syzygies of the gradient ideal:

$$\begin{pmatrix} 6y^2+xz \\ 3yz \\ -z^2 \end{pmatrix}, \quad \begin{pmatrix} 7x^2+18yz \\ 3xy \\ 6y^2-7xz \end{pmatrix}.$$

A straightforward calculation gives the right codimension.

Block (4) (Triple point)

(k) The discussion of this case is much like the one of (a) in that the parametric structure of $\tilde{f}$ allows to see that the syzygies of $I_{\tilde{f}(\mathbf{0})}$ are contained in the syzygies of $I_{\tilde{f}}$ evaluated at $\mathbf{0}$. It then suffices to check that $I_{\tilde{f}(\mathbf{0})}$ is of linear type. We do this by arguing that it is locally a complete intersection at its unique singular prime (x, y). For this, it suffices to consider the syzygy

$$\begin{pmatrix} x^2 - 2/3y^2 + 1/6xz \\ 1/6yz \\ -(3x + 1/2z)z \end{pmatrix}$$

where the last coordinate is invertible locally at (x, y).

(l) As in (k), the syzygies of $I_{\tilde{f}(\mathbf{0})}$ are contained in the syzygies of $I_{\tilde{f}}$ evaluated at $\mathbf{0}$. Here, it is elementary to guess the syzygy

$$\begin{pmatrix} 0 \\ xy \\ -4y^2 - 2xz \end{pmatrix}$$

Using further the generators of the gradient ideal of $\tilde{f}(\mathbf{0})$, one readily detects powers of the variables among the entries.

(m) This case is like the previous one, only more elementary. We argue that $\mathcal{I}(\mathbf{0})$ has codimension 3 as before by looking at the obvious syzygy of $I_{\tilde{f}(\mathbf{0})}$,

$$\begin{pmatrix} 0 \\ y \\ -3z \end{pmatrix}$$

which clearly tells us that the ideal is locally a complete intersection at (x, y). □

Remark 5.1.27. The above has drawn quite a bit on computer calculation [79] to verify all cases of the theorem. Using Theorem 5.1.21, to have a computation-free argument it would suffice to show that the contraction $(\mathcal{I} : \mathfrak{m}^\infty) \cap k[\mathbf{u}]$ coincides set-theoretically with the product $\Delta\mathfrak{a}$ of the discriminant and the strata degeneration locus $\mathfrak{a} \subset k[\mathbf{u}]$. A computational verification of this fact has been carried for all four blocks of families. Thus, morally, the conjecture for rational quartics is settled. Unfortunately, even the ideal inclusion $(\mathcal{I} : \mathfrak{m}^\infty) \cap k[\mathbf{u}] \subset \Delta$ still lacks a theoretical argument.

It seems natural to conjecture that any *irreducible* quartic has gradient ideal of linear type.

5.2 Free divisors

The concept of a free divisor f was introduced by Saito [169] in the realm of complex analytic logarithmic differential forms, in terms of requiring that a certain module be free. The intent of this section is to elaborate on the close relationship between this notion and the gradient ideal of f. In addition, we wish to give the transcription of this notion into a purely algebraic way, involving ideals and their properties.

Now, Saito's background is local analytic. As a natural counterpart, we ought to take a regular domain essentially of finite type over a field of characteristic zero. However, in order to turn the exposition as simple as possible, we will work with a polynomial ring over such a field. The interested reader will find no hardship in redoing the theory in the case of a regular ring of fractions, at least in its relevant contents.

5.2.1 General nonsense

In this part, we develop a few general facts of elementary module theory.

Proposition 5.2.1. *Let R be a Noetherian commutative ring, let $f \in R$ be a regular element and let $\varphi\colon G \to R$ be an R-homomorphism, with G a free module of finite rank. Set $I := \operatorname{Im}(\varphi)$, $Z := \ker\varphi$ and $Z_{[f]} := \ker\varphi_{[f]}$, where $\varphi_{[f]}$ is the composite map*

$$G \to R \twoheadrightarrow R/(f).$$

Then:

(i) *There are exact sequences of R-modules*

$$0 \to Z \to Z_{[f]} \to I\colon(f) \to 0, \tag{51}$$

and of $R/(f)$-modules

$$0 \to Z/fZ \to Z_{[f]}/fG \to \frac{I\colon(f)}{I} \to 0. \tag{52}$$

(ii) *If $\psi\colon= \varphi \oplus f\colon G \oplus R \longrightarrow (I,f)$, then $Z_{[f]} \simeq \ker(\psi)$ and $Z_{[f]}$ is reflexive.*

Proof. (i) Let $\mathfrak{z} \in Z_{[f]} \subset G$. By definition, $\varphi(\mathfrak{z}) = af$, for some $a \in R$, and this expression is unique since f is regular. On the other hand, $\varphi(\mathfrak{z}) \in I$. Therefore, $af \in I$, i. e., $a \in I : (f)$. This gives a well-defined map $Z_{[f]} \to I : (f)$, which is obviously surjective and the kernel is Z by construction. This establishes the first exact sequence.

The second exact sequence follows from the first. Indeed, since f is a regular element of R and $\ker(Z/fZ \to Z_f/fG) = (Z \cap fG)/fZ$, it follows readily that $\ker(Z/fZ \to Z_f/fG) = 0$. On the other hand, as explained in the proof of item (i), the homomorphism $Z_{[f]} \to I\colon(f)$ maps an element $\mathfrak{z} \in Z_{[f]}$ to $a \in I$, where $\varphi(\mathfrak{z}) = af$. In particular, if

$\mathfrak{z} = fe \in fG \subset Z_{[f]}$, with $e \in G$, then $\varphi(\mathfrak{z}) = \varphi(e)f$, hence $\mathfrak{z}$ is mapped to $\varphi(e)$. Therefore, the restriction to fG maps onto I. This explains the second exact sequence.

(ii) Let $\widetilde{Z_{[f]}} = \ker(\psi)$. Since f is a regular element, projecting $G \oplus R$ onto the first summand induces a bijection of $\widetilde{Z_{[f]}}$ onto $Z_{[f]}$. Moreover, the ideal (I,f) has grade at least one, hence it has a well-defined rank (one) as an R-module. Therefore, $\widetilde{Z_{[f]}}$ being a first syzygy of such a module, it too has a well-defined rank. But a finitely generated second syzygy with this property is reflexive. □

Corollary 5.2.2. *With the notation of* Proposition 5.2.1, *one has:*
(a) *If $f \in I$, then $Z_{[f]} \simeq Z \oplus R$.*
(b) *Suppose that projective R-modules of finite rank are free. If the ideal (I,f) is proper and has grade at least two, then $Z_{[f]}$ is a free module if and only (I,f) is a codimension two perfect ideal.*

Proof. (a) The first assertion is obvious since R is free. Note, for future reference, that a splitting map ρ will map 1 to a vector $\mathfrak{z}$ such that $\varphi(\mathfrak{z}) = f$.

(b) If $Z_{[f]}$ is free, then (I,f) has projective dimension one; hence, it must be a codimension two perfect ideal because it has grade at least two. Conversely, if (I,f) is a codimension two perfect ideal then it has projective dimension one. Since $Z_{[f]}$ is a first syzygy thereof, it must be a projective module (of finite rank) by a well-known device; hence, it is free. □

5.2.2 Algebraic logarithmic vector fields

We assume throughout that k is a field of characteristic zero.

Let $R := k[x_1, \dots, x_n]$ be a polynomial ring over k. Let $f \in R$ be nonzero, J_f its gradient ideal and $J := (J_f, f)$ the corresponding full Jacobian ideal. As in the preliminaries of the chapter, the first module of syzygies of J_f will be denoted by $\mathrm{Syz}(J_f)$, having in mind that we are talking about the syzygies of the partial derivatives.

We denote $\mathrm{Der}_f(R) := Z_{[f]}$, a reflexive module called the *differential idealizer*. Its complex analytic version, here denoted $\mathrm{Derlog}(f)$, is the module of logarithmic derivations of f. Note that $\mathrm{Derlog}(f) = \{\theta \in \mathrm{Der}(R) \mid \theta(f) \in (f)\}$, where $\mathrm{Der}(R)$ is the module of k-derivations of R. We now translate Proposition 5.2.1 in this environment.

The ideal $J_f : (f)$ is called the *Euler conductor* of f and f is called *Eulerian* if $f \in J_f$.

Proposition 5.2.3. *There are exact sequences of R-modules*

$$0 \to \mathrm{Syz}(J_f) \to \mathrm{Der}_f(R) \to J_f : (f) \to 0, \tag{53}$$

and of $R/(f)$-modules

$$0 \to \frac{\mathrm{Syz}(J_f)}{f\,\mathrm{Syz}(J_f)} \to \mathrm{Der}_k(R/(f)) \to \frac{J_f : (f)}{I} \to 0. \tag{54}$$

In particular, if f is Eulerian then the first exact sequence above gives a direct sum decomposition

$$\mathrm{Derlog}(f) = \mathrm{Der}_f(R) = \mathrm{Syz}(J_f) \oplus R\theta_E,$$

where $\theta_E = \sum_{i=1}^{n} x_i \frac{\partial}{\partial x_i}$ denotes the Euler derivation.

Definition 5.2.4. (According to Saito) f is said to be a *free divisor* if f is reduced and $\mathrm{Der}_f(R)$ is a free R-module.

Any smooth f (i. e., $J = R$) is a free. (Of course, no homogeneous f of degree ≥ 2 is smooth since it is a cone.) Any quasi-homogeneous divisor is Eulerian. A smooth f is Eulerian if and only if $I = R$. Also, if f is Eulerian but not smooth then $f + a$ is smooth non-Eulerian for every $a \in k$, $a \neq 0$. If $f \in R$ is an irreducible homogeneous free divisor of degree ≥ 2, then the singular locus of the associated projective hypersurface has codimension 2—this will readily follow from the proposition below.

For nonsmooth free divisors the characterization below, (Proposition 5.2.6) makes it easy to produce the square matrix whose determinant is the free divisor (up to a scalar). In the smooth case, one has to compute $\mathrm{Der}_f(R)$ directly. Here is a celebrated case.

Example 5.2.5. Let $f = x_1^2 + \cdots + x_n^2 - 1$ (over any perfect field of characteristic $\neq 2$). Then

$$f = (-1)^{n-1} \det \begin{pmatrix} x_1^2 - 1 & x_1x_2 & \cdots & x_1x_n \\ x_1x_2 & x_2^2 - 1 & \cdots & x_2x_n \\ \vdots & & \vdots & \\ x_1x_n & x_2x_n & \cdots & x_n^2 - 1 \end{pmatrix}.$$

The next result is a simple consequence of the previous account. For an earlier appearance, see [206, Proposition 2.4] for the case of a hyperplane arrangement and a quote of the result in general in [207, Proposition 3]. It seems like Saito (personal communication) was already aware of such a result.

Proposition 5.2.6. *Let $R/(f)$ be reduced.*
(i) *f is a free divisor if and only if either f is smooth or else the full Jacobian ideal J is a codimension two perfect ideal.*
(ii) *If f is a nonsmooth Eulerian divisor, then f is a free divisor if and only if J_f is a codimension two perfect ideal.*

Proof. (i) The smooth case has already been discussed above. For the nonsmooth case, since $R/(f)$ is reduced, $\mathrm{ht}\, J \geq 2$. Therefore, the result is an immediate consequence of Corollary 5.2.2, (b).

(ii) This is a particular case of (i). □

Let us make the above result explicit. If f is a nonsmooth free divisor then $\mathrm{Der}_f(R)$ as a submodule of R^n (identified with $\mathrm{Der}_k(R)$) has a free basis of column vectors, say, represented by an $n \times n$ matrix

$$M = \begin{pmatrix} g_{11} & \cdots & g_{1n} \\ \vdots & & \vdots \\ g_{n1} & \cdots & g_{nn} \end{pmatrix},$$

while its "lifting" $\widetilde{\mathrm{Der}_f(R)} = \ker(R^{n+1} \to J)$ as in Proposition 5.2.1, (ii), is generated by the column vectors of the $(n+1) \times n$ matrix obtained from M by stacking to it the row vector $(-h_1, \dots, -h_n)$, where $\sum_i g_{ij} \frac{\partial f}{\partial x_i} = h_j f$ for $1 \leq j \leq n$ (cf. the proof of Proposition 5.2.1). Then the other $n \times n$ signed minors of the latter Hilbert–Burch matrix are the partial derivatives of f. On the other hand, the result of part (i) of Proposition 5.2.6 reads to the effect that $\mathrm{Der}_f(R)$ is freely generated by the column vectors of an $n \times n$ matrix such as M, where one of the columns can be taken to have as coordinates the coefficients of an Euler vector and the remaining columns can be taken to be syzygies of the gradient ideal J. Then the $(n-1)\times(n-1)$ signed minors of these $n\times(n-1)$ columns will be the partial derivatives while the total determinant gives back f, as is seen by expansion along the Euler column.

Remark 5.2.7. The search for irreducible homogeneous free divisors in $k[x,y,z]$ goes far beyond the mere finding of 3×3 graded matrices with nonzero determinant f. The above teaches us that, in order that f be a free divisor, it is required in addition that such a matrix have the following form up to k-linear elementary transformations

$$\frac{1}{\deg(f)} \begin{pmatrix} x & p_1(x,y,z) & q_1(x,y,z) \\ y & p_2(x,y,z) & q_2(x,y,z) \\ z & p_3(x,y,z) & q_3(x,y,z) \end{pmatrix}$$

where further the partial derivatives of f are the signed cofactors of x, y, z, respectively.

A piece of notation: given an R-module E, an element $\mathfrak{z} \in E$ (resp., a submodule $Z \subset E$) gives rise to a 1-form $\mathfrak{z}(1)$ (resp., to an ideal $Z(1)$ generated by 1-forms) in the symmetric algebra $\mathcal{S}_R(E)$.

The following notion appears in [30] in the local analytic setup. With appropriate slight language changes, we have the following.

Definition 5.2.8. Let $f \in R$ be such that $R/(f)$ is reduced. One says that f is a *Koszul free divisor* if:

(1) f is a free divisor (i. e., $\mathrm{Der}_f(R)$ is free), and
(2) Locally everywhere on R, the ideal $\mathrm{Der}_f(R)(1) \subset \mathcal{S}_R(\mathrm{Der}(R))$ has grade n.

Since $\mathrm{Der}_f(R)$ is n-generated when it is free, then $\mathrm{Der}_f(R)(1)$ is globally generated by n elements when f is a free divisor. If, moreover, f is Koszul-free then this ideal is locally everywhere on $S_R(\mathrm{Der}(R))$ generated by a regular sequence of n elements. Since in a Noetherian ring (global) grade is attained locally somewhere, it follows that a free divisor f is a Koszul free divisor if and only if the ideal $\mathrm{Der}_f(R)(1)$ is globally generated by a regular sequence.

Example 5.2.9. The easiest example of a (reduced, nonsmooth) Koszul-free divisor is the square-free monomial $f = x_1 \cdots x_n$.

Note that the cokernel of the inclusion $\mathrm{Der}_f(R) \subset \mathrm{Der}(R)$ is $J/(f)$. Finally, since grade and height coincide in the polynomial ring $S_R(\mathrm{Der}(R))$, f is Koszul-free if and only if f is free and $\dim \mathcal{S}_R(J/(f)) = n$. We then arrive at the following characterization of a Koszul-free divisor.

Proposition 5.2.10. *Let $f \in R$ be a free divisor. Then f is Koszul-free if and only if $\dim \mathcal{S}_R(J) = n+1$ and $f(1)$ avoids all minimal primes of $\mathcal{S}_R(J)$ of maximal dimension.*

Proof. We note that since $\mathrm{Der}(R)/\mathrm{Der}_f(R) \simeq J/(f)$, our subject fits as the cokernel of the exact sequence $0 \to (f(1)) \longrightarrow \mathcal{S}_R(J) \longrightarrow \mathcal{S}_R(J/(f)) \to 0$, where $f(1)$ is f viewed as a 1-form in $\mathcal{S}_R(J)_1 = J$. Now, $\dim \mathcal{S}_R(J) \geq n+1$ and $\dim \mathcal{S}_R(J/(f)) \geq n$—these are the minimum possible dimensions of these symmetric algebras. Clearly then, $\dim \mathcal{S}_R(J/(f)) = n$ if and only if $\dim \mathcal{S}_R(J) = n+1$ and $f(1)$ is not contained in any minimal prime of $\mathcal{S}_R(J)$ of maximal dimension. Therefore, we are through. □

For nonsmooth divisors, one derives the following version.

Proposition 5.2.11. *Let $f \in R$ be a nonsmooth reduced divisor. The following conditions are equivalent:*
(i) *f is Koszul-free.*
(ii) *The full Jacobian ideal J is a codimension two perfect ideal with $\mathcal{S}_R(J)$ a Cohen–Macaulay ring and $f(1)$ a nonzero divisor.*

Proof. It is known that, over a Cohen–Macaulay ring, the symmetric algebra of a codimension two perfect ideal is of minimal dimension if and only it is Cohen–Macaulay as well (cf., e. g., [96]). □

The proposition has the following consequence, which is an analogue of a result of Calderón–Narváez [30, Proposition 3.2].

Corollary 5.2.12. *Let f be a reduced free divisor. If full Jacobian ideal J is a codimension two perfect ideal of linear type then f is Koszul-free. In particular, if f is Eulerian then it is Koszul-free if J_f is a codimension two perfect ideal of linear type.*

Proof. Since J is of linear type, the symmetric algebra $\mathcal{S}_R(J)$ is a Cohen–Macaulay domain (see [96, Corollary 10.8]). □

In dimension three, in the homogeneous case we have the following special result.

Proposition 5.2.13. *Let $f \in R = k[x,y,z]$ be a reduced homogeneous polynomial of degree ≥ 3. The following conditions are equivalent:*
(i) *f is a free divisor*
(ii) *f is a Koszul-free divisor*
(iii) $\operatorname{depth}(R/J) > 0$

Proof. The equivalence of (i) and (iii) follows from Proposition 5.2.6. It suffices to show that (i) implies (ii). For that, since f is Eulerian, according to Proposition 5.2.3, the Euler vector can always be taken as part of a free basis of $\operatorname{Der}_f(R)$. Since f is further homogeneous, the coordinates of the Euler vector generate an ideal of codimension three. This readily implies that $\operatorname{Der}(R)/\operatorname{Der}_f(R) \simeq J/(f)$ satisfies the property called $\mathcal{F}_0$, hence its symmetric algebra has minimal dimension [96]. □

Remark 5.2.14. It is clear that the equivalence of conditions (i) and (ii) above hold also for a quasi-homogeneous divisor.

5.2.3 Families of free divisors

It was shown in [30, Theorem 5.6] in the local analytic setup that the Jacobian ideal of a locally quasi-homogeneous free divisor is of linear type. This motivates introducing a notion of a family of free divisors (relative to the base of the family). In this extended context, there are examples of families of free divisors whose (relative) gradient ideal is not of linear type.

For this, let $k[\mathbf{u}]$ denote the parameter polynomial ring over k and $S = k[\mathbf{u}, \mathbf{x}]$ a standard graded polynomial ring over $k[\mathbf{u}]$. If $F \in S$ is a form with unit $\mathbf{u}$-content, then one speaks of F as a *family of divisors of linear type* if the relative gradient ideal $J_{F,\mathbf{x}} \subset B$ ($\mathbf{x}$-derivatives only) is an ideal of linear type.

One can also speak of families of free divisors. For this, let $\operatorname{Der}_{F,\mathbf{x}}(B) \subset \operatorname{Der}_{\mathbf{x}}(B)$ stand for the (F)-idealizer of $\operatorname{Der}_{\mathbf{x}}(B)$ (i. e., the analogue of the modules $\operatorname{Der}_f(R) \subset \operatorname{Der}(R)$ with respect to the $\mathbf{x}$-derivatives only). Then, by a similar token, F will be said to be a *family of free divisors* provided $\operatorname{Der}_{F,\mathbf{x}}(B)$ is a free B-module of rank n.

5.2.3.1 The Cayley family of divisors

Consider a cubic equation $T^3 - pT + q$ with parametric coefficients, where $p = x^2 + vy^2 + uxz$, $q = u(x^2 + vy^2)z \in B = k[u,v,x,y,z]$. Then $T^3 - pT + q$ is a quasi-homogenous polynomial of degree 6 with respect to the following weights:

$$\begin{array}{ll} \deg u = 1 & \deg x = 2 \\ \deg v = 2 & \deg y = 1 \\ \deg T = 2 & \deg z = 1 \end{array}$$

Then the discriminant of $T^3 - pT + q$ is a quasi-homogeneous polynomial of degree 12,

$$F = F(u,v,x,y,z) = 4(x^2 + vy^2 + uxz)^3 - 27u^2(x^2 + vy^2)^2z^2,$$

which gives a family of quasi-homogeneous surfaces in the affine space $\mathbb{A}^3$ with weights 2, 1, 1 over the affine plane $\mathbb{A}^2$ with weights 1, 2—called the *Cayley family*.

Now, for every $(a,b) \in \mathbb{A}^2$ such that $ab \neq 0$, the corresponding fiber gives a form $f = f_{a,b} \in k[x,y,z]$ of (standard) degree 6. Such an individualized form is called a *Cayley sextic*.

Let now F and f denote, respectively, the Cayley family and any Cayley sextic. Since F has unit u,v-content and f is irreducible over $k[x,y,z]$, F is irreducible over $S = k[u,v,x,y,z]$. Furthermore, we have the following.

Proposition 5.2.15. *Let F denote the Cayley family and let f stand for any of its Cayley sextics. Then:*

(i) *$J_{F,\mathbf{x}} \in S$ and $J_f \in k[x,y,z]$ are both codimension two perfect ideals.*
(ii) *F is a (quasi-homogeneous) family of free divisors.*
(iii) *f is a (standard homogeneous) free divisor of linear type.*
(iv) *$J_{F,\mathbf{x}}$ is not of linear type as an ideal in S.*

Proof. (i) A calculation with [79] yields that the relative gradient ideal $J_{F,\mathbf{x}}$ is a codimension two perfect ideal with Hilbert–Burch matrix

$$\mathcal{M} = \begin{pmatrix} ux^2 - 7uvy^2 - 2u^2xz & vx^2y + v^2y^3 - uvxyz \\ 8uxy - u^2yz & -x^3 - vxy^2 + 1/2ux^2z - 1/2uvy^2z \\ -2x^2 - 2vy^2 - 6uxz + 2u^2z^2 & uvyz^2 \end{pmatrix}$$

As for f, since it is the image of F by a ring homomorphism $k[u,v,x,y,z] \to k[x,y,z]$, its partial derivatives are specialized from the x, y, z-partial derivatives of F. Further, since the latter are the (maximal) minors of a matrix then the partial derivatives of f are the (maximal) minors of the specialized matrix. Since f is irreducible, the gradient ideal J_f has codimension two, hence it is a perfect ideal.

(ii) This follows from (i) by Proposition 5.2.6.

(iii) This follows from (i) by Proposition 5.2.6.

To see that f is a divisor of linear type, it suffices by Proposition 2.3.2 to check that the required specialization of the syzygy matrix $\mathcal{M}$ has ideal of entries of codimension 3. For this, a direct inspection of the entries of $\mathcal{M}$, pretending that u,v are units, is rather obvious since we see pure powers of x, y, z appearing each in a different entry of the matrix.

(iv) By the same token as in (iii), one now sees that $I_1(\mathcal{M})$ is contained in the ideal $(u, x^2 + vy^2)$. □

5.2.3.2 A family of monoid free divisors

This part shows that there are families in (at least) two parameters of ternary forms of arbitrary degree ≥ 5 whose general member is a free divisor, while the special member is not a free divisor.

The member of the family presented here is a plane higher cusp-like singularity, defined by a z-monoid (hence, rational). It can be extended to more parameters as in [182, Proposition 2.2]. Other results and/or generalizations can be found in [182, 22, 146, 50] (and its list of references).

Proposition 5.2.16. *Consider the two-parameter family $F = y^{d-1}z + ux^d + vx^2y^{d-2} \in k[u,v][x,y,z]$, with $d \geq 5$. Then:*
(i) *The members of F over $\mathbb{A}^2_{u,v} \setminus V(uv)$ are irreducible free divisors.*
(ii) *A member of F over $\mathbb{A}^2_{u,v} \setminus V(uv)$ is of linear type if and only if $d = 5$.*

Proof. (i) Let $a,b \in k$ with $a \neq 0$. Then $f_{a,b} = F(a,b)$ is irreducible because it is a z-monoid with $\gcd(y^{d-1}z, ax^d + bx^2y^{d-2}) = 1$.

To see that $f := f_{a,b}$ is a free divisor, we will show directly that it is the ideal of maximal minors of a 3×2 matrix. At the outset, one has

$$J_f = ((ad)x^{d-1} + (2b)xy^{d-2}, y^{d-3}(b(d-2)x^2 + (d-1)yz), y^{d-1}).$$

Claim. Consider the k-vector V_{d-3} subspace of $k[x,y,z]_{d-3}$ spanned by the monomials $x^{d-3}, x^{d-5}yz, x^{d-7}y^2z^2, \dots, x^{d-l}y^pz^p$, where $l = d-1$ (resp., $l = d$) and $p = (d-2)/2$ (resp., $p = (d-3)/2$) if d is even (resp., odd). Then there exists a form $\mathfrak{g} \in V_{d-3}$ and field elements $\lambda, \eta \in k$ satisfying the following relation:

$$\mathfrak{g}(b(d-2)x^2 + (d-1)yz) - y^2((2b)xy^{d-4} + \lambda x^{d-l}y^{p-1}z^{p+1}) = \eta((ad)x^{d-1} + (2b)xy^{d-2}).$$

To solve this equation, we write $\mathfrak{g} = \alpha_3x^{d-3} + \alpha_5x^{d-5}yz + \alpha_7x^{d-7}y^2z^2 + \dots + \alpha_lx^{d-l}y^pz^p$. Say, $l = d-1$ (the case $l = d$ is similar). Then we first take $\alpha_l = \alpha_{d-1}$ to be an arbitrary nonzero scalar; then take $\lambda = -\alpha_{d-1}$. Next, we solve sequentially for the rest of the α_i's by the rule $\alpha_{i+1} = \beta\alpha_i$, where $\beta = b(d-2)/(d-1)$. Finally, take $\eta = \alpha_3b(d-2)/ad$.

From the claim, we see that the maximal minors of the following matrix generate J_f:

$$\varphi = \begin{pmatrix} 0 & \gamma y^{d-3} \\ -y^2 & \mathfrak{g} \\ b(d-2)x^2 + (d-1)yz & 2b\,xy^{d-4} + \lambda\,x^{d-l}y^{p-1}z^{p+1} \end{pmatrix},$$

for suitable nonzero $\gamma \in k$.

(ii) By direct inspection, the ideal of entries of φ contains a power of the ideal (x,y). If moreover $d = 5$, then $d - l = 0$ and $p = (d-3)/2 = 1$. Therefore, the lower rightmost entry of φ contains a pure power of z as a term. This shows that $I_1(\varphi)$ is (x,y,z)-primary.

If $d \geq 6$, the same analysis shows that no pure power of z belongs to $I_1(\varphi)$. □

Remark 5.2.17. The above family for $d = 5$ should be compared with the family of monoid quintics of Example 5.1.22. Thus, a slight change of a term away from the singularity may give substantially different results. For an additional colorful collection of free divisors, some of linear type, see [182].

5.2.4 Weighted homogeneous divisors

Let $f \in R := k[x_1,\dots,x_n]$ be an arbitrary polynomial. One wonders what homological properties of the gradient ideal J_f of f are inherited by the gradient ideal J_F of the homogenization $F \in R[t]$ of f with respect to a new variable t. A leitmotiv of this part is a discussion of this matter in the case f is quasi-homogeneous.

5.2.4.1 Recap of weighted homogeneous polynomials

Definition 5.2.18. A polynomial $f \in R := k[x_1,\dots,x_n]$ over an infinite field k is called *weighted homogeneous* (or *quasi homogeneous*) if there exist positive rational numbers $0 < a_1,\dots,a_n \leq \frac{1}{2}$ (called *rational weights*) such that

$$f(s^{a_1}x_1,\dots,s^{a_n}x_n) = sf(x_1,\dots,x_n),$$

where s is an indeterminate over R.

Taking the partial derivative with respect to s and using the chain rule yields

$$\sum_{i=1}^{n} a_i s^{a_i-1} x_i f_{x_i}(s^{a_1}x_1,\dots,s^{a_n}a_n) = f(x_1,\dots,x_n).$$

Evaluating at $s = 1$ gives

$$f = a_1x_1f_{x_1} + \cdots + a_nx_nf_{x_n}.$$

Conversely, let $f \in R$ be such that there exist positive rational numbers $0 < a_1,\dots,a_n \leq \frac{1}{2}$ with $f = a_1x_1f_{x_1} + \cdots + a_nx_nf_{x_n}$. Then by mapping $x_i \mapsto s^{a_i}x_i$, one obtains

$$f(s^{a_1}x_1,\dots,s^{a_n}x_n) = s\frac{\partial}{\partial s}(f(s^{a_1}x_1,\dots,s^{a_n}x_n)).$$

This is possible only when

$$f(s^{a_1}x_1,\dots,s^{a_n}x_n) = sC(x_1,\dots,x_n),$$

for some $C(x_1,\dots,x_n) \in R$. Evaluating at $s = 1$, yields $C(x_1,\dots,x_n) = f(x_1,\dots,x_n)$. So f is a weighted homogeneous polynomial of rational weights $a_1,\dots,a_n$.

Instead of using rational weights it is common to use integer weights $w_1,\dots,w_n \geq 1$ with gcd $= 1$, in which case there exists a positive integer λ such that $f(s^{w_1}x_1,\dots,s^{w_n}x_n) = s^{\lambda}f(x_1,\dots,x_n)$ and $\lambda f = w_1x_1f_{x_1} + \cdots + w_nx_nf_{x_n}$. The rational weights in the first form are then $a_i := \frac{w_i}{\lambda}$, $i = 1,\dots,n$.

Frequently, one accepts a degenerated situation where some of the weights are zero, in order to deal with general Eulerian divisors. Naturally, for these some of the standing properties of the weighted homogeneous case may fail to hold.

The following plays a role in the subsequent results.

Lemma 5.2.19. *Let $f \in R$ be a weighted homogeneous polynomial of (standard) degree d and rational weights $a_1,\dots,a_n$. Let $F \in R[t]$ be the homogenization of f with respect to a new variable t. Then the partial derivatives of F admit the following linear syzygy:*

$$(da_1 - 1)x_1F_{x_1} + \cdots + (da_n - 1)x_nF_{x_n} - tF_t = 0.$$

Proof. Dehomogenizing the Euler relation $dF = x_1F_{x_1} + \cdots + x_nF_{x_n} + tF_t$ via $t \to 1$ gives

$$df = x_1f_{x_1} + \cdots + x_nf_{x_n} + \widetilde{F_t},$$

where $\tilde{F}_t$ denotes the dehomogenization of F_t.

Using the assumption $f = a_1x_1f_{x_1} + \cdots + a_nx_nf_{x_n}$ yields

$$(da_1 - 1)x_1f_{x_1} + \cdots + (da_n - 1)x_nf_{x_n} - \widetilde{F_t} = 0. \tag{55}$$

On the other hand, writing $f = h_1 + \cdots + h_d$, where $h_i \in R$ is homogeneous of degree i, one has $F = h_1t^{d-1} + \cdots + h_{d-1}t + h_d$. Then (55) implies the following relations:

$$\begin{aligned}
(da_1 - 1)x_1(h_1)_{x_1} + \cdots + (da_n - 1)x_n(h_1)_{x_n} &= (d-1)h_1\\
(da_1 - 1)x_1(h_2)_{x_1} + \cdots + (da_n - 1)x_n(h_2)_{x_n} &= (d-2)h_2\\
&\vdots\\
(da_1 - 1)x_1(h_{d-1})_{x_1} + \cdots + (da_n - 1)x_n(h_{d-1})_{x_n} &= h_{d-1}\\
(da_1 - 1)x_1(h_d)_{x_1} + \cdots + (da_n - 1)x_n(h_d)_{x_n} &= 0.
\end{aligned}$$

Multiplying each of these equations by a suitable power of t and adding up the resulting relations yields

$$(da_1 - 1)x_1F_{x_1} + \cdots + (da_n - 1)x_nF_{x_n} - tF_t = 0,$$

as was to be shown. □

5.2.4.2 Binary weighted homogeneous polynomials

The main properties of binary quasi-homogeneous are readily stated.

Proposition 5.2.20. *Let $f \in k[x,y]$ be a weighted homogeneous polynomial with positive integral weights. Then:*

(1) $f = c_x x^{sq} + c_y y^{sp} + \sum_{1\le r<s} c_r x^{(s-r)q} y^{rp}$, *for any $p,q,s \in \mathbb{N}_{>0}$, such that $\gcd(p,q) = 1$ and weights pe, qe, with $e \ge 1$ arbitrarily chosen.*
(2) *f is homogeneous if and only if $p = q = 1$.*
(3) *If f is irreducible, then $c_x \ne 0$, $c_y \ne 0$.*
(4) *Let $\deg(f)$ denote the standard degree of f. Then neither x nor y is a factor of f if and only if at least one of the rational weights of f is $1/\deg(f)$.*
(5) *f is smooth at $(0,0)$ if and only if it is irreducible and, in addition, $s = 1$ and either $p = 1$ or $q = 1$.*

Proof. The shape of f follows immediately from the definitions. The other items follow by close inspection of this shape and are left to the reader. □

For the case of divisors on $\mathbb{P}^2$ obtained by homogenization of weighted homogeneous polynomials in two variables, we obtain a complete homological characterization as follows.

Theorem 5.2.21. *Let $f \in R := k[x,y]$ be a (nonhomogeneous) weighted homogeneous polynomial of degree $d \ge 3$, with $f = axf_x + byf_y$, $a,b > 0$. Let $F \in S := R[z]$ be the homogenization of f with respect to a new variable z. Then:*

(1) *If $a \ne \frac{1}{d}$ and $b \ne \frac{1}{d}$, then J_F is perfect of linear type with graded minimal free resolution:*

$$0 \to S(-(2d-3)) \oplus S(-d) \to S^3(-(d-1)) \to J_F \to 0.$$

(2) *If $a = \frac{1}{d}$ or $b = \frac{1}{d}$, then J_F is an ideal of linear type. In this case, J_F has a graded minimal free resolution:*

$$0 \to S(-(2d-1)) \to S(-d) \oplus S(-2(d-1))^2 \to S(-(d-1))^3 \to J_F \to 0.$$

Summing up, if f is not standard homogeneous, then its homogenization is a free divisor if and only if f is divisible by x or by y.

Proof. (1) By Lemma 5.2.19, $((da-1)x, (db-1)y, -z)$ is a syzygy on J_F. Thus, on one hand, its coordinates form a regular sequence of length 3, hence it is known that J_F is perfect (see, e. g., [208, Theorem 2.1]). The shape of the graded free resolution follows suit. On the other hand, J_F is generically a complete intersection, hence is of linear type.

(2) Since the data are symmetric, we may assume that $a = 1/d$ (then $b \ne 1/d$, as we are assuming that f is not homogeneous in the standard grading of R). Then x^{d-1}

occurs as a pure power in F_x. Also, Lemma 5.2.19 gives

$$(db-1)yF_y = zF_z.$$

Denote $\mathbf{z} := (0,(db-1)y,-z)^t$ the corresponding linear syzygy.

Now, then $I_1(\varphi)$ contains y, z, where φ denotes a minimal graded presentation matrix of J_F. On the other hand, F_x contains a nonzero pure power in x, while its remaining terms belong to the ideal (y,z). Because of any of the Koszul syzygies involving F_x this forces $I_1(\varphi)$ to be (x,y,z)-primary. Therefore, J_F is generically a complete intersection, and since it is an almost complete intersection, it must be of linear type.

To see the resolution, consider the two Koszul syzygies $\mathbf{z}_{12} := (F_y,-F_x,0)^t$ and $\mathbf{z}_{13} := (F_z,0,-F_x)^t$. Clearly, $\mathbf{z},\mathbf{z}_{12},\mathbf{z}_{13}$ form a minimal set of generators of the S-submodule of S^3 that they generate. A direct calculation shows that the vector $(F_x,-z,(db-1)y)^t$ is a syzygy of the set $\{\mathbf{z},\mathbf{z}_{12},\mathbf{z}_{13}\}$. This way one has a complex of R-modules:

$$0 \to S(-(2d-1)) \overset{\psi_2}{\to} S(-d)\oplus S(-2(d-1))^2 \overset{\psi_1}{\to} S(-(d-1))^3 \overset{\psi_0}{\to} S,$$

where $\mathrm{Im}(\psi_2) = (F_x,\beta z,\alpha y)^t$, $\mathrm{Im}(\psi_1) = S\mathbf{z}+S\mathbf{z}_{12}+S\mathbf{z}_{13}$ and $\mathrm{Im}(F_x,F_y,F_z)$.

To see that it is a resolution of its cokernel J_F, we apply the well-known acyclicity criterion as formulated in [57, Theorem 20.9]. The ranks of the maps clearly add up. Thus, it suffices to check the heights of the appropriate Fitting ideals.

Since F_x contains a nonzero pure power in x, while its remaining terms belong to the ideal (y,z), then the height of $I_1(\psi_2)$ is 3.

As to ψ_1, it has the 2-minor F_x^2 which has a nonvanishing pure power term in x, while it has another 2-minor belonging to the ideal (y,z). The two cannot have a proper common factor; hence, $I_2(\psi_1)$ has height at least 2. □

5.2.4.3 Nonhomogeneous free divisors

The goal of this part is to look closer at a couple of examples discussed in [30, Example 1.11 and Example 6.2] that may throw further light in the algebraic side of the theory. The fact that they are quasi-homogeneous free divisors adds up to the complexity of the involved algebra.

Proposition 5.2.22. *Let* $f = 256z^3-128x^2z^2+16x^4z+144xy^2z-4x^3y^2-27y^4 \in R = k[x,y,z]$. *Then:*

(i) f *is a weighted homogenous Koszul free divisor with Euler equality* $12f = 2xf_x+3yf_y+4zf_z$.

(ii) *The gradient ideal* $I \subset R$ *of* f *is an ideal of linear type.*

(iii) *Let* $F \in k[x,y,z,t]$ *denote the homogenization of* f *relative to a new variable* t. *Then*

  (a) $V(F)$ *is not a free divisor of* $\mathbb{P}^3$.

  (b) J_F *is of linear type.*

Proof. (i) Using [79], one sees that J_f is the ideal generated by the 2-minors of the following matrix:

$$\begin{pmatrix} -3y & x^2 - 12z \\ x^2 - 4z & 3xy \\ 1/2xy & 9/4y^2 - 4xz \end{pmatrix} \tag{56}$$

By Proposition 5.2.13 and Remark 5.2.14, $V(f)$ is Koszul-free. By direct verification, f is weighted homogeneous with weights 2, 3, 4 for x, y, z, respectively. A calculation yields the coefficient of f in the Euler relation.

(ii) It is straightforward to see that the entries of the above matrix generate an (x, y, z)-primary ideal. Therefore, I is generically a complete intersection, hence of linear type.

(iii) (a) To show that $V(F)$ is not a free divisor, one could of course use computer assistance; however, we choose to give a conceptual argument that may be used elsewhere.

By Lemma 5.2.19, J_F admits the linear syzygy $(x, -3/2y, -4z, 6t)^t$. On the other hand, the leftmost syzygy of J_f in (56) gives by homogenization a syzygy of J_F of degree 2, which is visibly not a multiple of the linear one. Clearly, J_F cannot be generated the 2-minors of these two syzygies (either because it is 4-generated or by degree reason since $\deg(F) = 5$). Therefore, it admits at least one additional minimal syzygy independent of the first two. Showing that any of the additional ones has standard degree at least 2 will do it since the sum of the degrees of any 3 minimal syzygies would exceed the degree of the partial derivatives of F, which is 4. Equivalently, we show that J_F has no additional minimal linear syzygies.

To prove the latter claim, suppose that $\ell_x F_x + \ell_y F_y + \ell_z F_z + \ell_t F_t = 0$, with the ℓ's linear forms in $k[x, y, z, t]$. Mapping $t \mapsto 1$ gives the relation

$$\widetilde{\ell_x} f_x + \widetilde{\ell_y} f_y + \widetilde{\ell_z} f_z = -\widetilde{\ell_t}\widetilde{F_t}, \tag{57}$$

where $\widetilde{\ }$ indicates the resulting polynomial in $k[x, y, z]$. On the other hand, dehomogenizing the Euler relation $5F = xF_x + yF_y + zF_z + tF_t$ yields

$$5f = xf_x + yf_y + zf_z + \widetilde{F_t}. \tag{58}$$

Using the weighted homogeneous Euler relation $12f = 2xf_x + 3yf_y + 4zf_z$ and (58) furnishes $\widetilde{F_t} = (-5/6)xf_x + (5/4)yf_y + (10/3)zf_z$. Comparing with (57) leads to the relation

$$(\widetilde{\ell_x} - (5/6)\widetilde{\ell_t}x)f_x + (\widetilde{\ell_y} + (5/4)\widetilde{\ell_t}y)f_y + (\widetilde{\ell_z} + (10/3)\widetilde{\ell_t}z)f_z = 0.$$

Since f is weighted homogeneous, so are its partial derivatives (of weighted degrees 10, 9, 8, resp.). Therefore, the coefficients of the last relation must be weighted homogeneous as well. But since a polynomial of degree 1 can only be weighted homogeneous for distinct weights of the variables (as is the case here) if it is a k-multiple of a variable, a close scrutiny of the above coefficients shows that they are not weighted homogeneous unless they vanish. In this case, one gets the following equalities:

$$\begin{cases} \widetilde{\ell_x} = (5/6)\widetilde{\ell_t}x \\ \widetilde{\ell_y} = -(5/4)\widetilde{\ell_t}y \\ \widetilde{\ell_z} = -(10/3)\widetilde{\ell_t}z \end{cases}$$

Since these are equalities of polynomials with standard degrees, we must have $\widetilde{\ell_t} = \alpha$, for some $\alpha \in k$. It follows that $\widetilde{\ell_t} = \alpha t$, hence

$$(\ell_x, \ell_y, \ell_z, \ell_t)^t = \frac{6}{5}\alpha\,(x, -3/2y, -4z, 6t)^t,$$

showing that the assumed linear syzygy is a multiple of the linear syzygy already obtained.

(b) To show that J_F is an ideal of linear type, one uses that $\mu(J_F) = 4 = \operatorname{ht} J_F + 2$. For such ideals, which are moreover Cohen–Macaulay ideals in a Cohen–Macaulay ring, it is known that their Koszul homology modules are Cohen–Macaulay [9]. Therefore, by [96, Theorem 9.1] it suffices to show that J_F is locally at every prime ideal P generated by at most $\dim R_P$ elements. This is obvious if P is the maximal homogeneous ideal of $k[x, y, z, t]$. If $\dim R_P = 3$, the linear syzygy of J_F shows that $(J_F)_P$ is generated by at most 3 elements.

It remains to prove that J_F is generically a complete intersection. Now, the singular points of $V(F)$ in $\mathbb{P}^2$ are either those of $V(f)$ in $\mathbb{A}^2$ $(t = 1)$ or else lie on the plane $t = 0$. By [79], the radical of J_f is generated by the polynomials

$$x^3 + 9/2y^2 - 4xz, \quad y(x^2 + 12z), \quad xy^2 - 8/3x^2z + 32/3z^2.$$

Therefore, any minimal prime of J_f contains either y or $x^2 + 12z$. From this, one gets that the minimal primes are $P := (x^2 - 4z, y)$ and $Q := (x^2 + 12z, y^2 - 32/9xz)$. Now, the entry $x^2 - 12z$ of the matrix (56) does not belong to P; hence, locally at P the first partial derivative belongs to the ideal generated by the other two partial derivatives. Similarly, the entry $-3y$ does not belong to Q, hence locally at Q the ideal J_f is again a complete intersection.

It remains to check the minimal primes of J_F that contain t. Again [79] gives that the radical of J_F is generated by the homogenized generators of J_f and the additional polynomial $y^3 - 32/9xyz$. It follows from this that t does not belong to any minimal prime of J_F (i. e., $V(F)$ has no singularities at infinity).

This finishes the proof. □

Remark 5.2.23. From the examples obtained so far, it looks tempting to ask if there exists an irreducible free divisor $f \in k[x, y, z]$ (quasi-homogeneous or not) of leading standard degree 4.

Proposition 5.2.24. *Let $f = xy(x+y)(x+yz) \in R = k[x,y,z]$. Then:*

(i) *f is a reduced free Eulerian divisor, with a degenerate weight decomposition—namely, one has $f = 1/4xf_x + 1/4yf_y$.*

(ii) *J_f is not an ideal of linear type.*

(iii) *The symmetric algebra $\mathcal{S}_R(J_f)$ is a complete intersection on which the Euler equation is a zero-divisor, in particular, f is not Koszul-free.*

(iv) *Letting $F \in R[t]$ denote the homogenization of f relative to the variable t, one has:*
   (a) *F is a free divisor.*
   (b) *J_F is not syzygetic, hence not of linear type.*
   (c) *The symmetric algebra of J_F is a reduced complete intersection, but F is not Koszul-free.*
   (d) *F is a homaloidal form.*

Proof. (i) The Euler-like expression of f is a direct calculation with the partial derivatives. With [79] we find that the syzygy matrix of J_f is

$$\begin{pmatrix} xy & x^2+4xy \\ y^2 & -3xy \\ -4x-4yz & -8x+4xz-4yz \end{pmatrix} \tag{59}$$

(ii) As one readily checks, the ideal generated by the entries of the matrix is contained in (x,y). Therefore, J_f is not of linear type.

(iii) The assertion about $\mathcal{S}_R(J_f)$ being a complete intersection is clear. By the same token as in (ii), $(x,y)R[T,U,V]$ is a minimal prime of the ideal $(\mathcal{J}_1)$ of definition of $\mathcal{S}_R(J_f)$ over the polynomial ring $R[T,U,V]$. Since the Euler relation read in degree one is $e := 1/4(xT+yU) \in (x,y)R[T,U,V]$, it is a zerodivisor on $\mathcal{S}_R(J_f)$.

Note in passing that Proposition 5.2.13 and Remark 5.2.14 tell us nothing since the Euler relation has a zero weight.

(iv) Nearly everything in this item seems to require computer assistance.

(a) An exception are the two linear syzygies of J_F, which can be guessed by inspection due to the particular shape of f and its "nearly" weighted homogeneous structure. With [79], the syzygy matrix of J_F is

$$\begin{pmatrix} 0 & x & 0 \\ 0 & y & xy+y^2 \\ x & 4z & -3yz \\ -y & -4t & -2yz-xt-2yt \end{pmatrix}. \tag{60}$$

Items (b) through (d) are discussed in Exercise 5.3.6. □

5.3 Exercises

Exercise 5.3.1. Let $f = x^d y^{d-i} + y^d z^{d-i} + x^{d-i} z^d \in k[x,y,z]$, with $0 \leq i \leq d$, a homogeneous polynomial of degree $2d - i$.

(1) Show that the corresponding projective hypersurface is smooth if and only if $i \geq d - 1$.

(2) Suppose that $i \leq d - 2$. Prove that the gradient ideal J_f is of linear type. (*Hint*: argue that the partial derivatives of f, up to scalar factors, have the form $x^{d-i-1}f, y^{d-i-1}g, z^{d-i-1}h$, where $f = z^d + \alpha x^i y^{d-i}$ and so on; then prove that J_f admits syzygies of the form $(xg, \_, \_)^t, (\_, yh, \_)^t, (\_, \_, zf)^t$.)

(3) Is the symmetric algebra (i. e., the Rees algebra) of J_f Cohen–Macaulay?

Exercise 5.3.2 (char(k) $\neq 2$). Let $f = xyz(x + y + z) \in k[x,y,z]$.

(1) Show that the unmixed part of J_f is the ideal $(xyz, xy(x + y), xz(x + z), yz(y + z))$.

(2) Prove that J_f is generically a complete intersection.

(3) Deduce that J_f is an ideal of linear type.

Exercise 5.3.3 (char(k) = 2). Let $f = xyz(x + y + z) \in k[x,y,z]$. Show:

(1) $f \not\subset J_f$.

(2) J_f is a perfect ideal (i. e., f is a free divisor). (*Hint*: show that J_f has a linear syzygy whose coordinates form a regular sequence of grade 3.)

(3) The free resolution of J_f has the shape $0 \to R(-4) \oplus R(-5) \to R(-3)^3 \to J_f \to 0$.

(4) Derlog(f) does not split as in (47).

(5) Derlog(f) has homological dimension 1 (while Syz(J_f) is a free module).

(6) J_f is not a reduction of $\mathbb{I}$ since $J_f : \mathbb{I}^\infty \neq (1)$.

Exercise 5.3.4 (char(k) $\neq 2$). Let $f = xyz(x + y)(x + z)(y + z) \in k[x,y,z]$.

With a computer assistant, establish:

(1) The unmixed part of J_f.

(2) Whether J_f is generically a complete intersection

(3) The minimal graded free resolution of J_f.

Exercise 5.3.5. Let $f = x_1x_2x_3x_4(x_1 + x_2)(x_3 + x_4) \in k[x_1, x_2, x_3, x_4]$ and $J = J_f$ its gradient ideal.

(1) Give an explicit linear syzygy of J (i. e., of the partial derivatives of f).

(2) Argue that the paired partial derivatives f_{x_1}, f_{x_2} and f_{x_3}, f_{x_4} have both reduced Koszul relations of degree 2.

(3) Deduce that f is a free divisor. (*Hint*: show that the 4×3 syzygy matrix coming from the previous two items has maximal rank.)

Exercise 5.3.6. This exercise is intended to close the proof of Proposition 5.2.24. Computer assistance is welcome, but as frugally as possible.

(1) Prove that J_F is not of linear type by showing that it is not syzygetic.
(*Hint*: find a syzygy $\mathbf{z}$ of J_F with coefficients in J_F, not belonging to the submodule of Koszul syzygies.)
(2) Give a relation of J_F of bidegree $(1,2)$ drawing upon the hint of the previous item.
(3) Prove that J_F is not of linear type by showing that it fails for property (F_1).
(*Hint*: the ideal of 2-minors of the matrix in (60) has only height 2.)
(4) Let $\mathcal{S}(J_F)$ denote the symmetric algebra of J_F. Prove that $\dim \mathcal{S}(J_F) = 5$.
(*Hint*: J_F satisfies the property (F_0).)
(5) Let T, U, V, W be presentation variables for the symmetric algebra of J_F. Prove that the ideal $P := (x, y, zV + tW) \subset S := k[x, y, z, t, T, U, V, W]$ is a minimal prime of $\mathcal{S}(J_F)$.
(6) Prove that $\mathcal{J}$ is reduced.
(*Hint*: let $\mathcal{J} \subset S$ denote the presentation ideal of the Rees algebra of J_F. Show that the presentation ideal of $\mathcal{S}(J_F)$ is $P \cap \mathcal{J}$. It may be helpful verifying that $\mathcal{I} : e = \mathcal{J}$, where $\mathcal{S}(J_F) = S/\mathcal{I}$ and $e = xT + yU + zV + tW$ is the Euler form.)
(7) Prove that J_F is not Koszul-free.
(*Hint*: the Euler form $e = xT + yU + zV + tW$ belongs to P.)
(8) Prove that F is homaloidal.
(*Hint*: collect the two linear syzygies of J_F and the bidegree $(1,2)$ relation of item (2) to concoct a submatrix of the Jacobian dual matrix with maximal rank.)

Exercise 5.3.7. Let $f = x_n^{n-1} - x_1 \cdots x_{n-1} \in k[x_1, \dots, x_n]$.

(1) Prove that f is not a free divisor.
(2) Let $\mathbf{x} = x_{i_1} \cdots x_{i_r}$, with $r \geq n-2$. Prove that $\mathbf{x}f$ is a free divisor.
(3) What goes wrong with taking $\mathbf{x} = x_{i_1} \cdots x_{i_r}$, with $r \leq n-3$?

Exercise 5.3.8 (char$(k) = 0$). Consider the 3×3 generic Hankel matrix

$$H = \begin{pmatrix} x_1 & x_2 & x_3 \\ x_2 & x_3 & x_4 \\ x_3 & x_4 & x_5 \end{pmatrix}.$$

(1) Let $J \subset k[x_1, \dots, x_5]$ denote the gradient ideal of $\det H$ and let $P := I_2(H)$. Show that $J \subset P = (J, \Delta)$, for a suitable 2-minor and that $\operatorname{ht} J = \operatorname{ht} P$.
(2) Prove that $\mathfrak{m} := (x_1, \dots, x_5) \subset J : P$. Deduce that P is the minimal component of a primary decomposition of J and $J : \mathfrak{m}^\infty = P$.
(3) Prove that $\det H$ is not homaloidal.
(*Hint*: by a result of Pan–Russo, otherwise it would trigger the lower bound $\operatorname{indeg}(P/J) \geq 3$.)

Exercise 5.3.9 (char(k) = 0). Consider the following degeneration of the 3×3 generic matrix:

$$H_0 = \begin{pmatrix} x_1 & x_2 & x_3 \\ x_2 & x_3 & x_4 \\ x_3 & x_4 & 0 \end{pmatrix}.$$

(1) Prove that the partial derivatives of $\det H_0$ are algebraically independent. (*Hint*: verify the rank of the corresponding Hessian.)

(2) Show that the partial derivatives of $\det H_0$ have (at least) three independent linear syzygies and deduce that $\det H_0$ is homaloidal.

(3) Argue that the three linear syzygies found in the previous item lift to linear syzygies of the partial derivatives of the partial derivatives of $\det H$. So, what goes wrong for $\det H$?

Exercise 5.3.10 (char(k) = 0). Consider the following degeneration of the 3×3 generic matrix:

$$G = \begin{pmatrix} x_1 & x_2 & x_3 \\ x_2 & 0 & x_4 \\ x_3 & x_4 & x_5 \end{pmatrix}.$$

(1) Show that the image of the polar map of $\det G$ is the quadric cone hypersurface $V(y_3^2 - 4y_1y_5) \subset \mathbb{P}^4$.

(2) In what sense is G a degeneration of the 3×3 generic symmetric matrix, but not of the 3×3 generic Hankel matrix?

6 Cremona transformations, I

Cremona transformations of $\mathbb{P}^n$ constitute a celebrated, fascinating subject in algebraic geometry. Since they are birational maps of $\mathbb{P}^n$ onto itself, they can be composed as rational maps and the resulting object is a group, called the *Cremona group* of $\mathbb{P}^n$. This group is trivial for $n = 1$, since it coincides with the projective linear group, while pretty much understood for $n = 2$. In the latter case, the main structural result is the celebrated Noether–Castelnuovo theorem, saying that it is generated by the projective linear group and the simplest quadratic involution. The subgroup structure of the group is also reasonably understood, although quite involved. As for $n \geq 3$, nothing much is really known about the group structure.

Essentially, there are two basic platforms to stay on: first, to ask whether given rational maps of $\mathbb{P}^n$ to itself are Cremona transformations; second, to ask how to classify given Cremona transformations inside its containing group. Following the previous chapter, the typically chosen platform in this book is the first one, with a distinctive role played by algebraic actors of the kind that have a stable place in commutative algebra. Thus, a full accomplishment means not just good testing theorems but effective characterizations as well in terms of the underlying ideal theory and homology.

6.1 Plane Cremona transformations: homaloidal nets

This section is mostly about Cremona transformations of the projective plane $\mathbb{P}^2$.

In this part we focus on aspects of the classical material. Our basic reference for the classical theory is the excellent book [4].

6.1.1 Weighted cluster of base points

Throughout, k denotes an infinite ground field. A good deal of the first principles make sense in that abstract landscape. However, it is clear that for the pure geometric purpose, as the theory deepens, one may have to assume that k is algebraically closed and, often, of zero characteristic. Our main source [4] assumes that $k = \mathbb{C}$, but it is clear that quite a bit goes through for any algebraically closed field of characteristic zero, while having k of infinite transcendence degree over $\mathbb{Q}$ may only be required in very particular passages. Of course, the theory of Cremona maps in characteristic $p > 0$ has many pitfalls. For this reason, the whole chapter unfolds under the assumption that $\mathrm{char}(k) = 0$.

6.1.1.1 Basic definitions

Let $R := k[x, y, z]$ be a standard graded polynomial ring in three variables over k. By a *linear system* of plane curves of degree d, we mean a k-vector subspace L_d of the

https://doi.org/10.1515/9783110640694-006

vector space of forms of degree d in R. The *fixed part* of L_d is the greatest homogeneous polynomial F of degree ≥ 1 that divides every element of L_d. We say that the linear system L_d has no fixed part if there is no such polynomial F.

If the linear system L_d has no fixed part, then the ideal $I = (L_d)$ has codimension ≥ 2 and hence, the radical of I define a reduced finite set $V(I)$ of points. Clearly, the codimension of I is in this case exactly 2 if and only if I is not (x,y,z)-primary. The points of $V(I) \subset \mathbb{P}^2$ are called *proper base points* of L_d. Thus, if I has codimension 2 then it has a primary decomposition whose minimal primary components are associated to the minimal primes of R/I defining the proper base points $V(I) = \{p_1, \ldots, p_r\}$.

We next recall the notion of (effective) multiplicity. Namely, given a variety X a smooth point $p \in X$ and a hypersurface (divisor) D, then the *multiplicity* $e_p(D)$ of D at p is the order of vanishing of a local equation of D at p; algebraically, if g is a local equation of D at p, then $e_p(D) = \min\{r \geq 0 \mid g \notin \mathfrak{m}^{r+1}\}$, where $\mathfrak{m}$ is the maximal ideal of the local ring of X at p. For our purpose, X will always be a smooth projective surface; more particularly, X is either $\mathbb{P}^2$ or the resulting surface of repeatedly blowing-up a reduced finite set of points on the latter.

In $\mathbb{P}^2$, a divisor D is defined by one single form $f \in k[x_0, x_1, x_2]$. In this case, we use interchangeably, $e_q(D)$ or $e_q(f)$ to mean the same number.

Assuming that the linear system L_d has no nontrivial fixed part, for each proper base point $p \in V(I)$, with $I = (L_d)$, one introduces into the picture a *virtual multiplicity*

$$\mu_p = \mu_p(L_d) := \min\{e_p(f) \mid f \in L_d\}.$$

Note that the subset of the linear system whose elements have at p effective multiplicity equal to μ_p forms an open set U_p (in the set of parameters).

Besides the proper base points, we can define the *improper base points* of L_d. To define these points, fix a proper base point p of L_d and let $\mathcal{B}_p$ denote the surface obtained by blowing-up p on $\mathbb{P}^2$. Let $E_p \subset \mathcal{B}_p$ stand for the corresponding exceptional divisor. A point $q \in E_p$ is a *base point infinitely near* to p if it belongs to the proper transform of every divisor in the open set U_p as introduced above. We then define the *virtual multiplicity* of L_d at q to be the integer

$$\mu_q = \mu_q(L_d) := \{e_q(\tilde{D}) \mid D \in U_p\},$$

where $\tilde{}$ denotes proper transform. This procedure is to be repeated successively (see [4, Chapter 1] for the details).

Denote by $K = \{p_1, \ldots, p_r\}$ the total set of base points, proper and improper, of L_d and by $\boldsymbol{\mu} = \{\mu_1, \ldots, \mu_r\}$ the set of multiplicities corresponding to the base points.

Definition 6.1.1. The datum $\mathcal{K} = (K, \boldsymbol{\mu})$ is called the (weighted) *cluster of base points* of L_d. The *type* associated to L_d is the vector $(d; \boldsymbol{\mu})$, as a preamble to the idea of a homaloidal type to come next.

A linear system L_d defines a rational map with source $\mathbb{P}^2$ and target $\mathbb{P}^r$, where $r+1$ is the vector space dimension of L_d. We are particularly interested in the situation in which the linear system L_d it is spanned by three independent forms in $R_d = k[x,y,z]_d$ without a proper common factor. In this case, the linear system L_d is called a *net*.

We now introduce the main object of this chapter.

Definition 6.1.2. Let $n \geq 2$ be an integer. A *Cremona transformation* (or *Cremona map*) of order n (or dimension n) is a birational map of $\mathbb{P}^n_k$.

Clearly, for $n = 2$ such a map is given by a particular net, called a *homaloidal net*. As a birational map, a Cremona one has a base ideal. Introducing linear systems as we now have, its base ideal is the ideal generated by one such system. Requiring that the linear system has no common factor (i. e., has no fixed part in the classical terminology) makes the *defining degree* of the transformation uniquely defined. However, in the theory of linear systems, it is often the case that a linear system without fixed part has interesting subsystems with a nontrivial fixed part.

In this part, the focus is on the case where $n = 2$. If L_d is a homaloidal net, we call the *characteristic* of L_d the set of multiplicities $\boldsymbol{\mu} = \{\mu_1,\dots,\mu_r\}$ corresponding to the base points of L_d, preceded by the degree d—usually denoted $(d;\mu_1,\dots,\mu_r)$ or $(d;\mu_1^{m_1},\dots,\mu_s^{m_s})$ to account for the repetitions. The characteristic of a homaloidal net L_d satisfies the classical *equations of condition*

$$\sum_{i=1}^{r}\mu_i = 3d-d, \quad \sum_{i=1}^{r}\mu_i^2 = d^2-1 \tag{61}$$

An abstract configuration $(d;\mu_1,\dots,\mu_r)$ satisfying the equations of condition is called a *homaloidal type*, and is denoted in the same fashion as the characteristic of a homaloidal net. A homaloidal type is called *proper* if there exists a plane Cremona map whose characteristic coincides with it.

6.1.1.2 Arithmetic quadratic transformations, I: the Hudson test

One of the great achievements of the classical theory was an effective tool to test whether a given homaloidal type is proper—it is called *Hudson test*, first introduced informally in [111]. To state this test in a rigorous way, we recall the notion of (arithmetic) quadratic transformations, introduced by M. Nagata [144].

Let $\mathbf{v} := (d;\boldsymbol{\mu})$ be a vector in $\mathbb{N}\times\mathbb{N}^r = \mathbb{N}^{r+1}$, where $d > 0$ and $\boldsymbol{\mu} = (\mu_1,\dots\mu_r)$—to fix ideas think of d as a "degree" and $\boldsymbol{\mu}$ as "virtual multiplicities." As a minor abuse, we will often refer to $\boldsymbol{\mu}$ as a set rather than a vector. In addition, we will often call a vector such as $\mathbf{v}$ an (abstract) *type* as a reminiscent of the type associated to a linear system, as defined in Definition 6.1.1.

Definition 6.1.3. Fix three indices j, k, l among $\{1,\dots,r\}$. The *arithmetic quadratic transformation* based on the triple j, k, l is the map $\mathcal{Q}_{j,k,l} : \mathbb{N}^{r+1} \longrightarrow \mathbb{N}^{r+1}$ such that

$\mathcal{Q}_{j,k,l}(\mathbf{v}) = \tilde{\mathbf{v}}$ with $\tilde{\mathbf{v}} = (\tilde{d}; \tilde{\boldsymbol{\mu}})$, where

$$\begin{cases} \tilde{d} := 2d - \mu_j - \mu_k - \mu_l \\ \widetilde{\mu_j} = d - \mu_k - \mu_l \\ \widetilde{\mu_k} = d - \mu_j - \mu_l \\ \widetilde{\mu_l} = d - \mu_j - \mu_k \\ \widetilde{\mu_i} = \mu_i, \ i \notin \{j,k,l\}. \end{cases} \tag{62}$$

Note an important feature of this map: if $\mu_j \geq \mu_k \geq \mu_l$, then $\widetilde{\mu_j} \geq \widetilde{\mu_k} \geq \widetilde{\mu_l}$.

Whenever the indices j, k, l are fixed in the discussion, we denote the transformation simply by $\mathcal{Q}$. A triple j, k, l of indices such that $\mu_j \geq \mu_k \geq \mu_l \geq \mu_i$, for all i, is called a *major triple* of indices.

Definition 6.1.4. An (abstract) type $\mathbf{v} = (d; \mu_1, \ldots, \mu_r)$ *satisfies the Hudson test* if $\mu_i \geq 0$ for every i and, for every major triple of indices, every arithmetic quadratic transformation based on it returns a type with nonnegative virtual multiplicities.

Clearly, the main purpose of this test is to apply to homaloidal types and to understand what one gets. Next is the state-of-the-art about this question, as collected from [4], where the main results were rigorously proved.

Theorem 6.1.5 ([4, Theorem 5.3.4]). *The following conditions are equivalent for a homaloidal type* $\mathbf{v} = (d; \mu_1, \ldots, \mu_r)$:

(1) $\mathbf{v}$ *is a proper homaloidal type.*
(2) (Hudson) $\mathbf{v}$ *satisfies the Hudson test.*
(3) (Enriques) *There exists a cluster of r base points* $\{p_1, \ldots, p_r\}$ *and an irreducible form* $f \in k[x,y,z]_d$ *such that* $e_{p_i}(f) = \mu_i$, *for every i such that* $\mu_i \neq 0$.
(4) (Nagata) *there is a finite sequence* $\mathcal{Q}_1, \ldots, \mathcal{Q}_\sigma$ *of arithmetic quadratic transformations such that* $(\mathcal{Q}_\sigma \circ \cdots \circ \mathcal{Q}_1)(\mathbf{v}) = (1; 0, \overset{r}{\ldots}, 0)$.

Remark 6.1.6. In addition, condition (2) is an open condition of proper points in the parameter space. More precisely, there exists a Zariski open set $U \subset \mathbb{P}^2 \times \overset{r}{\cdots} \times \mathbb{P}^2$ such that, for every $(p_1, \ldots, p_r) \in U$, there is a plane Cremona map of degree d having p_i as a base point with corresponding multiplicity μ_i, for every i for which $\mu_i > 0$, and no other base points.

Example 6.1.7. For $d = 2$, the equations of condition readily imply that a homaloidal type has the form $\mathbf{v} = (2; 1, 1, 1, 0, \overset{r}{\ldots}, 0)$. Applying a quadratic transformation, (62) outputs the type $\tilde{\mathbf{v}} = (1; 0, \overset{r}{\ldots}, 0)$. To get an explicit Cremona map with homaloidal type $\mathbf{v}$, it suffices to pick three nonaligned points in $\mathbb{P}^2$, i. e., one can take the following open set:

$$U := \{((a_1 : a_2 : a_3), (b_1 : b_2 : b_3), (c_1 : c_2 : c_3)) \in \mathbb{P}^2 \times \mathbb{P}^2 \times \mathbb{P}^2\}$$

such that

$$\det\begin{pmatrix} a_1 & a_2 & a_3 \\ b_1 & b_2 & b_3 \\ c_1 & c_2 & c_3 \end{pmatrix} \neq 0$$

6.1.2 Relation to fat ideals

Let $L_d \subset R_d = k[x,y,z]_d$ be a homaloidal net and let $F : \mathbb{P}^2 \dashrightarrow \mathbb{P}^2$ denote the associated rational map. The ideal $I = (L_d)$ is called the *base ideal* of F. The core of this section is the intertwining between the algebraic properties of the base ideal I and the weighted cluster of base points $\mathcal{K} = (K, \boldsymbol{\mu})$ of L_d.

6.1.2.1 The ideal of forms virtually through a cluster

The objective is to introduce another ideal, closely associated to a given cluster $\mathcal{K} = (K, \boldsymbol{\mu})$. Since the cluster may have infinitely near points, the general concept will naturally rely on handling exceptional divisors and total transforms. We observe that the terminology below, inspired in [4], may not be necessarily universally accepted (see [136, Definition 1] for an alternative).

Let $\mathcal{B}_K$ stand the blowup of the set K, by which one understands the final composite of the successive blowups $\mathcal{B}_{p_i}$ of the individual points $p_i \in K$. It can be shown that $\mathcal{B}_K$ does not depends on the ordering of the points by proximity; we refer to [31, 32] for further details.

In addition, given a plane curve $C = V(f) \subset \mathbb{P}^2$, let $\overline{C}^K$ and $\overline{E}_p^K$ denote the total transform of C and the total exceptional p-component of the exceptional divisor of $\mathcal{B}_K$, respectively

One says that $C \subset \mathbb{P}^2$ *passes virtually through* $\mathcal{K}$ if the divisor

$$\overline{C}^K - \sum_{p\in K} \mu_p \overline{E}_p^K \tag{63}$$

on $\mathcal{B}_K$ is effective.

Given $m \geq 1$, let $\ell_{\mathcal{K}}(m) \subset R_m$ consist of all forms f of degree m such that the curve $C = V(f)$ passes virtually through $\mathcal{K}$.

Definition 6.1.8. The *ideal of curves virtually through* $\mathcal{K}$ is the homogeneous ideal

$$I_{\mathcal{K}} := \bigoplus_{m\geq 0} \ell_{\mathcal{K}}(m) \subset R.$$

This ideal is a generalized version of an ideal of fat points in that it admits infinitely near points. Its nature was investigated by various authors (see, e. g., [83, 136]).

However, we have the usual ideal of fat points

$$\mathfrak{F}_{\mathcal{K}} = I(p_1)^{\mu_1} \cap \cdots \cap I(p_r)^{\mu_r},$$

corresponding to the proper points of K, where $I(p_j)$ stands for the defining prime ideal of the proper point p_j. This ideal will be referred here as the ***μ**-fat ideal* associated to the given cluster.

We next establish the relationship between the two notions. A few auxiliary results will be needed. First, to complete the notation, denote by $\widetilde{C}^K$ the proper transform of C under the blowup $\mathcal{B}_K$. Similarly, let $\widetilde{E}_p^K$ be its proper exceptional p-component. We have the well-known additive formula

$$\overline{C}^K = \widetilde{C}^K + \sum_{p \in K} e_p(C)\overline{E}_p^K. \tag{64}$$

It follows from (64) that a form f passes virtually through $\mathcal{K}$ if and only if the divisor $\sum_{p \in K} (e_p(C) - \mu_p)\,\bar{E}_p^K$ is effective.

Now, the obvious way to describe the **μ**-fat ideal in terms of inequalities is

$$\mathfrak{F}_{\mathcal{K}} = \{f \in R : e_p(f) \geq \mu_p, \ \forall \text{ proper } p \in K\}.$$

In order to state a parallel characterization of $I_{\mathcal{K}}$, one wishes to translate effectiveness as above in terms of numerical inequalities. The tool is the following notion devised by Du Val [55].

Definition 6.1.9. The *proximity matrix* of the set K of a cluster with n points is the $n \times n$ matrix $\mathbf{P}_{\mathcal{K}} = (m_{q,p})$ with

$$m_{q,p} = \begin{cases} 1 & \text{if } p = q, \\ -1 & \text{if } q \text{ is proximate to } p, \\ 0 & \text{otherwise.} \end{cases}$$

Here, one says that q is *proximate* to p if q comes after p in the chosen order of points and if q belongs to the exceptional divisor of the blowup of p or is infinitely near a point there.

Now, the connection between this matrix and the various proper transforms is the notable fact that the (q, p) entry of the proximity matrix coincides with the intersection product of $\widetilde{E}_p^K$ and $\overline{E}_q^K$. Drawing upon the above results, one has the following.

Lemma 6.1.10. *A plane curve $C \subset \mathbb{P}^2$ passes virtually through $\mathcal{K}$ if and only if*

$$\mathbf{P}_{\mathcal{K}}^{-1} \cdot (\mathbf{e}_K(C) - \boldsymbol{\mu}_K)^t \geq 0, \tag{65}$$

where $\mathbf{e}_K(C)$ *(resp.,* $\boldsymbol{\mu}_K$*) denotes the vector of effective multiplicities at the points of* K *(resp., the vector of virtual multiplicities of* $\mathcal{K}$*), and* $^{-1}$, ^{t} *denote matrix inverse and matrix transpose, respectively.*

The above relation makes sense as $\det(\mathbf{P}_{\mathcal{K}}) \neq 0$, in fact, it is not difficult to see that any such matrix is unimodular, so its inverse is an integer matrix and, moreover, its inverse has only nonnegative entries.

Proposition 6.1.11. *Let* $F : \mathbb{P}^2 \dashrightarrow \mathbb{P}^2$ *be a rational map without fixed part, with cluster of base points* $\mathcal{K} = (K, \boldsymbol{\mu})$ *and base ideal* $I \subset R$*. Then*

$$I \subset I_{\mathcal{K}} \subset \mathfrak{F}_{\mathcal{K}}.$$

Moreover, if K *consists of proper points then* $I_{\mathcal{K}} = \mathfrak{F}_{\mathcal{K}}$*.*

Proof. Let d be the degree of the forms defining F. By the very nature of the definition of the associated weighted cluster $\mathcal{K}$ of F, any curve of the linear system in degree d defining F passes virtually through $\mathcal{K}$—note that the condition in (65) is here verified "on the nose" through the inequalities $e_p(C) \geq \mu_p$ for every $p \in K$. Therefore, $I_d \subset (I_{\mathcal{K}})_d$. It follows that $I_m \subset (I_{\mathcal{K}})_m$ for every degree $m \geq d$. If $m < d$, then $I_m = \{0\}$, hence $I_m \subset (I_{\mathcal{K}})_m$ holds trivially. Thus, $I \subset I_{\mathcal{K}}$.

For the right-most inclusion, we can order the points of K so that the proper ones come first and determine the upper left block of $\mathbf{P}_{\mathcal{K}}$. This block has to be the identity matrix by definition and splits $\mathbf{P}_{\mathcal{K}}$ in a block-diagonal shape. Then the inverse $\mathbf{P}_{\mathcal{K}}^{-1}$ will also contain an identity block that multiplies the proper part of the vector $(\mathbf{e}_K(C) - \boldsymbol{\mu}_K)^t$. On the other hand, as mentioned earlier, one has $\mathfrak{F}_{\mathcal{K}} = \{f \in R : e_p(f) \geq \mu_p, \ \forall \text{ proper } p \in K\}$. Therefore, the inclusion $I_{\mathcal{K}} \subset \mathfrak{F}_{\mathcal{K}}$ follows immediately.

The supplementary statement is pretty clear now. □

Remark 6.1.12. (1) If the rational map F above is birational (i. e., a Cremona transformation), then a classical result proves that there is actually an equality $I_d = (I_{\mathcal{K}})_d$ (see, e. g., [4, Proposition 2.5.2]). This knowledge will be used in Theorem 6.1.16.

(2) The history of the proximity matrix, first introduced rigorously by Du Val in [55], is quite interesting. It came out in the way of improving on results of Barber and Zariski of the mid 1930s. The latter, in turn, had in mind making rigorous the theory of Enriques on matters of infinitely near points. Actually, in a second paper [56], Du Val addresses directly some processes in the famous book by Enriques and Chisini, that the latter called "scaricamento" (hence, "unloading" in Du Val's) and "scorrimento."

6.1.2.2 Homology of the μ-fat ideal

A strategy to access the algebraic properties of the ideal I is to make use of the inclusions in Proposition 6.1.11. The more encompassing $I_{\mathcal{K}}$ does not seem to be well captured in an algebraic way so as to understand its ideal theoretic properties.

A clear advantage of the $\boldsymbol{\mu}$-fat ideal $\mathfrak{F}_{\mathcal{K}}$ at the outset is that some properties are readily guessed. For example, $R/\mathfrak{F}_{\mathcal{K}}$ is always Cohen–Macaulay and the degree of $R/\mathfrak{F}_{\mathcal{K}}$ is

$$\deg(R/\mathfrak{F}_{\mathcal{K}}) = \sum_{i=1}^{r} \frac{\mu_i(\mu_i+1)}{2}.$$

In this part, we discuss the homological nature of the ideal of the $\boldsymbol{\mu}$-fat ideal of a homaloidal type. The following are well-known basic facts about $\boldsymbol{\mu}$-fat ideals—a proof is provided for the reader's convenience.

Lemma 6.1.13. *Let $F : \mathbb{P}^2 \dashrightarrow \mathbb{P}^2$ be a Cremona map with homaloidal type $(d;\boldsymbol{\mu})$ whose base points $\mathbf{p}$ are proper (i. e., $\mathbf{p} \subset \mathbb{P}^2$) and let $J \subset R := k[x,y,z]$ denote the associated $\boldsymbol{\mu}$-fat ideal. Then:*
(i) *The initial degree of J is d.*
(ii) *The Hilbert function of R/J is maximal with $h_{R/J}(t) = e(R/J)$ for $t \geq d$ (i. e., the regularity index of R/J is d).*
(iii) *$n_t = 0$ for $t \geq d+2$, where n_t is the t-th graded Betti number of J.*

Proof. (i) Suppose that $0 \neq f \in J_{d-1}$. Then $\{xf, yf, zf\}$ is a k-linearly independent subset of J_d. Since F is a Cremona map, its underlying 2-dimensional linear system L_d is complete by [4, Proposition 2.5.2], i. e., $L_d = J_d$. Therefore, $\{xf, yf, zf\}$ is a k-basis of L_d, contradicting the fact that L_d has no fixed part.

(ii) For the second assertion we recall that the equations of condition satisfied by a Cremona map plus knowing that $\dim_k(J_d) = \dim_k(L_d) = 3$ imply the equality

$$\dim_k(J_d) = \frac{(d+2)(d+1)}{2} - \sum \frac{(\mu_i+1)\mu_i}{2}. \tag{66}$$

Since d is the initial degree of J, the Hilbert function of R/J is maximal.

(iii) This follows from (ii) (see, e. g., [49]). □

We will need the following structural result, which could not be found in this exact form in the literature. For the definition of de Jonquières maps, see Section 7.1.14 below.

Proposition 6.1.14. *Let $F : \mathbb{P}^2 \dashrightarrow \mathbb{P}^2$ be a Cremona map whose base ideal is perfect and has a linear syzygy. Then F is a de Jonquières map.*

Proof. Since any quadratic Cremona map is de Jonquières, we may assume at the outset that $\mathfrak{F}$ has degree ≥ 3. Consider the presentation matrix of its base ideal I:

$$L := \begin{pmatrix} \ell_1 & q_1 \\ \ell_2 & q_2 \\ \ell_3 & q \end{pmatrix},$$

where ℓ_i is a linear form and q_i, q are forms of degree ≥ 2. By Corollary 3.2.27, I is not an ideal of linear type. It follows that the codimension of the ideal generated by the entries of L is ≤ 2 since otherwise I would be generically a complete intersection; but, by Proposition 2.3.2, an almost complete intersection, which is generically a complete intersection must be of linear type. In particular, the three linear forms along the first column cannot be k-linearly independent. By suitable elementary operation, we may assume that $\ell_1 = x$, $\ell_2 = y$, $\ell_3 = 0$. Then one has $I = (xq, yq, \ell_1 q_2 - \ell_2 q_1)$. Moreover, q and $\ell_1 q_2 - \ell_2 q_1$ are relatively prime z-monoids, as will be shown in greater generality in Section 7.1.6. This means that $\mathfrak{F}$ is a plane de Jonquières map. □

Remark 6.1.15. There are examples of dominant rational maps $\mathbb{P}^2 \dashrightarrow \mathbb{P}^2$, which are not birational, whose base ideal I is perfect and has a unique linear syzygy (up to scalars). Such examples exist, in addition, where I is an ideal of linear type—equivalently, such that the three coordinates of the linear syzygy are k-linearly independent.

We are now ready for the main result of this part.

Theorem 6.1.16. *Let $F : \mathbb{P}^2 \dashrightarrow \mathbb{P}^2$ be a Cremona map of degree $d \geq 4$ with cluster $\mathcal{K} = (K, \boldsymbol{\mu})$, where K has only proper points. Let $I \subset R = k[x, y, z]$ denote its base ideal. Then*

(a) *The minimal graded free resolution of the $\boldsymbol{\mu}$-fat ideal $\mathfrak{F}_{\mathcal{K}}$ is of the form*

$$0 \to R(-(d+2))^{d-2} \oplus R(-(d+1))^{s} \to R(-(d+1))^{d-4+s} \oplus R(-d)^3 \to \mathfrak{F}_{\mathcal{K}} \to 0$$

where s is the number of independent linear syzygies of the ideal I. Furthermore, $0 \leq s \leq 1$, while $s = 1$ if and only if F is a de Jonquières map.

(b) *If F is not a de Jonquières map, the minimal graded free resolution of the ideal $([\mathfrak{F}_{\mathcal{K}}]_{d+1})$ is of the form*

$$0 \to R(-(d+3))^3 \longrightarrow R(-(d+2))^{d+7} \longrightarrow R(-(d+1))^{d+5} \longrightarrow ([\mathfrak{F}_{\mathcal{K}}]_{d+1}) \to 0.$$

Proof. Set $J := \mathfrak{F}_{\mathcal{K}}$ for lighter reading.

(a) By Lemma 8.4.25, the initial degree of J is d.

Next, by the equations of condition (61) we have

$$\deg(R/J) = \frac{d^2 + 3d - 4}{2}. \tag{67}$$

In this particular case, we know that

$$\operatorname{reg} J = \min\left\{s \mid h_J(t) = \binom{t+1}{2} - \deg(R/J) \quad \text{for all } t \geq s\right\} + 1$$

where h_J is the Hilbert function of J. By [4, Proposition 2.5.2], we have $h_J(d) = 3$. A straightforward calculation using (67) implies that

$$h_J(d) = 3 = \binom{d+2}{2} - \deg(R/J).$$

Thus, $\operatorname{reg} J \leq d+1$.

With these two facts, the minimal graded free resolution of J has the following shape:

$$0 \to R(-(d+2))^{\nu_{d+2}} \oplus R(-(d+1))^{\nu_{d+1}} \to R(-(d+1))^{n_{d+1}} \oplus R(-d)^3 \to J \to 0.$$

Claim. $n_{d+1} = d - 4 + s$, where $s = \dim_k(\ker(R_1 \otimes_k (J)_d \longrightarrow (J)_{d+1}))$.

Indeed, since $\operatorname{reg} J \leq d+1$ then

$$\dim_k(J)_{d+1} = \frac{(d+3)(d+2)}{2} - \deg(R/J) = d+5.$$

Let $U := \ker(R_1 \otimes_k (J)_d \longrightarrow (J)_{d+1})$ and $V := \operatorname{coker}(R_1 \otimes_k (J)_d \longrightarrow (J)_{d+1})$. From the exact sequence,

$$0 \to U \longrightarrow R_1 \otimes_k (J)_d \longrightarrow (J)_{d+1} \longrightarrow V \to 0,$$

and since $\dim_k(R_1 \otimes_k (J)_d) = 9$, we then get $n_{d+1} = d - 4 + s$ where $s := \dim_k(U)$, as required.

Next, one knows that $\nu_{d+1} = n_{d+1} - \Delta^3 h_J(d+1)$, where Δ denotes the difference function (see, e. g., [62]). A calculation then gives $\nu_{d+1} = s$; hence, $\nu_{d+2} = d - 2$. Thus, the minimal graded free resolution of J is as stated.

Claim. $s = \dim_k(U)$ coincides with the number of independent linear syzygies of the ideal I.

For this, let $\{D_1, D_2, D_3\}$ denote a k-vector basis of $(J)_d$. Letting $\{T_1, T_2, T_3\}$ denote indeterminates over R, one has $k[T_1, T_2, T_3]_1 = kT_1 + kT_2 + kT_3 \simeq (J)_d$ as k-vector spaces. Therefore, $R_1 \otimes_k (J)_d \simeq R_1 \otimes_k k[T_1, T_2, T_3]_1 \simeq R[T_1, T_2, T_3]_{(1,1)}$ as k-vector spaces, where the latter is the graded part in bidegree $(1, 1)$ of the standard bigraded polynomial ring $R[T_1, T_2, T_3]$. With this identification, it follows immediately that N coincides with the bigraded $(1, 1)$-part of the ideal generated by the bigraded relations of $\{D_1, D_2, D_3\}$—the so-called Rees ideal of $I = (D_1, D_2, D_3)$—which is identified with the k-vector space spanned by the R-linear syzygies of I. This settles the first statement in the proposition.

For the supplementary statement, note that the number $s = \dim_k(U)$ of independent linear syzygies of I is exactly the number of independent k-linear relations among the elements $\{xD_j, yD_j, zD_j\}$ spanning $R_1(J)_d$. Thus, to show that $0 \leq s \leq 1$ it suffices to prove that these forms span a k-vector space of dimension at least 8. For this purpose, consider a linear form $\ell \in R$, which is a nonzerodivisor on R/J. We claim that xD_1, xD_2,

$xD_3, \ell D_1, \ell D_2, \ell D_3$ are k-linearly independent. Indeed, from a relation

$$\alpha_1 xD_1 + \alpha_2 xD_2 + \alpha_3 xD_3 + \beta_1 \ell D_1 + \beta_2 \ell D_2 + \beta_3 \ell D_3 = 0,$$

with $\alpha_j, \beta_j \in k$, one deduces

$$x(\alpha_1 D_1 + \alpha_2 D_2 + \alpha_3 D_3) = \ell(-\beta_1 D_1 - \beta_2 D_2 - \beta_3 D_3);$$

and hence

$$(-\beta_1 D_1 - \beta_2 D_2 - \beta_3 D_3)/x = f \in k[x,y,z],$$

with f a form of degree $d-1$. Noting that $\ell f \in (D_1, D_2, D_3) \subset J$, since ℓ is regular modulo J we must have $f \in J$. But the initial degree of J being d, we conclude that $f = 0$. Bringing this back in the above relations, one gets the relations

$$\alpha_1 D_1 + \alpha_2 D_2 + \alpha_3 D_3 = 0 \quad \text{and} \quad \beta_1 D_1 + \beta_2 D_2 + \beta_3 D_3 = 0$$

thus forcing $\alpha_1 = \alpha_2 = \alpha_3 = \beta_1 = \beta_2 = \beta_3 = 0$ as $\{D_1, D_2, D_3\}$ is k-linearly independent.

We have proved that $\dim_k(R_1(J)_d) \geq 6$. We can do one better readily, to wit, force ℓ to avoid the prime (x,y) as well. We claim that for some $1 \leq j \leq 3$, one has $yD_j \notin k\,xD_1 + k\,xD_2 + k\,xD_3 + k\,\ell D_1 + k\,\ell D_2 + k\,\ell D_3$. Supposing the contrary, in particular, $yD_j \in (x,\ell)$ for $1 \leq j \leq 3$. But since $\ell \notin (x,y)$ by our choice, then $y \notin (x,\ell)$, and hence $D_j \in (x,\ell)$ for every $j \in \{1,2,3\}$. It follows that (x,ℓ) is a minimal prime of $k[x,y,z]/(D_1,D_2,D_3)$, thus a minimal prime of R/J as well, contradicting the choice of ℓ.

This gives $\dim_k(R_1(J)_d) \geq 7$.

We still have one to go. Thus, assume $\dim_k(R_1(J)_d) = 7$, i. e., the ideal I has 2 independent linear syzygies. Then the syzygy matrix of J would have the form

$$\varphi = \left(\begin{array}{cc|cccc}
\ell_{1,1} & \ell_{1,2} & q_{1,1} & q_{1,2} & \cdots & q_{1,d-2} \\
\ell_{2,1} & \ell_{2,2} & q_{2,1} & q_{2,2} & \cdots & q_{2,d-2} \\
\ell_{3,1} & \ell_{3,2} & q_{3,1} & q_{3,2} & \cdots & q_{3,d-2} \\
\hline
0 & 0 & h_{1,1} & h_{1,2} & \cdots & h_{1,d-2} \\
0 & 0 & h_{2,1} & h_{2,2} & \cdots & h_{2,d-2} \\
0 & 0 & h_{3,1} & h_{3,2} & \cdots & h_{3,d-2} \\
\vdots & \vdots & \vdots & \vdots & & \vdots \\
0 & 0 & h_{d-2,1} & h_{d-2,2} & \cdots & h_{d-2,d-2}
\end{array}\right)$$

Now, note that for each $1 \leq j \leq 3$, the generator D_j can be taken to be the determinant of the submatrix of φ obtained by omitting the jth row. This gives that the determinant of the right lower corner of size $(d-2) \times (d-2)$ is a factor of every D_j. But this is absurd as we are assuming $d \geq 5$ and the linear system $(J)_d$ has no fixed part. Therefore, $\dim_k(R_1(J)_d) \geq 8$ as contended.

To complete the proof of the supplementary assertions of this item, we explain the issue about the de Jonquières map, for which we need a later result in the book. Namely, by Section 7.1.14, the base ideal of the latter admits a unique linear syzygy up to scalars, i. e., $s = 1$ in the notation of (a).

Conversely, assume that $s = 1$. We now focus on the $(d-3) \times (d-1)$ submatrix $\mathfrak{L}$ of the matrix φ obtained by omitting the first three rows.

Note that $\mathfrak{L}$ has the form $(\mathbf{0} \mid \mathfrak{L}')$ where $\mathbf{0}$ is the $(d-3) \times 1$ zero matrix and $\mathfrak{L}'$ is a $(d-3) \times (d-2)$ matrix with linear entries. The ideal of maximal minors of the latter matrix has codimension ≥ 2 as it contains I. It follows readily that the following complex:

$$0 \to R \longrightarrow R^{d-2} \xrightarrow{\mathfrak{L}'} R^{d-3}$$

is exact, where the entries of the leftmost map are the (signed, ordered) maximal minors of $\mathfrak{L}'$.

On the other hand, the syzygies $\mathfrak{K}$ of $\mathfrak{L} = (\mathbf{0} \mid \mathfrak{L}')$ are $(1\,0\,\dots\,0)^t$ and the syzygies of $\mathfrak{L}'$. It follows that $\mathfrak{K}$ is a matrix with 2 columns. Therefore, by Proposition 6.1.18, (iv) below, whose proof is independent, the syzygies of the ideal I are generated by 2 syzygies, i. e., I is a perfect ideal.

Now apply Proposition 6.1.14.

(b) Note that $J = ([J]_d, B)$ and $([J]_{d+1}) = (R_1[J]_d, B)$ where B is a k-vector basis of the direct complement of $\mathrm{Im}(R_1 \otimes_k [J]_d \longrightarrow [J]_{d+1})$. By (a), it follows that $([J]_{d+1})$ is generated by $9 + d - 4 = d + 5$ forms of degree $d + 1$. Furthermore, again by (a) any minimal syzygy of J is the transpose of a vector $(q_1\, q_2\, q_3\, \ell_4\, \dots\, \ell_{d-1})$, where each q_i is a 2-form and each ℓ_j is a linear form.

Writing every q_i as an element of $\mathfrak{m} := (x, y, z)$, with linear coefficients, readily yields a linear syzygy of the above generators of $([J]_{d+1})$ (the expression of q_i as an element of $\mathfrak{m}$ is not unique, but it will turn out to be irrelevant which one is picked). This gives a total of $d - 2$ linear syzygies.

One has to add up the reduced Koszul relations of each one of the following pairs $\{xD_i, yD_i\}$, $\{xD_i, zD_i\}$, $\{yD_i, zD_i\}$, for $i = 1, 2, 3$. These contribute 9 additional syzygies, giving a total of $d - 2 + 9 = d + 7$ syzygies. Note that the latter syzygies are just the diagonal sum of three times the matrix of Koszul syzygies of $\mathfrak{m}$, in other words, they generate the syzygies of the R-module $\mathfrak{m} \oplus \mathfrak{m} \oplus \mathfrak{m}$.

On the other hand, one has the obvious free resolution:

$$0 \to R(-3)^3 \xrightarrow{\Psi} R(-2)^9 \longrightarrow R(-1)^9 \longrightarrow \mathfrak{m} \oplus \mathfrak{m} \oplus \mathfrak{m} \to 0. \tag{68}$$

Splicing, we get the following free complex:

$$0 \to R(-(d+3))^3 \xrightarrow{\Psi(-d)} R(-(d+2))^{9+d-2} \xrightarrow{\Phi} R(-(d+1))^{d+5} \longrightarrow ([J]_{d+1}) \to 0.$$

Now use the Buchsbaum–Eisenbud acyclicity criterion: the ranks are readily seen to add up; as to the heights of the required Fitting, one has $\operatorname{ht} I_3(\Psi(-d)) = \operatorname{ht} I_3(\Psi) = 3$, as is obvious from (68), while for Φ one notes its shape as a $(d+5)\times(d+7)$ matrix:

$$\Phi = \left(\begin{array}{ccc|c} \mathfrak{K}(\mathfrak{m}) & & & \\ & \mathfrak{K}(\mathfrak{m}) & & \widetilde{\mathcal{L}} \\ & & \mathfrak{K}(\mathfrak{m}) & \\ \hline & \mathbf{0} & & \mathcal{L} \end{array}\right),$$

where, to the left of the vertical line, $\mathfrak{K}(\mathfrak{m})$ stands for the first Koszul matrix of $\mathfrak{m}$ and the empty slots are all zeros, whereas the matrix to the right of the vertical line is the $(d-2)\times(d+5)$ matrix of linear syzygies as obtained above from the original syzygies of J, the lower matrix $\mathcal{L}$ coinciding with the one described in (a) and the upper $\widetilde{\mathcal{L}}$ being linear.

Note that the diagonal sum of $\mathfrak{K}(\mathfrak{m})$ has only rank 6, so it can only contribute this much to get a nonzero minor of size $(d+4)\times(d+4)$; forcefully, then one has to use a full maximal minor of the right side matrix. It is clear that this patching allows for the construction of at least two $(d+4)$-minors with no proper common factor. This shows that $\operatorname{ht} I_{d+4}(\Phi) \geq 2$, as required for the acyclicity criterion. □

Remark 6.1.17. (i) Note that, quite trivially, the linear system $[\mathfrak{F}_{\mathcal{K}}]_{d+1}$ in (b) defines a birational mapping of $\mathbb{P}^2$ onto the image. Indeed, more generally, for $R = k[x_1,\ldots,x_n]$ and a given integer $d \geq 0$, the rational map defined by $(x_1 f : \cdots : x_n f)$, with $\deg(f) = d$, is obviously birational onto the image (a "fat" version of the identity). Alternatively, the field extension $k(fR_1) \subset k(x_1^{d+1}, x_2/x_1,\ldots,x_n/x_1) = k(R_{d+1})$ is an equality because $x_i/x_1 = x_i f/x_1 f \in k(x_1 f,\ldots,x_n f)$, while $f^{d+1} = f(x_1 f,\ldots,x_n f) \in k(x_1 f,\ldots,x_n f)$, triggering $x_1^{d+1} \in k(x_1 f,\ldots,x_n f)$ as well via the equality $x_1^{d+1} = (x_1 f)^{d+1}/f^{d+1}$.

In particular, if $J \subset R = k[x_1,\ldots,x_n]$ is a homogeneous ideal and $d \geq \operatorname{indeg}(J)$, then $k([J]_{d+1}) = k(x_1^{d+1}, x_2/x_1,\ldots,x_n/x_1)$, which proves the contention.

(ii) The birational content of [77, Theorem 2.1(ii)] falls off the above approach because, as one observes in the notation there, $\sigma = d+1$ as follows from Lemma 6.1.13. Moreover, the condition on how a straight line intersects the fat scheme translates here, via Bézout theorem, into the inequality $\mu_1 + \mu_2 \leq d$, a condition that is always satisfied by a proper homaloidal type as is the present case.

By that result the image of the birational map is a smooth surface and, moreover, the surface is arithmetically Cohen–Macaulay [77, Proposition 2.4] (hence also projectively normal). Since the present situation is more special it would be of some interest to have a direct algebraic proof of the Cohen–Macaulayness assertion. For certain proper homaloidal types, this will be accomplished in Section 6.2.2.

(iii) Although a free resolution of the ideal $I_{\mathcal{K}}$ of virtual forms is not known in general, it is available for points in certain special position. This idea is due to Harbourne and an example is [136, Theorem A].

6.1.3 From fat ideals back to the Cremona base ideal

The next result explains how information about the $\boldsymbol{\mu}$-fat ideal $\mathfrak{F}_{\mathcal{K}}$ travels back to the base ideal I.

Proposition 6.1.18. *Keep the notation and hypotheses of* Theorem 6.1.16 *and let $\mathfrak{L}$ denote the $(d-4+s)\times(d-2+s)$ submatrix of φ omitting the first three rows. Then:*

(a) *The R-module $\mathfrak{F}_{\mathcal{K}}/I$ is linearly presented by $\mathfrak{L}$ if $s=0$ (resp., by its submatrix omitting the zero column if $s=1$).*

(b) $\mathfrak{F}_{\mathcal{K}} = I^{\mathrm{sat}} \Rightarrow \dim(\mathfrak{F}_{\mathcal{K}}/I) = 0 \Rightarrow \mu_i \leq 2$, *for all* $i \Rightarrow (d;\boldsymbol{\mu}) = (4;2^3,1^3)$ *or* $(d,\boldsymbol{\mu}) = (5;2^6) \Rightarrow \mathfrak{F}_{\mathcal{K}} = I^{\mathrm{sat}}$; *then all these conditions are mutually equivalent.*

(c) *If $\dim \mathfrak{F}_{\mathcal{K}}/I = 1$, then R/I is Cohen–Macaulay if and only if $\mathfrak{F}_{\mathcal{K}}/I$ is Cohen–Macaulay.*

(d) *Let $\mathfrak{K}$ denote the null matrix of $\mathfrak{L}$ – i. e., the matrix, in the canonical bases, representing the kernel of the map defined by $\mathfrak{L}$. Then the submatrix of $\varphi\cdot\mathfrak{K}$ consisting of its first three rows is a syzygy matrix of I.*

Proof. (a) By Theorem 6.1.16, we know that $s = 1$ means that one column of φ has linear entries in the first three rows and zeros for the remaining rows. Therefore, $\mathfrak{L}$ has the stated decomposition. Since the generators of I are minimal generators of $\mathfrak{F}_{\mathcal{K}}$, the result is clear.

(b) Clearly, $\dim(I^{\mathrm{sat}}/I) = 0$, so the first implication of the stated string is obvious.

To proceed, by (i) one has $\mathrm{ht}(I:\mathfrak{F}_{\mathcal{K}}) = \mathrm{ht}\, I_{d-4+s}(\mathfrak{L})$ since $\mathfrak{F}_{\mathcal{K}}/I$ and its zeroth Fitting ideal have the same radical. Therefore, $\dim(\mathfrak{F}_{\mathcal{K}}/I) = \dim(R/I:\mathfrak{F}_{\mathcal{K}}) = 3 - \mathrm{ht}\, I_{d-4+s}(\mathfrak{L})$, so we are to show that $\mathrm{ht}\, I_{d-4+s}(\mathfrak{L}) = 3 \Rightarrow \mu_i \leq 2$, $\forall i$. Assuming the contrary, one has $\mu_1 \geq 3$. Then $d-1+s-\mu_1 \leq d-1+s-3 = d-4+s$, hence for even more reason $\mathrm{ht}\, I_{d-1+s-\mu_1}(\mathfrak{L}) \geq 3$. But letting P_1 denote the prime ideal for the first multiplicity, we see that $I_{d-1+s-\mu_1}(\mathfrak{L}) \subset P_1$ since otherwise we could invert a minor in the localization R_{P_1}, and thus have $(\mathfrak{F}_{\mathcal{K}})_{P_1} = P_{1P_1}^{\mu_1}$ generated by $d-1+s-(d-1+s-\mu_1) = \mu_1$ elements—this is absurd as $P_{1P_1}^{\mu_1}$ is minimally generated by μ_1+1 elements.

We move on to the next implication, namely that $\mu_i \leq 2$, $\forall i$ can only happen for those two particular homaloidal types. This is an easy calculation using the equations of condition. Thus, letting $(d;2^{r_2},1^{r_1})$ be the assumed type, one has

$$\begin{cases} r_1 + 2r_2 = 3d-3 \\ r_1 + 4r_2 = d^2 - 1 \end{cases}$$

Manipulation of these equations yields $r_1 = d(6-d)-5$, from which (since $r_1 \geq 0$) the only possibilities are $d=4$, $r_1 = 3$ or else, $d=5$, $r_1 = 0$; correspondingly, $r_2 = 3$ or else $r_2 = 6$.

The last implication is well known—for $d=4$ the base ideal I is always saturated (cf. [89, Theorem 1.5(i)]), while $(5;2^6)$ is a classic (see, e. g., [89, Theorem 2.14(iii)] Supplement] and the preliminaries of the proof, where it is argued that $(x,y,z)f \in I$, where f is the unique minimal generator of degree 6 of $\mathfrak{F}_{\mathcal{K}}$).

(c) This follows readily chasing depth along the following structural exact sequences:

$$0 \to I \longrightarrow R \longrightarrow R/I \to 0 \quad \text{and} \quad 0 \to I \longrightarrow \mathfrak{F}_{\mathcal{K}} \longrightarrow \mathfrak{F}_{\mathcal{K}}/I \to 0,$$

using that $\dim(\mathfrak{F}_{\mathcal{K}}/I) = 1$ by hypothesis and that $\operatorname{depth}(\mathfrak{F}_{\mathcal{K}}) = 2$.

(d) Note that any syzygy of $\{D_1, D_2, D_3\}$ stacked on top of the $(d-4+s) \times 1$ zero matrix is a syzygy of $\mathfrak{F}_{\mathcal{K}}/I$. Since $I = (D_1, D_2, D_3)$, one can write

$$\left(\frac{\psi}{\mathbf{0}}\right) = \varphi \cdot \mathfrak{C}$$

for a suitable matrix $\mathfrak{C}$ ("content" matrix), where the matrix on the left-hand side is the matrix ψ of syzygies of I stacked on top of a zero matrix with $(d-4+s)$ rows and same number of columns as ψ—note that in the case where $s = 1$ this content matrix will involve the unit column $(1\,0\,\ldots\,0)^t$. Clearly, $\operatorname{Im}(\mathfrak{C}) \subset \mathfrak{K} := \ker(\mathfrak{L})$; in other words, every column of $\mathfrak{C}$ is a syzygy of $\mathfrak{L}$.

We claim the reverse inclusion as well. To see this, let ρ denote the submatrix of φ with the first three rows. Then we wish to show that

$$(D_1\, D_2\, D_3) \cdot \rho \cdot \mathfrak{K} = 0.$$

Recall that D_j is the minor of φ omitting row j, for $j = 1, 2, 3$. Thus, if $\tilde{\varphi}$ denotes the submatrix of φ omitting its last row and $\mathfrak{Q}$ denotes the matrix of cofactors of $\tilde{\varphi}$, one has $(D_1\, D_2\, D_3) = L \cdot \mathfrak{Q}$, where L denotes the last row of $\mathfrak{L}$.

Then one has

$$(D_1\, D_2\, D_3) \cdot \rho \cdot \mathfrak{K} = L \cdot \mathfrak{Q} \cdot \rho \cdot \mathfrak{K} = \det(\tilde{\varphi}) \cdot L \cdot \mathfrak{K} = 0$$

as was to be shown. □

6.1.4 Proper homaloidal types with large highest virtual multiplicity

Some classical inequalities for Cremona map stem from the consideration of the three highest virtual multiplicities of the corresponding homaloidal type. Not much however has appeared in the more recent literature stressing the role of the behavior of the map in the neighborhood of a single point with highest virtual multiplicity. The exception, as always, is the de Jonquières transformations.

If d denotes the degree and μ denotes the highest virtual multiplicity we mind about the interval $d-1 \geq \mu \geq \lceil \frac{d}{2} \rceil$, with a big question mark about the subinterval

$d-5 \geq \mu \geq \lceil \frac{d}{2} \rceil$. Here, we discuss the cases $\mu \geq d-4$. Applying an arithmetic quadratic transformation (Definition 6.1.3) to a homaloidal type belonging to this imprecise family will often yield a homaloidal type of the same shape (informally, a "telescopic" type), while in a few other instances will afford a homaloidal type belonging to the inverse of the former. Thereby, using repeatedly (or inductively) the Hudson test will show that these homaloidal types are proper.

The case where $\mu = d-1$ corresponds to a de Jonquières type, which is sufficiently discussed elsewhere in the book. The case $\mu = d-2$ is sufficiently elementary and is left to the reader (Exercise 6.3.1).

6.1.4.1 Highest virtual multiplicity *d* – 3

The typical proper homaloidal type that will emerge has as remaining virtual multiplicities 3, 2, 1, distinguished according to the values of d modulo 3.

Proposition 6.1.19. *Let T denote a proper homaloidal type of degree $d \geq 6$ and highest virtual multiplicity $d-3$. Then T is one of the following, where $t \geq 2$:*

(0) $d \equiv 0 \pmod 3$
Proper homaloidal types ($d = 3t$):
(0.1)$(d; d-3, 3^{2t-2}, 2, 1^4)$
(0.2)$(d; d-3, 3^{2t-3}, 2^4, 1)$

(1) $d \equiv 1 \pmod 3$
Proper homaloidal types ($d = 3t+1$):
(1.1)$(d; d-3, 3^{2t-1}, 2^0, 1^5)$
(1.2)$(d; d-3, 3^{2t-2}, 2^3, 1^2)$

(2) $d \equiv 2 \pmod 3$
Proper homaloidal types ($d = 3t+2$):
(2.1)$(d; d-3, 3^{2t-1}, 2^2, 1^3)$
(2.2)$(d; d-3, 3^{2t-2}, 2^5, 1^0)$

Conversely, any of the above homaloidal types is proper.

Proof. We first argue that any proper type T with highest virtual multiplicity $d-3$ is necessarily of the form $T = (d; d-3, 3^{r_3}, 2^{r_2}, 1^{r_1})$. Since T is proper and $\mu_2 \leq d-3$, then applying an arithmetic quadratic transformation based on the highest virtual multiplicities yields the virtual multiplicity $d-(d-3)-\mu_2 = 3-\mu_2$, which is nonnegative only if $\mu_2 \leq 3$. Thus, for a proper type necessarily $\mu_i \leq 3$ for every $i \geq 2$. Therefore, we can set $T = (d; d-3, 3^{r_3}, 2^{r_2}, 1^{r_1})$.

The equations of condition yield

$$\begin{cases} r_1 + 2r_2 + 3r_3 + d - 3 &= 3d - 3 \\ r_1 + 4r_2 + 9r_3 + (d-3)^2 &= d^2 - 1 \end{cases}$$

An easy manipulation of these two relations yields the following ones:

$$\begin{cases} r_1 + r_2 &= 5 \\ r_2 + 3r_3 &= 2d - 5 \end{cases} \tag{69}$$

The solutions are straightforward by taking a nonnegative integer partition of 5 and plugging into the second of these relations, thus producing a congruence modulo 3, whose solutions are pretty immediate.

To see that all these homaloidal types are proper we apply the Hudson algorithm in terms of arithmetic quadratic transformations based on the three highest multiplicities (cf. [4, Corollary 5.2.21]). As above, in the case of highest multiplicity $d-1$ or $d-2$, the result of applying such an arithmetic transformation is telescopic. Indeed, for $t \geq 2$ the arithmetic transformation is exactly the same in all cases, as the three highest multiplicities are $d-3, 3, 3$. Moreover, since one goes down by 3 in the degree every time an arithmetic quadratic transformation is applied, then the degree of the resulting type does not leave its class modulo 3. Explicitly, one obtains:

Case (0.1) $(d-3; d-6, 3^{2(t-2)}, 2, 1^4)$
Case (0.2) $(d-3; d-6, 3^{2t-5}, 2^4, 1)$ $(t \geq 3)$
Case (1.1) $(d-3; d-6, 3^{2t-3}, 2^0, 1^5)$
Case (1.2) $(d-3; d-6, 3^{2(t-2)}, 2^3, 1^2)$
Case (2.1) $(d-3; d-6, 3^{2t-3}, 2^2, 1^3)$
Case (2.2) $(d-3; d-6, 3^{2(t-2)}, 2^5, 1^0)$

To start the induction, one needs the bottom type in each case, along with one additional arithmetic quadratic transformations $\mathcal{Q}$:

Case (0.1) $(6; 3 = 6-3, 3^2, 2, 1^4) = (6; 3^3, 2, 1^4) \overset{\mathcal{Q}}{\rightsquigarrow} (3; 2, 1^4)$: de Jonquières
Case (0.2) $(6; 3 = 6-3, 3, 2^4, 1) = (6; 3^2, 2^4, 1) \overset{\mathcal{Q}}{\rightsquigarrow} (4; 2^3, 1^3)$: the non de Jonquières
Case (1.1) $(7; 4 = 7-3, 3^3, 2^0, 1^5) = (7; 4, 3^3, 1^5) \overset{\mathcal{Q}}{\rightsquigarrow} (4; 3, 1^6)$: de Jonquières
Case (1.2) $(7; 4 = 7-3, 3^2, 2^3, 1^2) = (7; 4, 3^2, 2^3, 1^2) \overset{\mathcal{Q}}{\rightsquigarrow} (4; 2^3, 1^3)$: the non de Jonquières
Case (2.1) $(8; 5, 3^3, 2^2, 1^3) \overset{\mathcal{Q}}{\rightsquigarrow} (5; 3, 2^3, 1^3)$: the saturated non de Jonquières
Case (2.2) $(8; 5, 3^2, 2^5) \overset{\mathcal{Q}}{\rightsquigarrow} (5; 2^6)$: the nonsaturated.

This shows the properness of all types. □

Remark 6.1.20. Note that the virtual multiplicity 2 is present in all cases above with the exception of (1.1). Thus, in all these cases one has $\mu_1 + 2 = d - 3 + 2 = d - 1$. By [15, Proposition 5.2] this means that when the highest multiplicity is $d-3$, except for case (1.1), the Cremona map belongs to the closure of the set of Cremona maps of degree $d+1$.

6.1.4.2 Highest virtual multiplicity $d - 4$

We first note that any proper homaloidal type T with highest virtual multiplicity $d-4$ is necessarily of the form

$$T = (d; d-4, 4^{r_4}, 3^{r_3}, 2^{r_2}, 1^{r_1}).$$

Indeed, since T is proper and $\mu_2 \le d-4$, then applying an arithmetic quadratic transformation based on the highest virtual multiplicities yields the virtual multiplicity $d-(d-4)-\mu_2 = 4-\mu_2$, which is nonnegative only if $\mu_2 \le 4$. Thus, for a proper type necessarily $\mu_i \le 4$ for every $i \ge 2$.

We will accordingly assume this form at the outset in the next proposition. In addition, the bulk of this case will have $d-4 \ge 4$, i. e., $d \ge 8$, though we harmlessly include $d = 7$ as a sort of degenerate case.

Proposition 6.1.21. *Let $T = (d; d-4, 4^{r_4}, 3^{r_3}, 2^{r_2}, 1^{r_1})$ denote a proper homaloidal type of degree d and highest virtual multiplicity $d-4 \ge 3$. One has:*
(i) *r_3 and r_4 cannot vanish simultaneously.*
(ii) *r_1 and r_2 cannot vanish simultaneously.*

Moreover:
(a) *If $r_4 = 0$ then T is one of the following types:*

$$(7; 3 = 7-4, 3^3, 2^3) = (7; 3^4, 2^3), \ (8; 4 = 8-4, 3^5, 1^2), \ (10; 6 = 10-4, 3^7, 1^0),$$

where the second is obtained from the last applying an arithmetic transformation on the three highest multiplicities.
(b) *If $r_3 = 0$, then $r_4 = \lfloor (d-3)/2 \rfloor \ge 2$ and*

$$T = \begin{cases} (d; d-4, 4^{\lfloor (d-3)/2 \rfloor}, 2^3, 1^3) & \text{if } d \text{ is even} \\ (d; d-4, 4^{(d-3)/2}, 1^7) & \text{if } d \text{ is odd and} \not\equiv 3 \bmod 4 \end{cases}$$

(c) *Suppose that neither r_3 nor r_4 vanishes.*
(c.1) *$r_1 = 0$, $r_2 \ne 0$: in this case, $T = (d; d-4, 4^{(d-7)/2}, 3^3, 2^3)$, where $d \ge 9$ is odd.*
(c.2) *$r_2 = 0$, $r_1 \ne 0$: in this case, $T = (d; d-4, 4^{(d-(r_3+3))/2}, 3^{r_3}, 1^{7-r_3})$, where d is even and $r_3 = 1, 3, 5$.*

Conversely, any of the above types is proper.

Proof. Consider the equations of condition for this type:

$$\begin{cases} r_1 + 2r_2 + 3r_3 + 4r_4 = 3d - 3 - (d-4) = 2d + 1 \\ r_1 + 4r_2 + 9r_3 + 16r_4 = d^2 - 1 - (d-4)^2 = 8d - 17 \end{cases}$$

An easy manipulation of these relations allows us to express r_1 and r_2 in terms of r_3 and r_4 and d:

$$\begin{cases} r_1 = 3r_3 + 8r_4 - 4d + 19 \\ r_2 = 3(d-3-(r_3+2r_4)) \end{cases} \tag{70}$$

Since r_1 and r_2 are nonnegative, we have the following inequalities:

$$3r_3 + 8r_4 \geq 4d - 19, \quad r_3 + 2r_4 \leq d - 3. \tag{71}$$

(i) This one is clear: $r_3 = r_4 = 0$ imply $r_1 = 19 - 4d < 0$ for $d \geq 5$.

(ii) Plugging $r_1 = r_2 = 0$ into their expressions above yields two linear equations in r_3, r_4 in terms of d which are readily solved, giving $2r_4 = d-10$, $r_3 = d-3-(d-10) = 7$ Then d must be even and $d \geq 12$ since $r_4 > 0$ for $r_1 = r_2 = 0$ as will be proven in the next item.

Note the telescopic behavior:

$$(d; d-4, 4^{\frac{d-10}{2}}, 3^7) \overset{Q}{\rightsquigarrow} (d-4; d-8, 4^{\frac{(d-4)-10}{2}}, 3^7),$$

while for $d = 12$ one has

$$(12; 8, 4, 3^7) \overset{Q}{\rightsquigarrow} (9; 5, 3^6, 1) \overset{Q}{\rightsquigarrow} (7; 3^5, 1^2) \overset{Q}{\rightsquigarrow} (5; 3^2.1^6) \overset{Q}{\rightsquigarrow} (3; -1, \ldots)$$

Therefore, this homaloidal type is not proper.

(a) If $r_4 = 0$, then $(4d-19)/3 \leq r_3 \leq d - 3$. Necessarily, $d - 3 \geq (4d-19)/3$ which means that $d \leq 10$. Now, for $8 \leq d \leq 10$ we then see that $r_3 = d - 3$. Substituting for $r_3 = d-3$, $r_4 = 0$ in the formulas of r_1, r_2 yields $r_1 = 10 - d$, $r_2 = 0$.

The case where $d = 10$ reduces to the case where $d = 8$ by applying an arithmetic quadratic transformation, as one readily sees. Therefore, it suffices to check the latter. Applying again such a transformation yields the case (0.1) of Proposition 6.1.19, with $d = 6$.

Now, as for $d = 7$, the interval for r_3 allows for two possibilities: $r_3 = 3$ or $r_3 = 4$. In the first case, one gets the homaloidal type $(7; 3 = 7-4, 3^3, 2^3, 1^0) = (7; 3^4, 2^3)$. Applying two successive arithmetic quadratic transformations yields the type $(3; 2, 1^4)$ – a de Jonquières type. As to the case $r_3 = 4$, we get the type $(7; 3 = 7-4, 3^4, 2^0, 1^3) = (7; 3^4, 1^3)$, then $(5; 3^2, 1^6)$; the latter is not a proper type since for $d = 5$ there are only three proper homaloidal types (de Jonquières, symmetric $(5; 2^6)$ and $(5; 3, 2^3, 1^3)$).

Finally, the case $d = 9$ reduces to the above nonproper type in degree 7 by an arithmetic quadratic transformation, hence is not proper either.

(b) With $r_3 = 0$, the inequalities (71) imply

$$\frac{4d-19}{8} \leq r_4 \leq \frac{d-3}{2}.$$

But the positive difference between these two bounds is $19/8 - 3/2 = 7/8 < 1$. This yields the stated value of r_4 as the floor of half $d - 3$.

On the other hand, the value of r_2 is now given by $r_2 = 3(d - 3 - 2r_4)$, from which follows that $r_2 = 3$ or $r_2 = 0$ according to whether d is even or odd, respectively.

Finally, plugging each of these values into the expression $r_1 = 8r_4 - 4d + 19 = \lfloor (d-3)/2 \rfloor - 4d + 19$ readily gives the two respective values of r_1, as stated.

To prove that both types are proper, one notes that since $d \geq 7$, then applying an arithmetic quadratic transformation to each of them yields a homaloidal type of the same shape by substituting $d - 4$ for d. Explicitly, since $r_4 \geq 2$ then $d \geq 8$ and further the three highest virtual multiplicities are $d - 4, 4, 4$. Thus, we get $(d - 4; d - 8, 4^{\lfloor (d-4)-3)/2 \rfloor}, \ldots)$.

By recursion, according to the Hudson test, the two homaloidal types will be proper if and only if the resulting types of this operation for $7 \leq d \leq 11$ are proper (this is because we jump 4 at the time).

One has

$(11; 7.4^4, 1^7) \overset{Q}{\rightsquigarrow} (7; 4^2, 3, 1^7) \overset{Q}{\rightsquigarrow} (3; -1, \ldots)$: nonproper

$(10; 6, 4^3, 2^3, 1^3) \overset{Q}{\rightsquigarrow} (6; 4, 2^4, 1^3)$: proper by Exercise 6.3.1

$(9; 5, 4^3, 1^7) \overset{Q}{\rightsquigarrow} (5; 4, 1^8)$: de Jonquières

$(8; 4^3, 2^3, 1^3) \overset{Q}{\rightsquigarrow} (4; 2^3, 1^3)$: the saturated non-de Jonquières.

This shows the statement of this item.

(c.1) Let $r_1 = 0$. Then the first of the basic relations in (70) yields $3r_3 + 6r_4 = 4d - 19 - 2r_4$; upon substitution in the second relation, one obtains $r_2 = 2(r_4 + 5) - d$. Since $r_2 \neq 0$, $d < 2(r_4 + 5)$. Back to the first relation gives $3r_3 + 8r_4 < 8(r_4 + 5) - 19 = 8r_4 + 21$, from which we deduce that $r_3 < 7$. But again the first of the relations in (70) forces 4 to divide $19 + 3r_3$. The only possibility in the interval $1 \leq r_3 \leq 6$ is $r_3 = 3$. This in turn implies that $r_4 = (d-7)/2$, so in particular d is odd. Also then $r_2 = 2((d-7)/2+5)-d = 3$.

In conclusion, $T = (d; d-4, 4^{(d-7)/2}, 3^3, 2^3)$. To show it is proper, apply an arithmetic quadratic transformation to the three highest multiplicities; assuming $d \geq 11$ yields the same sort of type. At the bottom, one gets $(11; 7; 4^2, 3^3.2^3)$, which in turns eventually transforms to $(5; 3, 2^3, 1^3)$, a well-known proper homaloidal type.

(c.2) Let $r_2 = 0$. From the second relation, $r_3 + 2r_4 = d - 3$ upon substitution on the first relation yields $r - 1 = 10 + 2r_4 - d$, and since $r_1 \neq 0$, $d < 10 + 2r_4$. Back to the second relation gives once more $r_3 < 7$. Also $r_4 = (d - (r_3 + 3))/2$, so r_3 and d must have opposite parities. Further, $r_1 = 7 - r_3$.

This gives the claimed form of T. Changing by an arithmetic quadratic transformation is again telescopic, for $d \geq r_3 + 7$. At the bottom, we find $T = (r_3 + 7; r_3 + 3, 4^2, 3^{r_3}, 1^{7-r_3})$. Letting r_3 run through 1 to 6 we readily test the obtained types for properness applying arithmetic quadratic transformations, thus finding that they are proper for r_3 odd and improper for r_3 even. □

Remark 6.1.22. By [15, Corollary 5.2], except for the two cases in item (b), all such Cremona maps lie on the closure of Cremona maps of degree $d+1$.

Of course, this is pretty general: a proper homaloidal type whose highest virtual multiplicity is, say, $\mu_1 = d - \ell$ divides into two classes of topological nature: the ones such that there exists a multiplicity $\mu_i = \ell - 1$ corresponding to Cremona maps of degree d lying on the closure of the set of Cremona maps of degree $d+1$, while the rest is the union of the two classes where either $\mu_2 = \ell$ and $\mu_i \leq \ell - 2$ for every $i \geq 3$ or else $\mu_i \leq \ell - 2$ for every $i \geq 2$, with neither case corresponding to Cremona maps of degree d lying on the closure of the set of Cremona maps of degree $d+1$.

Conjecture 6.1.23. *The base ideal of a Cremona map with proper points and one of the homaloidal types in Proposition* 6.1.19 *and Proposition* 6.1.21 *is perfect.*

The argument used in the cases where $\mu_1 \geq d-1$ does not readily extend all the way over.

6.2 Plane Cremona transformations: subhomaloidal types

6.2.1 Arithmetic quadratic transformations acting on fat points

Given an abstract type $\mathbf{v} := (d;\boldsymbol{\mu}) \in \mathbb{N} \times \mathbb{N}^r = \mathbb{N}^{r+1}$, where $d > 0$ and $\boldsymbol{\mu} = (\mu_1, \ldots \mu_r)$ (see Subsection 6.1.1.2), and a set $\mathbf{p} \subset \mathbb{P}^2$ of r points, then the arithmetic quadratic transformation $\mathcal{Q} := \mathcal{Q}_{jkl} : \mathbb{N}^{n+1} \to \mathbb{N}^{n+1}$ based on a triple j, k, l of indices in $\{1, \ldots, r\}$ transforms the type $\mathbf{v}$ into a type $(\tilde{d};\tilde{\boldsymbol{\mu}})$ as explained in Subsection 6.1.1.2.

Now, let $\mathfrak{Q} := \mathfrak{Q}_{jkl}$ stand for the plane quadratic Cremona map based on the three points p_i, p_j, p_k. Set

$$\tilde{p}_i = \begin{cases} p_i, & \text{if } \imath \in \{j,k,l\} \\ \mathfrak{Q}(p_i), & \text{if } i \notin \{j,k,l\} \end{cases}$$

Recall that, if j, k, l is a triple of major points then the corresponding points are not aligned.

Definition 6.2.1. The *transform* of the fat ideal $I = I(\mathbf{p},\boldsymbol{\mu}) \subset R = k[x,y,z]$ by $(\mathcal{Q},\mathfrak{Q})$ is the fat ideal $\tilde{I} := I(\tilde{\mathbf{p}},\tilde{\boldsymbol{\mu}}) \subset R = k[x,y,z]$.

The above transform commutes with taking rational maps, in a way to be made precise. Since we could not find an exact reference, we give a proof.

Recall that if the three base points are the coordinate points then $\mathfrak{Q} = (yz : xz : yz)$, the standard quadratic map. Also, for a major triple j, k, l, up to a change of variables, the corresponding points are the three coordinate points.

Lemma 6.2.2. *Assume that $\mu_1 \geq \cdots \geq \mu_r$ and $j = 1, k = 2, l = 3$. As above, p_1, p_2, p_3 are the coordinate points and $I = I(\mathbf{p}, \boldsymbol{\mu}) \subset R = k[x, y, z]$. Set $\Delta := x^{\mu_1}y^{\mu_2}z^{\mu_3}$. Then, for any given $d > 0$ one has:*

(a) *If $f(x, y, z) \in I_d$, then $f(yz, xz, yz)/\Delta \in \widetilde{I}_{\widetilde{d}}$.*

(b) *For any integer $n \geq 1$, the map $\eta^{(n)} : f(x, y, z) \mapsto f(yz, xz, yz)/\Delta^n$ induces an isomorphism of the k-vector spaces $(I^n)_{nd}$ and $(\widetilde{I}^n)_{n\widetilde{d}}$.*

Proof. (a) Since $f \in (y, z)^{\mu_1}$, then $f(yz, xz, xy) \in (x^{\mu_1})$. Similarly, $f(yz, xz, xy) \in (y^{\mu_2})$ and $f(yz, xz, xy) \in (z^{\mu_3})$. Thus, Δ is a factor of $f(yz, xz, xy)$.

Next, we argue that the multiplicity of $f(yz, xz, xy)/\Delta$ at each q_i is at least $\widetilde{\mu}_i$. Since f is homogeneous of degree d, $f(yx, xz, yz) \in (y, z)^d \cap (x, z)^d \cap (x, y)^d$. Thus,

$$e_{p_1}\left(\frac{f(yz, xz, xy)}{\Delta}\right) = e_{p_1}(f(yz, xz, xy)) - e_{p_1}(\Delta) \geq d - (\mu_2 + \mu_3) = \widetilde{\mu}_1,$$

$$e_{p_2}\left(\frac{f(yz, xz, xy)}{\Delta}\right) = e_{p_2}(f(yz, xz, xy)) - e_{p_2}(\Delta) \geq d - (\mu_1 + \mu_3) = \widetilde{\mu}_2$$

and

$$e_{p_3}\left(\frac{f(yz, xz, xy)}{\Delta}\right) = e_{p_3}(f(yz, xz, xy)) - e_{p_3}(\Delta) \geq d - (\mu_1 + \mu_2) = \widetilde{\mu}_3$$

Next, let $p = (\alpha_1 : \alpha_2 : \alpha_3)$ denote any of the remaining points p_i, with $4 \leq i \leq r$, and let μ denote the respective virtual multiplicity. Since the defining ideal of p is $(\alpha_3 x - \alpha_1 z, \alpha_3 y - \alpha_2 z)$, then that of $\Omega(p)$ is $(\alpha_1 x - \alpha_3 z, \alpha_2 y - \alpha_3 z)$. Then one must have

$$f = \sum_{i=1}^{\mu} a_i(x, y, z)(\alpha_3 x - \alpha_1 z)^i(\alpha_3 y - \alpha_2 z)^{\mu - i}$$

for certain $a_i \in k[x, y, z]$. It follows that

$$f(yz, xz, xy) = \sum_{i=1}^{\mu} a(yz, xz, xy) y^i x^{\mu - i}(\alpha_3 z - \alpha_1 x)^i(\alpha_3 z - \alpha_2 y)^{\mu - i}.$$

Since $(\alpha_3 z - \alpha_1 x)^i(\alpha_3 z - \alpha_2 y)^{\mu - i}$ is up to a coefficient the typical term in the binomial expansion of the power $(\alpha_1 x - \alpha_3 z, \alpha_2 y - \alpha_3 z)^{\mu}$, then $f(yz, xz, xy) \in (\alpha_1 x - \alpha_3 z, \alpha_2 y - \alpha_3 z)^{\mu}$, which implies that $e_p(f(yz, xz, xy)) \geq \mu$. Therefore,

$$e_p\left(\frac{f(yz, xz, xy)}{\Delta}\right) = e_p(f(yz, xz, xy)) - e_p(\Delta) \geq \mu$$

as claimed.

(b) Set $\widetilde{\Delta} := x^{\widetilde{\mu}_1}y^{\widetilde{\mu}_2}z^{\widetilde{\mu}_3}$ and consider the analogous map

$$\zeta^{(n)} : g(x, y, z) \mapsto g(yz, xz, xy)/\widetilde{\Delta}$$

coming the other way. By the same token, it maps $J_{\tilde{d}}$ to J_d. We claim that the two maps are inverse of each other.

First, consider the case $n = 1$, setting $\eta^{(1)} = \eta$ and $\zeta^{(1)} = \zeta$ for lighter reading. One has

$$\begin{aligned}\zeta \circ \eta(f(x,y,z)) = \zeta(f(yz,xz,xy)/\Delta) &= \frac{f((xyz)x,(xyz)y,(xyz)z)}{\Delta(yz,xz,xy)\tilde{\Delta}} \\ &= \frac{x^d y^d z^d f(x,y,z)}{(x^{d-\tilde{\mu}_1}y^{d-\tilde{\mu}_2}z^{d-\tilde{\mu}_3})(x^{\tilde{\mu}_1}y^{\tilde{\mu}_2}z^{\tilde{\mu}_3})} \\ &= f(x,y,z).\end{aligned}$$

The equality in the other composite direction is argued similarly.

Now, the arbitrary case $n \geq 1$. Let $\{f_1,\ldots,f_m\}$ denote a vector basis of J_d. Since any element of $(J^n)_{nd}$ is a k-linear combination of the forms $\{f_1^{\alpha_1}\cdots f_m^{\alpha_m} \mid \alpha_1+\cdots+\alpha_m = n\}$, it suffices to deal with one such form. Clearly,

$$\eta^{(n)}(f_1^{\alpha_1}\cdots f_m^{\alpha_m}) = \eta(f_1)^{\alpha_1}\cdots\eta(f_m)^{\alpha_m}$$

A similar rule works for $\zeta^{(n)}$ as well. Therefore,

$$\zeta^{(n)}(\eta^{(n)}(f_1^{\alpha_1}\cdots f_m^{\alpha_m})) = \zeta(\eta(f_1))^{\alpha_1}\cdots\zeta(\eta(f_m))^{\alpha_m} = (f_1)^{\alpha_1}\cdots(f_m)^{\alpha_m}$$

from the case $n = 1$.

The required equality in the other direction works the same way. □

Remark 6.2.3. Let $\mathbf{v} = (d;\boldsymbol{\mu})$ be an abstract type, with $\boldsymbol{\mu} = \{\mu_1,\ldots,\mu_r\}$ and $d \geq 1$. It is known that there is an integer $e(d;\boldsymbol{\mu})$ such that $\dim_k I(\mathbf{p},\boldsymbol{\mu})_d \geq e(d;\boldsymbol{\mu})$ for every set of points $\mathbf{p} = \{p_1,\ldots,p_r\} \subset \mathbb{P}^2$ and such that the set of points $\mathbf{p}$ for which the equality holds is parameterized by a dense Zariski open subset of $(\mathbb{P}^2)^r$ (see, e. g., [141]). The number $e(d;\boldsymbol{\mu})$ is the *generic dimension* of the linear system of curves of degree d having multiplicities $\boldsymbol{\mu}$ at given r points. It is bounded below by the so-called *expected dimension* $\max\{0, \binom{d+2}{2} - \sum_{i=1}^{r}\binom{\mu_i+1}{2}\}$. A fundamental question of linear systems is as to when the two dimensions coincide. A basic tool to address this question consists in transforming the given type $\mathbf{v}$ by successive quadratic transformations into a type where the relation between the two dimensions is easier to verify.

The transform of a fat ideal will be thoroughly discussed in Subsection 6.2.3.

6.2.2 Subhomaloidal multiplicity sets

In this part, we will focus on a set of virtual multiplicities $\boldsymbol{\mu}$ satisfying certain relations which are naturally related to the equations of condition of a Cremona map. Doubling these multiplicities will in fact satisfy the equations of condition. This is a purely arithmetic procedure, but it impacts the corresponding ideals of fat points.

Definition 6.2.4. Let $\boldsymbol{\mu} = \{\mu_1, \ldots, \mu_r\}$ be a set of nonnegative integers satisfying the following condition: there exists an integer $s \geq 2$ such that

$$\sum_{i=1}^{r} \mu_i = 3(s-1) \quad \text{and} \quad \sum_{i=1}^{r} \mu_i^2 = s(s-1). \tag{72}$$

Note that the first of the above relations implies that s is uniquely determined. We will say that $\boldsymbol{\mu}$ is a *subhomaloidal multiplicity set in degree s*.

We will say that a set of nonnegative integers $\mathbf{v} = \{v_1, \ldots, v_r\}$ is a *homaloidal multiplicity set* in degree t if the corresponding type $(t; \mathbf{v})$ is homaloidal, i. e., satisfies the classical equations of condition for a Cremona map. The reason to emphasize this seemingly pedantic terminology is to be able to treat both multiplicity notions on the same foot and so also corresponding ideals with respect to the same set of points, in case the respective multiplicity sets have the same number.

Quite generally, the arithmetic as well as the geometric relation between the two are as follows.

Lemma 6.2.5. *Let $\boldsymbol{\mu} = \{\mu_1, \ldots, \mu_r\}$ stand for a set of nonnegative integers and let $s \geq 2$ denote an integer. The following conditions are equivalent:*

(i) *$\boldsymbol{\mu}$ is a subhomaloidal multiplicity set in degree s.*
(ii) *$2\boldsymbol{\mu}$ is a homaloidal multiplicity set in degree $2s-1$.*

In addition, fixing a set of points $\mathbf{p}$ with same cardinality as $\boldsymbol{\mu}$, then $I(\mathbf{p}, 2\boldsymbol{\mu}) = I(\mathbf{p}, \boldsymbol{\mu})^{(2)}$ (second symbolic power).

Proof. The proof of the equivalence is immediate noting that $2(3s-3) = 3(2s-1) - 3$ and $4(s2-s) = (2s-1)^2 - 1$. The supplementary assertion follows since points in $\mathbb{P}^n$ are complete intersections and, therefore, the symbolic power of arbitrary order of the ideal of a point coincides with the respective ordinary power. □

The following result shows a stronger tie between the two types above.

Proposition 6.2.6. *Let $\boldsymbol{\mu} = \{\mu_1 \geq \cdots \geq \mu_r\}$ denote a subhomaloidal multiplicity set in degree $s \geq 3$ and let $\mathbf{p} \subset \mathbb{P}^2$ stand for a set of r general points. Write $J := I(\mathbf{p}, \boldsymbol{\mu}) \subset R = k[x,y,z]$. If the "doubled" homaloidal type $(2s-1; 2\boldsymbol{\mu})$ is proper then* $\operatorname{indeg}(J) = s$.

Proof. Since the points in $\mathbf{p}$ are general there is a Cremona map with homaloidal type $(2s-1; 2\boldsymbol{\mu})$ and these base points. Letting $\mathbb{J} := I(\mathbf{p}, 2\boldsymbol{\mu})$, by Lemma 6.2.5 the initial degree of $\mathbb{J}$ is $2s-1$, in the above notation.

First, it is clear that $\operatorname{indeg}(J) \geq s$, by using that $\mathbb{J} = J^{(2)}$ from Lemma 6.2.5. Indeed, if $\operatorname{indeg}(J) \leq s-1$ then $\operatorname{indeg}(J^2) \leq 2s-2 < 2s-1$, and hence also $\operatorname{indeg}(\mathbb{J}) < 2s-1$

as $J^2 \subset J^{(2)}$. But this would contradict the result of Theorem 6.1.16, which says that $\operatorname{indeg}(\mathbb{J}) \geq 2s-1$.

For the reverse inequality, an easy calculation gives $e(R/J) = (1/2)(s+3)(s-1)$, and hence $\binom{s+2}{2} - e(R/J) = (s+5)/2 > 0$. On the other hand, for any t one has

$$\dim_k J_t \geq \max\left\{\binom{t+2}{2} - e(R/J), 0\right\}.$$

Therefore, $\dim_k J_s \geq (s+5)/2 > 0$. In particular, $\operatorname{indeg}(J) \leq s$, hence equality must be the case, as stated. □

6.2.3 3-uniform subhomaloidal types

Setup 6.2.7. Fix the following general notation, as introduced in Subsection 6.2.1:

- $\boldsymbol{\mu} = \{\mu_1 \geq \cdots \geq \mu_r\}$: a subhomaloidal multiplicity set in degree $s \geq 3$.
- $S := (s; \boldsymbol{\mu})$: the corresponding subhomaloidal type.
- $S^{(2)} := (2s-1; 2\boldsymbol{\mu})$: the "doubled" homaloidal type.
- $\mathbf{p} = \{p_1, \dots, p_r\}$: a set of points in $\mathbb{P}^2$ (general most often).
- $\mathcal{Q}$ and $\mathfrak{Q}$: the arithmetic quadratic transformation and the quadratic Cremona, respectively, based on a major triple of virtual multiplicities $\mu_1 \geq \mu_2 \geq \mu_3$.
- $\widetilde{S} := (\widetilde{s}; \widetilde{\boldsymbol{\mu}})$: the transform of S by $\mathcal{Q}$.
- $J = I(\mathbf{p}, \boldsymbol{\mu})$, $\widetilde{J} = I(\widetilde{\mathbf{p}}, \widetilde{\boldsymbol{\mu}})$, $\mathbb{J} = I(\mathbf{p}, 2\boldsymbol{\mu})$, the corresponding fat ideals in $R = k[x,y,z]$, where $\widetilde{J}$ is the $(\mathcal{Q}, \mathfrak{Q})$-transform of J.

Note that we are using J for the fat ideal instead of the usual I, reserving the latter for the base ideal of the Cremona map corresponding to a proper $S^{(2)}$.

Assuming that the "doubled" homaloidal type $(2s-1; 2\boldsymbol{\mu})$ is proper, Proposition 6.2.6 yields that the initial degree of J is s. Thus, this seems like a good hypothesis to keep and we include it in the definition of an important class:

Definition 6.2.8. Let $\boldsymbol{\mu} = \{\mu_1 \geq \cdots \geq \mu_r\}$: a subhomaloidal multiplicity set in degree $s \geq 3$. We say that the subhomaloidal type $S = (s; \boldsymbol{\mu})$ is *3-uniform subhomaloidal type* if $S^{(2)}$ is proper and $\mu_1 = \mu_2 = \mu_3 = (s-1)/2$.

We also say, in this case, that $S^{(2)}$ is a 3-uniform homaloidal type (properness being subsumed). Note that in this class one has

$$\sum_{i=1}^{r} \mu_i - 3(s-1)/2 = 3(s+1)/2 - 1, \quad \text{and} \quad \sum_{i=1}^{r} \mu_i^2 - 3(s-1)^2/4 = ((s+1)/2)^2 - 1. \tag{73}$$

Thus, the room for the remaining virtual multiplicities is often tight.

6.2.3.1 Structure and resolution of the Cremona base ideal

The goal of this part is to look closer at the base ideal of the Cremona map defined by a 3-uniform homaloidal type $S^{(2)}$.

The generators and free resolution to be described unravels a deep geometric content, involving the so-called exceptional types introduced in the classical theory. A vector $\mathbf{v} = (t; v_1, \ldots, v_r)$ is an *exceptional type* if it satisfies the following modified equations of condition:

$$\sum_{i=1}^{r} v_i = 3t - 1, \quad \text{and} \quad \sum_{i=1}^{r} v_i^2 = t^2 + 1. \tag{74}$$

In order to define properness of an exceptional type, one can proceed pretty much as in the case of a homaloidal type. In the latter case, one introduces a so-called *characteristic matrix* of a Cremona transformation in which the corresponding homaloidal type is the 0th row of this matrix. One then says that the exceptional type is *proper* if its negative gives one of the last r rows of the characteristic of a Cremona map.

Working with these matrices is a tour de force of the classical theory. Fortunately, one has the following analogue of the Hudson test for properness of exceptional types:

Proposition 6.2.9 ([4, Proposition 5.5.5]). *An exceptional type $\mathbf{v} = (t; v_1, \ldots, v_r)$ is proper if and only if there is a finite sequence $\mathcal{Q}_1, \ldots, \mathcal{Q}_\sigma$ of arithmetic quadratic transformations such that $(\mathcal{Q}_\sigma \circ \cdots \circ \mathcal{Q}_1)(\mathbf{v}) = (0; -1, 0, \overset{r-1}{\ldots}, 0)$.*

The proof is based on associating a matrix in $\mathrm{GL}_{r+1}(\mathbb{Z})$ to an arithmetic quadratic transformation and dealing with a certain subgroup of $\mathrm{GL}_{r+1}(\mathbb{Z})$. The details are beyond the scope of this book, for which we refer to [4, Sections 5.4 and 5.5].

Corollary 6.2.10. *Let $J \subset R = k[x,y,z]$ denote the $\boldsymbol{\mu}$-fat ideal associated to an exceptional type $\mathbf{v}$ in degree t on a set of proper points. Then the k-vector space J_t is one-dimensional.*

Proof. First note that, by Proposition 6.2.2(b), the dimension of any transformed of a fat ideal by quadratic transformations is invariant. By Proposition 6.2.9 as applied to $\mathbf{v}$ yields the type $\mathbf{e} := (0; -1, 0, \overset{r-1}{\ldots}, 0)$. Let $(d; \boldsymbol{\mu})$ be a type transformed to $\mathbf{e}$, with μ_1, μ_2, μ_3 three highest multiplicities. From the equations of transformation in Definition 6.1.3, one readily derives that $d = 1$, $\mu_1 = \mu_2 = 1$, $\mu_i = 0$, for $i \geq 3$, so $(d; \boldsymbol{\mu}) = (1; 1, 1, 0, \overset{r-2}{\ldots}, 0)$. We may assume that the first two of the given points are (x,y) and (x,z). Then the corresponding exceptional $\boldsymbol{\mu}$-fat ideal is (x, yz), in which the degree one piece is spanned by x. □

Lemma 6.2.11. *Let $S = (s; \boldsymbol{\mu})$ ($\boldsymbol{\mu} = \{\mu_1, \ldots, \mu_r\}$) be a 3-uniform subhomaloidal type in degree s. Then*

$$V_1 := ((s-1); (s-3)/2, (s-1)/2, (s-1)/2, \mu_4, \ldots, \mu_r) \tag{75}$$

$$V_2 := ((s-1); (s-2)/2, (s-3)/2, (s-1)/2, \mu_4, \dots, \mu_r), \tag{76}$$

$$V_3 := ((s-1); (s-2)/2, (s-1)/2, (s-3)/2, \mu_4, \dots, \mu_r) \tag{77}$$

are exceptional types. Moreover, $-V_1$, $-V_2$ and $-V_3$ correspond to lines of the characteristic matrix of the proper homaloidal type $T := (2s-1; (s-1)^3, 2\mu_4, \dots, 2\mu_r)$.

Proof. One has $T := S^{(2)} = (2s-1; (s-1)^3, 2\mu_4, \dots, 2\mu_r)$, where one may harmlessly assume that $\mu_4 \geq \dots \geq \mu_r > 0$.

We first show that V_i $(i = 1, 2, 3)$ are exceptional types. Namely, since T is homaloidal then one has

$$\begin{cases} 3(s-1) + 2\sum_{i\geq 4}\mu_i = 6(s-1) \\ 3(s-1)^2 + 4\sum_{i\geq 4}\mu_i^2 = 4s(s-1), \end{cases}$$

from which we derive $\sum_{i\geq 4}\mu_i = 3(s-1)/2$ and $\sum_{i\geq 4}\mu_i^2 = (s^2+2s-3)/4$.

It follows that

$$\begin{cases} 2(s-1)/2 + (s-3)/2 + \sum_{i\geq 4}\mu_i = 3(s-1) - 1 \\ 2((s-1)/2)^2 + ((s-3)/2)^2 + \sum_{i\geq 4}\mu_i^2 = (s-1)^2 + 1, \end{cases}$$

as required.

Next, we look at properness. For $j \geq 0$, define T_j and V_{ij} $(i = 1, 2, 3)$ iteratively by $T_0 = T$, $V_{i,0} = V$ and for $j \geq 1$, T_j (resp., V_{ij}) is the type obtained from T_{j-1} (resp., from $V_{i,j-1}$) by applying an arithmetic quadratic transformation based on three major multiplicities.

Since we are assuming that T is a proper homaloidal type, then $T_u = (1; \underbrace{0, \dots, 0}_{r})$ for a certain $u \geq 1$. Let T_j be any transformed type, with $j < u$.

Claim 1. T_j is of the form $(2t; 1, 1, 1, 2\eta_1, \dots, 2\eta_\sigma, 0^{r-\sigma-3})$, for suitable $t \geq 1$, $\eta_1 \geq \dots \geq \eta_\sigma \geq 1$ and some $0 \leq \sigma \leq r-3$.

One proceeds by induction on $j \geq 1$. Applying an arithmetic quadratic transformation based on the first three multiplicities, it is readily seen that $T_1 = (s+1; 1, 1, 1, 2\mu_4, \dots, 2\mu_r)$. Note that now the degree is even.

Assume inductively that T_{j-1} has the form $(2t; 1, 1, 1, 2\eta_1, \dots, 2\eta_\sigma, 0^{r-\sigma-3})$ and let us show that necessarily $\sigma \geq 3$. Indeed, first it is obvious that $\sigma \neq 0$ as otherwise $T_{j-1} = (2; 1^3)$, hence $T_j = (1; 0, \dots, 0)$, contradicting the assumption that $j < u$. If $\sigma = 1$, then the first equation of condition gives $2\eta_1 + 3 = 3.2t - 3 = 6t - 3$. In addition, one knows that the sum of the highest multiplicities does not exceed the degree, hence $2\eta_1 + 1 \leq 2t$. Together this yields $t \leq 1$, thus implying $\eta_1 = 0$; this contradicts the assumption that $\eta_1 > 0$.

The case where $\sigma = 2$ is similar: $t \geq \eta_1 + \eta_2 = 3t - 3$ still yields $t = 1$, forcing $\eta + 1 + \eta_2 \leq 1$, i. e., $\eta_2 = 0$ – a contradiction.

Now, since $\sigma \geq 3$, one sees that the transformed type is

$$(2(2t - \eta_1 - \eta_2 - \eta_3); 2(t - \eta_2 - \eta_3), 2(t - \eta_1 - \eta_3), 2(t - \eta_1 - \eta_2), 2\eta_4, \ldots, 2\eta_\sigma, 1^3, 0 \ldots, 0),$$

where the first three multiplicities are in decreasing order, but need to be reordered as compared to the subsequent multiplicities $2\eta_4 \geq \cdots \geq 2\eta_\sigma$. In fact, some of the first three written above may turn out to vanish, thus increasing the number of zeros and decreasing σ. However, since for $i \geq 1$ both the degree and the multiplicities are even, the number of 1's will not change. The number of zeros is a consequence of the fact that the total number of multiplicities does not change under an arithmetic quadratic transformation.

This proves the claim.

Next, one has the following.

Claim 2. If $T_j = (2t; 1, 1, 1, 2\eta_1, \ldots, 2\eta_\sigma, 0^{r-\sigma-3})$, then

$$V_{1j} = (t; 0, 1, 1, \eta_1, \ldots, \eta_\sigma, 0^{r-\sigma-3}),$$
$$V_{2j} = (t; 1, 0, 1, \eta_1, \ldots, \eta_\sigma, 0^{r-\sigma-3}),$$
$$V_{3j} = (t; 0, 1, 1, \eta_1, \ldots, \eta_\sigma, 0^{r-\sigma-3}).$$

The proof is again by induction on i and similar to the previous one, so we only stress the main points. Thus, one readily sees that

$$\begin{aligned} V_{11} &= ((s+1)/2; 0, 1, 1, \mu_4, \ldots, \mu_r) \\ V_{21} &= ((s+1)/2; 1, 0, 1, \mu_4, \ldots, \mu_r) \\ V_{31} &= ((s+1)/2; 1, 1, 0, \mu_4, \ldots, \mu_r) \end{aligned} \tag{78}$$

Supposing that

$$V_{1i-1} = (t; 0, 1, 1, \eta_1, \ldots, \eta_\sigma, 0^{r-\sigma-3}),$$
$$V_{2i-1} = (t; 1, 0, 1, \eta_1, \ldots, \eta_\sigma, 0^{r-\sigma-3}),$$
$$V_{3i-1} = (t; 1, 1, 0, \eta_1, \ldots, \eta_\sigma, 0^{r-\sigma-3})$$

the next stage will have the form

$$(2t - \eta_1 - \eta_2 - \eta_3; 0, 1, 1, t - \eta_2 - \eta_3, t - \eta_1 - \eta_3, t - \eta_1 - \eta_2, \eta_4, \ldots, \eta_\sigma, 0, \ldots, 0),$$
$$(2t - \eta_1 - \eta_2 - \eta_3; 1, 0, 1, t - \eta_2 - \eta_3, t - \eta_1 - \eta_3, t - \eta_1 - \eta_2, \eta_4, \ldots, \eta_\sigma, 0, \ldots, 0),$$
$$(2t - \eta_1 - \eta_2 - \eta_3; 1, 1, 0, t - \eta_2 - \eta_3, t - \eta_1 - \eta_3, t - \eta_1 - \eta_2, \eta_4, \ldots, \eta_\sigma, 0, \ldots, 0).$$

So much for the claim.

To conclude, in particular, since in the one before the last stage of transforming T one gets the type $(2; 1^3, 0^{r-3})$, then the corresponding transformed type of V_1 (resp., V_2,

V_3) is $(1;0,1^2,0^{r-3})$ (resp., $(1;1,0,1,0^{r-3})$, $(1;1,1,0,0^{r-3})$)—the latter clearly transforms to $(0;-1,0^{r-1})$. Now apply Proposition 6.2.9.

For each $1 \leq j \leq u-1$, consider the matrix Q_1 corresponding to the arithmetic quadratic transformation that transforms T_{j-1} in T_j. The characteristic matrix of T_{u-1} is

$$C = \left(\begin{array}{c|ccc|ccc} 2 & 1 & 1 & 1 & 0 & \cdots & 0 \\ \hline -1 & 0 & -1 & -1 & 0 & \cdots & 0 \\ -1 & -1 & 0 & -1 & 0 & \cdots & 0 \\ -1 & -1 & -1 & 0 & 0 & \cdots & 0 \\ \hline 0 & 0 & 0 & 0 & -1 & \cdots & 0 \\ \vdots & \vdots & \vdots & \vdots & \vdots & \ddots & \vdots \\ 0 & 0 & 0 & 0 & 0 & \cdots & -1 \end{array}\right)$$

Thus,

$$C' = CQ_{u-1}Q_{u-2}\cdots Q_1$$

is also a characteristic matrix of a proper homaloidal type. By the above argument, the four first lines of C' are $T_1, -V_1, -V_2$ and $-V_3$. □

A main result of this part is as follows.

Proposition 6.2.12. *Let $S = (s;\boldsymbol{\mu})$ ($\boldsymbol{\mu} = \{\mu_1,\ldots,\mu_r\}$) be a 3-uniform subhomaloidal type in degree s. Suppose that the corresponding doubled type $(2s-1;2\boldsymbol{\mu})$ is proper. Then, for any bisimple Cremona map F having homaloidal type $(2s-1;2\boldsymbol{\mu})$, with base ideal $I \subset R = k[x,y,z]$, there are forms f_1, f_2, f_3 of degree $s-1$, and independent linear forms $\ell_1, \ell_2, \ell_3 \in R$ such that $I = (\ell_3 f_1 f_2, \ell_2 f_1 f_3, \ell_1 f_2 f_3)$.*

Proof. Since F is simple, its base points are proper. Let $\{P_1,\ldots,P_r\} \subset R$ denote the corresponding defining prime ideals. Since the first three points of highest virtual multiplicity are nonaligned, we may assume that the first three of the given points are the coordinate ones. By Theorem 6.1.16, the base ideal $I \subset R$ of F is generated by the homogeneous piece of degree $2s-1$ of the ideal of fat points

$$\mathbb{J} := J^{(2)} = P_1^{s-1} \cap P_2^{s-1} \cap P_3^{s-1} \cap P_4^{2\mu_4} \cap \cdots \cap P_r^{2\mu_r},$$

where $J := P_1^{(s-1)/2} \cap P_2^{(s-1)/2} \cap P_3^{(s-1)/2} \cap P_4^{\mu_4} \cap \cdots \cap P_r^{\mu_r}$.

Consider the following additional ideals of fat points:

$$J_1 = P_1^{(s-3)/2} \cap P_2^{(s-1)/2} \cap P_3^{(s-1)/2} \cap P_4^{\mu_4} \cap \cdots \cap P_r^{\mu_r}, \tag{79}$$

$$J_2 = P_1^{(s-1)/2} \cap P_2^{(s-3)/2} \cap P_3^{(s-1)/2} \cap P_4^{\mu_4} \cap \cdots \cap P_r^{\mu_r} \tag{80}$$

$$J_3 = P_1^{(s-1)/2} \cap P_2^{(s-1)/2} \cap P_3^{(s-3)/2} \cap P_4^{\mu_4} \cap \cdots \cap P_r^{\mu_r}. \tag{81}$$

By Lemma 6.2.11, the respective types of J_1, J_2, J_3 are proper exceptional in degree $s-1$, corresponding to lines of the characteristic matrix of F. Then, by Corollary 6.2.10, for every $i = 1,2,3$, $[J_i]_{s-1} = kf_i$, where f_i is the defining equation of a total principal curve Θ_i corresponding to a certain base point of the inverse of F. Since F is bisimple, the total principal curves are principal curves [4, Lemma 2.2.25]; hence, Θ_i $(i = 1,2,3)$ are irreducible curves. Since we are assuming that $P_1 = (x,y)$, $P_2 = (x,z)$, $P_3 = (y,z)$, the forms $xf_1f_2, zf_1f_3, yf_2f_3$ belong to $\mathbb{J}$. But they are of degree $2s-1$, hence $(xf_1f_2, zf_1f_3, yf_2f_3) \subset I$. On the other hand, no two $f_i, f_j (i \neq j)$ have a proper common factor as otherwise, since the curves Θ_i are irreducible, $f_i \in J$ while the latter has initial degree s by Proposition 6.2.6. Therefore, e. g., $\{f_1(xf_2 + yf_3), zf_2f_3\}$ have no proper common factors; hence, the ideal $(xf_1f_2, zf_1f_3, yf_2f_3)$ has codimension 2. By the same token, $\{xf_1f_2, zf_1f_3, yf_2f_3\}$ are k-linearly independent. The equality now follows. □

Remark 6.2.13. Regarding the assumptions of the proposition, see Exercise 6.3.7.

Theorem 6.2.14. *Let $I \subset R$ denote the base ideal of a Cremona map satisfying the conditions of* Proposition 6.2.12 *and let f_1, f_2, f_3 be forms as provided in its statement. Then*
(a) *I is a nonsaturated ideal with minimal graded resolution*

$$0 \to R(-3s) \xrightarrow{\psi} R^3(-(3s-1)) \xrightarrow{\varphi} R^3(-(2s-1)) \to I \to 0,$$

where

$$\varphi = \begin{pmatrix} \ell_2 f_3 & -\ell_3 f_3 & 0 \\ -\ell_1 f_1 & 0 & \ell_3 f_1 \\ 0 & \ell_1 f_2 & -\ell_2 f_2 \end{pmatrix} \quad \text{and} \quad \psi = \begin{pmatrix} \ell_3 \\ \ell_2 \\ \ell_1 \end{pmatrix}.$$

(b) $I^{\mathrm{sat}} = (I, f_1f_2f_3)$.

Proof. (a) Note that φ is just the obvious reduced Koszul syzygies. Then the maps are readily seen to give a complex. A simple application of the Buchsbaum–Eisenbud acyclicity criterion shows the contention.

(b) Since I is the base ideal of a Cremona map, $\mathrm{indeg}(I^{\mathrm{sat}}/I) \geq 2s-1$ [155]. Since R/I^{sat} is Cohen–Macaulay, part (a) and [89, Proposition 1.3 (ii)] imply that I^{sat} is minimally generated by 4 forms, three of which can be taken to be the minimal generators of I as in (a) and the fourth one some form of degree $3(s-1)$.

Consider the following matrix:

$$\begin{pmatrix} f_1 & 0 & 0 \\ 0 & f_2 & 0 \\ 0 & 0 & f_3 \\ \ell_3 & \ell_2 & \ell_1 \end{pmatrix}$$

Its 3-minors form a minimal set of generators of the ideal $(I, f_1f_2f_3)$, since for any $i \neq j$ one has $\gcd(f_i, f_j) = 1$. This ideal is contained in I^{sat} because $(x, y, z)f_1f_2f_3 = (\ell_1, \ell_2, \ell_3)f_1f_2f_3 \in I$. Since $\deg f_1f_2f_3 = 3(s-1)$, we are done. □

Remark 6.2.15. Theorem 6.2.14 is an algebraic rap up of a special geometric datum. Yet, it turns out that the base ideals of many plane Cremona transformations have a free resolution with those same Betti numbers. In particular, the hypothesis of bisimplicity may not be essential in many cases. It would be interesting to decide whether these transformations form a more visible slice of the Cremona group. Note that, for any plane Cremona transformation in degree d, the largest twist in the presentation map of its base ideal is at least $3d/2$, as it follows from the general [179, Proposition 6.2.30]. Thus, the above attains its minimum possible.

6.2.3.2 The structure of the transform

The following establishes the behavior of the transform $\widetilde{J}$ (Definition 6.2.1) when J is the fat ideal of a 3-uniform subhomaloidal type on a set of general points $\mathbf{p} = \{p_1, \dots, p_r\} \subset \mathbb{P}^2$.

Proposition 6.2.16. *Let $S = (s; \boldsymbol{\mu})$ denote a 3-uniform subhomaloidal type and let $\widetilde{J}$ denote the transformed fat ideal of the fat ideal J of S on a set $\mathbf{p}$ of general points. Then the minimal resolution of $\widetilde{J}$ has the form*

$$0 \to R^{(s+3)/2}(-((s+5)/2)) \longrightarrow R^{(s+5)/2}(-(s+3)/2) \to \widetilde{J} \to 0. \tag{82}$$

In particular, $\operatorname{reg}(\widetilde{J}) = (s+3)/2$.

Proof. For simplicity, write $N = (s+3)/2$. A calculation using the fact that $S = (s; \boldsymbol{\mu})$ is a 3-uniform subhomaloidal type gives $e(R/\widetilde{J}) = \binom{N+1}{2}$. In particular,

$$h_{\widetilde{J}}(N-1) = 0 \Leftrightarrow h_{\widetilde{J}}(N-1) = \binom{N+1}{2} - e(R/\widetilde{J}) \Leftrightarrow \operatorname{reg}(\widetilde{J}) = N$$

Any of these equivalent equalities implies that the minimal resolution of $\widetilde{J}$ has the shape as in (82).

Claim. $h_{\widetilde{J}}(N-1) = 0$.

Let $s = 3$. Then one readily verifies that the only 3-uniform subhomaloidal type is $S = (3; 1^6)$. Here, $\widetilde{S} = S$ as is also readily verified. Therefore, $\widetilde{J} = J$, while the latter is a well-known linearly presented reduced 4-generated ideal, hence is generated in degree $3 = N$.

Suppose now that $s \geq 5$ and consider the abstract type $S^* := (s-2; \boldsymbol{\mu}^*)$, where

$$\boldsymbol{\mu}^* = ((s-3)/2, (s-3)/2, (s-3)/2, \mu_4, \dots, \mu_r) \quad (s \geq 5).$$

Note that $\widetilde{S^*} = ((s+1)/2; \widetilde{\boldsymbol{\mu}}) = (N-1; \widetilde{\boldsymbol{\mu}})$, by applying an arithmetic quadratic transformation based on the first three multiplicities.

Set

$$T := S^{(2)} = (2s-1; (s-1)^3, 2\mu_4, \dots, 2\mu_r) \tag{83}$$

We draw upon parts of the argument in the proof of Lemma 6.2.11. Thus, for $j \geq 0$, define T_j and S_j^* iteratively setting $T_0 = T$, $S_0^* = S^*$ and for $j \geq 1$, setting T_j (resp., S_j^*) to be the type obtained from T_{j-1} (resp., from S_{j-1}^*) by applying an arithmetic quadratic transformation based on three major multiplicities of T_{j-1}.

Since T is proper, $T_u = (1; 0, \dots, 0)$ for some u. As in the proof of Lemma 6.2.11, we can conclude by induction on $1 \leq j < u$ that

$$T_j = (2t; 1, 1, 1, 2\eta_1, \dots, 2\eta_\sigma, 0^{r-\sigma-3}) \quad \text{and} \quad S_j^* = (t; 1, 1, 1, \eta_1, \dots, \eta_\sigma, 0^{r-\sigma-3})$$

for suitable $t \geq 1$, $\eta_1 \geq \cdots \geq \eta_\sigma \geq 1$ and some $0 \leq \sigma \leq r-3$. In particular,

$$T_{u-1} = (2; 1, 1, 1, 0^{r-3}) \quad \text{and} \quad S_{u-1}^* = (1; 1, 1, 1, 0^{r-3}).$$

But, for r general points $\mathbf{p} \subset \mathbb{P}^2$ and $\mathbf{v} = (1, 1, 1, 0^{r-3})$, the linear system $I(\mathbf{p}, \mathbf{v})_1$ has the expected dimension zero. Thus, as observed in Remark 6.2.3, the linear system $I(\mathbf{p}, \boldsymbol{\mu}^*)_{s-2}$ has also the expected dimension zero. Since $\widetilde{S^*} = (N-1; \widetilde{\boldsymbol{\mu}})$ then, by Lemma 6.2.2, $\dim_k \widetilde{J}_{N-1} = \dim_k I(\widetilde{\mathbf{p}}, \widetilde{\boldsymbol{\mu}})_{N-1} = \dim_k I(\mathbf{p}, \boldsymbol{\mu}^*)_{s-2} = 0$, as required. □

Keeping the notation $N := (s+3)/2$, one has the following.

Corollary 6.2.17. *For any integer $n \geq 1$, one has:*

(a) $\operatorname{reg}(\widetilde{J}^n) = nN$.

(b) $(\widetilde{J}^{(n)})_{nN} = (\widetilde{J}^n)_{nN}$.

(c) $\operatorname{reg}(\widetilde{J}^{(n)}) \leq nN$.

(d) $\dim_k \widetilde{J}^n_{ns} = \dim_k \widetilde{J}^{(n)}_{nN} = n(sn+3)/2 + 1$.

Proof. (a) By [77, Theorem 1.1], one has

$$\operatorname{reg}(\widetilde{J}^n) \leq n \cdot \operatorname{reg}(\widetilde{J}) = nN.$$

Since $nN = \operatorname{indeg}(\widetilde{J}^n) \leq \operatorname{reg}(\widetilde{J}^n)$ we are done.

(b) For a homogeneous ideal $I \subset k[x, y, z]$, let $\mathfrak{n}(I)$ denote its *saturation degree*, i. e., the least positive integer s such that $I_t = I_t^{\text{sat}}$ for every $t \geq s$. One has

$$\operatorname{reg}(I) = \max\{\mathfrak{n}(I), \operatorname{reg}(I^{\text{sat}})\}.$$

The result follows immediately from this and item (a).

(c) The proof is the same as in item (b).

(d) Using the hypothesis that $\boldsymbol{\mu}$ is subhomaloidal and that $\mu_1 = \mu_2 = \mu_3 = (s-1)/2$, a straightforward calculation yields the following relations:

$$\sum_{i=1}^{r}(n\widetilde{\mu}_i) = 3n(N-1) \quad \text{and} \quad \sum_{i=1}^{r}(n\widetilde{\mu}_i)^2 = n^2(N-2)N + 3n^2, \tag{84}$$

for every integer $n \geq 1$.

From this, $e(R/\widetilde{J}^{(n)}) = \frac{1}{2}(n^2(N-2)N + 3n^2 + 3n(N-1))$.

On the other hand, quite generally for an unmixed ideal $I \subset R$, the Hilbert function of I at any $D \geq \rho := \operatorname{reg}(I)$ has the expected value $\binom{D+2}{2} - e(R/I)$.

Applying this in the present case with $\operatorname{reg}(\widetilde{J}^{(n)}) \leq nN$ yields that $\widetilde{J}^{(n)}_{nN}$ has the expected dimension, hence

$$\begin{aligned}
\dim_k \widetilde{J}^{(n)}_{nN} &= \binom{nN+2}{2} - e(R/\widetilde{J}^{(n)}) \\
&= \frac{(nN+2)(nN+1)}{2} - \frac{n^2(N-2)N + 3n^2 + 3n(N-1)}{2} \\
&= \frac{2Nn^2 - 3n^2 + 3n + 2}{2} = \frac{2N-3}{2}n^2 + \frac{3}{2}n + 1 \\
&= \frac{s}{2}n^2 + \frac{3}{2}n + 1 = \frac{n(sn+3)}{2} + 1.
\end{aligned}$$

□

Proposition 6.2.18. *For any integer $n \geq 1$, one has:*

(a) *The arithmetic transform of the type $(ns; n\boldsymbol{\mu})$ by $\mathcal{Q}$ is the type $(nN; n\widetilde{\boldsymbol{\mu}})$.*

(b) $J^{(n)}_{ns} = J^n_{ns}$.

(c) $\dim_k J^n_{ns} = n(sn+3)/2 + 1$.

Proof. (a) This is an obvious calculation from the definitions:

$$\begin{aligned}
2ns - n\mu_1 - n\mu_2 - n\mu_3 &= n(2s - \mu_1 - \mu_2 - \mu_3) = nN, \\
ns - n\mu_2 - n\mu_3 &= n(s - \mu_2 - \mu_3) = n\widetilde{\mu}_1, \\
ns - n\mu_1 - n\mu_3 &= n(s - \mu_1 - \mu_3) = n\widetilde{\mu}_2, \\
ns - n\mu_1 - n\mu_2 &= n(s - \mu_1 - \mu_2) = n\widetilde{\mu}_3
\end{aligned}$$

and

$$n\mu_i = n\widetilde{\mu}_i, \quad \text{for } i \geq 4.$$

(b) Since $J^n_{ns} \subset J^{(n)}_{ns}$, it suffices to show that these are k-vector spaces of the same dimension. One has

$$\begin{aligned}
\dim_k (J^n)_{ns} &= \dim_k (\widetilde{J}^n)_{nN} \\
&= \dim_k (\widetilde{J}^{(n)})_{nN}
\end{aligned}$$

$$= \dim_k (J^{(n)})_{ns},$$

where the first equality comes from Lemma 6.2.2(d) and the last two come from Corollary 6.2.17(b).

(c) This is a straightforward consequence of Corollary 6.2.17(d) and from the equality $\dim_k (J^n)_{ns} = \dim_k (\tilde{J}^n)_{nN}$. □

Proposition 6.2.19. *Let $W \subset \mathbb{P}^N$ denote the image of the rational map defined by the linear system J_s. Then the Hilbert function of W is given by $H_W(n) = n(sn+3)/2 + 1$, for all $n \geq 0$. Thus, the Hilbert polynomial of W coincides with its Hilbert function for $n \geq 1$; in particular, the degree of W is s.*

Proof. The homogeneous coordinate ring C of W in its embedding is isomorphic by a degree preserving isomorphism to the homogeneous k-subalgebra $k[f_0t, \ldots, f_Nt] \subset R[t]$ generated in degree 1, where J_s is spanned by $\{f_0, \ldots, f_N\}$. Therefore, for any $n \geq 1$ one obtains k-vector space isomorphisms:

$$C_n \simeq k[f_0t, \ldots, f_Nt]_n \simeq \sum_{|\alpha|=n} k(\mathbf{f}t)^\alpha = t^n \sum_{|\alpha|=n} k(\mathbf{f})^\alpha$$
$$\simeq \sum_{|\alpha|=n} k(\mathbf{f})^\alpha \simeq k[f_0, \ldots, f_N]_n \simeq (J^n)_{ns}.$$

Thus, the result follows from Proposition 6.2.18(c). □

6.2.3.3 The plus/minus Harbourne criterion

Let $I := I(\mathbf{p}; \boldsymbol{\mu}) \subset R = k[x,y,z]$ be a fat ideal. Any such ideal is a perfect ideal, i. e., the ring R/I is Cohen–Macaulay. In addition, the degree of the corresponding scheme depends only on the number of points and their appended multiplicities:

$$e(R/I) = \sum_{j=1}^{n} \frac{\mu_j(\mu_j+1)}{2}. \tag{85}$$

For convenience, we switch to the Hilbert function of the ideal I. Here, for any $t \geq 0$, one has

$$h_I(t) := \dim_k I_t \geq \max\left\{0, \binom{t+2}{2} - \sum_{j=1}^{n} \frac{\mu_j(\mu_j+1)}{2}\right\}.$$

When the Hilbert function $h_I(t)$ achieves this lower bound in every degree t one says that it is *minimal* —this is the same as the more common designation in the literature saying that R/I has *maximal* Hilbert function.

A major question asks when the multiplication map

$$\mathfrak{m}_t : R_1 \otimes_k I_{t-1} \longrightarrow I_t, \quad \sum \ell_i \otimes f_j \mapsto \sum \ell_i f_j$$

has maximal rank. This question has much interest regardless of the assumption that I has minimal Hilbert function, and yet, in the presence of the latter, the crucial multiplication map is the one with $t = \text{indeg}(I)$. In this case saying that $\mathfrak{m}_{\text{indeg}(I)}$ is injective or surjective means, respectively, that I admits no nonzero linear syzygies or that I is generated in one single degree.

The philosophy of B. Harbourne (see [86, 84, 85]) is that, in order to obtain information about the maximal rank property of $\mathfrak{m}_t$ when I has minimal Hilbert function, one needs some elbow room. The recurrent idea behind this philosophy is to introduce the following additional ideals of fat points:

$$\boldsymbol{\mu}^- := \{\mu_1 - 1, \mu_2, \dots, \mu_r\}, \quad I^- := I(\mathbf{q}, \boldsymbol{\mu}^-), \tag{86}$$

if $\mu_1 \geq 1$, and

$$\boldsymbol{\mu}^+ := \{\mu_1 + 1, \mu_2, \dots, \mu_r\}, \quad I^+ := I(\mathbf{q}, \boldsymbol{\mu}^+). \tag{87}$$

We will make use of one of his results.

Lemma 6.2.20 ([86, Lemma 2.4 (d)]). *With the above notation, assume that:*
(a) *All three ideals I, I^- and I^+ have minimal Hilbert function.*
(b) $h_{I^-}(\text{indeg}(I) - 1) > 0$ *and* $h_{I^+}(\text{indeg}(I)) > 0$.

Then the multiplication map $\mathfrak{m}_{\text{indeg}(I)}$ is surjective; in particular, I is generated in one single degree.

We apply this result in the particular case of the ideal J of the Setup 6.2.7.

Proposition 6.2.21. *Notation as in the* Setup 6.2.7. *Let J^- and J^+ be as in* (86) *and* (87), *respectively. If $S^{(2)} = (2s - 1; 2\boldsymbol{\mu})$ is a proper homaloidal type then hypotheses* (a) *and* (b) *of* Lemma 6.2.20 *hold for J, J^-, J^+.*

Proof. Thus, we are to show that the three fat ideals J, J^-, J^+ have minimal Hilbert function and that $h_{J^-}(\text{indeg}(J) - 1) > 0$ and $h_{J^+}(\text{indeg}(J)) > 0$. As explained in the first section, in order to show that the Hilbert function of a fat ideal $I := I(\mathbf{p}, \boldsymbol{\mu})$ is minimal it suffices to show that

$$h_I(\alpha) = \binom{\alpha + 2}{2} - \sum_{i=1}^{r} \binom{\mu_i + 1}{2},$$

where $\alpha := \text{indeg}(J)$.

For J, it follows from the proof of Proposition 6.2.16 that $h_J(s)$ has the expected dimension; hence, J has minimal Hilbert function.

As for J^-, note that it coincides with the fat ideal on the virtual multiplicities of the type (75) based on the originally given points $p_i, 1 \leq i \leq r$. By the same token as argued about (75), the former is a proper exceptional type, in particular one has $\operatorname{indeg}(J^-) = s-1$ and $\dim h_{J^-}(s-1) = 1$. On the other hand, a direct calculation gives

$$\binom{s+1}{2} - \binom{\mu_1}{2} - \sum_{i=2}^{r} \binom{\mu_i+1}{2} = 1.$$

Since the left-hand side of the above equality is at most $\dim h_{J^-}(s-1)$, then we get equalities throughout. This shows both that J^- has minimal Hilbert function and that its value on $\operatorname{indeg}(J) - 1$ is positive, as required in item (b) of Lemma 6.2.20.

We are left with J^+. The easy piece is the equality

$$\binom{s+2}{2} - \binom{\mu_1+2}{2} - \sum_{i=2}^{r} \binom{\mu_i+1}{2} = 2,$$

again a direct calculation using the hypothesis that $S^{(2)} = (2s-1; 2\boldsymbol{\mu})$ is a proper homaloidal type.

To close the cycle as above, we now show that $h_{J^+}(s) = 2$ and $\operatorname{indeg}(J^+) = s$. For this, consider the type $(s; \boldsymbol{\mu}^+)$, where $\boldsymbol{\mu}^+$ are the virtual multiplicities of J^+. Then, as in the proof of Lemma 6.2.11, applying iterated arithmetic quadratic transformations based on the three highest virtual multiplicities we eventually reach the type $(1; 1, 0^{r-1})$. The fat ideal on the virtual multiplicities of the latter type, based on the current set of transformed points, clearly has vector dimension 2. Since the iterated procedure preserves the vector dimension of the graded part of the fat ideal in the degree of the type, we get $h_{J^+}(s) = \dim_k (J^+)_s = 2$, as desired.

To conclude, we now have $\operatorname{indeg}(J^+) \leq s$. If $h_{J^+}(s-1) > 0$ let $0 \neq f \in J^+_{s-1}$. Then $\{xf, yf, zf\} \subset J^+_s$ which is absurd for reasons of vector space dimension. Therefore, $\operatorname{indeg}(J^+) = s$, as desired. □

As a consequence of Lemma 6.2.20 we get the following.

Corollary 6.2.22. *If J is the fat ideal coming from a 3-uniform proper subhomaloidal type then the multiplication map $\mathfrak{m} : R_1 \otimes_k J_s \to J_{s+1}$ is surjective; in particular, J is generated in degree s.*

6.2.3.4 The second symbolic power of the transform

In this part, we focus mainly on the transform $\widetilde{J}$ of the fat ideal J of a 3-uniform subhomaloidal type, on general points. We will give additional details about second symbolic power of $\widetilde{J}$. The obtained structure will have impact on the search of a minimal reduction of J and the corresponding reduction number, a matter of crucial use in the proof of the main theorem in the next section.

In the sequel, we keep the notation already established in the subsection.

We will make use of the following elementary result.

Lemma 6.2.23. *Let $f, g \in R = k[x,y,z]$ denote forms of the same degree $\alpha \geq 2$. Then the linear system $L = \{xf, yf, xg, zg\}$ defines a birational map of $\mathbb{P}^2$ onto the image.*

Proof. Since $k([R]_{\alpha+1}) = k(x^{\alpha+1}, y/x, z/x)$, it suffices to argue with the latter. Clearly, $y/x = yf/xf$ and $z/x = zg/xg$ belong both to $k(xf, yf, xg, zg)$. Then, also $zf = xf \cdot (z/x) \in k(xf, yf, xg, zg)$. Finally,

$$x^{\alpha+1} = x^{\alpha+1}f^{\alpha+1}/f(xf, yf, zf) \in k(xf, yf, xg, zg).$$

This proves the required inclusion $k(x^{\alpha+1}, y/x, z/x) \subset k(xf, yf, xg, zg)$, as desired. □

Theorem 6.2.24. *Let $\mathfrak{F} : \mathbb{P}^2 \dashrightarrow \mathbb{P}^{\frac{s+3}{2}}$ denote the rational map defined by a set of minimal generators of $\widetilde{J}$. Then:*
(a) *$\mathfrak{F}$ is birational onto the image.*
(b) *The initial degree of $\widetilde{J}^{(2)}$ is $s + 2$.*
(c) *$\dim_k((\widetilde{J}^{(2)})_{s+2}) = s$ and $\operatorname{reg}(\widetilde{J}^{(2)}) = s + 3$.*

Proof. (a) Since $\widetilde{J} = (\widetilde{J}_{(s+3)/2})$ by Proposition 6.2.16, then the map is defined by a k-basis of the linear system $\widetilde{J}_{(s+3)/2}$. Still by the proof of Proposition 6.2.16, the vector $\widetilde{E} = ((s+1)/2; \mu_4, \ldots, \mu_r, 1^2, 0)$ as in (78) is an exceptional type for either version of the virtual multiplicities

$$\boldsymbol{\mu}_1 = (0, 1, 1, \mu_1, \ldots, \mu_r) \quad \text{and} \quad \boldsymbol{\mu}_2 = (1, 0, 1, \mu_1, \ldots, \mu_r).$$

Thus, by Corollary 6.2.10, the linear systems $I(\tilde{\mathbf{p}}; \boldsymbol{\mu}_1)_{(s+1)/2}$ and $I(\tilde{\mathbf{p}}; \boldsymbol{\mu}_2)_{(s+1)/2}$ are spanned by forms f and g, respectively. By Lemma 6.2.23, $\{xf, yf, xg, zg\}$ defines a birational map of $\mathbb{P}^2$ onto the image. Clearly, $\{xf, yf, xg, zg\} \subset \widetilde{J}_{(s+3)/2}$. Therefore, $\widetilde{J}_{(s+3)/2}$ also defines a birational map of $\mathbb{P}^2$ onto the image.

(b) We first argue that $\operatorname{indeg}(\widetilde{J}^{(2)}) \geq s + 2$, i. e., that $(\widetilde{J}^{(2)})_{s+1} = 0$. For this, we go back to the level of J and consider the transformed homaloidal type $\widetilde{S^{(2)}} = (s+1; 1, 1, 1, 2\mu_4, \ldots, 2\mu_r)$ of the homaloidal type in (83). By the proof of Lemma 6.2.11, this type is eventually transformed into the homaloidal type $(2; 1^3, 0^{r-3})$ by applying a finite number of arithmetic quadratic transformations based on three major multiplicities. Accordingly, similarly the vector $(s+1; 2, 2, 2, 2\mu_4, \ldots, 2\mu_r)$ is transformed into the vector $(2; 2^3, 0^{r-3})$. By an iterative application of Lemma 6.2.2, the corresponding linear systems in degree $s+1$ and 2, respectively, have the same vector dimension. But since $\{2\mu_4, \ldots, 2\mu_r, 2^3\}$ are the virtual multiplicities of $\widetilde{J}^{(2)}$ then $\dim_k(\widetilde{J}^{(2)})_{s+1} = \dim_k I_2$, where $I = p_1^2 \cap p_2^2 \cap p_3^2 = (x,y)^2 \cap (x,z)^2 \cap (y,z)^2$. Clearly, $\dim_k I_2 = 0$.

On the other hand, having in mind that

$$\widetilde{\mu}_i = \begin{cases} 1 & \text{if } i \in \{1,2,3\}, \\ \mu_i & \text{if } i \geq 4, \end{cases}$$

and drawing upon the subhomaloidal equations (72) of J, one obtains

$$h_{\tilde{J}^{(2)}}(s+2) \geq \binom{s+4}{2} - e(R/\tilde{J}) = s. \tag{88}$$

Hence, $\operatorname{indeg}(\tilde{J}^{(2)}) \leq s+2$.

(c) By Corollary 6.2.17(c), $\operatorname{reg}(\tilde{J}^{(2)}) \leq 2N = s+3$. Thus, since $h_{\tilde{J}^{(2)}}(t) = \binom{t+2}{2} - e(R/\tilde{J}^{(2)})$ for every $t \geq \operatorname{reg}(\tilde{J}^{(2)}) - 1$ then, in particular,

$$h_{\tilde{J}^{(2)}}(s+2) = \binom{s+4}{2} - e(R/\tilde{J}^{(2)}) = s.$$

We have $s+2 = \operatorname{indeg}(\tilde{J}^{(2)}) \leq \operatorname{reg}(\tilde{J}^{(2)}) \leq s+3$. If $\operatorname{reg}(\tilde{J}^{(2)}) = s+2$ then $\tilde{J}^{(2)}$ would be a linearly presented perfect ideal minimally generated by s forms of degree $s+2$, which is nonsense. Hence, $\operatorname{reg}(\tilde{J}^{(2)}) = s+3$. □

We now analyze closely the structure of the birational map $\mathfrak{F}$ of Theorem 6.2.24.

Set $N := (s+3/2)$ as before. By Proposition 6.2.16, the ideal $\tilde{J}$ is linearly presented. As in the proof of Theorem 6.2.24 (a), since $\{xf, yf, xg, zg\}$ is k-linearly independent we can assume that they form a subset of minimal generators of $\tilde{J}$. For the sake of dealing with the $(N+1) \times N$ presentation matrix φ of $\tilde{J}$, we will write xf, yf, xg, zg as the first four generators in this order. With this proviso, we may assume that φ has the shape

$$\varphi = (\varphi_1 \mid \varphi_2), \tag{89}$$

where φ_1 is the transpose of the matrix

$$\begin{pmatrix} y & -x & 0 & 0 & 0 & \cdots & 0 \\ 0 & 0 & z & -x & 0 & \cdots & 0 \end{pmatrix}$$

and φ_2 is an $(N+1) \times (N-2)$ block.

The following ought to be a sufficiently simple matrix calculation.

Conjecture 6.2.25. *Let $\mathbf{z}_3$ and $\mathbf{z}_1$ denote the transposes of*

$$\begin{pmatrix} 0 & 0 & 0 & z & 0 & \cdots & 0 \\ 0 & 0 & z & 0 & 0 & \cdots & 0 \end{pmatrix} \quad \text{and} \quad \begin{pmatrix} 0 & z & 0 & 0 & 0 & \cdots & 0 \\ z & 0 & 0 & 0 & 0 & \cdots & 0 \end{pmatrix},$$

respectively. Then no vector of the column space of φ_2 belongs to the column space of either of the matrices $V_1 := (x\mathbb{I}_{N+1} \mid y\mathbb{I}_{N+1} \mid \mathbf{z}_3)$ or $V_2 := (x\mathbb{I}_{N+1} \mid y\mathbb{I}_{N+1} \mid \mathbf{z}_1)$, where $\mathbb{I}_{N+1}$ stands for the identity matrix.

Proposition 6.2.26 (Conjectured). *With the notation of* Theorem 6.2.24 *one has:*

(i) *The inverse map of $\mathfrak{F}$ admits s representatives, each given by the 2-minors of a 2×3 matrix with linear entries.*

(ii) *The source inversion factors* $D_1, \ldots, D_s$ *corresponding to the representatives of* $\mathfrak{F}^{-1}$ *in* (i) *are k-linearly independent elements of a vector basis of* $(\tilde{J}^{(2)})_{s+2}$.
(iii) *If* $\tilde{J}^{(2)} = (D_1, \ldots, D_s)$ *then* $\tilde{J}^2 = (x, y, z)\tilde{J}^{(2)} = (xf, yf, xg, zg)\tilde{J}$.

Proof. We assume that φ has the shape as in (89) and that φ_2 satisfies the statement of Conjecture 6.2.25.

Let $\mathbf{t} := \{t_1, \ldots, t_{N+1}\}$ denote coordinates of the target space $\mathbb{P}^N$, where $N = (s+3)/2$. Consider the ideal $I_1(\mathbf{t}{\cdot}\varphi) \subset R[\mathbf{t}]$. Since φ is a linear matrix with φ_1 the initial submatrix, the Jacobian matrix of these biforms with respect to x, y, z has the following shape:

$$B = \begin{pmatrix} -t_2 & t_1 & 0 \\ -t_4 & 0 & t_3 \\ \ell_{3,1} & \ell_{3,2} & \ell_{3,3} \\ \ell_{4,1} & \ell_{4,2} & \ell_{4,3} \\ \vdots & \vdots & \vdots \\ \ell_{N,1} & \ell_{N,2} & \ell_{N,3} \end{pmatrix}, \tag{90}$$

where $\ell_{i,j}$ is a certain linear form in $\mathbf{t}$.

Claim. Each of the following two sets of 1-forms in $k[\mathbf{t}]$ is k-linearly independent

$$\{t_3, t_4, \ell_{3,3}, \ldots, \ell_{N,3}\} \quad \text{and} \quad \{t_1, t_2, \ell_{3,2}, \ldots, \ell_{N,2}\}.$$

This will follow from the specified shape of $\varphi = (\varphi_1|\varphi_2)$. Namely, for $1 \le i \le N+1$ and $3 \le j \le N$, the (i,j)-entry of φ_2 is

$$\frac{\partial \ell_{j,1}}{\partial t_i}x + \frac{\partial \ell_{j,2}}{\partial t_i}y + \frac{\partial \ell_{j,3}}{\partial t_i}z \tag{91}$$

Say we have

$$\alpha_1 t_3 + \alpha_2 t_4 = \sum_{u=3}^{N} \beta_u \ell_{u3} \tag{92}$$

for certain $\alpha_1, \alpha_2, \beta_3, \ldots, \beta_N \in k$.

Taking derivatives with respect to the variables $\mathbf{t}$ yields

$$\sum_{u=3}^{N} \beta_u \frac{\partial \ell_{u3}}{\partial t_i} = \begin{cases} \alpha_1, & \text{if } i = 3 \\ \alpha_2, & \text{if } i = 4 \\ 0, & \text{otherwise} \end{cases} \tag{93}$$

Therefore, it suffices to show that $\beta_u = 0$ for $3 \le u \le N$. Taking a k-linear combination of the columns of φ_2 with the β_i's as coefficients, then (91) and (93) yield a vector in the column space of the matrix V_1 of Conjecture 6.2.25, thus contradicting this conjecture.

The argument for the other set will be analogous, by applying Conjecture 6.2.25 as regards V_2 instead.

We now come to the proof proper of the statements.

(i) First, note that the image of the map cannot be cut by a linear form because the map is defined by a k-basis of a linear system. Then, by Theorem 3.2.22, the 2-minors of any one among the following matrices gives a representative of the inverse map:

$$\begin{pmatrix} -t_2 & t_1 & 0 \\ -t_4 & 0 & t_3 \end{pmatrix}, \quad \begin{pmatrix} -t_2 & t_1 & 0 \\ \ell_{i,1} & \ell_{i,2} & \ell_{i3} \end{pmatrix} \quad \text{or} \quad \begin{pmatrix} -t_4 & 0 & t_3 \\ \ell_{i,1} & \ell_{i,2} & \ell_{i3} \end{pmatrix} \tag{94}$$

where $3 \leq i \leq N$ and the $\ell_{i,j}$s are the linear forms in the matrix (90). The reason is that the rank of any of these matrices modulo the defining equations of the image is 2, as the minors are products of linear forms and the defining equations of the image generate a prime ideal with initial degree ≥ 2 by the above observation.

Each of these representatives is spanned in degree 2 and there are $2(N-2)+1 = 2N-3 = s$ of them.

(ii) Proving this item will imply that no two of the representatives in (i) are proportional, thus avoiding repetition. We will draw upon the notion of an inverse factor (see Section 4.2).

Let $D_1, \ldots, D_s \in R$ denote these factors. Note that $\deg(D_i) = 2N - 1 = s + 2$.

First, one observes that letting $\mathbf{h} := \{xf, yf, xg, zg, h_5, \ldots, h_{N+1}\}$ denote a k-basis of $\tilde{J}_N$, an easy calculation shows that the inversion factors corresponding to the three blocks of matrices in (94) have the respective forms xfg, $f\ell_{i,3}(\mathbf{h})$, $g\ell_{i2}(\mathbf{h})$, where $3 \leq i \leq N$.

Say,

$$\alpha xfg + f\sum_{i=3}^{N} \beta_i \ell_{i,3}(\mathbf{h}) + g\sum_{i=3}^{N} \gamma_i \ell_{i,2}(\mathbf{h}) = 0 \quad (\alpha, \beta_i, \gamma_i \in k).$$

From this, there are linear forms L and L' in $k[x, y, z]$ such that

$$Lf = \sum_{i=3}^{N} \gamma_i \ell_{i,2}(\mathbf{h}) \quad \text{and} \quad L'g = \sum_{i=3}^{N} \beta_i \ell_{i,3}(\mathbf{h}).$$

Therefore, $Lf, L'g \in \tilde{J}$, and hence $L = a_1x + b_1y$ and $L' = a_1'x + b_1'z$ for certain $a_1, a_1', b_1, b_1' \in k$. It then follows that

$$a_1t_1 + b_1t_2 - \sum_{i=3}^{N} \gamma_i \ell_{i,2}(\mathbf{t}) \quad \text{and} \quad a_1't_3 + b_1't_4 - \sum \beta_i \ell_{i,3}(\mathbf{t})$$

are among the defining equations of the image of $\mathfrak{F}$. However, as remarked earlier, the image is not cut by any linear form. Consequently,

$$a_1t_1 + b_1t_2 - \sum_{i=3}^{N} \gamma_i\ell_{i,2}(\mathbf{t}) = a_1't_3 + b_1't_4 - \sum \beta_i\ell_{i,3}(\mathbf{t}) = 0.$$

By the above claim, $\beta_u = 0$ and $\gamma_u = 0$, as was to be shown.

By Proposition 4.2.4 $D_1, \ldots, D_s$ belong to the second symbolic power of $\tilde{J}$.

(iii) For the leftmost equality, proceed as follows. By Corollary 6.2.17(b), one has $(\tilde{J}^2)_{2N} = (\tilde{J}^{(2)})_{2N}$. By (a), $\tilde{J}^{(2)} = (D_1, \ldots, D_s)$, hence $(\tilde{J}^{(2)})_{2N-1} = \sum_i kD_i$. It follows that $(\tilde{J}^2)_{2N}$ is the span of $(x,y,z)_1\,(\tilde{J}^{(2)})_{2N-1}$. But since both $\tilde{J}^2$ and $\tilde{J}^{(2)}$ are generated in one single degree, we get the desired equality.

As for the rightmost equality, by definition $(x,y,z)D_i$ $(1 \leq i \leq 2N-1 = s)$ is contained in the ideal of 2-minors of one among the matrices in (94), further evaluated on the generators $\mathbf{h} = \{xf, yf, xg, zg, h_5, \ldots, h_{N+1}\}$. These evaluated matrices have the form

$$\begin{pmatrix} -yf & xf & 0 \\ -zg & 0 & xg \end{pmatrix}, \quad \begin{pmatrix} -yf & xf & 0 \\ \ell_{i,1}(\mathbf{h}) & \ell_{i,2}(\mathbf{h}) & \ell_{i3}(\mathbf{h}) \end{pmatrix} \quad \text{and} \quad \begin{pmatrix} -zg & 0 & xg \\ \ell_{i,1}(\mathbf{h}) & \ell_{i,2}(\mathbf{h}) & \ell_{i3}(\mathbf{h}) \end{pmatrix}.$$

Clearly, any 2-minor above is contained in the product $(xf, yf, xg, zg)\tilde{J}$. Therefore, we obtain the following string of relations:

$$\tilde{J}^2 = (x,y,z)\tilde{J}^{(2)} \subset (xf, yf, xg, zg)\tilde{J} \subset \tilde{J}^2.$$

Thus, we are home. □

Remark 6.2.27. It is reasonable to suspect that the assumption of item (iii) above always takes place. As for Conjecture 6.2.25, it looks fairly at hand, albeit the flawed argument stated in [161] based on introducing exceptional curves, by assuming the classical mistake that these are always irreducible.

6.2.4 Main theorem

In this part, we prove a couple of results concerning the fat ideal J itself.

Proposition 6.2.28. *Let $(s;\boldsymbol{\mu})$ denote a 3-uniform subhomaloidal type in degree $s \geq 3$, whose doubled type is proper. Let $J := I(\mathbf{p}, \boldsymbol{\mu})$, where $\mathbf{p} \subset \mathbb{P}^2$ is a set of r general points. Then J is generated in degree s and its minimal graded free resolution of J has the form*

$$0 \to R^3(-(s+1)) \oplus R^{\frac{s-3}{2}}(-(s+2)) \xrightarrow{\varphi} R^{\frac{s+5}{2}}(-s) \to J \to 0. \tag{95}$$

Proof. Since the points in $\mathbf{p}$ are general there is a Cremona map with homaloidal type $(2s-1; 2\boldsymbol{\mu})$ and these base points. That the initial degree of J is s is the content of Proposition 6.2.6—note that this result depended on the fact that $\mathbb{J} := I(\mathbf{p}, 2\boldsymbol{\mu})$ is the second symbolic power of J and on the equations of subhomaloidness, but on no assumption about three major virtual multiplicities.

To see that J is actually generated in degree s under the 3-uniform hypothesis, we draw upon Corollary 6.2.22 and, moreover, from the proof of Proposition 6.2.16, we know that $\dim J_s = \frac{s+5}{2}$, the expected dimension.

Finally, J is a codimension 2 perfect ideal, the graded minimal free resolution of J has the form

$$0 \to \bigoplus_i R^{u_i}(-i) \to R^g(-s) \to J \to 0$$

with $g = (s+5)/2$, e $i > s$. Now, quite generally, for a perfect ideal with graded free resolution

$$0 \to \bigoplus_i R^{u_i}(-i) \to \bigoplus_i R^{v_i}(-i) \to J \to 0,$$

the Betti numbers u_i can be computed as

$$u_i = v_i - \Delta^3 h_J(i),$$

where Δ denotes the difference function and h_J the Hilbert function of J [62]. Then

$$u_i = v_i - h_J(i) + 3h_J(i-1) - 3h_J(i-2) + h_J(i-3)$$

Since J_s has the expected dimension (Proposition 6.2.21), a calculation using the equations of condition of $(2s-1; \boldsymbol{\mu})$ give

$$u_{s+1} = -h_J(s+1) + 3h_J(s) - 3h_J(s-1) + h_J(s-2) = -3\frac{(s+3)}{2} + 3\frac{s+5}{2} = 3$$

and

$$u_{s+2} = -h_J(s+2) + 3h_J(s+1) - 3h_J(s) + h_J(s-1) = -5\frac{s+3}{2} + 9\frac{s+3}{2} - 3\frac{(s+5)}{2} = \frac{s-3}{2}.$$

Since $u_{s+1} + u_{s+2} = g - 1$, it obtains

$$\oplus R^{u_i}(-i) = R^3(-(s+1)) \oplus R^{g-4}(-(s+2)),$$

as claimed. □

For the next result, we will use a theorem of Cortadellas–Zarzuela. For convenience, we state the part of their result in the form that best suits our present application.

Theorem 6.2.29 ([43, Theorem 5.12]). *Let R denote a Cohen–Macaulay local ring and let $J \subset R$ stand for an ideal satisfying the following properties:*

(1) *J is unmixed, and R/J and R/J^2 have different depths.*

(2) $\max\{r(J_P) \mid J \subset P, \mathrm{ht}(P) = \mathrm{ht}(J)\} \leq 1$.
(3) $\ell(J) = \mathrm{ht}(J) + 1 \geq 3$.
(4) $r(J) \leq 2$.

Then $\operatorname{depth} \mathcal{R}_R(J) = \min\{\operatorname{depth} R/J, \operatorname{depth} R/J^2 + 1\} + \mathrm{ht}(J) + 1$.

Recall that $\ell(\_)$ denotes analytic spread, while $r(\_)$ stands for the minimum of all reduction numbers with respect to minimal reductions.

We also make use of the following basic result.

Proposition 6.2.30. *Suppose that $I \subset R = k[x,y,z]$ is a homogeneous ideal such that its special fiber $\mathcal{F}(I)$ is Cohen–Macaulay and has dimension 3. The following conditions are equivalent:*
(i) *The reduction number $r(I)$ is at most 2.*
(ii) *The Hilbert function $H_{\mathcal{F}(I)}(t)$ of $\mathcal{F}(I)$ coincides with its Hilbert polynomial for all $t \geq 0$.*

Proof. Since the degree of the numerator of the Hilbert series of $\mathcal{F}(I)$ is at most 2 by [217, Proposition 1.85], the equivalence of (i) and (iii) follows by [20, Proposition 4.1.12] or [20, Proposition 4.3.5 (c)]. □

The required estimate for the reduction number now follows from Proposition 6.2.19.

Now apply the theorem with the present values, thus getting

$$\operatorname{depth} \mathcal{R}_R(J) = \min\{1, 0 + 1\} + 2 + 1 = 4.$$

Since $\dim \mathcal{R}_R(J) = \dim R + 1 = 4$, we have proved that the Rees algebra of J is Cohen–Macaulay as claimed. □

Theorem 6.2.31. *Let $(s;\boldsymbol{\mu})$ denote a 3-uniform subhomaloidal type in degree $s \geq 3$, whose doubled type is proper, and let $J := I(\mathbf{p},\boldsymbol{\mu})$ denote the corresponding fat ideal on a set $\mathbf{p} \subset \mathbb{P}^2$ of general points. Letting $\mathfrak{F}$ denote the rational map defined by the linear system J_s, with image $W \subset \mathbb{P}^{\frac{s+3}{2}}$, one has:*
(i) *$\mathfrak{F}$ is a birational map of $\mathbb{P}^2$ onto W.*
(ii) *The following properties are satisfied:*
 (a) *W has degree s.*
 (b) *W is arithmetically Cohen–Macaulay.*
 (c) *W is ideal theoretically defined by quadrics and cubics.*

Proof. (i) By Lemma 6.2.2, the rational map defined by the linear system J_s is obtained from the rational map defined by the linear system of $\widetilde{J}_N$, where $N = (s+3)/2$. Therefore, the result follows from Theorem 6.2.24 (a).

(ii) (a) This follows from Proposition 6.2.19.

(b) The assertion means that the homogeneous coordinate ring of W is a Cohen–Macaulay ring. The latter is identified with the k-subalgebra $k[J_s] \subset R$ up to degree rescaling, while $k[J_s] \simeq k[\tilde{J}_N]$ as graded k-algebras as a consequence of the material in Subsection 6.2.1. Thus, we may assume that W is the image of the birational map defined by $\tilde{J}_N$.

Let $Z \subset \mathbb{P}^2$ denote the projective scheme defined by the ideal $\tilde{J}$.

Claim. $\deg(Z \cap L) < \operatorname{reg}(\tilde{J})$.

For this recall that since $(2s-1; s-1, s-1, s-1, 2\mu_4, 2\mu_5, \dots, 2\mu_r)$ is a proper homaloidal type, hence so is its arithmetic transform $(s+1; 2\mu_4, 2\mu_5, \dots, 2\mu_r, 1^3)$. But for a proper homaloidal type, one must have $2\mu_4 + 2\mu_5 \le s+1$, and so $\mu_4 + \mu_5 \le (s+1)/2 < (s+3)/2 = \operatorname{reg}(\tilde{J})$.

Thus, it suffices to show that $\deg(Z \cap L) \le \mu_4 + \mu_5$. But, since these multiplicities are major ones, it suffices to show that L meets Z in at most two distinct base points of $\tilde{J}$. Then we will be done if we show that no three of the base points of $\tilde{J}$ are aligned—an obvious datum for the base points of J since they are general, but not immediately obvious for the transformed points as they may not be general anymore. They do nevertheless satisfy the weaker property of no three distinct points being aligned, as we now verify.

Now, for this, we may harmless assume that the first three points of $\mathbf{p}$ are the coordinate points $p_1 = (0:0:1)$, $p_2 = (0:1:0)$, $p_3 = (1:0:0)$. Set in addition $p_i = (\alpha_1 : \beta_i : \gamma_i)$, for $4 \le i \le r$, with a general choice of all coordinates. The quadratic transformation being given by $(yz : xz : xy)$, we get $\tilde{\mathbf{p}} = \{p_1, p_2, p_3, (\gamma_i\beta_i : \alpha_i\gamma_i : \alpha_i\beta_i)\ (4 \le i \le r)\}$. Thus, the following finitely many determinants do not vanish:

$$\begin{vmatrix} 1 & 0 & 0 \\ 0 & 1 & 0 \\ \gamma_i\beta_i & \alpha_i\gamma_i & \alpha_i\beta_i \end{vmatrix}, \quad \begin{vmatrix} 1 & 0 & 0 \\ 0 & 0 & 1 \\ \gamma_i\beta_i & \alpha_i\gamma_i & \alpha_i\beta_i \end{vmatrix}, \quad \begin{vmatrix} 0 & 1 & 0 \\ 0 & 0 & 1 \\ \gamma_i\beta_i & \alpha_i\gamma_i & \alpha_i\beta_i \end{vmatrix} \quad (4 \le i \le r)$$

$$\begin{vmatrix} 1 & 0 & 0 \\ \gamma_i\beta_i & \alpha_i\gamma_i & \alpha_i\beta_i \\ \gamma_j\beta_j & \alpha_j\gamma_j & \alpha_j\beta_j \end{vmatrix}, \quad \begin{vmatrix} 0 & 1 & 0 \\ \gamma_i\beta_i & \alpha_i\gamma_i & \alpha_i\beta_i \\ \gamma_j\beta_j & \alpha_j\gamma_j & \alpha_j\beta_j \end{vmatrix}, \quad \begin{vmatrix} 0 & 0 & 1 \\ \gamma_i\beta_i & \alpha_i\gamma_i & \alpha_i\beta_i \\ \gamma_j\beta_j & \alpha_j\gamma_j & \alpha_j\beta_j \end{vmatrix} \quad (4 \le i < j \le r)$$

$$\begin{vmatrix} \gamma_i\beta_i & \alpha_i\gamma_i & \alpha_i\beta_i \\ \gamma_j\beta_j & \alpha_j\gamma_j & \alpha_j\beta_j \\ \gamma_k\beta_k & \alpha_k\gamma_k & \alpha_k\beta_k \end{vmatrix} \quad (4 \le i < j < k \le r).$$

This means that no three distinct points in $\mathbf{p}$ are aligned.

Granted the claim, we apply [77, Theorems 2.1 and 2.4] by taking I in these theorems to be $\tilde{J}$. In the notation there, the divisor D_σ is then very ample, where $\sigma = \operatorname{reg}(\tilde{J})$. But the image of the embedding it defines is the same as the image W of the birational map defined by $\tilde{J}_N$. Therefore, we are through.

(c) In order to argue about the generating degrees of the defining ideal of the image we will prove that the Rees algebra $\mathcal{R}_R(J) = R[Jt] \subset R[t]$ is Cohen–Macaulay. Once this is granted, we apply [3, Corollary 6.3] by which the relation type (Section 2.3.4) of J is at most $\dim R = 3$. But since the defining ideal of the image in the embedding is contained in $\mathcal{J}$, its minimal generators must be minimal generators of the latter. This implies that the image is ideal theoretically generated in degree ≤ 3, and hence is generated by quadrics and cubics.

To show the Cohen–Macaulayness of $\mathcal{R}_R(J)$, we will apply Theorem 6.2.29. In order to apply it to our present ideal J, we take $R = k[x,y,z]_{(x,y,z)}$ as usual and go through the list of hypotheses of the theorem. Condition (1) has two parts. The first is obvious since R/J is even Cohen–Macaulay; in particular, R/J has depth 1. The second statement follows from the fact that the symbolic power $J^{(2)} = \mathbb{J} \neq J^2$ as it contains elements of degree $2s-1$, while J^2 is generated in degree $\geq 2s$. Therefore, R/J^2 has depth zero.

Condition (2) goes as follows: locally at a minimal prime P of R/J the ideal J is a power of certain order μ of a complete intersection which, up to a change of coordinates, can be thought of as $(x,y)^\mu$ in $k[x,y]_{(x,y)}$. Clearly, (x^μ, y^μ) is a minimal reduction of the power with reduction number 1. Thus, the required maximum is attained at 1.

Condition (3) is equivalent to having $\dim k[J_s] = 3$, which follows immediately from the birationality assertion of item (i).

Finally, condition (4) is a consequence of the Cohen–Macaulayness of the image via Proposition 6.2.30. □

6.2.5 Relation to Bordiga–White parametrizations

The notion of a Bordiga–White surface has been introduced in [77, Definition 3.1].

Definition 6.2.32. Let $\mathbf{q} = \{q_1, \dots, q_r\} \subset \mathbb{P}^2$ be a set of r points and let $\mathbf{v} = (v_1, \dots, v_r)$ be a r-tuple of nonnegative integers such that:

(i) For the fat ideal $I = I(\mathbf{q}, \mathbf{v}) \subset R = k[x,y,z]$, $\deg R/I = \binom{d+1}{2}$ and $I_{d-1} = (0)$.

(ii) For any line L in $\mathbb{P}^2$, $\deg(L \cap Z) \leq d-1$, where Z is the 0-dimensional subscheme of $\mathbb{P}^2$ corresponding to the fat ideal I.

It is possible to show that the linear system I_d define a birational map $\mathbb{P}^2 \dashrightarrow \mathbb{P}^d$ onto its image (see [77, Theorem 2.1]). The closed image of this map is called a *Bordiga–White* surface.

We wish to relate this class of surfaces to the previous results.

As in the previous section, let $\boldsymbol{\mu} = (\mu_1, \dots, \mu_r)$ denote a subhomaloidal multiplicity set in degree $s \geq 3$ (s necessarily odd) and such that the associated doubled homaloidal type $(2s-1; 2\boldsymbol{\mu})$ is proper. Letting $\mathbf{p} \subset \mathbb{P}^2$ denote a set of r general points, fix a plane Cremona map $\mathfrak{F}$ with this type and further write $J := I(\mathbf{p}, \boldsymbol{\mu})$.

Let us assume that J is generated in degree s—as is the case, e. g., under the hypotheses of Proposition 6.2.28. Let as well $\tilde{J} \subset R$ as in Subsection 6.2.1 denote the ideal of the transformed points $\{p_1, p_2, p_3, \mathcal{Q}(p_4), \ldots, \mathcal{Q}(p_r)\}$ with the virtual multiplicities of $S(\mathcal{Q})$.

A general partial result away from the strict hypothesis of Proposition 6.2.28 goes as follows.

Proposition 6.2.33. *Assume that the linear system J_s has the expected dimension and that $\tilde{J}$ is generated in degree $d := 2s - (\mu_1 + \mu_2 + \mu_3)$, where $\mu_1 \geq \mu_2 \geq \mu_3$ are the three highest multiplicities. The following conditions are equivalent:*
(i) *$\tilde{J}$ is minimally generated by $d + 1$ forms.*
(ii) $e(R/\tilde{J}) = \binom{d+1}{2}$.
(iii) $\mu_1 + \mu_2 + \mu_3 = 3(s - 1)/2$.

Moreover, under any of these equivalent conditions $\tilde{J}_d$ defines a birational map to $\mathbb{P}^d$ whose image is a Bordiga–White surface.

Proof. (i) ⇒ (ii) Since $\tilde{J}$ is generated in degree d and $R/\tilde{J}$ is Cohen–Macaulay, it is generated by the maximal minors of a $(d+1)\times d$ matrix with homogeneous entries. It follows that all entries must be linear forms. This implies that $\tilde{J}$ has minimal free resolution of the form

$$0 \to R^d(-(d+1)) \longrightarrow R^{d+1}(-d) \to \tilde{J} \to 0. \tag{96}$$

Computing the degree $e(R/\tilde{J})$ out of the Hilbert series read on (96), one readily gets the stated value in (ii).

(ii) ⇒ (iii) We express the degree $e(R/\tilde{J})$ from its known value in terms of the virtual multiplicities. Thinking of $\tilde{J}$ as $J^{\mathcal{Q}}$, one has the transformed multiplicities

$$m_1 = s - (\mu_2 + \mu_3), \quad m_2 = s - (\mu_1 + \mu_3), \quad m_3 = s - (\mu_1 + \mu_2),$$

yielding $m_1 + m_2 + m_3 = 3s - 2\underline{\mu} = (2s - \underline{\mu}) + (s - \underline{\mu}) = d + (s - \underline{\mu})$, where we have written $\underline{\mu} := \mu_1 + \mu_2 + \mu_3$ for lighter reading.

Since $m_i = \mu_i$ for $i \geq 4$, it obtains

$$\begin{aligned}\sum_{i\geq 1} m_i = m_1 + m_2 + m_3 + \sum_{i\geq 4} m_i &= d + (s - \underline{\mu}) + \sum_{i\geq 4} \mu_i \\ &= d + (s - \underline{\mu}) + 3(s-1) - \underline{\mu} = d + 2(2s - \underline{\mu}) - 3 \\ &= d + 2d - 3 = 3(d-1).\end{aligned}$$

The computation of the sum of squares is slightly more involved but follows the same pattern. As $m_1^2 = s^2 - 2s(\mu_2 + \mu_3) + \mu_2^2 + \mu_3^2 + 2\mu_2\mu_3$, and m_2^2, m_3^3 have similar expressions, we derive

$$\begin{aligned}\sum_{i\geq 1} m_i^2 &= m_1^2+m_2^2+m_3^2+\sum_{i\geq 4}\mu_i^2 = m_1^2+m_2^2+m_3^2+s(s-1)-(\mu_1^2+\mu_2^2+\mu_3^2)\\ &= 3s^2-4s\underline{\mu}+2(\mu_1\mu_2+\mu_1\mu_3+\mu_2\mu_3)+2(\mu_1^2+\mu_2^2+\mu_3^2)+s^2-s-(\mu_1^2+\mu_2^2+\mu_3^2)\\ &= 4s^2-s(1+4\underline{\mu})+2(\mu_1\mu_2+\mu_1\mu_3+\mu_2\mu_3)+\mu_1^2+\mu_2^2+\mu_3^2\\ &= 4s^2-4s\underline{\mu}+\underline{\mu}^2-s=(2s-\underline{\mu})^2-s=d^2-s=d^2-(d+\underline{\mu})/2.\end{aligned}$$

Assembling yields

$$\frac{1}{2}(m_i(m_i+1))=\frac{1}{2}\left(3(d-1)+d^2-\frac{d+\underline{\mu}}{2}\right)=\frac{1}{4}(d(2d+5)-(\underline{\mu}+6)).$$

By assumption, this value is also $\binom{d+1}{2}$. This forces, upon an easy calculation, that $\underline{\mu}=3(d-2)$. Since $d=s-\underline{\mu}$, this gives in turn $\underline{\mu}=(3/2)(s-1)$.

(iii) $\Rightarrow$ (i) The expected dimension of J_s is $\dim_k(J_s)=(s+5)/2=(s+3)+1$, while $d=2s-(\mu_1+\mu_2+\mu_3)=2s-3(s-1)/2=(s+3)/2$. Since $\tilde{J}$ is generated in degree $d=(s+3)/2$ and by construction, we must have $\dim_k(J_s)=\dim_k(\tilde{J}_{(s+3)/2})$ then the minimal number of generators of $\tilde{J}$ is $d+1$.

The birationality of the map follows from the criterion of Theorem 3.2.26, provided one shows that the dimension of the k-algebra $k[\tilde{J}_d]$ is 3 (maximal possible). For this, recall once more the nature of $\tilde{J}$ which came from the linear system J_s evaluated at the three algebraically independent 2-forms defining $\mathcal{Q}$, further canceling a common factor of the resulting forms (Lemma 6.2.2). Since $\dim k[J_s]=3$, clearly then $\dim k[\tilde{J}_d]=3$ as well.

In order to conclude that the image is a Bordiga–White surface, we check the present situation against the ingredients of [77, Definition 3.1], whereby the ideal I in this reference is taken to be $\tilde{J}$. First, from (96) we see that the regularity of $R/\tilde{J}$ is d. Second, the degree of $R/\tilde{J}$ is the assumed combinatorial number. Finally, the condition on how a line intersects the corresponding ideal of fat points translates into the inequality $m_1+m_2<d$. The latter holds since $m_1+m_2=2s-(\mu_1+\mu_2+\mu_3)-\mu_3=d-\mu_3<d$ as $\mu_3>0$ due to Noether's inequality [4, Proposition 2.6.4]. □

Corollary 6.2.34. *Fix the notation and hypotheses of* Proposition 6.2.28. *Set $d:=(s+3)/2$ and let $\tilde{J}$ be as in the previous theorem. Then statements* (i) *through* (ii) *and the birationality statement of the previous theorem hold true and, moreover, the image of the Bordiga–White map defined by $\tilde{J}_d$ has degree $2d-3$.*

Proof. The assertions are clear since the hypotheses of the proposition are implied by Proposition 6.2.28. To see the degree of the image, note that composing with a quadratic transformation does not change the properties of the map. Therefore, the image under $\tilde{J}_d$ has the same degree as the image under J_s, which is $s=2d-3$ by Theorem 6.2.31(b). □

Remark 6.2.35. The map defined by $\widetilde{J}_d$ lifts to a well-known embedding of $\mathbb{P}^2$ blown-up on the base points of $\widetilde{J}_d$, induced by the corresponding very ample divisor, and the embedded surface is one of the Bordiga–White surfaces, studied in [77], where it is shown in addition that its homogeneous defining ideal is generated by the 2-minors of a suitable $2 \times (d-1)$ matrix.

6.3 Exercises

(As granted at the beginning of the chapter here, too, one assumes that $\operatorname{char}(k) = 0$, unless otherwise stated.)

Exercise 6.3.1. Prove that a proper homaloidal type T of degree $d \geq 4$ and highest virtual multiplicity $d-2$ has the shape $(d; d-2, 2^{d-2}, 1^3)$.

(*Hint*: (i) Write $T = (d; d-2, \mu_2 \geq \cdots \geq \mu_{r+1})$ and apply an arithmetic quadratic transformation Q, getting in $Q(T)$ a virtual multiplicity $d-(d-2)-\mu_2 = 2-\mu_2$, which implies that $\mu_2 \leq 2$ since $Q(T)$ is proper; (ii) Thus, $T = (d; d-2, 2^s, 1^{r-s})$ for suitable r, s; use the equations of condition to solve for r, s a linear system of two equations.)

Exercise 6.3.2. Prove that the type $(d; d-2, 2^{d-2}, 1^3)$ of the previous exercise is indeed proper.

(*Hint*: apply the Hudson test telescopically (since, at each step, the type has the same shape); finalize with $(4; 2^3, 1^3) \rightsquigarrow ((2; 1^3))$.)

Exercise 6.3.3. Let $J \subset R = k[x,y,z]$ denote the fat ideal with proper points $\{p_1, \dots, p_{d+2}\}$ and type as in the previous exercise. Let $I = ([J]_d)$ denote the base ideal of the corresponding Cremona map.

1. Give the graded free resolution of J.
2. Letting φ denote the syzygy matrix of J in (1), show that $I_1(\varphi) \subset P_1$, where P_1 is the prime ideal of p_1.
3. Assuming that $p_1 = (0:0:1)$, show that the entries of the linear submatrix $\mathfrak{L}$ of φ belong to the subring $k[x,y] \subset R$.
4. Deduce that I is a perfect ideal. (*Hint*: use Proposition 6.1.18.)

Exercise 6.3.4. Let $I \subset R = k[x,y,z]$ be the ideal of 2-minors of a 3×2 rank 2 matrix whose entries are 2-forms. Suppose that I has maximal analytic spread and that I^2 is also perfect. Show that I is the base ideal of a Cremona transformation of $\mathbb{P}^2$.

(*Hint*: Show that I^2 must have two independent linear syzygies.)

Exercise 6.3.5. Let $T = (s; \boldsymbol{\mu})$ be a subhomaloidal type with $\boldsymbol{\mu} \subset \{1,2,4\}$.

1. Prove that T is one of the following types:
$$(3; 1^6),\ (5: 2^4, 1^4),\ (7; 4, 2^6, 1^2),\ (7; 4^2, 1^{10}),\ (9; 4^3, 2^6),\ (9; 4^4, 1^8).$$
2. For which of these is the doubled type proper? (*Hint*: Apply the Hudson test to the doubled type.)

(3) By computer assistance, for each of the improper doubled types choose a corresponding set of general proper points and:
   - Give the degrees of the minimal generators of the fat ideal of the corresponding subhomaloidal type.
   - Prove that the fat ideal of the corresponding doubled type has height one.

Exercise 6.3.6. Let T denote any one of the types $(9; 4^3, 2^6)$, $(9; 4^3, 3, 2^3, 1^3)$ and $(7; 3^3, 2^3, 1^3)$ and let $\mathbf{p}$ denote a corresponding set of general points, the first three of which are the coordinate points. Let $\tilde{J}$ be the transform of the fat ideal of T on $\mathbf{p}$.

(1) Prove that T is subhomaloidal.
(2) Compute the initial degree d of $\tilde{J}$ and show that $\tilde{J}_d$ defines a rational map whose image is a Bordiga–White surface.

Exercise 6.3.7. (A. Doria) Consider the type $S = (27; 13^2, 10, 7^5, 3^2, 1)$ and its derived types $V' = (26; 12, 13, 10, 7^5, 3^2, 1)$ and $V'' = (26; 13, 12, 10, 7^5, 3^2, 1)$ as in the proof of Theorem 6.2.12. Prove:

(1) S is subhomaloidal (but only 2-uniform).
(2) V' and V'' are not exceptional types.
(3) Let J' and J'' denote the respective corresponding fat ideals of the types V' and V'' on a set of general proper points. Then $\dim_k J'_{s-1} = \dim_k J''_{s-1} = 1$.
(4) Letting $J'_{s-1} - kf'$ and $J''_{s-1} - kf''$, show that $\gcd(f', f'') \neq 1$.
(*Hint*: The last two will possibly require computer assistance.)
(5) Let J denote the fat ideal of S. Does either J or its transform in the initial degree induce a birational map onto the image?

Exercise 6.3.8. Let $J \subset k[x, y, z]$ denote the ideal of general fat points corresponding to the type $(s = 27; 13^2, 10, 7^5, 3^2, 1)$.

(1) Does J_s define a birational map onto the image?
(2) Let $\tilde{J}$ stand for the (first) transform of J, with initial degree $\tilde{s}$. Does $\tilde{J}_{\tilde{s}}$ define a birational map onto the image?

Exercise 6.3.9. Let $I \subset k[x, y, z]$ denote the reduced ideal of a set of points of $\mathbb{P}^2$ in linear general position. Show that I_2 is spanned as a vector space by an irreducible quadratic form f.

Exercise 6.3.10. Prove:

(1) $(5; 2^6)$ is a proper homaloidal type.

Let $F : \mathbb{P}^2 \dashrightarrow \mathbb{P}^2$ be a Cremona map with characteristic $(5; 2^6)$ on a set of sufficiently general proper base points, and let $I \subset k[x, y, z]$ denote the corresponding base ideal.

(2) Neither I nor the base ideal of the inverse map to F is perfect.
(3) The shape and properties of the free resolution of I as stated in Proposition 6.2.12 and Theorem 6.2.14 directly, with similar strategy, using the previous exercise.

7 Cremona transformations, II

Whereas the previous chapter dealt with the details of plane Cremona transformations, this chapter is devoted to such transformations in arbitrary dimension. As before, the goal is to give an algebraic treatment throughout. As regards the Cremona group, we envisage some special classes that include a generalization of the plane de Jonquières transformations and monomial transformations, along with properties preserved by the action of the Newton duality.

7.1 Generalized de Jonquières transformations

The notion of a Cremona transformation in arbitrary dimension that parallels the plane de Jonquières transformation was stated in [154] under the name *stellar transformations*. A group theoretic approach subsequently took place in [156]. Here, the objective is a thorough study of such maps from the purely algebraic angle, thus obtaining structural results and invariants.

As a matter of notation, we will use $\mathfrak{F}$ for certain Cremona maps, instead of the usual F—this will raise no confusion regarding the notation $\mathfrak{F}_K$ for an ideal of fat points from the previous chapter as the latter will rarely appear in the context. Since a Cremona map of $\mathbb{P}^n$ is usually assumed to have no fixed part, we will throughout accordingly focus on rational maps F defined by forms $f_1, \ldots, f_{m+1}$ having no proper common factor. In an imprecise way, we write $F = (f_1 : \cdots : f_{m+1})$, where the colon signals for homogeneous coordinates in $\mathbb{P}^n_{k(x_1,\ldots,x_{n+1})}$

Fix a geometric dimension n and let $\mathfrak{J} : \mathbb{P}^n \dashrightarrow \mathbb{P}^n$ and $\mathfrak{F} : \mathbb{P}^{n-1} \dashrightarrow \mathbb{P}^{n-1}$ be two such rational maps. For clarity in the discussion, we will distinguish $\mathbb{P}^n_x = \mathrm{Proj}(k[x_1, \ldots, x_{n+1}])$ (source) from $\mathbb{P}^n_y = \mathrm{Proj}(k[y_1, \ldots, y_{n+1}])$ (target), and, similarly, $\mathbb{P}^{n-1}_x = \mathrm{Proj}(k[x_1, \ldots, x_n])$ (source) from $\mathbb{P}^{n-1}_y = \mathrm{Proj}(k[y_1, \ldots, y_n])$ (target).

Let $\pi_x : \mathbb{P}^n_x \twoheadrightarrow \mathbb{P}^{n-1}_x$ and $\pi_y : \mathbb{P}^n_y \twoheadrightarrow \mathbb{P}^{n-1}_y$ denote the natural coordinate projections, themselves rational maps.

Definition 7.1.1. We will say that $\mathfrak{J}$ and $\mathfrak{F}$ are *confluent* if

$$\mathfrak{F} \circ \pi_x = \pi_y \circ \mathfrak{J} \tag{97}$$

as maps from $\mathbb{P}^n_x$ to $\mathbb{P}^{n-1}_y$.

Since we deal quite a bit with inverses, we will obviously allow for a similar notion reversing variables.

Lemma 7.1.2. *Let $\{g_1, \ldots, g_n\} \subset k[x_1, \ldots, x_n]$ be forms of the same degree with no proper common factor and let $\mathfrak{J}$ and $\mathfrak{F}$ be confluent rational maps as above. Then $\mathfrak{F} = (g_1 : \cdots : g_n)$ if and only if $\mathfrak{J} = (fg_1 : \cdots : fg_n : g)$ for suitable forms $f, g \in k[x_1, \ldots, x_{n+1}]$ satisfying* $\gcd(f, g) = 1$.

https://doi.org/10.1515/9783110640694-007

Proof. Let $\mathfrak{F} = (g_1 : \cdots : g_n)$ and set $\mathfrak{J} = (h_1 : \cdots : h_n : h_{n+1})$, with $\gcd(h_1,\ldots,h_n,h_{n+1}) = 1$. Then

$$\mathfrak{F} \circ \pi_x = (g_1(x_1,\ldots,x_n) : \cdots : g_n(x_1,\ldots,x_n)) = (g_1 : \cdots : g_n)$$

and

$$\pi_y \circ \mathfrak{J} = (y_1(h_1,\ldots,h_n,h_{n+1}) : \cdots : y_n(h_1,\ldots,h_n,h_{n+1})) = (h_1 : \cdots : h_n).$$

Therefore, $(h_1 : \cdots : h_n) = (g_1 : \cdots : g_n)$ define the same rational map, hence the corresponding vectors are proportional. This means that there exist nonzero forms $p,q \in k[x_1,\ldots,x_{n+1}]$ with the right degrees such that $ph_i = qg_i$ for $1 \le i \le n$. But, since $\gcd(g_1,\ldots,g_n) = 1$, then $p = 1$. Thus,

$$(h_1 : \cdots : h_n; h_{n+1}) = (qg_1 : \cdots : qg_n : h_{n+1}).$$

Now, take $f := q$ and $g := h_{n+1}$. Finally, since f divides every $h_i (1 \le i \le n)$ and $\gcd(h_1,\ldots,h_n,h_{n+1}) = 1$, then $\gcd(f,g) = 1$.

Conversely, let $\mathfrak{J} = (fg_1 : \cdots : fg_n; g)$ with $\gcd(f,g) = 1$. Then

$$\begin{aligned}\mathfrak{F} \circ \pi_x = \pi_y \circ \mathfrak{J} &= (y_1(fg_1,\ldots,fg_n,g) : \cdots : y_n(fg_1,\ldots,fg_n,g)) = (fg_1 : \cdots : g_n)\\ &= (g_1 : \cdots : g_n)\end{aligned}$$

as a rational map. Therefore, $\mathfrak{F}$ is defined by $g_1,\ldots,g_n$. □

Lemma 7.1.3. *With the above notation, suppose that $g_1,\ldots,g_n$ are algebraically independent over k and that $\gcd(f,g) = 1$. Then $fg_1,\ldots,fg_n,g$ are algebraically independent over k.*

Proof. The argument is an elementary calculation using the fact that $\gcd(f,g) = 1$. Namely, suppose $F(y_1,\ldots,y_{n+1}) \in k[y_1,\ldots,y_{n+1}]$ is a nonzero homogeneous polynomial such that $F(fg_1,\ldots,fg_n,g) = 0$. Write F as a polynomial in y_{n+1} with coefficients in $k[y_1,\ldots,y_n]$, say, $F = p_m y_{n+1}^m + \cdots + p_1 y_{n+1} + p_0$, with $p_i \in k[y_1,\ldots,y_n]$ homogeneous polynomials of degree $\deg F - i$ with $p_m, p_0 \ne 0$. Thus, the vanishing of $F(fg_1,\ldots,fg_n,g)$ translates into

$$f^{\deg F - m}[p_m(g_1,\ldots,g_n)g^m + \cdots + p_i(g_1,\ldots,g_n)f^{m-i}g^i + \cdots + p_0(g_1,\ldots,g_n)f^m] = 0,$$

or, equivalently,

$$p_m(g_1,\ldots,g_n)g^m + \cdots + p_i(g_1,\ldots,g_n)f^{m-i}g^i + \cdots + p_0(g_1,\ldots,g_n)f^m = 0.$$

Hence, $[p_m(g_1,\ldots,g_n) \ \ldots \ p_0(g_1,\ldots,g_n)]^t$ is a syzygy of the ideal power $(f,g)^m$, where $p_m(g_1,\ldots,g_n) \ne 0$ and $p_0(g_1,\ldots,g_n) \ne 0$ since $g_1,\ldots,g_n$ are algebraically independent over k. But the syzygy matrix of $(f,g)^m$ is well known to be the following $(m+1)\times m$ matrix:

$$\begin{pmatrix} -g & & & \\ f & -g & & \\ & f & \ddots & \\ & & \ddots & \\ & & & -g \\ & & & f \end{pmatrix}.$$

Thus, $p_m(g_1,\dots,g_n) = pf$ and $p_0(g_1,\dots,g_n) = qg$ for certain $p,q \in k[x_1,\dots,x_{n+1}]$. Since $p_m(g_1,\dots,g_n) \neq 0$ and $p_0(g_1,\dots,g_n) \neq 0$ and neither depends on x_{n+1}, while either f or g does, we arrive at a contradiction. □

7.1.1 Fully downgraded sequences

The idea of introducing certain downgraded forms to get insight into the structure of generators of the blowup equations has appeared in many places (see [180, 12, 89, 183, 24] for a few recent ones).

Introduce additional notation, as follows: $R' := k[\mathbf{x}'] = k[x_1,\dots,x_n]$ and $R := k[\mathbf{x}] = k[x_1,\dots,x_n,x_{n+1}]$. As in the previous section, $g_1,\dots,g_n \in R'$ are forms of the same degree d' with no proper common factor and $f,g \in R := k[\mathbf{x}] = k[x_1,\dots,x_n,x_{n+1}]$ are additional forms of degrees $d-d'$ and $d > d'$, respectively, such that $\gcd(f,g) = 1$ and, moreover, one at least involves x_{n+1} effectively. Set, in addition, $\mathbf{g} := \{g_1,\dots,g_n\}$ and $f\mathbf{g} := \{fg_1,\dots,fg_n\}$.

In the sequel, we will emphasize f, g as polynomials in x_{n+1} with coefficients in $k[\mathbf{x}']$, in particular, the respective x_{n+1}-degree will come up quite often.

Let $I \subset R$ denote the base ideal generated by $fg_1,\dots,fg_n,g$. We are interested in the Rees algebra $\mathcal{R}(I) = R[fg_1t,\dots,fg_nt,gt] \subset R[t]$ and, more precisely, in its presentation ideal defined as the kernel $\mathcal{I}$ of the homomorphism of bigraded k-algebras:

$$S := R[y_1,\dots,y_{n+1}] \twoheadrightarrow \mathcal{R}(I), \quad x_i \mapsto x_i,\ y_j \mapsto fg_jt\ (1 \le j \le n),\ y_{n+1} \mapsto gt.$$

Consider the extended ideal $J := (g_1,\dots,g_n)R \subset R$. Since the ideals $J, (fg_1,\dots,fg_n) \subset R$ have the same syzygies, there is an isomorphism of the respective Rees algebras. On the other hand, $\mathcal{R}_R(fg_1,\dots,fg_n) \subset \mathcal{R}_R(I)$ induces an inclusion of the respective presentation ideals.

Assume, moreover, that $g_1,\dots,g_n$ define a Cremona map $\mathfrak{F}$ of $\mathbb{P}^{n-1}$.

Let $\mathbf{h} = h_1,\dots,h_n \in k[y_1,\dots,y_n]$ be forms of degree d' defining the inverse map $\mathfrak{F}^{-1}$ and let Δ be the corresponding source inversion factor, i. e., the uniquely defined form of R' of degree $dd' - 1$ such that

$$h_i(\mathbf{g}) = \Delta \cdot x_i \quad (i = 1, \ldots, n) \tag{98}$$

Now let $\mathfrak{z}$ be a syzygy of $f\mathbf{g}, g$ of degree e whose last coordinate is nonzero, e. g., any Koszul syzygy involving g. Say, $\mathfrak{z}$ has standard degree e. Denote by δ the largest integer such that $(x_1, \ldots, x_n)^\delta$ contains every entry of $\mathfrak{z}$—clearly, $\delta \leq e$. Then $F_1 := [y_1 \cdots y_{n+1}] \cdot \mathfrak{z} \in \mathcal{I}$ is a form of bidegree $(e, 1)$ and, moreover, $F_1 \in (x_1, \ldots, x_n)^\delta S$. Thus, write $F_1 = F_{1,1}x_1 + \cdots + F_{1,n}x_n$, where $F_{1,i}$ are bihomogeneous polynomials in $(x_1, \ldots, x_n)^{\delta-1}S$ of bidegree $(e-1, 1)$. Define

$$F_2 = F_{1,1}h_1 + \cdots + F_{1,n}h_n.$$

In particular, F_2 has bidegree $(e-1, d'+1)$. Moreover,

$$\begin{aligned} F_2(\mathbf{x}, f\mathbf{g}, g) &= F_{1,1}(\mathbf{x}, f\mathbf{g}, g)h_1(f\mathbf{g}) + \cdots + F_{1,n}(\mathbf{x}, f\mathbf{g}, g)h_n(f\mathbf{g}) \\ &= f^{d'}\Delta[F_{1,1}(\mathbf{x}, f\mathbf{g}, g)x_1 + \cdots + F_{1,n}(\mathbf{x}, f\mathbf{g}, g)x_n] \quad \text{(by (98))} \\ &= f^{d'}\Delta \cdot F_1(\mathbf{x}, f\mathbf{g}, g) \\ &= 0 \quad (\text{because } F_1 \in \mathcal{I}) \end{aligned}$$

Thus, $F_2 \in \mathcal{I}_{(e-1,d'+1)}$.

Note that F_2 is obtained from F_1 via replacing the template of $x_1, \ldots, x_n$ by the template of $h_1, \ldots, h_n$. Moreover, $h_1, \ldots, h_n$ are algebraically independent, as they define (the inverse of) a Cremona map. Therefore, F_2 is a nonzero y_{n+1}-monoid.

Again, $F_{1,i} \in (x_1, \ldots, x_n)^{\delta-1}S$, so we can write $F_2 = F_{2,1}x_1 + \cdots + F_{2,n}x_n$ where $F_{2,i}$ are bihomogeneous polynomials in $(x_1, \ldots, x_n)^{\delta-2}S$ of bidegree $(e-2, d'+1)$. Define

$$F_3 := F_{2,1}h_1 + \cdots + F_{2,n}h_n.$$

As above, F_3 is a form of bidegree $(e-2, 2d'+1)$ and

$$F_3(\mathbf{x}, f\mathbf{g}, g) := f^{d'}\Delta \cdot F_2(\mathbf{x}, f\mathbf{g}, g) = 0.$$

Thus, $F_3 \in \mathcal{I}_{(e-2,2d'+1)}$. We also have, similarly for F_2, that F_3 is a nonzero y_{n+1}-monoid. Iterating this process eventually gives a nonzero form $F_\delta \in \mathcal{I}$ of bidegree $(e-\delta, \delta d'+1)$.

Definition 7.1.4. The set $\{F_1, F_2, \ldots, F_\delta\}$ is a *fully downgraded sequence* associated to the syzygy $\mathfrak{z}$ of $I = (f\mathbf{g}, g)$, where $\mathbf{g}$ defines a Cremona map of $\mathbb{P}^{n-1}$.

This sort of sequence, obtained by suitably downgrading the degrees of the $\mathbf{x}$-terms in benefit of upgrading the ones of the $\mathbf{y}$-terms will be invoked in other parts (see Section 8.3.3).

Corollary 7.1.5. *Keeping the above notation, suppose that* $\mathbf{g} := \{g_1, \dots, g_n\}$ *define a Cremona map of* $\mathbb{P}^n$ *and that* f, g *are* x_{n+1}*-monoids with no proper common factor, one of which at least involves* x_{n+1} *effectively. Then the ideal* $I = (f\mathbf{g}, g)$ *admits a nonzero Rees equation* F *of bidegree* $(1, D)$*, for suitable* $D \geq 1$*, depending effectively on* x_{n+1}*.*

Proof. Since f, g are x_{n+1}-monoids then there is a syzygy $\mathfrak{z}$ of degree, say, $\delta + 1$ such that δ is the largest integer for which the ideal $(x_1, \dots, x_n)^\delta$ contains the entries of $\mathfrak{z}$. Hence, the term F_δ of the fully downgraded sequence associated to $\mathfrak{z}$ has bidegree $(\delta + 1 - \delta, \delta d' + 1) = (1, D)$, with $D = \delta d' + 1$. Now, F_δ depends effectively on x_{n+1}, as otherwise, we could produce a term $0 \neq F_{\delta+1} \in \mathcal{I}$ of bidegree $(0, \delta d' + 2)$, which is an absurd by Lemma 7.1.3. □

7.1.2 Generalized de Jonquières transformations

With the notation and content of Lemma 7.1.2, we give an algebraic proof of the basic structural result of [154].

Proposition 7.1.6. *Suppose that* $\mathfrak{F}$ *and* $\mathfrak{J}$ *are confluent, where* $\mathfrak{F} = (g_1 : \cdots : g_n)$*, with* $g_1, \dots, g_n$ *having no proper common factor, and* $\mathfrak{J} = (fg_1 : \cdots : fg_n : g)$*, with* $\gcd(f, g) = 1$*. The following are equivalent:*

(i) $\mathfrak{J}$ *is birational.*
(ii) $\mathfrak{F}$ *is birational and, moreover,* f, g *are* x_{n+1}*-monoids one of which at least involves* x_{n+1} *effectively.*

Proof. (i) $\Rightarrow$ (ii) Let $\mathfrak{i}_n$ denote the identity map of $\mathbb{P}^n$ – regardless as to whether it is source or target since the details of calculation will readily identify which is which. Similarly, $\mathfrak{i}_{n-1}$ is the identity map of $\mathbb{P}^{n-1}$.

By the present assumption, $\mathfrak{J} \circ \mathfrak{J}^{-1} = \mathfrak{J}^{-1} \circ \mathfrak{J} = \mathfrak{i}_n$. The standing assumption implies

$$\mathfrak{F} \circ \pi_x \circ \mathfrak{J}^{-1} = \pi_y \circ \mathfrak{J} \circ \mathfrak{J}^{-1} = \pi_y \circ \mathfrak{i}_n = \pi_y. \tag{99}$$

On the other hand, letting $L_y = V(y_{n+1}) \subset \mathbb{P}^n$, one has an isomorphism $\pi_y|_{L_y} : L_Y \simeq \mathbb{P}^{n-1}$ induced by restriction of the coordinate projection π_y. Let $(\pi_y|_{L_y})^{-1}$ denote its inverse. Applying (99), we find

$$(\mathfrak{F} \circ \pi_x \circ \mathfrak{J}^{-1})|_{L_y} = \pi_y|_{L},$$

which further implies

$$\mathfrak{F} \circ (\pi_x \circ \mathfrak{J}^{-1}|_{L_y} \circ (\pi_y|_{L_y})^{-1}) = \pi_y|_{L_y} \circ (\pi_y|_{L_y})^{-1} = \mathfrak{i}_{n-1}.$$

Therefore, $\mathfrak{F}$ has an inverse.

For the additional assertion, we argue the following.

Claim 1. If $\deg_{x_{n+1}}(f) = \deg_{x_{n+1}}(g) = 0$, the map $\mathfrak{J}$ is not dominant.

This is clear as the defining forms of $\mathfrak{J}$ do not involve x_{n+1}, triggering

$$\dim k[fg_1, \ldots, fg_n, g] \leq n.$$

Claim 2. $\deg_{x_{n+1}}(f) \leq 1$ and $\deg_{x_{n+1}}(g) \leq 1$.

Set $\deg_{x_{n+1}}(f) = r$ and $\deg_{x_{n+1}}(g) = s$.

Let $\boldsymbol{\alpha} = (\alpha_1 : \cdots : \alpha_n : \alpha_{n+1}) \in \mathbb{P}^n_x$ denote a general point in the source and let $\boldsymbol{\alpha}' = (\alpha_1 : \cdots : \alpha_n) \in \mathbb{P}^{n-1}_{x'}$. Since f and g have no proper common factor and $\boldsymbol{\alpha}$ has general coordinates, then $f(\boldsymbol{\alpha}', x_{n+1})$ and $g(\boldsymbol{\alpha}', x_{n+1})$ have no common root as polynomials in x_{n+1}. In particular, the polynomial

$$p = p(x_{n+1}) := g(\boldsymbol{\alpha})f(\boldsymbol{\alpha}', x_{n+1}) - f(\boldsymbol{\alpha})g(\boldsymbol{\alpha}', x_{n+1}) \in k[x_{n+1}]$$

has no common root with either $f(\boldsymbol{\alpha}', x_{n+1})$ or $g(\boldsymbol{\alpha}', x_{n+1})$.

Moreover, one has $\deg p = \max\{r, s\}$. Indeed, clearly, $\deg p \leq \max\{r, s\}$. If equality fails, one must have $r = s$. Let $\mathfrak{f}, \mathfrak{g} \in k[\mathbf{x}']$, respectively, denote the rth coefficients of f, g as polynomials in x_{n+1} over $k[\mathbf{x}']$. Then the rth coefficient of p in $k[x_{n+1}]$ is $g(\boldsymbol{\alpha})\mathfrak{f} - f(\boldsymbol{\alpha})\mathfrak{g}$. Now, $g\mathfrak{f} - f\mathfrak{g} \neq 0$ since $\gcd(f, g) = 1$ entails that any syzygy of $\{f, g\}$ is a multiple of its Koszul syzygy, while $\mathfrak{f}$, $\mathfrak{g}$ do not depend on x_{n+1}. Since $\boldsymbol{\alpha}$ is a general point then also $g(\boldsymbol{\alpha})\mathfrak{f} - f(\boldsymbol{\alpha})\mathfrak{g} \neq 0$, which is a contradiction.

To proceed, it will suffice to show that $\deg p = 1$.

For this, let $\boldsymbol{\beta} = (\beta_1 : \cdots : \beta_{n+1}) \in \mathbb{P}^n_y$ be the point in the target which is the image by $\mathfrak{J}$ of the point $\boldsymbol{\alpha}$, a well-defined value as $\boldsymbol{\alpha}$ is general. According to [130, Definition 3.3], the fiber (inverse image of $\boldsymbol{\beta}$ by $\mathfrak{J}$) is defined by the ideal

$$\mathbb{J}(\boldsymbol{\alpha}) := I_2\begin{pmatrix} fg_1 & \cdots & fg_n & g \\ \beta_1 & \cdots & \beta_n & \beta_{n+1} \end{pmatrix} : I^{\infty}.$$

Since $\mathfrak{J}$ is assumed to be birational, this fiber consists of a single proper point of multiplicity one. Thus, $\mathbb{J}(\alpha)$ is one-dimensional unmixed, hence regular. Therefore, it coincides with the prime ideal of the point $\boldsymbol{\alpha}$. Since the latter is generated by (n independent linear forms among) the 2-minors of $I_2\left(\begin{smallmatrix} x_1 & \cdots & x_n & x_{n+1} \\ \alpha_1 & \cdots & \alpha_n & \alpha_{n+1} \end{smallmatrix}\right)$, in particular one has $l := \alpha_1 x_{n+1} - \alpha_{n+1}x_1 \in \mathbb{J}(\boldsymbol{\alpha})$. But $g \in I$ implies that

$$lg^t \in I_2\begin{pmatrix} fg_1 & \cdots & fg_n & g \\ \beta_1 & \cdots & \beta_n & \beta_{n+1} \end{pmatrix}, \tag{100}$$

for a certain $t \geq 1$.

Now recall that $\beta_i = f(\boldsymbol{\alpha})g_i(\boldsymbol{\alpha}')$, for $1 \leq i \leq n$, while $\beta_{n+1} = g(\boldsymbol{\alpha})$. Thus, by evaluating $\mathbf{x}' \mapsto \boldsymbol{\alpha}'$ all 2-minors of the right side of (100) involving the first n columns vanish, while the ones involving the last column are all multiple of $p = p(x_{n+1})$. Therefore, the result of so evaluating (100) is that $lg(\boldsymbol{\alpha}', x_{n+1})^t$ is a multiple of $p(x_{n+1})$ in the polynomial ring $k[x_{n+1}]$. Since $p(x_{n+1})$ has no common root with $g(\boldsymbol{\alpha}', x_{n+1})$, then $\deg p(x_{n+1}) = 1$, as was to be shown.

(ii) $\Rightarrow$ (i) Let B an $n \times (n-1)$ submatrix of the Jacobian dual matrix of $g_1, \ldots, g_n$ having rank $n-1$. By Corollary 7.1.5, there is a form F of bidegree $(1, D)$ in the defining ideal of the Rees algebra $\mathcal{R}(I)$ such that $\partial F_d / \partial x_{n+1} \neq 0$. Hence, the matrix

$$\left(\begin{array}{c|c} & \partial F/\partial x_1 \\ B & \vdots \\ & \partial F/\partial x_n \\ \hline \mathbf{0} & \partial F/\partial x_{n+1} \end{array}\right)$$

is an $(n+1) \times n$ submatrix of the Jacobian dual matrix of $fg_1, \ldots, fg_n, g$ having rank n. Hence, by Theorem 3.2.22, $fg_1, \ldots, fg_n, g$ define a Cremona map of $\mathbb{P}^n$. □

Definition 7.1.7. Under the equivalent conditions of Proposition 7.1.6, the Cremona transformation $\mathfrak{J}$ is called a *generalized de Jonquières transformation*, with *Cremona support* $\mathfrak{F}$. In particular, every generalized de Jonquières transformation is confluent with its Cremona support.

Proposition 7.1.8. *The inverse transformation of a generalized de Jonquières transformation $\mathfrak{J}$ with Cremona support $\mathfrak{F}$ is a generalized de Jonquières transformation with Cremona support $\mathfrak{F}^{-1}$*

Proof. Multiply both members of the basic equation (97) by $\mathfrak{F}^{-1}$ on the left and by $\mathfrak{J}^{-1}$ on the right, to get $\mathfrak{F}^{-1} \circ \pi_y = \pi_x \circ \mathfrak{J}^{-1}$. Therefore, $\mathfrak{J}^{-1}$ and $\mathfrak{F}^{-1}$ are confluent. Then, by Lemma 7.1.2, one may set

$$\mathfrak{F}^{-1} = (g_1' : \cdots : g_n') \quad \text{and} \quad \mathfrak{J}^{-1} = (f'g_1' : \cdots : f'g_n' : g'),$$

for certain forms $g_1', \ldots, g_n', f', g'$ in $k[y_1, \ldots, y_n, y_{n+1}]$, with $\gcd(f', g') = 1$. Thus, since $\mathfrak{J}^{-1}$ is birational, Proposition 7.1.6 implies that f', g' are y_{n+1}-monoids with at least one of them effectively involving y_{n+1}. This shows that $\mathfrak{J}^{-1}$ is a generalized de Jonquières transformation with Cremona support $\mathfrak{F}^{-1}$. □

7.1.3 Generalized de Jonquières transformations under Newton duality

It can be seen that the composite of two generalized de Jonquières transformations is one of the same kind. Therefore, by Proposition 7.1.8, the subset of the Cremona group whose elements are generalized de Jonquières transformations is a subgroup. Due to a characteristic geometric property of these transformations, this group is denoted $J_o(1; \mathbb{P}^n)$, a notation borrowed from [156].

In this part, we focus on the properties of this group under the Newton dual transform. We refer to Section 4.3 for this notion and for the notation introduced there.

Proposition 7.1.9. *If $\mathfrak{F} \in J_o(1; \mathbb{P}^n)$, then $\widehat{\mathfrak{F}} \in J_o(1; \mathbb{P}^n)$.*

Proof. As in the previous subsection, an element of $J_o(1; \mathbb{P}^n)$ can be characterized as having a representative of the form

$$\{fg_1 : \ldots : fg_n : g\},$$

where $(g_1 : \ldots : g_n)$ defines a Cremona map of $\mathbb{P}^{n-1}$ and $f, g \in k[\mathbf{x}] = k[x_1, \ldots, x_n, x_{n+1}]$ are relatively prime x_{n+1}-monoids one of which at least has positive x_n-degree. Set $\mathfrak{f} = \{fg_1, \ldots, fg_n : g\}$ and $\mathfrak{g} = \{g_1, \ldots, g_n\}$.

Set

$$\boldsymbol{\alpha} := (\alpha_1, \ldots, \alpha_{n+1})^t, \quad \boldsymbol{\beta} := (\beta_1, \ldots, \beta_{n+1})^t, \quad \boldsymbol{\gamma} := (\gamma_1, \ldots, \gamma_{n+1})^t \quad \text{and} \quad \boldsymbol{\delta} := (\delta_1, \ldots, \delta_{n+1})^t$$

for the directrix vectors of $N(\mathfrak{f})$, $N(\mathfrak{g})$, $N(f)$ and $N(g)$, respectively.

By Theorem 4.3.6, one has

$$\begin{aligned}
\widehat{\mathfrak{f}} &= \left(x_1^{d-\alpha_1} \cdots x_{n+1}^{d-\alpha_{n+1}}\right)^{-1} \mathfrak{f}(\widehat{\mathbf{x}}) \\
&= \left(x_1^{d-\alpha_1} \cdots x_{n+1}^{d-\alpha_{n+1}}\right)^{-1} \{f(\widehat{\mathbf{x}})\mathfrak{g}(\widehat{\mathbf{x}}), g(\widehat{\mathbf{x}})\} \\
&= \left(x_1^{d-\alpha_1} \cdots x_{n+1}^{d-\alpha_{n+1}}\right)^{-1} \{(x_1^{r-\gamma_1} \cdots x_{n+1}^{r-\gamma_{n+1}})\,(x_1^{s-\beta_1} \cdots x_{n+1}^{s-\beta_{n+1}})\,\widehat{f}\,\widehat{\mathfrak{g}}, (x_1^{d-\delta_1} \cdots x_{n+1}^{d-\delta_{n+1}})\,\widehat{g}\} \\
&= \{(x_1^{\alpha_1-\gamma_1-\beta_1} \cdots x_{n+1}^{\alpha_{n+1}-\gamma_{n+1}-\beta_{n+1}})\,\widehat{f}\,\widehat{\mathfrak{g}}, (x_1^{\alpha_1-\delta_1} \cdots x_{n+1}^{\alpha_{n+1}-\delta_{n+1}})\,\widehat{g}\},
\end{aligned}$$

where $d := \deg(\mathfrak{f}) = \deg(g)$, $s := \deg(\mathfrak{g})$ and $r := \deg(f)$. By Corollary 4.3.9, $\widehat{\mathfrak{g}}$ is a Cremona map of $\mathbb{P}^{n-1}$.

In addition, since g and f are relatively prime x_{n+1}-monoids one of which at least has positive x_{n+1}-degree, then so are $\widehat{g}$ and $\widehat{f}$. Note that $\beta_{n+1} = 0$, $\alpha_{n+1} = 1$ and $\alpha_i = \max\{\gamma_i + \beta_i, \delta_i\}$, for every $i = 1, \ldots, n+1$. Therefore, $(x_1^{\alpha_1-\gamma_1-\beta_1} \cdots x_{n+1}^{\alpha_{n+1}-\gamma_{n+1}-\beta_{n+1}})\,\widehat{f}$ and $(x_1^{\alpha_1-\delta_1} \cdots x_{n+1}^{\alpha_{n+1}-\delta_{n+1}})\,\widehat{g}$ are relatively prime x_{n+1}-monoids one of which at least has positive x_{n+1}-degree. □

The question as to which Cremona maps have the property that taking the Newton dual commutes with taking the inverse map is quite puzzling. By Corollary 4.3.10(ii), this is always the case for any monomial Cremona map, and, in general, this property is equivalent to having the map commute with the reciprocal involution $\mathfrak{M}$ (Corollary 4.3.10(ii)). The next proposition gives a slight generalization of the monomial case for generalized de Jonquières maps.

Proposition 7.1.10. *Let $\mathfrak{F} \in J_0(1; \mathbb{P}^n)$ be defined by the forms $fg_1, \ldots, fg_n, g$, where the underlying Cremona map $(g_1 : \ldots : g_n)$ is monomial and $f, g \in k[\mathbf{x}]$ are relatively prime forms. The following conditions are equivalent:*

(i) *$\mathfrak{F}$ commutes with the reciprocal involution $\mathfrak{M}$.*
(ii) $\widehat{f'} = g'$,

where $f = f'M_f$ and $g = g'M_g$ with M_f and M_g monomials of highest degrees dividing f and g, respectively.

Proof. By Corollary 4.3.10(i), $\mathfrak{F} \circ \mathfrak{M} = \widehat{\mathfrak{F}}$. In addition, by the proof of Proposition 7.1.9, the Newton dual $\widehat{\mathfrak{F}}$ is defined by the forms $\{\widehat{f}\,\widehat{\mathfrak{g}}\,T_1, \widehat{g}\,T_2\}$, where T_1 and T_2 are suitable monomials and this representative has no fixed part. On the other hand, the map $\mathfrak{M} \circ \mathfrak{F}$ admits as representative

$$\{gg_2\cdots g_n : gg_1g_3\cdots g_n : \ldots : gg_1\cdots g_{n-1} : fg_1\cdots g_n\}.$$

Let T be the fixed part of representative above, i. e., the gcd of its terms. Since $\{g_1, \ldots, g_n\}$ are monomials and $\gcd(f, g) = 1$, T is a monomial.

(i) $\Rightarrow$ (ii) Drawing upon the format of the preceding representatives, the assumed equality $\mathfrak{M} \circ \mathfrak{F} = \mathfrak{F} \circ \mathfrak{M}$ implies the equality $\widehat{g}\,TT_2 = fg_1\cdots g_n$ and the latter implies that $\widehat{\widehat{g}} = \widehat{f}$ by Corollary 4.3.5, and hence $g' = \widehat{f'}$, by Newton duality.

(ii) $\Rightarrow$ (i) We argue that the above respective representatives of $\mathfrak{M} \circ \mathfrak{F}$ and $\mathfrak{F} \circ \mathfrak{M}$ are proportional and, therefore, they define the same map.

In other words, we claim that the following matrix has rank 1:

$$A := \begin{pmatrix} T_1\widehat{f}\,h_1 & T_1\widehat{f}\,h_2 & \cdots & T_1\widehat{f}\,h_n & T_2\widehat{g} \\ gg_2\cdots g_n & fg_1g_3\cdots g_n & \cdots & fg_1\cdots g_{n-1} & fg_1\cdots g_n \end{pmatrix},$$

where $\widehat{\mathfrak{g}} = \{h_1, \ldots, h_n\}$. Let $A_{i,j}$ denote the 2×2 submatrix of A with the ith and jth columns $(1 \le i < j \le n+1)$. For $j \le n$, one has

$$\begin{aligned} \det(A_{i,j}) &= (T_1\widehat{f}\,h_i)\Big(g\,\frac{g_1\cdots g_n}{g_j}\Big) - (T_1\widehat{f}\,h_j)\Big(g\,\frac{g_1\cdots g_n}{g_i}\Big) \\ &= T_1\widehat{f}\,g\,\frac{g_1\cdots g_n}{g_ig_j}(h_ig_i - h_jg_j) \\ &= T_1\widehat{f}\,g\,\frac{g_1\cdots g_n}{g_ig_j}(x_1^{\beta_1}\cdots x_{n+1}^{\beta_n} - x_1^{\beta_1}\cdots x_{n+1}^{\beta_n}) \\ &= 0, \end{aligned}$$

where $\boldsymbol{\beta} := (\beta_1, \ldots, \beta_n)^t$ is the directrix vector of $N(\mathfrak{g})$. For $j = n+1$, one finds

$$\begin{aligned} \det(A_{i,n}) &= (T_1\widehat{f}\,h_i)(fg_1\cdots g_n) - (T_2\widehat{g})\Big(g\,\frac{g_1\cdots g_n}{g_i}\Big) \\ &= \Big(\frac{g_1\cdots g_n}{g_i}\Big)(T_1 f\widehat{f}\,h_i\,g_i - T_2\,g\,\widehat{g}). \end{aligned}$$

By the proof of Proposition 7.1.9, one has

$$T_1 = x_1^{\alpha_1-\gamma_1-\beta_1} \cdots x_{n+1}^{\alpha_{n+1}-\gamma_{n+1}-\beta_{n+1}} \quad \text{and} \quad T_2 = x_1^{\alpha_1-\delta_1} \cdots x_{n+1}^{\alpha_{n+1}-\delta_{n+1}},$$

where $\boldsymbol{\alpha} := (\alpha_1, \dots, \alpha_{n+1})^t$, $\boldsymbol{\gamma} := (\gamma_1, \dots, \gamma_{n+1})^t$ and $\boldsymbol{\delta} := (\delta_1, \dots, \delta_{n+1})^t$ are the directrix vectors of $N(\mathfrak{f})$, $N(f)$ and $N(g)$, respectively. Therefore, we get

$$\begin{aligned} T_1 f \hat{f} h_i g_i &= x_1^{\alpha_1-\gamma_1} \cdots x_{n+1}^{\alpha_{n+1}-\gamma_{n+1}} f \hat{f} \\ &= x_1^{\alpha_1-\gamma_1} \cdots x_{n+1}^{\alpha_{n+1}-\gamma_{n+1}} M_f f' \widehat{M_f f'} \\ &= x_1^{\alpha_1-\gamma_1} \cdots x_{n+1}^{\alpha_{n+1}-\gamma_{n+1}} M_f f' \widehat{f'} \\ &= x_1^{\alpha_1-\gamma_1} \cdots x_{n+1}^{\alpha_{n+1}-\gamma_n} M_f f' g' \end{aligned}$$

and

$$\begin{aligned} T_2 g \hat{g} &= x_1^{\alpha_1-\delta_1} \cdots x_{n+1}^{\alpha_{n+1}-\delta_{n+1}} g \hat{g} \\ &= x_1^{\alpha_1-\delta_1} \cdots x_{n+1}^{\alpha_{n+1}-\delta_{n+1}} M_g g' \widehat{M_g g'} \\ &= x_1^{\alpha_1-\delta_1} \cdots x_{n+1}^{\alpha_{n+1}-\delta_{n+1}} M_g g' \widehat{g'} \\ &= x_1^{\alpha_1-\delta_1} \cdots x_{n+1}^{\alpha_{n+1}-\delta_{n+1}} M_g f' g'. \end{aligned}$$

Write $M_f := x_1^{b_0} \cdots x_{n+1}^{b_n}$ and $M_g := x_1^{c_0} \cdots x_{n+1}^{c_n}$. Since $\widehat{f'} = g'$, $\widehat{\widehat{f'}} = f'$ and $\widehat{\widehat{g'}} = g'$, then f' and g' have the same directrix vector $\boldsymbol{\eta} := (\eta_1, \dots, \eta_{n+1})^t$. In addition, $f = M_f f'$ and $g = M_g g'$, hence $\boldsymbol{\eta} = \boldsymbol{\gamma} - (b_1, \dots, b_{n+1})$ and $\boldsymbol{\eta} = \boldsymbol{\delta} - (c_1, \dots, c_n)$. Therefore, $\det(A_{i,n})$ vanishes. □

Remark 7.1.11. Note that the condition of the proposition implies that if f (resp., g) is a monomial then so is g (resp., f).

7.1.4 Generalized de Jonquières transformations with identity support

Changing a bit our original notation, let $\mathfrak{F}$ denote a generalized de Jonquières transformation of $\mathbb{P}^n$, with Cremona support the identity $(x_1 : \dots : x_n)$. Let $I = (fx_1, \dots, fx_n, g) \subset R = k[x_1, \dots, x_n, x_{n+1}]$ stand for the corresponding base ideal.

7.1.5 The minimal free resolution of the base ideal

A free (not necessarily minimal) resolution of the base ideal I is available for arbitrary n and arbitrary Cremona support by a straightforward adaptation of the result in [90, Proposition 2.3]. As a matter of fact, this resolution is kosher regardless of the monoid condition. We give a proof in the case of identity support for the reader's convenience.

We assume throughout that $g \in (x_1, \dots, x_n)R$, where $R = k[x_1, \dots, x_n, x_{n+1}]$, i. e., the x_{n+1}-coefficient of g is not a scalar.

Proposition 7.1.12. *Let $f, g \in R = k[x_1, \dots, x_n, x_{n+1}]$ be relatively prime forms of respective degrees $d-1$ and $d \geq 2$ and such that $g \in (x_1, \dots, x_n)R$. Set $I := (x_1 f, \dots, x_n f, g)$. Then the minimal graded free resolution of R/I is*

$$0 \to R(-(n+d-1)) \to \cdots \to R(-(d+2))^{\binom{n}{3}} \to \begin{array}{c} R(-(d+1))^{\binom{n}{2}} \\ \oplus \\ R(-(2d-1)) \end{array} \xrightarrow{\psi} R(-d)^{n+1} \to R,$$

where

(a) *The syzygy matrix Ψ has the form*

$$\psi = \left(\begin{array}{ccc|c} & & & -q_1 \\ & \mathcal{K} & & \vdots \\ & & & -q_n \\ \hline 0 & \cdots & 0 & f \end{array} \right)$$

with $\mathcal{K}$ denoting the first Koszul matrix of $x_1, \dots, x_n$ and $g = q_1 x_1 + \cdots + q_n x_n$.

(b) *The $n-2$ tail terms are the $n-2$ tail terms of the Kozul complex of $\{x_1, \dots, x_n\}$ shifted by $d-1$.*

Proof. Set $\mathfrak{m} = (x_1, \dots, x_n)$. The minimal graded free resolution of $R/f\mathfrak{m}$ is the Koszul complex of the generators of $\mathfrak{m}$ shifted by $\deg f = d-1$:

$$0 \to R(-(n+d-1)) \to \cdots \to R(-(d+1))^{\binom{n}{2}} \xrightarrow{\psi} R(-d)^n \to R \to R/f\mathfrak{m} \to 0. \tag{101}$$

Now, $f\mathfrak{m} : g = (f)$. Shifting by d, we get the exact sequence

$$0 \to R(-(2d-1)) \to R(-d) \to (R/f\mathfrak{m} : g)(-d) \to 0. \tag{102}$$

On the other hand, multiplication by g on R induces the exact sequence

$$0 \to (R/f\mathfrak{m} : g)(-d) \xrightarrow{\cdot g} R/f\mathfrak{m} \to R/I \to 0.$$

Extending to a map from (101) to (102), the resulting mapping cone

$$0 \to R(-(n+d-1)) \to \cdots \to R(-(d+2))^{\binom{n}{3}} \to \begin{array}{c} R(-(d+1))^{\binom{n}{2}} \\ \oplus \\ R(-(2d-1)) \end{array} \xrightarrow{\psi} R(-d)^{n+1} \to R$$

is a graded free resolution of R/I. Due to its shifts, this resolution is actually minimal. □

Corollary 7.1.13. *With the notation of* Proposition 7.1.12, *one has:*
(i) *I is saturated.*
(ii) *$(x_1, \dots, x_n)$ is an associated prime of R/I. In particular, R/I is Cohen–Macaulay if and only if $n = 2$.*

Proof. (i) It follows from the resolution which gives depth $R/I = 1$.

(ii) Since $(x_1, \dots, x_n) \subset I : f$, then $(x_1, \dots, x_n)$ is contained in an associated prime of R/I. If the containment were proper, it would contradict (i). □

Example 7.1.14 (Plane de Jonquières transformation). The generalized de Jonquières transformation of dimension 2 with identity support is the classical de Jonquières transformation with homaloidal type $(d; d-1, 1^{2d-2})$, for any given $d \geq 1$. It is in fact the only generalized de Jonquières transformation for $n = 2$. This is because the only Cremona map of $\mathbb{P}^1$ is the identity.

The base ideal $I \subset k[x, y, z]$ of a plane de Jonquières transformation of degree $d \geq 2$ has the following properties:
- $k[x, y, z]/I$ is a Cohen–Macaulay ring of multiplicity $d(d-1) + 1$.
- I is an ideal of linear type if and only if $d = 2$.

7.2 The graph (blowup) of a Cremona map

Quite generally, if $I \subset R = k[x_1, \dots, x_n]$ is the base ideal of a Cremona map in $\mathbb{P}_k^{n-1}$ then I is generated by algebraically independent elements. This means that the special fiber of I is but a copy of R. Thus, typically the interest lies in studying the ring R/I itself and the Rees algebra $\mathcal{R}(I) = \mathcal{R}_R(I)$.

7.2.1 The graph of a de Jonquières map with identity support

Recall from Subsection 7.1.4 that a generalized de Jonquières transformation has base ideal $I = (x_1 f, \dots, x_n f, g)$, where $\deg(g) = d \geq 2$ and f, g are x_{n+1}-monoids satisfying an additional condition. In particular, $g \in (x_1, \dots, x_n)R$, with $R = k[x_1, \dots, x_{n+1}]$.

By Proposition 7.1.12(a), the syzygy matrix of I has the following shape:

$$\psi = \left(\begin{array}{ccc|c} & & & -q_1 \\ & \mathcal{K} & & \vdots \\ & & & -q_n \\ \hline 0 & \cdots & 0 & f \end{array} \right)$$

where $\mathcal{K}$ is the first Koszul matrix of $x_1, \dots, x_n$ and $g = q_1 x_1 + \cdots + q_n x_n$.

The forms

$$p_{i,j} := x_j y_i - x_i y_j \quad (1 \le i < j \le n) \quad \text{and} \quad F := f\, y_{n+1} - \sum_{i=1}^{n} q_i \cdot y_i \tag{103}$$

associated to the column of ψ belong to the presentation ideal $\mathcal{I} \subset R[y_1, \dots, y_{n+1}]$ of the Rees algebra $\mathcal{R}_R(I)$ of I, obtained by mapping $y_i \mapsto x_i f$, for $1 \le i \le n$, and $y_{n+1} \mapsto g$.

7.2.1.1 Main structure result

A fully downgraded sequence is much simpler when the Cremona support is the identity. There is an obvious syzygy to play the role of $\mathfrak{z}$ in the notation of Subsection 7.1.1, and that is F as in (103); it has standard degree $\deg(f) = \deg(g) - 1 = d - 1$. Besides, the inverse map to the identity is defined by the variables $y_1, \dots, y_n$, hence most of the calculation becomes obvious.

Let then $\{F_0 = F, F_1, \dots, F_{d-2}\} \subset \mathcal{I}$ stand for a fully downgraded sequence, where for convenience we have shifted the indices by one to start at zero. Recall that, by construction, in this case F_{d-2} has bidegree $(1, d-1)$, since $d-1$ is the standard degree of the right most column of ψ. We state the main highlights about these forms in the next remark.

Remark 7.2.1. The forms F_i, $0 \le i \le d-2$, have the following properties:
(a) F_i is a strict y_{n+1}-monoid in $k[x_1, \dots, x_{n+1}, y_1, \dots, y_{n+1}]$ that depends effectively on y_{n+1}, as was already remarked.
(b) If $d \ge 3$, then $F_i \in (x_1, \dots, x_n)^{d-(i+2)} R[y_1, \dots, y_{n+1}]$, for $0 \le i \le d-2$.
(c) $x_j F_i - y_j F_{i-1} = \sum (x_j y_k - x_k y_j) F_{i-1,k} \quad (j = 1, \dots, n)$, for $1 \le i \le d-2$.

Only (a) needs a bit of an argument. Namely, for any chosen decomposition $F_0 = F_{0,1}x_1 + \cdots + F_{0,n}x_n$ as displayed in Subsection 7.1.1, there is a unique $1 \le i \le n$ such that $F_{0,i} = f\, y_{n+1}/x_i$. This nonzero form $F_{0,i}$ is then the only entry of the corresponding "content matrix" having a term in y_{n+1}. Therefore, the determinant has degree one in y_{n+1}. The argument for the subsequent F_i's is the same.

Using F_{d-2}, we can prove the following naturally expected property.

Proposition 7.2.2. *Let $\mathfrak{J}$ denote a generalized de Jonquières map of degree $d \ge 2$ with identity support. Then the inverse map is defined by the forms*

$$\frac{\partial F_{d-2}}{\partial x_{n+1}} y_1, \dots, \frac{\partial F_{d-2}}{\partial x_{n+1}} y_n, \sum_{i=1}^{n} \frac{\partial F_{d-2}}{\partial x_i} y_i.$$

In particular, the inverse map is a generalized de Jonquières map of degree d with identity support.

Proof. An $(n+1)\times n$ submatrix of the Jacobian dual matrix of $\mathfrak{F}$ is

$$\Psi = \left(\begin{array}{ccc|c} & & & \partial F_{d-2}/\partial x_1 \\ & \mathcal{K}(\mathbf{y}') & & \vdots \\ & & & \partial F_{d-2}/\partial x_n \\ \hline 0 & \cdots & 0 & \partial F_{d-2}/\partial x_{n+1} \end{array}\right), \qquad (104)$$

where $\mathcal{K}(\mathbf{y}')$ denotes the first Koszul matrix of the variables $\mathbf{y}' = y_1,\dots,y_n$. Set

$$\mathfrak{f} := \frac{\partial F_{d-2}}{\partial x_{n+1}} \quad \text{and} \quad \mathfrak{g} := \sum_{i=1}^{n} \frac{\partial F_{d-2}}{\partial x_i} y_i.$$

By Theorem 3.2.22(i), the ordered signed n-minors of Ψ define the inverse map $\mathfrak{F}^{-1}$. The latter are $y_n^{n-1}\mathfrak{f}y_1,\dots,y_n^{n-1}\mathfrak{f}y_n, y_n^{n-1}\mathfrak{g}$. Hence, $\mathfrak{f}y_1,\dots,\mathfrak{f}y_n,\mathfrak{g}$ also define $\mathfrak{F}^{-1}$. To conclude, we prove the following.

Claim. $\mathfrak{f}$ and $\mathfrak{g}$ have no proper common factor.

Set $\mathfrak{h} = \gcd(f,g)$. Then, for $\mathfrak{f}' = \mathfrak{f}/\mathfrak{h}$ and $\mathfrak{g}' = \mathfrak{g}/\mathfrak{h}$, the forms $\mathfrak{f}'y_1,\dots,\mathfrak{f}'y_n,\mathfrak{g}'$ define also a representative to $\mathfrak{F}^{-1}$. Say that $d' := \deg \mathfrak{f}'y_i = \deg \mathfrak{g}'$. By definition, $d' \le d$. Let $\mathcal{I}'$ be presentation ideal of the Rees algebra of the ideal $(\mathfrak{f}'y_1,\dots,\mathfrak{f}'y_n,\mathfrak{g}') \subset k[y_1,\dots,y_n,y_{n+1}]$. By the above discussion, there is a biform $F'_{d'-2} \in \mathcal{I}'_{(d'-1,1)}$ effectively depending on y_{n+1}. But, $\mathcal{I}' = \mathcal{I}$. Hence, $F'_{d'-2}$ gives rise to a syzygy of $f\mathbf{g}$, g of degree $d'-1$ whose $(n+1)$th coordinate is nonzero. Thus, by Proposition 7.1.12, $d' \ge d$. Hence, $d = d'$. Thus, the equality $\mathfrak{g}' = \mathfrak{g}/\mathfrak{h}$ implies $\mathfrak{h} = 1$. Therefore, the claim follows. □

Let again $f,g \in R = k[x_1,\dots,x_n,x_{n+1}]$ be relatively prime forms of degree $d-1$ and $d \ge 2$, respectively, and set $I := (x_1f,\dots,x_nf,g)$. By Corollary 7.1.13, $(x_1,\dots,x_n)$ is an associated prime of R/I, while $\mathfrak{m} = (x_1,\dots,x_{n+1})$ is not. Therefore, R admits a linear form of the shape $\ell = x_{n+1}-\lambda$, regular on R/I, where $\lambda \in k[x_1,\dots,x_n]$. Then the k-algebra surjection

$$R = k[x_1,\dots,x_{n+1}] \twoheadrightarrow k[x_1,\dots,x_n], \quad x_i \mapsto x_i\,(1 \le i \le n), \quad x_{n+1} \mapsto \lambda$$

yields an identification $R/(\ell) \simeq S := k[x_1,\dots,x_n]$.

Now, set $J = (I,\ell)/(\ell) \subset R/(\ell) = S$. Since ℓ is regular both on R and on R/I, the ring S/J is a specialization of R/I. In particular, J is an ideal of height 2 minimally generated by $n+1$ forms of degree d whose syzygy matrix is the corresponding specialization of ψ.

Denote by $\mathcal{J} \subset S[y_1,\dots,y_{n+1}]$ the presentation ideal of the Rees algebra $\mathcal{R}_S(J)$. One has $(\mathcal{I},\ell)/(\ell)$ upon the identification $R[y_1,\dots,y_{n+1}]/(\ell) = S[y_1,\dots,y_{n+1}]$. For a form $P \in R[y_1,\dots,y_{n+1}]$ denote by $\overline{P}$ its image in $S[y_1,\dots,y_{n+1}]$.

Drawing upon the images of the downgraded sequence $F_0,F_1,\dots,F_{d-2}$, one knows by [12, Theorem 3.1] that $\mathcal{J} = (\overline{p}_{i,j},\ 1 \le i < j \le n, \overline{F}_0,\overline{F}_1,\dots,\overline{F}_{d-2},\mathfrak{H})$ where $\mathfrak{H}$ is a form of bidegree $(0,d)$ in $S[y_1,\dots,y_{n+1}]$ (the implicit equation).

Theorem 7.2.3. *Let $I \subset R = k[x_1, \ldots, x_n, x_{n+1}]$ denote the base ideal of a generalized de Jonquières map $\mathfrak{J}$ with identity support and let $\mathcal{I} \subset R[y_1, \ldots, y_{n+1}]$ be the presentation ideal of the Rees algebra $\mathcal{R}_R(I)$ of I obtained by mapping $y_i \mapsto x_i f$, for $1 \le i \le n$, and $y_{n+1} \mapsto g$. Then*

$$\{p_{i,j},\ 1 \le i < j \le n,\ F_0, F_1, \ldots, F_{d-2}\} \tag{105}$$

is a minimal set of generators of $\mathcal{I}$.

Proof. Let $P \in \mathcal{I}_{(i,j)}$ be a form of bidegree (i,j). Consider two cases:

($j < d$) Induct on i.

There is nothing to prove if $i = 0$ since $\mathfrak{J}$ is in particular a dominant map. Let $i \ge 1$. Then $\overline{P} \in \mathcal{J}_{(i,j)}$ and, since $j < d$, it cannot involve effectively the form $\mathfrak{H}$, hence we can write

$$\overline{P} = \sum_{1 \le u < v \le n} \overline{A}_{u,v} \overline{p}_{u,v} + \overline{B} \overline{F}_0 + \sum_{u=1}^{d-2} \overline{C}_u \overline{F}_u$$

for certain forms $\overline{A}_{u,v}$, $\overline{B}$, $\overline{C}_u$ of appropriate bidegrees in $S[y_1, \ldots, y_{n+1}]$. Lifting back to $R[y_1, \ldots, y_{n+1}]$ yields

$$P - \sum_{1 \le u < v \le n} A_{u,v} p_{u,v} - BF_0 - \sum_{u=1}^{d-2} C_u F_u = \ell Q$$

for a form $Q \in R[y_1, \ldots, y_{n+1}]$ of bidegree $(i - 1, j)$. Clearly, then $\ell Q \in \mathcal{I}_{(i,j)}$, hence $Q \in \mathcal{I}_{i-1,j}$ since $\mathcal{I}$ is prime. Thus, by induction $Q \in (p_{i,j},\ 1 \le i < j \le n,\ F, F_1, \ldots, F_{d-2})$. Therefore $P \in (p_{i,j},\ 1 \le i < j \le n,\ F_0, F_1, \ldots, F_{d-2})$.

($j \ge d$) Let $\mathfrak{J}^{-1} = (\mathfrak{g}_1 : \ldots : \mathfrak{g}_{n+1}) : \mathbb{P}^n \dashrightarrow \mathbb{P}^n$ be the inverse to $\mathfrak{J}$. By Proposition 7.2.2, $\mathfrak{g}_i$ is a form in $k[y_1, \ldots, y_{n+1}]$ of degree d.

Claim. $\ell(\mathfrak{g}_1, \ldots, \mathfrak{g}_{n+1})$ and $\mathfrak{H}$ coincide up to a scalar multiplier.

To see this, let $\delta \in R$ denote the source inversion factor of $\mathfrak{J}$. That is, δ is a homogeneous polynomial of R of degree $d^2 - 1$ such that

$$\mathfrak{g}_i(x_1 f, \ldots, x_n f, g) = x_i \delta, \quad (1 \le i \le n+1).$$

Hence,

$$\ell(\mathfrak{g}_1, \ldots, \mathfrak{g}_{n+1})(x_1 f, \ldots, x_n f, g) = \ell(\delta x_1, \ldots, \delta x_{n+1}) = \delta \ell(x_1, \ldots, x_{n+1}).$$

Reducing modulo ℓ gives

$$\ell(\mathfrak{g}_1, \ldots, \mathfrak{g}_{n+1})(\overline{x_1 f}, \ldots, \overline{x_n f}, \overline{g}) = 0.$$

This means that $\ell(\mathfrak{g}_1,\dots,\mathfrak{g}_{n+1})$ is a multiple of the implicit equation $\mathfrak{H}$. But $\deg\mathfrak{H} = d = \deg\ell(\mathfrak{g}_1,\dots,\mathfrak{g}_{n+1})$. Hence, $\ell(\mathfrak{g}_1,\dots,\mathfrak{g}_{n+1})$ and $\mathfrak{H}$ are the same up to a scalar multiplier as required.

Again induct on i. As before, we can assume that $i \geq 0$. Since $j \geq d$, then

$$\overline{P} = \sum_{1\leq u<v\leq n} \overline{A}_{u,v}\overline{p}_{u,v} + \overline{BF}_0 + \sum_{u=1}^{d-2} \overline{C}_u\overline{F}_u + \overline{D}\ell(\mathfrak{g}_1,\dots,\mathfrak{g}_{n+1}),$$

for appropriate coefficients. Hence,

$$P - \sum_{1\leq u<v\leq n} A_{u,v}p_{u,v} - BF_0 - \sum_{u=1}^{d-2} C_uF_u = D\ell(\mathfrak{g}_1,\dots,\mathfrak{g}_{n+1}) + \ell Q,$$

for some $Q \in R[y_1,\dots,y_{n+1}]$ of bidegree $(i-1,j)$. Then

$$D\ell(\mathfrak{g}_1,\dots,\mathfrak{g}_{n+1}) + \ell(x_1,\dots,x_{n+1})Q \in \mathcal{I}_{i,j}. \tag{106}$$

Now, letting $I' := (\mathfrak{g}_1,\dots,\mathfrak{g}_{n+1})$, there is a k-algebra isomorphism $\mathcal{R}_R(I) \simeq \mathcal{R}_R(I')$ induced by the identity map of the bigraded polynomial ring $R[y_1,\dots,y_{n+1}]$ (Proposition 3.2.14). There follows the equality $\mathcal{I} = \mathcal{I}'$ of the respective presentation ideals. Thus,

$$D\ell(\mathfrak{g}_1\dots,\mathfrak{g}_{n+1}) + \ell(x_1,\dots,x_{n+1})Q \in \mathcal{I}_{(i,j)} \subset \mathcal{I}'$$

as well. This means that, by evaluating $x_i \mapsto \mathfrak{g}_i$ $(1\leq i\leq n+1)$, (106) gives

$$D(\mathfrak{g}_1,\dots,\mathfrak{g}_{n+1},y_1,\dots,y_{n+1})\ell(\mathfrak{g}_1,\dots,\mathfrak{g}_{n+1}) + \ell(\mathfrak{g}_1,\dots,\mathfrak{g}_{n+1})Q(\mathfrak{g}_1,\dots,\mathfrak{g}_{n+1},y_1,\dots,y_{n+1}) = 0.$$

Since $\ell \notin \mathcal{I}$, $\ell(\mathfrak{g}_1,\dots,\mathfrak{g}_{n+1}) \neq 0$; canceling yields

$$D(\mathfrak{g}_1,\dots,\mathfrak{g}_{n+1},y_1,\dots,y_{n+1}) + Q(\mathfrak{g}_1,\dots,\mathfrak{g}_{n+1},y_1,\dots,y_{n+1}) = 0,$$

which means similarly that

$$D + Q = D(x_1,\dots,x_{n+1},y_1,\dots,y_{n+1}) + Q(x_1,\dots,x_{n+1},y_1,\dots,y_{n+1}) \in \mathcal{I}.$$

But, D and Q have respective bidegrees $(i,j-d) \neq (i-1,j)$. Hence, $D, Q \in \mathcal{I}$. By the inductive assumption $Q \in (p_{i,j},\ 1\leq i<j\leq n,\ F_0,F_1,\dots,F_{d-2})$. As to D, if $j-d<d$ then by the previous argument $D \in (\overline{p}_{i,j},\ 1\leq i<j\leq n,\ \overline{F}_0,\overline{F}_1,\dots,\overline{F}_{d-2},\mathfrak{H})$. Otherwise, we replace P by D and repeat the above argument. Since the second component of the bidegree goes down at each such step, we are through.

The assertion about minimality follows from [12], where it is proved that $\mathcal{J}$ is minimally generated by the set $\{\overline{p}_{i,j},\ 1\leq i<j\leq n,\ \overline{F}_0,\overline{F}_1,\dots,\overline{F}_{d-2},\mathfrak{H}\}$. □

7.2.1.2 The depth of the Rees algebra

We now look more closely at the structure of the Rees algebra $\mathcal{R}_R(I)$ of the base ideal of a generalized de Jonquières map. At the outset, consider the ideal $I_2(\mathbf{x}|\mathbf{y}) = (p_{i,j}) \subset k[\mathbf{x}, \mathbf{y}]$ of 2-minors of the $2 \times n$ matrix $(\mathbf{x}|\mathbf{y})$ with rows $\mathbf{x} = x_1 \dots x_n$ and $\mathbf{y} = y_1 \dots y_n$. The next result seems to be known, but we have not been able to find a precise reference. It is a certain analogue of the well-known property saying that the associated graded ring of the irrelevant maximal ideal of a standard graded residue ring $k[\mathbf{x}]/J$ is (isomorphic to) $k[\mathbf{x}]/J$ itself.

Lemma 7.2.4. *Let* $\mathfrak{n}$ *denote the residue ideal in* $k[\mathbf{x}, \mathbf{y}]/I_2(\mathbf{x}|\mathbf{y})$ *of the ideal* $(\mathbf{x}) \subset k[\mathbf{x}, \mathbf{y}]$. *Then the associated graded ring of* $\mathfrak{n}$ *is isomorphic to* $k[\mathbf{x}, \mathbf{y}]/I_2(\mathbf{x}|\mathbf{y})$. *In particular, all powers of* $\mathfrak{n}$ *are* $\mathfrak{n}$*-primary.*

Proof. It suffices to show the main assertion because, as is well known, the associated graded ring being a domain implies that the symbolic powers are ordinary powers.

Set $A := k[\mathbf{x}, \mathbf{y}]/I_2(\mathbf{x}|\mathbf{y})$. The assertion follows from the following.

Claim. Let $\mathcal{R}_A(\mathfrak{n}) \simeq A[\mathbf{z}]/\mathcal{L}$ be a presentation of the Rees algebra of $\mathfrak{n}$ over A, where $\mathbf{z} = \{z_1, \dots, z_n\}$ are indeterminates over A. Then $\mathcal{L}$ is the residue of the ideal $I_2(G) \subset k[\mathbf{x}, \mathbf{y}, \mathbf{z}]$, where

$$G = \begin{pmatrix} x_1 & \cdots & x_n \\ y_1 & \cdots & y_n \\ z_1 & \cdots & z_n \end{pmatrix}.$$

To prove the claim, as is easily seen, $I_2(G)$ is contained in the inverse image of $\mathcal{L}$ in $k[\mathbf{x}, \mathbf{y}, \mathbf{z}]$. But the first is a prime ideal of height $(3-2+1)(n-2+1) = 2(n-1)$ in $k[\mathbf{x}, \mathbf{y}][\mathbf{z}]$. Therefore, its height in $A[\mathbf{z}]$ is $2(n-1) - (n-1) = n-1$, because the height of $I_2(\mathbf{x}|\mathbf{y})$ in $k[\mathbf{x}, \mathbf{y}]$ is $n-1$. But we know that the height of $\mathcal{L}$ is one less the number of generators of $\mathfrak{n}$, hence is also $n-1$. This triggers the required equality. □

Now, $I_2(\mathbf{x}|\mathbf{y})$ fits in the following increasing inclusion sequence of ideals in the presentation ideal $\mathcal{I}$:

$$\mathcal{P}_0 := (I_2(\mathbf{x}|\mathbf{y})) \subset \mathcal{P}_1 := (\mathcal{P}_0, F_0) \subset \mathcal{P}_2 := (\mathcal{P}_1, F_1) \subset \cdots \subset \mathcal{P}_{d-1} := (\mathcal{P}_{d-2}, F_{d-2}).$$

Here, the round brackets indicate an ideal of $R[y_1, \dots, y_{n+1}]$, where $R = k[x_1, \dots, x_{n+1}]$. Note that, by Theorem 7.2.3, $\mathcal{P}_{d-1} = \mathcal{I}$.

The following lemma clarifies the role of these ideals.

Lemma 7.2.5. *With the above notation, one has:*

(i) *Let* $\mathfrak{f} \in \mathcal{I}$ *be a form of bidegree* (r, s). *If* $\mathfrak{f} \in (\mathcal{P}_0, (x_1, \dots, x_n)^r) \subset R[y_1, \dots, y_{n+1}]$ *then* $\mathfrak{f} \in \mathcal{P}_0$.

(ii) $F_i \notin (\mathcal{P}_0, (x_1, \dots, x_n)^{d-i})R[y_1, \dots, y_{n+1}]$, *for every* $1 \le i \le d-2$.

(iii) $\mathcal{P}_0 : F_0 = \mathcal{P}_0$ *and* $\mathcal{P}_i : F_i = (x_1, \dots, x_n)$ $(1 \le i \le d-2)$.

Proof. (i) Writing $\mathfrak{f} = \mathfrak{f}_1 + \mathfrak{f}_2$ where $\mathfrak{f}_1 \in \mathcal{P}_0$ and $\mathfrak{f}_2 \in (x_1, \ldots, x_n)^r$ we have that $\mathfrak{f}_2$ is a form of bidegree (r, s) in $\mathcal{I}$ that depends only on the variables $x_1, \ldots, x_n, y_1, \ldots, y_{n+1}$. Using an exchange principle based on Proposition 3.2.14(b), one sees that $\mathfrak{f}_2$ is a Rees relation of the ideal $(y_1, \ldots, y_n)$, hence $\mathfrak{f}_2 \in \mathcal{P}_0$.

(ii) If $F_i \in (\mathcal{P}_0, (x_1, \ldots, x_n)^{d-i})R[y_1, \ldots, y_{n+1}]$ then, by (i), we have $F_i \in \mathcal{P}_0$, which contradicts the fact that, by Theorem 7.2.3, $\{p_{i,j}, 1 \le i < j \le n, F_0, F_1, \ldots, F_{d-2}\}$ is a minimal set of generators of $\mathcal{I}$.

(ii) The first equality is clear since $\mathcal{P}_0$ is a prime ideal and $F_0 \notin \mathcal{P}_0$, by Remark 7.2.1(a). Now, take $1 \le i \le d-2$. By Remark 7.2.1(c), $(x_1, \ldots, x_n) \subset \mathcal{P}_i : F_i$. For the other inclusion, consider $a \in \mathcal{P}_i : F_i$. Then

$$aF_i \in \mathcal{P}_i \subset (\mathcal{P}_0, (x_1, \ldots, x_n)^{d-i})R[y_1, \ldots, y_{n+1}], \tag{107}$$

where the inclusion of ideals is consequence of Remark 7.2.1(b). Thus,

$$\overline{aF_i} \in (\overline{x}_1, \ldots, \overline{x}_n)^{d-i}$$

where "–" denotes the residual class of a element module $\mathcal{P}_0 = I_2(\mathbf{x}|\mathbf{y})$. By Lemma 7.2.4, $(\overline{x}_1, \ldots, \overline{x}_n)^{d-i}$ is $(\overline{x}_1, \ldots, \overline{x}_n)$-primary and, by part (i), $\overline{F_i} \notin (\overline{x}_1, \ldots, \overline{x}_n)^{d-i}$. Therefore, $\overline{a} \in (\overline{x}_1, \ldots, \overline{x}_n)$. That is, $a \in (x_1, \ldots, x_n)$ since $\mathcal{P}_0 \subset (x_1, \ldots, x_n)$. □

Proposition 7.2.6. *Let $R = k[x_1, \ldots, x_n, x_{n+1}]$, with $n \ge 2$, and let $I \subset R$ denote the base ideal of a generalized de Jonquières map of degree $d \ge 2$ with identity support. Then the Rees algebra $\mathcal{R}(I)$ is almost Cohen–Macaulay.*

Proof. Set $A = R[y_1, \ldots, y_{n+1}]$. We depart from two well-known free resolutions. The first is the 2-linear resolution of $A/\mathcal{P}_0$:

$$0 \to G_{n-1,0} \to \cdots \to G_{2,0} \to G_{1,0} \to A \to A/\mathcal{P}_0 \to 0 \tag{108}$$

where $G_{i,0} = A(-(i+1))^{i\binom{n}{i+1}}$, for $i \ge 1$ and $G_{0,0} = A$. The second is the shifted Koszul complex resolving the complete intersection $(A/(x_1, \ldots, x_n))(-d)$:

$$0 \to K_n(-(d+n)) \to \cdots \to K_1(-(d+1)) \to A(-d) \to (A/(x_1, \ldots, x_n))(-d) \to 0. \tag{109}$$

To start, consider the exact sequence of multiplication by F_0 on $A/\mathcal{P}_0$:

$$A/\mathcal{P}_0 \xrightarrow{\cdot F_0} A/\mathcal{P}_0 \longrightarrow A/(\mathcal{P}_0, F_0) = A/\mathcal{P}_1 \to 0,$$

using the definition of $\mathcal{P}_1$. This map extends to a map of (108) to itself and the resulting mapping cone [179, Section 6.2.3.2] has the form

$$0 \to G_{n,1} \to G_{n-1,1} \to \cdots \to G_{1,1} \to A \to A/\mathcal{P}_1 \to 0, \tag{110}$$

where $G_{i,1} = G_{i,0} \oplus G_{i-1,0}$, for $0 \le i \le n$. This a graded free resolution of $A/\mathcal{P}_1$ of length n with $G_{n,1} = G_{n-1,0} = A(-n)^{n-1}$.

Next, consider the homogeneous map

$$(A/(x_1,\ldots,x_n))(-d) = (A/\mathcal{P}_1 : F_1)(-d) \to A/\mathcal{P}_1$$

induced by multiplication by F_1, with cokernel $A/(\mathcal{P}_1, F_1) = A/\mathcal{P}_2$. It extends to a map from (109) to (110), with resulting mapping cone given by

$$0 \to G_{n+1,2} \to G_{n,2} \to \cdots \to G_{1,2} \to A \to A/\mathcal{P}_2 \to 0, \qquad (111)$$

where $G_{i,2} = G_{i,1} \oplus K_{i-1}(-(d+i-1))$. This resolution has length $n+1$.

Repeat the last argument with multiplication by F_2 on $A/\mathcal{P}_2$ and the extended map from (109) to (111) above to get a graded free resolution of $A/\mathcal{P}_3$ of length $n+1$.

Proceeding this way, we eventually reach a graded free resolution of $A/\mathcal{P}_{d-1} = A/\mathcal{I}$ of length $n+1$. Therefore, proj.dim $A/\mathcal{I} \le n+1$, i. e., depth $A/\mathcal{I} \ge n+1$, showing that $\mathcal{R}_R(I)$ is almost Cohen–Macaulay. □

Conjecture 7.2.7. *With the same notation as in the previous theorem, $\mathcal{R}_R(I)$ is Cohen–Macaulay if and only if $d \le n+1$.*

Note that if $\mathcal{R}(I)$ is Cohen–Macaulay then the relation type of I is at most the analytic spread of I (see, e. g., [3, Corollary 6.3]). Therefore, $d-1 \le n+1$, i. e., $d \le n+2$, it remaining to exclude the case where $d = n+2$. For the converse, one follows the same line of proof above, thereby arguing that, if $d \le n+1$, there is a canceling at the tail of (111) and the subsequent iterated mapping cones, resulting a minimal resolution of length n at the end.

7.2.2 Cremona maps with a Cohen–Macaulay graph

Let $\mathfrak{F} : \mathbb{P}^n_k \dashrightarrow \mathbb{P}^n_k$ be a Cremona transformation defined by forms of degree d. A natural question is whether conditions on the Rees algebra $\mathcal{R}_R(I)$ of the base ideal $I \subset R = k[x_0,\ldots,x_n]$ of $\mathfrak{F}$ have any meaningful impact back on its structure. The question we wish to address is the case where $\mathcal{R}_R(I)$ is Cohen–Macaulay and wonder whether this has any imprint on the defining degree of $\mathfrak{F}$ and/or of its inverse. The material is essentially lifted from [91].

The case of arbitrary order n will require some ingenuity by moving across several homological-like numerical invariants. For $n=2$, a direct approach gives a very precise account.

Theorem 7.2.8. *Let $I \subset R := k[x,y,z]$ denote the base ideal of a plane Cremona transformation defined by forms of degree $d \ge 1$ and admitting at least three proper non-aligned base points. The following two conditions are equivalent:*

(a) *I is a saturated ideal and the Rees algebra $\mathcal{R}_R(I)$ of I is Cohen–Macaulay.*
(b) *$d \leq 3$ or else $d = 4$ and the map is not a de Jonquières map.*

Proof. (a) ⇒ (b) First, since $\mathcal{R}_R(I)$ has codimension 2, the Cohen–Macaulayness condition implies that the defining ideal $\mathcal{J}$ of $\mathcal{R}_R(I)$ on the polynomial ring $R[t, u, v]$, where t, u, v are variables for the target projective space $\mathbb{P}^2$, is a codimension two perfect ideal; as such it is generated by the maximal minors of an $(m + 1) \times m$ matrix Ψ with bigraded entries.

Among the minimal generators of $\mathcal{J}$ are included the ones of bidegree $(\_, 1)$ coming from a complete set of minimal syzygies of I. Since we are assuming that I is saturated, then I is generated by the 2-minors of a rank two 3×2 matrix φ. Letting r, r' denote the (standard) degrees of the two columns of φ (hence, $d = r + r'$), the ideal $\mathcal{J}$ has two minimal generators of bidegrees $(r, 1)$ and $(r', 1)$. As these are maximal minors Δ and Δ' of Ψ, we may assume that they are, respectively, the minor omitting the last row and the minor omitting the first row. By suitably permuting columns, one may assume that Ψ has the form

$$\Psi = \begin{pmatrix} a_1 & & & & \\ & a_2 & & b_1 & \\ & b_2 & a_3 & & \\ & & \ddots & \ddots & \\ & & & b_{m-1} & a_m \\ & & b_m & & \end{pmatrix}, \tag{112}$$

where $\Delta = \overbrace{a_1 a_2 \cdots a_m}^{\neq 0} + \cdots$ and, similarly, $\Delta' = \overbrace{b_1 b_2 \cdots b_m}^{\neq 0} + \cdots$ (note the random positioning of the b_i's as one may not be able to similarly dispose them along a subdiagonal without disrupting the diagonal arrangement of the a_i's, but this will turn out to be irrelevant).

Setting $\operatorname{bideg}(a_i) := (r_i, s_i)$, $1 \leq i \leq m$, and similarly, $\operatorname{bideg}(b_i) := (r'_i, s'_i)$, $1 \leq i \leq m$, a minute thought, keeping in mind that Δ (resp., Δ') has bidegree $(r, 1)$ (resp., $(r', 1)$), yields $r = \sum_i r_i$ and $s_2 = \cdots = s_m = 0$ (resp., $r' = \sum_i r'_i$ and $s'_1 = \cdots = s'_{m-1} = 0$). For further visualization, one can draw a corresponding matrix whose entries are the bidegrees of the entries of Ψ:

$$\operatorname{bidegs}(\Psi) = \begin{pmatrix} (r_1, 1) & & & & \\ & (r_2, 0) & & & (r'_1, 0) \\ & (r'_2, 0) & (r_3, 0) & & \\ & & \ddots & \ddots & \\ & & & (r'_{m-1}, 0) & (r_m, 0) \\ & & (r'_m, 1) & & \end{pmatrix}. \tag{113}$$

Moreover, since the total degree is fixed along any row, we have $r'_i = r_{i+1}$ for $1 \le i \le m-1$. Therefore, one also has $r - r' = r_1 - r'_m$.

Now, any other minimal generator of $\mathcal{J}$ being still a maximal minor thereof has to include the first and last rows of Ψ. This leaves no choice for the bidegrees of the remaining minimal generators of $\mathcal{J}$ except to have the form $(y_j, 2)$, for some $y_j \ge 2$, $2 \le j \le m$. Moreover, necessarily, $y_j = r_1 + r_2 + \cdots + \widehat{r_j} + \cdots + r_m + r'_m$.

On the other hand, since I is the base ideal of a Cremona map, Theorem 3.2.22 implies that $\mathcal{J}$ admits at least two minimal generators of bidegree $(1, s)$, for some $s \ge 1$.

Say, $r' \le r$. If $r' = 1$, and since one is assuming that I is saturated, then the map is a de Jonquières map. A full proof of this assertion is in [160, Proposition 3.4], but the main line of argument is as follows: if $\mathrm{cod}(I_1(\varphi)) = 3$ then I is a generically a complete intersection and, hence of linear type by Proposition 2.3.2; but this is not possible by Proposition 3.2.27. Thus, $I_1(\varphi)$ has codimension ≤ 2 and up to a change of variables one may assume that the first column of φ has coordinates x, y, 0. Since the map is assumed to be a Cremona map, the coordinates of the second column can be taken to be forms of degree at most 1 in z (a particular case of Proposition 7.1.6).

Now, by the "only if" part of [89, Theorem 2.7(iii)], one has $d \le 3$.

Thus, assume that $r' \ge 2$. Then there are two distinct minimal generators of bidegree $(1, 2)$. Letting Δ_1 be one of them, we must have $1 = y_j = r_1 + r_2 + \cdots + \widehat{r_j} + \cdots + r_m + r'_m$, for some $2 \le j \le m$. It follows that $r = (r_1 + r_2 + \cdots + \widehat{r_j} + \cdots + r_m) + r_j \le r_j + 1$. Since the row where the entry a_j appears is omitted in the minor Δ_1, it must appear in the other minor Δ_2 of bidegree $(1, 2)$. By a similar token, $1 = y_{j'}$, with $j' \ne j$, and hence we must have $r_j \le 1$. We conclude that $r \le r_j + 1 \le 2$. Since we are assuming that $2 \le r' \le r$, it follows finally that $r = r' = 2$, in which case the map has degree 4 and is necessarily non-de Jonquières.

(b) $\to$ (a) Assume first that $d - 2$, i. e., the map is a quadratic Cremona map. This case is sufficiently well known: the ideal is generated by the 2-minors of a rank two 2×3 matrix with linear entries. Therefore, R/I is a Cohen–Macaulay ring—equivalently, in this situation, I is saturated—and, moreover, I is an ideal of linear type, i. e., $\mathcal{R}_R(I)$ coincides with the symmetric algebra of I, the latter being a complete intersection.

If $d = 3$, the map is a de Jonquières map as a consequence of the well-known equations of condition satisfied by a Cremona map. In this case, saturation follows from Corollary 7.1.13(i) and Cohen–Macaulayness follows from the "if" part of [89, Theorem 2.7(iii)].

Finally, assume that $d = 4$. By Corollary 7.1.13(i) again, the base ideal is saturated. Since one is assuming that the map is not a de Jonquières transformation, there is only one additional proper homaloidal type for this degree, namely, $(4; 2^3, 1^3)$ (this follows readily from the Hudson test (Theorem 6.1.5)). Since we are assuming that three of the base points are proper and nonaligned then Cohen–Macaulayness is proved in [45, Proposition A.1 and Remark A.3, Appendix]. □

Remark 7.2.9. It will be shown in Theorem 7.2.11 that, in arbitrary dimension, saturation of the base ideal I is a consequence of $\mathcal{R}_R(I)$ being Cohen–Macaulay. Note that this fails for arbitrary ideals, the simplest example being perhaps the ideal of $k[x,y,z]$ generated by the partial derivatives of $y^2z - x^3$.

At the other end, although the implication (b) $\Rightarrow$ (a) is not the high point of the theorem, arguing that the base ideal of a plane Cremona map of degree 4, which is not de Jonquières, has Cohen–Macaulay Rees algebra is not entirely a trivial matter as would perhaps be expected at this early degree. Though a minimal set of bihomogeneous generators is easy to guess in shape and number ($= 4$) thanks to the inverse map, proving that these are maximal minors of a 4×3 matrix with the expected rank requires some toil.

To proceed, we need the following result, where as previously, $\ell(I)$ denote the analytic spread of the ideal I. The proof is essentially an extraction of [3, Corollary 4.11] by bringing up a tighter bound. One wonders if there is a simpler argument.

Lemma 7.2.10. *Let $(A, \mathfrak{m})$ be a Cohen–Macaulay Noetherian local ring and let $I \subset \mathfrak{m}$ be an ideal such that $\operatorname{ht}(I) < \ell(I) = \mu(I) = \dim A$. If $\mathcal{R}_A(I)$ is Cohen–Macaulay then $\operatorname{reg} \mathcal{R}_A(I) \leq \dim A - 2$.*

Proof. Since I is generated by analytically independent elements, it is basic, i. e., its reduction number with respect to itself is zero. Since $\operatorname{reg} \mathcal{R}_A(I) = \operatorname{reg} \operatorname{gr}_I(A)$ (see, [152, Lemma 4.8]) then [3, Corollary 4.11] gives $\operatorname{reg} \mathcal{R}_A(I) \leq r_{\ell-1}(I)$, where $\ell = \ell(I)$ and

$$r_{\ell-1}(I) := \max_{\substack{\operatorname{ht}(\mathfrak{p}) < \ell(I) \\ \ell(I_{\mathfrak{p}}) = \operatorname{ht}(\mathfrak{p})}} \{r(I_{\mathfrak{p}}) - \ell(I_{\mathfrak{p}})\} + \ell(I) - 1 \quad ([3, \text{p. } 415]) \tag{114}$$

Recall the meaning of the invariant a_d^* from Subsection 2.3.4, (20).

Claim 1. Let $d = \dim A$. Then

$$a_d^*(\operatorname{gr}_I(A)) = \max\{r(I_{\mathfrak{p}}) - \ell(I_{\mathfrak{p}}) \mid \mathfrak{p} \in V(I)\}. \tag{115}$$

This follows recursively from [92, Corollary 4.2].

Claim 2. $a_d^*(\operatorname{gr}_I(A)) \leq -1$.
This is well-known (see [209, Theorem 1.1]).

The two claims together imply that $\max\{r(I_{\mathfrak{p}}) - \ell(I_{\mathfrak{p}}) \mid \mathfrak{p} \in V(I)\} \leq -1$. Since $\operatorname{ht}(I) < \ell(I)$ by assumption, by definition (114) this bound implies

$$r_{\ell-1}(I) = \max_{\substack{\operatorname{ht}(\mathfrak{p}) < \ell(I) \\ \ell(I_{\mathfrak{p}}) = \operatorname{ht}(\mathfrak{p})}} \{r(I_{\mathfrak{p}}) - \ell(I_{\mathfrak{p}})\} + \ell(I) - 1 \leq \ell(I) - 2 = \dim A - 2,$$

as was to be shown. □

Theorem 7.2.11. *Let $\mathfrak{F} : \mathbb{P}^n_k \dashrightarrow \mathbb{P}^n_k$ be a Cremona transformation with base ideal $I \subset R = k[x_0, \ldots, x_n]$. Suppose that the Rees algebra $\mathcal{R}_R(I)$ is Cohen–Macaulay. Then:*

(i) *Both I and the base ideal of the inverse transformation are generated in degree at most n^2.*

(ii) *The saturation exponent of I is $n-2$. Consequently, the base ideal I of a plane Cremona map such that $\mathcal{R}_R(I)$ is Cohen–Macaulay is saturated.*

Proof. For clarity, let $S := k[y_0, \ldots, y_n]$ denote the homogeneous coordinate of the target. We highlight the following general facts:

- Let $I' \subset S$ stand for the base ideal of the inverse map to $\mathfrak{F}$. Then there is an isomorphism $\mathcal{R}_R(I) \simeq \mathcal{R}_S(I')$ of k-algebras (Proposition 3.2.14).
- Let $\mathcal{R}_S(I') \simeq S[x_0, \ldots, x_n]/\mathcal{J}'$ be a presentation, with $\mathcal{J}'$ a standard bigraded ideal. Let $\mathrm{rel}(I')$ denote the relation type of I', i. e., the maximum degree in $\mathbf{x} = \{x_0, \ldots, x_n\}$ of a minimal generator of $\mathcal{J}'$. Then $\mathrm{rel}(I') \leq \mathrm{reg}(\mathcal{R}_S(I')) + 1$, where the regularity is computed with respect to the maximal irrelevant ideal $\mathcal{R}_S(I')_+$ (this is clear by taking resolution over $S[\mathbf{x}]$, with S in degree zero).
- Let I be generated in degree d. Then $d \leq n\,\mathrm{rel}(I')$. This follows from Theorem 3.2.22, Supplement (ii).
- $\mathrm{reg}(\mathcal{R}_S(I')) = \mathrm{reg}_{\mathfrak{m}}(\mathcal{R}_R(I))$, where $\mathfrak{m} = (\mathbf{x})$. To see this, by a previous highlight, $\mathcal{R}_R(I) \simeq \mathcal{R}_S(I')$ based on the standard bigraded structure of the polynomial ring $R[\mathbf{y}] = k[\mathbf{x}, \mathbf{y}]$. If one changes this grading by setting $\deg(\mathbf{y}) = 0$ and $\deg(\mathbf{x}) = 1$, it follows that $H^i_{\mathfrak{m}}(\mathcal{R}_S(I')) \simeq H^i_{\mathfrak{m}}(\mathcal{R}_R(I))$ for every i. Since $\mathrm{reg}(\mathcal{R}_S(I'))$ is computed with respect to the irrelevant ideal $\mathcal{R}_S(I')_+$, it is defined in terms of local cohomology based on the ideal $S[\mathbf{x}]_+ = \mathfrak{m}$. Then the above isomorphisms of local cohomology modules imply the required equality.

Now, for the proof proper.

(i) Assume in addition that $\mathcal{R}_R(I)$ is Cohen–Macaulay. Then the above highlights, alongside with Lemma 7.2.10, yield:

$$\begin{aligned} d &\leq n\,\mathrm{rel}(I') \leq n\,(\mathrm{reg}(\mathcal{R}_S(I')) + 1) = n\,(\mathrm{reg}(\mathcal{R}_R(I)) + 1) \\ &\leq n\,(\dim R - 2 + 1) = n^2, \end{aligned}$$

as was to be shown. The argument for the inverse base ideal is the same.

(ii) As seen above, $\mathrm{reg}(\mathcal{R}_R(I)) \leq n-1$. By [35], it then follows that $\mathrm{reg}(I) - d \leq n-1$, i. e., $\mathrm{reg}(R/I) \leq d + n - 2$. In particular, $\mathrm{end}(H^0_{\mathfrak{m}}(R/I)) = \mathrm{end}(I^{\mathrm{sat}}/I) \leq d + n - 2$. On the other hand, by [155, Proposition 1.2(i)], one has $\mathrm{indeg}(I^{\mathrm{sat}}/I) \geq d + 1$. Therefore, $\mathfrak{m}^{n-2}(I^{\mathrm{sat}}/I) = 0$, thus proving the assertion. □

7.2.3 Cremona maps and symbolic Rees algebras

Let $R = k[\mathbf{x}] = k[x_0, \ldots, x_n]$ denote a standard graded polynomial ring over an infinite field. Given an ideal $I \subset R$, we denote by $\mathcal{R}^{(I)} \subset R[t]$ the symbolic Rees algebra of I.

To move deeper into the theory, one way is to use the nature of symbolic powers in terms of ideal transforms. Namely, given an ideal $\mathfrak{a} \subset A$, where A is a Noetherian domain with field of fractions K, the *ideal transform* of A relative to $\mathfrak{a}$ is the A-subalgebra $T_A(\mathfrak{a}) := A :_K \mathfrak{a}^\infty \subset K$. We will draw on the following fundamental principle.

Proposition 7.2.12 (The Ideal transform principle [215, Proposition 7.1.4]). *If $C \subset T_A(\mathfrak{a})$ is a finitely generated A-algebra such that* $\operatorname{depth}_{\mathfrak{a}C}(C) \geq 2$, *then $C = T_A(\mathfrak{a})$.*

Next is an application of this principle, where $A = \mathcal{R}_R(I) = R[It]$ is the Rees algebra of an ideal $I \subset R$ in a domain R with fraction field k, and $K = k(t)$.

Proposition 7.2.13. *Let $(R, \mathfrak{m})$ be a local domain and $I \subset \mathfrak{m}$ an ideal with* $\operatorname{depth} R/I > 0$, *satisfying the following properties:*

(i) *For every $r \geq 1$, the R-module $I^{(r)}/I^r$ is either zero or $\mathfrak{m}$-primary.*
(ii) *$C \subset \mathcal{R}_R^{(I)}$ is a finitely generated graded R-subalgebra containing $R[It]$ and such that* $\operatorname{depth}_{\mathfrak{m}C}(C) \geq 2$.

Then $\mathcal{R}_R^{(I)} = C$.

Proof. The condition (i) implies that for every $r \geq 1$, there is a integer $s \geq 1$ such that $y\mathfrak{m}^s \subset \mathcal{R}_R(I)$. Hence, the symbolic Rees algebra $\mathcal{R}_R^{(I)}$ is contained in the ideal transform $T_{\mathcal{R}_R(I)}(\mathfrak{m}\mathcal{R}_R(I))$. In particular, we have

$$\mathcal{R}_R(I) \subset C \subset \mathcal{R}_R^{(I)}t \subset T_{\mathcal{R}_R(I)}(\mathfrak{m}\mathcal{R}_R(I)).$$

Hence, the condition (ii) and Proposition 7.2.12 imply the desired equality. □

Remark 7.2.14. The hypothesis (i) of the above proposition is automatic in the following case: $(R, \mathfrak{m})$ is a Cohen–Macaulay local domain and $I \subset R$ a prime ideal of height 2 such that R/I is Cohen–Macaulay and $\mu(I_P) \leq \max\{\operatorname{ht} P - 1, \operatorname{ht} I\}$ for every prime ideal $P \supsetneq I$ with $P \neq \mathfrak{m}$.

Theorem 7.2.15. *Let $\mathfrak{F} : \mathbb{P}^n \dashrightarrow \mathbb{P}^n$ stand for a Cremona map with base ideal $I \subset R = k[\mathbf{x}] = k[x_0, \dots, x_n]$ such that* $\operatorname{depth}(R/I) > 0$. *Suppose that:*

(i) *For every $\ell \geq 1$, the module $I^{(\ell)}/I^\ell$ is either zero or $(\mathbf{x})$-primary.*
(ii) *The Rees algebra $R[It]$ has the property (S_2) of Serre.*

Then

(a) *The symbolic algebra $\mathcal{R}^{(I)} = R[It, Ft^{d'}]$, where F denotes the source inversion factor of $\mathfrak{F}$ and d' is the coordinate degree of its inverse map.*
(b) *A presentation ideal of $\mathcal{R}^{(I)}$ on $R[\mathbf{y}, z]$ has the form $(\mathcal{I}, \{x_i z - g_i(\mathbf{y}), 0 \leq i \leq n\})$, where $\mathcal{I} \subset R[\mathbf{y}]$ denotes a presentation ideal of $R[It]$, $g_0, \dots, g_n$ denote coordinate forms of the inverse map and z denotes an indeterminate over $R[\mathbf{y}]$.*

Proof. (a) Let $I = (f_0,\dots,f_n)$ and $I' = (g_0,\dots,g_n) \subset S$ denote the respective base ideals of the map and of its inverse, both of height ≥ 2. By Proposition 3.2.14, there is an identification of Rees algebras $R[It] = S[I'u]$ inside $k[\mathbf{x},\mathbf{y},t,u]$. Therefore, $x_i = g_i u$, $y_i = f_i t$. Now consider the element $Ft^{d'} \in R[t] = k[\mathbf{x},t]$, where F is the source inversion factor of $\mathfrak{F}$ and d' is the degree of the g_i's. By definition, say, $F = g_1(\mathbf{f})/x_1$. Therefore,

$$\begin{aligned} Ft^{d'} &= g_1(\mathbf{f})t^{d'}/x_1 = g_1(\mathbf{f}t)/x_1 = g_1(\mathbf{y})/g_1(\mathbf{y})u \\ &= u^{-1}, \end{aligned}$$

always by the above identification inside the field $k(\mathbf{x},\mathbf{y},t,u)$. This gives $\mathcal{R}_R(I)[Ft^{d'}] = \mathcal{R}_S(I')[u^{-1}]$, the latter being the extended Rees algebra of the ideal $I' \subset S$.

Now, by (ii) $\mathcal{R}_S(I') = \mathcal{R}_R(I)$ satisfies (S_2), hence the associated graded ring $\mathrm{gr}_{I'}(S)$ satisfies (S_1) by the usual exact sequences

$$\begin{array}{ccccccccc} 0 & \to & I'\mathcal{R}_S(I') & \longrightarrow & \mathcal{R}_S(I') & \longrightarrow & \mathrm{gr}_{I'}(S) & \to & 0 \\ & & \wr\| & & & & & & \\ 0 & \to & \mathcal{R}_S(I')_+ & \longrightarrow & \mathcal{R}_S(I') & \longrightarrow & S & \to & 0 \end{array}$$

Since $\mathrm{gr}_{I'}(S) \simeq \mathcal{R}_S(I')[u^{-1}]/(u^{-1})$, with u^{-1} a nonzero divisor, it follows that $\mathcal{R}_S(I')[u^{-1}]$ satisfies (S_2) and so does $\mathcal{R}_R(I)[Ft^{d'}]$ by the above identification. This implies that the extended ideal $(\mathbf{x})\mathcal{R}_R(I)[Ft^{d'}]$ has depth at least 2. Finally, use condition (i) to conclude by applying Proposition 7.2.13.

(b) By the proof of (a), $\mathcal{R}^{(I)} = \mathcal{R}_S(I')[u^{-1}] = \mathcal{R}_R(I)[u^{-1}]$. Therefore, the assertion follows from the well-known presentation of the extended Rees algebra (see, e. g., [118, Proposition 5.5.7]). □

The next result is based on Theorem 7.2.15, reshaped for the case of ambient dimension 2. The notation and details are the same as in that theorem. However, it does not apply directly because the Rees algebra of a base ideal of a Cremona map does not in general satisfy Serre's property (S_2). However, in dimension two we can bypass this deficiency.

Theorem 7.2.16. *Let $\mathfrak{F} : \mathbb{P}^2 \to \mathbb{P}^2$ be a Cremona map of degree d. Suppose that the base ideal I of F is saturated. Then:*
(a) $\mathcal{R}^{(I)} = R[It, Dt^d]$, *where D denotes the source inversion factor of F.*
(b) *A presentation ideal of $\mathcal{R}^{(I)}$ on $R[\mathbf{y},z]$ has the form*

$$(\mathcal{I}, \{x_i z - g_i(\mathbf{y}), 0 \leq i \leq n\}),$$

where $\mathcal{I}$ denotes a presentation ideal of the ordinary Rees algebra $\mathcal{R}_R(I)$ and z is an indeterminate over $R[\mathbf{y}]$, while $\{g_0, g_1, g_2\} \subset k[\mathbf{y}]$ denote defining forms of the inverse map.

Proof. We are missing property (S_2), but the argument in Theorem 7.2.15 remains valid if one can prove directly that the extended ideal

$$(\mathfrak{m}, \mathfrak{J})/\mathfrak{J} \subset T := R[\mathbf{y}, z]/\mathfrak{J}$$

has grade ≥ 2, where $\mathfrak{J} = (\mathcal{I}, \{x_i z - g_i(\mathbf{y}), 0 \leq i \leq 2\})$. As a side note, since the generators of I are analytically independent, one has

$$(\mathfrak{m}, \mathfrak{J}) = (\mathfrak{m}, \{g_i(\mathbf{y}), 0 \leq i \leq 2\}),$$

an ideal of codimension 5. Since $\mathfrak{J}$ has codimension 3 (one more than the codimension of the defining ideal of the Rees algebra of I), we would be done if codimension ≥ 2 were the issue at stake. Alas, we need to check the grade instead which, in principle, could be different as the Rees algebra might not satisfy (S_2). For this, since R/I has positive depth, up to a change of coordinates we may assume that x_2 is a regular element modulo I.

Letting $I \subset k[x_0, x_1, x_2]$ denote the base ideal of $\mathfrak{F}$, set $J := (I, x_2)/(x_2) \subset k[x_0, x_1]$, by an obvious identification. Let $\mathcal{J}$ denote the presentation ideal of the Rees algebra of J.

Claim 1. The rational map $\eta : \mathbb{P}^1 \dashrightarrow \mathbb{P}^2$ defined by the generators of J is birational onto its image, a curve of degree d.

Since $\mathfrak{F}$ is birational, one can use the criterion of Theorem 3.2.22 in both directions to deduce the claim.

Claim 2. g_2 is up to a nonzero scalar the defining equation $F \subset k[\mathbf{y}]$ of the image curve of η.

We leave this to the reader.

From the claim also follows that the inverse map to η is defined by the residue classes of g_0, g_1 on the k-algebra $k[y_0, y_1, y_2]/(g_2)$. Since $\mathcal{J}$ is the presentation ideal of η, then the extended Rees algebra of this base ideal is

$$\begin{aligned} \frac{k[x_0, x_1][\mathbf{y}, z]}{(\mathcal{J}, x_0 z - g_0, x_1 z - g_1)} &\simeq \frac{R[\mathbf{y}, z]}{(\mathcal{J}, x_0 z - g_0, x_1 z - g_1, x_2)} \\ &\simeq R[\mathbf{y}, z]/(\mathfrak{J}, x_2) \\ &\simeq T/x_2 T, \end{aligned}$$

where the following equalities have been used:

$$(\mathcal{J}, x_2) = (\mathcal{A}, E, x_2) = (\mathcal{A}, g_2, x_2) = ((\mathcal{A}, x_2), g_2) = ((\mathcal{I}, x_2), g_2) = (\mathcal{I}, x_2 z - g_2, x_2),$$

with $\mathcal{A} \subset k[x_0, x_1][\mathbf{y}]$ being such that $(\mathcal{I}, x_2) = (\mathcal{A}, x_2)$.

Claim. The symbolic Rees algebra of I is isomorphic to T.

As observed before, it suffices to show that the ideal $\mathfrak{m}T$ has grade at least 2. Since T is a domain, x_2 is a regular element on T. Now, we have seen that the defining ideal of T/x_2T is also prime. Since, e. g., x_0 does not belong to this ideal by a degree argument, then $\{x_2, x_0\}$ is a regular sequence on T, as was to be shown. □

7.3 Monomial Cremona transformations

7.3.1 The group of monomial Cremona transformations

Among monomial birational maps, the Cremona ones form a distinguished class with many special properties Drawing on the criterion of birationality of Theorem 3.2.22 and the fact that the Fitting ideals of graded matrices under general gradings are homogeneous in the grading makes it possible to prove that the inverse of a monomial Cremona map is also monomial. More is true, as these maps will form a subgroup of the whole Cremona group. However, such general approach does not reveal the combinatorial details hidden in these processes.

Here, we set up a direct way to convert geometric results into numeric or combinatorial data in a characteristic free way. In particular, it is shown that the monomial plane Cremona group is completely understood without appealing to the known results about arbitrary plane Cremona maps. Thus, one shows that this group is generated, up to reordering of variables in the source and the target, by two basic monomial quadratic maps. The result is not a trivial consequence of Noether's theorem since the latter requires composing with projective transformations, which is out of the picture here. Moreover, the known proofs of Noether's theorem (see, e. g., [4]) reduce to various especial situations, passing through de Jonquières maps, which are rarely monomial.

Monomial Cremona maps have been dealt with in [75] and in [126], but the methods and some of the goals are different. In particular, [75] proves that the monomial Cremona group of arbitrary dimension is also generated by quadratic monomial such maps up to the action of the integer general linear group.

Most of the material of this part is based on [197].

7.3.1.1 The combinatorial mirror

We need the following lemma, for which one could not find an appropriate source in the classical matrix literature.

Lemma 7.3.1. *Let $\Gamma = (y_{i,j})$ be an $n \times n$ square matrix over an arbitrary commutative ring, and let $D = \operatorname{diag}(d_1, \ldots, d_n)$ be a diagonal matrix over the same ring. Then*

$$\det(\Gamma + D) = \det(\Gamma) + \sum_i d_i \Delta_{[n]\setminus\{i\}} + \sum_{1\le i_1<i_2\le n} d_{i_1} d_{i_2} \Delta_{[n]\setminus\{i_1,i_2\}}$$
$$+ \cdots + \sum_{1\le i_1<\cdots<i_{n-1}\le n} d_{i_1}\cdots d_{i_{n-1}} \Delta_{[n]\setminus\{i_1,\ldots,i_{n-1}\}} + \det D,$$

where $[n] = \{1, \ldots, n\}$ *and* $\Delta_{[n]\setminus\{i_1,\ldots,i_k\}}$ *denotes the principal* $(n-k)\times(n-k)$*-minor of* Γ *with rows and columns* $[n]\setminus\{i_1,\ldots,i_k\}$.

In particular, if Γ *has rank at most one and* D *is the identity matrix, one gets* $\det(\Gamma + I) = \mathrm{trace}(\Gamma) + 1$.

Proof. Due to the multilinearity of the determinant, the following equality of determinants holds:

$$\det\begin{pmatrix} y_{1,1}+d_1 & y_{1,2} & \cdots & y_{1,n} \\ y_{2,1} & y_{2,2}+d_2 & \cdots & y_{2,n} \\ \vdots & \vdots & & \vdots \\ y_{n,1} & y_{n,2} & \cdots & y_{n,n}+d_n \end{pmatrix} = \det\begin{pmatrix} y_{1,1} & y_{1,2} & \cdots & y_{1,n} \\ y_{2,1} & y_{2,2}+d_2 & \cdots & y_{2,n} \\ \vdots & \vdots & & \vdots \\ y_{n,1} & y_{n,2} & \cdots & y_{n,n}+d_n \end{pmatrix} + \det\begin{pmatrix} d_1 & y_{1,2} & \cdots & y_{1,n} \\ 0 & y_{2,2}+d_2 & \cdots & y_{2,n} \\ \vdots & \vdots & & \vdots \\ 0 & y_{n,2} & \cdots & y_{n,n}+d_n \end{pmatrix}$$

The recurrence for the second determinant on the right-hand side is obvious, while for the first determinant we repeat the procedure, thus writing

$$\det\begin{pmatrix} y_{1,1} & y_{1,2} & y_{1,3} & \cdots & y_{1,n} \\ y_{2,1} & y_{2,2}+d_2 & y_{2,3} & \cdots & y_{2,n} \\ y_{3,1} & y_{3,2} & y_{3,3}+d_3 & \cdots & y_{3,n} \\ \vdots & \vdots & \vdots & & \vdots \\ y_{n,1} & y_{n,2} & y_{n,3} & \cdots & y_{n,n}+d_n \end{pmatrix} = \det\begin{pmatrix} y_{1,1} & y_{1,2} & y_{1,3} & \cdots & y_{1,n} \\ y_{2,1} & y_{2,2} & y_{2,3} & \cdots & y_{2,n} \\ y_{3,1} & y_{3,2} & y_{3,3}+d_3 & \cdots & y_{3,n} \\ \vdots & \vdots & \vdots & & \vdots \\ y_{n,1} & y_{n,2} & y_{n,3} & \cdots & y_{n,n}+d_n \end{pmatrix} + \det\begin{pmatrix} y_{1,1} & 0 & y_{1,3} & \cdots & y_{1,n} \\ y_{2,1} & d_2 & y_{2,3} & \cdots & y_{2,n} \\ y_{3,1} & 0 & y_{3,3}+d_3 & \cdots & y_{3,n} \\ \vdots & \vdots & \vdots & & \vdots \\ y_{n,1} & 0 & y_{n,3} & \cdots & y_{n,n}+d_n \end{pmatrix}$$

and so forth. □

Next is a main combinatorial assessment of the monomial Cremona theory. Set $e_i := (0 \ldots 010 \ldots 0)^t \in \mathbb{N}^n$ for the canonical basis vector.

Theorem 7.3.2. *Let $v_1,\dots,v_n$ be a set of vectors in $\mathbb{N}^n$ such that $|v_i| = d \geq 1$ for every $1 \leq i \leq n$ and $\det(A) = \pm d$, where A is the $n \times n$ matrix with column vectors $v_1,\dots,v_n$. Then there are unique vectors $\beta_1,\dots,\beta_n,\gamma \in \mathbb{N}^n$ such that the following two conditions hold:*

(a) *$A\beta_i = \gamma + e_i$ for all i, where β_i, γ and e_i are regarded as column vectors.*
(b) *The matrix B whose columns are $\beta_1,\dots,\beta_n$ has at least one zero entry in every row.*

Moreover, $\det(B) = \pm(|\gamma| + 1)/d = \pm|\beta_i|$ for all i.

Proof. First, we show the uniqueness. Assume that $\beta_1',\dots,\beta_n',\gamma'$ is a set of vectors in $\mathbb{N}^n$ such that: (a') $A\beta_i' = \gamma' + e_i$ for all i, and (b') The matrix B' whose column vectors are $\beta_1',\dots,\beta_n'$ has at least one zero entry in every row. Let $\Delta = (\Delta_i)$ and $\Delta' = (\Delta_i')$ be nonnegative vectors such that $A^{-1}(\gamma - \gamma') = \Delta' - \Delta$. Then from (a) and (a'), we get

$$\beta_i - \beta_i' = A^{-1}(\gamma - \gamma') = \Delta' - \Delta, \quad \forall i \implies \beta_{ik} - \beta_{ik}' = \Delta_k' - \Delta_k, \ \forall i, k, \tag{116}$$

where $\beta_i = (\beta_{i1},\dots,\beta_{in})$ and $\beta_i' = (\beta_{i1}',\dots,\beta_{in}')$. It suffices to show that $\Delta = \Delta'$. If $\Delta_k' > \Delta_k$ for some k, then, by (116), we obtain $\beta_{ik} > 0$ for $i = 1,\dots,n$, which contradicts (b). Similarly if $\Delta_k' < \Delta_k$ for some k, then again by (116), we obtain $\beta_{ik}' > 0$ for $i = 1,\dots,n$, which contradicts (b'). Thus, $\Delta_k = \Delta_k'$ for all k, i. e., $\Delta = \Delta'$.

Next, we prove the existence of $\beta_1,\dots,\beta_n$ and γ. By Proposition 3.3.5, for $i \geq 2$ we can write

$$0 \neq \alpha_i = A^{-1}(e_1 - e_i) = \alpha_i^+ - \alpha_i^-$$

where α_i^+ and α_i^- are in $\mathbb{N}^n$. Notice that $\alpha_i^+ \neq 0$ and $\alpha_i^- \neq 0$. Indeed the sum of the entries of $A^{-1}(e_i)$ is equal to $1/d$. Thus, $|\alpha_i| = |\alpha_i^+| - |\alpha_i^-| = 0$, and consequently the positive and negative parts of α_i are both nonzero for $i \geq 2$. The vector α_i^+ can be written as $\alpha_i^+ = (\alpha_{i1}^+,\dots,\alpha_{in}^+)$ for $i \geq 2$. For $1 \leq k \leq n$, consider the integers given by

$$m_k = \max_{2 \leq i \leq n}\{\alpha_{ik}^+\}$$

and set $\beta_1 = (m_1,\dots,m_n)$. Since $\beta_1 \geq \alpha_i^+$, for each $i \geq 2$ there is $\theta_i \in \mathbb{N}^n$ such that $\beta_1 = \theta_i + \alpha_i^+$. Therefore,

$$\alpha_i = A^{-1}(e_1 - e_i) = \alpha_i^+ - \alpha_i^- = \beta_1 - (\theta_i + \alpha_i^-).$$

We set $\beta_i = \theta_i + \alpha_i^-$ for $i \geq 2$. Since we have $A\beta_1 - e_1 = A\beta_i - e_i$ for $i \geq 2$, it follows readily that $A\beta_1 - e_1 \geq 0$ (make $i = 2$ in the last equality and compare entries). Thus, making $\gamma = A\beta_1 - e_1$, we obtain that $\beta_1,\dots,\beta_n$ and γ satisfy (a). If each row of B has a zero entry the proof of the existence is complete. If a row of B has all its entries positive we subtract the vector $\mathbf{1} = (1,\dots,1)$ from that row and modify the γ accordingly so that

(a) is still satisfied. Applying this argument repeatedly, we get a sequence $\beta_1, \ldots, \beta_n, \gamma$ satisfying (a) and (b).

To see the last part of the assertion, note that condition (a) is equivalent to the equality $AB = \Gamma + \mathbb{I}$, where Γ is the matrix all of whose columns are equal to γ and $\mathbb{I}$ denotes the identity matrix of order n. Since $A\beta_i = \gamma + e_i$ for all i and $\det(B) = \pm\det(\Gamma + \mathbb{I})/d$, it suffices to show that $\det(\Gamma + I) = |\gamma| + 1$, which follows from the above lemma. □

The fundamental matrix equation in item (a) of this theorem will be referred to as the *inversion matrix equation* of the matrix A, while the matrix B is called the *Cremona inverse matrix* of A, a terminology justified by Theorem 7.3.4 below. The set $\{v_1, \ldots, v_n\}$ of defining monomials will sometimes be referred to as a *Cremona set*.

Remark 7.3.3. Note that even if the vectors $v_1, \ldots, v_n$ in the theorem give rise to monomials with a nontrivial common factor, the resulting column vectors of B can be taken so that the corresponding monomials have no nontrivial common factor. Thus, e. g., if

$$A = \begin{pmatrix} 2 & 1 \\ 0 & 1 \end{pmatrix}$$

then B is the identity matrix.

7.3.1.2 The geometric side

Let $R = k[x_1, \ldots, x_n]$ be a polynomial ring over a field k and set $x^\alpha := x_1^{a_1} \cdots x_n^{a_n}$ for $\alpha = (a_1, \ldots, a_n) \in \mathbb{N}^n$. In the sequel, we consider a finite set of distinct monomials $F = \{x^{v_1}, \ldots, x^{v_q}\} \subset R$ of the same degree $d \geq 1$ and having no nontrivial common factor. We also assume throughout that every x_i divides at least one element of F, a harmless condition. The set F defines a rational (monomial) map $\mathbb{P}^{n-1} \dashrightarrow \mathbb{P}^{n-1}$, which will also be denoted F and written as a tuple $F = (x^{v_1}, \ldots, x^{v_q})$.

Note that a rational monomial map is defined everywhere if and only if the defining monomials are pure powers of the variables, in which case it is a Cremona map if and only if $d = 1$ (the identity map).

Theorem 7.3.4. *Let $F : \mathbb{P}^{n-1} \dashrightarrow \mathbb{P}^{n-1}$ stand for a rational map defined by monomials. If F is a Cremona map then its inverse is also defined by monomials. Moreover, the degree as well as a set of monomials defining the inverse can be obtained explicitly in terms of the given set of monomials defining F.*

Proof. Let F be a Cremona map defined by a set of monomials $f_1, \ldots, f_n$ of the same degree $d \geq 1$. By Proposition 3.3.1, the log-matrix A of these monomials has determinant $\pm d$. Therefore, Theorem 7.3.2 implies the existence of an $n \times n$ matrix B such that $AB = \Gamma + I$, where Γ is a matrix with repeated column γ throughout. Letting $g_1, \ldots, g_n$ denote the monomials whose log-matrix is B and letting x^γ denote the monomial whose exponents are the coordinates of y, the above matrix equality translates into the equality

$$(f_1(g_1,\dots,g_n),\dots,f_n(g_1,\dots,g_n)) = (x^y \cdot x_1,\dots,x^y \cdot x_n).$$

Thus, the left-hand side is proportional to the vector $(x_1,\dots,x_n)$, which means that the composite map $F \circ G$ is the identity map, where G is defined by $g_1,\dots,g_n$, wherever the two are defined (see Section 3.1). On the other hand, since B is the log-matrix of $g_1,\dots,g_n$, Theorem 7.3.2 applied in the opposite direction says that G is also a Cremona map. Therefore, G has to be inverse of F, as required. Finally, notice that the proof of Theorem 7.3.2 provides a means of computing B and y. □

The theorem allows us to introduce the following group.

Definition 7.3.5. The *monomial Cremona group* of order $n-1$ is the subgroup of the Cremona group of $\mathbb{P}^{n-1}$ whose elements are Cremona monomial maps.

Here, we will not distinguish between rational maps $F, G : \mathbb{P}^{n-1} \dashrightarrow \mathbb{P}^{m-1}$ defined, respectively, by forms $f_1,\dots,f_m$ of the same degree and their multiples $g_1 = gf_1,\dots,g_m = gf_m$ by a fixed form g of arbitrary degree.

There is a potential confusion in this terminology. For instance, in [126] one allows for more general maps by considering the free product with a certain Klein subgroup. Since our goal is combinatorial we will stick to the above definition of the monomial Cremona group, but will allow for source and target permutation of variables.

As a matter of notation, the composite of two Cremona maps F, G will be indicated by FG (first G, then F). Likewise, the power composite $FF\cdots F$ of m factors will be denoted F^m.

We shift our attention to *plane* monomial Cremona maps, i. e., we will study the structure of the monomial Cremona group for $n = 3$. Consider the maps $H = (x_1^2 : x_1x_2 : x_2x_3)$ and $S = (x_1x_2 : x_1x_3 : x_2x_3)$.

- For any $d \geq 1$, one has $H^{d-1} = (x_1^d : x_1^{d-1}x_2 : x_2^{d-1}x_3)$.
  This is a straightforward composite calculation: by induction, we assume that H^{d-2} is defined by x_1^{d-1}, $x_1^{d-2}x_2$, $x_2^{d-2}x_3$, hence $H^{d-1} = H^{d-2}H$ is defined by the monomials $x_1^{2(d-1)}$, $x_1^{2d-3}x_2$, $x_1^{d-2}x_2^{d-1}x_3$, which is the same rational map as the one defined by x_1^d, $x_1^{d-1}x_2$, $x_2^{d-1}x_3$ (by canceling the gcd x_1^{d-2}).
- For any $d \geq 1$, up to permutation of the variables both in the source and the target, H^{d-1} is involutive, i. e., coincides with its own inverse.
  Namely, consider the Cremona map $G = (x_1x_2^{d-1} : x_2^d : x_1^{d-1}x_3)$. Again, one finds straightforwardly that $H^{d-1}G$ is defined by $x_1^dx_2^{d(d-1)}$, $x_1^{d-1}x_2^{(d-1)^2+d}x_2^d$, $x_1^{d-1}x_2^{d(d-1)}x_3$; canceling the gcd $x_1^{d-1}x_2^{d(d-1)}$ yields the identity map (x_1, x_2, x_3).
- The monomial maps S and H "nearly" commute; actually they are conjugate by a transposition.

This is once more straightforward: the general composites SH^{d-1} and $H^{d-1}S$ are defined, respectively, by x_1^d, $x_1x_2^{d-2}x_3$, $x_2^{d-1}x_3$ and $x_1x_2^{d-1}$, $x_1x_2^{d-2}x_3$, x_3^d and these are the same up to transposing x_1 and x_3 and the extreme terms. This means that they are conjugate by a transposition. It has as a consequence that the subgroup of the Cremona group of $\mathbb{P}^2$ generated by S and H is Abelian up to free products with suitable permutation subgroups.

Lemma 7.3.6. *Up to permutation of the source (variables) and the target (monomials), a plane Cremona map of degree d, defined by monomials $x_1^{a_1}x_2^{a_2}$, $x_2^{b_2}x_3^{b_3}$, $x_1^{c_1}x_3^{c_3}$ is one of the following two kinds: x_1^d, $x_2x_3^{d-1}$, $x_1^{d-1}x_3$ or x_1x_2, x_2x_3, x_1x_3.*

Proof. By hypothesis, $d = a_1 + a_2 = b_2 + b_3 = c_1 + c_3 = a_1b_2c_3 + c_1a_2b_3$. Then one has

$$a_1(b_2c_3 - 1) = a_2(1 - c_1b_3). \tag{117}$$

The cases below can be readily verified using these equations.

Case (I): $a_1 \geq 1$, $a_2 = 0$. Then $a_1 = d$, $b_2 = c_3 = 1$ and $F = (x_1^d : x_2x_3^{d-1} : x_1^{d-1}x_3)$.
Case (II): $a_1 = 0$, $a_2 \geq 1$. Then $a_2 = d$, $b_3 = c_1 = 1$ and $F = (x_2^d : x_2^{d-1}x_3 : x_1x_3^{d-1})$.
Case (III)(a): $a_1 \geq 1$, $a_2 \geq 1$, $b_2c_3 = 0$, $b_2 = 0$. Then $F = (x_1^{d-1}x_2 : x_3^d : x_1x_3^{d-1})$.
Case (III)(b): $a_1 \geq 1$, $a_2 \geq 1$, $b_2c_3 = 0$, $c_3 = 0$. Then $F = (x_1^{d-1}x_2 : x_2^{d-1}x_3 : x_1^d)$.
Case (III)(c): $a_1 \geq 1$, $a_2 \geq 1$, $c_1b_3 = 0$, $c_1 = 0$. Then $F = (x_1x_2^{d-1} : x_2x_3^{d-1} : x_3^d)$.
Case (III)(d): $a_1 \geq 1$, $a_2 \geq 1$, $c_1b_3 = 0$, $b_3 = 0$. Then $F = (x_1x_2^{d-1} : x_2^d : x_1^{d-1}x_3)$.
Case (III)(e): $a_1 \geq 1$, $a_2 \geq 1$, $b_2c_3 - 1 \geq 0$, $1 - c_1b_3 \leq 0$. From equation (117), we get $b_2c_3 = 1$ and $c_1b_3 = 1$. Then $F = (x_1x_2 : x_2x_3 : x_1x_3)$. □

We will next give a purely combinatorial proof of the following result (see [126] for similar dealings).

Proposition 7.3.7. *Up to permutation of the variables (both in the source and the target), the plane monomial Cremona group is generated by the maps S, H, where S is the quadratic map of first kind defined by x_1x_2, x_1x_3, x_2x_3 (reciprocal involution) and H is the quadratic map of second kind defined by x_1^2, x_1x_2, x_2x_3 ("hyperbolism").*

Proof. Let $F = (x_1^{a_1}x_2^{a_2}x_3^{a_3} : x_1^{b_1}x_2^{b_2}x_3^{b_3} : x_1^{c_1}x_2^{c_2}x_3^{c_3})$ be a plane Cremona map of degree d. We will prove that, up to a permutation of the variables and the monomials, every plane Cremona monomial map is of the form $E \cdots SH^{d_{i_k}}SH^{d_{i_{k+1}}} \cdots G$, where E, G are in $\{S, H^d \mid d \geq 1\}$. The proof is by induction on the degree. Consider the corresponding log-matrix

$$A = \begin{pmatrix} a_1 & b_1 & c_1 \\ a_2 & b_2 & c_2 \\ a_3 & b_3 & c_3 \end{pmatrix}.$$

Up to permutation of variables and monomials, there are essentially two cases to consider:

$$A = \begin{pmatrix} a_1 & b_1 & 0 \\ a_2 & b_2 & 0 \\ a_3 & 0 & d \end{pmatrix} \quad \text{or} \quad A = \begin{pmatrix} a_1 & 0 & c_1 \\ a_2 & b_2 & 0 \\ 0 & b_3 & c_3 \end{pmatrix}.$$

By Lemma 7.3.6, we may assume that $F = (x_1^{a_1}x_2^{a_2}x_3^{a_3}, x_1^{b_1}x_2^{b_2}, x_3^d)$, corresponding to the matrix on the left.

Case (I): $a_1 \geq b_1$. From $a_1 + a_2 + a_3 = b_1 + b_2 = d$, we get that $b_2 \geq a_2$. Consider the reciprocal involution given by $S = (x_1x_2 : x_2x_3 : x_1x_3)$. Then

$$\begin{aligned} SF &= (x_1^{a_1+b_1}x_2^{a_2+b_2}x_3^{a_3} : x_1^{b_1}x_2^{b_2}x_3^{d} : x_1^{a_1}x_2^{a_2}x_3^{d+a_3}) \\ &= x_1^{b_1}x_2^{a_2}x_3^{a_3}(x_1^{a_1}x_2^{b_2} : x_2^{b_2-a_2}x_3^{d-a_3} : x_1^{a_1-b_1}x_3^{d}). \end{aligned}$$

Thus, by Lemma 7.3.6 we get $SF = H^{a_1+b_2-1}$, for some hyperbolism H or else SF is the reciprocal involution. Thus multiplying by the inverse of S, we obtain that F has the required form.

Case (II)(a): $b_1 > a_1$ and $\det(A) = d$. From the equality $d = \det(A) = d(a_1b_2 - a_2b_1)$, we get that $b_2 > a_2$. Consider the hyperbolism $H = (x_1^2 : x_1x_2 : x_2x_3)$. In this case, we have

$$\begin{aligned} HF &= (x_1^{2a_1}x_2^{2a_2}x_3^{2a_3} : x_1^{a_1+b_1}x_2^{a_2+b_2}x_3^{a_3} : x_1^{b_1}x_2^{b_2}x_3^{d}) \\ &= x_1^{a_1}x_2^{a_2}x_3^{a_3}(x_1^{a_1}x_2^{a_2}x_3^{a_3} : x_1^{b_1}x_2^{b_2} : x_1^{b_1-a_1}x_2^{b_2-a_2}x_3^{d-a_3}) \\ &= x_1^{a_1}x_2^{a_2}x_3^{a_3}F_1. \end{aligned}$$

Since we can factor x_1 from F_1 and F_1 has degree d, we have lowered the degree of F. Thus by induction F has the required form.

Case (II)(b): $b_1 > a_1$ and $\det(A) = -d$. We may assume that $a_2 \geq b_2$, otherwise we may proceed as in Case (II)(a). By Lemma 7.3.6, we may also assume that $a_1 \geq 1$. We claim that F has the form $F = (x_1^{d-2}x_2x_3 : x_1^{d-1}x_2 : x_3^d)$. The condition on the determinant is equivalent to the equality $a_2b_1 - a_1b_2 = 1$. Thus, $a_2 \geq 1$. Hence, using that $b_1 \geq a_1 + 1$, one has

$$a_2b_1 \geq a_2a_1 + a_2 \geq b_2a_1 + 1 = a_2b_1.$$

Consequently, $a_2 = b_2 = 1$. The condition $\det(A) = -d$, becomes $a_1 = b_1 - 1 = d - 2$. Therefore, F has the asserted form. Consider the hyperbolism $H = (x_1^2 : x_1x_2 : x_2x_3)$. It is easy to see that HF is equal to $x^\gamma(x_1^{d-3}x_2x_3 : x_1^{d-2}x_2 : x_3^{d-1})$ for some monomial x^γ. Thus, we have lowered the degree of F and we may apply induction. □

The next result is classically well known. We give a direct combinatorial proof in the case of Cremona monomial maps. The present proof shows explicit simple formulae for the Cremona inverse in the case of 3 variables.

Proposition 7.3.8. *If $n = 3$, then a Cremona monomial map and its inverse have the same degree.*

Proof. The proof is based on the method employed in the proof of Theorem 7.3.2. Let A be the matrix defining a Cremona map of degree d:

$$A = \begin{pmatrix} a_1 & b_1 & c_1 \\ a_2 & b_2 & c_2 \\ a_3 & b_3 & c_3 \end{pmatrix}$$

For simplicity, we assume that $\det(A) = d$, the case $\det(A) = -d$ can be shown similarly. Up to permutation of variables, as in the proof of Proposition 7.3.7, there are essentially two cases to consider:

$$A = \begin{pmatrix} a_1 & b_1 & 0 \\ a_2 & b_2 & 0 \\ a_3 & 0 & d \end{pmatrix} \quad \text{or} \quad A = \begin{pmatrix} a_1 & 0 & c_1 \\ a_2 & b_2 & 0 \\ 0 & b_3 & c_3 \end{pmatrix}.$$

It is readily seen that the inverse of A is given by

$$A^{-1} = \begin{pmatrix} b_2 & -b_1 & 0 \\ -a_2 & a_1 & 0 \\ -a_3b_2/d & b_1a_3/d & 1/d \end{pmatrix} \quad \text{or} \quad A^{-1} = \frac{1}{d}\begin{pmatrix} b_2c_3 & b_3c_1 & -b_2c_1 \\ -a_2c_3 & a_1c_3 & a_2c_1 \\ a_2b_3 & -a_1b_3 & a_1b_2 \end{pmatrix},$$

respectively. By the argument given in the proof of Theorem 7.3.2 we get that

$$\beta_1 = (d, 0, 0),\ \gamma = (da_1 - 1, da_2, da_3) \quad \text{or} \quad \beta_1 = (b_2, 0, b_3),\ \gamma = (a_1b_2 + b_3c_1 - 1, a_2b_2, b_3c_3),$$

respectively. Therefore,

$$B = \begin{pmatrix} d & 0 & b_1 \\ 0 & a_1 + a_2 & a_2 \\ 0 & a_3 & (a_3b_2 + 1)/d \end{pmatrix} \quad \text{or} \quad B = \begin{pmatrix} b_2 & c_1 & 0 \\ 0 & c_3 & a_2 \\ b_3 & 0 & a_1 \end{pmatrix},$$

respectively. Using that the sum of the entries in every column of A is equal to d and $\det(A) = d$, it is seen that the sum of the entries in each column of B is equal to d and $\det(B) = d$. □

7.3.2 Monomial Cremona transformations of degree 2

Monomial Cremona maps in this degree have a distinct combinatorial flavor coming from graph theory. This part will exploit this relationship in some detail.

Keep the notation of Subsection 3.3.2.

7.3.2.1 Main criterion

Proposition 7.3.9. *Let $F \subset k[x_1, \dots, x_n]$ be a cohesive set of n monomials of degree two having no nontrivial common factor and let $\mathcal{G} \subset \tilde{\mathcal{G}}$ denote the corresponding graphs as above. Let A denote the $n \times n$ incidence matrix of $\tilde{\mathcal{G}}$. The following conditions are equivalent:*

(i) $\det A \neq 0$.

(ii) *F defines a Cremona transformation of $\mathbb{P}^{n-1}$.*

(iii) *Either*

  (a) *$\tilde{\mathcal{G}} = \mathcal{G}$ (i. e., no loops), $\mathcal{G}$ has a unique cycle and this cycle has odd length; or else*

  (b) *$\tilde{\mathcal{G}}$ is a tree with exactly one loop.*

(iv) *The ideal $(F) \subset k[x_1, \dots, x_n]$ is of linear type.*

Proof. The equivalence of (i) through (iii) follows immediately from Theorem 3.3.9, by noticing that if the underlying simple graph $\mathcal{G}$ is bipartite and $\tilde{\mathcal{G}}$ has exactly n edges and loops, then the latter has to be a tree with exactly one loop. To see that the first three conditions are also equivalent to (iv), notice that (iv) implies (i) since the generators of an ideal of linear type are analytically independent, hence algebraically independent as they are forms of the same degree. Now, when $\tilde{\mathcal{G}} = \mathcal{G}$, the implication (iii)(a) $\Rightarrow$ (iv) is part of [195, Corollary 3.3] but has really been noticed way before in [218, Corollary 3.2] (see also [219, Corollary 8.2.4]). Thus, it remains to see that (iii)(b) $\Rightarrow$ (iv) in the case where $\tilde{\mathcal{G}}$ effectively has loops. This follows from Lemma 7.3.10 below by induction on n. □

Lemma 7.3.10. *Let $F = \{f_1, \dots, f_q\} \subset R = k[x_1, \dots, x_n]$ be a set of monomials of degree two. Given $1 \leq i \leq n$, let $f_{q+1} = x_i x_{n+1} \in R' = R[x_{n+1}]$, where x_{n+1} is a new variable. If $(F) \subset R$ is an ideal of linear type, then $(F, f_{q+1}) \subset R'$ is an ideal of linear type.*

Proof. Set $I = (F)$ and $I' = (I, f_{q+1})$. Let $R'[I't]$ be the Rees algebra of I' over R'. Let $J \subset S := R[y_1, \dots, y_q]$ and $J' \subset S' := R'[y_1, \dots, y_q, y_{q+1}]$ denote respective homogeneous presentation ideals of the Rees algebras $R[It]$ over R and $R'[I't]$ over R'. The natural inclusion $R[It] \subset R'[I't]$ induces an homogeneous inclusion $J \subset J'$ for the standard grading of the two Rees algebras. It suffices to show that, for any $s \geq 1$, the graded piece J'_s lives in $S'J'_1$.

We induct on s, the result being vacuous for $s = 1$. Thus, let $s \geq 2$. Since I' is a monomial ideal, J' is generated by binomials homogeneous in the y-variables. By the inductive hypothesis, it suffices to show that these belong to $S'J'_{s-1}$. Let then

$$h = x^{\alpha} y_{i_1}^{a_1} \cdots y_{i_k}^{a_k} - x^{\beta} y_{j_1}^{b_1} \cdots y_{j_r}^{b_r}$$

be such a binomial in J' binomial in J'_s, where $i_1, \dots, i_k, j_1, \dots, j_r$ are distinct integers between 1 and $n+1$, $a_i > 0$, $b_i > 0$ for all i, j and $a_1 + \cdots + a_k = b_1 + \cdots + b_r = s$. We may assume that $f_{i_k} = f_{q+1} = x_i x_{n+1}$, otherwise $h \in SJ_1 \subset S'J'_1$ because I is of linear type. By definition,

$$x^{\alpha} f_{i_1}^{a_1} \cdots f_{i_k}^{a_k} = x^{\beta} f_{j_1}^{b_1} \cdots f_{j_r}^{b_r},$$

from which follows that x_{n+1} divides x^{β}, since no f_j on the right side of this equality involves the variable x_{n+1}. Thus, there is a relation

$$x^{\alpha} f_{i_1}^{a_1} \cdots f_{i_{k-1}}^{a_{k-1}} f_{q+1}^{a_{q+1}-1} = x^{\delta} f_{j_1}^{c_1} \cdots f_{j_r}^{c_r} \tag{118}$$

where one of the c_i's may be zero and $c_1 + \cdots + c_r = s - 1 \geq 1$. We may assume that $c_1 > 0$ because not all c_i's are zero. Write $h = y_{i_{q+1}} F_1 + y_{j_1} F_2$, where

$$F_1 = x^{\alpha} y_{i_1}^{a_1} \cdots y_{i_{k-1}}^{a_{k-1}} y_{i_{q+1}}^{a_{q+1}-1} - x^{\delta} y_{j_1}^{c_1} \cdots y_{j_r}^{c_r} \quad \text{and} \quad F_2 = x^{\delta} y_{i_{q+1}} y_{j_1}^{c_1-1} y_{j_2}^{c_2} \cdots y_{j_r}^{c_r} - x^{\beta} y_{j_1}^{b_1-1} y_{j_2}^{b_2} \cdots y_{j_r}^{b_r}.$$

Since $F_1 \in J'$ because of (118), then $y_{j_1} F_2 \in J'$, hence $F_2 \in J'$ as J' is prime. Therefore, h is an element of $S'J'_{s-1}$, as required. □

Example 7.3.11. $\{x_1x_2, x_1x_3, x_2x_3\}$ and $\{x_1x_2, x_1x_3, x_3^2\}$ are examples of each of the subcases (a) and (b) in Proposition 7.3.9. They respectively define the standard Cremona plane maps with 3 distinct base points and with 2 base points and one infinitely near point. The third type of standard Cremona map is a double structure on one single point, hence is not monomial.

7.3.2.2 The degree of the inverse map

This part is lifted from [46] and [47].

Let $V = \{v_1, \dots, v_q\} \subset \mathbb{N}^n$ be a set of distinct vectors and consider the corresponding *log*-set of monomials $F^V = \{\mathbf{x}^{v_1}, \dots, \mathbf{x}^{v_q}\} \subset R = k[x_1, \dots, x_n]$. The following *canonical restrictions* will be assumed throughout:

- For every $i \in \{1, \dots, n\}$, there is a $j \in \{1, \dots, q\}$ such that $v_{ij} = 0$, where $v_j = (v_{1j}, \dots, v_{nj})$ (i. e., the monomials in F^V have no nontrivial common factor).
- For every $i \in \{1, \dots, n\}$ there is at least one $j \in \{1, \dots, q\}$ such that $v_{ij} \neq 0$, where $v_j = (v_{1j}, \dots, v_{nj})$.

Recall once more that the log-matrix A_V of V as above is the integer matrix whose columns are the (transposes of the) vectors in V. Also we say that V is a *d-stochastic* set if $|v_1| = \cdots = |v_q| = d$, for some fixed integer $d \geq 1$, i. e., if the monomials have the same degree d. This is of course the typical environment when considering rational maps of projective varieties.

Drawing upon the structural result of Proposition 7.3.9, one has the following.

Lemma 7.3.12. *Let $F^V = \{\mathbf{x}^{v_1}, \ldots, \mathbf{x}^{v_n}\} \subset R$ be a 2-stochastic set. If F^V defines a Cremona map of $\mathbb{P}^n_k$, then up to permutations of the variables and the monomials of F^V, the latter has a log-matrix of the following shape:*

$$\mathcal{N}_F = \left(\begin{array}{c|c|c|c|c} \mathcal{N}_{F_r} & M_1 & & & \\ \hline & I_{s_1} & M_2 & & \\ \hline & & I_{s_2} & & \\ \hline & & & \ddots & M_p \\ \hline & & & & I_{s_p} \end{array}\right),$$

where:

(i) *$\mathcal{N}_{F_r}$ denotes either the incidence matrix of the unique odd circuit of length r, written in the form*

$$\begin{pmatrix} 1 & 0 & & 0 & 1 \\ 1 & 1 & & 0 & 0 \\ 0 & 1 & & 0 & 0 \\ & & \ddots & & \\ 0 & 0 & & 1 & 0 \\ 0 & 0 & & 1 & 1 \end{pmatrix}$$

or the 1×1 matrix (2) corresponding to the unique loop.

(ii) *I_{s_j} denotes the identity matrix of size s_j, where j runs through the ordered list of the neighborhoods of the unique circuit and s_j denotes the number of edges in the jth neighborhood of the circuit, with $j = 1, \ldots, p$.*

(iii) *M_j is a certain matrix with 0, 1 entries distributed in such a way as to have exactly one 1 on every column.*

(iv) *The empty slots are filled with zero entries.*

Proof. Take the corresponding graph as a rooted tree, where the root is either the unique circuit or else the unique loop and the successive neighborhoods of the circuit (or loop). The disposition on the matrix (left to right) obeys the order of the neighborhoods from the root. Finally, along every single neighborhood we order the vertices in an arbitrary way. □

According to Proposition 7.3.9, the graph whose edges define a monomial Cremona map in degree 2 has either a unique circuit and this circuit has odd length $r \geq 3$ or else degenerates into a loop of length $r = 1$. For convenience, we will call *root circuit* this basic unique subgraph in both cases. Also, when talking about the log-matrix in this case we always mean the normal form $\mathcal{N}_F$ as established in Lemma 7.3.12.

We will make strong use of Theorem 7.3.2 and refer to the terminology right after it. In the notation there, in order to emphasize that A comes from a d-stochastic set F of monomials, we set $A = A_F$.

Theorem 7.3.13. *Let $F = \{\mathbf{x}^{v_1}, \ldots, \mathbf{x}^{v_n}\} \subset k[x_1, \ldots, x_n] (n \geq 2)$ define a monomial Cremona map in degree 2 and let G_F stand for the corresponding graph. Let r denote the length of the root circuit of G_F and let s stand for the number of vertices of G_F off the root circuit whose vertex degree is ≥ 2. Then:*
(a) *The degree of the Cremona inverse F^{-1} is $(r+1)/2 + s$.*
(b) *The entries of the Cremona inverse matrix $A_{F^{-1}}$ satisfy the following conditions:*
   (i) *The entries on the main diagonal are all nonzero.*
   (ii) *All entries belong to $\{0, 1, 2\}$; moreover, 2 is an entry on the ith row if and only if the corresponding vertex x_i does not belong to the root circuit of G_F and has degree ≥ 2; in particular, the inverse map is defined by square-free monomials if and only if the root circuit of G_F has empty second neighborhood.*
(c) *Write $\widetilde{G_F}$ for the subgraph obtained from G_F by omitting the vertices of degree 1 and the corresponding incident edges and let $\{f_{i_1}, \ldots, f_{i_m}\}$ denote the monomials corresponding to the edges of $\widetilde{G_F}$ not belonging to the root circuit $\{x_1, \ldots, x_r\}$ of G_F. Then the inversion monomial factor is $x_1 \cdots x_r f_{i_1} \cdots f_{i_m}$, where the order of the vertices of the circuit and of the other edges is as in the normal form $\mathcal{N}_F$.*

Proof. Let us first focus on the square-free case (i. e., when G_F is a simple graph). The degenerate case of a loop will need just a few adjustments to be explained at the end.

Drawing on Proposition 7.3.9, one considers the family of all connected graphs on n vertices, having a unique circuit of a fixed length $3 \leq r \leq n$ and proceed by induction on $n \geq r$.

For $n = r$, the graph reduces to a circuit of length r. The defining monomials of the inverse map F^{-1} are well known to come from the set of minimal coverings of the circuit by taking, for each such covering, the product of the variables corresponding to the vertices of the covering. Clearly, the degree is $(n+1)/2$ and the entries on the main diagonal of $A_{F^{-1}}$ are all equal to 1. Further, the entries of the Cremona inverse matrix are 0, 1 and the monomial inversion factor is $x_1 \cdots x_n$ as is seen directly. This takes care of statements (a) through (c) in this particular situation.

Now assume that $n \geq r+1$. Then there exists a vertex off the circuit whose degree in G_F is 1—any vertex lying on the last nonempty neighborhood will do. Moreover, by suitably reordering the indices corresponding to the vertices on this neighborhood, one may assume that the vertex corresponds to x_n and the unique edge issuing from it corresponds to the monomial $\mathbf{x}^{v_n} = x_j x_n$, where $j \in \{1, \ldots, n-1\}$ is such that x_j corresponds to the unique vertex adjacent to x_n, a vertex on a neighborhood next to last neighborhood of the circuit in the established order.

Clearly, the set $F' = F \setminus \mathbf{x}^{v_n} = \{\mathbf{x}^{v_1}, \dots, \mathbf{x}^{v_{n-1}}\} \subset k[\mathbf{x} \setminus x_n] = k[x_1, \dots, x_{n-1}]$ is still a Cremona set in degree 2 satisfying the canonical restrictions. By the inductive hypothesis, its Cremona inverse set $(F')^{-1}$ and inversion vector y' satisfy statements (a) through (c). In particular, the degree of $(F')^{-1}$ is $(r+1)/2 + s'$, where s' is the number of vertices of $G_{F'}$ off the circuit whose vertex degree in $G_{F'}$ is ≥ 2.

Since we have not changed the normal arrangement of the earlier neighborhoods of the circuit, the normal form of the log-matrix of F has the shape

$$\mathcal{N}_F = \left(\begin{array}{c|c} \mathcal{N}_{F'} & M \\ \hline 0 & 1 \end{array} \right),$$

where $\mathcal{N}_{F'}$ is a normal form of the log-matrix of F' and M denotes an $(n-1) \times 1$ matrix whose only nonzero entry is its jth entry.

We will now argue that the Cremona inverse matrix of $\mathcal{N}_F$ inherits a similar shape, with $\mathcal{N}_{F'}$ replaced by the Cremona inverse matrix of $\mathcal{N}_{F'}$. From this shape, we will read off the degree of F^{-1} and the remaining assertions in the statements.

For this, one has to analyze the role of the special vertex corresponding to x_j, where $j \in \{1, \dots, n-1\}$. We accordingly divide this analysis into three distinct cases.

Case 1. The vertex corresponding to x_j belongs to the common circuit of G_F and $G_{F'}$.

In this case, since x_j belongs to the common circuit of G_F and $G_{F'}$, one has $s' = s$ and, clearly, $\widetilde{G_F} = \widetilde{G_{F'}}$ since x_j has degree ≥ 2.

Further, since $r \geq 3$, one can rewrite a normal form of F' in which $j = 2$ and $\mathbf{x}^{v_1} = x_1x_2$, $\mathbf{x}^{v_2} = x_2x_3$ are the two adjacent edges of the circuit. The claim is that the following matrix has the properties in the statement of Theorem 7.3.2 with regard to $\mathcal{N}_F$, i. e., is the Cremona inverse matrix of $\mathcal{N}_F$:

$$B = \left(\begin{array}{c|c} A_{(F')^{-1}} & N \\ \hline 0 & 1 \end{array} \right)$$

where $A_{(F')^{-1}}$ is the log-matrix of a uniquely defined inverse to F' as in [loc. cit.], and N is the $(n-1) \times 1$ matrix in which

$$N_l = \begin{cases} (A_{(F')^{-1}})_{l,1} - 1, & \text{if } l = 1 \\ (A_{(F')^{-1}})_{l,1}, & \text{otherwise.} \end{cases}$$

Note that subtracting 1 in the first alternative above makes sense since, by the inductive hypothesis, no entry along the main diagonal of $A_{(F')^{-1}}$ is null.

Let $\mathcal{N}_{F'} \cdot A_{(F')^{-1}} = \Gamma' + I_{n-1}$ be the fundamental matrix equation of inversion as in Theorem 7.3.2 relative to F', with $\Gamma' = \underbrace{[y' | \cdots | y']}_{n-1}$. The usual block multiplication then yields

$$\mathcal{N}_F \cdot B = \left(\begin{array}{c|c} \mathcal{N}_{F'} \cdot A_{(F')^{-1}} & \mathcal{N}_{F'} \cdot N + M \\ \hline 0 & 1 \end{array}\right) = \left(\begin{array}{c|c|c|c} y' & \cdots & y' & \mathcal{N}_{F'} \cdot N + M \\ \hline 0 & \cdots & 0 & 0 \end{array}\right) + I_n,$$

where M is $(n-1) \times 1$ and $y' = [(y')_1, \ldots, (y')_{n-1}]^t$. A straightforward calculation now yields $(\mathcal{N}_{F'} \cdot N + M)_l = (y')_l$, for $l = 1, \ldots, n-1$. Indeed:

– If $l \neq 1, 2$ one has $M_l = 0$ and $(\mathcal{N}_{F'})_{l,1} = 0$, hence

$$(\mathcal{N}_{F'} \cdot N + M)_l = \sum_{k=1}^{n-1} (\mathcal{N}_{F'})_{l,k} n_k = \sum_{k=1}^{n-1} (\mathcal{N}_{F'})_{l,k} (A_{(F')^{-1}})_{k,1} = (y')_l.$$

– If $l = 2$, it is the case that $M_2 = 1$ and $(\mathcal{N}_{F'})_{2,1} = 1$, therefore,

$$(\mathcal{N}_{F'} \cdot N + M)_2 = \sum_{k=1}^{n-1} (\mathcal{N}_{F'})_{2,k} n_k + M_2 = \sum_{k=1}^{n-1} (\mathcal{N}_{F'})_{2,k} (A_{F'^{-1}})_{k,1} - (\mathcal{N}_{F'})_{2,1} + M_2 = (y')_2.$$

– If $l = 1$, then $M_1 = 0$ and $(\mathcal{N}_{F'})_{1,1} = 1$, and hence

$$\begin{aligned}(\mathcal{N}_{F'} \cdot N + M)_1 &= \sum_{k=1}^{n-1} (\mathcal{N}_{F'})_{1,k} n_k = \sum_{k=1}^{n-1} (\mathcal{N}_{F'})_{1,k} (A_{F'^{-1}})_{k,1} - (\mathcal{N}_{F'})_{1,1} \\ &= (y')_1 + 1 - (\mathcal{N}_{F'})_{1,1} = (y')_1.\end{aligned}$$

Hence, this will yield an inversion equation $\mathcal{N}_F \cdot B = \Gamma + I_n$, with $\Gamma = [y|\cdots|y]$, where $y_l = (y')_l$ for $l = 1, \ldots, n-1$ and $y_n = 0$. Moreover, since $(F')^{-1}$ satisfies the canonical restrictions, so does B by construction. By Theorem 7.3.2, B is the uniquely defined matrix giving the inverse to F, as stated. Moreover, by the explicit format of B with the nature of the entries of N, the inductive hypothesis implies that the entries of B are 0, 1, 2, while 2 appears exactly on the ith row if and only if x_i is a vertex off the root circuit with degree ≥ 2. Since $|y| = |y'|$ and, besides, no entry on the main diagonal of B is null, we are through in this case for all statements of the theorem.

Case 2. The vertex corresponding to x_j does not belong to the circuit and its degree in $G_{F'}$ is ≥ 2

In this case, since $G_{F'}$ only misses the vertex x_n and its unique adjacent edge $x_j x_n$, and since s only counts vertices of degree ≥ 2, it follows that $s' = s$ in this case, too, and further, $\widetilde{G_F} = \widetilde{G_{F'}}$ since x_j has degree ≥ 2.

By hypothesis, there are at least two indices $1 \leq i' < j < i \leq n-1$ such that $x_{i'} x_j$ and $x_j x_i$ belong to the set F. Note that the vertex $x_{i'}$ might belong to the circuit. We claim that the following matrix has the properties in the statement of Theorem 7.3.2

with regard to $\mathcal{N}_F$, i. e., is its Cremona inverse matrix:

$$B = \left(\begin{array}{c|c} A_{(F')^{-1}} & N \\ \hline 0 & 1 \end{array} \right)$$

where $A_{(F')^{-1}}$ is the Cremona inverse matrix of $\mathcal{N}_{F'}$ as in [loc. cit.], and N is the $(n-1)\times 1$ matrix in which

$$N_l = \begin{cases} (A_{(F')^{-1}})_{l,i} - 1, & \text{if } l = i \\ (A_{(F')^{-1}})_{l,i}, & \text{otherwise,} \end{cases}$$

where i is the largest of the two indices taken above. Let $\mathcal{N}_{F'} \cdot A_{(F')^{-1}} = \Gamma' + I_{n-1}$ be the fundamental matrix equation of inversion as in Theorem 7.3.2 relative to F', with $\Gamma' = [\underbrace{y'|\cdots|y'}_{n-1}]$. Block multiplication then yields

$$\mathcal{N}_F \cdot B = \left(\begin{array}{c|c} \mathcal{N}_{F'} \cdot A_{(F')^{-1}} & \mathcal{N}_{F'} \cdot N + M \\ \hline 0 & 1 \end{array} \right) = \left(\begin{array}{c|c|c|c} y' & \cdots & y' & \mathcal{N}_{F'} \cdot N + M \\ \hline 0 & \cdots & 0 & 0 \end{array} \right) + I_n,$$

where M is $(n-1)\times 1$ and $y' = [(y')_1, \ldots, (y')_{n-1}]^t$. A straightforward calculation now yields $(\mathcal{N}_{F'} \cdot N + M)_l = (y')_l$, for $l = 1, \ldots, n-1$:

- If $l \neq j, i$ one has $M_l = 0$ and $(\mathcal{N}_{F'})_{l,i} = 0$; hence,

$$(\mathcal{N}_{F'} \cdot N + M)_l = \sum_{k=1}^{n-1} (\mathcal{N}_{F'})_{l,k} n_k = \sum_{k=1}^{n-1} (\mathcal{N}_{F'})_{l,k} (A_{F'^{-1}})_{k,i} = (y')_l.$$

- If $l = j$, it is the case that $M_j = 1$ and $(\mathcal{N}_{F'})_{j,i} = 1$, therefore,

$$(\mathcal{N}_{F'} \cdot N + M)_j = \sum_{k=1}^{n-1} (\mathcal{N}_{F'})_{j,k} n_k + M_j = \sum_{k=1}^{n-1} (\mathcal{N}_{F'})_{j,k} (A_{F'^{-1}})_{k,i} - (\mathcal{N}_{F'})_{ji} + M_j = (y')_j.$$

- If $l = i$, then $M_i = 0$ and $(\mathcal{N}_{F'})_{i,i} = 1$, and hence

$$\begin{aligned} (\mathcal{N}_{F'} \cdot N + M)_i &= \sum_{k=1}^{n-1} (\mathcal{N}_{F'})_{i,k} n_k = \sum_{k=1}^{n-1} (\mathcal{N}_{F'})_{i,k} (A_{(F')^{-1}})_{k,i} - (\mathcal{N}_{F'})_{i,i} \\ &= (y')_i + 1 - (\mathcal{N}_{F'})_{i,i} = (y')_i. \end{aligned}$$

Hence, this will yield an inversion equation $\mathcal{N}_F \cdot B = \Gamma + I_n$, with $\Gamma = [y|\cdots|y]$, where $y_l = (y')_l$ for $l = 1, \ldots, n-1$ and $y_n = 0$. The conclusion is identical to the one in the previous case for all three statements (a) through (c), so we are done in this case as well.

Case 3. The vertex corresponding to x_j does not belong to the circuit and its degree in $G_{F'}$ is 1.

In this case, one readily sees that $s = s' + 1$ and, by a similar token, $\widetilde{G_F} = \widetilde{G_{F'}} \cup \{x_i x_j\}$, where x_i is the unique vertex of $G_{F'}$ adjacent to x_j. Note that, in this situation, we need to prove that the vector y that appears in the inversion equation of F has modulo $|y'|+2$, where y' is the corresponding vector for the inversion equation of F'; more precisely, we need $\mathbf{x}^y = \mathbf{x}^{y'} x_i x_j$.

Now, let again $i < j$ denote the unique index such that $x_i x_j$ belongs to F'—here x_i may or may not belong to the circuit. In this case, we find it appropriate to express the Cremona inverse matrix of $\mathcal{N}_F$ in the form

$$B = \left(\begin{array}{c|c} A_{(F')^{-1}} & N \\ \hline 0 & 1 \end{array} \right) + E,$$

where N is the ith column of $A_{(F')^{-1}}$ and $E = (e_{k,l})$ is the $n \times n$ matrix defined by

$$e_{k,l} = \begin{cases} 1, & \text{if } k = j, l \neq n \\ 0, & \text{otherwise.} \end{cases}$$

Multiplying we find

$$\begin{aligned} \mathcal{N}_F \cdot B &= \left(\begin{array}{c|c} \mathcal{N}_{F'} \cdot A_{(F')^{-1}} & \mathcal{N}_{F'} \cdot N + M \\ \hline 0 & 1 \end{array} \right) + \left(\begin{array}{ccc|c} \alpha_j & \cdots & \alpha_j & \mathbf{0} \\ \hline 0 & \cdots & 0 & 0 \end{array} \right) \\ &= \left(\begin{array}{c|c|c|c} y' + \alpha_j & \cdots & y' + \alpha_j & \mathcal{N}_{F'} \cdot N + M \\ \hline 0 & \cdots & 0 & 0 \end{array} \right) + I_n, \end{aligned}$$

where α_j is the jth column of $\mathcal{N}_{F'}$.

We now assert that:

(1) $\mathcal{N}_{F'} \cdot N + M = y' + \alpha_j$.

(2) Every row of B has a zero entry.

The first of these statements gives an inversion equation for F with unique vector $y := y' + \alpha_j$ and, since α_j is a column of a Cremona matrix in degree 2, it follows that $\mathbf{x}^y = \mathbf{x}^{y'+\alpha_j} = \mathbf{x}^{y'} x_i x_j$ – in particular, $|y| = |y'| + 2$ – as was to be shown. It is also clear from the form of B that there is no null entry along its main diagonal. Moreover, since x_j has degree 1 on $G_{F'}$ the entries on the jth row of $A_{(F')^{-1}}$ are 0, 1 and not all are zero. Once more, the explicit format of B shows that its entries are 0, 1, 2, while 2 appears exactly on the ith row if and only if x_i is a vertex off the root circuit with degree ≥ 2.

To prove (1), we proceed again along several cases:

- If $l \neq j, i$ one has $M_l = 0$ and $(\mathcal{N}_{F'})_{l,j} = 0$, hence

$$(\mathcal{N}_{F'} \cdot N + M)_l = \sum_{k=1}^{n-1} (\mathcal{N}_{F'})_{l,k} (A_{(F')^{-1}})_{k,i} = (y')_l + \alpha_{l,j}.$$

- If $l = j$, it is the case that $M_j = 1$ and $(\mathcal{N}_{F'})_{j,j} = 1$, therefore,

$$(\mathcal{N}_{F'} \cdot N + M)_j = \sum_{k=1}^{n-1} (\mathcal{N}_{F'})_{j,k} (A_{(F')^{-1}})_{k,i} + 1 = (y')_j + 1 = (y')_j + \alpha_{j,j}.$$

- If $l = i$, then $M_i = 0$ and $(\mathcal{N}_{F'})_{i,j} = 1$; hence,

$$(\mathcal{N}_{F'} \cdot N + M)_i = \sum_{k=1}^{n-1} (\mathcal{N}_{F'})_{i,k} (A_{(F')^{-1}})_{k,i} = (y')_i + 1 = (y')_i + \alpha_{i,j}.$$

As for the second assertion, it is obvious for any row except possibly for the jth row. For the latter, it suffices to show that $(A_{F^{-1}})_{j,n} = (A_{(F')^{-1}})_{j,i} = 0$. First, observe that $(\mathcal{N}_{F'})_{j,j} = 1$ and $(\mathcal{N}_{F'})_{j,k} = 0$ for every $k \in \{1, \dots n-1\} \setminus \{j\}$ since we are in the normal form and j is the index of a vertex of degree 1 off the circuit. Then the inversion equation of F' implies that $(A_{(F')^{-1}})_{j,k} = y'_j + \delta_{j,k}$ for $k \in \{1, \dots n-1\}$. Since $A_{(F')^{-1}}$ satisfies the canonical restrictions, some entry along its jth row is null. Therefore, $(A_{(F')^{-1}})_{j,k} = 0$ for all $k \neq j$ and $(A_{(F')^{-1}})_{j,j} = 1$; in particular, $(A_{(F')^{-1}})_{j,i} = 0$.

To conclude, we explain the adjustment in the case where the circuit degenerates into a loop. This concerns only the initial step in the induction process. Since F satisfies the canonical restrictions, with $n \geq 2$ by assumption, its constituents have no proper common factor. Therefore, $n \geq s_1 + 2$ where, we recall, $s_1 \geq 1$ stands for the set of edges of G_F in the first neighborhood of the loop. Thus, the initial step could be vacuous or, alternatively, would start from $n = s_1 + 2$, while in the inductive step one would then assume that $n > s_1 + 2$. Taking up the second alternative, the initial step has $F = \{x_1^2, x_1x_2, \dots, x_1x_{n-1}, x_{n-1}x_n\}$, with $n \geq 3$. Consider the following set in degree 2:

$$\{x_1x_{n-1}, x_2x_{n-1}, \dots, x_{n-2}x_{n-1}, x_{n-1}^2, x_1x_n\}, \tag{119}$$

where the roles of x_1 and x_{n-1} have been interchanged and the loop has moved to another slot in the sequence. A direct calculation show that this set is the Cremona inverse set of F – one can compose the two sets on the nose or else pass to the respective log-matrices and multiply them out to get the inversion equation, with inversion vector $y = (2, 0, \dots, 0, 1, 0)^t$.

The rest of the argument stays unchanged, as far as the inductive step goes. □

7.3.2.3 A role of Hilbert bases

The goal of this part is to prove that, among the generators of an equigenerated normal monomial ideal in n variables, a subset will define a Cremona transformation of $\mathbb{P}^{n-1}$.

The plan is to approach a proof of the theorem by drawing upon the combinatorics of the so-called Hilbert bases. We now expand on these matters by reviewing a few elementary notions of polyhedral combinatorics. As a general reference, we use the beautiful exposé in [219].

Recall the partial order on $\mathbb{R}^n$ defined by $a = (a_1, \ldots, a_n) \le c = (c_1, \ldots, c_n)$ if $a_i \le c_i$ for every i. Given vectors $a, b \in \mathbb{R}^n$, their inner product will be denoted by $\langle a, b\rangle$. A hyperplane $H = H(a,c) = \{x \in \mathbb{R}^n \mid \langle x, a\rangle = c\} \subset \mathbb{R}^n$ determines two closed half-spaces:

$$H^+(a,c) = \{x \in \mathbb{R}^n \mid \langle x, a\rangle \ge c\} \quad \text{and} \quad H^-(a,c) = \{x \in \mathbb{R}^n \mid \langle x, a\rangle \le c\}.$$

We follow common usage of writing $H_a^+ = H^+(a,0)$ and $H_a^- = H^-(a,0)$ when the hyperplane goes through the origin.

A (*convex*) *cone* in $\mathbb{R}^n$ is a nonempty set $C \subset \mathbb{R}^n$ such that, for all $x, y \in C$ and all real $\lambda, \mu \ge 0$, one has $\lambda x + \mu y \in C$.

A cone C is *polyhedral* if it there exists a finite subset $S = \{s_1, \ldots, s_q\} \subset C$ such that $C = \mathbb{R}_+ S := \{\lambda_1 s_1 + \cdots + \lambda_q s_q, \lambda_j \in \mathbb{R}_+\}$. We then refer to S as a generating set of C.

The *dual* to a polyhedral cone $C \subset \mathbb{R}^n$ is defined as

$$C^* := \{u \in \mathbb{R}^{n*} = \mathrm{Hom}_{\mathbb{R}}(\mathbb{R}^n, \mathbb{R}) \mid \langle v, u\rangle \le 0, \forall v \in C\},$$

where $\langle\, ,\, \rangle$ denotes the ordinary pairing on $\mathbb{R}^n \times \mathbb{R}^{n*}$. It is immediate to see that it suffices to take the pairing over a finite set of cone generators of C.

Common practice identifies $\mathbb{R}^n$ with its dual space by identifying a vector basis with its dual basis. In this way, the pairing can be seen as the usual inner product and the dual to a cone can be considered in the same space. As such, one has a representation $C^* = \cap_{s \in S} H_s^-$ as intersection of half-spaces through the origin. Moreover, a fundamental result going back to Farkas, Minkowski and Weyl (see [219, Theorem 1.1.31]) says that a cone is polyhedral if and only if it is the intersection of a finite set of closed half-spaces through the origin. From this follows that the dual to a polyhedral cone is also a polyhedral cone and $(C^*)^* = C$ for every polyhedral cone.

This well-known dichotomy of representing a polyhedral cone, both as the intersection of closed half-spaces through the origin and as the set of nonnegative linear combinations of a finite set of vectors is very useful. The first of these representation allows to write a polyhedral cone in the form $\{v \in \mathbb{R}^n \mid Av \le 0\}$, for some real matrix A.

In this vein, we say that C is *pointed* if the linear system $Ax = 0$ has only the trivial solution $x = 0$, i. e., A has maximal rank. In other words, a pointed polyhedral cone contains no straight lines. Since this condition also means that $C \cap (-C) = \{0\}$, where $-C$ has the obvious meaning, a pointed cone is also called *strongly convex*. To free ourselves from the matrix representation in the notion, we can use the characterization in [219, Proposition 1.1.56] to the effect that a polyhedral cone is pointed if and only if its dual has maximal dimension (i. e., n).

We now come to the main concept of this part. As a matter of further notation, given a subset $\mathcal{A} \subset \mathbb{R}^n$, denote by $\mathbb{Z}\mathcal{A}$ (resp., $\mathbb{N}\mathcal{A}$) the integer lattice generated by $\mathcal{A}$ (resp., the set of lattice elements with nonnegative coefficients).

Definition 7.3.14. A finite subset $H \subset \mathbb{R}^n$ is a *Hilbert base* if $\mathbb{Z}^n \cap \mathbb{R}_+H = \mathbb{N}H$. A polyhedral cone is said to admit a Hilbert base if it contains a Hilbert base and is generated by it.

Note that the definition implies that a Hilbert base is contained in $\mathbb{Z}^n$.

The fundamental results regarding Hilbert bases are as follows. The first tells us that Hilbert bases are pretty ubiquitous and often uniquely defined.

Theorem 7.3.15 ([171, Theorem 16.4]). *A rational polyhedral cone C admits a Hilbert base. If, moreover, C is pointed then it contains a unique minimal such base in the sense that no proper subset is a generating Hilbert base.*

The second result even gives a hint as to the nature of a minimal such base.

Theorem 7.3.16 ([72, proof of Theorem 1]). *Let H denote a Hilbert base of a pointed polyhedral cone and let r stand for the rank of the lattice generated by H. Then H has a subset of r linearly independent vectors forming a Hilbert base.*

Recall that an ideal I of a ring R is *normal* if all its powers are integrally closed in R.

Our first main result of this section is the following.

Theorem 7.3.17. *Let $v_1, \dots, v_q \in \mathbb{N}^n$ $(q \geq n)$ be given such that the associated monomials $\mathbf{x}^{v_1}, \dots, \mathbf{x}^{v_q}$ have the same degree $d \geq 1$. If the ideal $(\mathbf{x}^{v_1}, \dots, \mathbf{x}^{v_q}) \subset k[\mathbf{x}]$ is normal then $\{(v_1, 1), \dots, (v_q, 1)\}$ is a Hilbert base.*

Proof. We first note that, since the ring $k[\mathbf{x}]$ is normal, the ideal $(\mathbf{x}^{v_1}, \dots, \mathbf{x}^{v_q}) \subset k[\mathbf{x}]$ is normal if and only if the Rees algebra of this ideal is normal. But the Rees algebra is isomorphic to the semigroup ring $k[\mathbf{x}, \mathbf{x}^{v_1}T, \dots, \mathbf{x}^{v_q}T] \subset k[\mathbf{x}, T]$. Applying the well-known combinatorial criterion (see, e. g., [219, Corollary 7.2.29]), we find that this algebra is normal if and only if

$$\mathbb{Z}H' \cap \mathbb{R}_+H' = \mathbb{N}H',$$

where $H' = \{e_1, \dots, e_n\} \cup H$, $H = \{(v_1, 1), \dots, (v_q, 1)\} \subset \mathbb{N}^{n+1}$, and $\{e_1, \dots, e_n, e_{n+1}\}$ stands for the canonical basis of $\mathbb{Z}^{n+1}$.

On the other hand, it is well known that, as a consequence of the so-called Farkas Lemma ([219, Corollary 1.1.29], also [171, Corollary 7.1d]), for any subset $\mathcal{A} \subset \mathbb{Z}^m$ the natural inclusions $\mathbb{Z}\mathcal{A} \cap \mathbb{Q}_+\mathcal{A} \subset \mathbb{Z}\mathcal{A} \cap \mathbb{R}_+\mathcal{A}$ and $\mathbb{Z}^m \cap \mathbb{Q}_+\mathcal{A} \subset \mathbb{Z}^m \cap \mathbb{R}_+\mathcal{A}$ are equalities. Therefore, we are to prove that

$$\mathbb{Z}H' \cap \mathbb{Q}_+H' \subset \mathbb{N}H' \Rightarrow \mathbb{Z}^{n+1} \cap \mathbb{Q}_+H \subset \mathbb{N}H.$$

Now, $\mathbb{Z}H' = \mathbb{Z}^{n+1}$ because, e. g., $e_{n+1} = (v_1, 1) - v_{1,1}e_1 + \cdots + v_{n,1}e_n$, where $v_1 = (v_{1,1}, \dots, v_{n,1})$. Thus, let $z \in \mathbb{Z}^{n+1} \cap \mathbb{Q}_+H$, say, $z = (z_1, \dots, z_{n+1}) = \lambda_1(v_1, 1) + \cdots + \lambda_q(v_q, 1)$.

Then

$$z_{n+1} = \sum_{i=1}^{q} \lambda_i \quad \text{and} \quad |z| = \left(\sum_{i=1}^{q} \lambda_i \right)(d+1) = z_{n+1}(d+1).$$

Since $H \subset H'$, also $z \in \mathbb{Z}^{n+1} \cap \mathbb{Q}_+ H'$, hence by the assumption and the above remark, one can write

$$z = \alpha_1(v_1, 1) + \cdots + \alpha_q(v_q, 1) + \beta_1(e_1, 0) + \cdots + \beta_n(e_n, 0),$$

for suitable $\alpha_1, \ldots, \alpha_q, \beta_1, \ldots, \beta_n \in \mathbb{N}$. But then $z_{n+1} = \sum_{i=1}^{q} \alpha_i$ and

$$|z| = \left(\sum_{i=1}^{q} \alpha_i \right)(d+1) + \sum_{j=1}^{n} \beta_j = z_{n+1}(d+1) + \sum_{j=1}^{n} \beta_j,$$

hence $\sum_{j=1}^{n} \beta_j = 0$, that is, $\beta_j = 0$ for every j. Consequently, $z \in \mathbb{N}H$ as was required to show. □

The following result is pretty elementary, but we give a proof for the sake of completeness.

Lemma 7.3.18. *Let $H = \{v_1, \ldots, v_q\} \subset \mathbb{Z}^n$ be an arbitrary subset. Then the subset $\{(v_1, 1), \ldots, (v_q, 1)\} \subset \mathbb{Z}^{n+1}$ generates a pointed cone. If, moreover, the associated monomials of the vectors in H have the same degree then this cone has the same dimension as the cone generated by H.*

Proof. To prove that the polyhedral cone generated by $(H, 1) := \{(v_1, 1), \ldots, (v_q, 1)\}$ is pointed one argues that it is strongly convex, namely, let

$$\sum_j a_j(v_j, 1) = \sum_j b_j(v_j, 1),$$

with $a_j \geq 0$, $\forall j$ and $b_j \leq 0$, $\forall j$. Looking at the last coordinate, we get $\sum_j a_j = \sum_j b_j$. Forcefully, $a_j = 0$, $\forall j$.

Assuming now that the matrix whose columns are the vectors in H is d-stochastic for some $d \geq 1$, the matrix

$$\begin{pmatrix} v_1 & \cdots & v_n \\ d & \cdots & d \end{pmatrix}$$

has the same rank. But this is also the rank over $\mathbb{Q}$ (hence, over $\mathbb{Z}$) of the matrix whose columns are $(v_1, 1), \ldots, (v_q, 1)$.

Thus, we are done. □

Remark 7.3.19. Regarding pointedness, one notes that, more generally, a finite set of vectors in $\mathbb{Z}^n$ generates a pointed polyhedral cone if, for some $i \in \{1, \ldots, n\}$, their ith coordinates are positive natural numbers.

The main combinatorial result of this part now follows.

Theorem 7.3.20. *Let $H = \{v_1, \ldots, v_q\} \subset \mathbb{N}^n$ be such that the associated monomials have the same degree $d \geq 1$ and $\mathbb{Z}H$ has rank n. If $(H, 1) = \{(v_1, 1), \ldots, (v_q, 1)\} \subset \mathbb{Z}^{n+1}$ is a Hilbert base then there exists an $n \times n$ submatrix of $[v_1 | \ldots | v_q]$ whose determinant is d.*

Proof. By Lemma 7.3.18, the cone generated by $(H, 1)$ is pointed. Therefore, by Theorem 7.3.16, $(H, 1)$ admits a subset $(H', 1)$ of n linearly independent vectors which is a Hilbert base.

We claim that the matrix whose columns are the vectors in H' gives the required result. For this, it suffices to show an isomorphism of $\mathbb{Z}$-modules $\mathbb{Z}^n/\mathbb{Z}H' \simeq \mathbb{Z}/d\mathbb{Z}$. First, note that $\mathbb{Z}^n/\mathbb{Z}H'$ is a torsion $\mathbb{Z}$-module since $\mathbb{Z}H'$ has rank n. On the other hand, one has an exact sequence of $\mathbb{Z}$-modules

$$0 \to T(\mathbb{Z}^{n+1}/\mathbb{Z}H) \to T(\mathbb{Z}^n/\mathbb{Z}H') \to \mathbb{Z}/d\mathbb{Z} \to 0,$$

where T denotes $\mathbb{Z}$-torsion (see, e. g., [195, the proof of Theorem 1.1]).

Now, $T(\mathbb{Z}^{n+1}/\mathbb{Z}H) = \mathbb{R}H \cap \mathbb{Z}^{n+1}/\mathbb{Z}H$ by [219, Section 1.3]. Since H is a Hilbert base, $\mathbb{R}_+H \cap \mathbb{Z}^{n+1} = \mathbb{N}H$. But the latter equality implies the containment $\mathbb{R}H \cap \mathbb{Z}^{n+1} \subset \mathbb{Z}H$, a fact that is readily checked by writing every coefficient $a \in \mathbb{R}$ in the form $\lfloor a \rfloor + b$, with $b \in \mathbb{R}_+$. □

Corollary 7.3.21. *Let $H := \{v_1, \ldots, v_q\} \subset \mathbb{N}^n$ be such that the associated monomials have the same degree $d \geq 1$. If $\mathbb{Z}H$ has rank n and $\{(v_1, 1), \ldots, (v_q, 1)\} \subset \mathbb{Z}^{n+1}$ is a Hilbert base, then there exist n vectors in H such that the associated monomials define a Cremona transformation of $\mathbb{P}^{n-1}$.*

Proof. By Theorem 7.3.20, there exists an $n \times n$ submatrix of $[v_1 | \ldots | v_q]$ whose determinant is d. Therefore, the criterion of Proposition 3.3.1 applies. □

Now, for the main promised result.

Theorem 7.3.22. *Let $\mathbf{x}^{v_1}, \ldots, \mathbf{x}^{v_q} \subset k[\mathbf{X}] = k[x_1, \ldots, x_n]$ ($q \geq n$) be monomials of the same degree generating a normal ideal. Then there exist n among these monomials defining a Cremona transformation of $\mathbb{P}^{n-1}$.*

Proof. It follows immediately from the previous corollary and Theorem 7.3.17. □

Remark 7.3.23. Note that, by Corollary 3.3.4, in any case the rational map defined by the given monomials is birational onto the image. The theorem has a subtlety in the sense that it may happen that the Cremona transformations so obtained are defined by monomials with a nontrivial gcd. For a simplest example, take $\{x^2, xy, y^2\} \subset k[x, y]$, generating a well-known normal ideal. These define the birational map $\mathbb{P}^1_k \dashrightarrow \mathbb{P}^2_k = \mathrm{Proj}(k[t,u,v])$ of the projective line onto the quadric $V(tv-u^2)$. The Cremona submaps are defined by $\{x^2, xy\}$ or $\{xy, y^2\}$. Clearly, as elements of the trivial Cremona group of $\mathbb{P}^1$, both give the identity transformation.

7.3.3 A duality principle

We say that two Cremona sets are *permutable* if they coincide up to a permutation of the source and of the target variables.

Obviously, for a given pair n, d, where n is the number of variables and d is the degree of the monomials, there are a finite number of mutually nonpermutable square-free Cremona sets with these values. Classifying means finding this complete list.

Classifying square-free Cremona sets looks within grasp since necessarily $d \le n-1$. Up to permutability, the only Cremona transformation of degree $d = n-1$ in n variables whose terms are square-free monomials is the analogue of the classical reciprocal involution, given by $F = \{x_1 \cdots x_{n-1}, x_1 \cdots x_{n-2}x_n, \dots, x_2 \cdots x_n\}$. For degrees $d \le n-2$, the classification becomes more involved. Our purpose in this part is to convey the impact of combinatorics on birationality by examining some scattered examples for low values of n and degrees $d \le n-2$. The case where $d = 2$ was completely covered by case (a) of Proposition 7.3.9.

7.3.3.1 Dual complement

Recall the following notion of combinatorial nature. If F is a set of monomials of the same degree with log-matrix $A = (a_{ij})$, its *dual complement* is the set $\hat{F}$ of monomials whose log-matrix is $\hat{A} = (1 - a_{ij})$.

Clearly, this is a special case of the Newton complementary dual as exposed in Section 4.3, but we deal with it here ab initio in an elementary way. For the time being, the following basic principle guides us into further simplification.

Proposition 7.3.24 (Duality principle). *Let F be a set of monomials in an odd number n of variables, of the same degree d, with no common factor. Then F is a Cremona set if and only if $\hat{F}$ is a Cremona set.*

Proof. First, there is a known equality that works for all n and d: $(n-d)\det(A) = (-1)^{n-1} d \det(\hat{A})$. The simple proof consists in adding the rows of A to get $\det(A) = d\det(A')$, where

$$A' = \begin{bmatrix} a_{1,1} & \dots & a_{1,n} \\ \vdots & & \vdots \\ a_{n-1,1} & \dots & a_{n-1,n} \\ 1 & \dots & 1 \end{bmatrix}.$$

Similarly, adding the rows of $\hat{A}$ we get $\det(\hat{A}) = (n-d)\det(\hat{A}')$, where

$$\hat{A}' = \begin{bmatrix} b_{1,1} & \dots & b_{1,n} \\ \vdots & \vdots & \vdots \\ b_{n-1,1} & \dots & b_{n-1,n} \\ 1 & \dots & 1 \end{bmatrix}.$$

Then $\hat{A}'$ is obtained from A' by subtracting the row $\mathbf{1} = (1,\dots,1)$ from each of the first $n-1$ rows of A' and making a change of sign at each step. Thus, the determinants of A' and $\hat{A}'$ differ by (at most) a sign given by $(-1)^{n-1}$. Thus,

$$\det(A) = d\det(A') = (-1)^{n-1} d\det(\hat{A}') = (-1)^{n-1}\frac{d}{n-d}\det(\hat{A}),$$

which yields the required formula.

Now, if F is Cremona then $\det(A) = d$ by $\mathcal{DP}$. The formula yields in this case $\det(\hat{A}') = (-1)^{n-1}(n-d) = n-d$, as n is odd. The reverse implication is obtained by a symmetrical argument. □

Proposition 7.3.25. *Up to permutation of the variables, the complete list of distinct squarefree Cremona sets of degree 3 in 5 variables is as follows:*
- $F = \{x_3x_4x_5, x_1x_4x_5, x_1x_2x_5, x_1x_2x_3, x_2x_3x_4\}$ ($\mathcal{DB}$)
- $F = \{x_3x_4x_5, x_1x_4x_5, x_1x_2x_5, x_1x_3x_5, x_1x_2x_4\}$ (p-involutive)
- $F = \{x_3x_4x_5, x_1x_4x_5, x_1x_2x_5, x_1x_3x_5, x_1x_3x_4\}$ (p-involutive)
- $F = \{x_3x_4x_5, x_1x_4x_5, x_1x_2x_5, x_2x_4x_5, x_1x_2x_3\}$ (apocryphal)

Proof. According to Proposition 7.3.24, the required complete list is the complete list of the dual-complements. The latter is the list of all squarefree degree 2 Cremona sets obtained from Proposition 7.3.9(a). Their corresponding graphs are shown below:

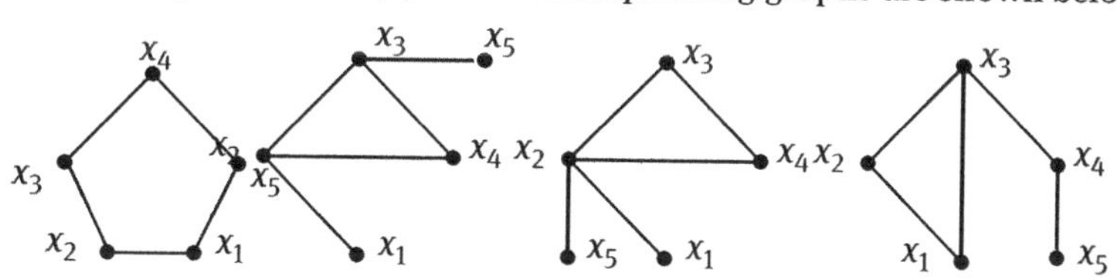

This concludes the proof. □

Remark 7.3.26. We explain the appended terminology. A set F of square-free monomials is called *d-doubly-stochastic* (short: $\mathcal{DB}$) if its log-matrix $A = (a_{ij})$ is doubly-stochastic, i. e., the entries of each column sum up to d (i. e., the monomials have fixed degree d) and so do the entries of each row (i. e., no variable is privileged or, the "incidence" degrees of the variables is also d). A Cremona set is called *p-involutive* if it coincides with its inverse set up to permutability. Finally, a Cremona set is called *apocryphal* if its inverse set has at least one nonsquare-free monomial.

It may be easier to classify $\mathcal{DB}$ Cremona sets. For instance, the following simple result considerably restricts the possibilities.

Proposition 7.3.27. *If $A = (a_{ij})$ is a doubly stochastic integer matrix and $|\det(A)| = d$, then $\gcd\{n, d\} = 1$.*

Proof. Adding the first $n-1$ rows of A to its last row and factoring out d, we get

$$\det(A) = d\det\begin{bmatrix} a_{11} & \cdots & a_{1n-1} & a_{1n} \\ \vdots & & \vdots & \vdots \\ a_{n-11} & \cdots & a_{n-1n-1} & a_{n-1n} \\ 1 & \cdots & 1 & 1 \end{bmatrix}.$$

Next, we add the first $n-1$ columns of the matrix occurring in the right-hand side of this equality to its last column to get

$$\det(A) = d\det\begin{bmatrix} a_{11} & \cdots & a_{1n-1} & d \\ \vdots & & \vdots & \vdots \\ a_{n-11} & \cdots & a_{n-1n-1} & d \\ 1 & \cdots & 1 & n \end{bmatrix}.$$

Hence, since $\det(A) = \pm d$ we obtain that n and d are relatively prime. □

Another useful tool is the following.

Lemma 7.3.28 (Inductive principle for $\mathcal{DB}$). *Let $F = u_1, \ldots, u_n \subset k[x_1, \ldots, x_n]$ be a $\mathcal{DB}$ set of square-free monomials of degree d. Then, given a permutation $\{i_1, \ldots, i_n\}$ of the indices such that the set $u_1/x_{i_1}, \ldots, u_n/x_{i_n}$ has no repeated monomials, then this set is a $\mathcal{DB}$ set of square-free monomials of degree $d-1$.*

Proof. The proof follows immediately from a close inspection of the corresponding log-matrices. □

Of course, the result of the lemma can be read backwards, i. e., from degree $d-1$ up to degree d by multiplying by variables out of $\{x_{i_1}, \ldots, x_{i_n}\}$.

Corollary 7.3.29. *For $n = 6$, the only $\mathcal{DB}$ square-free Cremona set of degree d is the reciprocal involution, given by*

$$x_1x_2x_3x_4x_5,\ x_1x_2x_3x_4x_6,\ x_1x_2x_3x_5x_6,\ x_1x_2x_4x_5x_6,\ x_1x_3x_4x_5x_6,\ x_2x_3x_4x_5x_6$$

Proof. It follows readily from Proposition 7.3.27. In particular, the inductive principle above does not preserve the rank of the log-matrix. □

To classify the square-free Cremona sets with $n = 6$ we only need to look at degree $d = 3$, since $d = 2$ follows from Proposition 7.3.9 and $d = 4$ goes by duality.

We give in the exercises examples of Cremona sets with $n = 6$ and $d = 3$, with special care for their linear syzygy behavior. These examples will hopefully give some measure of the theoretical hardship in classifying square-free Cremona sets for $n \geq 6$. In this regard, a recent announcement has been out where the complete classification of the case $n = 6$ is obtained [44].

7.4 Exercises

Exercise 7.4.1. Consider the de Jonquières map F of $\mathbb{P}^2$ defined by x^4, x^3y, y^3z. Prove by a direct approach that the Rees algebra of the base ideal of F is not Cohen–Macaulay.

(*Hint*: Show that the presentation ideal has at least 4 minimal generators.)

Exercise 7.4.2. Let $I \subset k[x_0, x_1, x_2]$ denote the base ideal of a plane Cremona map in degree 4, which is not a de Jonquières map. Let y_0, y_1, y_2 denote the target variables. Prove:

(1) The Rees algebra $\mathcal{R}_R(I)$ of I is defined by the maximal minors of a 4×3 matrix with two rows of linear forms in x_0, x_1, x_2 and two rows of linear forms in y_0, y_1, y_2.
(*Hint*: By Theorem 7.2.8, $\mathcal{R}_R(I)$ is Cohen–Macaulay and I is Cohen–Macaulay; then argue that the inverse map is in degree 4 and not de Jonquières either and get the defining degrees of $\mathcal{R}_R(I)$ by the Jacobian dual criterion.)

(2) $\mathcal{R}_R(I)$ is normal.
(*Hint*: Show that $\mathcal{R}_R(I)$ is regular locally in codimension one by inspecting the Jacobian matrix of the generators in (1).)

Exercise 7.4.3. Work out the details of the previous exercise for the Cremona map of [4, Example 2.1.8], defined by the forms

$$x_0x_1(x_1 - x_2)(x_0 - x_2),\ x_0x_2(x_1 - x_2)(x_0 - 2x_1),\ x_1x_2(x_0 - x_2)(x_0 - 2x_1).$$

In addition, prove:

(1) Its base ideal is the fat ideal of the six points with the following coordinates:

$$\begin{pmatrix} 1 & 0 & 0 & 1 & 2 & 2 \\ 0 & 1 & 0 & 1 & 1 & 1 \\ 0 & 0 & 1 & 1 & 1 & 2 \end{pmatrix}$$

and respective multiplicities 2, 2, 2, 1, 1, 1.

(2) The inverse map has only three proper points.

Exercise 7.4.4. Let $F : \mathbb{P}^n \dashrightarrow \mathbb{P}^n$ denote a monomial Cremona map. Show that the inverse map is also monomial arguing directly with the corresponding Jacobian dual matrix as in Section 3.2.3.

Exercise 7.4.5. Consider the set

$$F = \{x_1x_2x_6, x_2x_3x_6, x_1x_3x_6, x_1x_3x_4, x_1x_4x_5, x_3x_4x_6\}.$$

Argue (or compute):

(1) F is a Cremona set.
(*Hint*: Both the log matrix and the linear syzygy matrix of F have maximal rank.)

(2) F is p-involutive.
(*Hint*: Use the recipe in the proof of Theorem 7.3.4.)

(3) The dual complement of F is also p-involutive, but not permutably equivalent to F.

Exercise 7.4.6. Consider the set

$$F = \{x_1x_2x_3, x_2x_3x_4, x_3x_4x_5, x_1x_3x_6, x_2x_5x_6, x_4x_5x_6\}.$$

Argue (or compute):

(1) F is a Cremona set.

(2) F is apocryphal with inverse in degree 4
(*Hint*: for example, $y_1^2y_6^2$ belong to the inverse set.)

Exercise 7.4.7. Consider the set:

$$F = \{x_1x_2x_4, x_2x_3x_5, x_3x_4x_6, x_1x_4x_5, x_1x_4x_6, x_2x_5x_6\}.$$

Argue (or compute):

(1) F is a Cremona set.
(*Hint*: The log matrix of F has maximal rank; the linear syzygy matrix has one rank less—so, what is happening?)

(2) F is apocryphal with inverse in degree 5: show a nonsquare-free term in the inverse set.

Exercise 7.4.8. Let $F : \mathbb{P}^2 \dashrightarrow \mathbb{P}^2$ denote a monomial Cremona map and let $I \subset R$ stand for its base ideal. Prove that up to permutation of the variables (source) and the defining monomials (target) F is a de Jonquières transformation exactly in the following cases:

(1) $I = (xy, xz, yz)$.

(2) $I = (x^d, x^{d-1}y, y^{d-1}z)$, with $d \geq 2$.

(3) $I = (x^d, xyz^{d-2}, yz^{d-1})$, with $d \geq 3$.

Exercise 7.4.9. Consider the following monomials:

$$\{x_0x_1, x_1x_2, x_0x_2, x_2x_3, x_3x_4\} \subset k[x_0, x_1, x_2, x_3, x_4].$$

(1) Show that they define a Cremona map F of $\mathbb{P}^4$.
(2) Show that the inverse map is defined by monomials of degree 3, with source inversion factor $f := x_0x_1x_2^2x_3$.
(3) Show that, although $f \in I^{(3)} \setminus I^3$, it is not a minimal generator of the symbolic Rees algebra of the base ideal of F.

Exercise 7.4.10. Let $I = (xy, xz, yz) \subset k[x, y, z]$ and let $J \subset k[x, y, z]$ be the ideal of the reduced points $(0 : 1 : 0)$, $(1 : 0 : 0)$, $(1 : 1 : 1)$.
(1) (Warm up) Prove that the generators of J define a quadratic Cremona map by finding an inverse.
(2) Letting $\mathfrak{F}$ and $\mathfrak{G}$ stand for the respective quadratic Cremona maps defined by I and J, show that the composite $\mathfrak{F} \circ \mathfrak{G}$ is a quadratic Cremona map with only two distinct proper points.
(3) (Harder) As a generalized converse to (ii), show that any quadratic Cremona map with only two proper points is the composite of two quadratic Cremona maps with three distinct proper points each.

8 Steps in elimination

Elimination is a classical topic both in commutative algebra and algebraic geometry. Never mind that it has been subtracted from subsequent editions of van der Waerden's *Moderne Algebra*. For an overview of the underlying geometry, see the beautiful [57, Chapter 14]. By and large, in commutative algebra, elimination has to do with contraction of ideals in a ring inclusion, while in applied theory the main object envisaged is the effective discussion of the structure of the so-called implicit equations. All three setups are maneuvers on an essentially the same question.

8.1 Elimination algebras

The variant of elimination in this book requires the idea of saturation. The latter is of fundamental importance in both commutative algebra and algebraic geometry. Besides allowing to obtaining the homogeneous defining ideal of a projective scheme out of any of its affine pieces, saturation yields the components of a projective scheme off a certain locus and the defining equations of a projection into projective space. From a more algebraic angle, it also allows for obtaining the torsion of certain algebras over a base ring.

With so many applications, we will be willing to approach this notion as soon as possible.

Definition 8.1.1. Let B be a Noetherian ring and let F a free B-module of finite rank. Let $M \subset F$ be a submodule and $J \subset B$ an ideal. One defines the *saturation of M with respect to J* to be the B-submodule of F,

$$(M : J^{\infty}) := \bigcup_{t \geq 1} (M : J^t) \subset F.$$

Here, $(M : J^t) = \{f \in F \mid fJ^t \subset M\}$.

A major landscape occurs when $F = B$ and M is an ideal.

The next simple result tells how saturation yields torsion.

Lemma 8.1.2. *Let A be a Noetherian ring and let B be a Noetherian A-algebra. Let $\mathfrak{n} \subset A$ be an ideal containing a non-zero-divisor such that B is A-torsionfree locally on* $\operatorname{Spec}(A) \setminus V(\mathfrak{n})$ *(i. e., B_p is A_p-torsion-free for every $p \in \operatorname{Spec}(A)$, $p \not\supset \mathfrak{n}$). Then the A-torsion submodule of B is the ideal* $(0) :_B \mathfrak{n}^{\infty}$.

Proof. Let $\mathcal{A} \subset B$ denote the A-torsion submodule of B. Since $\mathfrak{n}$ (hence any of its powers) contains a regular element, it is clear that $(0) :_B \mathfrak{n}^{\infty} \subset \mathcal{A}$. Conversely, since $\mathcal{A}_p = 0$ for every $p \in \operatorname{Spec}(A)$, $p \not\supseteq \mathfrak{n}$, it follows that some power of $\mathfrak{n}$ annihilates $\mathcal{A}$. This shows the reverse inclusion $\mathcal{A} \subset (0) :_B \mathfrak{n}^{\infty}$. ◻

https://doi.org/10.1515/9783110640694-008

Notation 8.1.3. $(A, \mathfrak{m})$ denotes a Noetherian local ring and its maximal ideal or an $\mathbb{N}$-graded ring of finite type and its maximal graded ideal. Let $k \subset A$ be a Noetherian subring (mostly, k will be a field and A will be a localization of an algebra of finite type over k or a standard graded ring with $A_0 = k$). Let $\mathbf{T}$ be a set of indeterminates over A and let $\mathcal{J} \subset (\mathfrak{m}, \mathbf{T})$ be an ideal of $A[\mathbf{T}]$. Set $\mathcal{A} = A[\mathbf{T}]/\mathcal{J}$. Finally, let $\mathfrak{n} \subset \mathfrak{m}$ be an ideal (homogeneous in the graded case).

Definition 8.1.4. The *T-elimination algebra of $\mathcal{A}$ off* $\mathfrak{n}$ is the k-algebra,

$$\mathbb{E}_{\mathfrak{n}}(\mathcal{A}) := k[T]/k[T] \cap (\mathcal{J} :_{A[T]} \mathfrak{n}^{\infty}).$$

If $\mathfrak{n} = \mathfrak{m}$, we write simply $\mathbb{E}(\mathcal{A})$ for $\mathbb{E}_{\mathfrak{m}}(\mathcal{A})$ and omit "off $\mathfrak{m}$" from the terminology. One also considers the *conic elimination algebra* $\mathbb{F}(\mathcal{A}) := k[T]/\mathcal{J} \cap k[T]$. There is a canonical surjective homomorphism

$$\mathbb{F}(\mathcal{A}) \to \mathbb{E}(\mathcal{A}).$$

Clearly, this is an isomorphism if and only if $\mathfrak{m}$ is not contained in any associated prime of $\mathcal{A}$.

In the early literature, a polynomial $f \in A[T]$ was called an *inertia form ("Trägesheitform")* of $\mathcal{J}$ (relative to $\mathfrak{m}$) if $\mathfrak{m}^N f \in \mathcal{J}$ for some $N \geq 0$. Thus, the ideal of inertia forms of $\mathcal{J}$ in $A[T]$ is the lifting of the local cohomology module $H^0_{\mathfrak{m}}(\mathcal{A}) = (0) :_{\mathcal{A}} \mathfrak{m}^{\infty} = \mathcal{J} :_{A[T]} \mathfrak{m}^{\infty}/\mathcal{J}$.

It is not at all clear that $\mathbb{E}(\mathcal{A})$ depends only on $\mathcal{A}$. However, in many interesting cases it will be independent of the presentation ideal $\mathcal{J}$.

Example 8.1.5. Let $\mathfrak{m} = (x, y) \subset k[x, y]$, $T = T_1, T_2$ and $\mathcal{J} = (xT_1, yT_2)$. Then $\mathcal{J} : \mathfrak{m}^{\infty}$ is readily seen to be generated by T_1T_2, xT_1, yT_2; hence, the elimination algebra is $k[T_1, T_2]/(T_1T_2)$. Observe, however, that $k[T] \cap \mathcal{J} = (0)$.

In this book, we will mainly look at the case where $\mathcal{A} = \mathcal{R}_A(E)$ is the Rees algebra of an appropriate A-module E. In this case, $\mathcal{J}$ is a presentation ideal of $\mathcal{R}_A(E)$ so that $A[T]/\mathcal{J} \simeq \mathcal{R}_A(E)$ with T_j mapping to a generator of E (viewed in degree one)—often, if A is a residue ring of a polynomial ring $R = k[\mathbf{x}]$, $\mathcal{J}$ will denote the lifting of the presentation ideal all the way back to $R[T]$. Although $\mathcal{J}$ is not uniquely determined by either $\mathcal{R}_A(E)$ or E, we shall work from such a fixed presentation. Thus, whenever we talk about the elimination algebra $\mathbb{E}(\mathcal{R}_A(E))$, we mean that we are eliminating toward $k[T]$ off the maximal (resp., graded maximal) ideal $\mathfrak{m} \subset A$, where T are presentation variables taken as explained (however, see Example 8.5 where the elimination procedure is toward a subring in degree zero).

Of course, an important case is when $E = \mathfrak{a} \subset A$ is an ideal (homogeneous in the graded case). Here, besides $\mathbb{E}(\mathcal{R}_A(\mathfrak{a}))$, one can also look at the elimination algebra $\mathbb{E}(\mathcal{G}(\mathfrak{a}))$ coming from the associated graded ring $\mathcal{G}(\mathfrak{a}) = \mathrm{gr}_{\mathfrak{a}}(A) \simeq A[T]/(\mathfrak{a}, \mathcal{J})$ and off $\mathfrak{m}/\mathfrak{a}$.

Moreover, if $\mathfrak{a}$ is locally of linear type at its associated primes (e. g., generically a complete intersection) then a third level can be stacked on to the list by Proposition 2.3.18. Therefore, under this hypothesis, there is a commutative diagram of surjections:

$$\begin{array}{ccc} \mathbb{F}(\mathcal{R}_A(\mathfrak{a})) & \rightarrow & \mathbb{E}(\mathcal{R}_A(\mathfrak{a})) \\ \downarrow & & \downarrow \sigma \\ \mathbb{F}(\mathcal{G}(\mathfrak{a})) & \rightarrow & \mathbb{E}(\mathcal{G}(\mathfrak{a})) \\ \downarrow & & \downarrow \tau \\ \mathbb{F}(\mathcal{R}_{A/\mathfrak{a}}(\mathfrak{a}/\mathfrak{a}^2)) & \twoheadrightarrow & \mathbb{E}(\mathcal{R}_{A/\mathfrak{a}}(\mathfrak{a}/\mathfrak{a}^2)) \end{array} \qquad \begin{array}{c} \\ (120) \\ \\ (121) \\ \\ \end{array}$$

Often the fiber cone $\mathcal{F}_A(\mathfrak{a}) = \mathcal{R}_A(\mathfrak{a})/\mathfrak{m}\mathcal{R}_A(\mathfrak{a}) = \mathcal{G}(\mathfrak{a})/\mathfrak{m}\mathcal{G}(\mathfrak{a})$ comes into the picture. The composition map $k \subset A \to A/\mathfrak{m}$ induces a map $\mathbb{F}(\mathcal{G}(\mathfrak{a})) \to \mathcal{F}_A(\mathfrak{a})$ which, in general, is neither injective nor surjective.

Two major problems regarding this diagram are as follows, where the parenthetical designations are a geometric mnemonic for a later development.

Question 8.1.6 (The abstract span problem). When is σ injective (resp., injective at the level of the corresponding reduced structures)?

Question 8.1.7 (The abstract tangential problem). When is τ injective (resp., injective at the level of the corresponding reduced structures)?

A couple of elementary examples to bear in mind is as follows.

Example 8.1.8. Let $A = k[x,y]/(x^2,xy)$ and let $\mathfrak{a} = (x) \subset A$ be the unique non-embedded associated prime of A. Here, $\dim \mathcal{R}_A(\mathfrak{a}) = \dim A = 1$ and

$$\mathcal{R}_A(\mathfrak{a}) \simeq k[x,y,T]/(x^2,xy,xT,yT,T^2).$$

In particular, depth $\mathcal{R}_A(\mathfrak{a}) = 0$. By a direct calculation, $\mathbb{F}(\mathcal{R}_A(\mathfrak{a})) = k[T]/(T^2)$ while $\mathbb{E}(\mathcal{R}_A(\mathfrak{a})) = \mathbb{E}(\mathcal{G}(\mathfrak{a})) = k[T]/(T) = k$.

Example 8.1.9. Let $A = k[x,y,z]/(xy,xz)$ and let $\mathfrak{a} := (y,z) \subset A$ be the unique minimal prime of A of least dimension. Then $\dim \mathcal{R}_A(\mathfrak{a}) = \dim A + 1 = 3$ and

$$\mathcal{R}_A(\mathfrak{a}) \simeq k[x,y,z,T_1,T_2]/(xy,xz,xT_1,xT_2,zT_1 - yT_2),$$

a reduced ring. A calculation yields $\mathbb{F}(\mathcal{R}_A(\mathfrak{a})) = \mathbb{E}(\mathcal{R}_A(\mathfrak{a})) = k[T_1,T_2]$ while $\mathbb{E}(\mathcal{G}(\mathfrak{a})) = k[T_1,T_2]/k[T_1,T_2] \cap (xT_1,xT_2) : (x)^\infty = k[T_1,T_2]/(T_1,T_2) = k$.

The preceding examples show that some of the maps in (120) can behave quite erratically. In a special nonetheless sufficiently important case, the fundamental relation among the maps on the first two stacks of (120) is within reach.

Proposition 8.1.10. *Let A be a standard graded k-algebra with $[A]_0 = k$ and $\mathfrak{m} = [A]_1 A$. Let $\mathfrak{a} \subset A$ be a homogeneous ideal minimally generated by elements of the same degree $d \geq 1$. Then:*

(i) *$k[\mathfrak{a}_d] \simeq \mathcal{F}_A(\mathfrak{a}) \simeq \mathbb{F}(\mathcal{R}_A(\mathfrak{a}))$ as graded k-algebras, where $k[\mathfrak{a}_d]$ is the subalgebra of A generated by the d-graded part of $\mathfrak{a}$.*

(ii) *If $\operatorname{depth}_{\mathfrak{m}} A > 0$ (resp., $\operatorname{depth}_{\mathfrak{m}} \mathcal{G}(\mathfrak{a}) > 0$) then $\mathbb{E}(\mathcal{R}_A(\mathfrak{a})) = \mathbb{F}(\mathcal{R}_A(\mathfrak{a}))$ (resp., $\mathbb{E}(\mathcal{G}(\mathfrak{a})) = \mathbb{F}(\mathcal{G}(\mathfrak{a}))$).*

Proof. (i) This follows from (2).

(ii) Since $\mathcal{R}_A(\mathfrak{a})$ is torsion-free over A, the contraction to A of any of its associated primes is contained in an associated prime of A. Therefore, the hypothesis implies that $\mathfrak{m}$ also contains a regular element on $\mathcal{R}_A(\mathfrak{a})$, and hence $\mathcal{J} = \mathcal{J} : \mathfrak{m}^\infty$. Similarly, for the associated graded ring. □

Remark 8.1.11. It is important to understand the difference of nature between the span problem, which by and large bears a dimension-theoretic obstruction, and the tangential problem which has to do with the existence of certain torsion components coming from the normal cone. We will consider variations of these problems in a later section.

8.2 Implicitization

In this section, one assumes that $R = k[\mathbf{x}] = k[x_1, \ldots, x_n]$ $(n \geq 2)$ is a standard graded polynomial ring over an infinite field k and that $\mathfrak{a} \subset R$ is an equigenerated ideal in degree $d \geq 1$. By Proposition 8.1.10, the elimination algebra of $\mathcal{R}_R(\mathfrak{a}) \simeq R[T]/\mathcal{J}$ toward $k[T]$ is isomorphic to the k-subalgebra $k[\mathfrak{a}_d] \subset k[R_d]$. Recall that, in this landscape, $k[\mathfrak{a}_d]$ gets identified with the fiber cone of $\mathfrak{a}$.

Definition 8.2.1. One says that the elimination is *implicitization* if $\operatorname{tr.deg.}_k k(\mathfrak{a}_d) = \dim_k \mathfrak{a}_d - 1$, where $k(\mathfrak{a}_d)$ denotes the field of fractions of $k[\mathfrak{a}_d]$.

In implicitization, since k is infinite, one can assume that $\mathfrak{a} = (f_1, \ldots, f_m, f_{m+1})$, where $f_i \in \mathfrak{a}_d$ and $\{f_1, \ldots, f_m\}$ is an algebraically independent set over k. Up to a nonzero scalar, the uniquely defined homogeneous polynomial $F = F(T) \subset k[T]$ such that $k[\mathfrak{a}_d] \simeq k[T]/(F)$ is called the *implicit equation* of the (rational) parametrization defined by $f_1, \ldots, f_m, f_{m+1}$.

This parametrization defines a rational map $\mathfrak{F} : \mathbb{P}_k^{n-1} \dashrightarrow \mathbb{P}_k^m$ whose image is the hypersurface $V(F) \subset \mathbb{P}_k^m$, which is a rational hypersurface when the map is birational onto it. A necessary condition for birationality is that the map be generically finite, i. e., that $n = m = \dim_k \mathfrak{a}_d - 1$.

The typical modern view of implicitization, both theoretic and applied, carries along two main aspects: a description of the implicit equation and an understanding of the remaining defining equations (or *moving curve ideal*). The first has to do with the fiber cone and the second with the entire defining ideal of the Rees algebra of the given parametrization. As an extension of considering the implicit equation, one is interested in other equations living in $R[T]$, whose coefficients are not elements of k. While in the applied theory, some of these appear under the terminology of *moving lines* and *moving quadrics*, in algebra they mean other generators of the defining ideal $\mathcal{J}$ of the Rees algebra $\mathcal{R}_R(\mathfrak{a})$.

8.3 De Jonquières like implicitization

8.3.1 Algebraic prelims

In this part, we focus on properties of an ideal of the form $(If, g) \subset R = k[x_1, \dots, x_n]$, where $I \subset R$ is a (not necessarily homogeneous) ideal and f, g are polynomials. The material of this section is lifted from [90].

Lemma 8.3.1. *If* $\gcd(f, g) = 1$, *then:*
(i) $If : (g) = (I : (g))f$.
(ii) *Multiplication by* g *induces an isomorphism* $R/(I : (g))f \simeq (If, g)/If$ *of* R*-modules.*

Proof. (i) The inclusion $If : (g) \supset (I : (g))f$ is obvious regardless of any relative assumption about f, g. Conversely, let $b \in R$ be such that $bg \in If$. Then f divides bg and, since $\gcd(f, g) = 1$, then f divides b. Say, $b = af$, with $a \in R$. Then $(ag)f \in If$, hence $ag \in I$, i. e., $a \in I : (g)$. Therefore, $b \in (I : (g))f$.

(ii) One has $(If, g)/If \simeq (g)/(g) \cap If = (g)/(If : (g))g \simeq R/If : (g)$, where the last isomorphism is multiplication by g. Now apply (i). □

Let $\pi : R^s \twoheadrightarrow I : (g)$ be the surjective R-map induced by choosing a set of generators $\{c_1, \dots, c_s\}$ of $I : (g)$, so that $\pi(v_j) = c_j$, where $\{v_1, \dots, v_s\}$ is the canonical basis of R^s. Now, let $\{g_1, \dots, g_r\}$ be a set of generators of I. Then π naturally induces another R-map $c(g) : R^s \longrightarrow R^r$; namely, letting $\{e_1, \dots, e_r\}$ denote the canonical basis of R^r, write $c_j g = \sum_{i=1}^r h_{ij} g_i$, with $h_{ij} \in R$. Then $c(g)(v_j) = \sum_{i=1}^r h_{ij} e_i$, for $j = 1, \dots, s$.

The map $c(g)$ is called a *content map* induced by the given surjection π.

This construction is used in the following mapping cone like result.

Lemma 8.3.2. *Let* $\mathfrak{R}$ *and* $\mathfrak{S}$ *denote finite-free resolutions of* R/I *and* $R/(I : (g))f$, *respectively. Then multiplication by* g *lifts to a map* $\mathfrak{S} \to \mathfrak{R}$ *whose associated mapping cone is a free resolution of* $R/(If, g)$. *In particular, a syzygy matrix of the generators of* (If, g) *has the form*

$$\Psi = \begin{pmatrix} \varphi & \widetilde{c(g)} \\ \mathbf{0} & -f\tilde{\pi} \end{pmatrix},$$

where φ denotes a syzygy matrix of a given set of generators of I, and $\widetilde{c(g)}$ and $\tilde{\pi}$ denote, respectively, the matrices representations of the maps $c(g)$ and π.

Proof. As in Lemma 8.3.1(ii), multiplication by g induces an injective R-module homomorphism $R/(I:(g))f \hookrightarrow R/If$ with image $(If,g)/If$. This homomorphism lifts to a map of complexes (free resolutions)

$$\begin{array}{llllllllll}
\mathfrak{R}: & \cdots & \longrightarrow & R^{r_1} & \xrightarrow{\varphi} & R^{r} & \xrightarrow{f\mathbf{g}} & R & \longrightarrow & R/If & \to & 0 \\
 & & & \uparrow & & c(g)\uparrow & & \cdot g\uparrow & & \cdot g\uparrow & & \\
\mathfrak{N}: & \cdots & \longrightarrow & R^{s_1} & \xrightarrow{\psi} & R^{s} & \xrightarrow{f\pi} & R & \longrightarrow & R/(I:(g))f & \to & 0
\end{array},$$

where $I = (g_1,\dots,g_r)$, $\mathbf{g} = (g_1\dots g_r)$ and $c(g)$ is the above content map. Then the corresponding mapping cone is an R-free resolution of $(R/If)/((If,g)/If) \simeq R/(If,g)$ (see [57, Exercise A3.30]). □

8.3.2 The (If, g)-parametrization

We go back to the original graded situation. Namely, let $I = (g_1,\dots,g_n) \subset R = k[x_1,\dots,x_n]$, where the g_i's are algebraically independent forms of the same degree $d \geq 1$ minimally generating I, and f, g are forms with $\deg(g) = d + \deg(f)$ such that $\gcd(f,g) = 1$.

Note that the hypothesis on I says that one has an implicitization case, parameterized by the natural generators $g_1f,\dots,g_nf,g$ of the ideal $\mathfrak{a} := (If,g)$. Interesting behavior unfolds if this parametrization is birational onto the implicit equation. Now, for arbitrarily chosen $g_1,\dots,g_n$, there is no reason to expect this property. For example, if $I = (x^2,y^2,z^2) \subset R = k[x,y,z]$, $f = x$, $g = yz^2$, then the parametrization $(x^3 : xy^2 : xz^2 : yz^2)$ is not birational onto the cubic hypersurface of $\mathbb{P}^3$ defined by the implicit equation.

To account for this deficiency, we have the following.

Lemma 8.3.3. *Let $g_1,\dots,g_n$ define a Cremona transformation of $\mathbb{P}^{n-1}$. Then the parametrization $g_1f,\dots,g_nf,g$ defines a birational map of $\mathbb{P}^{n-1}$ onto the hypersurface of $\mathbb{P}^n$ defined by the implicit equation.*

Proof. Let $\mathfrak{G}$ denote the given Cremona transformation, with inverse defined by forms $\{g'_1\dots,g'_n\} \subset k[\mathbf{y}] := k[y_1,\dots,y_n]$, where $(\mathbb{P}^{n-1})^* = \mathrm{Proj}(k[\mathbf{y}])$. Let a bar over an element of $k[\mathbf{y},y_{n+1}]$ denote its class module the ideal of the implicit equation F. Then $\{\overline{g'_0},\dots,\overline{g'_n}\}$ defines a map $\mathbb{P}^n = \mathrm{Proj}(k[\mathbf{y},y_{n+1}]/(F)) \dashrightarrow \mathbb{P}^{n-1}$ which is inverse to the parametrization map. (Note that this representative of $\mathfrak{F}^{-1}$ does not involve the last variable y_{n+1}.) □

Due to its close behavior to the details of the generalized de Jonquières transformations of Chapter 7, the above parametrization will be called a *de Jonquières parametrization* based on the given underlying Cremona transformation $\mathfrak{G}$.

8.3.2.1 The expected shape of the implicit equation

The idea of a de Jonquières Cremona transformation is very closely handled in terms of the so-called monoids, in entire similarity to the material of Chapter 7. Recall once more that a form F of the shape $F = G + Hy_{n+1}$, where G, H are forms in $k[y_0,\dots,y_n]$, is called a *monoid*. In other words, we are dealing with a polynomial of degree 1 in the one variable polynomial ring $B[y_{n+1}]$, with homogeneous coefficients in $B = k[y_0,\dots,y_n]$ – calling it an y_{n+1}-monoid stresses the privileged variable y_{n+1}.

The typical syzygy generator in Ψ as in Lemma 8.3.2, having nonzero last coordinate, has the following polynomial 1-form stunt:

$$\sum_{i=1}^{n} h_{ij}y_i - fc_jy_{n+1} \in R[\mathbf{y}, y_{n+1}], \tag{122}$$

for some element c_j of a chosen set of generators of $I : (g)$ and $c_jg = \sum_{i=0}^{n} h_{ij}g_i$.

Definition 8.3.4. Let $(g_1 : \cdots : g_n)$ define a Cremona transformation of $\mathbb{P}^{n-1}$ and let $(g'_1 : \cdots : g'_n)$ define its inverse map, where $g'_i \in k[\mathbf{y}]$. A *syzygetic polynomial* is one of the shape

$$P_j := \sum_{i=1}^{n} h_{ij}(g'_1,\dots,g'_n)y_i - f(g'_1,\dots,g'_n)c_j(g'_1,\dots,g'_n)y_{n+1} \in k[\mathbf{y}, y_{n+1}], \tag{123}$$

obtained from (122) by evaluating $x_i \mapsto g'_i$, for $i = 0,\dots,n$.

Note that P_j depends on the choice of a generator of $I : g$.

Its relevant property is that it is a multiple of the implicit equation. To see this, one evaluates it directly by $y_i \mapsto fg_i$ and $y_{n+1} \mapsto g$ and use the fact that

$$g'_1(fg_1,\dots,fg_n),\dots,g'_n(fg_1,\dots,fg_n)$$

retrieves $x_1,\dots,x_n$ up to a suitable factor, by the inverse map property. The resulting vanishing is then an expression of the original syzygy nature of P_j.

A more conceptual, but essentially the same, way is to first show that P_j belongs to the presentation ideal of the Rees algebra of the ideal $\mathfrak{a} = (If, g)$, and since it is a polynomial form to start with, necessarily it is a multiple of the implicit equation.

This procedure draws from the fact that, since the rational map defined by the generators of $\mathfrak{a}$ is birational onto $V(F)$, one has a k-algebra isomorphism of the Rees algebras

$$\mathcal{R}_R((If,g)) \simeq \mathcal{R}_S(I'), \tag{124}$$

where $I' := (\overline{g'_1},\ldots,\overline{g'_n})S$, with $S := k[\mathbf{y},y_{n+1}]/(F)$, where the isomorphism is induced by the identity map of $k[\mathbf{x}][\mathbf{y},y_{n+1}] = k[\mathbf{y},y_{n+1}][\mathbf{x}] = k[\mathbf{x},\mathbf{y},y_{n+1}]$ (see Proposition 3.2.14).

The reader can fill in the required details.

Thus, a syzygetic polynomial is a fair candidate for the implicit equation F and it will be so if it happens to be irreducible.

8.3.2.2 The degree of the implicit equation

In this part, we show that the existence of an underlying Cremona transformation $\mathfrak{G}$ goes beyond the mere birationality of the parametrization, namely, the details of its structure will be intrinsic to the expression of the implicit equation.

For this, recall the notion of a target inverse factor (Section 4.2).

Proposition 8.3.5. *Keeping the notation of the previous part, let* $\mathbf{g}' = (g'_1,\ldots,g'_n)$ *define the inverse map* $\mathfrak{G}^{-1}$, *with target inverse factor D. One has:*

(i) *F is the* y_{n+1}*-monoid*

$$\frac{g(\mathbf{g}') - y_{n+1}f(\mathbf{g}')\,D}{\gcd(g(\mathbf{g}'),f(\mathbf{g}')\,D)}. \tag{125}$$

(ii) *Its degree is*

$$\begin{aligned}\deg(F) &= \deg(g)\deg(\mathfrak{G}^{-1}) - \deg(\gcd(g(\mathbf{g}'),f(\mathbf{g}')\,D))\\ &= \deg(f)\deg(\mathfrak{G}^{-1}) + \deg(D) + 1 - \deg(\gcd(g(\mathbf{g}'),f(\mathbf{g}')\,D)),\end{aligned}$$

where $\deg(\mathfrak{G}^{-1})$ *stands for the common degree of the forms in* $\mathbf{g}'$. *In particular,* $\deg(F) \le \deg(g)\deg(\mathfrak{G}^{-1})$.

Proof. (i) Since the de Jonquières parametrization is birational onto the implicit equation F, the two vectors $((fg_0)(\mathbf{g}'),\ldots,(fg_n)(\mathbf{g}'),\, g(\mathbf{g}'))$ and $(y_1,\ldots,y_n,y_{n+1})$ are proportional modulo (F), i. e., the 2×2 minors of

$$\begin{pmatrix}(fg_0)(\mathbf{g}') & \cdots & (fg_n)(\mathbf{g}') & g(\mathbf{g}')\\ y_0 & \cdots & y_n & y_{n+1}\end{pmatrix}$$

vanish modulo (F). In particular, each of the minors $Q_i := y_i g(\mathbf{g}') - y_{n+1}f(\mathbf{g}')g_i(\mathbf{g}')$ fixing the last column is a multiple of F.

On the other hand, by definition, $g_i(\mathbf{g}') = y_i D$, for $i = 1,\ldots,n$. Since y_i is not a factor of F, it follows that F divides the nonzero y_{n+1}-monoid $g(\mathbf{g}') - y_{n+1}f(\mathbf{g}')\,D$. Clearly,

F is not a factor of $\gcd(g(\mathbf{g}'), f(\mathbf{g}')\,D)$ as the latter lives in $k[\mathbf{y}] = k[y_1,\ldots,y_n]$ while F involves effectively the variable y_{n+1}.

Now

$$\frac{g(\mathbf{g}') - y_{n+1}f(\mathbf{g}')\,D}{\gcd(g(\mathbf{g}'), f(\mathbf{g}')\,D)}$$

is an y_{n+1}-monoid with relatively prime components, hence is irreducible. Therefore, it must coincide with F.

(ii) Taking degrees, the first equality in the stated formula follows readily, while the subsequent equality follows from the standing assumption that $\deg(g) = \deg(f) + \deg(\mathfrak{G})$ and from the equality $\deg(\mathfrak{G})\deg(\mathfrak{G}^{-1}) = \deg(D) + 1$ by definition of the inversion factor D. □

Under a certain restriction, one can state upper and lower bounds for $\deg(F)$:

Corollary 8.3.6. *If* $\gcd(f(\mathbf{g}'), g(\mathbf{g}')) = 1$ *then*

$$\deg(F) = \deg(g)\deg(\mathfrak{G}^{-1}) - \deg(\gcd(g(\mathbf{g}'), D)).$$

In particular, $\deg(f)\deg(\mathfrak{G}^{-1}) + 1 \leq \deg(F) \leq \deg(g)\deg(\mathfrak{G}^{-1})$.

Proof. The displayed equality follows immediately from the formula in Proposition 8.3.5. Thus, the only issue is the stated lower bound. For that, writing $\deg(g) = \deg(f)+\deg(\mathfrak{G})$ yields $\deg(F) = \deg(f)\deg(\mathfrak{G}^{-1})+\deg(\mathfrak{G})\deg(\mathfrak{G}^{-1})-\deg(\gcd(g(\mathbf{g}'), D))$. But $\deg(\mathfrak{G})\deg(\mathfrak{G}^{-1}) = \deg(D) + 1$, while obviously $\deg(\gcd(g(\mathbf{g}'), D)) \leq \deg(D)$. □

8.3.2.3 The implicit equation as a syzygetic monoid

In this part, we search for reasonable conditions under which the implicit equation is given by a syzygetic polynomial. The analysis will be carried in two extreme cases.

Inclusion case. $g \in I$.

Then $I : (g) = (1)$, which in the notation of Subsection 8.3.1 tells us that $\pi : R \longrightarrow R$ can be taken to be the identity map and the content map $c(g) : R \longrightarrow R^n$ picks up only one syzygy. Here, the syzygy matrix of the given generators of (If, g) is of the form

$$\Psi = \begin{pmatrix} \varphi & \widetilde{c(g)} \\ \mathbf{0} & -f \end{pmatrix}, \tag{126}$$

where $\widetilde{c(g)}$ is the column vector whose entries are the coefficients h_i of g as a combination of $g_1,\ldots,g_n$.

Thus, there is only one syzygetic polynomial $P := \sum_{i=1}^{n} h_i(\mathbf{g}')y_i - f(\mathbf{g}')y_{n+1}$. Keeping the assumptions of Proposition 8.3.5, one has the following.

Proposition 8.3.7. *If* $\gcd(f(\mathbf{g}'), g(\mathbf{g}')) = 1$ *then the following conditions are equivalent:*

(i) $g \in I$.
(ii) $\deg(F) = \deg(f)\deg(\mathfrak{G}^{-1}) + 1$ *and* $(F) = (P)$.

Proof. (i) ⇒ (ii) Quite generally, when $g \in I$ one has $\deg(P) = \deg(f)\deg(\mathfrak{G}^{-1}) + 1$. On the other hand, by Corollary 8.3.6, $\deg(F) \geq \deg(f)\deg(\mathfrak{G}^{-1}) + 1$. Since F is a factor of P, we are through.

(ii) ⇒ (i) Since $\deg(P) = \deg(f)\deg(\mathfrak{G}^{-1})+1$, confronting with the general shape of P as in Definition 8.3.4 yields $\deg(c_j(g'_1, \dots, g'_n)) = 0$ for a generator c_j of the conductor $I : g$. This forces c_j to be invertible, so $g \in I$. □

A special case is a result of [12].

Corollary 8.3.8. *If $\mathfrak{G}$ is the identity map of $\mathbb{P}^{n-1}$ and $\gcd(f,g) = 1$, then*

$$F = g(y_1, \dots, y_n) - f(y_1, \dots, y_n)\, y_{n+1}. \tag{127}$$

In particular, $\deg(F) = \deg(f) + 1$.

Remark 8.3.9. Having chosen the underlying Cremona transformation, with base ideal I, and a form $g \in I$, if f is sufficiently general then $\gcd(f(\mathbf{g}'), g(\mathbf{g}')) = 1$ since $\mathbf{g}'$ depends only on I. Therefore, $(F) = (P)$ for such a choice. Of course, easy examples show that this conclusion fails if f is not sufficiently general.

Another special case of $g \in I$ is worth isolating, where the data are somewhat twisted around. Recall that a form $g \in R = k[\mathbf{x}]$ is called homaloidal if its partial derivatives define a (polar) Cremona map (Definition 5.1.2).

Corollary 8.3.10 (char$(k) = 0$). *Let $g \in R = k[\mathbf{x}]$ denote a reduced homaloidal form of degree $d + 1$ and let $\{g'_1, \dots, g'_n\} \subset k[y_1, \dots, y_n]$ define the inverse map of the polar map of g. If $f = \sum_{i=1}^{n} \lambda_i x_i$ is a general linear form, then*

$$F = \sum_{i=1}^{n} (y_i - (d+1)\lambda_i\, y_{n+1})\, g'_i(y_1, \dots, y_n).$$

In particular, $\deg(F) = \deg(g'_i) + 1$.

The above implicit equation might be called the *general Eulerian form* of a polar Cremona map.

Nonzero-divisor case. g is a regular element on R/I.

Thus, $I : (g) = I$, hence $\pi : R^n \longrightarrow I$ maps to the generators $g_1, \dots, g_n$ of I. Accordingly, the content map $c(g)$ reduces to g times the identity map of R^n. Therefore, a presentation matrix of $J = (If, g)$ has now the form

$$\Psi = \begin{pmatrix} \varphi & g \cdot \mathbb{I}_n \\ \mathbf{0} & -f\mathbf{g} \end{pmatrix},$$

where φ is a syzygy matrix of I, $\mathbf{g} = (g_1 \dots g_n)$ and $\mathbb{I}_n$ is the $n \times n$ identity matrix.

Proposition 8.3.11. *Keeping the notation of the previous part, let* $\mathbf{g}' = (g'_1, \dots, g'_n)$ *define the inverse map* $\mathfrak{G}^{-1}$, *with target inverse factor D. Suppose that g is a nonzero divisor on R/I. Then the implicit equation F is a factor of*

$$P := g(g'_0, \dots, g'_n) - f(g'_0, \dots, g'_n)\, D\, y_{n+1}.$$

In particular, one has $\deg(F) \leq \deg(f)\deg(\mathfrak{G}^{-1}) + \deg(D) + 1$ *and, moreover, the following conditions are equivalent:*
(a) $(P) = (F)$.
(b) $\gcd(g(\mathbf{g}'), f(\mathbf{g}')D) = 1$.
(c) $\deg(F) = \deg(f)\deg(\mathfrak{G}^{-1}) + \deg(D) + 1$.

Proof. Consider the 1-form corresponding to a Koszul syzygy in Ψ as above $Q_i(\mathbf{x}, \mathbf{y}) := gy_i - fg_i y_{n+1}$, $1 \leq i \leq n$, and take the corresponding syzygetic $\mathbf{y}$-polynomial,

$$P_i := g(g'_0, \dots, g'_n) y_i - f(g'_0, \dots, g'_n)\, g_i(g'_0, \dots, g'_n)\, y_{n+1}.$$

Note that P_i is the numerator in the expression of F as obtained in the proof of Proposition 8.3.5(i).

Clearly, then (a) through (c) are equivalent assertions. □

8.3.3 The equations of the Rees algebra

So far, the focus has been solely on the implicit equation of a de Jonquières parametrization. In this part, we aim at describing ways to get to the defining equations of the Rees algebra $\mathcal{R}_R(\mathfrak{a})$, where $\mathfrak{a} = (fI, g)$ in the notation of the previous sections.

8.3.3.1 The downgrading method

Versions of this notion have been considered before in different contexts (Section 7.1, [28, 96, 108]). Here, we emphasize the birationality context. We will work in the context of the bigraded ring $k[\mathbf{x}, \mathbf{y}, y_{n+1}]$ of the previous sections, but the notion works in any bigraded size.

Given a bihomogeneous polynomial $Q \in k[\mathbf{x}, \mathbf{y}, y_{n+1}]$, a decomposition of the form $Q = \sum_{i=1}^{n} x_i Q_i$, where Q_i is bihomogeneous, is called an $\mathbf{x}$-*framing* of Q. The classical downgrading consists in trading y_i for the multiplier x_i, for every $1 \leq i \leq n$. Here, we fix a sequence of forms of the same degree $\mathfrak{H} := \{\mathfrak{h}_1, \dots, \mathfrak{h}_n\} \subset k[\mathbf{y}]$ and trade each $\mathfrak{h}_i$ for the multiplier x_i.

The resulting bihomogeneous polynomial $\sum_{i=1}^{n} \mathfrak{h}_i Q_i(\mathbf{x}, \mathbf{y})$ is called an $\mathfrak{H}$-*downgraded polynomial* of Q. We use the notation $D_{\mathfrak{H}}(Q)$ for an $\mathfrak{H}$-downgraded polynomial even though it is not well-defined since the **x**-*framing* is only stable modulo the trivial (Koszul) relations of **x**.

Our present interest is when $\mathfrak{H} \subset k[\mathbf{y}]$ is the set of forms defining the inverse of the underlying Cremona transformation, in which case, we talk informally about a *birational downgrading*. Note that our previous syzygetic polynomials are among the fully downgraded $D_{\mathfrak{H}}^{(\ell)}(Q)$, for Q a syzygy of $\mathfrak{a}$ with nonzero last coordinate. The classical downgrading is then the case where the Cremona transformation is the identity.

We introduce an iterated version by setting

$$D_{\mathfrak{H}}^{0}(Q) = Q, \quad D_{\mathfrak{H}}^{(\ell)}(Q) := D_{\mathfrak{H}}(D_{\mathfrak{H}}^{(\ell-1)}(Q)), \quad \forall \ell \geq 1.$$

We say that $D_{\mathfrak{H}}^{(\ell)}(Q)$ is *fully downgraded* when it eventually lands in $k[y_0, \dots, y_{n+1}]$, that is, when $\ell = \deg_{\mathbf{x}}(Q)$.

Theorem 8.3.12. *Let $\mathcal{J} \subset k[\mathbf{x}, \mathbf{y}, y_{n+1}]$ denote the presentation ideal of the Rees algebra of the ideal $\mathfrak{a} = (If, g) \subset k[\mathbf{x}]$ and let*

$$\mathfrak{D} := (\mathcal{I}, \{D_{g'_0,\dots,g'_n}^{(\ell)}(Q),\ 0 \leq \ell \leq \deg_{\mathbf{x}}(Q)\}) \subset k[\mathbf{x}, \mathbf{y}, y_{n+1}],$$

where $\mathcal{I} \subset k[\mathbf{x}, \mathbf{y}]$ stands for the presentation ideal of $\mathcal{R}_R(I)$ and $Q \in k[\mathbf{x}, \mathbf{y}]$ runs through the biforms corresponding to the syzygies of $\mathfrak{a}$ with nonzero last coordinate. Then:

(i) *$\mathcal{J}$ is a minimal prime ideal of $\mathfrak{D}$.*

(ii) *Suppose in addition the following conditions:*
   (a) *f is a general form.*
   (b) *I is saturated.*
   (c) $\operatorname{depth} R/(I{:}g) \geq \operatorname{depth} R/I + 1$.

   Then $\mathcal{J} = \mathfrak{D} : C^{\infty}$, where C is the source inversion factor of the underlying Cremona transformation defined by the generators of I.

Proof. (i) Let $\mathcal{J} \subset k[\mathbf{x}, \mathbf{y}, y_{n+1}]$ denote the presentation ideal of $\mathcal{R}_{k[\mathbf{x}]}(\mathfrak{a})$ induced by the natural generators $\{fg_1, \dots, fg_n, g\}$.

Claim. $\mathfrak{D} \subset \mathcal{J}$.

It suffices to show that $D_{g'_0,\dots,g'_n}(Q) \subset \mathcal{J}$ for every such Q.

Recall that $Q = \sum_{i=1}^{n} h_i y_i - fcy_{n+1} \in R[\mathbf{y}, y_{n+1}]$, for some element c of a chosen set of generators of $I : (g)$ and $cg = \sum_{i=1}^{n} h_i g_i$. Then $D_{g'_0,\dots,g'_n}(Q)$ is obtained by first **x**-framing Q, followed by trading g'_i for the multiplier x_i of the framing.

On the other hand, we have the k-algebra isomorphism of (124),

$$\mathcal{R}_R((If, g)) \simeq \mathcal{R}_S(I'),$$

where $I' := (\overline{g'_1}, \dots \overline{g'_n})$. This says that fully evaluating Q or $D_{g'_0,\dots,g'_n}(Q)$ by $x_i \mapsto g'_i$ yields a form vanishing at I', hence $D_{g'_0,\dots,g'_n}(Q) \subset \mathcal{J}$.

This proves the claim. Now, $\mathcal{I}$ has height $n-1$ because I has n generators. It is a prime ideal contained in the ideal $(\mathbf{x})k[\mathbf{x},\mathbf{y},y_{n+1}]$ because I is generated by algebraically independent elements over k. Since the fully downgraded elements of $\mathfrak{D}$ belong to $k[\mathbf{y},y_{n+1}]$, this ideal has codimension at least one more.

This completes the proof since $\mathcal{J}$ has height n.

(ii) Since $\mathcal{J}$ is prime and $C\notin\mathcal{J}$, the inclusion $\mathfrak{D}\colon C^{\infty}\subset\mathcal{J}$ is obvious. Conversely, by construction $\mathfrak{D}$ contains the presentation ideal $(\mathfrak{J}_1)$ of the symmetric algebra of $\mathfrak{a}$. Therefore, $\mathfrak{D}\colon\mathfrak{a}^{\infty}=\mathcal{J}$.

Claim. $\operatorname{depth} R/\mathfrak{a}>0$.

To see this, consider the map $R/(I:g)f\to R/If$ given by multiplication by g. This map is injective and its image is $(If,g)/If=\mathfrak{a}/If$. Therefore, the associated mapping cone is a resolution of the cokernel $R/\mathfrak{a}$. Now, up to shifts R/I (resp., $R/(I{:}g)$) and R/If (respectively, $R/(I{:}g)f$) have the same homological dimension. Thus, (c) implies that the homological dimension of $R/(I{:}g)f$ is at most the one of R/If minus 1. Then the mapping cone has length at most the homological dimension of R/I. This implies that $\operatorname{depth} R/f\mathfrak{a}\geq\operatorname{depth} R/I>0$, by (b).

So much for the claim. Now, since the inverse map to the parametrization defined by $\mathfrak{a}$ is defined by the residues modulo F of $g_1',\dots,g_n'$, an easy calculation yields that the source inversion factor of the parametrization is $D:=f^{d'}C$. Since $\operatorname{depth} R/\mathfrak{a}>0$, the form D belongs to the symbolic power $\mathfrak{a}^{(d')}$ (Proposition 4.2.4(a)). Therefore, $\mathfrak{D}:{\mathfrak{a}^{(d')}}^{\infty}\subset\mathfrak{D}:D^{\infty}$. But $\mathfrak{a}$ and $\mathfrak{a}^{(d')}$ share the same radical, hence $\mathfrak{D}:{\mathfrak{a}^{(d')}}^{\infty}=\mathfrak{D}:\mathfrak{a}^{\infty}$. Therefore, $\mathfrak{D}:D^{\infty}=\mathcal{J}$. By (a), f is regular on $\mathfrak{D}$. Recalling that $D=f^{d'}C$, this implies that $\mathfrak{D}:C^{\infty}=\mathcal{J}$, as was to be shown. □

Remark 8.3.13. (1) Conditions (b) and (c) above are only needed to guarantee that $R/\mathfrak{a}$ has positive depth. As to condition (a) it is not clear if one can do with much less.

(2) The ideal $\mathfrak{D}$ is seldom generated by a subset of minimal generators of the full Rees ideal $\mathcal{J}$, even in the case where $g\in I$. The case where $\mathfrak{G}$ is the identity—a so-called monoid parametrization—is known to have $\mathfrak{D}=\mathcal{J}$ (see [12, Theorem 3.1]), but it is not a consequence of item (ii) above. To follow in this case a similar path to the above argument, if $\operatorname{char}(k)$ does not divide $\deg(f)$ then $f^2\in\mathfrak{a}$, hence by a same token we get $\mathfrak{D}:f^{\infty}$; thus, it would suffice to prove that f is regular on $\mathfrak{D}$ (even if f is not general).

8.3.3.2 The associated monoid method

A de Jonquières parametrization with underlying Cremona transformation the identity is often called a *monoid*—possibly, a bad terminology choice. Thus, a monoid parametrization has the shape $(fx_1:\cdots:fx_n:g)$, for arbitrary forms $f,g\in k[x_1,\dots,x_n]$ of suitable degrees such that $\gcd(f,g)=1$.

It has the following properties:

- $\operatorname{depth} R/\mathfrak{a}=0$, where $\mathfrak{a}=(fx_1,\dots,fx_n,g)\subset R=k[x_1,\dots,x_n]$.

- The implicit equation is $g(\mathbf{y}) - f(\mathbf{y})y_{n+1}$ and coincides with the unique syzygetic polynomial derived from the polynomial stunt $fy_{n+1} - \sum_{i=1}^{n} h_i y_i$, where $g = \sum_{i=1}^{n} h_i x_i$.
- The presentation ideal of the Rees algebra $\mathcal{R}_R(\mathfrak{a})$ is generated by the polynomial stunts of the Koszul syzygies of $x_1, \ldots, x_n$ and the birational downgradings of $fy_{n+1} - \sum_{i=1}^{n} h_i y_i$ [12, Theorem 3.1].

Now let $\mathfrak{a} \subset R = k[x_1, \ldots, x_n]$ denote the base ideal of an arbitrary de Jonquières parametrization, with implicit equation F. By Proposition 8.3.5(i), F is an y_{n+1}-monoid, say, $F = F_\delta - y_{n+1}F_{\delta-1}$, where $\delta = \deg(F)$, and $F_\delta, F_{\delta-1} \in k[\mathbf{y}]$ are forms of degrees δ, $\delta - 1$, respectively, such that $\gcd(F_\delta, F_{\delta-1}) = 1$.

Set $h_\delta := F_\delta(\mathbf{x})$, $h_{\delta-1} := F_{\delta-1}(\mathbf{x})$ and consider the standard monoid parametrization $(h_{\delta-1}x_0 : \cdots : h_{\delta-1}x_n : -h_\delta)$. Let $\mathfrak{b} \subset R$ denote the corresponding base ideal. Finally, set $\mathfrak{J}_\mathfrak{a}$ and $\mathfrak{J}_\mathfrak{b}$ for the respective presentation ideals of $\mathcal{R}_R(\mathfrak{a})$ and $\mathcal{R}_R(\mathfrak{b})$.

We will make use of the following construction. Let $\Psi_\mathbf{g} : k[\mathbf{x}, \mathbf{y}, y_{n+1}] \to k[\mathbf{x}, \mathbf{y}, y_{n+1}]$ denote the k-algebra map such that $\Psi_\mathbf{g}(x_i) = g_i$, $\Psi_\mathbf{g}(y_i) = y_i$, $\Psi_\mathbf{g}(y_{n+1}) = y_{n+1}$, for $1 \le i \le n$. Set $\mathfrak{J}_\mathfrak{b}(\mathfrak{G})$ for the ideal generated by the image of $\mathfrak{J}_\mathfrak{b}$ by $\Psi_\mathbf{g}$.

Let $\Psi_{\mathbf{g}'(\mathbf{x})}$ denote the similar map with the evaluated inverse forms $\mathbf{g}'(\mathbf{x})$ and set $\mathfrak{J}_\mathfrak{a}(\mathfrak{G}^{-1})$ for the ideal generated by the image of $\mathfrak{J}_\mathfrak{a}$ by $\Psi_{\mathbf{g}'(\mathbf{x})}$.

The notation is a bit convoluted, but it pays for the simplicity of the following result.

Theorem 8.3.14. *With the above notation, one has:*

(a) *The implicit equations of $\mathfrak{a}$ and $\mathfrak{b}$ coincide.*

(b) *Let $\mathfrak{F}_\mathfrak{a}$ and $\mathfrak{F}_\mathfrak{b}$ denote the respective birational maps of the parametrizations $\mathfrak{a}$ and $\mathfrak{b}$.*

Then $\mathfrak{F}_\mathfrak{a} = \mathfrak{F}_\mathfrak{b} \circ \mathfrak{G}$.

(c) *Let $C \in k[\mathbf{x}]$ (resp., $D \in k[\mathbf{y}]$) denote the source (resp., target) inversion factor of the underlying Cremona $\mathfrak{G}$. Then*

$$\mathfrak{J}_\mathfrak{a} = \mathfrak{J}_\mathfrak{b}(\mathfrak{G}) : C^\infty$$

and

$$\mathfrak{J}_\mathfrak{b} = \mathfrak{J}_\mathfrak{a}(\mathfrak{G}^{-1}) : D(\mathbf{x})^\infty.$$

Proof. (a) This follows immediately from Corollary 8.3.8 and the definition of the forms h_δ, $h_{\delta-1}$.

(b) Let $\boldsymbol{\alpha} \in \mathbb{P}^{n-1} \setminus V(\mathfrak{a})$. Then

$$\boldsymbol{\beta} := \mathfrak{F}_\mathfrak{a}(\boldsymbol{\alpha}) = (f(\boldsymbol{\alpha})g_1(\boldsymbol{\alpha}) : \cdots : f(\boldsymbol{\alpha})g_n(\boldsymbol{\alpha}) : g(\boldsymbol{\alpha})) \in V(F) \subset \mathbb{P}^n,$$

hence

$$\begin{aligned} \mathfrak{F}_\mathfrak{b}^{-1}(\boldsymbol{\beta}) &= (f(\boldsymbol{\alpha})g_1(\boldsymbol{\alpha}) : \cdots : f(\boldsymbol{\alpha})g_n(\boldsymbol{\alpha})) \\ &= (g_1(\boldsymbol{\alpha}) : \cdots : g_n(\boldsymbol{\alpha})) = \mathfrak{G}(\boldsymbol{\alpha}). \end{aligned}$$

(c) We first show the inclusions $\mathfrak{J}_a(\mathfrak{G}^{-1}) \subset \mathfrak{J}_b$ and $\mathfrak{J}_b(\mathfrak{G}) \subset \mathfrak{J}_a$.

For the first of these, let $\mathfrak{h}(\mathbf{x};\mathbf{y},y_{n+1}) \in \mathfrak{J}_a$ be a biform. By definition, one has $\mathfrak{h}(\mathbf{x};fg_0,\dots,fg_n,g) = 0$, while we wish to show that $\mathfrak{h}(\mathbf{g}'(\mathbf{x});h_{\delta-1}x_0,\dots,h_{\delta-1}x_n,h_\delta) = 0$. By construction and Proposition 8.3.5, one has $h_\delta = \frac{g(\mathbf{g}'(\mathbf{x}))}{\mathfrak{f}}$ and $h_{\delta-1} = \frac{f(\mathbf{g}'(\mathbf{x}))\,D(\mathbf{x})}{\mathfrak{f}}$, where $\mathfrak{f} = \gcd(g(\mathbf{g}'(\mathbf{x})),f(\mathbf{g}'(\mathbf{x}))\,D(\mathbf{x}))$. Since $\mathfrak{h}$ is bihomogeneous, we can pull out a power of $\mathfrak{f}$ as a factor, hence the assertion is equivalent to showing that

$$\mathfrak{h}(\mathbf{g}'(\mathbf{x});f(\mathbf{g}'(\mathbf{x}))D(\mathbf{x})x_1,\dots,f(\mathbf{g}'(\mathbf{x}))D(\mathbf{x})x_n,g(\mathbf{g}'(\mathbf{x}))) = 0. \tag{128}$$

By definition, $D(\mathbf{x})x_i = g_i(\mathbf{g}'(\mathbf{x}))$, $\forall i$. Therefore, (128) is equivalent to the vanishing of

$$\mathfrak{h}(\mathbf{g}'(\mathbf{x});f(\mathbf{g}'(\mathbf{x}))g_1(\mathbf{g}'(x)),\dots,f(\mathbf{g}'(\mathbf{x}))g_n(\mathbf{g}'(x)),g(\mathbf{g}'(\mathbf{x}))) = \mathfrak{h}(\mathbf{x};fg_1,\dots,fg_n,g)(\mathbf{g}'(\mathbf{x})).$$

The rightmost polynomial is the resulting of evaluating the null polynomial, so itself is null.

To argue for the second inclusion above, let likewise $\mathfrak{h}(\mathbf{x};\mathbf{y},y_{n+1}) \in \mathfrak{J}_b$ be a biform. By definition, $\mathfrak{h}(\mathbf{x};h_{\delta-1}\mathbf{x},h_\delta) = 0$ whereas one wishes to show that

$$H := \mathfrak{h}(\mathbf{g}(\mathbf{x});f(\mathbf{x})\mathbf{g}(\mathbf{x}),g(\mathbf{x})) = 0.$$

For this, we first prove that substituting $\mathbf{g}'(\mathbf{x})$ for $\mathbf{x}$ in H gives zero; namely, by a similar token as above, using the characteristic property of the target factor D, there are suitable integers s, r such that

$$\begin{aligned} H(\mathbf{g}'(\mathbf{x})) &= \mathfrak{h}(\mathbf{g}(\mathbf{g}'(\mathbf{x}));f(\mathbf{g}'(\mathbf{x}))g_1(\mathbf{g}'(\mathbf{x})),\dots,f(\mathbf{g}'(\mathbf{x}))g_n(\mathbf{g}'(\mathbf{x})),g(\mathbf{g}'(\mathbf{x}))) \\ &= \mathfrak{f}^r D^s \mathfrak{h}\left(\mathbf{x};\frac{f(\mathbf{g}'(\mathbf{x}))}{\mathfrak{f}}\mathbf{g}(\mathbf{g}'(\mathbf{x})),\frac{g(\mathbf{g}'(\mathbf{x}))}{\mathfrak{f}}\right) \\ &= \mathfrak{f}^r D^s \mathfrak{h}(\mathbf{x};h_{\delta-1}\mathbf{x},h_\delta) = 0. \end{aligned}$$

Consider now the source inversion factor $C \in k[\mathbf{x}]$, whose characteristic property is that $g_i'(\mathbf{g}) = Cx_i$, $\forall i$. Then, for a suitable exponent t, one has

$$\begin{aligned} C^t H = C^t \mathfrak{h}(\mathbf{g}(\mathbf{x});f(\mathbf{x})\mathbf{g}(\mathbf{x}),g(\mathbf{x})) &= (\mathfrak{h}(\mathbf{g}(\mathbf{x});f(\mathbf{x})\mathbf{g}(\mathbf{x}),g(\mathbf{x})))(\mathbf{g}'(\mathbf{g}(\mathbf{x}))) \\ &= ((\mathfrak{h}(\mathbf{g}(\mathbf{x});f(\mathbf{x})\mathbf{g}(\mathbf{x}),g(\mathbf{x})))(\mathbf{g}'(\mathbf{x})))(\mathbf{g}(\mathbf{x})) = H(\mathbf{g}'(\mathbf{x}))(\mathbf{g}(\mathbf{x})) = 0, \end{aligned}$$

which proves the assertion.

To complete the proof of the theorem, we show the equality $\mathfrak{J}_a = \mathfrak{J}_b(\mathfrak{G}) : C^\infty$—the other equality being proved in the same fashion, by keeping in mind that D is to be replaced by the evaluated $D(\mathbf{x})$. Since $\mathfrak{J}_b(\mathfrak{G}) \subset \mathfrak{J}_a$ and the ideal $\mathfrak{J}_a$ is prime, the inclusion $\mathfrak{J}_a \supset \mathfrak{J}_b(\mathfrak{G}) : C^\infty$ is clear. Conversely, let $\mathfrak{h}(\mathbf{x};\mathbf{y}) \in \mathfrak{J}_a$. By what we have proved above, $\mathfrak{h}(\mathbf{g}'(\mathbf{x});\mathbf{y}) \in \mathcal{I}_{\mathfrak{M}}$, and hence $\mathfrak{h}(\mathbf{g}'(\mathbf{g}(\mathbf{x}));\mathbf{y}) \in \mathcal{I}_{\mathfrak{M}}(\mathfrak{G})$. Again, $\mathbf{g}'(\mathbf{g}(\mathbf{x})) = C\mathbf{x}$ and $\mathfrak{h}(\mathbf{x};\mathbf{y})$ is bihomogeneous. Therefore, for a suitable exponent u, one has $C^u\mathfrak{h}(\mathbf{x};\mathbf{y}) =$

$\mathfrak{h}(\mathbf{g}'(\mathbf{g}(\mathbf{x})); \mathbf{y}) \in \mathfrak{J}_{\mathfrak{b}}(\mathfrak{G})$, which says that $\mathfrak{h}(\mathbf{x}; \mathbf{y}) \in \mathfrak{J}_{\mathfrak{b}}(\mathfrak{G}) : C^{\infty}$. This proves the reverse inclusion. □

Question 8.3.15. What are the applications of Theorem 8.3.14?

8.4 Implicitization in dimension ≤ 1

A special case of implicitization occurs very often in the applications and that is the case where $\mathfrak{a} \subset R$ is a d-equigenerated ideal of height $\geq n-1$, i. e., $\dim R/\mathfrak{a} \leq 1$. Geometrically, one is in the landscape of a finite (possibly, empty) set points in $\mathbb{P}^n$. There are two cases: either $\mathfrak{a}$ has height n, i. e., is $(\mathbf{x})$-primary, or else it has analytic deviation one, that is to say, $\ell(\mathfrak{a}) = \operatorname{ht} \mathfrak{a} + 1$.

In the first case, one can assume that the first n generators of $\mathfrak{a}$ form a maximal regular sequence, automatically algebraically independent over k. In both cases, one can assume that the first n generators of $\mathfrak{a}$ form a reduction.

By and large, the goal is to obtain the implicit equation by theoretic methods not resorting to computer calculation or at least to describe some of its relevant properties.

8.4.1 Dimension zero

Unless explicitly specified otherwise, we assume throughout this part that $\mathfrak{a} \subset k[x_1, \ldots, x_n]$ is an $\mathfrak{m}$-primary ideal, where $\mathfrak{m} = (x_1, \ldots, x_n)$.

8.4.1.1 Recap of Hilbert coefficients

When $(R, \mathfrak{m})$ is a Noetherian local ring of dimension n and $\mathfrak{a} \subset \mathfrak{m}$ is an $\mathfrak{m}$-primary ideal, one has the classical Hilbert–Samuel function and polynomial associated to $\mathfrak{a}$ by looking at the length of the graded component $\mathfrak{a}^m/\mathfrak{a}^{m+1}$ of the ring $\operatorname{gr}_{\mathfrak{a}}(R)$. This length is a polynomial for $m \gg 0$. Because of the sliding exact sequences

$$0 \to \mathfrak{a}^m/\mathfrak{a}^{m+1} \longrightarrow R/\mathfrak{a}^{m+1} \longrightarrow R/\mathfrak{a}^m \to 0,$$

the length of $R/\mathfrak{a}^{m+1}$ will also be a polynomial for $m \gg 0$. As any of the polynomials coming out this way (see, e. g., [179, Section 7.4]), one has

$$\lambda(R/\mathfrak{a}^{m+1}) = e_0(\mathfrak{a})\binom{n+m}{d} - e_1(\mathfrak{a})\binom{n+m-1}{d-1} + \text{lower degree terms of } m, \qquad (129)$$

where $m \gg 0$ has the role of a variable. The integers $e_i(\mathfrak{a})$ are called the *Hilbert coefficients* of $\mathfrak{a}$, while $e_0(\mathfrak{a})$ is the *multiplicity* of the ideal $\mathfrak{a}$ (or of R with respect to $\mathfrak{a}$).

Now, the same ideas go through ab initio in the graded case, namely, when $(R, \mathfrak{m})$ is a standard graded ring over an infinite field and its maximal irrelevant ideal, and $\mathfrak{a} \subset$

$\mathfrak{m}$ is an equigenerated irrelevant homogeneous ideal. Alternatively, one can localize at $\mathfrak{m}$ all visible rings and ideals, thus reducing to the local case.

Thus, $e_0(\mathfrak{a})$ is again the *multiplicity* of $\mathfrak{a}$ and $e_1(\mathfrak{a})$ its *first Hilbert coefficient*. Moreover, since $\mathfrak{a}$ is, say, d-equigenerated, the integral closure of any power $\mathfrak{a}^m$ is $(\mathfrak{m}^d)^m = \mathfrak{m}^{md}$. Since $\lambda(R/(\mathfrak{m}^d)^m)$ has a similar asymptotic polynomial as in (129), comparison yields

$$e_0(\mathfrak{a}) = e_0(\mathfrak{m}^d) = d^n$$
$$e_1(\mathfrak{a}) \leq e_1(\mathfrak{m}^d) = \frac{n-1}{2}(d^n - d^{n-1}).$$

The above inequality of first Hilbert coefficients is geometrically meaningful.

The following mix of algebra and geometry has been given in [107, Proposition 3.3] and [108, Proposition 3.3] (see also [131, Section 5] for a largely improved version).

Proposition 8.4.1. *Let $\mathfrak{F} : \mathbb{P}^{n-1} \dashrightarrow \mathbb{P}^n$ be the rational map defined by the above generators of $\mathfrak{a}$. The following are equivalent:*
(i) *$\mathfrak{F}$ is birational onto the image.*
(ii) $\deg k[\mathfrak{a}_d] = d^{n-1}$.
(iii) $e_1(\mathfrak{a}) = \frac{n-1}{2}(d^n - d^{n-1})$.
(iv) *The Rees algebra $\mathcal{R}_R(\mathfrak{a})$ is regular locally in codimension one.*

8.4.1.2 The syzygetic flagship

The following is a souped-up version of [108, Section 3.3].

Quite generally, let $\mathfrak{a} \subset R$ be an ideal in a Noetherian ring R with finite free presentation

$$F_1 \xrightarrow{\varphi} F_0 \longrightarrow \mathfrak{a} \to 0 \tag{130}$$

on a given set of generators of $\mathfrak{a}$. Let $Z \subset F_0$ denote the image of φ. Tensoring with $R/\mathfrak{a}$ over R yields the so-called syzygetic exact sequence

$$0 \to \delta(\mathfrak{a}) \longrightarrow H_1 \longrightarrow (F_0/\mathfrak{a}F_0) \longrightarrow \mathfrak{a}/\mathfrak{a}^2 \to 0, \tag{131}$$

where $H_1 = Z/B$ denote the first Koszul homology module on the chosen set of generators of $\mathfrak{a}$ and $\delta(\mathfrak{a}) = Z \cap \mathfrak{a}F_0/B$.

Definition 8.4.2. The ideal $\delta(\mathfrak{a})$ is called the *syzygetic component* of $\mathfrak{a}$ (or of H_1). The ideal $\mathfrak{a}$ is said to be *syzygetic* if its syzygetic component vanishes.

One essence of $\delta(\mathfrak{a})$ is that it does not depend on choice of set of generators of I (hence the notation). For the details on this notion and related questions, see [192] (also [96]).

Now assume the following.

Notation 8.4.3. $\mathfrak{a} = (f_1, \dots, f_n, f_{n+1}) \subset R = k[x_1, \dots, x_n]$ is an almost complete intersection of forms of the same degree, where $\{f_1, \dots, f_n\}$ is a regular sequence generating a minimal reduction $J \subset \mathfrak{a}$. As usual, set $\mathfrak{m} = R_+R = (x_1, \dots, x_n)$.

Since $\mathfrak{a}$ is an almost complete intersection, one knows that H_1 is isomorphic to the canonical module of $R/\mathfrak{a}$; hence, will be safely denoted $H_1(\mathfrak{a})$.

Proposition 8.4.4. *Let $\mathfrak{a} \subset R$ be as in (8.4.3). Then:*
(i) *$\delta(\mathfrak{a}) \simeq J{:}I_1(\varphi)/J$, as submodules of $H_1(\mathfrak{a})$.*
(ii) *$\mu(\delta(\mathfrak{a})) = \mu(I_1(\varphi) : \mathfrak{m}/I_1(\varphi))$. In particular, if $I_1(\varphi) = \mathfrak{m}^s$ for certain s, then $\mu(\delta(\mathfrak{a})) = \binom{n+s-2}{s-1}$.*

Proof. (i) Split (131) into two short exact sequences and dualize with respect to $H_1(\mathfrak{a})$. Since the latter is a canonical module of $R/\mathfrak{a}$, one has

$$\operatorname{Ext}^1_{R/\mathfrak{a}}(\mathfrak{a}/\mathfrak{a}^2, H_1(\mathfrak{a})) = \operatorname{Ext}^1_{R/\mathfrak{a}}(H_1(\mathfrak{a})/\delta(\mathfrak{a}), H_1(\mathfrak{a})) = 0.$$

Writing $\operatorname{Hom}((R/\mathfrak{a})^{n+1}, H_1(\mathfrak{a})) = H_1(\mathfrak{a}) \oplus \cdots \oplus H_1(\mathfrak{a}) := H_1(\mathfrak{a})^{n+1}$, one obtains an exact complex

$$H_1(\mathfrak{a})^{n+1} \longrightarrow \operatorname{Hom}_{R/\mathfrak{a}}(H_1(\mathfrak{a}), H_1(\mathfrak{a})) \simeq R/\mathfrak{a} \longrightarrow \operatorname{Hom}_{R/\mathfrak{a}}(\delta(\mathfrak{a}), H_1(\mathfrak{a})) \to 0.$$

A moment reflection tells us that the image of $H_1(\mathfrak{a})^{n+1}$ in $R/\mathfrak{a}$ gets identified with the ideal $I_1(\varphi)/\mathfrak{a}$, by noting that $\mathfrak{a} \subset I_1(\varphi)$ since $I_1(\varphi)$ contains the Koszul relations.

This gives $R/I_1(\varphi) \simeq \operatorname{Hom}_{R/\mathfrak{a}}(\delta(\mathfrak{a}), H_1(\mathfrak{a}))$, hence duality yields

$$\delta(\mathfrak{a}) \simeq \operatorname{Hom}_{R/\mathfrak{a}}(R/I_1(\varphi), H_1(\mathfrak{a})). \tag{132}$$

Since $H_1(\mathfrak{a})$ is a canonical module, $H_1(\mathfrak{a}) \simeq J : \mathfrak{a}/J$. Therefore, again because $\mathfrak{a} \subset I_1(\varphi)$, (132) finally yields the desired isomorphism $\delta(\mathfrak{a}) \simeq J{:}I_1(\varphi)/J \subset J : \mathfrak{a}/J \simeq H_1(\mathfrak{a})$.

(ii) By (132), $\delta(\mathfrak{a})$ gets identified with the canonical module of the ring $R/I_1(\varphi)$. It follows that $\mu(\delta(\mathfrak{a}))$ is the Cohen–Macaulay type of $R/I_1(\varphi)$. In other words, it has the same number of generators as its socle $I_1(\varphi) : \mathfrak{m}/I_1(\varphi)$.

The supplementary assertion is clear as $\mathfrak{m}^s : \mathfrak{m} = \mathfrak{m}^{s-1}$. □

In order to acquire further insight in the isomorphism of Proposition 8.4.4, one recalls a second interpretation of the syzygetic component $\delta(\mathfrak{a})$.

Consider a presentation of the Rees algebra $\mathcal{R}_R(\mathfrak{a}) \simeq R[T]/\mathcal{J}$. Here $\mathcal{J} \subset S := R[T]$ is a homogeneous ideal in the T-variables, so let $L_i, i \geq 1$ denote its ith graded R-module. We are often interested in the subideal $(L_i) \subset \mathcal{J}$ generated by L_i and the distinction may get blurred, but it becomes clear from the context. Note that (L_1) is generated by the syzygies of $\mathfrak{a}$ taken in their natural polynomial vest. Another important feature is that, quite generally, $\mathcal{J} = (L_1) : \mathfrak{a}^\infty$, as follows from the descent of the symmetric algebra to the Rees algebra. Moreover, if $\mathfrak{a}$ is a finite colength ideal then also $\mathcal{J} = (L_1) : \mathfrak{m}^\infty$ and the least saturation exponent thereof has some useful virtues.

As to L_2, one has the following.

Lemma 8.4.5 (The syzygetic lemma ([193, 1.2], [108, Lemma 3.6])). *Let $\mathfrak{a}$ be an ideal with presentation as in* (130). *Then*

$$\delta(\mathfrak{a}) \overset{\eta}{\simeq} L_2/S_1L_1. \tag{133}$$

Proof. The mapping φ is given as follows: given $z = \sum a_i T_i$, $(a_1, \ldots, a_n) \in Z \cap \mathfrak{a}R^n$, write $a_i = \sum_{j=1}^n c_{i,j} f_j$, for certain $c_{i,j} \in R$. Then set

$$\eta(z \bmod B) = \sum_{i,j} c_{i,j} T_i T_j \in L_2/S_1L_1.$$

We leave to the reader the required calculations showing that this association is a well-defined $(R/\mathfrak{a})$-module isomorphism. □

Some examples will be in exercises (see also Proposition 8.4.24 below). The overall philosophy is that L_1 and L_2 by themselves provide a good chunk of information about the whole presentation ideal $\mathcal{J}$. More on these modules as the theory develops on.

8.4.1.3 Threshold degree

We next look more closely at the relationship among the ideals $J \subset \mathfrak{a} \subset I_1(\varphi)$. First, one has $J{:}\,\mathfrak{a} = J{:}\,f_{n+1}$ with the notation of (8.4.3). Since $H_1(\mathfrak{a}) \simeq J{:}\,f_{n+1}/J$, the algebraic analogue of the homological definition $H_1(\mathfrak{a}) = Z/B$ (Z = cycles, B = boundaries) is that the syzygies of $\mathfrak{a}$ are generated (not necessarily minimally) by the syzygies lifted from $J{:}\,f_{n+1}$ modulo the Koszul syzygies of $f_1, \ldots, f_{n+1}$.

Let $\operatorname{indeg}(I)$ denote the initial degree of a homogeneous ideal $I \subset R$ with respect to the standard grading of R, i. e., the degree of a nonzero form of smallest standard degree. One difficulty of handling an appropriate notion of threshold degree is that the initial degrees of $H_1(\mathfrak{a})$ and the Gorenstein ideal $J : \mathfrak{a}$ are often unrelated. Clearly, $\operatorname{indeg}(J : \mathfrak{a}) \le d$, but $\operatorname{indeg}(H_1(\mathfrak{a}))$ can be larger than d.

Example 8.4.6. Let $n = 6$, $J = (x_1^2, \ldots, x_6^2)$ and $f = (\sum_{i=1}^6 x_i)^2$. A calculation with [79] yields $J : \mathfrak{a} = (J, \mathcal{H})$, with $\mathcal{H}$ a set of forms of degree 3; hence, $\operatorname{indeg}(H_1(\mathfrak{a})) = 3$.

One positive result in this regard is the following.

Proposition 8.4.7. *Let $\mathfrak{a} \subset R = k[x_1, \ldots, x_n]$ be as in* (8.4.3). *Then*

$$\dim_k [J : f_{n+1}]_{d-1} \ge \max\{(n+1)\dim_k R_{d-1} - \dim_k R_{2d-1}, 0\}.$$

Proof. Since J is generated by a regular sequence of n forms of degree d, one has $\dim_k (R/J)_{2d-1} = \dim_k R_{2d-1} - n \dim_k R_{d-1}$ (apply [179, Proposition 2.7.15] recursively).

Thus, $\dim_k J_{2d-1} = n \dim_k R_{d-1}$. Therefore,

$$\begin{aligned}\dim_k R_{2d-1} &\geq \dim_k(f_{n+1}R_{d-1} + J_{2d-1})\\ &= \dim_k f_{n+1}R_{d-1} + \dim_k J_{2d-1} - \dim_k(f_{n+1}R_{d-1} \cap J_{2d-1})\\ &= (n+1)\dim_k R_{d-1} - \dim_k(f_{n+1}R_{d-1} \cap J_{2d-1})\end{aligned}$$

Hence, $\dim_k(f_{n+1}R_{d-1} \cap J_{2d-1}) \geq (n+1)\dim_k R_{d-1} - \dim_k R_{2d-1}$.

Since multiplication by f_{n+1} induces an isomorphism $[J : f_{n+1}]_{d-1} \simeq f_{n+1}R_{d-1} \cap J_{2d-1}$ of k-vector spaces, we are done. □

Corollary 8.4.8. *Let $\mathfrak{a} \subset R$ be as in (8.4.3). If $n = 3$ then* $\operatorname{indeg}(H_1(\mathfrak{a})) \leq d - 1$.

Proof. It is enough to show that $\operatorname{indeg}(J : f_4) \leq d - 1$. By the previous proposition, we have $\dim_k[J : f_4]_{d-1} \geq \max\{4\dim_k R_{d-1} - \dim_k R_{d-1}, 0\} = \max\{0, d\} = d > 0$. Therefore, $\operatorname{indeg}(J : f_4) \leq d - 1$, and hence, $\operatorname{indeg}(H_1(\mathfrak{a})) \leq d - 1$ as well. □

The initial degree of $\delta(\mathfrak{a})$ is even more elusive. In general, all one can say is the following.

Proposition 8.4.9. $\operatorname{indeg}(\delta(\mathfrak{a})) \leq n(d-1) - s + 1$, *where* $s = \operatorname{indeg}(I_1(\varphi))$.

Proof. Invoke the fact that $J{:}\mathfrak{m}/J$ – as is the case of any nonzero submodule of R/J—shares the same socle as that of R/J; the latter is the degree $n(d-1)$ of the determinant of the Jacobian matrix of the generators of J. In particular, $(\mathfrak{m}/J)^{e-s+1}(\mathfrak{m}/J)^s = (\mathfrak{m}/J)^{e+1} = 0$, i. e., $(\mathfrak{m}/J)^{e-s+1} \in J : \mathfrak{m}^s/J$. Now, since $\delta(\mathfrak{a}) = J : I_1(\varphi)/J \supset J : \mathfrak{m}^s/J$, the assertion follows. □

The case of equality above is obtained by the following condition.

Definition 8.4.10. Let $\mathfrak{a}$ be an ideal as in (8.4.3) and let φ denote the corresponding presentation map φ. One says that the ideal $\mathfrak{a}$ is *s-balanced* if $I_1(\varphi) = \mathfrak{m}^s$.

Lemma 8.4.11. *If the ideal $\mathfrak{a}$ as above is s-balanced then the R-coefficients of all minimal generators of L_2/S_1L_1 live in degree $(n-1)(d-1) - s$.*

Proof. It suffices to prove that $\delta(\mathfrak{a}) = J : \mathfrak{m}^s/J$ is an equigenerated ideal in R/J, generated in degree $n(d-1) - s + 1$, because the required degree is then obtained by subtracting d.

We actually prove more.

Claim. $J : \mathfrak{m}^s/J = \mathfrak{n}^{n(d-1)-s+1}$, where $\mathfrak{n} = \mathfrak{m}/J$.

The proof is by induction on s. For $s = 1$, the assertion is clear since $J{:}\mathfrak{m}/J$ is the socle of R/J, which is the degree of the determinant of the Jacobian matrix of the generators of J.

We have already proved earlier that $\mathfrak{n}^{e-s+1} \in J : \mathfrak{m}^s/J$. Conversely, we show that the initial degree of $J : \mathfrak{m}^s/J$ is indeed $e - s + 1$. If not, let $h \in J : \mathfrak{m}^s/J$ denote a nonzero homogeneous element of degree $i < e - s + 1$. Then $h\mathfrak{n}^{s-1}\mathfrak{n} = hf\mathfrak{n}^s = 0$, hence, $h\mathfrak{n}^{s-1} \subset J : m/J = \mathfrak{n}^e$. On the other hand, $h\mathfrak{n}^{s-1} \subset \mathfrak{n}^{i+s-1}$. Since $i + s - 1 < e$, then $h\mathfrak{n}^{s-1} = 0$. Therefore, $h \in J : \mathfrak{m}^{s-1}/J = \mathfrak{n}^{e-s+2}$, where the inductive hypothesis has been applied. But $\deg(h) = i < e - s + 2$, thus implying that $h = 0$. □

As for the initial degree of $H_1(\mathfrak{a})$, the following is a special case of [168, Theorem 5] (see also [200]).

Theorem 8.4.12 (char(k) = 0). *Given an integer $d \geq 1$, let $J = (x_1^d, \dots, x_n^d)$, $f_{n+1} = (x_1 + \cdots + x_n)^d$ and $\mathfrak{a} = (J, f_{n+1})$. Then*

$$\operatorname{indeg}(H_1(\mathfrak{a})) \geq (n-1)(d-1)/2.$$

The proof in [168] is by induction and differentiation, requiring for it a larger landscape where d and the exponent of $x_1 + \cdots + x_n$ are not necessarily equal.

As seen above, in the balanced case of exponent s, the integer $(n-1)(d-1) - s$ plays a distinctive role. However, in general, s is different from the initial degree of $H_1(\mathfrak{a})$, so it is instead the number $\mu_0(\mathfrak{a}) := (n-1)(d-1) - \operatorname{indeg}(H_1(\mathfrak{a}))$ that has interest in itself regardless of the balancedness assumption, having been named the *threshold degree* in [28]. Its meaningfulness also comes from the employ of the approximation complexes in order to access the implicit equation.

Corollary 8.4.13 (char(k) = 0). *With the notation of Theorem* 8.4.12, *one has*

$$\mu_0(\mathfrak{a}) \leq (n-1)(d-1)/2.$$

On the other hand, if $\mathfrak{a}$ is generated by general forms, then one actually gets an equality

$$\mu_0(\mathfrak{a}) = \left\lfloor \frac{(n-1)(d-1)}{2} \right\rfloor. \tag{134}$$

See [28] for an argument.

A more detailed analysis in the setup of Theorem 8.4.12 may depend on the parity of either n or d (or both). It has a direct bearing on the associated Gorenstein ring $J : f_{n+1}$. We next give a partial account for $n = 3$.

Proposition 8.4.14 (char(k) = 0). *Let $R = k[x,y,z]$, $f = (x+y+z)^d$ and $J = (x^d, y^d, z^d)$ ($d \geq 2$). Denote $\mathfrak{a} := (J, f)$ and φ the syzygy matrix of $\mathfrak{a}$. Then:*

(i) *$\mathfrak{a}$ is $(d-1)$-balanced if and only if $d \leq 4$.*
(ii) $\operatorname{indeg}(J : f) = d - 1$.
(iii) *The resolution of $J : f$ has the shape*

$$0 \to R(-2d) \to R(-(d+1))^d \oplus R(-d)^\beta \xrightarrow{\rho} R(-(d-1))^d \oplus R(-d)^\beta \to J : f \to 0, \tag{135}$$

where $r = d + \beta$.

Proof. (i) By Exercise 8.7.10, if $\mathfrak{a}$ is $(d-1)$-balanced then $d \leq 7$. A straightforward computation with [79] yields:

- $d = 2$: $\mu(I_1(\varphi)) = 3 = \binom{3}{2}$
- $d = 3$: $\mu(I_1(\varphi)) = 6 = \binom{4}{2}$
- $d = 4$: $\mu(I_1(\varphi)) = 10 = \binom{5}{2}$
- $d = 5$: $\mu(I_1(\varphi)) = 14 < 15 = \binom{6}{2}$
- $d = 3$: $\mu(I_1(\varphi)) = 18 < 21 = \binom{7}{2}$
- $d = 3$: $\mu(I_1(\varphi)) = 22 < 28 = \binom{8}{2}$

Reading off these numbers yields the stated result.

(ii) This follows from Corollary 8.4.8 and Theorem 8.4.12.

(iii) By (ii), the graded minimal free resolution of the Gorenstein ideal $J : f$ has the following shape:

$$0 \to R(-(s+3)) \to \bigoplus_{i=1}^{r} R(-D_i) \xrightarrow{\rho} \bigoplus_{i=1}^{r} R(-d_i) \to J : f \to 0 \tag{136}$$

where s is the socle degree of $R/J : f$, $r = \mu(J : f)$ and $d - 1 = d_1 \leq \cdots \leq d_r$. Since this resolution is self-dual, $D_i = s + 3 - d_i$ for every $1 \leq i \leq r$. By [134, Proposition 3.2], $s = 3(d-1)-d = 2d-3$. Thus, $D_i = 2d - d_i$ for every $1 \leq i \leq r$. If $d_u > d$ for some $1 \leq u \leq r$, then the degree of the ith entry of the uth column of ρ is $D_u - d_i = 2d - (d_u + d_i) \leq 0$, contradicting the minimality of the resolution. Therefore, one has $d_i \leq d$ for every $1 \leq i \leq r$. Thus, (136) becomes

$$0 \to R(-2d) \to R(-(d+1))^{\alpha} \oplus R(-d)^{\beta} \xrightarrow{\rho} R(-(d-1))^{\alpha} \oplus R(-d)^{\beta} \to J : f \to 0$$

for certain integers $\alpha, \beta \geq 0$ such that $\alpha + \beta = r$. From this, we get the Hilbert series of $R/J : f$

$$H(t) = \frac{1 - \alpha t^{d-1} + \alpha t^{d+1} - t^{2d}}{(1-t)^3}.$$

Since the first derivative of the numerator vanishes at 1, we get $\alpha = d$. Therefore, the resolution becomes

$$0 \to R(-2d) \to R(-(d+1))^{d} \oplus R(-d)^{\beta} \xrightarrow{\rho} R(-(d-1))^{d} \oplus R(-d)^{\beta} \to J : f \to 0,$$

where $r = d + \beta$, as was to be shown. □

Conjecture 8.4.15. *In the notation of (135), one has: $\beta \leq 1$; moreover, $\beta = 0$ (i. e., no 'ghost terms') if and only if d is odd.*

Remark 8.4.16. Note that the conjecture depends on the shape of the defining colon data as one can easily concoct a Gorenstein ideal I with resolution

$$0 \to R(-6) \to R(-4)^3 \oplus R(-3)^2 \xrightarrow{\rho} R(-2)^3 \oplus R(-3)^2 \to I \to 0.$$

Clearly, here $\mu(I) = 5 = 2d - 1 >> d$ and I is not of the assumed colon form.

A consequence of the above conjecture is as follows.

Proposition 8.4.17 (Conjectured). *The following are equivalent:*
(a) *$J : f$ is equigenerated (in degree $d - 1$).*
(b) *The skew-symmetric matrix of Pfaffians defining the Gorenstein ideal $J{:}f$ has entries of degree 2.*
(c) *$\mu(J : f) = d$.*
(d) *d is odd.*

Moreover, if d is even then $\mu(J : f) = d + 1$ with d generators in degree $d - 1$ and one in degree d.

For the argument, any of the conditions (a), (b) or (c) of the statement is equivalent to having $\beta = 0$.

8.4.1.4 Binary case ($n = 2$): the implicit equation

Much work has been done in binary elimination, both by theoretic and applied methods, so one can say that the state of affairs is quite satisfactory. We reproduce one of these procedures based on elementary linkage theory, as obtained in [108].

Write $R = k[x, y]$ for lighter reading. Let $\mathfrak{a} \subset R = k[x, y]$ be an (x, y)-primary ideal generated by three forms of degree $d \geq 1$ and let

$$0 \to R^2 \xrightarrow{\varphi} R^3 \longrightarrow \mathfrak{a} \to 0$$

be a minimal-free resolution of $\mathfrak{a}$. Let $r, s \geq r$ be the standard degrees of the columns of φ.

Set $S = R[T_1, T_2, T_3]$. Let f and g denote the forms generating the defining ideal L_1 of the symmetric algebra of $\mathfrak{a}$, i. e.,

$$[f, g] = [T_1, T_2, T_3] \cdot \varphi.$$

Consider the subideal $N = (L_1) : (x, y)^r \subset \mathcal{J}$, where $\mathcal{J}$ designates the defining ideal of the Rees algebra $\mathcal{R}_R(\mathfrak{a})$.

Proposition 8.4.18. *The following hold:*
(i) *N is a perfect ideal of codimension two.*

(ii) *The canonical module of S/N is generated by $(x,y)^r S/(f,g)$; in particular, its Cohen–Macaulay type is $r+1$.*
(iii) *N is the ideal of maximal minors of an $(r+2)\times(r+1)$ matrix of the shape*

$$\zeta = \left[\frac{\sigma}{\tau}\right],$$

with σ a two rowed submatrix whose entries are, respectively, biforms of bidegree $(s-r,1)$ and $(0,1)$; and τ is an $(r+1)$ rowed submatrix whose entries are biforms of bidegree $(1,0)$.

Proof. (i) This is because N is a direct link of the Cohen–Macaulay ideal $(x,y)^r$.
(ii) This follows from the preliminaries of linkage [157].
(iii) Ibid as in (ii). □

Clearly, $N \subset \mathcal{J}$ and from (iii) above, the elements of N have T-degree at most 2. More precisely, $N = (f,g,\mathbf{h}_1,\dots,\mathbf{h}_r)$, where the forms $\mathbf{h}_i \in \mathcal{J}$ have T-degree 2 and R-coefficients in $(x,y)^{s-1}$.

Theorem 8.4.19. *Let $\mathbf{B}$ stand for the following content matrix with coefficients in $k[T_1,T_2,T_3]$:*
If $s=r$, write

$$[\mathbf{h}_1 \ \dots \ \mathbf{h}_r] = [x^{r-1} \ x^{r-2}y \ \dots \ xy^{r-2} \ y^{r-1}]\cdot \mathbf{B}. \tag{137}$$

If $s>r$, collect the $s-r$ forms $\mathbf{f} = \{x^{s-r-1}f, x^{s-r-2}yf,\dots,y^{s-r-1}f\}$ and write

$$[\mathbf{f}\,\mathbf{h}_1 \ \dots \ \mathbf{h}_r] = [x^{s-1} \ x^{s-2}y \ \dots \ xy^{s-2} \ y^{s-1}]\cdot \mathbf{B}. \tag{138}$$

Then
(a) $\det \mathbf{B}$ *is a nonzero polynomial of degree d.*
(b) $\det \mathbf{B}$ *is a power of the implicit equation of the parametrization given by the generators of $\mathfrak{a}$.*

Proof. (a) $(s=r)$. Suppose that $\det\mathbf{B}=0$. Let $[\mathbf{g}_1\dots\mathbf{g}_r]^t$ be a nonzero vector over $k[T_1,T_2,T_3]$ such that $\mathbf{B}\cdot[\mathbf{g}_1\dots\mathbf{g}_r]^t=0$. Then $[\mathbf{g}_1\dots\mathbf{g}_r]^t$ is a syzygy of $\mathbf{h}_1,\dots,\mathbf{h}_r$. Since the syzygies of the latter are S-linear combinations of the columns of ζ, we get a contradiction.

The assertion on the degree follows since the degree of $\det\mathbf{B}$ is $2r=r+s=d$.

$(s>r)$. Suppose that $\det\mathbf{B}=0$. Then, as above, there exists a nonzero vector $[\mathbf{g}_1\dots\mathbf{g}_r]^t\, k[T_1,T_2,T_3]$, which is a syzygy of $[\mathbf{f}\,\mathbf{h}_1 \ \dots \ \mathbf{h}_r]$. Write this relation as follows:

$$\sum_{i=1}^{s-r}\mathbf{g}_i x^{s-r-i}y^{i-1}f + \sum_{j=1}^{r}\mathbf{g}_{s-r+j}\mathbf{h}_j = \mathbf{g}f + \sum_{j=1}^{r}\mathbf{g}_{s-r+j}\mathbf{h}_j = 0,$$

where

$$\mathbf{g} = \sum_{i=1}^{s-r} \mathbf{g}_i x^{s-r-i} y^{i-1}.$$

Since the relations of $f, \mathbf{h}_1, \ldots, \mathbf{h}_r$ are S-linear combinations of the columns of ζ, one derives that $\mathbf{g} \in (x,y)^{s-r}S$ and $\mathbf{g}_{s-r+j} \in (x,y)S$, for $1 \le j \le r$. Therefore, $\mathbf{g}_i \in (x,y)S$, for all i. This gives a contradiction.

The assertion on the degree follows since the degree of $\det \mathbf{B}$ is $(s-r)+2r = r+s = d$.

(b)

Claim 1. One has a primary decomposition $N = \mathcal{J} \cap Q$, where Q is an $(x,y)S$-primary ideal.

To see this, let $P \subset S$ be a prime ideal containing L_1 but not (x,y). Then $N_P = (L_1)_P : (1) = (L_1)_P$. Similarly, $\mathcal{J} = (L_1) : \mathfrak{a}^\infty = (L_1) : (x,y)^\infty$ since $\mathfrak{a}$ is (x,y)-primary. It follows by the same token that $\mathcal{J}_P = (L_1)_P$. Therefore, $\mathcal{J}_P = N_P$. In particular, taking $P = \mathcal{J}$, one gets that the $\mathcal{J}$-primary component of N is $\mathcal{J}$ itself.

The remaining primary component must be $(x,y)S$-primary since N is unmixed.

Claim 2. In the notation of the first claim, one has $(x,y)^{s-1} \subset Q$.

By Claim 1, $N : \det \mathbf{B} = Q$, since $\det \mathbf{B} \in \mathcal{J}$ and is a form in $k[T]$. On the other hand, $(x,y)^{s-1} \subset N : \det \mathbf{B}$ by construction of $\mathbf{B}$ in Theorem 8.4.19.

Now, let $F \in \mathcal{J}$ denote the implicit equation. Using both claims above yields the inclusion $(x,y)^{s-1}F \subset N$. Since f, g are of bidegrees $(r,1)$ and $(s,1)$, it must be the case that each polynomial $x^i y^{s-1-i}F$ lies in the span of $\{f(x,y)^{s-r-1}, \mathbf{h}\}$. This yields a representation

$$F\left[(x,y)^{s-1}\right] = [\mathbf{f}; \mathbf{h}] \cdot \mathbf{A},$$

where $\mathbf{A}$ is an $s \times s$ matrix with entries in $k[T]$. Replacing $[\mathbf{f}; \mathbf{h}]$ by $[(x,y)^{s-1}] \cdot \mathbf{B}$, gives the matrix equation

$$[(x,y)^{s-1}](\mathbf{B} \cdot \mathbf{A} - F\mathbb{I}_s) = 0,$$

where $\mathbb{I}_s$ is the $s \times s$ identity matrix. But, since the minimal syzygies of $(x,y)^{s-1}$ have coefficients in (x,y), we must have $\mathbf{B} \cdot \mathbf{A} = F\mathbb{I}_s$, so that $\det \mathbf{B} \cdot \det \mathbf{A} = F^s$, as desired. □

Remark 8.4.20. We note that $\det \mathbf{B}$ as above is exactly the implicit equation if and only if the rational map $\mathbb{P}^1 \dashrightarrow \mathbb{P}^2$ is birational onto its image (see, e. g., [107, Proposition 3.3]). In particular, the degree of the image is the degree of the forms generating $\mathfrak{a}$ if and only if the map is birational onto it. Examples of non-birational such maps are the so-called improper parametrizations (see [107, Section 3.1] for a discussion of this notion.)

8.4.1.5 Binary case: the remaining equations

("Moving curve ideal")

We have already met some of these equations as generators of the subideal $N \subset \mathcal{J}$ of T-degree at most 2. Moreover, $R[T]/N$ is a Cohen–Macaulay ring. It seems natural to ask:

(1) Can one extend N to $\mathcal{J}$ so as to grant enough information about the remaining equations?
(2) Is $\mathcal{R}_R(\mathfrak{a})$ Cohen–Macaulay as well?

As to (1), there is a more or less algorithmic procedure to accomplish it by means of the Sylvester forms (already encountered in Chapter 7).

The answer to (2) is that it is rarely affirmative. The reason behind this phenomenon is a theorem by Goto and Shimoda [78, Theorem 3.1] in terms of reduction numbers. In terms of the Hilbert–Burch matrix representing the syzygies of $\mathfrak{a}$, the obstruction is as follows:

Proposition 8.4.21. *Let $I \subset R = k[x,y]$ be an (x,y)-primary ideal minimally generated by 3 forms of the same degree. Let*

$$\varphi = \begin{bmatrix} g_1 & h_1 \\ g_2 & h_2 \\ g_3 & h_3 \end{bmatrix}$$

be the syzygy matrix of I. Then $\mathcal{R}_R(\mathfrak{a})$ is Cohen–Macaulay if and only if there are forms $u, v \in R$ of the same degree such that

$$(g_1, g_2, g_3) = (h_1, h_2, h_3) = (u, v), \tag{139}$$

as ideals in R.

Proof. Suppose that $\mathcal{R}_R(\mathfrak{a})$ is Cohen–Macaulay. By [3, Corollary 6.3], the relation type of $\mathfrak{a}$ is at most $\dim R = 2$. It follows that the implicit equation F must be of degree 2. On the other hand, $\mathcal{R}_R(\mathfrak{a})$ has codimension 2, so it must be defined by the maximal minors of an $(m+1) \times m$ matrix whose entries are biforms in $R[T]$; to get F as one of these minors forces $m = 2$. Thus, $\mathcal{J}$ is generated by 3 biforms with syzygy matrix of the shape

$$\varphi = \begin{bmatrix} u & v \\ p_1 & q_1 \\ p_2 & q_2 \end{bmatrix},$$

where $u, v \in R$ are forms of the same degree and p_i, q_i are linear forms in $k[T]$. Here, $F = p_1q_2 - p_2q_1$, while $uq_1 - vp_1 = g_1T_1 + g_2T_2 + g_3T_3$ and $uq_2 - vp_2 = h_1T_1 + h_2T_2 + h_3T_3$. Expanding the last two relations in terms of T, one finds syzygies of T whose terms lie in R, hence they must vanish. This yields the inclusions $(g_1, g_2, g_3) \subset (u, v)$ and

$(h_1, h_2, h_3) \subset (u, v)$. To see the inclusion in the reverse direction, one uses that, since the generators of $\mathfrak{a}$ are assumed to be k-linearly independent, then at least a pair g_i, $g_j (i \neq j)$ is k-linearly independent. Similarly, for the column of the h_i's.

Now, conversely, suppose that the syzygy matrix of $\mathfrak{a}$ has the shape as in (139) Note that g_i, h_i, u, v have all the same degree as forms in R. Write each g_i, h_i as k-linear combinations of u, v, take the two syzygy polynomials $g, h \in R[T]$ and collect terms with respect to u, v. Then one obtains a content representation (Sylvester form)

$$[g\,h] = \mathbf{B}[u\,v]^t,$$

where $\mathbf{B}$ is a 2×2 matrix with linear entries in $k[T]$. Clearly, $F = \det \mathbf{B}$ is a nonzero form of degree 2. By a change of variables in $k[T]$—corresponding to an elementary transformation of the generators of $\mathfrak{a}$—one can assume that F has a pure power in, say, T_3. This gives that the third generator of $\mathfrak{a}$ is integral of degree 2 over the subideal J generated by the first two generators, which we are assuming to be a reduction of $\mathfrak{a}$. This implies that the reduction exponent is 1. Therefore, by [78, Theorem 3.1], $\mathcal{R}_R(\mathfrak{a})$ is Cohen–Macaulay. □

8.4.1.6 Ternary case ($n = 3$): the implicit equation

Here, it is assumed that $R = k[x, y, z]$, $\mathfrak{m} = (x, y, z)$, and $\mathfrak{a} = (f_1, f_2, f_3, f_4) \subset R$ is an $\mathfrak{m}$-primary ideal, where (f_1, f_2, f_3) is a minimal reduction generated by a regular sequence.

The difficulties are increasingly higher, both as to guessing the implicit equation or as to devising the structure of the other defining equations. By and large, one searches for the implicit equation as the determinant of a conveniently obtained square matrix or a square submatrix of certain larger matrices. From the applied angle, effective methods have been developed by the theoretic and applied communities. In particular, the use of the *raison d'être* of [94] and [95] has been proved very successful by the French school. Here, instead we will give but a taste of a still quite classic method, based on bringing out certain content square matrices.

Theorem 8.4.22. *Let $\mathfrak{a} \subset R = k[x, y, z]$ as above, generated in degree d, and let $\mathcal{J} \subset k[T] = k[T_1, T_2, T_3, T_4]$ denote a presentation ideal of the Rees algebra of $\mathfrak{a}$. Suppose that $\mathfrak{a}$ is $(d-1)$-balanced. Then:*

(i) *There is an equality $[R]_{2d-1} = [\mathfrak{a}]_{2d-1}$ of k-vector spaces.*
(ii) *$\mathcal{J}$ contains d independent forms $\{\mathfrak{f}_1, \dots, \mathfrak{f}_d\}$ of bidegree $(d-1, 1)$.*
(iii) *$\mathcal{J}$ contains $\binom{d}{2}$ independent forms $\{\mathfrak{h}_1, \dots, \mathfrak{h}_{\binom{d}{2}}\}$ of bidegree $(d-1, 2)$.*
(iv) *With the notation of (i) and (ii), there is a $\binom{d+1}{2} \times \binom{d+1}{2}$ matrix $\mathfrak{B}$ over $k[T]$ such that the following matrix equality holds:*

$$(\mathfrak{f}_1, \dots, \mathfrak{f}_d, \mathfrak{h}_1, \dots, \mathfrak{h}_{\binom{d}{2}}) = (\mathfrak{m}^{d-1})\mathfrak{B}, \tag{140}$$

where $(\mathfrak{m}^{d-1})$ denotes the row vector of a basis of the k-vector space $[\mathfrak{m}^{d-1}]_{d-1}$.

(v) *Any matrix* $\mathfrak{B}$ *as in* (iv) *has nonzero determinant. Moreover, if*

$$(L_1) : \mathfrak{m}^{d-1} = (\mathfrak{f}_1, \dots, \mathfrak{f}_d, \mathfrak{h}_1, \dots, \mathfrak{h}_{\binom{d}{2}}),$$

where $L_1 = [\mathcal{J}]_1$, *then* $\det \mathfrak{B}$ *is a power of the implicit equation.*

Proof. (i) Using the above notation, set $J := (f_1, f_2, f_3)$ and $f := f_4$. The socle of the canonical module $H_1(\mathfrak{a}) = J{:}f/J \subset R/J$ is the same as the one of its containing ring R/J. Since J is generated by a regular sequence of d-forms, the socle of R/J is the determinant of the Jacobian matrix of three d-forms, i. e., $3(d-1)$.

On the other hand, the $(d-1)$-balancedness assumption implies that $\operatorname{indeg}(H_1(\mathfrak{a})) = \operatorname{indeg}(J : f) = d - 1$. Then the number of nonzero homogeneous components of $H_1(\mathfrak{a})$ is given by $3(d-1) - (d-1) + 1 = 2d - 1$. Setting $\mathfrak{m} = (x, y, z)$, it follows that $\mathfrak{m}^{2d-1}$ annihilates $H_1(\mathfrak{a}) = J{:}f/J$. Since $J : f$ is Gorenstein, $J{:}(J{:}f) = (J, f) = \mathfrak{a}$. Then $(J{:}f)\mathfrak{a} \subset J$, hence $J{:}f/J$ annihilates $R/\mathfrak{a}$ as well. This means that $\mathfrak{m}^{2d-1} \subset \mathfrak{a}\mathfrak{m}^{d-1}$, hence equality holds, which proves the assertion.

(ii) Reading a surjective homogeneous R-map $R^4(-d) \twoheadrightarrow \mathfrak{a}$ in degree $d-1$ and using (i) yields a vector space surjection

$$[R]^4_{d-1} \xrightarrow{\psi} [\mathfrak{a}]_{2d-1} = [R]_{2d-1} \to 0.$$

Then $\dim \ker(\psi) = 4\binom{d+1}{2} - \binom{2d+1}{2} = d$. But, clearly, by assumption, $\ker(\psi)$ is the span of the syzygies of $\mathfrak{a}$ in (standard) degree $d-1$ based on the chosen map $R^4(-d) \twoheadrightarrow \mathfrak{a}$.

(iii) Since $\mathfrak{a}$ is $(d-1)$-balanced, this follows immediately from Proposition 8.4.4, via Lemma 8.4.5.

(iv) The existence of $\mathfrak{B}$ is clear, by expressing the R-coefficients of $\mathfrak{f}_i$s and $\mathfrak{h}_j$s in the chosen basis $(\mathfrak{m}^{d-1})$ ($\mathfrak{B}$ if often called an *R-content matrix*).

(v) Consider both $\mathfrak{f}_i$s and $\mathfrak{h}_j$s as polynomials in $R[T_1, T_2, T_3][T_4]$ and let $c(\mathfrak{f}_i)$ and $c(\mathfrak{h}_j)$ denote their respective R-coefficients in T_4 and T_4^2. Set $\mathfrak{c}$ for the ideal of R generated by these coefficients. Clearly, $\mathfrak{c} \subset \mathfrak{m}^{d-1}$. But, more surprisingly, we will have the following.

Claim. $\mathfrak{c} = \mathfrak{m}^{d-1}$.

To prove this, assume otherwise. Note that the $c(\mathfrak{f}_i)$s are k-linearly independent because a k-linear dependence would imply a syzygy of J of degree $d-1$, which is absurd since J is generated by a regular sequence of degree d. Therefore, we may replace one of the $\mathfrak{h}_j$ by a form of the shape

$$\begin{aligned}\mathfrak{h} &= T_4\,\mathfrak{f}(T_1, T_2, T_3) + \mathfrak{g}(T_1, T_2, T_3)\\ &= T_4(r_1T_1 + r_2T_2 + r_3T_3) + T_1\mathfrak{g}_1 + T_2\mathfrak{g}_2 + T_3\mathfrak{g}_3,\end{aligned}$$

where the r_i are $(d-1)$-forms in R and the $\mathfrak{g}$'s are T-linear involving only T_1, T_2, T_3. Evaluate now T_i at the corresponding generator of $\mathfrak{a}$ to get

$$(fr_1 + \mathfrak{g}_1(f_1,f_2,f_3))f_1 + (fr_2 + \mathfrak{g}_2(f_1,f_2,f_3))f_2 + (fr_3 + \mathfrak{g}_3(f_1,f_2,f_3))f_3 = 0,$$

a syzygy of the subideal $J = (f_1,f_2,f_3) \subset \mathfrak{a}$. Since J is a complete intersection, $fr_i + \mathfrak{g}_i(f_1,f_2,f_3) =: \mathfrak{u}_i(f_1,f_2,f_3) \in J$ for $i = 1,2,3$, with $\mathfrak{u}_i$ a linear form in T_1, T_2, T_3 with coefficients in R. These are syzygies of the generators of $\mathfrak{a}$, so lifting back to 1-forms in T and substituting yields $\mathfrak{h} = \mathfrak{h}' + \mathfrak{k}$, where

$$\begin{aligned}\mathfrak{h}' = {} & (r_1T_4 + \mathfrak{g}_1(T_2,T_2,T_3) - \mathfrak{u}_1(T_1,T_2,T_3))\,T_1\\ & + (r_2T_4 + \mathfrak{g}_2(T_1,T_2,T_3) - \mathfrak{u}_2(T_1,T_2,T_3))\,T_2\\ & + (r_3T_4 + \mathfrak{g}_3(T_1,T_2,T_3) - \mathfrak{u}_3(T_1,T_2,T_3))\,T_3\end{aligned}$$

is an element of (L_1) and $\mathfrak{k}$ involves only T_1, T_2, T_3. Since $\mathfrak{h}$ is a relation then so is the term $\mathfrak{k}$. But the latter belongs to the presentation ideal of the Rees algebra of the complete intersection J, hence also to the one of the symmetric algebra of J. But the latter is certainly contained in (L_1). Summing up, $\mathfrak{h} \in (L_1)$, which is a contradiction since $\mathfrak{h}$ is a minimal generator of $L_2/\mathbf{S}_1L_1$.

So much for the claim. It is clear that, since $\mathfrak{c} = \mathfrak{m}^{d-1}$, there is a noncanceling term in T_4^{2d} in the expansion of $\det \mathbf{B}$.

(v) Let $\mathbf{F}$ be the implicit equation.

Claim. $\mathcal{J} = (L_1) : \mathfrak{m}^{2d-2}$.

To see this, set $(L_1) : \mathfrak{m}^{d-1} = \mathcal{J} \cap Q$, where Q is an $\mathfrak{m}R[T]$-primary ideal. Since $\det \mathfrak{B} \neq 0$, it follows that

$$\mathfrak{m}^{d-1}R[T] \subset (\mathfrak{f},\mathfrak{h}) : \det \mathfrak{B} \subset ((L_1) : \mathfrak{m}^{d-1}) : \det \mathfrak{B} = Q.$$

This implies that

$$\mathcal{J} \cdot \mathfrak{m}^{d-1} \subset \mathcal{J}Q \subset (L_1) : \mathfrak{m}^{d-1}.$$

Hence $\mathcal{J} = (L_1) : \mathfrak{m}^{2d-2}$, as claimed.

Then

$$\mathbf{F} \in (L_1) : \mathfrak{m}^{2d-2} = ((L_1) : \mathfrak{m}^{d-1}) : \mathfrak{m}^{d-1}.$$

Therefore, $\mathbf{F}\mathfrak{m}^{d-1} \subset (\mathfrak{f},\mathfrak{h})$, which gives rise to a matrix representation

$$\mathbf{F}(\mathfrak{m}^{d-1}) = (\mathfrak{f},\mathfrak{h}) \cdot \mathfrak{A},$$

where $\mathfrak{A}$ is a square matrix with entries in $R[T]$. Replacing $(\mathfrak{f}, \mathfrak{h})$ by $(\mathfrak{m}^{d-1}) \cdot \mathfrak{B}$, gives the matrix equation

$$(\mathfrak{m}^{d-1}) \cdot (\mathfrak{B} \cdot \mathfrak{A} - \mathbf{F}\mathbb{I}) = 0,$$

where $\mathbb{I}$ is the identity matrix. Since the minimal syzygies of $\mathfrak{m}^{d-1}$ have coefficients in $\mathfrak{m}$, we must have

$$\mathfrak{B} \cdot \mathfrak{A} = \mathbf{F}\mathbb{I},$$

so that $\det \mathfrak{B} \cdot \det \mathfrak{A} = \mathbf{F}^m$, for some integer m. □

Remark 8.4.23. The above is one of the various theoretical effective ways of getting a power of the implicit equation. Unfortunately, there are no strictly effective ones to grab the actual implicit equation, except in some particular cases. This is a real thorn in the applied community as the computer programs based on current technology will not manage sufficiently large data.

8.4.1.7 Ternary case: the remaining equations

("Moving curve ideal")

Already in degree $d = 2$, the Rees algebra has an erratic behavior:

Proposition 8.4.24 (char$(k) \neq 2$). *Let $\mathfrak{a} \subset R = k[x,y,z]$ be an ideal of finite colength, minimally generated by 4 quadrics. Then $\mathfrak{a}$ has reduction number 1 or 3. In addition:*

(i) *If the reduction number is 1, $\mathcal{R}_R(\mathfrak{a})$ is Cohen–Macaulay, but not regular locally in codimension one. Moreover, this case takes place if and only if the k-vector space $\mathfrak{a}_d$ admits three algebraically dependent forms over k.*

(ii) *If the reduction number is 3, $\mathcal{R}_R(\mathfrak{a})$ is strictly almost Cohen–Macaulay and regular locally in codimension one. In particular, this case takes place if and only if no three forms in $\mathfrak{a}_d$ are algebraically dependent over k.*

Proof. As before, we may assume that $\mathfrak{a} = (J, f)$, where J is a minimal reduction. Recall that in the case of finite colength, the reduction number is independent of the chosen minimal reduction, so it will be denoted red$(\mathfrak{a})$. Let λ denote length as before.

Consider the homogeneous integral extension of graded k-algebras $k[\mathfrak{a}_2] \subset k[\mathfrak{m}_2^2]$. Then $4 = \deg k[\mathfrak{m}_2^2] = \deg k[\mathfrak{a}_2]\,[k(\mathfrak{m}_2^2) : k(\mathfrak{a}_2)]$, where round brackets denote the corresponding fraction fields (see, e. g., [190, Proposition 6.1 (b)]). Since $\deg k[\mathfrak{a}_2] \geq 2$, it follows that: either $\deg k[\mathfrak{a}_2] = [k(\mathfrak{m}_2^2) : k(\mathfrak{a}_2)] = 2$; or else $\deg k[\mathfrak{a}_2] = 4$ and $[k(\mathfrak{m}_2^2) : k(\mathfrak{a}_2)] = 1$. Let $\mathfrak{F}$ denote the rational map defined by the generators of $\mathfrak{a}$.

(i) In the first case, as already argued before, the implicit equation implies red$(\mathfrak{a}) = 1$ – hence, $\mathcal{R}_R(\mathfrak{a})$ is Cohen–Macaulay—and $\mathfrak{F}$ is not birational—i. e., $\mathcal{R}_R(\mathfrak{a})$ is not regular locally in codimension one by Proposition 8.4.1.

To see the supplementary characterization, note that $\mu(J\mathfrak{a}) = 3\mu(\mathfrak{a}) - 3 = 9$ [163, Proposition 2.22]. But then $\mathfrak{a}^2 = J\mathfrak{a}$ if and only if $\mu(\mathfrak{a}^2) = 9$ and this happens if and only if, up to a coordinate change, three of the minimal generators of $\mathfrak{a}$ are algebraically dependent.

(ii) In the second case, $\operatorname{red}(\mathfrak{a}) = 3$ and $\mathfrak{F}$ is birational—i. e., $\mathcal{R}_R(\mathfrak{a})$ is regular locally in codimension one. It remains in this case to see that depth $\mathcal{R}_R(\mathfrak{a}) = 3$.

Claim 1. $\lambda(\mathfrak{a}/J) = 2$.

As one readily sees, $\lambda(R/J) = 8$ and $\lambda(R/\mathfrak{a}) = 6$. Therefore, the exact sequence

$$0 \to \mathfrak{a}/J \longrightarrow R/J \longrightarrow R/\mathfrak{a} \to 0$$

yields the result.

Claim 2. $\lambda(\mathfrak{a}^2/J\mathfrak{a}) = 1$.

By the first line of the display in [109, Proposition 3.12], one has

$$\lambda(\mathfrak{a}^2/J\mathfrak{a}) = \lambda(\mathfrak{a}/J) - \lambda(R/I_1(\varphi)) = 2 - \lambda(R/I_1(\varphi)),$$

using Claim 1, where φ denotes the matrix of syzygies of $\mathfrak{a}$. But $\lambda(\mathfrak{a}^2/J\mathfrak{a}) \neq 0$ since $\operatorname{red}(\mathfrak{a}) \neq 1$. Clearly, $I_1(\varphi) \subset \mathfrak{m}$. Therefore, necessarily $\lambda(R/I_1(\varphi)) = 1$, so the result follows.

Claim 3. $\lambda(\mathfrak{a}^3/J\mathfrak{a}^2) = 1$.

By Claim 2, one can choose a pure product $fg \in \mathfrak{a}^2$, with $\{f, g\} \subset \mathfrak{a} \setminus J$, such that $\mathfrak{a}^2 = (fg, J\mathfrak{a})$. We contend that $\mathfrak{a}^3 = (f^2g, J\mathfrak{a}^2)$. Thus, letting $h_1, h_2, h_3 \in \mathfrak{a}$ one has $h_1h_2 = rfg + \mathfrak{f}$, $\mathfrak{f} \in J\mathfrak{a}$. Therefore, $h_1h_2h_3 = rfgh_3 + h_3\mathfrak{f}$. Writing $gh_3 = pfg + \mathfrak{g}$, where $\mathfrak{g} \in J\mathfrak{a}$, yields $h_1h_2h_3 = rpf^2g + \mathfrak{h}$, where $\mathfrak{h} = h_3\mathfrak{f} + rf\mathfrak{g} \in J\mathfrak{a}^2$, as was to be shown.

Finally, note that since $\lambda(\mathfrak{a}^3/J\mathfrak{a}^2) = 1$ then it must be the case that $\mathfrak{m}fg \subset \mathfrak{m}J\mathfrak{a}$; consequently, $\mathfrak{m}f^2g \subset \mathfrak{m}J\mathfrak{a}^2$ as well, thus showing that $\mathfrak{a}^3/J\mathfrak{a}^2$ has no nonzero element in degree ≥ 7. This concludes the proof of the claim.

Adding up the values found in the above three claims yields 4. But $e_1(\mathfrak{a}) = 4$ by Proposition 8.4.1. Therefore, by [110, Theorem 3.1] depth $\operatorname{gr}_{\mathfrak{a}}(R) \geq 2$, hence depth $\mathcal{R}_R(\mathfrak{a}) \geq 3$ and this is an equality as otherwise $\mathcal{R}_R(\mathfrak{a})$ would be Cohen–Macaulay. But this coupled with being regular locally in codimension one would make it normal, and hence equal to $\mathcal{R}_R(\mathfrak{m}^2)$. □

8.4.2 Dimension one

This is now the case where $\mathfrak{a} \subset R = k[x_1, \ldots, x_n]$ has analytic deviation one and height $n - 1$. As before, $\mathfrak{a}$ is d-equigenerated, say, $\mathfrak{a} = (f_1, \ldots, f_{n+1})$ and set $f := f_{n+1}$.

In this part it is more convenient to take the initial degree with respect to the shifted grading. Thus, for example, in the case of the Koszul homology modules

$H_i(\mathfrak{a}) = H_i(\mathfrak{a}; R)$ the initial degree is taken with respect to the grading of the free module of which cycles (and boundaries) are graded submodules. In dimension zero, the initial degree with respect to the standard grading was more appropriate.

A general theory in dimension one, with clearly stated guiding theorems, seems still off hand. Here, we assume moreover that the ideal $\mathfrak{a}$ satisfies condition (F_0), that is to say, $\mu(\mathfrak{a}_\wp) \leq \operatorname{ht}\wp + 1$ for every prime $\wp \supset \mathfrak{a}$. One advantage in this case is that one can, by taking n general k-linear combinations of the original generators of $\mathfrak{a}$, assume that $\mathfrak{J} = (f_1, \ldots, f_n)$ is a minimal reduction of $\mathfrak{a}$ such that $\mathfrak{a}_\wp = \mathfrak{J}_\wp$ for every prime $\wp \supset \mathfrak{a}$ such that $\wp \neq \mathfrak{m}$. It follows that these two ideals have the same saturation with respect to $\mathfrak{m}$.

Our purpose is to establish a few guiding principles, without going into the details of applied elimination proper. For the latter, the reader is referred to [29, 26, 28] and [27], where an important breakthrough has been given using the approximation complexes of [94].

We need the following result on the mobility of the initial degree.

Proposition 8.4.25. *With the above notation and assumption, one has:*

(i) $\operatorname{indeg}(H_2(\mathfrak{a})) \geq \operatorname{indeg}(H_1(\mathfrak{a})) + d$.

(ii) $\operatorname{indeg}(H^0_{\mathfrak{m}}(H_1(\mathfrak{a}))) \geq d + \operatorname{indeg}(\mathfrak{a}^{\mathrm{sat}})$, *where* $\mathfrak{a}^{\mathrm{sat}} = \mathfrak{a} : \mathfrak{m}^\infty$.

Proof. (i) Since $\mathfrak{J}$ is a reduction of $\mathfrak{a}$, its minimal set of generators can be extended to one of $\mathfrak{a}$. Say, $\mathfrak{a} = (\mathbf{f}, f)$, where $\mathbf{f} = \{f_1, \ldots, f_n\}$ and $f = f_{n+1}$. Consider the exact sequence of Koszul complexes with coefficients in R,

$$0 \to K_\bullet(\mathbf{f}) \xrightarrow{\iota} K_\bullet(\mathfrak{a}) \xrightarrow{\pi} K_{\bullet-1}(\mathbf{f})[-d] \to 0$$

induced by natural inclusion and projection. It induces in turn a long exact sequence in homology

$$\begin{aligned} \cdots \to H_2(\mathbf{f}) \xrightarrow{\iota_2} H_2(\mathfrak{a}) \xrightarrow{\overline{\pi_2}} H_1(\mathbf{f})[-d] \xrightarrow{\cdot(-f)} H_1(\mathbf{f}) \\ \xrightarrow{\overline{\iota_1}} H_1(\mathfrak{a}) \xrightarrow{\overline{\pi_1}} R/\mathfrak{J}[-d] \xrightarrow{\cdot f} R/\mathfrak{J} \xrightarrow{\overline{\iota_0}} R/\mathfrak{a} \to 0 \end{aligned} \tag{141}$$

Now, since $\mathfrak{J} = (\mathbf{f})$ has height $n-1$, then $H_1(\mathbf{f})$ is isomorphic to the canonical module $\omega_{R/\mathfrak{J}}$, which is annihilated by $\mathfrak{J}$. Since $\mathfrak{J}$ and $\mathfrak{a}$ coincide locally off $\mathfrak{m}$ then f annihilates $H_1(\mathfrak{a})$ locally off $\mathfrak{m}$, hence also annihilates $\omega_{R/\mathfrak{J}}$ locally off $\mathfrak{m}$. Since the canonical module is unmixed, it is annihilated by f, i. e., $H_1(\mathbf{f})$ is annihilated by f. Thus, the map

$$H_1(\mathbf{f})[-d] \xrightarrow{\cdot(-f)} H_1(\mathbf{f})$$

is zero. Then from (141), it follows that $H_2(\mathfrak{a}) \simeq H_1(\mathbf{f})[-d]$ and that $H_1(\mathbf{f})$ is a submodule of $H_1(\mathfrak{a})$, thus implying the claimed statement.

(ii) From the proof of item (i), one has an exact sequence

$$0 \to H_1(\mathbf{f}) \longrightarrow H_1(\mathfrak{a}) \longrightarrow (\mathfrak{J} : f)/\mathfrak{J}[-d] \to 0.$$

As seen above, $H_1(\mathbf{f})$ is unmixed, hence $H^0_{\mathfrak{m}}(H_1(\mathbf{f})) = 0$. Therefore, one has

$$\begin{aligned}\operatorname{indeg} H^0_{\mathfrak{m}}(H_1(\mathfrak{a})) &\geq \operatorname{indeg}(H^0_{\mathfrak{m}}((\mathfrak{J} : f)/\mathfrak{J})[-d]) = d + \operatorname{indeg}(\mathfrak{J} :_{H^0_{\mathfrak{m}}(R/\mathfrak{J}} f)/\mathfrak{J} \\ &= \operatorname{indeg}(\mathfrak{J} : f) \cap (\mathfrak{J} : \mathfrak{m}^{\infty})/\mathfrak{J} \geq d + \operatorname{indeg}(\mathfrak{J}^{\mathrm{sat}}/\mathfrak{J}) \\ &\geq d + \operatorname{indeg}(\mathfrak{J}^{\mathrm{sat}}) = d + \operatorname{indeg}(\mathfrak{a}^{\mathrm{sat}}).\end{aligned}$$

Above one has used that $\mathfrak{J}^{\mathrm{sat}} \neq \mathfrak{J}$, which holds since $\mathfrak{J} \neq \mathfrak{a}$ and both have the same saturation as mentioned earlier. □

In analogy to the zero-dimensional case, one can also introduce here a *threshold degree*:

$$\mu_0(\mathfrak{a}) := (n-1)(d-1) - \min\{\operatorname{indeg} H_1(\mathfrak{a}), \operatorname{indeg} H^0_{\mathfrak{m}}(H_1(\mathfrak{a})) - d\}. \tag{142}$$

Note that this definition is also kosher for the zero-dimensional case, that is, one recovers the expression introduced in (8.4.1.3). In fact, then $H_1(\mathfrak{a})$ is $\mathfrak{m}$-primary, hence $H^0_{\mathfrak{m}}(H_1(\mathfrak{a})) = H_1(\mathfrak{a})$, implying that the minimum above is attained at $\operatorname{indeg} H_1(\mathfrak{a}) - d$, which is the standard initial degree of $\mathfrak{J} : f/\mathfrak{J}$.

Corollary 8.4.26. *Assume as above that* $\mathfrak{a}$ *satisfies condition* (F_0). *Then*

$$\mu_0(\mathfrak{a}) \leq (n-1)(d-1) - \operatorname{indeg} \mathfrak{a}^{\mathrm{sat}}.$$

Proof. Clearly, $\operatorname{indeg} H_1(\mathfrak{a}) \geq d \geq \operatorname{indeg} \mathfrak{a}^{\mathrm{sat}}$. Therefore, the statement is a consequence of Proposition 8.4.25(ii). □

Remark 8.4.27. Lower bounds are more difficult to get, but see [28, Proposition 6] for a lower bound in terms of n, d analogous to the one in (134).

8.5 The classical resultant

Our treatment will be very brief. We refer to the modern account of [120] (see also [27], and [29] for a more applied view). Classically, one first defines the resultant of general forms and then specialize. This procedure works well at least when the number of forms coincides with the number of variables, in which case the so called resultant ideal is indeed principal. A generator of this ideal is called a resultant of the forms. The original ideas seem to go back to Hurwitz and Macaulay.

In order to see the generic resultant in terms of elimination algebras, one proceeds as follows. A set of r general forms $\{F_1, \ldots, F_r\}$ over k of respective degrees $d_1, \ldots, d_r$ is given. Here, k denotes a commutative ring and one takes F_j to be the form

$$F_j = \sum_{|a|=d_j} T_{j,a}\mathbf{X}^a,$$

where $T := \{T_{j,a}, 1 \le j \le r, |a| = d_j\}$ are mutually independent indeterminates over k and $\mathbf{X} = \{X_1, \dots, X_n\}$ is a set of n variables over $k[T]$. One eliminates toward $A := k[T]$ rather than to $k[\mathbf{X}]$. Thus, in the notation of Definition 8.1.4, $\mathcal{A} = k[\mathbf{X}, T]/(F_1, \dots, F_r)$ and $\mathbb{E}(\mathcal{A}) = k[T]/k[T]\cap((F_1, \dots, F_r) : (\mathbf{X})^\infty)$. The ideal $\mathfrak{R} = \mathfrak{R}(n,r) := k[T]\cap((F_1, \dots, F_r) : (\mathbf{X})^\infty)$ is called the (generic) *resultant ideal* (of r forms in n variables) and the k-algebra $k[T]/\mathfrak{R}$ is, accordingly, the (generic) *resultant algebra* of such forms.

By definition, one has an injection of k-algebras

$$\mathbb{E}(\mathcal{A}) = k[T]/\mathfrak{R} \hookrightarrow k[T, \mathbf{X}]/(F_1, \dots, F_r) : (\mathbf{X})^\infty \tag{143}$$

Proposition 8.5.1. [120] *With the preceding notation, one has*

(i) *If $r < n$, $\mathfrak{R} = (0)$.*

(ii) *If $r \ge n$ and k is a domain, the map* (143) *induces a birational morphism*

$$\mathrm{Proj}_S(k[T, \mathbf{X}]/(F_1, \dots, F_r) : (\mathbf{X})^\infty \to S = \mathrm{Spec}(\mathbb{E}(\mathcal{A})).$$

Proof. (Sketch) (i) Consider the (graded) Koszul complex on the polynomials $\{F_1, \dots, F_r\}$. In any case the homology of this complex has support on $V((\mathbf{X}))$ since making any $X_i = 1$ the resulting $F's$ form a regular sequence. Now, assume $r < n$. By induction, we may assume that $\{F_1, \dots, F_{r-1}\}$ is a regular sequence. Clearly, $k[T, \mathbf{X}]/(F_1, \dots, F_{r-1})$ is a polynomial ring on the variables $T_{r,a}$, $|a| = d_r$ over the ring $B' = k[T_{j,a} \mid j < r][\mathbf{X}]/(F_1, \dots, F_{r-1})$. The content ideal of F_r considered inside the latter ring is $(\mathbf{X})^{d_r}$ by the Dedekind–Mertens lemma. This in turns is not contained in any associated prime of B' since, by the inductive hypothesis, $\mathrm{depth}_{(\mathbf{X})}(B') = n-(r-1) > 0$. Therefore, F_r is indeed a non zero divisor modulo $(F_1, \dots, F_{r-1})$. Since $\{F_1, \dots, F_r\}$ is a regular sequence, any associated prime of $k[T, \mathbf{X}]/(F_1, \dots, F_r)$ has grade r, while $(\mathbf{X})$ has grade $n > r$. This shows that $(F_1, \dots, F_r) : (\mathbf{X})^\infty = (F_1, \dots, F_r)$, i. e., the elimination is conic in the sense of Definition 8.1.4. The final blow is to argue that this conic elimination ideal is null.

(ii) (Case $r = n$) In this situation, the resultant ideal $\mathfrak{R} \subset k[T]$ is principal. Let R be a generator therein. By reducing to the case where k is a prime field (hence perfect) one sees that there exists some variable $T_{j,a}$ such that $\partial R/\partial T_{j,a} \neq 0$. In fact, there is such an index j such that if ϵ_{jl}, with $l \in \{1, \dots, n\}$, denotes the T-coefficient of the term $X_l X_n^{d_j-1}$ in F_j, then $\partial R/\partial \epsilon_{jn} \neq 0$. Fixing one such j, introduce the ideal $J = (\partial R/\partial \epsilon_{j1}, \dots, \partial R/\partial \epsilon_{jn})k[T]/\mathfrak{R}$. The rest of the argument consists in showing that $k[T, \mathbf{X}]/((F_1, \dots, F_r) : (\mathbf{X})^\infty)$ is isomorphic to the Rees algebra of the ideal J. The birationality follows immediately from this. □

Remark 8.5.2. In (ii) the proof of the isomorphism for $r = n$ (or of the birationality for $r \geq n$) is not trivial. It depends on a key result proved in [120, Lemme 3.2.2] showing how to produce, out of a resultant $R \in \mathfrak{R}$ and any given $j \in \{1, \ldots, r\}$, an element of the saturation $((F_1, \ldots, F_r) : (\mathbf{X})^{\infty})$ of degree d_j in $\mathbf{X}$. Applying this suitably, one also finds that the Koszul relations of the generators of the above carefully chosen ideal J belong indeed to the saturation. Note that it is not clear from the definition of the saturation that there are at all 1-forms in $\mathbf{X}$ belonging to it. Another aspect of the Rees algebra $\mathcal{R}(J)$ intrusion is that, though one is still eliminating from it to obtain the algebra $k[T]/\mathfrak{R}$, the latter is not what we called $\mathbb{E}(\mathcal{R}(J))$.

8.6 Eliminating from correspondences

In this part, we focus on a biprojective variety $\mathcal{V} \subset \mathbb{P}^{n-1} \times_k \mathbb{P}^{n-1}$ (often the closure of an incidence correspondence). Some of the important varieties obtained from $\mathcal{V}$ by elimination require the intervention of the diagonal. Its geometric nature has been introduced in Section 1.3. Here, we wish to give an algebraic overture.

Setting $S := k[\mathbf{X}] \otimes_k k[\mathbf{X}]$ for the bihomogeneous coordinate ring of $\mathbb{P}^{n-1} \times_k \mathbb{P}^{n-1}$, let $\widetilde{\mathbb{D}} := (\{X_i \otimes 1 - 1 \otimes X_i \mid 0 \leq i \leq n\})$. Note that, quite generally, for any algebra R essentially of finite type over a Noetherian ring k, the kernel of the multiplication map $R \otimes_k R \to R$ is generated by the differences $t \otimes 1 - 1 \otimes t$, with $t \in R$ [179, Section 4.4.1]. In particular, if $R = k[\mathbf{X}]$ is a polynomial ring over k then this kernel is generated by the differences $X_i \otimes 1 - 1 \otimes X_i$. Therefore, our "diagonal" $\widetilde{\mathbb{D}}$ is the kernel of the multiplication map $k[\mathbf{X}] \otimes_k k[\mathbf{X}] \to k[\mathbf{X}]$. Note that $\widetilde{\mathbb{D}}$ is not bihomogeneous in S but it is homogeneous when considered as an ideal in the standard graded ring S^* in the notation of Subsection 1.2.2.

Now let B stand for the bihomogeneous coordinate ring of $\mathcal{V}$, along with the structural surjection $S^* \twoheadrightarrow B^*$ of $\mathbb{N}$-graded rings and set $\mathbb{D} := \widetilde{\mathbb{D}} B^*$.

Note that $\mathbb{D}$ is not the usual diagonal ideal since the ring B^* is not in general a tensor product of two standard graded rings, although it will be so in many interesting cases.

For easier reading, set $A := B^*$ and $\bar{A} := A/\mathbb{D}$.

Though slightly convoluted, this is the stage scene needed for an approach to intersection theory using the diagonal.

We will be concerned with the following diagram of algebra maps over A and $\overline{A}$:

$$\begin{array}{ccccc} \mathcal{S}_A(\mathbb{D}) & \to & \mathcal{S}_{\bar{A}}(\mathbb{D}/\mathbb{D}^2) & & \\ \downarrow & & \downarrow & & . \\ \mathcal{R}_A(\mathbb{D}) & \to & \mathcal{G}(\mathbb{D}) & \to & \mathcal{R}_{\bar{A}}(\mathbb{D}/\mathbb{D}^2) \end{array}$$

Here, $\mathcal{G}(\mathbb{D})$ is a lighter notation for the associated graded ring $\mathrm{gr}_{\mathbb{D}}(A)$, while in the remaining algebras the base ring is self-understood and may be often excluded. Similarly, one has a diagram for the respective elimination algebras:

$$\begin{array}{ccccc} \mathbb{E}(\mathcal{S}(\mathbb{D})) & \to & \mathbb{E}(\mathcal{S}(\mathbb{D}/\mathbb{D}^2)) & & \\ \downarrow & & \downarrow & & \\ \mathbb{E}(\mathcal{R}(\mathbb{D})) & \to & \mathbb{E}(\mathcal{G}(\mathbb{D})) & \to & \mathbb{E}(\mathcal{R}(\mathbb{D}/\mathbb{D}^2)) \end{array}$$

For the geometric purpose, the latter induces a corresponding diagram of inclusions of Proj's – some of the geometric constructions may even require to take first the associated reduced structures.

Then, drawing upon the terminology in [69], one has the following dictionary:

$\mathrm{Proj}(\mathbb{E}(\mathcal{R}(\mathbb{D})))$	*the span of the correspondence* $\mathcal{V}$
$\mathrm{Spec}(\mathcal{G}(\mathbb{D}))$	(the cone over) *the scheme of principal tangents of* $\mathcal{V}$
$\mathrm{Proj}(\mathbb{E}(\mathcal{G}(\mathbb{D})))$	*the span of the cone of principal tangents* of $\mathcal{V}$

A special case is when $\mathcal{V}$ is a product of two varieties, say, with bihomogeneous coordinate ring $k[\mathbf{X}]\otimes_k k[\mathbf{X}]/(I\otimes 1, 1\otimes J)$. In this case, the scheme $\mathrm{Proj}(\mathbb{E}(\mathcal{R}(\mathbb{D})))$ is called the *embedded join* of the schemes $\mathrm{Proj}(k[\mathbf{X}]/I)$ and $\mathrm{Proj}(k[\mathbf{X}]/J)$ in $\mathrm{Proj}(k[\mathbf{X}]) = \mathbb{P}^{n-1}$. The collapsing situation $I = J$ embodies many interesting features. The embedded join in this case becomes the *secant variety* to $\mathrm{Proj}(k[\mathbf{X}]/I)$ in $\mathbb{P}^{n-1}$. Both will be largely considered in Chapter 11.

On the other hand, $\mathrm{Spec}(\mathcal{G}(\mathbb{D}))$, called the *tangent star cone* in [188], will be treated at length in Chapter 10. Its geometric importance is at least the fact that the fibers of the structural projection $\mathrm{Spec}(\mathcal{G}(\mathbb{D})) \longrightarrow \mathrm{Spec}(A)$ at closed points are the so-named *tangent stars* in [119].

Let us collect a few facts about some of the maps in the above diagrams. Suppose that $\mathbb{D}$ is generically of linear type (e. g., $\bar{A}$ reduced). Then one has the following.

Proposition 8.6.1. *Suppose that k is a perfect field and $\bar{A}$ is as assumed above. Then*

$$\mathcal{R}_{\bar{A}}(\mathbb{D}/\mathbb{D}^2) = \mathcal{S}_{\bar{A}}(\mathbb{D}/\mathbb{D}^2)/((0):\mathfrak{J}_{\bar{A}}^{\infty}) = \mathcal{G}(\mathbb{D})/((0):\mathfrak{J}_{\bar{A}}^{\infty}),$$

where $\mathfrak{J}_{\bar{A}} \subset A$ is the Jacobian ideal of $\bar{A}$.

Proof. This follows from Lemma 8.1.2 by taking $\mathfrak{n} = \mathfrak{J}_A$ and from Corollary 2.3.19. □

The above says, in particular, that the Rees algebra of the module of differentials $\Omega(A/k)$ of A/k is obtained by killing the torsion of the so-called tangent star algebra $\mathbb{T}_{A/k}$, to be introduced in Section 10.2. As will be seen in detail in that section, the ring $A = k[\mathbf{X}]/I$ has an expected dimension (called the *expected star dimension*) expressed in terms of the potential equality $\dim \mathcal{S}(\mathbb{D}/\mathbb{D}^2) = \dim \mathcal{G}(\mathbb{D}) (= 2\dim A)$. One would like to introduce, in a similar way, an expected tangential dimension. For that, one has the following result.

translating the relevant incidence in terms of the Rees algebra of a suitable module E. It will be essentially the proof that the Rees algebra of E is the bigraded coordinate ring of the conormal variety, hence its projection of the latter on the second factor is given by the corresponding fiber cone—just as in the usual case of an incidence given by the graph of a rational map.

Throughout, assume that k is a perfect field. Let $I \subset (\mathbf{X})^2 \subset k[\mathbf{X}] = k[X_1, \ldots, X_n]$ denote a radical equidimensional homogeneous ideal of height g. Fix a regular sequence of forms $\mathbf{f} = f_1, \ldots, f_g \in I$ that generates I generically on $A = k[\mathbf{X}]/I$. Let $d_1 \geq \cdots \geq d_g \geq 2$ denote the degrees of these forms, respectively. Let $\Theta = \Theta(\mathbf{f})$ denote the Jacobian matrix of $\mathbf{f}$ viewed in A. Then Θ represents a map of free A-modules $A^n \to A^g$ whose image has rank g – it could be called an *essential Jacobian module* of I since the regular sequence $\mathbf{f}$ could in principle be taken at random making this module slightly more invariant on the ideal I than the Jacobian module on the given generators of I.

Remark 9.1.9. Let Θ_{tot} denote the Jacobian matrix of a whole set of generators of the ideal I. Then there is a natural surjective homomorphism $\operatorname{Im}(\Theta_{\text{tot}}) \twoheadrightarrow \operatorname{Im}(\Theta)$. Since $\operatorname{Im}(\Theta_{\text{tot}})$ is torsion-free and the modules have the same rank, this map is an isomorphism. Therefore, there is an isomorphism of the respective Rees algebras and this induces an isomorphism of k-algebras between the respective special fibers.

Let $\mathbf{t} = \{t_1, \ldots, t_g\}$ be weighted indeterminates over A with $\deg(t_i) = -d_i + 2$. As a result, the entries of the product matrix $\mathbf{t}\,\Theta$ are biforms of fixed total degree 1. Recall the identifications

$$\mathcal{R}_A(\operatorname{Im}(\Theta)) = A[\mathbf{t}\,\Theta] \subset A[\mathbf{t}] = \mathcal{R}_A(\oplus_{i=1}^{g} A(d_i)) = \mathcal{S}(\oplus_{i=1}^{g} A(d_i)),$$

as discussed in (2.3.2.4).

Theorem 9.1.10. *Let $I \subset (\mathbf{X})^2 \subset k[\mathbf{X}] = k[X_0, \ldots, X_N]$ denote a radical equidimensional homogeneous ideal of height g and let E denote a general essential Jacobian module of I. Set $A := k[\mathbf{X}]/I$ and let and let $V^* \subset \mathbb{P}^{N^*}$ stand for the dual variety of $V = \operatorname{Proj}(A) \subset \mathbb{P}^N$. Then the homogeneous coordinate ring of V^* is isomorphic to the fiber cone $\mathcal{F}(E)$ as graded k-algebras.*

Proof. The main argument in the proof is to show that the defining ideal as above actually comes from the Buchsbaum–Rim relations of the essential module $\operatorname{Im}(\Theta)$ which, in this setup, are given by the $(g+1)$-minors of $(\Theta^t|\mathbf{T})$.

Namely, consider the Jacobian matrix Ψ of these polynomials with respect to the $\mathbf{T}$-variables. More precisely, letting $A^N = \sum_{j=0}^{N} A\frac{\partial}{\partial x_j}$ in the usual way, one has

$$\bigwedge^{g+1}\left(\sum_{j=0}^{N} A\frac{\partial}{\partial x_j}\right) \xrightarrow{\Psi} \sum_{j=0}^{N} A\frac{\partial}{\partial x_j}, \tag{151}$$

where

$$\Psi\left(\frac{\partial}{\partial x_{j_1}} \wedge \cdots \wedge \frac{\partial}{\partial x_{j_{g+1}}}\right) = \sum_{l=1}^{g+1} (-1)^l \det\left(\frac{\partial \mathbf{f}}{\partial x_{j_1}} \wedge \cdots \wedge \widehat{\frac{\partial \mathbf{f}}{\partial x_{j_l}}} \wedge \cdots \wedge \frac{\partial \mathbf{f}}{\partial x_{j_{g+1}}}\right) \frac{\partial}{\partial x_{j_l}}.$$

Quite generally, the above are relations of Θ hence we get a surjective map $\operatorname{coker}(\Psi) \twoheadrightarrow \operatorname{Im}(\Theta)$. But the kernel of $\operatorname{coker}(\Psi) \twoheadrightarrow \operatorname{Im}(\Theta)$ is torsion since, by construction, A is generically a complete intersection of fixed codimension g and $\operatorname{Im}(\Theta)$ is generically the true Jacobian module of A, in which case one knows that the Buchsbaum–Rim relations generate $\ker(\Theta)$.

It follows that $\operatorname{coker}(\Psi) \twoheadrightarrow \operatorname{Im}(\Theta)$ induces an isomorphism between the corresponding Rees algebras. Therefore, $\mathcal{R}_A(\operatorname{Im}(\Theta))$ is isomorphic to

$$\mathcal{S}(\operatorname{coker}(\Psi)) \simeq k[\mathbf{X}, \mathbf{T}]/(I, I_{g+1}(\Theta^t|\mathbf{T}))$$

modulo its A-torsion. Finally, since Θ has rank g it is easy to see that the A-torsion of the latter symmetric algebra is the saturation $0 : I_g(\Theta)^\infty$. Therefore, we get

$$\mathcal{R}_A(\operatorname{Im}(\Theta)) \simeq k[\mathbf{X}, \mathbf{T}]/(I, I_{g+1}(\Theta^t|\mathbf{T}) : I_g(\Theta)^\infty).$$

To conclude note that $E = \operatorname{Im}(\Theta)$ is generated in the same degree as a submodule of $\oplus_{i=1}^{g} A(d_i)$, hence $\mathcal{F}(E)$ is isomorphic to the k-subalgebra of $\mathcal{R}_A(\operatorname{Im}(\Theta))$ generated in degree one, i. e., $\mathcal{F}(E) \simeq k[\mathbf{t}\Theta] = k[E_1]$. Clearly, the defining ideal of $k[E_1]$ is the contraction of $(I, I_{g+1}(\Theta^t|\mathbf{T}) : I_g(\Theta)^\infty)$ to the subring $k[\mathbf{T}] \subset k[\mathbf{X}, \mathbf{T}]$. Therefore, $k[\mathbf{t}\,\Theta]$ is indeed isomorphic to the homogeneous coordinate ring of the dual variety, as was to be shown. □

Remark 9.1.11. It is possible to introduce, in some sort of dual way, the homogeneous coordinate ring of the mth Gauss image in the Plücker embedding into $\mathbb{P}^{\binom{N+1}{m+1}-1}$. As it turns, the equations of the incidence in this formulation will involve vector fields (i. e., the module of derivations) rather than the module of Kähler differentials.

9.1.3 Relation to the tangential variety

Let V be a reduced and irreducible projective variety with a given nondegenerate embedding in $\mathbb{P}^n$. Set $d = \dim V$. Recall from the previous part (Remark 9.1.7) the Gauss map $\Gamma_V : V \to \mathbb{G}(d, n)$ associated to this embedding and (the closure of) its image, here denoted $\Gamma(V)$ for lack of an appropriate notation. Fixing an embedding, we talk in a slightly inappropriate way about the Gauss image of V.

Consider the incidence correspondence with its natural projections π_1 and π_2:

$$\Sigma = \{(p, \Lambda) \mid p \in \Lambda\} \subset \mathbb{P}^n \times \mathbb{G}(d, n).$$

The subvariety $T(V) := \pi_1(\pi_2^{-1}(\Gamma(V))) \subset \mathbb{P}^n$ is called the *tangential variety* to V (in the given embedding). Note that the closure of the image of V_{sm} by the Gauss map usually contains points that correspond to d-spaces which are not tangent to V. To get $T(V)$, one may first take the inverse image by π_2 of the points corresponding to tangent spaces to V_{sm}, then project to $\mathbb{P}^n$ by π_1 and close the image of this projection. It follows that $T(V)$ is as a set the closure of the union of all tangent spaces to V and that $V \subset T(V)$ (because $V_{\text{sm}} \subset T(V)$).

We observe that this notion can be approached from the point of view of the diagonal map (Chapter 10).

Remark 9.1.12. To describe the Gauss map in coordinates one looks at the abstract tangent space $T_{p,V}$ as the kernel of the linear map given by the Jacobian matrix at p of a set of generators of the defining ideal of V. However, there is a "dual" approach which consists in taking the kernel of the dual map $k^{n+1^*} \to T^*_{p,V} \to 0$, which is spanned by the column vectors of the transposed Jacobian matrix at p. If one searches for the Gauss image in terms of Plücker embeddings, then the difference amounts to choosing between the $(d+1) \times (d+1)$ minors of the null space of the Jacobian matrix, and the $(n-d) \times (n-d)$ minors of the Jacobian matrix.

We are interested in finding the defining equations of the tangential variety $T(V)$ by elimination. For this, first consider the Plücker embedding $\mathbb{G}(d,n) \subset \mathbb{P}^N$, with $N = \binom{n+1}{d+1} - 1$, letting $\mathbf{T} = \{T_{i_0 \dots i_d} \mid 0 \le i_0 < \dots < i_d \le n\}$ be the target coordinates. Then take the image of $\Gamma(V)$ under this embedding and let $\mathfrak{J} \subset k[\mathbf{T}]$ denote its homogeneous defining ideal.

Proposition 9.1.13. *Let $Q \subset k[\mathbf{X}, \mathbf{T}]$ be the bigraded ideal generated by the bidegree $(1,1)$ polynomials $\sum_{j=0}^{d} (-1)^{d+j} X_{i_j} T_{i_0 \dots \hat{i_j} \dots i_d}$, one for each choice of a subset $\{i_0, \dots, i_d\} \subset \{0, \dots, n\}$. Then the homogeneous defining ideal of the tangential variety $T(V)$ is the contraction to $k[\mathbf{X}]$ of the ideal $(\mathfrak{J}, Q) \subset k[\mathbf{X}, \mathbf{T}]$.*

Proof. The main point of the proof is to obtain the bihomogeneous ideal of the incidence correspondence Σ in the biprojective space $\mathbb{P}^N \times \mathbb{P}^n$, using the Plücker embedding $\mathbb{G}(d,n) \hookrightarrow \mathbb{P}^N$. The rest follows from the definition of $T(V)$ and from a translation of the elimination process.

For this, use the fact that a d-space $\Lambda \subset \mathbb{P}^n$ is defined by $d+1$ linearly independent vectors $v_0, \dots, v_d \in k^{n+1}$. The condition that $p \in \Lambda$ translates into $p \wedge v_0 \wedge \dots \wedge v_d = 0$. Using coordinates for p and for the v's, the condition amounts to the vanishing of the maximal minors of a $(n+1) \times (d+2)$ matrix over k.

Thus, let $\mathbf{X}$ be generic coordinates for a point of $\mathbb{P}^n$ and let Z_{ij}, $0 \le i \le n$, be generic coordinates for $d+1$ other vectors, with $j = 0, \dots, d$. Consider the augmented $(n+1) \times (d+2)$ matrix $(\mathbf{X} \mid (Z_{ij}))$. Recall that the homogeneous coordinate ring of $\mathbb{G}(d,n)$ is given (up to degree normalization) by the subalgebra $k[\Delta_{i_0 \dots i_d} \mid 0 \le i_0 < \dots < i_d \le n] \subset k[Z_{ij} \mid$

$0 \le i \le n$, $0 \le j \le d]$ generated by the maximal minors of the $(n+1) \times (d+1)$ matrix (Z_{ij}), with defining ideal on $k[\mathbf{t}]$ generated by the Grassmann–Plücker relations.

Now, expanding the maximal minors of the matrix $(\mathbf{X} \mid (Z_{ij}))$ by the Laplace rule along the first column and replacing the $(d+1)$-minors of (Z_{ij}) by the corresponding $\mathbf{T}$ variables, we arrive at the form of the generators for the bihomogeneous defining ideal of Σ. Namely, they are the Grassmann–Plücker relations (all of bidegree $(0,2)$) and the polynomials of bidegree $(1,1)$ obtained from lifting the Laplace expansions above.

Also, as a slight check on the final form of the proposition note that the ideal of the Gauss image under the Plücker embedding already contains the ideal of the Grassmann–Plücker relations, hence the bihomogeneous ideal of $\pi_2^{-1}(\Gamma(V))$ as a biprojective subvariety of $\mathbb{P}^n \times_k \mathbb{P}^N$ is indeed $(I(\Gamma(V)), Q)$. □

9.2 The Gauss image of a unirational variety

We focus on the situation where $V \subset \mathbb{P}^n$ is a unirational variety. We actually assume that a rational map $\mathbb{P}^d \dashrightarrow \mathbb{P}^n$ is given, where $d = \dim V$. This is indeed the classical case envisaged both in algebraic geometry and differential geometry.

9.2.1 The algebraic structural data

To elaborate on the algebraic side of the picture, let k be a field of characteristic zero and let $B = k[\mathbf{t}] = k[t_0, \dots, t_d]$ be a polynomial ring. Let $\mathbf{g} := \{g_0, \dots, g_n\} \subset (\mathbf{t})B$ be such that the k-subalgebra $A = k[\mathbf{g}] \subset B$ has dimension $d+1$. Then some subset $\{g_{i_0}, \dots, g_{i_d}\}$ is algebraically independent over k. By passing to the localization $B_{(\mathbf{t})}$, the determinant of the Jacobian matrix of $\{g_{i_0}, \dots, g_{i_d}\}$ generates the socle of the zero dimensional ring $B_{(\mathbf{t})}/(g_{i_0}, \dots, g_{i_d})$; hence, in particular, it is a nonzero element of B. It follows that the Jacobian matrix $\Theta = \Theta(\mathbf{g})$ of $\mathbf{g}$ has rank $d+1$. One denotes by $\mathbf{\Delta} = \mathbf{\Delta}(\mathbf{g})$ the set of $(d+1) \times (d+1)$ minors of Θ.

To recast the algebraic status of the previous subsection, we introduce the following additional notation. Let $k[\mathbf{X}]/P := k[X_0, \dots, X_n]/P \simeq k[\mathbf{g}]$ be a k-algebra presentation of A. From our standing assumption, P is a prime ideal of codimension $n-d$. Since P_P is generated by a regular sequence, one can choose $f_{d+1}, \dots, f_n \in P$ such that $P_P = (f_{d+1}, \dots, f_n)_P$. Setting $\mathbf{f} := \{f_{d+1}, \dots, f_n\}$, we let $\Theta(\mathbf{f})(\mathbf{g}) = \Theta(\mathbf{f})(g_0, \dots, g_d)$ denote the Jacobian matrix of $\mathbf{f}$ with respect to $X_0, \dots, X_n$ evaluated at $\mathbf{g}$ (or, which is the same, taken modulo the ideal P). Finally, $\mathbf{\Delta}(\mathbf{f})(\mathbf{g})$ will denote the set of $(n-d) \times (n-d)$ (i. e., maximal) minors of $\Theta(\mathbf{f})(\mathbf{g})$ evaluated at $\mathbf{g}$. This set generates a certain subideal of the Jacobian ideal of A, but we will rather be interested in the k-subalgebra $k[\mathbf{\Delta}(\mathbf{f})(\mathbf{g})] \subset A$. By the previous subsection, this algebra is (isomorphic to) the homogeneous coordinate ring of the Gauss image $\Gamma(V)$ in its dual Plücker embedding.

We need a lemma on matrices over a ring that can possibly be found in a classical book on the theory of matrices, but we have not been able to find a precise such source.

Let $0 < d < n$ be integers. To simplify the notation, an ordered subset $\{i_0 < \cdots < i_d\} \subset \{0, \ldots, n\}$ will be denoted by α, β, etc. For each $\alpha \subset \{0, \ldots, n\}$, $\sigma(\alpha)$ will stand for the ordered complementary subset of α in $\{0, \ldots, n\}$. If $\mathcal{M}$ is an $(n+1) \times (d+1)$ matrix with entries in A, a typical $(d+1) \times (d+1)$ (i. e., maximal) minor of $\mathcal{M}$ will be denoted by $\Delta_\alpha(\mathcal{M})$.

Lemma 9.2.1. *Let A be any commutative ring and let $\mathcal{Y}$ and $\mathcal{Z}$ be matrices over A of sizes $(n-d) \times (n+1)$ and $(n+1) \times (d+1)$, respectively, such that $\mathcal{Y}\mathcal{Z} = 0$. Then, for any two $\alpha, \beta \subset \{0, \ldots, n\}$ with $|\alpha| = |\beta| = d+1$ and $|\alpha \cap \beta| = d$, one has*

$$\Delta_\alpha(\mathcal{Z}) \cdot \Delta_{\sigma(\beta)}(\mathcal{Y}) = \pm \Delta_\beta(\mathcal{Z}) \cdot \Delta_{\sigma(\alpha)}(\mathcal{Y}).$$

Proof. Without loss of generality, assume that $\alpha = \{0, \ldots, d\}$ and $\beta = \{1, \ldots, d, d+1\}$. Write the matrices in block form

$$\mathcal{Y} = (Y' \mid Y), \quad \mathcal{Z} = \begin{pmatrix} Z \\ \hline Z' \end{pmatrix},$$

where Y has $n-d$ columns and Z has $d+1$ rows. Then $\Delta_{\sigma(\alpha)}(\mathcal{Y}) = \det(Y)$ and $\Delta_\alpha(\mathcal{Z}) = \det(Z)$. By hypothesis, $Y'Z + YZ' = \mathcal{Y}\mathcal{Z} = 0$. Multiplying on the left by $\operatorname{adj}(Y)$ and on the right, by $\operatorname{adj}(Z)$, one gets

$$\det(Z) \cdot \operatorname{adj}(Y) \cdot Y' = -\det(Y) \cdot Z' \cdot \operatorname{adj}(Z).$$

On the other hand, by an easy verification, one has $(\operatorname{adj}(Y) \cdot Y')_{1,1} = \Delta_{\{0,d+2,\ldots,n\}}(\mathcal{Y}) = \Delta_{\sigma(\beta)}(\mathcal{Y})$ and, similarly, $(Z' \cdot \operatorname{adj}(Z))_{1,1} = \Delta_{\{1,\ldots,d+1\}}(\mathcal{Z}) = \Delta_\beta(\mathcal{Z})$. This shows the contention. □

With the notation introduced prior to the above lemma, here is the main structural result of this part.

Theorem 9.2.2. *Let $\mathbf{g} \in k[\mathbf{t}]$ and $\mathbf{f} \in k[\mathbf{X}]$ as above. Then:*
(i) $\Theta(\mathbf{f})(\mathbf{g}) \cdot \Theta(\mathbf{g}) = 0$.
(ii) *Pairing each maximal minor of $\Theta(\mathbf{g})$ to its cofactor in the matrix obtained by concatenating $\Theta(\mathbf{g})$ and the transpose of $\Theta(\mathbf{f})(\mathbf{g})$, the (ordered) sets $(\mathbf{\Delta}(\mathbf{g}))$ and $(\mathbf{\Delta}(\mathbf{f})(\mathbf{g}))$ yield corresponding points in the natural duality*

$$\mathbb{P}(\Lambda^{d+1}(k(\mathbf{t})^{n+1})) \simeq \mathbb{P}(\Lambda^{n-d}(k(\mathbf{t})^{n+1^{\check{}}})).$$

(iii) *If in addition $\mathbf{g}$ are forms of fixed degree, there is an isomorphism $k[\mathbf{\Delta}(\mathbf{g})] \simeq k[\mathbf{\Delta}(\mathbf{f})(\mathbf{g})]$ of k-algebras.*

Proof. (i) Since $\mathbf{f}(g_0, \ldots, g_d) = 0$, this part follows from the chain rule for derivatives.

(ii) Because of part (i), Lemma 9.2.1 is applicable with $Y = (\Theta(\mathbf{f})(\mathbf{g}))$ and $Z = \Theta(\mathbf{g})$. The result is the first inductive step in proving the following generalization of Lemma 9.2.1: for *any* $\alpha, \beta \subset \{0, \dots, n\}$ with $|\alpha| = |\beta| = d+1$, one has

$$\Delta_\alpha(\Theta(\mathbf{g})) \cdot \Delta_{\sigma(\beta)}(\Theta(\mathbf{f})(\mathbf{g})) = \pm\Delta_\beta(\Theta(\mathbf{g})) \cdot \Delta_{\sigma(\alpha)}(\Theta(\mathbf{f})(\mathbf{g})).$$

Once this is proved, then by fixing α and letting β vary arbitrarily, we will be through.

Again, we assume with no loss of generality that $\alpha = \{0, \dots, d\}$. By descending induction on $r := |\alpha \cap \beta|$, assume $r < d$ and the result to be true for any $\beta' \subset \{0, \dots, n\}$ such that $|\beta'| = d+1$ and $|\alpha \cap \beta'| \geq r+1$.

One can assume that $\Delta_\beta(\Theta(\mathbf{g})) \neq 0$. Otherwise, as we now show, $\Delta_{\sigma(\beta)}(\Theta(\mathbf{f})(\mathbf{g})) = 0$ which would make the proposed equality a trivial equality. For that there is no harm in assuming that $\beta = \{0, \dots, d\}$ (this part of the argument will not touch our old α). The classical socle result says that if $\Delta_\beta(\Theta(\mathbf{g})) = 0$ then the set $\{g_0, \dots, g_d\}$ is algebraically dependent over k.

Claim. Letting $k[X_0, \dots, X_n]/P \simeq k[\mathbf{g}]$, there exists of a form $F \in P \cap k[X_0, \dots, X_d] \setminus P^{(2)}$, where $P^{(2)}$ denotes the symbolic power of P of order 2.

For this, since $P \cap k[X_0, \dots, X_d] \neq 0$, one may take a nonzero homogeneous polynomial in this contraction which contains, say, the variable X_d and is of least possible degree in X_d. It is well known, or easy to verify, that $\frac{\partial}{\partial X_d}(P^{(2)}) \subset P$. Hence, if $F \in P^{(2)}$, we get $\frac{\partial F}{\partial X_d} \in P$. Since the latter has smaller degree in X_d than F, it must vanish. Therefore, $F \in P \cap k[X_0, \dots, X_{d-1}]$. Repeating the argument, one eventually gets, say, $P \cap k[X_0] \neq (0)$. But this is impossible as no element of $\mathbf{g}$ belongs to k. This contradiction thus shows the existence of a homogeneous polynomial $F \in P \cap k[X_0, \dots, X_d] \setminus P^{(2)}$.

Fix a form F provided by the above claim.

To continue, since $P_P = (\mathbf{f})_P$, there exists an $H \in k[\mathbf{X}] = k[X_0, \dots, X_n] \setminus P$ such that $HF = \sum_{i=d+1}^n h_i f_i \in (\mathbf{f})k[\mathbf{X}]$. Taking partial derivatives with respect to the variables X_j, $j > d$ and using the chain rule, one finds

$$(h_{d+1}(\mathbf{g}), \dots, h_n(\mathbf{g})) \cdot \begin{pmatrix} \frac{\partial f_{d+1}}{\partial X_{d+1}}(\mathbf{g}) & \cdots & \frac{\partial f_{d+1}}{\partial X_n}(\mathbf{g}) \\ \vdots & \ddots & \vdots \\ \frac{\partial f_n}{\partial X_{d+1}}(\mathbf{g}) & \cdots & \frac{\partial f_n}{\partial X_n}(\mathbf{g}) \end{pmatrix} = (0, \dots, 0)$$

where the determinant of the Jacobian matrix in the middle is precisely $\Delta_{\sigma(\beta)}(\Theta(\mathbf{f})(\mathbf{g}))$. Therefore, if the latter is nonzero we must have $h_i(\mathbf{g}) = 0$, $\forall i$, i. e., $h_i(\mathbf{X}) \in P$, $\forall i$. But this would imply that $HF \in P^2$, hence $F \in P^{(2)}$, which contradicts our choice of F.

Thus, since we can assume $\Delta_\beta(\Theta(\mathbf{g})) \neq 0$, if we write $\beta = \{i_0, \dots, i_d\}$ (now we do not want to confuse it with α), then the set $\{g_{i_0}, \dots, g_{i_d}\}$ is algebraically independent over k. Fix an index $j \in \alpha \setminus \beta$. A slight modification of the classical exchange argument (cf. e. g., [225, Chapter II, p. 97]) shows that there exists an index $i_l \in \beta \setminus \alpha$ such that if $\beta' := (\beta \setminus \{i_l\}) \cup \{j\}$ then the corresponding $\mathbf{g}_{\beta'}$ is algebraically independent over k.

It follows that $\Delta_{\beta'}(\Theta(\mathbf{g})) \neq 0$. We now pull the pieces together, observing that $|\alpha \cap \beta'| = r+1$ and $|\beta \cap \beta'| = d$ by construction. By the first inductive step, one has

$$\Delta_{\beta}(\Theta(\mathbf{g})) \cdot \Delta_{\sigma(\beta')}(\Theta(\mathbf{f})(\mathbf{g})) = \pm\Delta_{\beta'}(\Theta(\mathbf{g})) \cdot \Delta_{\sigma(\beta)}(\Theta(\mathbf{f})(\mathbf{g})),$$

while the inductive assumption yields

$$\Delta_{\alpha}(\Theta(\mathbf{g})) \cdot \Delta_{\sigma(\beta')}(\Theta(\mathbf{f})(\mathbf{g})) = \pm\Delta_{\beta'}(\Theta(\mathbf{g})) \cdot \Delta_{\sigma(\alpha)}(\Theta(\mathbf{f})(\mathbf{g})).$$

Since $\Delta_{\beta'}(\Theta(\mathbf{g})) \neq 0$, together the two imply the required relation between α and β.

(iii) If $\mathbf{g}$ consists of forms of fixed degree then so does $\boldsymbol{\Delta}(\mathbf{g})$. In this case, a homogeneous polynomial $F(\mathbf{T}) \in k[\mathbf{T}]$, where $\#\mathbf{T} = \binom{n+1}{d+1}$, vanishes at $(\boldsymbol{\Delta}(\mathbf{g}))$ if and only it does at any proportional vector thereof, in particular, at $(\boldsymbol{\Delta}(\mathbf{f})(\mathbf{g}))$ by part (ii). This shows that the two algebras are isomorphic. □

Corollary 9.2.3. *If* $\mathbf{g}$ *are forms of fixed degree generating the k-subalgebra* $A \subset k[\mathbf{t}]$, *then* $\mathbb{G}(A)$ *is isomorphic to the homogeneous coordinate ring of the Gauss image of* $\mathrm{Proj}(A)$ *in its dual Plücker embedding. Moreover, there is an injective homomorphism of k-algebras* $\mathbb{G}(A) \hookrightarrow A$ *inducing the rational map from* $\mathrm{Proj}(A)$ *to its Gauss image.*

Remark 9.2.4. It is important to observe that the injective homomorphism of k-algebras $\mathbb{G}(A) \hookrightarrow A$ is not a natural inclusion induced by the respective inclusions in the polynomial ring $k[\mathbf{t}]$. It is rather given by an appropriate multiplier, which may vary from one situation to the other.

9.2.2 The Gauss image of toric varieties

We will give an application of the theory to the Veronese embeddings of a projective space. Actually, we deal with the corresponding homogeneous coordinate rings which are well understood. These are instances of the so-called toric rings. A *toric algebra* is a subalgebra of $k[\mathbf{t}]$ generated by a finite set of monomials (eventually with negative exponents). Let us agree to momentarily call *positive* a toric algebra generated by monomials with nonnegative exponents.

Lemma 9.2.5. *If* $A \subset k[\mathbf{t}]$ *is a positive toric algebra, then the subalgebra of* $k[\mathbf{t}]$ *generated by the minors of any size of the Jacobian matrix of the generators of A is a (positive) toric algebra.*

Proof. In fact, we prove the following assertion: if $\mathbf{f} = \{f_1, \dots, f_m\}$ is a set of monomials in $k[\mathbf{t}]$, then every minor of the Jacobian matrix of $\mathbf{f}$ is of the form $\alpha\mathbf{t}^{\mathbf{a}}$, where α belongs to the prime ring of k.

To see this, one endows the ring $k[\mathbf{t}]$ with the $\mathbb{N}^n$-grading in which $k[\mathbf{t}]_{\mathbf{a}} = k\,\mathbf{t}^{\mathbf{a}}$. Elements and matrices having this grading will be referred to as *multihomogeneous*.

One claims that the Jacobian matrix is also multihomogeneous, i. e., that any 2×2 minor is a multi-homogeneous element of $k[\mathbf{t}]$. This will imply that any minor is multihomogeneous, hence must be of the form $\alpha\mathbf{t}^{\mathbf{a}}$, for some $\mathbf{a} \in \mathbb{N}^n$ and some $\alpha \in k$. To conclude, the definition of derivatives readily implies that α actually belongs to the prime ring of k. Thus, say, the 2×2 minor in question is that of columns 1, 2 and rows 1, 2. Letting $f_1 = \mathbf{t}^{\mathbf{a}}$ and $f_2 = \mathbf{t}^{\mathbf{b}}$, the minor is a binomial with terms of multidegrees $(a_1 - 1 + b_1, a_2 + b_2 - 1, \dots)$ and $(a_1 + b_1 - 1, a_2 - 1 + b_2, \dots)$, respectively. Since the latter are equal, the minor is multihomogeneous. □

Thus, a variety that has a projective embedding whose homogeneous coordinate ring is toric is such that the homogeneous coordinate ring of its Gauss image in the Plücker embedding is again toric.

9.2.3 Algebras of Veronese type

This particular class of toric algebras was treated in [201]. The definition depends on the following set of data: $r \geq 1$ is an integer and $\mathbf{a} := \{a_1, \dots, a_d\}$ is a sequence of integers satisfying $1 \leq a_1 \leq \cdots \leq a_d \leq r$ and $\sum_{i=0}^{d} a_i > r$. Then the *Veronese algebra of type* $(\mathbf{a}; r)$ is

$$A(\mathbf{a}; r) := k\left[t_0^{e_0} \cdots t_d^{e_d} \mid \sum_i e_i = r,\ 0 \leq e_i \leq a_i,\ 0 \leq i \leq d\right] \subset k[\mathbf{t}].$$

It is known that $\dim A(\mathbf{a}; r) = d + 1$, for any $d, r, \mathbf{a}$ (cf. [147, (2.2) Corollary]).

The case $d = 1$ is rather familiar, since up to rescaling, $A(\mathbf{a}; r)$ is the homogeneous coordinate ring of a smooth projective monomial curve. We will see below that the Gauss algebra associated to the homogeneous coordinate ring of a projective monomial curve is (up to rescaling) again the homogeneous coordinate ring of a projective monomial curve.

It so happens that the Gauss algebra $\mathbb{G}(A(\mathbf{a}; r))$ is in general not an algebra of Veronese type. One obstruction is the introduction of nonisolated singularities, which can already be seen in the case of the cone over a smooth nonarithmetically Cohen–Macaulay monomial curve in $\mathbb{P}^3$ lying on the quadric surface $x_0x_3 - x_1x_2 = 0$ (the simplest case being that of the cone over the nonarithmetically normal rational quartic).

We need the following numerical matrix result.

Lemma 9.2.6. *For any positive integers $n, r, b_1, \dots, b_n$ such that $\sum_{i=1}^{n} b_i = nr$ there exists an integer matrix (e_{ij}) such that $\sum_{i=1}^{n} e_{ij} = b_j$ for all j, $\sum_{j=1}^{n} e_{ij} = r$ for all i and* $\det(e_{i,j}) \neq 0$.

Proof. We actually show that there exists in fact an upper triangular square matrix (e'_{ij}) and a permutation σ of $\{1, \dots, n\}$ satisfying $\sum_{i=1}^{n} e'_{ij} = b_{\sigma(j)}$ for all j, $\sum_{j=1}^{n} e'_{ij} = r$ for all i and $\det(e'_{i,j}) \neq 0$.

Proceed by induction on n, noting there is nothing to prove for $n = 1$. Let $n \geq 2$. If $b_j \leq r$ for every j, then r times the identity matrix will do it as $b_j = r$ for all j. Thus, we may assume that $b_j > r$ for at least one index. Say, up to a permutation, $b_1 > r$ and $b_n < r$. Set $b_1' = b_1 - (r - b_n)$ and $b_j' = b_j$ for $j = 2, \ldots, n-1$, so $b_1' > 0$. Applying the inductive hypothesis, there is an $(n-1) \times (n-1)$ upper triangular matrix (e_{ij}'') and a permutation σ of $\{1, \ldots, n-1\}$ with the required properties. Let $1 \leq k \leq n-1$ be such that $\sigma(k) = 1$. Then one can take for $(e_{i,j}')$ the bordered matrix

$$\begin{pmatrix} b_n & 0 & \cdots & 0 & r - b_n & 0 & \cdots & 0 \\ 0 & & & & & & & \\ \vdots & & & & (e_{ij}'') & & & \\ 0 & & & & & & & \end{pmatrix},$$

where $r - b_n$ sits on the $(k+1)$th position. It is now straightforward to check the required properties of $(e_{i,j}')$. □

Below we let $(\mathbf{t})_r$ denote the set of monomials of degree r.

The most celebrated instance of this class is given by the usual Veronese algebras. For given $r \geq 0$, the rth Veronese algebra is $k[(\mathbf{t})_r]$, the k-subalgebra generated by the monomials of degree r in $\mathbf{t}$. Clearly, $k[(\mathbf{t})_r] = A(\mathbf{a}; r)$, where $\mathbf{a} = \{r, \ldots, r\}$.

Proposition 9.2.7. *The Gauss algebra associated to a Veronese algebra is a Veronese algebra; more precisely,* $\mathbb{G}(k[(\mathbf{t})_r]) = k[(\mathbf{t})_{(d+1)(r-1)}]$.

Proof. One has to establish that the set of nonzero $(d+1) \times (d+1)$ minors of the Jacobian matrix of the set of all monomials of degree r coincides, up to multiplication by a nonzero element of k, with the set of all monomials of degree $(d+1)(r-1)$. Now, quite generally by Lemma 9.2.5, any such $(d+1) \times (d+1)$ minor is a monomial (with some coefficient from k). In the present case, we show that one can take $t_0 \cdots t_d$ as a common multiplier for these minors into $k[\mathbf{t}]$ (see Exercise 9.5.9) generated by a subset of terms of degree $(d+1)(r-1)$. Indeed, say, after renumbering, $\mathbf{g} = \{g_0, \ldots, g_d\}$ is a set of $d+1$ terms of degree r. Set $g_i = \mathbf{t}^{a_i}$, where $a_i = (a_{i0}, \ldots, a_{id})$ is such that $|a_i| = r$ for $i = 0, \ldots, d$. Then one has $t_0 \cdots t_d \det(\Theta(\mathbf{g})) = g_0 \cdots g_d \det(a_{ij})$, where $\Theta(\mathbf{g})$ denotes the Jacobian matrix of the elements $\mathbf{g}$. But any variable t_i divides at least one term in the set of any $d+1$ (distinct) terms of degree r. Therefore, $t_0 \cdots t_d$ divides $g_0 \cdots g_d$ and the result of the division, which is $\det(\Theta(\mathbf{g}))$ up to a constant from k is a term of degree $(d+1)r - (d+1) = (d+1)(r-1)$.

Conversely, let $\mathbf{t}^c$ be any term of degree $(d+1)(r-1)$, with $c = (c_0, \ldots, c_d)$. Applying the above lemma with $b_i = c_i + 1$ and $n = d+1$ to find a matrix such as (e_{ij}), one next sets $g_i = t_0^{e_{i0}} \cdots t_d^{e_{id}}$ and $\mathbf{g} = \{g_0, \ldots, g_d\}$. Then g_i has degree r. Moreover, one gets $t_0 \cdots t_d \det(\Theta(\mathbf{g})) = \det(e_{ij}) g_0 \cdots g_d = \det(e_{ij}) t_0 \cdots t_d \mathbf{t}^c$ (the latter because $\sum_{i=0}^d e_{ij} = b_j$ for all j). It follows that $\mathbf{t}^c = (\det(e_{ij}))^{-1} \det(\Theta(\mathbf{g}))$, as was to be shown. □

For additional recent results on the Gauss algebra of special toric varieties, see [93].

9.3 Abstract Gauss maps

In order to understand how abstract Gauss maps come about let us review the usual Gauss map as understood from Corollary 9.1.6 and Remark 9.1.7.

Fix a standard graded domain $R = k[R_1] = k[x_0,\dots,x_n]$ over an algebraically closed field k and let $\mathfrak{m} = (R_1) \subset R$ denotes its irrelevant maximal ideal. Write $R \simeq k[\mathbf{X}]/P$, with $P \subset k[\mathbf{X}] = k[X_0,\dots,X_n]$ a homogeneous prime ideal. Recall the conormal exact sequence of the module of k-differentials of R [179, Proposition 4.2.8]:

$$R^m \xrightarrow{\Theta^t} \sum_{i=0}^{n} R dX_i \longrightarrow \Omega_{R/k} \to 0, \tag{152}$$

where Θ^t denotes the transposed Jacobian matrix of a set of homogeneous generators of P, read on R. Here, $\operatorname{rank}\Omega_{R/k} = \dim R = n+1-g$, where $g = \operatorname{ht} P = \operatorname{rank}\Theta^t$. One can choose a set of homogeneous generators of P in such a way that the first g of these form a regular sequence (see Section 9.2). Note that, in particular, the image of the map represented by the first g columns of Θ^t coincides generically with the image of the map represented by the entire Θ^t.

Then Corollary 9.1.6 and Remark 9.1.7 say that the Gauss map of $X = \operatorname{Proj}(R)$ composed with the Plücker embedding has a representative whose coordinates are the maximal minors of the first g columns of Θ^t.

With this in mind, it becomes quite obvious what ought to be an abstract Gauss map. Namely, let M be a finitely generated graded R-module of rank r. As is the case of $\Omega_{R/k}$, assume that M is generated by homogeneous elements of the same degree.

Consider a graded-free presentation $0 \to Z \to F \to M \to 0$, where it is understood that $F \to M$ is a degree-preserving map. Say, $n = \operatorname{rank} F = \mu_R(M)$. By a k-linear combination of the minimal generators of Z, one can find a free submodule $Z' \subset Z$ generated by $g = \operatorname{rank} Z = n-r$ elements coinciding generically on R (i. e., at the level of the fraction field of R). Note that the restricted map $Z' \longrightarrow F$ is injective. By abuse, a minor of the matrix defining this inclusion will be called a minor of Z'.

Definition 9.3.1. The *(abstract) M-Gauss map* is the rational map

$$\Gamma_M : X = \operatorname{Proj}(R) \dashrightarrow \mathbb{P}^{\binom{n}{g}-1} = \mathbb{P}^{\binom{n}{r}-1}$$

given by a representative whose coordinates are the g-minors of Z' as above.

Proposition 9.3.2. *With the above notation, the homogeneous coordinate ring of the image of γ_M in $\mathbb{P}^{\binom{n}{r}-1}$ is the fiber cone*

$$\mathcal{R}_R\Big(\bigwedge^r M\Big) \otimes_R R/\mathfrak{m}$$

Proof. Consider the above graded free presentation $0 \twoheadrightarrow Z \twoheadrightarrow F \twoheadrightarrow M \twoheadrightarrow 0$ and the free submodule $Z' \subset Z$. The surjection $M' := F/Z' \twoheadrightarrow F/Z \simeq M$ is obtained by factoring out a torsion R-module, so similarly $\bigwedge^r M' \twoheadrightarrow \bigwedge^r M$ has an R-torsion kernel, thus implying that the induced R-module map $\mathcal{R}_R(\bigwedge^r M') \twoheadrightarrow \mathcal{R}_R(\bigwedge^r M)$ at the level of Rees algebras is an isomorphism. Thus without loss of generality, we may assume that $Z' = Z$.

Let $I \subset R$ denote the base ideal of Γ_M corresponding to the representative whose coordinate are the minors of Z'. Then the homogeneous coordinate ring of the image of Γ_M is the fiber cone of $\mathcal{R}_R(I)/\mathfrak{m}\mathcal{R}_R(I)$ of I (up to degree renormalization). To conclude, it remains only to show that $\mathcal{R}_R(I) \simeq \mathcal{R}_R((\bigwedge^r M))$. But the injective map $\varphi : Z \longrightarrow F$ induces a map

$$\bigwedge^r F \simeq \Big(\bigwedge^g F\Big)^* \stackrel{(\wedge^g \varphi)^*}{\longrightarrow} \Big(\bigwedge^g Z\Big)^* \simeq R$$

whose image is the ideal I. This map factors through $\bigwedge^r M$, and thus induces a surjection $\bigwedge^r M \twoheadrightarrow I$ of R-modules of same rank. The latter induces a surjective map $\mathcal{R}_R((\bigwedge^r M)) \twoheadrightarrow \mathcal{R}_R(I)$ of R-algebras of the same dimension, hence an isomorphism since R is a domain and $\mathcal{R}_R((\bigwedge^r M))$ is R-torsion-free. □

Thus, the dimension of the Gauss image of γ_M is controlled by the analytic spread of the top rank module $\bigwedge^r M$. This gives a motivation to look closer at the analytic spread with its built in geometric facet. The subsequent sections are dedicated to this objective.

9.4 The geometry of the analytic spread

9.4.1 Maximal analytic spread

For the preliminaries and notation regarding the analytic spread, see Section 2.3.2.3.

The relevance of the analytic spread cannot be exaggerated. In the local case, it measures the dimension of the fiber cone (central fiber in complex analytic geometry), while in the equigenerated graded case it gives one plus the dimension of the image of a rational map of projective varieties. In strict commutative algebra, it gives the minimal number of generators of a minimal reduction of an ideal in a local ring with infinite residue field.

In this section, we focus on aspects of the virtual maximality of the analytic spread. The material is based on [186].

9.4.1.1 Geometric approach

Throughout A denotes a Noetherian ring (as a change from previous parts, where we used the letter R instead of A; here, A is a slight mnemonic for "analytic" (spread)).

Suppose that A is a standard graded k-algebra, where k is an algebraically closed field. Set $X = \mathrm{Proj}(A) \subset \mathbb{P}^n_k$. An important case of a rational map $X \dashrightarrow \mathbb{P}^m$ is one defined by the sections of a line bundle on X. Algebraically, in the setup of Chapter 3, such a map is given by a representative $\mathbf{f} \subset R$ whose corresponding base ideal $I = (\mathbf{f}) \subset A$ is a codimension one (i. e., height one) ideal locally principal on the punctured spectrum of A.

Under this environment, the following question arises.

Problem 9.4.1. When does the image of such a map attains its maximum possible dimension $\min\{\dim X, m\}$?

The answer to the present question is negative in general. We thank Fancesco Russo for suggesting taking for X the projection of the 2-Veronesean embedding of $\mathbb{P}^n$ from a point on it and, for $I \subset A$, a suitable ideal generated by linear forms.

Next is the precise algebraic counterpart, with a couple of additional features.

Proposition 9.4.2. *Let k a field and let $n \geq 1$ be a fixed integer. Let B denote the determinantal ring $k[\mathbf{x}] = k[\mathbf{X}]/I_2(M_{n+1})$, where $M_{n+1} = (X_{ij})$ denotes the $(n+1) \times (n+1)$ generic symmetric matrix in the variables $\mathbf{X} = \{X_{ij} \mid 1 \leq i \leq j \leq n+1\}$ and $I_2(M_{n+1})$ denotes its ideal of 2-minors. Let $A \subset B$ denote the k-subalgebra generated by the residues of all variables except $X_{n+1,n+1}$. Then:*

(a) *There is a k-algebra presentation*

$$A \simeq k[\mathbf{X} \setminus \{X_{n+1,n+1}\}]/I_2(N_{n+1}),$$

where N_{n+1} denotes the semisymmetric matrix obtained from M_{n+1} by omitting its last row and $I_2(N_{n+1})$ denotes its ideal of 2-minors.

(b) *A is a normal Cohen–Macaulay isolated singularity of dimension $n+1$, and it is Gorenstein if and only if $n = 1$ or $n = 3$.*

(c) *Let $I \subset A$ denote the ideal generated by the residues of the $\binom{n+1}{2}$ entries X_{ij} for $1 \leq i \leq j \leq n$. If $n \geq 2$, then I is a nonprincipal codimension one ideal, locally principal on the punctured spectrum of A such that A/I is a regular ring. Moreover, $\mu(I) = \binom{n+1}{2}$ and $\ell(I) = n < \min\{\dim A, \mu(I)\} = n+1$.*

Proof. (a): We take the degree lexicographic term order on $k[\mathbf{X}]$ with the ordering of the variables given by $X_{ij} < X_{hl}$ if $j < l$ or if $j = l$ and $i < h$. By [40, 2.9], the 2-minors of M_{n+1} form a Gröbner basis of the ideal they generate. Thus the 2-minors not involving $X_{n+1,n+1}$ are a Gröbner basis of the elimination ideal $k[\mathbf{X} \setminus \{X_{n+1,n+1}\}] \cap I_2(M_{n+1})$ and it follows that this ideal coincides with $I_2(N_{n+1})$.

(b) and (c): First, note that the inclusion $A \subset B$ is birational since, for example, in the field of fractions of B one has $x_{n+1,n+1} = x_{n,n+1}^2/x_{n,n}$. In particular, $\dim A = \dim B = n+1$.

Next, recall that B is a Veronese subring of the polynomial ring over k in $n+1$ variables and notice that A is a monomial subring of B, hence can be regarded as a monomial subring of the same polynomial ring. But the proof of part (a) showed that the initial ideal of the toric ideal of A is generated by square-free monomials. By [202, 13.15], A is then normal, hence also Cohen–Macaulay according to [102, Theorem 1].

To argue that A is Gorenstein if and only if $n = 1$ or $n = 3$, we may obviously assume that $n \geq 2$. The result is then given in [42, 2.4(b)] as applied to the case when the matrix has size $n \times (n+1)$ and $t = 2$ (notation as in *loc. cit.*).

Next, note that we chose $I \subset A$ to be generated by the (residues of the) variables—call them $\widetilde{\mathbf{X}}$—of the initial submatrix M_n of M_{n+1}. The ideal generated by these variables clearly contains $I_2(N_{n+1})$, hence A/I is a regular ring of dimension n and, from part (b), the ideal I has codimension $(n+1) - n = 1$.

Now we prove that I is principal locally on the punctured spectrum of A. To this end, let $P \subset A$ be a prime ideal containing I, but not the homogeneous maximal ideal. We may assume that, say, $x_{1,n+1} \not\in P$ and then some of the determinantal relations in the ring A readily show that $I_P = (x_{11}, \dots, x_{1n})_P$. Again using some of the relations one sees that $x_{1j} \in x_{11}A_P$ for $1 \leq j \leq n$. Thus I_P is indeed principal.

It will now readily follow that A is an isolated singularity. In fact, let P be a prime ideal of A not coinciding with the homogeneous maximal ideal. If $I \subset P$, then by the above I_P is generated by a single regular element. As A_P/I_P is a regular ring, A_P is regular as well. If $I \not\subset P$, the determinantal relations on the Veronese ring B show that $x_{n+1,n+1} \in A_P$, hence $A_P = B_P$. Since $(\mathbf{X})B_P = B_P$ and B has an isolated singularity, the result follows in this case, too.

It remains to argue for the value of the analytic spread. But the latter is the dimension of the k-subalgebra,

$$C := \frac{k[\widetilde{\mathbf{X}}]}{I_2(N_{n+1}) \bigcap k[\widetilde{\mathbf{X}}]} \subset \frac{k[\mathbf{X} \setminus \{X_{n+1,n+1}\}]}{I_2(N_{n+1})},$$

and, we of course claim that it is just $k[\widetilde{\mathbf{X}}]/I_2(M_n)$ again. By the discussion in the proof of (a), we know that the 2-minors of N_{n+1} form a Gröbner basis of the ideal $I_2(N_{n+1})$. Therefore, the contraction of this ideal to the variables $\widetilde{\mathbf{X}}$ is generated by those 2-minors of N_{n+1} involving those sole variables. Thus indeed $\dim C = n$. □

Remark 9.4.3. We note that because of the presentation in (a), the ring A falls within the class of determinantal rings of semisymmetric matrices studied by Conca [42]. In particular, it is a normal Cohen–Macaulay domain. Since A is but a special case in this class, we thought a direct argument to show these two properties was desirable. We note that the divisor class group of A is $\mathbb{Z}$ and the class of the above ideal I is the canonical class of A [42, 2.4(b)]. This strongly corroborates the idea mentioned in Remark 9.4.7 below.

9.4.1.2 Analytic spread of a height one ideal

On the algebraic side, the image of the rational map as above is the subvariety whose homogeneous coordinate ring is the fiber cone $k[\mathbf{f}] \simeq \mathcal{F}(I) = k \otimes_A \mathcal{R}_A(I)$ up to degree normalization. Since the analytic spread $\ell(I)$ of I is $\dim \mathcal{F}(I)$, one is asking when $\ell(I)$ attains the maximum value $\min\{\dim A, \mu(I)\}$.

Asking when the analytic spread is maximal is natural and challenging.

Problem 9.4.1 admits the following purely algebraic version.

Problem 9.4.4. Let A be a Noetherian local ring and let $I \subset A$ be an ideal of height one, containing a regular element, and principal locally on the punctured spectrum. When is $\ell(I) = \min\{\dim A,\ \mu(I)\}$?

Next are some affirmative answers.

The case where $\dim A = 2$ is pretty much under control.

Proposition 9.4.5. *Let A be a Cohen–Macaulay local ring of dimension at most 2. Then* Problem 9.4.4 *has an affirmative answer.*

Proof. We may assume that I is not principal. Suppose that $\ell(I) < \dim A = 2$. Then $\ell(I) = \operatorname{ht} I$, i. e., I is equimultiple, and since I is in particular principal locally at its minimal primes of minimal height, it is generated by a regular sequence according to the main theorem of [48]. Therefore, I is principal, a contradiction. □

It would be interesting to study the non-Cohen–Macaulay case, as there are such examples for which Problem 9.4.4 has an affirmative answer (see Exercise 9.5.13).

In higher dimensions, one has the following results.

Theorem 9.4.6. *Let A be a Gorenstein local ring of dimension 3 with infinite residue field and let $I \subset A$ be an ideal of height one. Assume that I is principal locally on the punctured spectrum, and that there exists a regular element $x \in I$ and a positive integer n such that:*

(i) *x generates I at its minimal primes and is part of a minimal reduction of I.*
(ii) *the unmixed component of $A/(x, I^n)$ is Cohen–Macaulay.*

Then Problem 9.4.4 *has an affirmative answer for I.*

Proof. Let J^{unm} denote the unmixed part of an ideal J and let " $\bar{\ }$ " indicate residues modulo $(x){:}I^n$. We may assume that the latter ideal is proper as otherwise $I^n \subset (x)$ would imply that I is principal by the choice of x as in (i). Now, since A is a Gorenstein ring and $A/(x, I^n)^{\mathrm{unm}}$ is Cohen–Macaulay by (ii), linkage theory shows that $\bar{A} = A/(x){:}I^n = A/(x){:}(x, I^n)^{\mathrm{unm}}$ is again a Cohen–Macaulay ring (of dimension two). By (i), the ideal $I + ((x){:}I^n)$ has height at least two, hence $\operatorname{ht} \bar{I} \geq 1$.

Now, suppose that $\ell(I) < \dim A = 3$. Then $1 \leq \operatorname{ht} \bar{I} \leq \ell(\bar{I})$ and by (i), $\ell(\bar{I}) \leq \ell(I) - 1 \leq 1$, hence in fact $\ell(I) = 2$ and $\bar{I}$ is an equimultiple ideal of height one in $\bar{A}$. Since $\bar{I}$ is principal locally in codimension one and $\bar{A}$ is Cohen–Macaulay, by [48, main theorem] the ideal $\bar{I}$ is principal. In particular, its epimorphic image $I/(x)$ is cyclic and then $\mu(I) \leq 2$. Thus, indeed $\ell(I) = 2 = \min\{\dim A, \mu(I)\}$. □

Remark 9.4.7. Theorem 9.4.6 is best possible if $\dim A = 3$ even if A is a quadric hypersurface with an isolated singularity. This is a general principle that can be used whenever there is an element in the Picard group of the punctured spectrum of A having sufficiently small analytic spread (see Remark 9.4.3). Exercise 9.5.14 illustrates this behavior.

In dimension ≥ 4, very little seems to be known. Here is one affirmative result.

Proposition 9.4.8. *Let $(A, \mathfrak{m})$ be a local complete intersection of dimension ≥ 4. Then* Problem 9.4.4 *has an affirmative answer.*

Proof. Let I be a nonprincipal ideal of grade ≥ 1, locally principal in the punctured spectrum of A. By the so-called Grothendieck–Lefschetz theorem [81], A is parafactorial. Therefore, $I = (x) \cap Q = x(Q : x)$, for some element $x \in A$ and an $\mathfrak{m}$-primary ideal Q. Since I is not principal, $x \notin Q$, and hence $Q : x$ is also $\mathfrak{m}$-primary. But then $\ell(I) = \ell(Q : x) = \dim A$. □

Finally, we include an affirmative result of a different sort.

Theorem 9.4.9. *Let $(A, \mathfrak{m})$ be an equidimensional and universally catenary Noetherian local ring and let $I \subset A$ be an ideal of grade ≥ 1. If I is not a reduction of the saturation $I : (\mathrm{Fitt}_1(I))^\infty$ then* Problem 9.4.4 *has an affirmative answer for I.*

Proof. We are assuming that I is locally principal off $\mathfrak{m}$ and wish to show that $\ell(I) = \min\{\dim A, \mu(I)\}$. Now let

$$A^m \xrightarrow{\varphi} A^n \longrightarrow I \rightarrow 0$$

stand for a minimal presentation of I. Since $\mathrm{Fitt}_1(I) = I_{n-1}(\varphi)$, then $\mathrm{Fitt}_1(I)_P = A_P$ locally on the punctured spectrum of A. Thus, $\operatorname{ht} \mathrm{Fitt}_1(I) = \dim A$. It therefore suffices to show that $\ell(I) \geq \operatorname{ht}(\mathrm{Fitt}_1(I))$. Supposing to the contrary, we prove that J is integral over I, where $J = I : (\mathrm{Fitt}_1(I))^\infty$. For this, it suffices to show integrality locally at any associated prime P of $A/\bar{I}$, where $\bar{I}$ = denotes the integral closure of I. By [139, 4.1], $\operatorname{ht} P \leq \ell(I)$, and since $\ell(I) < \operatorname{ht}(\mathrm{Fitt}_1(I))$, we have $\operatorname{ht} P < \operatorname{ht}(\mathrm{Fitt}_1(I))$. Therefore, $J_P = I_P \subset \bar{I}_P$, as was to be shown. □

Remark 9.4.10. Note that the saturation $I : (\mathrm{Fitt}_1(I))^\infty$ is the largest ideal of A that coincides with I locally on the free locus of I. In the setting of Proposition 9.4.9, it can be replaced by either $I^{**} = \mathrm{Hom}(\mathrm{Hom}(I, A), A)$ or, by $I^{\vee\vee} = \mathrm{Hom}(\mathrm{Hom}(I, \omega_A), \omega_A)$ when

A satisfies condition (S_2) and has a canonical module ω_A, since either module embeds into $I : (\mathrm{Fitt}_1(I))^\infty$.

9.4.2 Analytic spread versus the Fitting ideal

The previous section has enough elements to allow for an extension to modules. And indeed, this extended theory is related to some important theorems.

Let $(A, \mathfrak{m})$ be a Noetherian local ring or a standard graded ring over a field, and let E be a finitely generated A-module having rank r. One deals with the expected inequality $\ell(\bigwedge^r E) \geq \mathrm{ht}(\mathrm{Fitt}_r(E))$, where $\mathrm{ht}(\mathrm{Fitt}_r(E))$ is the codimension of the rth Fitting ideal of E and $\ell(M)$ stands for the analytic spread of a module M. While the first is classically understood as a measure of the size of the nonfree locus of E, the second gives somewhat subtler information regarding the maximal number of analytically independent elements of $\bigwedge^r E$. Establishing lower bounds for analytic spreads is an important task not only in commutative algebra, but also in algebraic geometry, where in the graded case they often give the dimensions of fundamental varieties such as secant and tangential varieties, dual varieties and Gauss images, which are considered in this book.

Problem 9.4.11. Let E be a finitely generated A-module having rank r. If $E/\tau(E)$ is not free, when does the inequality $\ell(\bigwedge^r E) \geq \mathrm{ht}(\mathrm{Fitt}_r(E))$ hold?

Here, $\tau(M)$ denotes the torsion submodule of an A-module M. We observe that if $E/\tau(E)$ is free, the problem has no chance in general—this is clear already in Problem 9.4.4.

Our first task is to put in evidence that Problem 9.4.11 is indeed a natural extension of Problem 9.4.4 in the previous part. We do this by showing that the present module theoretic question eventually boils down to the terms of Problem 9.4.4. Here are the main steps:

- **Torsion killing reduction principle.** If Problem 9.4.11 has an affirmative answer for $E/\tau(E)$, then it is affirmative for E.

Proof. First, note that the assumption of the problem that $E/\tau(E)$ is not free remains unchanged. As for the analytic spreads, the surjective map $E \twoheadrightarrow E/\tau(E)$ induces a surjective map $\bigwedge^r E \twoheadrightarrow \bigwedge^r(E/\tau(E))$ of modules of the same rank. Therefore, $\mathcal{R}(\bigwedge^r E) \simeq \mathcal{R}(\bigwedge^r(E/\tau(E)))$, hence $\ell(\bigwedge^r E) = \ell(\bigwedge^r(E/\tau(E)))$. Finally, the surjection $E \twoheadrightarrow E/\tau(E)$ yields an inclusion of the respective Fitting ideals, which implies the inequality $\mathrm{ht}(\mathrm{Fitt}_r(E/\tau(E))) \geq \mathrm{ht}(\mathrm{Fitt}_r(E))$. □

- **Ideal reduction principle.** If 9.4.11 has an affirmative answer for $\bigwedge^r E/\tau(\bigwedge^r E)$ then it is affirmative for E.

Proof. Let E be a finitely generated A-module with rank r such that $E/\tau(E)$ is not free. Since taking Fitting ideals and forming wedges commute with localization, by [179,

Proposition 2.3.5] the two ideals $\mathrm{Fitt}_r(E)$ and $\mathrm{Fitt}_1(\bigwedge^r E)$ are equal up to radical, hence have the same height. On the other hand, by the same lemma, $\bigwedge^r E/\tau(\bigwedge^r E)$ is not free. Since we are assuming the statement of the problem valid for this module, we get $\ell(\bigwedge^r E/\tau(\bigwedge^r E)) \geq \mathrm{ht}(\mathrm{Fitt}_1(\bigwedge^r E/\tau(\bigwedge^r E))) \geq \mathrm{ht}(\mathrm{Fitt}_1(\bigwedge^r E))$, the latter inequality simply following again from the surjection $\bigwedge^r E \twoheadrightarrow \bigwedge^r E/\tau(\bigwedge^r E)$. Therefore, $\ell(\bigwedge^r E) \geq \mathrm{ht}(\mathrm{Fitt}_r(E))$. □

• **Dimension reduction principle.** If locally on the punctured spectrum of A, Problem 9.4.11 has an affirmative answer for the torsion free rank one module $\bigwedge^r E/\tau(\bigwedge^r E)$, one may suppose that $\bigwedge^r E/\tau(\bigwedge^r E)$ embeds as a nonprincipal height one ideal that contains a regular element and is principal locally on the punctured spectrum.

Proof. Let $\bigwedge^r E/\tau(\bigwedge^r E) \simeq I \subset A$. We may assume that I is not free. Thus, since I has rank one, it cannot be a principal ideal; in particular, $I \subsetneq A$. Again as I has rank one it contains a regular element a. Multiplying I by a we may assume that $\mathrm{ht}\, I = 1$.

Now let $P \in \mathrm{Supp}_A(A/\mathrm{Fitt}_1(I))$ be such that $P \neq \mathfrak{m}$. Then I_P is not A_P-free and, by the assumption on lower dimension, $\ell(I) \geq \ell(I_P) \geq \mathrm{ht}(\mathrm{Fitt}_1(I_P)) = \mathrm{ht}((\mathrm{Fitt}_1(I))_P) \geq \mathrm{ht}(\mathrm{Fitt}_1(I))$, yielding an affirmative answer to the problem. Thus, one may assume that $\mathrm{Supp}\,(A/\mathrm{Fitt}_1(I)) = \mathfrak{m}$. □

Summing up, if one is looking for an affirmative answer to Problem 9.4.11, one may and will assume that the A-module E is a nonprincipal ideal $I \subset A$ of height one, containing a regular element, and principal locally on the punctured spectrum. In this setup, an affirmative answer means that $\ell(I) = \dim A$. As $\ell(I) \leq \mu(I)$, the best one can hope for is that $\ell(I) = \min\{\dim A, \mu(I)\}$. Thus, 9.4.11 boils down to 9.4.4.

9.4.3 Case study: the module of differentials

The goal of this subsection is a detailed study of the previous problem in the case where E is the module of Kähler differentials. For this purpose, we must readjust our lens and focus on the graded case. Throughout, unless otherwise stated, we go back to the older notation with $(R, \mathfrak{m})$ denoting a standard graded ring over a field and its irrelevant maximal ideal.

9.4.3.1 Basic notions

In this part, we briefly recall the intervening concepts.

For the elementary parts of differentials, our source is [179, Chapter 4, Section 4.2].

Let $\Omega_{R/k}$ denote the module of Kähler k-differentials of R. Numerical invariants come at first through the following.

Proposition 9.4.12 ([179, Proposition 4.2.13]). *If R is finitely generated domain over a perfect field k, then* $\mathrm{rank}\ \Omega_{R/k} = \dim R$.

The well-known Jacobian criterion of regularity comes in various forms, one of which is through the freeness of $\Omega_{R/k}$.

Proposition 9.4.13 ([179, Corollary 4.3.5]). *Let R denote a finitely generated reduced equidimensional algebra over a perfect field k. Then R is locally regular everywhere if and only if $\Omega_{R/k}$ is a projective (i. e., locally-free everywhere) R-module.*

Since the nonfree locus of a finitely generated M module with rank r is its Fitting ideal $\mathrm{Fitt}(M) := \mathrm{Fitt}_r(M)$ of order r, one has immediately the following.

Corollary 9.4.14. *Let R be a finitely generated domain over a perfect field k. Then the singular locus of R is* $\mathrm{Fitt}(\Omega_{R/k})$.

Recall the notion of the a-invariant from Section 2.3.4: let $R = \oplus_{i\geq 0} R_i$ be a Noetherian $\mathbb{N}$-graded ring of dimension d with R_0 local, and let $\mathfrak{m}$ denote the irrelevant maximal ideal of R. The *a-invariant* of R is the integer

$$a(R) = \max\{n \mid [H^d_M(R)]_n \neq 0\} \tag{153}$$

where $H^d_{\mathfrak{m}}(R)$ denotes the top local cohomology module of R with respect to $\mathfrak{m}$.

Since $H^d_{\mathfrak{m}}(R)$ is Artinian, $a(R)$ is well-defined. If ω_R is the canonical module of R, then local duality affords the following characterization:

$$a(R) = -\min\{n \mid [\omega_R]_n \neq 0\}.$$

9.4.3.2 The canonical class

The references for this short subsection are [127], [128] and [133]. Never mind if one barely survives through the details of those beautiful manuscripts—here we will exploit but some of the formal behavior of the next concept.

We assume that $(R, \mathfrak{m})$ is a standard graded domain of dimension d over a field and its irrelevant maximal ideal. Assume in addition that k is an infinite field and that the fraction field L of R, is separable over k.

Consider the natural map of R-modules $\mathfrak{c} : \bigwedge^d \Omega_{R/k} \longrightarrow \bigwedge^d \Omega_{L/k}$ induced by localization. We first explain how the graded canonical module ω_R of R can be identified with a submodule $\omega_{R/k}$ of $\bigwedge^d \Omega_{L/k}$ in such a way that the image $\mathfrak{C}_{R/k}$ of $\mathfrak{c}$ is contained in $\omega_{R/k}$, and the inclusion $\mathfrak{C}_{R/k} \subset \omega_{R/k}$ is an isomorphism locally at any prime in the nonsingular locus of R. The resulting map

$$\bigwedge^d \Omega_{R/k} \longrightarrow \omega_{R/k}$$

is called the *canonical class* or *fundamental class*.

When R is normal, ω_R can be defined as the global sections $\omega_{R/k}$ of the sheaf $\bigwedge^d \Omega_{U/k}$ on the smooth locus U of $\mathrm{Spec}(R)$, or equivalently, as the unique reflexive submodule of $\bigwedge^d \Omega_{L/k}$ whose corresponding sheaf on $\mathrm{Spec}(R)$ agrees with $\bigwedge^d \Omega_{U/k}$ on U.

Obviously, each element of $\bigwedge^d \Omega_{R/k}$ maps to a global section of $\bigwedge^d \Omega_{U/k}$, so $\mathfrak{C}_{R/k} \subset \omega_{R/k}$. Clearly, this inclusion is an isomorphism at primes of U. Note also that $\mathfrak{C}_{R/k}$ sits inside $\omega_{R/k}$ as graded R-submodule.

Even when R is not normal, one can naturally interpret the image $\mathfrak{C}_{R/k}$ as a graded R-submodule of ω_R. For example, one can take for ω_R the complementary module $\omega_{R/k} \subset \bigwedge^d \Omega_{L/k}$. The complementary module is still a graded canonical module in the sense that its graded k-dual (i. e., its Matlis dual) is isomorphic as a graded R-module to the top local cohomology module $H^d_{\mathfrak{m}}(R)$, and the map $\mathfrak{C}_{R/k} \subset \omega_{R/k}$ is still an isomorphism locally on the nonsingular locus of R.

This discussion hopefully makes the idea of a canonical class acceptable, if not totally transparent.

We now consider the inclusion $\mathfrak{C}_{R/k} \subset \omega_{R/k}$ of submodules of $\bigwedge^d \Omega_{L/k}$. Fixing a separating transcendence basis $\{y_0, \dots, y_d\}$ for L over k, one has $\bigwedge^d \Omega_{L/k} = L dy_0 \wedge \cdots \wedge dy_d$. Thus, the modules $\mathfrak{C}_{R/k} \subset \omega_{R/k}$ are identified with fractional ideals of R.

9.4.4 Main theorem of abstract Gauss maps

The goal in this part is the following result proved in [181].

Theorem 9.4.15. *Let R be a standard graded domain of dimension d over a field k. Assume that k is algebraically closed in the quotient field of R. If R is not a polynomial ring, then*

$$\ell\Big(\bigwedge^d(\Omega_{R/k})\Big) \geq \operatorname{ht} \operatorname{Fitt}(\Omega_{R/k}).$$

Since $\dim R = \operatorname{rank} \Omega_{R/k}$ by Proposition 9.4.12, the above theorem answers affirmatively Problem 9.4.11 for the module of differentials. Moreover, since $\operatorname{Fitt}(\Omega_{R/k})$ defines the singular locus of R, the asserted inequality is equivalent to the following one: $\ell(\bigwedge^d(\Omega_{R/k})) \geq \operatorname{codim} \operatorname{Sing}(R)$.

Remark 9.4.16. A first proof of this theorem, based on discussion of the senior author with Bernd Ulrich, consisted in using the a-invariant in two situations: one, when R is Cohen–Macaulay instead of being a domain; the other, when R is an isolated singularity in characteristic zero, by drawing upon the Kodaira vanishing theorem. The proof below will be valid without those restrictions thanks to a characteristic $p > 0$ intervention by Karen Smith. The example to follow shows that at least equidimensionality is necessary for the main result to hold in the form of Proposition 9.4.21 below.

Example 9.4.17. Let $J \subset k[X_1,\dots,X_6]$ be the ideal generated by the 2×2 minors fixing the first column of the Hankel matrix

$$\mathbf{X} = \begin{pmatrix} X_1 & X_2 & X_3 & X_4 & X_5 \\ X_2 & X_3 & X_4 & X_5 & X_6 \end{pmatrix}.$$

This ideal is well known to be reduced but not equidimensional, with $I_2(\mathbf{X})$ a prime component. The free resolution of $R = k[X_1,\dots,X_6]/J$ is known [6], from which its regularity is 2. Therefore, $a(R) = -4 = -\dim R$.

In order to prove the main theorem above, we need a couple of lemmas. The first of these concerns the integrality of the extension of fractional ideals $\mathfrak{C}_{R/k} \subset \omega_{R/k}$ explained in the previous part.

Lemma 9.4.18. *Under the present assumptions on R, if $\bigwedge^d \Omega_{R/k} <$ codim Sing(R) then $\omega_{R/k}$ is integral over its submodule $\mathfrak{C}_{R/k}$.*

Proof. Multiplying the inclusion $\mathfrak{C}_{R/k} \subset \omega_{R/k}$ by a nonzero form of R yields an inclusion $J \subset I$ of homogeneous ideals of R. In this setup, one has to prove that I is contained in the integral closure $\bar{J}$ of J. Clearly, it is enough to verify the latter inclusion after localization at each associated prime of $R/\bar{J}$. Since R is universally catenary, [139, 4.1] shows that $\dim R_P \le \ell(\bar{J}) = \ell(J)$, for any associated prime P of $R/\bar{J}$. Since the assumption $\bigwedge^d \Omega_{R/k} < \operatorname{ht}\operatorname{Sing}(R)$ means that $\ell(J) < \operatorname{ht}\operatorname{Sing}(R)$, we then have the inequality $\dim R_P < \operatorname{ht}\operatorname{Sing}(R)$ for any associated prime P of $R/\bar{J}$. This forces any such P to be in the smooth locus of Spec(R). By the previous discussion on the canonical class, $J_P = I_P$ for all such primes, thus implying that $I_P \subset \bar{J}_P$ for any associated prime P of $R/\bar{J}$, as was claimed. □

Lemma 9.4.19. *Under the present assumptions on R, if $\omega_{R/k}$ is integral over $\mathfrak{C}_{R/k}$ then $a(R) \le -d$.*

Proof. Since R is standard graded, $\Omega_{R/k}$ is generated in degree one, hence $\bigwedge^d \Omega_{R/k}$ is generated in degree d. Since the inclusion $\mathfrak{C}_{R/k} \subset \omega_{R/k}$ preserves degrees, if $a(R) > -d$ then by definition of $a(R)$ (9.4.3.1), the canonical module $\omega_{R/k}$ has nontrivial elements of degree less than d. Therefore, $\mathfrak{C}_{R/k} \subset \omega_{R/k}$ cannot be integral. □

As for the proof of Theorem 9.4.15, we may further suppose that k is infinite and that codim Sing$(R) > 0$, in which case k is algebraically closed in the quotient field of R and the latter is separable over k. We further recall the following field theoretic data. For any field and any finitely generated k-algebra R, one says that R is *geometrically reduced* if $k^\infty \otimes_k R$ is reduced, where k^∞ denotes the perfect closure of k. If, moreover, R is a domain then $k^\infty \otimes_k R$ is a domain as well.

The following result is well known (see [225]).

Proposition 9.4.20. *Let R denote a domain of finite type over a field k, with fraction field $Q(R)$. The following are equivalent:*

(i) *R is geometrically reduced and k is algebraically closed in $Q(R)$.*
(ii) *$Q(R)$ is separable over k and k is algebraically closed in $Q(R)$.*
(iii) *$K \otimes_k R$ is a domain for every field extension $K|k$.*

On the face of these equivalences and the previous lemmas, the proof of Theorem 9.4.15 boils down to the following result.

Proposition 9.4.21. *Let $(R, \mathfrak{m})$ denote a standard graded domain of dimension d over a field k. Assume that k is a algebraically closed in the fraction field of R, and that R is geometrically reduced. If R is not a polynomial ring, then $a(R) > -d$.*

The following argument is due to K. Smith. It evolves around [198, Section 2], which assumes characteristic $p > 0$ throughout. It requires a few additional preliminaries.

Let R be a Noetherian ring of characteristic $p > 0$. For an ideal $I \subset R$ and an integer $e > 0$, let $I^{[p^e]}$ denote the ideal of R generated by the p^eth powers of the elements of I.

Let $\mathcal{S} : R \setminus \bigcup \mathfrak{p} \subset R$, where $\mathfrak{p}$ runs through the minimal primes of R, and let $\iota : R \to \mathcal{S}^{-1}R$ denote the canonical map of fractions.

Definition 9.4.22. The *tight closure* of I is the ideal

$$I^* := \bigcap_{e \gg 0} (\iota^{-1}(\mathcal{S}^{-1}I^{[p^e]})).$$

In connection to the above conditions, one has the following.

Theorem 9.4.23 ([198, Theorem 2.4]). *Let R be a geometrically reduced graded domain over a field $k = R_0$ such that k is algebraically closed in $Q(R)$. If $I \subset R$ is a homogeneous ideal whose initial degree is d, then $I^* \cap I_d \subset I$.*

We are now ready for the following.

Proof of Proposition 9.4.21. The module ω_R is the graded Matlis dual of the top local cohomology module $H^d_{\mathfrak{m}}(R)$. Thus, it is sufficient to show that $H^d_{\mathfrak{m}}(R)$ has a nonzero element of degree $-d+1$. After the faithfully flat base change $k \hookrightarrow k(t)$, where t is an indeterminate, we may assume without loss of generality that the ground field of R is infinite. In this case, R admits a homogeneous system of parameters $x_1, \ldots, x_d$ of degree one.

Recall that $H^d_{\mathfrak{m}}(R)$ can be identified with the cokernel of the map

$$\begin{array}{ccc} R_{x/x_1} \oplus \cdots \oplus R_{x/x_d} & \longrightarrow & R_x \\ \left(\dfrac{s_1x_1^t}{x^t}, \ldots, \dfrac{s_dx_d^t}{x^t}\right) & \mapsto & \dfrac{\sum_{i=1}^d (-1)^i s_i x_i^t}{x^t} \end{array}$$

where x denotes the product $x_1 \cdots x_d$. Thus, we represent an element η of $H^d_{\mathfrak{m}}(R)$ by an equivalence class of fractions

$$\eta = \left[\frac{w}{x^t}\right].$$

One can check that the element η is zero if and only if there exists a natural number s such that $x^s w \in (x_1^{t+s}, \ldots, x_d^{t+s})R$.

Since R is not a polynomial ring over k, $\mathfrak{m}$ is not generated by $x_1, \ldots, x_d$, so we can find an element $z \in R \setminus (x_1, \ldots, x_d)$ of degree one. Consider the element

$$\eta = \left[\frac{z}{x}\right]$$

in $H^d_{\mathfrak{m}}(R)$. This element has degree $-d$, provided it is nonzero. So we need only show that $\eta \neq 0$.

Supposing the contrary, there exists s such that $x^s z \in (x_1^{s+1}, \ldots, x_d^{s+1})R$. By the colon-capturing property of tight closure [103, 7.9] this implies that $z \in (x_1, \ldots, x_d)^*$, hence $z \in (x_1, \ldots, x_d)$ by Theorem 9.4.23, which contradicts our choice of z.

This takes care of the case where char $R > 0$. The result now follows in characteristic zero by a reduction device to positive characteristic as described in [105]. □

9.4.4.1 Application: Zak's inequality

The inequality of Theorem 9.4.15 will now be translated in geometric terms. Let R be as in Theorem 9.4.15, with given embedding dimension $n+1$, over an algebraically closed field k, and let $\mathfrak{m} = R_1 \subset R$ denote its maximal homogeneous ideal. Set $X := \operatorname{Proj}(R) \subset \mathbb{P}^n_k$, a reduced and irreducible variety of dimension $d-1$.

Proposition 9.4.24. *The homogeneous coordinate ring of the Gauss image of X in its Plücker embedding is isomorphic to $\mathcal{R}_R(\bigwedge^d \Omega_{R/k}) \otimes_R R/\mathfrak{m}$ as graded k-algebras.*

Proof. By Proposition 9.1.5, it suffices to show that the stated k-algebra is graded isomorphic to the k-subalgebra of R generated by the residues of the maximal minors of the Jacobian matrix of a sufficiently general regular sequence $\{f_1, \ldots, f_c\}$ in the homogeneous defining ideal of X, where $c = \operatorname{codim} R$. This is a special case of Proposition 9.3.2 where the module is $\Omega_{R/k}$. □

Recall the definition of the Gauss image from (9.1.1) in the case of d-planes, where d is the dimension of the given projective variety.

Corollary 9.4.25 ([224, Chapter I, 2.8. Corollary]). *Let $X \subset \mathbb{P}^n_k$ be a reduced and irreducible projective variety which is not a linear subspace, and let $\Gamma(X)$ denote its Gauss image. Then*

$$\dim \Gamma(X) \geq \dim X - \dim \operatorname{Sing}(X) - 1.$$

9.5 Exercises

Exercise 9.5.1. Show that the dual curve to a smooth conic $\mathfrak{C} \subset \mathbb{P}^2$ is again a smooth conic.

(*Hint*: First, note that in this ambient dimension there is only one variety of planes. In particular, the dual curve to $\mathfrak{C}$ is the image of the ordinary Gauss map which is given by the partial derivatives of the irreducible degree two equation defining $\mathfrak{C}$.)

Exercise 9.5.2 (P. Brumatti). Let $f \in R = k[x_0, \dots, x_n]$ be an irreducible form such that:

(i) The partial derivatives of f are algebraically independent.
(ii) f is integral over the k-subalgebra $k[f_{x_1}, \dots, f_{x_n}] \subset R$ generated by the partial derivatives.

Then the dual variety in $\mathbb{P}^n$ of the hypersurface defined by f is nondeficient (i. e., is a hypersurface).

Exercise 9.5.3. Show that the conic $X^2 + Y^2 + Z^2$ is self-dual.

(*Hint*: To avoid computer assistance, use the result of Proposition 9.1.8—the incidence ideal $(I, I_{g+1}(\Theta^t|\mathbf{T}))$ is generated by the equations $TY - UX, TZ - VX, UZ - VY, X^2 + Y^2 + Z^2$ and $I_g(\Theta) = (X, Y, Z)$. Manipulate directly the equations to get $T^2 + U^2 + V^2$ to multiply suitable powers of X, Y, Z inside the incidence ideal.)

Exercise 9.5.4. Let $S(a, b) \subset \mathbb{P}^{a+b+1}$ denote the rational normal scroll surface defined by integers a, b, i. e., the subvariety of $\mathbb{P}^{a+b+1}$ whose homogeneous defining ideal is generated by the 2-minors of the matrix

$$S_{a,b} = \left(\begin{array}{cccc|cccc} x_0 & x_1 & \cdots & x_{a-1} & x_{a+1} & x_{a+2} & \cdots & x_{a+b} \\ x_1 & x_2 & \cdots & x_a & x_{a+2} & x_{a+3} & \cdots & x_{a+b+1} \end{array} \right).$$

Let $a = 1$, $b = 3$ and $I = I_2(S_{a,b})$.

(1) Find a regular sequence of three forms in I that generates I generically.
(*Hint*: Prove that minors with columns [12], [23], [34] works.)
(2) Use the formula of Proposition 9.1.8 to prove that the dual variety to $S(1, 3)$ is a hypersurface F of degree 4.
(3) Show that the Hessian matrix of F has rank 4.
(4) Prove that the polar image of F is a cone with 2-dimensional vertex over the rational normal curve.

Exercise 9.5.5. In the notation of the previous exercise, project $S(2, 4)$ from the point $(0 : 1 : 0 : \cdots :) \notin S(2, 4)$ and let $T(2, 4) \subset \mathbb{P}^6$ denote the projected surface.

With computer assistance:

(1) Find a minimal set of generators of $T(2, 4)$.
(2) Show that $T(2, 4)$ is arithmetically Gorenstein, but not arithmetically normal.

(3) Use the formula of Proposition 9.1.8 to prove that the dual variety to $T(2,4)$ is a hypersurface F of degree 6.
(4) Prove that the homogeneous defining ideal of the polar image of F is the ideal of 2-minors of the catalecticant

$$\begin{pmatrix} x_2 & x_3 & x_4 \\ x_4 & x_5 & x_6 \end{pmatrix}.$$

Exercise 9.5.6. Show that, in the alternative definition of the tangential variety given in this chapter, there is an explicit surjective graded homomorphism from the homogeneous coordinate ring of $T(V)$ to the homogeneous coordinate ring of V (cf. Exercise 10.3.2).

Exercise 9.5.7. Consider the definition of the Gauss map by the first approach. Let Ψ denote the matrix representing the kernel of the map $(k[\mathbf{X}]/I(V))^{n+1} \to (k[\mathbf{X}]/I(V))^{d+1}$ defined by the Jacobian matrix modulo $I(V)$ (note that the submodule generated by the column vectors of Ψ coincides with the module $\mathrm{Der}_k(k[\mathbf{X}]/I(V))$ of k-derivations of the algebra $k[\mathbf{X}]/I(V)$). Let $\delta_0, \ldots, \delta_d$ general vectors in Ψ. Then the defining ideal of $\Gamma(V)$ is the kernel of the surjective homomorphism $k[\mathbf{T}] \to k[\boldsymbol{\Psi}]/k[\boldsymbol{\Psi}] \cap I(V)$, where $\#\mathbf{T} = \binom{n+1}{d+1} - 1$ and $\boldsymbol{\Psi}$ denotes the set of $(d+1)$-minors of $\delta_0, \ldots, \delta_d$. Clarify the meaning of "general" used here.

Exercise 9.5.8. Show that if $V \subset \mathbb{P}^n$ is a smooth variety then the tangential variety to V is isomorphic to $\mathrm{Proj}(\mathbb{E}(\mathcal{S}(\mathbb{D}/\mathbb{D}^2)))$.

(*Hint*: First, show that in this case one can define the tangential variety by using directly a natural incidence correspondence in $\mathbb{P}^n \times V$, rather than in $\mathbb{P}^n \times \Gamma(V)$ as we defined in general. Next, show that this incidence is actually the biprojective variety whose bihomogeneous coordinate ring is $\mathcal{S}(\mathbb{D}/\mathbb{D}^2)$. Finally, change the $\mathbf{T}$ variables into the $\mathbf{X}$ variables to strictly fall back in $\mathbb{P}^n$.)

Exercise 9.5.9. Show that if A is generated by square-free monomials of the same degree in $k[\mathbf{t}]$ then a multiplier giving the injective homomorphism in Remark 9.2.4 can be taken to be the product of all $\mathbf{t}$ variables.

Exercise 9.5.10 (Gauss image of monomial curves). Let $0 < a_1 < \cdots < a_n$ be a finite increasing sequence of integers with $\gcd(a_1, \ldots, a_n) = 1$ and let

$$A = k[t_0^{a_n}, t_0^{a_n - a_1} t_1^{a_1}, \ldots, t_0^{a_n - a_{n-1}} t_1^{a_{n-1}}, t_1^{a_n}] \subset k[t_0, t_1],$$

the homogeneous coordinate ring of a so-called *monomial curve* on parameters t_0, t_1; $\{a_1, \ldots, a_n\}$ is called the *structure sequence* of A.

(1) Show that A has multiplicity a_n and that it is of Veronese type (if and) only if it is the coordinate ring of the rational normal curve in $\mathbb{P}^{a_n}$.

(2) Show that

$$\mathbb{G}(A) = k[\{t_0^{a_n+a_{n-1}-(a_j+a_k)}\, t_1^{a_j+a_k-a_1} \mid 0 \le j < k \le n\}],$$

with the convention that $a_0 = 0$.

(3) Show that the multiplicity of $\mathbb{G}(A)$ is $a_n + a_{n-1} - a_1$.
(*Hint*: Deduce from (2) that $\mathbb{G}(A)$ admits a structure sequence with of the shape $\{a_2 - a_1, \dots, a_n + a_{n-2} - a_1, a_n + a_{n-1} - a_1\}$.)

Exercise 9.5.11. The structure sequence $\{a_1, \dots, a_n\}$ of the previous exercise is *centered* if $a_n = a_{n-i} + a_i$, $1 \le i \le n$.

(1) Show that, for a centered structure sequence, one has

$$2n - 1 \le \operatorname{edim}(\mathbb{G}(A)) \le \binom{n+1}{2} - \left\lfloor \frac{n+1}{2} \right\rfloor + 1,$$

where edim denotes embedding dimension.

(2) Show that the lower bound is attained for the structure sequence $\{1, 2, \dots, n\}$.

Exercise 9.5.12. With the notation of Exercise 9.5.10, let $n = 3$ and consider the structure sequence $\{c < a - c < a\}$, with $\gcd(c, a) = 1$.

(1) Show that $\mathbb{G}(A)$ has multiplicity $2(a - c)$ and embedding dimension 5.
(2) A is Cohen–Macaulay if and only if $\mathbb{G}(A)$ is an isolated singularity.
(*Hint*: Prove that $\mathbb{G}(A)$ is an isolated singularity if and only if $a - c = c + 1$ and use the properties of such monomials curves lying on a smooth quadric.)

Exercise 9.5.13. Let A stand for the localization of the homogeneous coordinate ring $S = k[s^a, s^b t^{a-b}, s^c t^{a-c}, t^a] \simeq k[x, y, z, w] = k[X, Y, Z, W]/P$ of a monomial curve in $\mathbb{P}^3$ at its irrelevant ideal. Show:

(1) S has at most two singular primes of height one, namely, (x, y, z) and (y, z, w).
(2) Assuming that (x, y, z) is a regular prime, the localization $I := (x, y, z)A$ is a non-principal ideal of height one, locally principal on the punctured spectrum of A, having $\ell(I) = 2$.

(*Hint*: Since the prime ideal $I = (x, y, z)$ is regular on A, the parameters of the curve can be taken to be of the form $(s^a, s^b t^{a-b}, s t^{a-1}, t^a)$. By the same token, I is generically a complete intersection. To see the analytic spread, one argues that the special fiber ring of I is $k[T, U, V]/(U^{a-1} - T^{b-1} V^{a-b})$ as all the remaining generators of the presentation ideal of the Rees algebra of I belong to the extension of the maximal ideal of A.)

Exercise 9.5.14. Let A be the localization of $k[x_1, x_2, x_3, x_4] = k[X_1, X_2, X_3, X_4]/(X_1X_4 - X_2X_3)$ at the maximal ideal (x_1, x_2, x_3, x_4), and let $I = (x_1, x_2)^2$. Clearly, $\dim A = 3$ and $\mu(I) = 3$.

(1) Prove that I is principal locally on the punctured spectrum and that $\ell(I) = \dim k[x_1^2, x_1x_2, x_2^2] = 2$.
(2) Justify the assertion that there is no element $x \in I$ satisfying both conditions (i) and (ii) of Proposition 9.4.6.

(*Hint*: $x = x_1^2$ is an element satisfying condition (i), but for no value of n is condition (ii) satisfied—note that for $n = 1$, $A/(x_1, x_2)^2 = k[X_1, X_2, X_3, X_4]/(X_1X_4 - X_2X_3, X_1^2, X_1X_2, X_2^2)$ is unmixed, but not Cohen–Macaulay.)

10 Tangent varieties

10.1 Algebraic Fulton–Hansen theorems

In this section, we prove a far reaching generalization of a result known as the Fulton–Hansen theorem. This remarkable geometric result was proved in [65] (cf. also [66, Corollary 5.5 and ff.] for a more thorough account of its history). Actually, we prove two different purely algebraic theorems that we call Fulton–Hansen-like theorems. Some of it has possibly been superseded in [17], but since they were obtained much earlier we thought it plentifully justified to have them exposed here. The results themselves are unpublished joint work with B. Ulrich. For a recent coverage of the geometric side, as appeared originally, we refer to [167].

The theorem relates the dimension of the secant variety $\mathrm{Sec}(V)$ to an irreducible variety $V \subset \mathbb{P}^n$ to that of the (embedded) variety $\mathrm{Tan}(V)$ of limiting positions of secant lines (in the sense of Whitney and Johnson).

The theorem was subsequently extended by Flenner–Gastel–Vogel (cf. [63, Corollary (3.6)]) to the case of the embedded join of two subvarieties of $\mathbb{P}^n$ connected in codimension one, and to the case of the span of a general irreducible correspondence by van Gastel [69, Proposition 4.1]. We remark that in both cases, one requires the nonemptiness of the locus in question; moreover, a suitable generalization of the variety $\mathrm{Tan}(V)$ is needed ("the span of the cone of principal tangents").

To see how the problem fits our setup, we go back to the notation established at the beginning of Subsection 8.1.4: let $(R, \mathfrak{m}, k)$ be a Noetherian local ring and let $I \subset R$ be a proper ideal. We have seen earlier that there are surjective graded homomorphisms

$$\mathbb{E}(\mathcal{R}_R(I)) \twoheadrightarrow \mathbb{E}(\mathrm{gr}_I(R)) \twoheadrightarrow \mathbb{E}(\mathcal{R}_{R/I}(I/I^2)). \tag{154}$$

(The second map needs a mild condition on I such as requiring that it be generically of linear type).

As we have seen, if $\operatorname{depth} R \geq 1$, then $\mathbb{E}(\mathcal{R}_R(I)) \simeq \mathcal{R}_R(I) \otimes_R k$, and if $\operatorname{depth} R/I \geq 1$ then $\mathbb{E}(\mathcal{R}_{R/I}(I/I^2)) \simeq \mathcal{R}_{R/I}(I/I^2) \otimes_R k$ and $\mathbb{E}(\mathrm{gr}_I(R)) \simeq (\mathrm{gr}_I(R))/(0 : \mathfrak{m}^\infty) \otimes_R k$, where $0 : \mathfrak{m}^\infty$ is short for $0 :_{\mathrm{gr}_I(R)} (\mathfrak{m}\,\mathrm{gr}_I(R))^\infty$. Thus, in the formulation of the following theorems we will stick to the fiber cone versions.

10.1.1 Algebraic preliminaries

For any plausible extension of the Fulton–Hansen theorem, some sort of topological condition on $\mathrm{Spec}(R)$ is required. We introduce this algebraically in the following format.

Definition 10.1.1. A Noetherian ring R is said to be an *Abhyankar–Hartshorne* ring if given any two nonnilpotent ideals $I, J \subset R$ such that IJ is nilpotent, then $\mathrm{ht}(I, J) \leq 1$

https://doi.org/10.1515/9783110640694-010

(equivalently, for every pair of nonzero radical ideals I, J whose intersection is zero, (I,J) has height at most one).

We leave to the reader the translation of the above definition in terms of Spec(R) being a connected topological space in codimension one, that is, such that $\mathrm{Spec}(R) \setminus V(J)$ is connected for every closed subset $V(J)$ with $\mathrm{ht}\, J \geq 2$. This condition is vacuously satisfied if R happens to be a domain. Also, being connected in codimension one is obviously stronger than being connected. The so called Abhyankar–Hartshorne connectedness lemma implies that if R is a ring satisfying Serre's condition S_2, then R is connected in codimension one if and only if it is connected.

Recall that a Noetherian local ring R is said to analytically have a certain property if the completion $\hat{R}$ of R at its maximal ideal has that property. Thus, R is analytically irreducible if $\hat{R}$ is a domain (this is of course imprecise as $\mathrm{Spec}(\hat{R})$ being an irreducible space is weaker than $\hat{R}$ being a domain, but long time usage has beaten exact terminology). It is known that if A is a $\mathbb{Z}_+$-graded ring with A_0 an analytically irreducible local ring and if $R = A_M$, where M is the maximal homogeneous ideal of A, then R is analytically irreducible.

We will often assume that a ring R is analytically connected in codimension one, a stronger condition that plays a relevant role in the main theorems of this section. In this connection, one can state the following general result.

Lemma 10.1.2. *Let A be a $\mathbb{Z}$-graded ring with A_0 a complete local ring and let M be the ideal of A generated by all homogeneous nonunits of A (a prime ideal by [20, 1.5.7]). If the local ring A_M is connected in codimension one, then so is the completion $\widehat{A_M}$.*

Proof. Notice that, since A_0 is local, $M \neq A$ is the unique maximal homogeneous ideal of A making the latter a **local ring* in the sense of [20, 1.5.13]. Furthermore, A is a finitely generated A_0-algebra [20, 1.5.5]. Since A_M is catenary and, by assumption, connected in codimension one, it is equidimensional. Consider a homogeneous presentation

$$0 \to I \to R[\mathbf{X}] \to A \to 0,$$

with R a local Gorenstein ring. Let N denote the maximal homogeneous ideal of $R[\mathbf{X}]$ and set $g = \mathrm{ht}\, I_N$, $W = \mathrm{Ext}^g_{R[\mathbf{X}]}(A, R[\mathbf{X}])$. Then W is a finitely generated $\mathbb{Z}$-graded A-module and we claim that the endomorphism ring $B = \mathrm{End}_A(W)$ is commutative. To see this, notice that the A_M-module W_M is isomorphic to the canonical module ω_{A_M} of A_M and the natural map $A \to B$ is a finite morphism of $\mathbb{Z}$-graded (possibly noncommutative) rings. However, $B \otimes_A A_M \simeq \mathrm{End}_{A_M}(W_M) \simeq \mathrm{End}_{A_M}(\omega_{A_M})$ is commutative and B is a finitely generated graded module over the *local ring (A, M). It follows that B is commutative as well [20, 1.5.15.a].

Next, we claim that B is a *local ring, too. Indeed, since A_M is connected in codimension one, it follows from [104, the proof of 3.3] that $\mathrm{Spec}(B \otimes_A A_M)$ is connected, hence so is $\mathrm{Spec}(B_0)$. Then, being a finitely generated module over the complete local ring A_0, B_0 has to be a local ring, as required. To see what its unique maximal homogeneous ideal of B is, let P be any prime of B contracting to M. Then P is a minimal prime of B/MB, hence is homogeneous. But, since its contraction is maximal homogeneous in A, P has to be, too. Therefore, P is the unique maximal homogeneous ideal of B. It follows that $B \otimes_A A_M$ is a local ring, hence $B \otimes_A \widehat{A_M} \simeq \mathrm{End}_{\widehat{A_M}}(\omega_{\widehat{A_M}})$ is a local ring as well. Then $\widehat{A_M}$ is connected in codimension one by [104, 3.3]. □

Now, another way of rephrasing connectedness in codimension one, which resembles the obstruction in order that two planes meet only in a point, is the following criterion.

Proposition 10.1.3. *A Noetherian ring R is connected in codimension one if and only if for any two minimal primes p, q of R, there exists a sequence of minimal primes $p_1, \ldots, p_n$ of R such that $p_1 = p$, $p_n = q$ and $\mathrm{ht}(p_i, p_{i+1}) \leq 1$ for $1 \leq i \leq n-1$.*

Proof. See [104, the proof of 3.3]. □

Using this criterion, one can readily establish the following result.

Lemma 10.1.4. *Let R be a Noetherian ring and let $I \subset R$ be an ideal. If R is connected in codimension one, then so is the extended Rees algebra $R[It, t^{-1}]$.*

Proof. As is well known, there is a bijection between the set of minimal primes of R and the set of minimal primes of $R[It, t^{-1}]$. Precisely, to a prime $p \in \mathrm{Min}(R)$ corresponds the prime $p' = pR[t, t^{-1}] \cap R[It, t^{-1}] \in \mathrm{Min}(R[It, t^{-1}])$ and the inverse is given by contraction [137, pp. 121–122]. Then, by Proposition 10.1.3, it suffices to show that given $p, q \in \mathrm{Min}(R)$ with $\mathrm{ht}(p, q) \leq 1$, then $\mathrm{ht}(p', q') \leq 1$. But this is obvious since for any prime $P \subset R$ of height ≤ 1, one has $\mathrm{ht}\, P' \leq 1$. □

We are now ready for the main technical result of this section, which extends [16, 2.5].

Proposition 10.1.5. *Let R be a Noetherian local ring and let $I \subset R$ be a proper ideal. If R is analytically connected in codimension one, then the associated graded ring $\mathrm{gr}_I(R)$ is connected in codimension one.*

Proof. Set $G = \mathrm{gr}_I(R)$ and observe that, since $\dim R = \dim G$, the result is vacuous if $\dim R = 1$. Thus, assume that $\dim R \geq 2$. Let N denote the maximal homogeneous ideal of G. Using the obvious implications $\widehat{G_N}$ connected in codimension one $\Rightarrow G_N$ connected in codimension one $\Rightarrow G$ connected in codimension one, we are reduced to show the required statement for $\widehat{G_N}$. In particular, one may assume that R is a complete equidimensional local ring.

Set $A = R[It, t^{-1}]$ and let $M \subset A$ denote its unique maximal homogeneous ideal. By Proposition 10.1.4, A_M is connected in codimension one, hence so is $\widehat{A_M}$ by Lemma 10.1.2. But the latter ring is a complete local ring of dimension $\dim R + 1 \geq 3$. Since $\widehat{G_N} \simeq \widehat{A_M}/(t^{-1})$, with t^{-1} a nonzero divisor on $\widehat{A_M}$, Grothendieck's connectedness theorem [81, Exp. XIII, 2.1] applies to show that $\widehat{G_N}$ is connected in codimension one, as required. □

We now proceed to state a fairly general algebraic formulation of the Fulton–Hansen theorem.

The first result deals with the first of the maps in (154).

Theorem 10.1.6. *Let* $(R, \mathfrak{m}, k)$ *be a Noetherian local ring of dimension* d *and let* $I \subset \mathfrak{m}$ *be an ideal which is not* $\mathfrak{m}$*-primary. Write* $G = \mathrm{gr}_I(R)$ *and* $G' = G/0 : \mathfrak{n}^\infty$*, where* $\mathfrak{n} := \mathfrak{m}\,\mathrm{gr}_I(R)$*. If* R *is analytically connected in codimension one, then one (and only one) of the following two statements hold:*

(i) $G_{\mathrm{red}} = (G')_{\mathrm{red}}$*, in which case* $\dim G \otimes_R k = \dim G' \otimes_R k \leq d - 1$.

(ii) $\dim G \otimes_R k = d$ *and* $\dim G' \otimes_R k = d - 1$.

Proof. The hypotheses and the statements are preserved by passing to the completion of R, hence one may assume that R is complete. Then G is equidimensional, a condition that is known to hold quite generally if R is equidimensional and universally catenary. By definition of G', $\mathrm{grade}\, \mathfrak{n}G' \geq 1$, hence $\dim G' \otimes_R k \leq d - 1$. Now, suppose that statement (i) does not hold. Then the ideal $0 : \mathfrak{n}^\infty$ is not nilpotent. So, let P be a minimal prime of G not containing $0 : \mathfrak{n}^\infty$. But then $\mathfrak{n}G \subset P$. Since G is equidimensional, one has $\dim G \otimes_R k (= \dim G/\mathfrak{n}G) \geq \dim G/P = \dim G = d$. Therefore, $\ell(I) = d$, which is the first assertion of (ii).

For the second assertion, notice that $\mathfrak{n}G$ is not nilpotent either as I is not $\mathfrak{n}$-primary. Therefore, $\mathfrak{n}G$ and $0 : \mathfrak{n}^\infty$ are nonnilpotent ideals whose product is nilpotent. Now, by Proposition 10.1.5, G is connected in codimension one. Thus, $\mathrm{ht}(\mathfrak{n}G + 0 : \mathfrak{n}^\infty) \leq 1$. But $G' \otimes_R k = G/(\mathfrak{n}G + 0 : \mathfrak{n}^\infty)$. It follows that $\dim G' \otimes_R k \geq \dim G - 1 = d - 1$, hence necessarily $\dim G' \otimes_R k = d - 1$ since G is equidimensional. This establishes the second assertion of (ii). □

Theorem 10.1.7. *Let* $(R, \mathfrak{m}, k)$ *be a Noetherian local ring of dimension* d *and let* $I \subset \mathfrak{m}$ *be an ideal that is generically of linear type and satisfies* $\ell(I_p) < \dim R_p$ *for every* $p \in V(I) \setminus (\{\mathfrak{m}\} \cup \mathrm{Ass}\, R/I)$*. Write* $G = \mathrm{gr}_I(R)$ *and* $\mathcal{R} = \mathcal{R}_{R/I}(I/I^2)$*. If* R *is analytically connected in codimension 1, then one (and only one) of the following two statements hold:*

(i) $G_{\mathrm{red}} = \mathcal{R}_{\mathrm{red}}$*, in which case* $\dim G \otimes_R k = \dim \mathcal{R} \otimes_R k \leq d - 1$.

(ii) $\dim G \otimes_R k = d$ *and* $\dim \mathcal{R} \otimes_R k = d - 1$.

Proof. The proof is similar to the one of Theorem 10.1.6, using the ideals $\mathfrak{m}G$ and the R/I-torsion ideal $\tau(G)$. As in [loc. cit.], one may assume that R is complete. Also, since I is assumed to be generically of linear type, we may skip the case where I is $\mathfrak{m}$-primary as in this situation trivially $G = \mathcal{S}_{R/I}(I/I^2) = \mathcal{R}$. Therefore, $\mathfrak{m}G$ is not nilpotent. By Corollary 2.3.19, $\mathcal{R} \simeq G/\tau(G)$. So suppose that statement (i) is false. Then $\tau(G)$ is not nilpotent either. We claim that the product $(\mathfrak{m}G)(\tau(G))$ is nevertheless nilpotent. This will follow from the hypothesis on the local analytic spreads. To wit, let P be any minimal prime of G and let p be its contraction to R. We have to prove that if $p \neq \mathfrak{m}$

then $\tau(G) \subset P$. Setting $A = R/I$, we will actually show that $p \in \operatorname{Ass} A$, which implies $\tau_{A_p}(G \otimes_A A_p) = 0$. To see this, recall that $G \otimes_A A_p$ is equidimensional since R_p is equidimensional and universally catenary. Therefore, $\dim R_p = \dim G \otimes_A A_p = \dim(G/P) \otimes_A A_p \leq \ell(I_p)$. By assumption, $\ell(I_p) < \dim R_p$ if $p \notin \operatorname{Ass} A$. Thus, we must have $p \in \operatorname{Ass} A$, as claimed. □

In most applications it will suffice to assume that R itself has the Abhyankar–Hartshorne property. Theorem 10.1.7 seems to yield new results even in the geometric context.

10.1.2 The underlying geometry

For the rest of this section, k is an algebraically closed field and we will typically focus on a reduced equidimensional standard graded k-algebra $A = k[x_0, \dots, x_n]$ of positive dimension. Some of the considerations below do not require all these restrictions, as one may realize. There is a corresponding scheme $X = \operatorname{Proj}(A) \subset \mathbb{P}^n = \mathbb{P}^n_k$ and we wish to translate the previous algebraic objects into meaningful geometric constructions attached to X. At the end of the section, we also consider the case of two subschemes $X, Y \subset \mathbb{P}^n$ and, more generally, arbitrary correspondences in $\mathbb{P}^n \times_k \mathbb{P}^n$.

The following geometric notion is very useful in the theory of projections.

Definition 10.1.8. Let $V \subset \mathbb{P}^n$ be a projective reduced algebraic set. The *tangential variety* to V (for the given embedding) is the Zariski closure of the set $\cup_p T_{p,V} \subset \mathbb{P}^n$, where p runs through the smooth points of V.

Here, $T_{p,V}$ denotes the embedded projective tangent space to V at p. If the embedding of V is clear from the context, we denote the tangential variety to V by $T(V)$.

It is known (and relatively easy to check) that $T(V)$ is indeed an algebraic set. The next proposition shows this with slightly more precision.

Proposition 10.1.9. *Let $A = k[\mathbf{X}]/I$ be a reduced standard graded k-algebra (k algebraically closed). If $V = \operatorname{Proj}(A)$, then $T(V) = \operatorname{Proj}(\mathcal{R}_A(\mathbb{D}/\mathbb{D}^2) \otimes_A k)$, where $\mathbb{D} = \ker(A \otimes_k A \to A)$.*

Proof. Let $V_{\mathrm{sm}} \subset V$ denote the smooth locus of V. By the Jacobian criterion, $V \setminus V_{\mathrm{sm}}$ is a closed set defined by the Jacobian ideal J_A. Consider the incidence

$$\mathfrak{T} := \{(p, q) \in V_{\mathrm{sm}} \times_k \mathbb{P}^n \mid q \in T_{p,V}\}.$$

Clearly, $q = (q_0 : \cdots : q_n) \in T_{p,V}$ if and only if the vector of these coordinates belongs to the kernel of the $k = k(p)$-linear map defined by the Jacobian matrix $\Theta = (\partial f_j / \partial X_i)$ evaluated at the coordinates of p, where the f_j's are polynomials generating the ideal I of the variety V. This condition then means that $(q_0, \dots, q_n) \cdot \Theta(p)^t = 0$. The Zariski closure Σ of $\mathfrak{T}$ in $V \times_k \mathbb{P}^n$ is a correspondence with bihomogeneous defining ideal

$(I, I_1(\mathbf{T}\cdot\Theta^t)) \subset k[\mathbf{X},\mathbf{T}]$. On the other hand, the factor ring $k[\mathbf{X},\mathbf{T}]/(I, I_1(\mathbf{T}\cdot\Theta^t))$ is isomorphic to the symmetric algebra $\mathcal{S}_A(\Omega_{A/k}) = \mathcal{S}_A(\mathbb{D}/\mathbb{D}^2)$. In order to disregard the component that creeps in Σ due to the singular locus of V, we must saturate with respect to the Jacobian ideal of J_A. The result, by Proposition 8.6.1, is the Rees algebra $\mathcal{R}_A(\mathbb{D}/\mathbb{D}^2)$.

Finally, note that the tangential variety $T(V)$ is obtained by projecting to $\mathbb{P}^n$. This means that the homogeneous coordinate ring of $T(V)$ is the $\mathbf{T}$-elimination algebra of $\mathcal{R}_A(\mathbb{D}/\mathbb{D}^2)$. But the latter is isomorphic to the fiber cone of $\mathcal{R}_A(\mathbb{D}/\mathbb{D}^2)$. This shows the contention. □

Corollary 10.1.10. *Let $A = k[\mathbf{X}]/I$ be a reduced standard graded k-algebra (k algebraically closed). Set $V = \mathrm{Proj}(A)$. Then $T(V)$ is a subvariety of the secant variety* $\mathrm{Sec}(V)$.

Proof. This follows from the considerations in (8.6), using the previous result and the fact that the homogeneous coordinate ring of $\mathrm{Sec}(V)$ is graded-isomorphic to the algebra $\mathbb{E}(\mathcal{R}_{A\otimes_k A}(\mathbb{D}))$. The composition of maps gives a surjective map between the respective homogeneous coordinate rings, hence the required inclusion of projective embeddings. □

Remark 10.1.11. In general, the map $\mathbb{E}(\mathcal{S}_A(\mathbb{D}/\mathbb{D}^2)) \twoheadrightarrow \mathbb{E}(\mathcal{R}_A(\mathbb{D}/\mathbb{D}^2))$ is not injective (not even at the reduced level) and the first of these elimination algebras lacks any interesting geometric meaning. However, if V is smooth then the map is an isomorphism—this is clear as in this case the radical of the Jacobian ideal $\mathfrak{J}_A$ is $\mathfrak{m} = (\mathbf{X})/I$.

The standard graded reduced k-algebra $A = k[\mathbf{X}]/I$ has the *expected star dimension* if

$$\dim \mathcal{S}_A(\mathbb{D}/\mathbb{D}^2) = \dim \mathrm{gr}_{\mathbb{D}}(A\otimes_k A)(= 2\dim A). \tag{155}$$

This notion extends to the case of a k-algebra essentially of finite type. As such, a criterion for when it takes place can be given in terms of A itself.

Proposition 10.1.12. *Assume that the ring $A = k[\mathbf{X}]/I$ is reduced equidimensional. Then $A = k[\mathbf{X}]/I$ has the expected star dimension if and only if* $\mathrm{edim}\, A_p \le 2\dim A_p$ *for every* $p \in \mathrm{Spec}(A)$, *where* edim *stands for embedding dimension.*

Proof. We show that both assertions are equivalent to the inequality

$$\mu(\mathbb{D}_\wp) \le \dim(A\otimes_k A)_\wp, \quad \text{for every } \wp \in V(\mathbb{D}). \tag{156}$$

Claim 1. $\mathrm{edim}\, A_p \le 2\dim A_p$ for every $p \in \mathrm{Spec}(A) \Leftrightarrow$ (156).

Let $p \in \mathrm{Spec}(A)$ denote the prime ideal image of $\wp$ by the canonical multiplication map $A\otimes_k A \longrightarrow A$. By [73], one has

$$\mu(\mathbb{D}_\wp) = \mu(\mathbb{D}_\wp/\mathbb{D}_\wp^2) = \mu(\Omega(A/k)_p) = \mathrm{edim}\, A_p + \mathrm{trdeg}_k A_p/pA_p.$$

It suffices to show that

$$\dim(A \otimes_k A)_{\wp} = 2\dim A_p + \operatorname{trdeg}_k A_p/pA_p. \tag{157}$$

Under the present hypothesis, $\dim A = \dim A_p + \dim A/p = \dim A_p + \operatorname{trdeg}_k A_p/pA_p$, thus yielding

$$\dim A + \dim A_p = 2\dim A_p + \operatorname{trdeg}_k A_p/pA_p. \tag{158}$$

Since $\mathbb{D}_{\wp} = \ker((A \otimes_k A)_{\wp} \longrightarrow A_p)$ and A is assumed to be equidimensional, one has $\dim(A \otimes_k A)_{\wp} = \dim A_p + \operatorname{ht} \mathbb{D}_{\wp}$. Writing $A \otimes_k A = k[\mathbf{X}] \otimes_k k[\mathbf{X}]/(I \otimes 1, 1 \otimes I)$ and P for the lifting of $\wp$, one has $\mathbb{D}_{\wp} = \Delta_P/(I \otimes 1, 1 \otimes I)_P$, where $\Delta = \{X_i \otimes 1 - 1 \otimes X_i \mid 1 \leq i \leq n\}$.

Clearly, $\operatorname{ht}(I \otimes 1)_P = \operatorname{ht}(1 \otimes I)_P = \operatorname{ht} I_P = \operatorname{ht} I$, since A is assumed to be equidimensional. Besides, by [172, Chapitre V, B-6, Théorème 3] (extended to the nonprime case as in [179, Section 2.7.5.1]) one has $\operatorname{ht}(I \otimes 1, 1 \otimes I)_P \leq 2\operatorname{ht} I_P = 2\operatorname{ht} I$. It follows that

$$\begin{aligned}
\operatorname{ht} \mathbb{D}_{\wp} &\geq \operatorname{ht} \Delta_P - 2\operatorname{ht} I = \operatorname{ht} P + n + 1 - 2\operatorname{ht} I \\
&= n + 1 - \operatorname{ht} I + \operatorname{ht} P - \operatorname{ht} I \\
&= \dim A + \operatorname{ht} P - \operatorname{ht} I \geq \dim A.
\end{aligned}$$

Carrying it over in (158), yields $\dim(A \otimes_k A)_{\wp} \geq 2\dim A_p + \operatorname{trdeg}_k A_p/pA_p$.

The reverse inequality $\dim(A \otimes_k A)_{\wp} \leq \dim A + \dim A_p$ is clear.

Claim 2. $\dim \mathcal{S}_A(\mathbb{D}/\mathbb{D}^2) = \dim \operatorname{gr}_{\mathbb{D}}(A \otimes_k A) \Leftrightarrow$ (156).

Letting $p \in \operatorname{Spec}(A)$ denote the prime ideal as above, the Huneke–Rossi formula ([114], also [179, Theorem 7.2.20]) gives

$$\dim \mathcal{S}_A(\mathbb{D}/\mathbb{D}^2) = \sup_{p \in \operatorname{Spec}(A)} \{\mu(\mathbb{D}_{\wp}) + \dim A/p\}.$$

Since $\dim \operatorname{gr}_{\mathbb{D}}(A \otimes_k A) = 2\dim A$ and $\dim \mathcal{S}_A(\mathbb{D}/\mathbb{D}^2) \geq \dim \operatorname{gr}_{\mathbb{D}}(A \otimes_k A)$, the equality is equivalent to having $\mu(\mathbb{D}_{\wp}) \leq 2\dim A - \dim A_p$ for every $\wp$. But by (157) and (158), the right side is $\dim(A \otimes_k A)_{\wp}$, so one is done. □

One would like to introduce, in a similar way, an expected tangential dimension. For that, one has the following result.

Proposition 10.1.13. *If k is an algebraically closed field and A is a reduced standard graded ring over k, then*

$$\dim \mathcal{R}_A(\mathbb{D}/\mathbb{D}^2) \otimes_A k \leq \min\{\operatorname{edim} A_{A_+}, 2\dim A - 1\}.$$

Proof. One can assume that $\dim A \geq 1$. Since $\mathcal{R}_A(\mathbb{D}/\mathbb{D}^2) \otimes_A k$ is the fiber cone of $\mathbb{D}/\mathbb{D}^2$, its dimension is the analytic spread of the module $\Omega_{A/k}$ of Kähler differentials of A. But the latter has rank $\dim A$ under the hypothesis. The result now follows from the expected dimension bounds for the analytic spread (see Corollary 2.3.21). □

One says that A has the *expected tangential dimension* if the above inequality is an equality. In this case, the tangential variety $\mathrm{Proj}(\mathcal{R}_A(\mathbb{D}/\mathbb{D}^2)\otimes_A k)$ is said to have no *deficiency*. Alas, no similar characterization as above is known in this case.

One other important construction attached to $X \subset \mathbb{P}^n$ is its secant variety $\mathrm{Sec}(X)$, defined to be the closure of the union of all secant lines to X in $\mathbb{P}^n$. This is a special case of the embedded join of two subschemes of $\mathbb{P}^n$, to be studied in detail in Chapter 11. In particular, it will be shown that $\mathrm{Sec}(X)$ is identified with $\mathrm{Proj}(\mathcal{R}_{A\otimes_k A}(\mathbb{D})\otimes_A k)$, i. e., with the projective variety whose homogeneous coordinate ring is the fiber cone of the diagonal ideal.

There is a third important construction attached to the geometry of the embedding $X \subset \mathbb{P}^n$ and that is the variety $\mathrm{Tan}(X)$, defined to be the union of all limiting secant lines $\overline{xy}$ to X as the two distinct points $x, y \in X$ approach a point of X. Now, another way to look at this construction is to take the *span* of the cone $\mathrm{Spec}(G)$ of the self-correspondence defined by X in $\mathbb{P}^n \times \mathbb{P}^n$, where $G = \mathrm{gr}_{\mathbb{D}}(A\otimes_k A)$. This is obtained by projecting the cone to the second factor. Therefore, the homogeneous coordinate ring of $\mathrm{Tan}(X)$ is the residue ring $G' = G/0:\mathfrak{n}G^{\infty}$, where $\mathfrak{n}$ is the irrelevant ideal of $A\otimes_k A$, the latter being considered as a graded ring by total degree.

Theorem 10.1.14 (Fulton–Hansen). *If $X \subset \mathbb{P}^n$ is reduced and connected in codimension one then either* $\mathrm{Tan}(X) = \mathrm{Sec}(X)$ *or else* $\dim(\mathrm{Tan}(X)) = \dim(\mathrm{Sec}(X)) - 1 = 2\dim X$.

Proof. One applies Theorem 10.1.6 with the following setup: $R = (A\otimes_k A)_{\mathfrak{n}}$, where A is the homogeneous coordinate ring of X and $\mathfrak{n}$ denotes the irrelevant ideal of $A\otimes_k A$ as explained above. By [104, 3.6 and 3.12], R is analytically connected in codimension one. For the ideal I one takes the ideal $\mathbb{D}_{\mathfrak{n}}$, where $\mathbb{D}$ is the diagonal ideal $\mathbb{D} \subset A\otimes_k A$, which is homogeneous in the total degree grading. The rest consists in recalling that, in this setup, by construction, one has the dimension equalities $\dim(\mathrm{Sec}(X)) = \dim(G\otimes_R k)-1$ and $\dim(\mathrm{Tan}(X)) = \dim(G'\otimes_R k)-1$, where $G = \mathrm{gr}_I(R)$ and $G' = G/0:\mathfrak{m}^{\infty}$, with $\mathfrak{m} = \mathfrak{n}_{\mathfrak{n}}$. One also needs to know that $G\otimes_R k$ is reduced and $G\otimes_R k \simeq G'\otimes_R k$. But this follows because $G\otimes_R k$ embeds into $A\otimes_k A$ (since A is standard graded) and A is reduced. □

10.1.3 The tangential variety to a unirational variety

Proposition 10.1.15. *Let $\mathbb{P}^d \dashrightarrow \mathbb{P}^n$ be a rational map defined by a set $\mathbf{g} \subset k[\mathbf{t}]$ of $n+1$ forms of the same degree and assume that the image $V \subset \mathbb{P}^n$ has dimension d. Let $\mathbf{X}$ be coordinates for $\mathbb{P}^n$. Let $\Delta(\Theta(\mathbf{g}))$ denote the set of maximal minors of the Jacobian matrix $\Theta(\mathbf{g})$ of the elements of $\mathbf{g}$ and let $(\mathbf{X} \mid \Theta(\mathbf{g}))$ denote the augmented matrix. Then the homogeneous defining ideal of the tangential variety $T(V) \subset \mathbb{P}^n$ is*

$$(I_{d+2}(\mathbf{X} \mid \Theta(\mathbf{g}) : \Delta(\Theta(\mathbf{g}))^{\infty})\bigcap k[\mathbf{X}].$$

Proof. The proof is parallel to that of Proposition 9.1.13, taking in account the present parametric data. Thus, e. g., the ideal $I_{d+2}(\mathbf{X} \mid \Theta(\mathbf{g}))$ of $k[\mathbf{X}, \Theta(\mathbf{g})]$ is in fact the bihomogeneous ideal of the incidence correspondence Σ in $\mathbb{P}^n \times \Gamma(V)$ and its extension to the ring $k[\mathbf{X}, \mathbf{t}]$ is the bihomogeneous ideal of the inverse image of Σ in $\mathbb{P}^n \times \mathbb{P}^d$ via the composition of the fibered parametrization $\mathbb{P}^n \times \mathbb{P}^d \to \mathbb{P}^n \times V$ with the fibered Gauss map $\mathbb{P}^n \times V \to \mathbb{P}^n \times \Gamma(V)$.

Algebraically, recall that, in the notation of Proposition 9.2.3, $\Delta(\Theta(\mathbf{f})(\mathbf{g}))$ is proportional to $\Delta(\Theta(\mathbf{g}))$ by an element of the field $k(\mathbf{t})$. Therefore, also the set of $(d+2)$-minors of the matrix $(\mathbf{X} \mid {}^t\Theta(\mathbf{f})(\mathbf{g}))$ is proportional to the set of $(d+2)$-minors of the matrix $(\mathbf{X} \mid \Theta(\mathbf{g}))$. At the outset the respective saturations are proportional by a factor in $k(\mathbf{t})$, which shows that we get the same contraction in $k[\mathbf{X}]$. □

Caveat 10.1.16. • Since one is going to contract all the way to $k[\mathbf{X}]$, it does not matter whether the saturated is taken in the ring $k[\mathbf{X}, \Theta(\mathbf{g})]$ or in its overring $k[\mathbf{X}, \mathbf{t}]$.

• As regards the computational side, there are advantages and disadvantages to the present form as compared to the one in Proposition 9.1.13. We have now replaced relatively simple $(1,1)$-biforms (those that generate Q in [loc. cit.]) by the computationally complex $(d+2)$-minors of $(\mathbf{X} \mid {}^t\Theta(\mathbf{g}))$—though the number of $\mathbf{t}$ variables is substantially smaller than that of the $\mathbf{T}$ variables. On the bright side, we now do not need to compute the defining ideal of the Gauss image which can be a hassle. Finally, what it matters for the saturation is the radical of the ideal of maximal minors of $\Theta(\mathbf{g})$ and this is sometimes a help—though it is in general preferable *not* to compute the radical but rather saturate directly.

10.2 Refined properties of tangent algebras

In the previous section, we have seen the prominence of the diagonal ideal $\mathbb{D}$ in the geometry of tangents and secants, via the introduction of graded algebras associated to $\mathbb{D}$. In this section, we develop further the underlying algebraic constructions involving $\mathbb{D}$.

The following notation will prevail throughout.

Notation 10.2.1. The following will be fixed throughout:

k	a perfect field
R	a ring of fractions of a polynomial ring $k[\mathbf{X}] = k[X_1, \ldots, X_n]$
A	a reduced residue k-algebra of R
$\mathbb{D}$	the kernel of the multiplication map $A \otimes_k A \to A$ (*diagonal ideal*)
$\Omega_{A/k}$	the module of Kähler k-differentials of A
$\mathbb{S}_{A/k}$	the symmetric algebra of $\Omega_{A/k}$
$\mathbb{R}_{A/k}$	the Rees algebra of $\Omega_{A/k}$
$\mathbb{T}_{A/k}$	the associated graded ring $\operatorname{gr}_{\mathbb{D}}(A \otimes_k A)$ of the ideal $\mathbb{D}$

The scheme defined by the algebra $\mathbb{S}_{A/k}$ is called the *Zariski tangent cone* of (the variety defined by) A, while the one defined by $\mathbb{T}_{A/k}$ will be named the *tangent star cone* of (the variety defined by) A. Although we are mostly interested in the case where A is a standard graded k-algebra, it is convenient for various purposes to keep A more flexible.

Remark 10.2.2. The latter terminology is akin to [119] who used the name *tangent star* for the fiber of $\mathbb{T}_{A/k}$ over a closed point $p \in \mathrm{Spec}(A)$. There is a slight potential confusion in terminology since the tangent star is a variant of the more usual Zariski tangent space at p, having originally been introduced by H. Whitney [221] who called it a *tangent cone* at p and named it C_5. He moreover introduced a sequence of six tangent cones at p, along with the somewhat surprising assertion that the present tangent star C_5 is of a lesser importance.

The algebras $\mathbb{S}_{A/k}$ and $\mathbb{T}_{A/k}$ will be called the *Zariski tangent algebra* and the *tangent star algebra*, respectively, according to the terminology introduced in [188].

Using the well-known identification $\Omega_{A/k} \simeq \mathbb{D}/\mathbb{D}^2$, the universal property of the symmetric algebra yields a surjective homomorphism

$$\mathbb{S}_{A/k} \twoheadrightarrow \mathbb{T}_{A/k}. \tag{159}$$

In addition, if $\mathbb{D}$ is an ideal of linear type locally at its associated primes, Corollary 2.3.19 implies that this surjection factors in a commutative diagram of surjective A-algebra maps

$$\begin{array}{ccc} \mathbb{S}_{A/k} & \longrightarrow & \mathbb{T}_{A/k} \\ & \searrow \quad \swarrow & \\ & \mathbb{R}_{A/k} & \end{array}$$

where both downward arrows are induced by killing torsion—the leftmost one since $\Omega_{A/k}$ is generically free because A is reduced.

Though $\mathbb{S}_{A/k}$ and $\mathbb{T}_{A/k}$ have a clear geometric source, for the algebra $\mathbb{R}_{A/k}$ the geometric picture is subtler. Let $\mathfrak{T}_V$ denote the scheme defined by the Zariski algebra $\mathbb{S}_{A/k}$ along with the natural projection $\pi : \mathfrak{T}_V \to V$, where V is the scheme defined by A. Since the singular locus of V is defined by the Jacobian ideal J_A of A, Proposition 8.6.1 implies that the “blowup” scheme $\mathfrak{B}_V$ defined by $\mathbb{R}_{A/k}$ is identified with the closure of $\pi^{-1}(\mathrm{Reg(V)})$ in $\mathfrak{T}_V$.

In particular, $\mathfrak{T}_V$ and $\mathfrak{B}_V$ are the same if V is smooth. In general, the inclusion $\mathfrak{B}_V \subset \mathfrak{T}_V$ has been the object of commutative algebra since long from the point of view of the corresponding surjective algebra map $\mathbb{S}_{A/k} \twoheadrightarrow \mathbb{R}_{A/k}$. Recall that, according to Definition 2.3.17, to say that this surjection is an equality means that the module $\mathbb{D}/\mathbb{D}^2$ of differentials is of linear type.

Definition 10.2.3. (i) A has the *expected star dimension* if

$$\dim \mathbb{S}_{A/k} = \dim \mathbb{T}_{A/k}.$$

(ii) A is *set theoretically starlike linear* if the map (159) induces an equality

$$(\mathbb{S}_{A/k})_{\mathrm{red}} = (\mathbb{T}_{A/k})_{\mathrm{red}}.$$

(iii) A is *starlike linear* if the map (159) induces an equality

$$\mathbb{S}_{A/k} = \mathbb{T}_{A/k}.$$

Obviously, (iii) $\Rightarrow$ (ii) $\Rightarrow$ (i). These relations were discussed in some detail for arbitrary Noetherian rings and ideals (Chapter 2, Subsection 2.3.1.4).

10.2.1 Expected star dimension

The notion in (i) has been introduced earlier in the case where A is a standard graded reduced k-algebra (155). As a preamble, it implies that A is generically reduced, hence reduced if it is also unmixed (in particular, an Artinian algebra A having the expected star dimension must be a product of fields). If, moreover, $A \simeq k[\mathbf{X}]/I$, with I a perfect ideal of codimension two, then (i) can be simply restated by saying that A is reduced and, locally in codimension one, (isomorphic to) a hypersurface ring—in higher codimension, the condition required by (i) is automatically satisfied in this case.

Having the expected star dimension property may be quite elusive, as the following example shows.

Example 10.2.4. Let $A = k[X,Y,Z]/(XY,XZ,YZ)$, a Cohen–Macaulay reduced ring of dimension one. The presentation ideal J of $\mathbb{T}_{A/k}$ over the polynomial ring $A[T,U,V]$ contains the element TUV which is a nonzero divisor on the subideal $J(1)$ generated in degree one (by the linear relations coming from the transposed Jacobian matrix of the generators XY, XZ, YZ). Therefore, $\dim \mathbb{S}_{A/k} > \dim \mathbb{T}_{A/k}$, so A fails to have the expected star dimension. And yet the "cylinder" $A[T] = k[X,Y,Z,T]/(XY,XZ,YZ)$ over A does have the expected star dimension, a geometrically intriguing situation.

The next result gives criteria for the expected star dimension to hold, based on the general development in Chapter 2 and Proposition 10.1.12. For simplicity, we give the proof in the case of finitely generated domain over an algebraically closed field.

Theorem 10.2.5. *Let $A = k[\mathbf{X}]/I$ be a finitely generated domain over an algebraically closed field k. The following conditions are equivalent:*

(i) *A has the expected star dimension.*

(i') $\operatorname{edim} A_p \leq 2\dim A_p$ *for every* $p \in \operatorname{Spec} A$.

(ii) $\mathbb{D}$ *satisfies has the property* (F_1).
(iii) $\operatorname{ht} I_t(\Theta) \geq \operatorname{ht} I - t + 1$, *for* $1 \leq t \leq \operatorname{ht} I$, *where* Θ *denotes the transposed Jacobian matrix of a set of generators of* I *read modulo* I.

Proof. (i) ⇔ (i') This is Proposition 10.1.12.

(i) ⇒ (ii). Since $A = k[\mathbf{X}]/I$ is a finitely generated domain over an algebraically closed field, then $A \otimes_k A$ is also a domain (and finitely generated over k). Therefore, the requirements for $R = A \otimes_k A$ in Proposition 2.3.6 are satisfied and, consequently, (i) and (ii) are equivalent.

(ii) ⇒ (iii). This follows from Lemma 2.3.8 since the property (F_0) is an invariant property of a module E with a rank because it is stated in terms of Fitting ideals. In fact, one knows that it is equivalent to having $\mu(E_P) \leq \operatorname{ht} P + \operatorname{rank} E$, for every $P \in \operatorname{Spec}(R)$. Therefore, one can start with any free presentation of $\mathbb{D}$ and get out with any free presentation of $\mathbb{D}/\mathbb{D}^2$.

(iii) ⇒ (i). Since the assumption means that $\Omega_{A/k}$ has property (F_0), then [194, Theorem 1.1.3] implies that

$$\dim \mathbb{S}_{A/k} = \dim \mathcal{S}_A(\Omega_{A/k}) = \dim A + \operatorname{rank} \Omega_{A/k} = 2 \dim A,$$

since $\operatorname{rank} \Omega_{A/k} = \dim A$ in the present condition on A. But, likewise

$$\dim \mathbb{T}_{A/k} = \dim \operatorname{gr}_{A \otimes_k A}(\mathbb{D}) = \dim A \otimes_k A = 2 \dim A,$$

since k is algebraically closed. □

Remark 10.2.6. The above equivalent conditions also hold under weaker restrictions on A, requiring instead properties such as equidimensionality and equicodimensionality (see also Proposition 10.1.12 in the graded case).

Regarding the implication (ii) ⇒ (iii) above, it is curious to ask if there is a natural set of syzygies of the diagonal ideal $\mathbb{D}$ whose residues modulo $\mathbb{D}$ recover the conormal sequence of the Kähler module $\mathbb{D}/\mathbb{D}^2$. In other words, is there a presentation ideal of $\mathcal{S}_{A \otimes_k A}(\mathbb{D})$ that maps "down" to the usual presentation $\mathbb{S}_{A/k} \simeq A[\mathbf{T}]/I_1(\mathbf{T} \cdot \Theta)$?

To engage, let A be as in Notation 10.2.1 and set $A = R/I = k[\mathbf{X}]/I$. Identify $A \otimes_k A$ with the ring $k[\mathbf{X}, \mathbf{Y}]/(I(\mathbf{X}), I(\mathbf{Y}))$, where $I(\mathbf{X}) = I$ and $I(\mathbf{Y})$ is obtained from I by the substitution $\mathbf{X} \mapsto \mathbf{Y}$. Let $\mathbf{T} = T_1, \ldots, T_n$ denote new variables over $k[\mathbf{X}, \mathbf{Y}]/(I(\mathbf{X}), I(\mathbf{Y}))$ and define a surjective homomorphism

$$\rho : k[\mathbf{X}, \mathbf{Y}, \mathbf{T}]/(I(\mathbf{X}), I(\mathbf{Y})) \to \mathcal{S}_{A \otimes_k A}(\mathbb{D})$$

by the assignment $T_i \mapsto X_i - Y_i \pmod{I(\mathbf{X}), I(\mathbf{Y})}$, where $X_i - Y_i$ is taken in degree 1.

It is well known that the presentation ideal $\ker \rho$ is generated by $\mathbf{T}$-linear polynomials with coefficients in $k[\mathbf{X}, \mathbf{Y}]/(I(\mathbf{X}), I(\mathbf{Y}))$ corresponding to the syzygies of $\mathbb{D}$. Now, collect among these $\mathbf{T}$-linear polynomials those whose $\mathbf{X}$, $\mathbf{Y}$-coefficients actually belong to $\mathbb{D}$ and take their liftings to $k[\mathbf{X},\mathbf{Y},\mathbf{T}]$. Denote by $\mathcal{D}(\mathbf{X}-\mathbf{Y}) \subset k[\mathbf{X},\mathbf{Y},\mathbf{T}]$ the ideal generated by these liftings. We informally call the latter the *syzygetic generators* of the presentation ideal $\ker\rho$.

We now introduce fresh generators, to be called the *Taylor generators* of $\ker\rho$. Namely, for each generator $f_j(\mathbf{X})$ of $I(\mathbf{X})$, consider the Taylor expansion of $f_j(\mathbf{X})$ at the point $\mathbf{Y}$ and collect the polynomial coefficients of the linear terms $X_i - Y_i$ in an arbitrary fashion to get an expression of the form $\sum g_{ij}(X_i - Y_i)$ – modulo $\mathcal{D}(\mathbf{X}-\mathbf{Y})$, it will be immaterial the way one collects them, due to the Koszul relations. Let $\mathcal{T}(I)$ denote the ideal of $k[\mathbf{X},\mathbf{Y},\mathbf{T}]$ generated by the corresponding $\mathbf{T}$-linear forms $\sum g_{ij}T_i$.

The next result says that, at least in characteristic zero, together the two kinds not only generate but are also natural generators.

Proposition 10.2.7. (char $k=0$). *With the above notation:*

(1) *There is a presentation*

$$\mathcal{S}_{A\otimes_k A}(\mathbb{D}) \simeq k[\mathbf{X},\mathbf{Y},\mathbf{T}]/(I(\mathbf{X}), I(\mathbf{Y}), \mathcal{D}(\mathbf{X}-\mathbf{Y}), \mathcal{T}(I)).$$

(2) *The above presentation of $\mathcal{S}_{A\otimes_k A}(\mathbb{D})$ modulo $\mathbb{D}$ yields the conormal presentation of the Zariski tangent algebra by means of the transposed Jacobian matrix.*

Proof. (1) Quite generally, for any ring B and any ideal J, from a presentation

$$0 \longrightarrow Z \longrightarrow B^n \longrightarrow J \longrightarrow 0,$$

one obtains a presentation of the conormal module

$$0 \longrightarrow Z/Z\cap JB^n \longrightarrow B^n/JB^n \longrightarrow J/J^2 \longrightarrow 0.$$

This says that $Z = Z\cap JB^n + L$, where L is generated by lifted generators. Applying to the present situation, with $B = A\otimes_k A$ and $J = \mathbb{D}$, it will suffice to show that the Taylor expansions at $\mathbf{Y}$ of the generators $f_1,\dots,f_m$ of I, read modulo the ideal $(I(\mathbf{X}), I(\mathbf{Y}))$, are liftings of the differentials

$$\sum_{i=1}^{n} \frac{\partial f_1}{\partial X_i} dX_i, \dots, \sum_{i=1}^{n} \frac{\partial f_m}{\partial X_i} dX_i$$

read modulo $I = I(\mathbf{X})$.

Now, for that, in the Taylor expansion

$$f_j(\mathbf{X}) - f_j(\mathbf{Y}) = \sum_{i=1}^{n} \frac{\partial f_m}{\partial X_i}(\mathbf{Y})(X_i - Y_i) + \frac{1}{2} \sum_{1\le i\le k\le n} \frac{\partial^2 f_j}{\partial X_i \partial X_j}(\mathbf{Y})(X_i - Y_i)(X_k - U_k) + \cdots$$

one collects terms as follows:

$$\begin{aligned} f_j(\mathbf{X}) - f_j(\mathbf{Y}) &= \left(\frac{\partial f_j}{\partial X_1} + \sum_{k\geq 2} \frac{\partial f_j}{\partial X_1 \partial X_k}(X_k - U_k) + \cdots \right)(X_1 - U_1) \\ &+ \left(\frac{\partial f_j}{\partial X_2} + \sum_{k\geq 3} \frac{\partial f_j}{\partial X_2 \partial X_k}(X_k - U_k) + \cdots \right)(X_2 - U_2) \\ &+ \cdots . \end{aligned}$$

It is clear that, modulo $\mathbb{D}$, these yield the above differentials.

(2) The second part follows immediately from the preceding proof. □

Before going on to further properties of the tangent algebras, here are a few features of $\mathbb{S}_{A/k}$:

- (Relation to deformation theory) Under suitable conditions, the presentation ideal of $S_{A\otimes_k A}(\mathbb{D})$ is actually generated by the Koszul relations on the generators of $\mathbb{D}$ and the Taylor relations $\mathcal{T}(I)$. This requires that the ideal $\mathbb{D}$ be syzygetic (Definition 8.4.2) or, in the language of deformation functors [132], that $T_2(A|k, A) = 0$.
- (Relation to embedding dimension of affine varieties) Let A stand for a finitely generated algebra over an infinite field k. In [199, Theorem 1], the following estimate is obtained:

$$\operatorname{edim} A \leq \dim S_{A\otimes_k A}(\mathbb{D}).$$

(This is not stated as such but it turns out to be the same statement if grade $\mathbb{D} > 0$). Here, edim A stands for the affine embedding dimension of A, i. e., the least $n \geq 0$ such that there is a surjective k-homomorphism

$$k[X_1, \ldots, X_n] \longrightarrow A.$$

The proof given in [199] hinges on a rather involved geometric argument, so one would naturally wonder if there is a simpler algebraic proof.

We note incidentally that, if A is an equidimensional k-algebra of finite type having the expected star dimension, then the above estimate reduces to $\operatorname{edim} A \leq 2 \dim A + 1$ (Proposition 10.1.12 and [179, Proposition 7.2.15]), which is a classical result in the case A is smooth. Thus, it seems even more natural to ask for a direct algebraic proof of this inequality for algebras A having the expected star dimension. It would suffice to show that, for such algebras, the following upper bound always works:

$$\operatorname{edim} A \leq \sup_{\mathfrak{m}\in \operatorname{Max} A} \{\operatorname{edim} A_{\mathfrak{m}}\} + 1.$$

10.2.2 Set theoretic starlike linearity

As a special case off the diagram (5), one knows that A is set theoretically starlike linear if and only if $\mathbb{D}$ is locally everywhere generated by analytically independent elements.

We ask: if the diagonal ideal $\mathbb{D} \subset A \otimes_k A$ satisfies property (F_1) then when is it locally everywhere generated by analytically independent elements? This sort of question has been brought up earlier in greater generality (Question 2.3.5). In order words, when expected star dimension implies set theoretical starlike linearity? This is was a question by van Gastel in [69], in the geometric landscape.

The overall question in this part is to what extent expected star dimension already triggers starlike equality at the set level. Since this is such a general question about any surjective homomorphism of two algebras of the same dimension, one wonders what specifics could be brought over in the present context.

If $A = k[\mathbf{X}]/I$ (of finite type) but I not required to be a domain, counterexamples are easy, many of which are Cohen–Macaulay. Some of these are in the exercises. By a process of double linkage, introducing new variables, it is possible to get counterexamples with A a domain as well. Instead we wish to bring up some more natural examples upon developing a little more the theory.

10.2.2.1 The projecting cone of the Veronese surface

As is well known, the projecting cone of this surface is ideal-theoretically defined by the 2×2 minors of the generic symmetric matrix

$$\begin{pmatrix} X_1 & X_2 & X_3 \\ X_2 & X_4 & X_5 \\ X_3 & X_5 & X_6 \end{pmatrix}.$$

Let $R := k[\mathbf{X}] = k[X_1, \ldots, X_6]$ and $A := R/I$, where I is the ideal generated by these minors. Also, A is a normal Cohen–Macaulay isolated singularity of dimension 3.

First, by Theorem 10.2.5(ii), A has the expected star dimension since it is an isolated singularity (alternatively, by a computation with [79] using (iii) of the same proposition).

Let $\mathbf{T} = T_1, \ldots, T_6$ be presentation variables for $\mathbb{S}_{A/k}$ (or $\mathbb{T}_{A/k}$) over A and let $D(\mathbf{T})$ stand for the determinant of the above matrix evaluated on the variables $\mathbf{T}$. A classical argument shows that $D(\mathbf{T})$ belongs to the presentation ideal of $\mathbb{T}_{A/k}$ over A, thus yielding a relation of analytic dependence.

The Veronese falls within a way more general intriguing phenomenon.

Proposition 10.2.8. *Let k be a field, let $R = k[\mathbf{X}]$ and let $\mathfrak{M}$ be an $r \times r$ matrix of linear forms with nonzero determinant. Then, for any ideal $I \subset (\mathbf{X})^2$ containing the ideal of minors $I_{\lfloor (r+1)/2 \rfloor}(\mathfrak{M})$, the ring R/I is not set theoretically starlike linear.*

Proof. Expanding the determinant of the $r \times r$ matrix $\mathfrak{M} \otimes 1 - 1 \otimes \mathfrak{M}$ by Laplace, using multilinearity and the assumption that the $[(r+1)/2] \times [(r+1)/2]$ minors of $\mathfrak{M}$ belong to the given ideal I, one finds

$$\det(\mathfrak{M} \otimes 1 - 1 \otimes \mathfrak{M}) \subset I_{\lfloor (r+1)/2 \rfloor}(\mathfrak{M}) \otimes R + R \otimes I_{\lfloor (r+1)/2 \rfloor}(\mathfrak{M}) \subset I \otimes R + R \otimes I,$$

which yields a relation of analytic dependence. □

10.2.3 Starlike linearity

From the strict geometric side, the interest lies in the crossing over from expected star dimension to set theoretical starlike linearity. On the algebraic side, it often comes as easy to cross straight to starlike linearity due to the homological tool in the background. This was the guiding principle in [188]. For the reader's convenience, we give here an encapsulated view of the main results.

The basic actors are the approximation complexes and the sliding depth property. Acyclicity of the first typically follow by applying the second in conjunction with the properties (F_i), with $i = 1, 2$. Sliding depth was introduced in the early 1980s. Since it is going to be used in other parts, it seems reasonable to state its definition.

Thus, let R denote a Cohen–Macaulay Noetherian ring and let $I = (\mathbf{a}) \subset R$ be an ideal, with $\mathbf{a} = \{a_1, \ldots, a_n\}$.

Definition 10.2.9 ([96, 97]). I is said to satisfy the *sliding depth condition* if

$$\operatorname{depth} H_i(\mathbf{a}; R) \geq \dim R - n + i, \tag{160}$$

for all $i \geq 0$.

Here, by $H_i(\mathbf{a}; R)$ we mean the ith Koszul homology module of $\mathbf{a}$, with the convention that the depth of the null module is ∞.

For the definition to make sense, one has to prove that it is independent of a chosen set of generators of I. Second, for technical reasons, one wishes the notion to localize—the latter is where the assumption that R is Cohen–Macaulay intervenes. We refer to [97, Section 6] for the details of these assertions.

Next, since we are dealing mainly with the case where I is a version of the diagonal ideal, sliding depth can be converted in depth conditions of the Tor's of the original ring $A = R/I$. This is done by means of the often called *Cartan–Eilenberg–Serre principle*, reading as follows.

Theorem 10.2.10 ([172, V-5]). *Let k be a Noetherian ring and let R stand for a ring of fractions of the polynomial ring $k[X_1, \ldots, X_n]$. Given R-modules M and N, one of which is* *flat over k, then there is an isomorphism of graded R-modules*

$$H_\bullet(\Delta; M \otimes_k N) \simeq \operatorname{Tor}_\bullet^R(M,N),$$

where Δ *denotes the sequence* $\{X_i \otimes 1 - 1 \otimes X_i \mid 1 \le i \le n\} \subset R \otimes_k R$.

Then the criterion for having starlike linearity in terms of approximation complexes becomes the same as the one for the linear type property of the diagonal ideal $\mathbb{D}$. Namely, one has the following.

Theorem 10.2.11. *Let k be a Noetherian ring and let A be a flat k-algebra with a presentation* $A \simeq R/I$, *where R is a ring of fractions of the polynomial ring of dimension n over k. Assume that the following conditions hold:*
(i) $\operatorname{depth} \operatorname{Tor}_i^R(A,A)_{\mathfrak{m}} \ge \dim A_{\mathfrak{m}} - \operatorname{ht} I_{\mathfrak{M}} + i$, *for every integer* $i \ge 0$ *and every maximal ideal* $\mathfrak{M} \supseteq I$, *with* $\mathfrak{m} = \mathfrak{M}/I$.
(ii) $\mu(\Omega(A/k)_{\wp}) \le \dim A_{\wp} + n - \operatorname{ht} I_P$, *for every prime ideal* $P \supseteq I$, *where* $\wp = P/I$.

Then A is starlike linear. Moreover, if A is Cohen–Macaulay the so is $\mathbb{S}_{A/k} = \mathbb{T}_{A/k}$.

Remark 10.2.12. Note that $\Omega_{A/k} \simeq \mathbb{D}/\mathbb{D}^2$ and that $\dim A_{\wp} + n - \operatorname{ht} I_P = 2\dim A_{\wp} + \operatorname{trdeg}_k A_{\wp}/\wp A_{\wp} = \dim(A \otimes_k A)_{\wp}$ in the case where k is a perfect field. Therefore, if k is a perfect field, the assumption in (ii) is that $\mathbb{D}$ satisfies the property F_1. Thus, over a perfect ground field this is but the stated game where F_1 plus sliding depth imply that the pertinent approximation complex is acyclic, and hence, the ideal $\mathbb{D} \subset A \otimes_k A$ is of linear type.

We next give an application. For simplicity, we take the case of a finitely generated algebra.

Corollary 10.2.13. *Let k be a perfect field and let R be a polynomial ring over k. Assume that* $I \subset R$ *is a perfect ideal of codimension two. If* $A := R/I$ *has the expected star dimension, then it is starlike linear and* $\mathbb{T}_{A/k}$ *is Cohen–Macaulay.*

Proof. Since $A := R/I$ has the expected star dimension, by the above remark, $\mathbb{D}$ satisfies the property F_1. This says that condition (ii) of Theorem 10.2.11 is satisfied. Since I has codimension two, to show condition (i) in the theorem it will suffice to deal with $\operatorname{Tor}_i^R(A,A)$, for $i = 1, 2$.

Now, I is generated by the maximal minors of a certain $n \times (n-1)$ matrix φ. Taking the Koszul homology of these generators, it has been proved in [9, Theorem 2.1(a)] that the corresponding Koszul homology modules H_i are maximal Cohen–Macaulay. From the exact sequence $0 \to I \longrightarrow R \longrightarrow A \to 0$, one has $\operatorname{Tor}_1^R(A,A) \simeq I/I^2$. But I is syzygetic since H_1 is torsion-free, yielding an exact sequence $0 \to H_1 \longrightarrow A^n \longrightarrow I/I^2 \to 0$. From this, it follows that $\operatorname{depth} \operatorname{Tor}_1^R(A,A) \ge \operatorname{depth} A - 1 = \dim A - 1$, as required in (i).

As for $\operatorname{Tor}_2^R(A,A)$, by décalage we know that $\operatorname{Tor}_2^R(A,A) = \operatorname{Tor}_1^R(I,A)$. Applying the Tor functor to the Hilbert–Burch resolution of I yields the exact sequence

$$0 \to \operatorname{Tor}_1^R(I,A) \longrightarrow A^{n-1} \xrightarrow{\varphi} A^n \longrightarrow I/I^2 \to 0.$$

By the above, H_1 has maximal depth. Therefore, the exact sequence $0 \to \operatorname{Tor}_1^R(I,A) \longrightarrow A^{n-1} \longrightarrow H_1 \to 0$ yields $\operatorname{depth} \operatorname{Tor}_1^R(I,A) = \operatorname{depth} A = \dim A$, hence $\operatorname{Tor}_1^R(I,A)$ is Cohen–Macaulay. □

For perfect ideals of codimension three, assuming that the first Koszul homology module H_1 is Cohen–Macaulay, one can still go through, although the proofs require a higher degree of sophistication [188, Proposition 3.7]. Examples of this environment are the codimension three perfect ideals generated by at most 5 generators. Thus, true emotion starts when the number of generators is ≥ 6. A first question is as to what happens if I is not syzygetic? Here is an example to bear in mind.

Example 10.2.14. Let $I \subset R = k[X_1, \ldots, X_6]$ be the ideal generated by the following quadrics:

$$X_2X_4 + X_3X_6, \quad X_3X_5 + X_1X_6, \quad X_1X_2 - X_2X_5 + X_3X_5 - X_5X_6,$$
$$X_2X_3 + X_2X_6 - X_3X_6 + X_6^2, \quad X_3^2 + X_3X_4 + X_3X_6 - X_4X_6,$$
$$X_1X_3 + X_1X_4 + X_4X_5 + X_1X_6.$$

This ideal is perfect of codimension three and has analytic spread $= 5$ (in particular, it is not of linear type). Here, $H_1(I;R)$ on the given generators is not Cohen–Macaulay as I is not even syzygetic: a quadratic relation off $(\mathbf{X})R[\mathbf{T}]$ is

$$T_1T_2 + T_2T_4 - T_2T_5 + T_3T_5 + T_1T_6 - T_4T_6,$$

with $\mathbf{T}$ presentation variables. And yet, $A = R/I$ is starlike linear (see Exercise 10.3.12).

We close this part with a partial result in the case of determinantal ideals.

Proposition 10.2.15. *Let R be a regular domain, let $1 \leq t \leq r \leq s$ and let $A = R[\mathbf{X}]/I_t(\mathbf{X})$, where $\mathbf{X} = (X_{i,j})_{1\leq i\leq r, 1\leq j\leq s}$ is an $r \times s$ matrix of indeterminates over R. The following conditions are equivalent:*
(a) $\operatorname{edim} A_\wp \leq 2 \dim A_\wp$ *for every* $\wp \in \operatorname{Spec}(A)$.
(b) $t = 1$, *or* $t = r$, *or* $r \leq t+1$ *and* $s \leq t+2$.

Proof. (a) ⇒ (b): Suppose that $2 \leq t < r$ and that either $r \geq t+2$ or $s \geq t+3$. By the well-known Eagon–Northcott localization trick, one has $A[X_{1,1}^{-1}] \simeq S[\mathbf{Z}]/I_{t-1}(\mathbf{Z})$, where S is a localization of a polynomial ring over R and $\mathbf{Z}$ is an $(r-1) \times (s-1)$ matrix of indeterminates over S. This procedure preserves the given data and hypotheses. By iterating, we can assume that $t = 2$, $r \geq 3$, and either $r \geq 4$ or $s \geq 5$.

Now, take $\wp = I_1(\mathbf{X})$. Then $\operatorname{edim} A_\wp = rs > 2(s + r - 1) = 2\dim A_\wp$, which yields a contradiction.

(b) $\Rightarrow$ (a): Induct on t. For $t = 1$, the assertion is clear. Let $t > 1$ and let $\wp \in \operatorname{Spec}(A)$, say, $\wp = Q/I_t(\mathbf{X})$. If $I_1(\mathbf{X}) \not\subset Q$, then we are done again the above localization trick and the induction hypothesis. If, on the other hand, $I_1(\mathbf{X}) \subset Q$ then

$$\dim R[\mathbf{X}]_Q \geq rs \geq 2(r - t + 1)(s - t + 1) = 2 \operatorname{ht} I_t(\mathbf{X})_Q.$$

Therefore,

$$\dim A_\wp = \dim R[\mathbf{X}]_Q - \operatorname{ht} I_t(\mathbf{X}) \geq \operatorname{ht} I_t(\mathbf{X})_Q \geq \operatorname{edim} A_\wp - \dim A_\wp,$$

thus yielding the required inequality. □

10.2.4 Behavior under flat deformation

Throughout this part, we fix the following sufficiently typical setup of flat deformations: k is a perfect field, A is a regular local ring with residue field k and B is a flat A-algebra essentially of finite type.

As a matter of additional notation, quite generally, for any A-algebra B we denote by $\mathbb{D}_{B/A}$ the kernel of the multiplication map $B \otimes_A B \longrightarrow B$.

Proposition 10.2.16. *Let* $(A, \mathfrak{n}, k)$ *and* B *be as above.*
Suppose that
(i) $\mathbb{T}_{B/A}$ *is Cohen–Macaulay.*
(ii) $\operatorname{edim} \bar{B}_\wp \leq 2\dim \bar{B}_\wp$ *for every* $\wp \in \operatorname{Spec}(\bar{B})$, *where* $\bar{B} = B \otimes_A k$.

Then:
(a) $\mathbb{T}_{B/A}$ *is A-flat.*
(b) $\mathbb{T}_{B/A} \otimes_A k \simeq \mathbb{T}_{\bar{B}/k}$.
(c) $\mathbb{T}_{\bar{B}/k}$ *is Cohen–Macaulay.*

Proof. We apply [179, Theorem 7.3.47] with the following settings: $R = B \otimes_A B$, $\mathcal{J} = \mathbb{D}_{B/A}$, $\mathcal{N} = \mathfrak{n}R$, so that $R/\mathcal{N} = \bar{B} \otimes_k \bar{B}$, $(\mathcal{J}, \mathcal{N})/\mathcal{N} = \mathbb{D}_{\bar{B}/k}$ and $\operatorname{gr}_{\mathcal{J}} R = \mathbb{T}_{B/A}$.

Let $P \subset B \otimes_A B$ be a prime ideal containing $(\mathbb{D}_{B/A}, \mathfrak{n}(B \otimes_A B))$. Its direct image in $B \simeq (B \otimes_A B)/\mathbb{D}_{B/A}$ will be denoted by the same letter P. Let $\wp = P/\mathbf{m}(B \otimes_A B)$. By a similar token, the latter is identified with a prime in $\bar{B} \otimes_k \bar{B}$ which will be denoted by the same letter $\wp$. Then

$$\begin{aligned} \ell(\mathcal{J}_P) &\leq \mu(\mathcal{J}_P) = \mu(\Omega(B/A)_P) = \mu(\Omega(\bar{B}/k)_\wp) \\ &= \operatorname{edim} \bar{B}_\wp + \operatorname{trdeg}_k \bar{B}_\wp/\wp\bar{B}_\wp, && \text{(cf., e. g., [73])} \\ &\leq 2\dim \bar{B}_\wp + \operatorname{trdeg}_k \bar{B}_\wp/\wp\bar{B}_\wp, && \text{by condition (ii)} \end{aligned}$$

$$= \dim(\bar{B} \otimes_k \bar{B})_{\wp} = \dim R_{\wp}. \qquad \text{by (157)}$$

Thus, one of the statements of [179, Theorem 7.3.47] is satisfied, hence so are the others. Therefore, (b) and (c) hold.

It remains to verify (a). For this, one may assume that the homomorphism $A \to B$ is local (possibly, at the expense of k being nonperfect). But the argument on the vanishing of $\mathrm{Tor}_i^R(\mathrm{gr}_{\mathcal{J}} R, R)$ in the proof of [179, Theorem 7.3.47] actually shows that a minimal generating set of $\mathfrak{n}$ is a regular sequence on $\mathrm{gr}_{\mathcal{J}} R = \mathbb{T}_{B/A}$ as well. Therefore, the latter is indeed A-flat as required. □

Remark 10.2.17. If k is a coefficient field of A then condition (i) of Proposition 10.2.16 can be replaced by the requirement that $\mathbb{T}_{B/k}$ be Cohen–Macaulay—more precisely, this requirement implies condition (i) provided (ii) holds. To see this, first note that condition (iii) of Proposition 10.2.16 implies condition (iii) of [179, Proposition 7.3.49]. Then apply the latter with the settings $R = B \otimes_k B$, $\mathcal{J} = \mathbb{D}_{B/k}$ and $\mathcal{N} = (x \otimes 1 - 1 \otimes x \mid x \in A) \subset R$ (so that $R/\mathcal{N} = B \otimes_A B$). The conclusion is that $\mathbb{T}_{B/A}$ is Cohen–Macaulay, as desired.

The following consequence is archetypal of one-parameter deformations.

Corollary 10.2.18. *Let $A = k[t]_{(t)}$ and let B as before.*
Suppose that:
(i) $\mathbb{T}_{B_{\mathfrak{n}}/k}$ *is Cohen–Macaulay for every maximal ideal* $\mathbf{n} \supseteq tB$.
(ii) $\mathrm{edim}\, \bar{B}_{\wp} \leq 2 \dim \bar{B}_{\wp}$ *for every* $\wp \in \mathrm{Spec}(\bar{B})$, *where* $\bar{B} = B \otimes_A k$.

Then:
(a) $\mathbb{T}_{B/A}$ *is A-flat.*
(b) $\mathbb{T}_{B/A} \otimes_A k \simeq \mathbb{T}_{\bar{B}/k}$.
(c) $\mathbb{T}_{\bar{B}/k}$ *is Cohen–Macaulay.*

Proof. The proof is clear and left to the reader. □

Example 10.2.19. Let k and A be as in Corollary 10.2.18 and let $B = A[\{X_{ij}\}]/I_2(\mathbf{X}_t)$, where

$$\mathbf{X}_t = \begin{pmatrix} X_{1,1} & X_{1,2} & X_{1,3} & X_{1,4} \\ X_{2,1} + t & X_{2,2} & X_{2,3} & X_{2,4} \end{pmatrix}.$$

The reader can check that B is A-flat. By Proposition 10.2.15, condition (ii) of Corollary 10.2.18 holds true. As for condition (i), in this case we only have to prove that $\mathbb{T}_{B/k}$ is Cohen–Macaulay. For this, one can use [79].

The following result puts in evidence the full nature of the preceding deformation landscape.

Theorem 10.2.20. *Let $(A, \mathfrak{m}, k)$ be as before, let B moreover be local with local structural map $A \to B$. Suppose that:*
(i) $\mathbb{T}_{B/A}$ *is Cohen–Macaulay.*
(ii) $\bar{B} = B \otimes_A k$ *is an isolated singularity with* $\operatorname{edim} \bar{B} = 2\dim \bar{B} + 1$.

Then one and only one of the following conditions hold:
(a) $\mathbb{T}_{B/A} \otimes_A k \simeq \mathbb{T}_{\bar{B}/k}$ *and* $\mathbb{T}_{\bar{B}/k}$ *is Cohen–Macaulay.*
(b) *B is set theoretically starlike linear over A.*

Proof. Let P be the preimage of the maximal ideal $\mathfrak{n}$ of B via the diagonal map $B \otimes_A B \to B$. As in the proof of Proposition 10.2.16, we set $\wp = P/\mathfrak{m}(B \otimes_A B) = \mathfrak{n}/\mathfrak{m}B$ and obtain, similarly:

$$\begin{aligned}\mu((\mathbb{D}_{B/A})_P) = \mu(\Omega(B/A)_P) &= \mu(\Omega(\bar{B}/k)_\wp)\\ &= \operatorname{edim} \bar{B} + \operatorname{trdeg}{}_k \bar{B}_\wp/\wp\bar{B}_\wp\\ &= \operatorname{edim} \bar{B} + \operatorname{trdeg}{}_k B/\mathfrak{n}\\ &= 2\dim \bar{B} + \operatorname{trdeg}{}_k B/\mathfrak{n} + 1\\ &= \dim(\bar{B} \otimes_k \bar{B})_\wp + 1.\end{aligned}$$

Thus, we have two mutually exclusive cases:

(1) $\ell((\mathbb{D}_{B/A})_\wp) \le \dim(\bar{B} \otimes_k \bar{B})_\wp$.
In this case, [179, Theorem 7.3.47] applies with $\mathcal{J} = \mathbb{D}_{B/A}$ and yields (a).

(2) $\ell((\mathbb{D}_{B/A})_\wp) = \dim(\bar{B} \otimes_k \bar{B})_\wp + 1$.
In this case, $\ell((\mathbb{D}_{B/A})_\wp) = \mu((\mathbb{D}_{B/A})_\wp)$ holds. By assumption, $\bar{B}$ is an isolated singularity. Moreover, there is natural bijection between the set of primes of $\bar{B}$ and the set of primes of $B \otimes_k B$ containing $\mathbb{D}_{B/k}$ and the inverse image of $\mathfrak{m}B$ or, equivalently, the set of primes of $B \otimes_A B$ containing $\mathbb{D}_{B/A}$. This means that the ideal $\mathbb{D}_{B/A}$ is locally everywhere generated by analytically independent elements which is equivalent to the required assertion in (b) as has been seen earlier in the section. □

The next example is a deformation of a generic 3×3 Hankel matrix to a symmetric matrix. It was extensively considered by Pinkham [158, Chapter II] and became very popular in singularity theory. Geometrically, it gives a deformation of the projecting cone of the rational normal quartic.

Example 10.2.21. Let k be a perfect field of characteristic $\neq 2$ and let $A = k[t]_{(t)}$. Set $B = A[X_{1,1}, X_{1,2}, X_{1,3}, X_{2,3}, X_{3,3}]/I_2(\mathbf{X})$ where

$$\mathbf{X} = \begin{pmatrix} X_{1,1} & X_{1,2} & X_{1,3} \\ X_{1,2} & X_{1,3} + t & X_{2,3} \\ X_{1,3} & X_{23,} & X_{3,3} \end{pmatrix},$$

Here, $\mathbb{T}_{B/k}$ is a normal Cohen–Macaulay domain. Therefore, $t \otimes 1 - 1 \otimes t$ is a non-zero-divisor on $\mathbb{T}_{B/k}$, hence

$$\begin{aligned}\mathbb{T}_{B/k}/(t \otimes 1 - 1 \otimes t) &= \mathrm{gr}_{\mathbb{D}_{B/k}/(t\otimes 1 - 1\otimes t)}((B \otimes_k B)/(t \otimes 1 - 1 \otimes t)) \\ &\simeq \mathrm{gr}_{\mathbb{D}_{B/A}}(B \otimes_A B) = \mathbb{T}_{B/A}\end{aligned}$$

is Cohen–Macaulay as well. Next, since B is not set theoretically starlike linear over k by Proposition 10.2.8, one can apply Theorem 10.2.20 to conclude that $\mathbb{T}_{B/A}$ is A-flat and specializes, and the specialized algebra $\mathbb{T}_{\bar{B}/k}$ is Cohen–Macaulay.

A second example in [158] is the rational normal scroll deformation

Example 10.2.22. Let $B = A[X_{1,1}, X_{1,2}, X_{1,3}, X_{1,4}, X_{1,5}]/I_2(\mathbf{X})$, where

$$\mathbf{X} = \begin{pmatrix} X_{1,1} & X_{1,2} & X_{1,3} & X_{1,4} \\ X_{1,5} & X_{1,1} + t & X_{1,2} & X_{1,3} \end{pmatrix}.$$

Again $\mathbb{T}_{B/A}$ is Cohen–Macaulay. We may therefore apply Theorem 10.2.20. By [188, Theorem 4.10] and its proof, B is set theoretically starlike linear both as a k-algebra and as an A-algebra. Also, $\mathbb{T}_{B/A}$ is A-flat if and only if it specializes, in which case the specialized algebra $\mathbb{T}_{\bar{B}/k}$ would be Cohen–Macaulay. Thus, Theorem 10.2.20 shows that $\mathbb{T}_{B/A}$ is not A-flat and does not specialize. More exactly, the polynomial

$$T_3^3 - 2T_2T_3T_4 + T_2^2T_5 + T_4^2T_6 - T_3T_5T_6,$$

for suitable presentation variables, gives a relation of analytic dependence of $\mathbb{D}_{\bar{B}/k}$. Of course, $\bar{B}$ does not have the expected star dimension. On the other hand, $\mathbb{T}_{\bar{B}/k}$ is nevertheless Cohen–Macaulay.

10.2.5 A glimpse of the blowup of the Kähler differentials

We go back to the absolute case, where k is a perfect field and A is an algebra essentially of finite type over k.

Recall that if, for example, A is Cohen–Macaulay then A has the expected star dimension if and only if $\mathbb{D}$ satisfies the condition (F_1) (Proposition 10.2.5). By Proposition 2.3.8 and the remark following it, this is equivalent to requiring that the module of Kähler differentials $\Omega(A/k) \simeq \mathbb{D}/\mathbb{D}^2$ satisfy condition (F_0).

Our basic quest is about the assumption that $\mathbb{D}$ satisfies (F_1), but not (F_2) and what is lost in this landscape. The answer is not clear in that generality, but assuming that A has a more particular nature, one has the quite surprising result.

Theorem 10.2.23. *Suppose that A is locally a complete intersection. If $\mathbb{D}$ satisfies (F_1), but not (F_2) then the torsionfree tangent star algebra $\mathbb{T}_{A/k}/A$-torsion is not Cohen–Macaulay.*

By drawing upon Proposition 8.6.1, the theorem can be rephrased in the language of differentials.

Theorem 10.2.24 ([191, Theorem 3.1]). *Suppose that A is locally a complete intersection. If $\Omega(A/k)$ satisfies (F_0), but not (F_1) then the Rees algebra $\mathcal{R}_A(\Omega(A/k))$ is not Cohen–Macaulay.*

This result tells us, perhaps over again, that certain properties of the differentials hide in between the sheets of Fitting ideals. Note, however, that the hypothesis on A means that the module $\Omega(A/k)$ has homological dimension one, which is a very special case in the study of the module.

Nevertheless, the theorem has a geometric flavor. To see this, first note that condition (F_1), pretty similarly to the case of (F_0), translates into the inequalities

$$\operatorname{edim} A_{\wp} \leq 2 \dim A_{\wp} - 1, \tag{161}$$

for every nonminimal prime $\wp$ of A.

Now, suppose that V is a smooth projective variety of dimension d over an algebraically closed field k. It is classical that X can be embedded in $\mathbb{P}_k^{2d+1}$. Let $A \subset k[X_0, \ldots, X_{2d}]$ denote the homogeneous coordinate ring of X in this embedding. Suppose, moreover, that $\operatorname{char}(k) = 0$ and that A is a complete intersection locally at the maximal irrelevant ideal. If the torsionfree tangent star algebra $\mathbb{T}_{A/k}/A$-torsion is Cohen–Macaulay then $\mathbb{D} = \mathbb{D}_{A/k}$ satisfies (F_2) as in (161); hence, in particular, globally,

$$\operatorname{edim}(A) \leq 2 \dim(A) - 1.$$

Translating back to the geometric setup, we find that $\operatorname{edim}(V) \leq 2 \dim(V)$. Thus, X is degenerate in the original embedding.

It would be interesting having a purely geometric interpretation of the assumption to the effect that the torsion-free tangent star algebra $\mathbb{T}_{A/k}/A$-torsion is Cohen–Macaulay.

The proof of Theorem 10.2.24 is quite technical and is beyond the present objectives. The reader is referred to the original paper [191] for the details and for further results on the nature of the Rees algebra $\mathcal{R}_A(\Omega(A/k)) \simeq \mathbb{T}_{A/k}/A$-torsion, including the normality property and applications to smooth projective varieties whose homogeneous coordinate ring is quasi-Gorenstein. For lack of a better designation, the authors of [191] called them varieties of Calabi–Yau type, perhaps an unfortunate choice as recent notions of a variety of Calabi–Yau type may go a little askew (see, e. g., [11, Definition 2.1]).

10.3 Exercises

Exercise 10.3.1. With the same assumption as in Proposition 8.6.1, describe a suitable ideal $\mathfrak{n} \subset A \otimes_k A$ for which $\mathcal{R}_{A\otimes_k A}(\mathbb{D}) = \mathcal{S}_{A\otimes_k A}(\mathbb{D})/((0) : \mathfrak{n}^\infty)$.

Exercise 10.3.2. Let $A = k[\mathbf{X}]/I$ be a reduced standard graded k-algebra (k algebraically closed). Show that $V = \mathrm{Proj}(A)$ is a subvariety of its tangential variety by showing, more strongly, that there is a surjective graded homomorphism

$$\mathcal{R}_A(\mathbb{D}/\mathbb{D}^2) \otimes_A k \twoheadrightarrow A.$$

(*Hint*: The crucial step is to observe that, as a consequence of Euler's formula, one has an inclusion $(I_1(\mathbf{T} \cdot \Theta^t), \mathbf{\Delta}) \subset (I(\mathbf{T}), \mathbf{\Delta})$, of ideals in the doubled polynomial ring $k[\mathbf{X}, \mathbf{T}]$, where $I(\mathbf{T})$ is the given ideal I read in the $\mathbf{T}$-variables and Θ is the Jacobian matrix of a set of homogeneous generators of I. Having this, use the composite map $\mathbb{E}(\mathcal{R}_A(\mathbb{D}/\mathbb{D}^2)) \hookrightarrow \mathcal{R}_A(\mathbb{D}/\mathbb{D}^2) \twoheadrightarrow k[\mathbf{X}, \mathbf{T}]/(I, I(\mathbf{T}), I_1(\mathbf{T} \cdot \Theta^t), \mathbf{\Delta})$.)

Exercise 10.3.3. Let $\mathbb{P}^d \to \mathbb{P}^n$ be a rational *uniformizing* map (in the sense that among its coordinate functions there are powers of each coordinate of $\mathbb{P}^d$; so, in particular, it is a regular map). What is the form of the saturation in this case?

Exercise 10.3.4. Consider the monomial curve $C \subset \mathbb{P}^4$ parameterized by $(t_0^5 : t_0^3t_1^2 : t_0^2t_1^3 : t_0t_1^4 : t_1^5)$. Argue (or compute):

(1) The tangential variety $T(C)$ is ideal theoretically cut out by the following 3 hypersurfaces:

$$x_2^2x_3 - 4/3x_1x_3^2 + 2/9x_1x_2x_4 + 1/9x_0x_4^2,$$
$$x_2^3 - 4/3x_1x_2x_3 + 1/4x_1^2x_4 + 1/12x_0x - 3x_4,$$
$$x_1x_2^2 - 9/8x_1^2x_3 - 3/8x_0x_3^2 + 1/2x_0x_2x_4.$$

(2) $T(C)$ is an arithmetically Cohen–Macaulay variety of degree 7.

(3) $T(C)$ is not smooth locally in codimension one.
(*Hint*: Show that tangential variety is singular along the curve; for this, let $\mathfrak{J}$ be the Jacobian ideal of the homogeneous coordinate ring of $T(C)$ lifted to $k[\mathbf{X}]$. Show that the ideal $(\mathfrak{J}, I(T(C)))$ is contained in the radical of the ideal $I(C)$. For the last verification, one could do the saturation of $I(C)$ by $(\mathfrak{J}, I(T(C)))$.)

Exercise 10.3.5. Let $\mathbb{P}^2 \to \mathbb{P}^9$ be the Veronese embedding of order 3, a smooth surface V of degree 9. Argue (or compute):

(1) The tangential variety $T(V)$ is a 4-dimensional variety of degree 24.

(2) The homogeneous defining ideal of $T(V)$ is minimally generated by quadrics and cubics. How many? (Many!)

(3) $T(C)$ is not arithmetically Cohen–Macaulay.

(4) $T(V)$ is singular along V (There is a prize for those that succeed in proving this by computer calculation).
(5) Does the homogeneous defining ideal of $T(C)$ satisfy the property S_2? (a prize for this as well)

Exercise 10.3.6. Let $I = (x_1x_2, x_2x_3, x_3x_1, x_1x_4) \subset R = k[x_1, x_2, x_3, x_4]$. Setting $A = R/I$, prove that $\dim \mathbb{S}_{A/k} = \dim \mathbb{T}_{A/k} (= 4)$, but the generators of $\mathbb{D}$ are analytically dependent.

Exercise 10.3.7. Let $I = (x_1x_2, x_2x_3, x_3x_1, x_1x_4, x_2x_5, x_3x_6) \subset R = k[x_1, x_2, x_3, x_4, x_5, x_6]$. Setting $A = R/I$, show:
(1) A is a 3-dimensional unmixed reduced ring.
(2) A has the expected star dimension (what is it?)
(3) The generators of $\mathbb{D}$ admit a nonzero monomial dependence relation.

Exercise 10.3.8. Let $R = k[x_1, \dots, x_5]$ and let $I \subset R$ denote the ideal of maximal minors fixing the first column of the Hankel matrix

$$H := \begin{pmatrix} x_1 & x_2 & x_3 & x_4 \\ x_2 & x_3 & x_4 & x_5 \end{pmatrix}.$$

Show:
(1) $I = (x_1, x_2) \cap D$, where $D = I_2(H)$.
(2) A is set theoretically starlike linear.
(*Hint*: The minimal primes of either $\mathbb{S}_{A/k}$ or $\mathbb{T}_{A/k}$ are the respective extended ideals of (x_1, x_2, T_1, T_2), $(\mathbf{x})$, $(x_1, x_2, x_3, x_4, T_1)$.)
(3) A is not starlike linear.
(*Hint*: The presentation ideal of $\mathbb{S}_{A/k}$ is minimally generated by forms of bidegrees $(1,1)$, while $\mathbb{T}_{A/k}$ admits a minimal relation of bidegree $(1,3)$.)

Exercise 10.3.9. Show that in the proof of Proposition 10.2.8 one can assume that the matrix $\mathfrak{M}$ is generic. Find $\dim \mathbb{S}_{A/k}$ and $\dim \mathbb{T}_{A/k}$ in the generic case when $I = I_{\lfloor (r+1)/2 \rfloor}(\mathfrak{M})$. Are there other relations of analytic dependence besides the determinantal one?

Exercise 10.3.10. Let M denote either one of the following matrices:
(i) The 3×3 generic matrix over a field.
(ii) The 3×3 generic symmetric matrix over a field.
(iii) The 3×3 generic Hankel matrix over a field.

If $I = I_2(M)$ is the ideal of 2-minors in the polynomial ring R of the entries over k, prove that R/I has the expected star dimension but is not set theoretic starlike linear.

Exercise 10.3.11 (2-edge paths in a pentagon). Consider the monomial ideal $I = (x_1x_2x_3, x_2x_3x_4, x_3x_4x_5, x_4x_5x_1, x_5x_1x_2) \subset R = k[x_1, x_2, x_3, x_4, x_5]$ and set $A := R/I$.

(1) Prove that A is set theoretic starlike linear.
(*Hint*: Show that A is locally everywhere generated by analytically independent elements. For this, localize separately at a prime ideal in the punctured spectrum and the irrelevant maximal ideal; in the first case, it reduces to edge-ideals of certain trees, while the secant ideal of I vanishes.)

(2) Argue (or compute) that $\mathbb{D}$ is not of linear type, by showing that it is not even syzygetic.
(*Hint*: Verify that the Rees algebra of $\mathbb{D}$ admits a binomial minimal generator of bidegree $(3, 2)$.)

Exercise 10.3.12. Consider the ideal $I \subset R$ of Example 10.2.14. Set $A := R/I$. Argue (or compute):

(1) I is a prime ideal.
(*Hint*: Show that A is a normal ring by the Serre criterion.)

(2) I has analytic spread 5. What is a relation of analytic dependence?

(3) A is Cohen–Macaulay, but not Gorenstein.

(4) A is starlike linear. Is $\mathbb{T}_{A/k}$ Cohen–Macaulay?

Exercise 10.3.13. Let G denote a simple graph on n vertices and let $I \subset R = k[X_1, \dots, X_n]$ stand for its ideal of edges over a field k. The *suspension* over G is the graph $S(G)$ on $2n$ vertices obtained by adding to each vertex of G an edge, so that the ideal of edges of $S(G)$ is $\tilde{I} = (I, X_1Y_1, \dots, X_nY_n) \subset k[X_1, \dots, X_n, Y_1, \dots, Y_n]$, where the Y_i's are new indeterminates. Set $A := k[\mathbf{X}, \mathbf{Y}]/\tilde{I}$. Prove:

(1) A has the expected star dimension.

(2) If A is set theoretically starlike linear then G is bipartite.
(*Hint*: if G admits an odd cycle of order m, then $T_1 \cdots T_m$ is a relation of analytic dependence of $\mathbb{D}$, for suitable ordering of the presentation variables.)

Exercise 10.3.14. Provide an argument for all assertions in Example 10.2.21 and Example 10.2.22. It is meant to use the computer for some of these assertions. At the other end, can one give a simpler proof of [188, Theorem 4.10] for a generic $2 \times s$ matrix?

11 Embedded joins and secants

11.1 Joins

In previous chapters we focused on the diagonal ideal $\mathbb{D}$ as being the kernel of the multiplication map $A \otimes_k A \to A$, where A is a given k-algebra. The nature of $\mathbb{D}$ depended on one single algebra A. In this chapter, we extend the definition to a pair of k-algebras with the same embedding dimension.

11.1.1 Algebraic side

The main material of this part is extracted from [185].

Let R be a ring of fractions of a polynomial ring $k[\mathbf{X}] = k[X_1, \ldots, X_n]$ over a Noetherian ring k, and let A, B be two k-algebras, admitting structural surjections $R \twoheadrightarrow A$ and $R \twoheadrightarrow B$, with respective kernels I, J.

Definition 11.1.1. The kernel of the multiplication map $A \otimes_k B \to R/(I,J)$ is the *diagonal ideal* associated to the maps $R \twoheadrightarrow A$ and $R \twoheadrightarrow B$.

Instead of a notation that involves the two structural maps, we use one that enhances the ideals I, J: $\mathbb{D}(I,J)$. Letting x_i (resp., y_i) denotes the image of X_i in $A = R/I$ (resp., in $B = R/J$), one has

$$\mathbb{D}(I,J) = (\mathbf{x} \otimes 1 - 1 \otimes \mathbf{y}) := (x_1 \otimes 1 - 1 \otimes y_1, \ldots, x_n \otimes 1 - 1 \otimes y_n) \subset (R/I \otimes_k R/J).$$

The proof of this fact is the same as in [179, Section 4.4.1].

Besides the diagonal ideal, one considers also its ring counterpart. For this purpose, assume that either R is a local ring which is a ring of fractions of $k[\mathbf{X}]$ or else, that $R = k[\mathbf{X}]$ and I, J are homogeneous ideals in the standard grading of $k[\mathbf{X}]$. Let $\mathfrak{m}$ denote, respectively, the maximal ideal of R or the maximal ideal $(\mathbf{X})$.

Definition 11.1.2. (i) When $(R, \mathfrak{m})$ is local, let

$$\mathfrak{M} := ((\mathfrak{m}/I) \otimes_k 1, 1 \otimes_k (\mathfrak{m}/J)) \subset R/I \otimes_k R/J$$

and consider the Rees algebra $\mathcal{R} := \mathcal{R}(\mathbb{D}(I,J)_{\mathfrak{M}})$. The *join algebra* of the pair I, J is the fiber cone algebra $\mathcal{R}/\mathfrak{M}\mathcal{R}$ (note that this is a finitely generated K-algebra, where K is the residue field of R).

(ii) When $R = k[\mathbf{X}]$, let $(\mathfrak{m}/I) \otimes_k 1 - 1 \otimes_k (\mathfrak{m}/J) := \{x_i \otimes 1 - 1 \otimes y_i \mid 1 \le i \le n\}$ in the above notation. The *join algebra* of I, J is the k-subalgebra

$$k[(\mathfrak{m}/I) \otimes_k 1 - 1 \otimes_k (\mathfrak{m}/J)] \subset R/I \otimes_k R/J.$$

https://doi.org/10.1515/9783110640694-011

Remark 11.1.3. Note that the two definitions coincide when k is a field, I, J are homogeneous and for the local ring in (i) one takes the localization $k[\mathbf{X}]_{(\mathbf{X})}$.

If no confusion arises, we will denote the join algebra in both situations by $\mathbb{J}(I,J)$. In the homogeneous case, one has

$$\mathbb{J}(I,J) \simeq k[\mathbf{X} \otimes_k 1 - 1 \otimes_k \mathbf{X}]/\mathcal{D}(I,J),$$

where $\mathbf{X} \otimes_k 1 - 1 \otimes_k \mathbf{X} = \{X_i \otimes 1 - 1 \otimes X_i \mid 1 \leq i \leq n\}$ and

$$\mathcal{D}(I,J) = k[\mathbf{X} \otimes_k 1 - 1 \otimes_k \mathbf{X}] \cap (I \otimes_k 1, 1 \otimes_k J).$$

We call $\mathcal{D}(I,J)$ the *(nonembedded) join ideal* of I, J. Note that the subring $k[\mathbf{X} \otimes_k 1 - 1 \otimes_k \mathbf{X}]$ is naturally $\mathbb{N}$-graded and $\mathcal{D}(I,J)$ is a homogeneous ideal in this grading. We will have more to say on the join algebra (and the join ideal) in the subsequent parts.

The important situation in intersection theory is that when $I, J \subset R = k[\mathbf{X}]$ are homogeneous ideals in the standard grading of R (with $R_0 = k$) and $\mathcal{J} = (I(\mathbf{X}), J(\mathbf{Y})) \subset S$. In this case, the span algebra is called the *join algebra* of I and J. Accordingly, in this case, we shift the notation to $\mathbb{J}(I,J)$. The ideal $\mathcal{D}(I,J) = k[\mathbf{X} \otimes_k 1 - 1 \otimes_k \mathbf{X}] \cap (I \otimes_k 1, 1 \otimes_k J)$ is the *(nonembedded) join ideal* of I and J. Note that, by regrading $k[\mathbf{X} \otimes_k 1, 1 \otimes_k \mathbf{X}]$ by total degree, the subring $k[\mathbf{X} \otimes_k 1 - 1 \otimes_k \mathbf{X}]$ is naturally $\mathbb{N}$-graded and $\mathcal{D}(I,J)$ is a homogeneous ideal in this grading.

11.1.2 The geometric background

The geometric version of the structural exact sequence

$$0 \to \mathbb{D}(I,J) \longrightarrow A \otimes_k B \longrightarrow R/(I,J) \to 0$$

is classically known as "reduction to the diagonal." Here, assume that k is a field, that $R = k[\mathbf{X}] = k[X_0, \ldots, X_n]$ and that A, B are standard graded quotients of R. Set $\mathbb{P}^n = \mathrm{Proj}(k[\mathbf{X}])$ and let $V, W \subset \mathbb{P}^n$ denote the varieties defined by A and B, respectively. Then the above exact sequence transcribes the existence of an isomorphism of the varieties $\mathcal{D} \cap (V \times W)$ and $V \cap W$, where $\mathcal{D} \subset \mathbb{P}^n \otimes \mathbb{P}^n$ denotes the biprojective diagonal (Section 1.3).

The geometric transcription of the above definition of the join algebra is a bit more delicate. Actually, there are various ways of introducing the embedded join. We next explain two of these.

- (VIA INCIDENCE THEORY)

Let $V, W \subset \mathbb{P}^n$ be two reduced algebraic sets and let $\mathcal{D} \subset \mathbb{P}^n \times \mathbb{P}^n$ denote the biprojective diagonal. Consider the rational map $\eta : V \times W \dashrightarrow \mathbb{G}(1,n)$ mapping a bipoint (p, q) to the line $[p, q]$ in $\mathbb{P}^n$ joining $p \in \mathbb{P}^n$ and $q \in \mathbb{P}^n$, considered as a point of the Grassmannian $\mathbb{G}(1, n)$. This is a regular map off $\mathcal{D} \cap (V \times W)$.

Definition 11.1.4. The closure $\mathcal{J}(V, W)$ of the image of η in $\mathbb{G}(1, n)$ is the *variety of joining lines* of V and W.

If $V = W$, this is the *variety of secant lines* of V.

Now, consider the incidence correspondence $\mathcal{I} := \{(\ell, p) \in \mathbb{G}(1, n) \times \mathbb{P}^n \mid p \in \ell\}$ and let $\pi_1 : \mathcal{I} \to \mathbb{G}(1, n)$ and $\pi_2 : \mathcal{I} \to \mathbb{P}^n$ denote the restrictions of the two projections, respectively.

Set $VW := \pi_2(\pi_1^{-1}(\mathcal{J}(V, W)))$.

• (VIA RULED JOINS)

Consider the composite rational map $\mathbb{P}^{2n+1} \dashrightarrow \mathbb{P}^n \times \mathbb{P}^n \dashrightarrow \mathbb{P}^n$ defined by

$$(a_0 : \ldots : a_n : b_0 : \ldots : b_n) \mapsto (a_0 - b_0 : \ldots : a_n - b_n),$$

and let κ denote its restriction to ${}^{\mathcal{Y}}(V \times W) \subset \mathbb{P}^{2n+1}$ (see Remark 1.3.2). This a regular map off ${}^{\mathcal{Y}}(V \times W) \cap \Delta$, where Δ denotes the projective diagonal.

Let $\mathfrak{J}(V, W) \subset \mathbb{P}^n$ denote the closure of the image of κ.

Proposition 11.1.5. *Let $V, W \subset \mathbb{P}^n$ be two reduced algebraic sets. Then $\mathfrak{J}(V, W) = VW$ as sets.*

Proof. Note that the rational map $(x_0 : \ldots : x_n : y_0 : \ldots : y_n) \mapsto (x_0 - y_0 : \ldots : x_n - y_n)$ is regular off the projective diagonal $\Delta \subset \mathbb{P}^{2n+1}$ as introduced in Section 1.3. We show that $VW \subset \mathfrak{J}(V, W)$, the reverse inclusion being completely similar.

The argument is readily read off the principles in Section 1.3. For that, consider a point $p = (\mathbf{a} : \mathbf{b}) \in^{\mathcal{Y}} (V \times W) \setminus (L_{\mathbf{X}} \cup L_{\mathbf{Y}})$. As has been argued, a point q on the fiber is $\mathcal{Y}^{-1}(\mathcal{Y}(p))$ is of the form $(\lambda\mathbf{a} : \mu\mathbf{b})$, for some $\lambda, \mu \in k \setminus \{0\}$. By construction, the image of q under η is the point $q' = (\lambda a_0 - \mu b_0 : \ldots : \lambda a_n - \mu b_n) \in \mathbb{P}^n$. Consider the line $\ell = [\mathbf{a}, \mathbf{b}] \in \mathbb{G}(1, n)$. Then $q' \in \ell$ since the vector with coordinates $(\lambda a_0 - \mu b_0, \ldots, \lambda a_n - \mu b_n)$ is linearly dependent on the vectors $\mathbf{a}, \mathbf{b}$. Therefore, $(\ell, q') \in \mathcal{L}$ and, clearly, $\pi_2(\ell, q') = q'$. This means that $q' \in \mathfrak{J}(V, W)$. By taking closure, we find that $VW \subset \mathfrak{J}(V, W)$, as required. □

This justifies the following definition.

Definition 11.1.6. Let k be an algebraically closed field and let $V, W \subset \mathbb{P}^n$ be reduced projective algebraic sets. The *embedded join* of V and W is $\mathfrak{J}(V, W) = VW$.

It is time to harmonize the algebraic and the geometric notions.

Proposition 11.1.7. *Let k be an algebraically closed field and let $V, W \subset \mathbb{P}^n$ be reduced projective algebraic sets. Then the homogeneous coordinate ring of the embedded join of V and W is graded isomorphic to the join algebra $\mathbb{J}(I, J)$, where $I, J \subset k[\mathbf{X}]$ stand for the respective homogeneous ideals of V, W in the given embeddings.*

Proof. We take the second construction $\mathfrak{J}(V,W)$. It says that $\mathfrak{J}(V,W)$ is the locus of the parametric equations $t_i = x_i - y_i$, for $i = 0,\dots,n$. This gives in principle Cartesian equations in $\mathbf{T}$; then one replaces back the original $\mathbf{X}$ coordinates.

To see what is going on behind the scenes on the algebraic side, one proceeds further as follows: consider the k-algebra homomorphisms $p_1 : k[\mathbf{X} \otimes_k 1, 1 \otimes_k \mathbf{X}] \to k[\mathbf{X} \otimes_k 1]$ such that $p_1(X_i \otimes_k 1) = X_i \otimes_k 1$ and $p_1(1 \otimes_k X_i) = 0$, for all i, and $p_2 : k[\mathbf{X} \otimes_k 1, 1 \otimes_k \mathbf{X}] \to k[1 \otimes_k \mathbf{X}]$ such that $p_2(X_i \otimes_k 1) = 0$ and $p_2(1 \otimes_k X_i) = 1 \otimes_k X_i$, for all i. Since the subring $k[\mathbf{X} \otimes_k 1 - 1 \otimes_k \mathbf{X}]$ is isomorphically mapped to its image under either p_1 or p_2, the same is true of the ideal $\mathcal{D}(I,J)$. Moreover, since $\mathcal{D}(I,J) \subset (I \otimes_k 1, 1 \otimes_k J)$, it is clear that $p_1(\mathcal{D}(I,J)) \subset I \otimes_k 1$ and $p_2(\mathcal{D}(I,J)) \subset 1 \otimes_k J$.

Next, consider the isomorphisms $\tau_1 : k[\mathbf{X} \otimes_k 1] \simeq k[\mathbf{X}]$ and $\tau_2 : k[1 \otimes_k \mathbf{X}] \simeq k[\mathbf{X}]$ such that $\tau_1(X_i \otimes_k 1) = X_i$ and $\tau_2(1 \otimes_k X_i) = X_i$, respectively. Clearly, $I \otimes_k 1$ (resp., $1 \otimes_k J$) is isomorphically mapped to I (resp., J). Since $\mathcal{D}(I,J) \subset k[\mathbf{X} \otimes_k 1 - 1 \otimes_k \mathbf{X}]$, the images by $\tau_1 \circ p_1$ and by $\tau_2 \circ p_2$ of a homogeneous element of $\mathcal{D}(I,J)$ coincide in $k[\mathbf{X}]$. Thus, $\mathcal{D}(I,J)$ is isomorphic to this common image in $k[\mathbf{X}]$ and it is contained in both I and J. □

The ideal of $k[\mathbf{X}]$ explicitly constructed is the *embedded join ideal* of I and J; it will be denoted $\mathfrak{J}(I,J)$. Thus, with this notation, $\mathfrak{J}(I,J) \subset I \cap J$ and $\mathbb{J}(I,J) \simeq k[\mathbf{X}]/\mathfrak{J}(I,J)$. Of course, the difference between the nonembedded version and the present one is merely explained by a change of variables and we use this whenever appropriate. However, the embedded version clarifies the algebraic side of the inclusion $V \cup W \subset VW$.

The following proposition sums up a few elementary properties of the ideal of the embedded join.

Proposition 11.1.8. *Let k be a field and let $I, J \subset R = k[\mathbf{X}]$ be homogeneous ideals.*
(i) *If $I = I_1 \cap I_2$ is a decomposition of I into homogeneous ideals, then $\mathfrak{J}(I,J) = \mathfrak{J}(I_1,J) \cap \mathfrak{J}(I_2,J)$.*
(ii) *If k is a perfect field, then $\sqrt{\mathfrak{J}(I,J)} = \mathfrak{J}(\sqrt{I}, \sqrt{J})$; in particular, if I and J are radical then so is $\mathfrak{J}(I,J)$.*
(iii) *If k is perfect and I and J are absolutely prime, then $\mathfrak{J}(I,J)$ is a prime ideal.*
(iv) *If k is perfect and I and J are primary ideals whose respective radicals are absolutely prime, then $\mathfrak{J}(I,J)$ is a primary ideal.*

Proof. (i) Since in the realm of ideals intersection commutes with contraction, it suffices to check that $((I_1 \cap I_2) \otimes_k 1, 1 \otimes_k J) = (I_1 \otimes_k 1, 1 \otimes_k J) \cap (I_2 \otimes_k 1, 1 \otimes_k J)$. Identifying $R \otimes_k R$ with $k[\mathbf{X},\mathbf{Y}]$ and $I \otimes_k 1$, $I_1 \otimes_k 1$, $I_2 \otimes_k 1$ and $1 \otimes_k J$ with I, I_1, I_2 and $J(\mathbf{Y})$, respectively, it suffices to see that $(I_1 \cap I_2, J(\mathbf{Y})) = (I_1, J(\mathbf{Y})) \cap (I_2, J(\mathbf{Y}))$. This follows immediately if we show that $k[\mathbf{X},\mathbf{Y}]/(J(\mathbf{Y}))$ is a free $k[\mathbf{X}]$-module. But the latter is clear as $k[\mathbf{X},\mathbf{Y}]/(J(\mathbf{Y})) \simeq k[\mathbf{Y}] \otimes_k k[\mathbf{X}]$ and $k[\mathbf{Y}]/J(\mathbf{Y})$ is a k-vector space.

(ii) Since k is a perfect field, $(\sqrt{I} \otimes_k 1, 1 \otimes_k \sqrt{J})$ is a radical ideal. Therefore, one obtains $\sqrt{(I \otimes_k 1, 1 \otimes_k J)} = (\sqrt{I} \otimes_k 1, 1 \otimes_k \sqrt{J})$. Finally, contracting to the diagonal subalgebra $k[\mathbf{X} \otimes_k 1 - 1 \otimes_k \mathbf{X}]$, one reads off $\sqrt{\mathcal{D}(I,J)} = \mathcal{D}(\sqrt{I}, \sqrt{J})$ for the nonembedded join ideals, hence the same holds for the corresponding embedded join ideals.

(iii) Let $\bar{k}$ be the algebraic closure of k. Setting $\bar{R} = R \otimes_k \bar{k} = \bar{k}[\mathbf{X}]$, then $I\bar{R}$, $J\bar{R}$ are prime ideals by hypothesis, hence $\bar{R}/I\bar{R} \otimes_{\bar{k}} \bar{R}/J\bar{R}$ is a domain. Therefore, its subring $R/I \otimes_k R/J$ is a domain, too. It follows that the join algebra $\mathbb{J}(I,J)$ is a domain.

(iv) Say, $\operatorname{Ass} R/I = \{P\}$, $\operatorname{Ass} R/J = \{Q\}$. Since $R/I \to R/I \otimes_k R/J$ is flat, $\operatorname{Ass}(R/I \otimes_k R/J) = \operatorname{Ass}(R/P \otimes_k R/J)$ [137, Theorem 23.2]. By the same token, $\operatorname{Ass}(R/P \otimes_k R/J) = \operatorname{Ass}(R/P \otimes_k R/Q)$. On the other hand, since P and Q are absolutely prime, it follows by part (ii) that $\operatorname{Ass}(R/P \otimes_k R/Q)$ consists of the single element. Thus, $R/I \otimes_k R/J$ has only one associated prime, and the same holds for the subring $k[\mathbf{x} \otimes_k 1 - 1 \otimes_k \mathbf{x}]$. □

11.1.3 Relation to the Bayer–Mumford deformation

We briefly recall the Bayer deformation [10] in the form that suits our purpose, drawing upon the contents of [57, Section 15.8].

Throughout, k is assumed to be a field. Let $R = k[\mathbf{X}] = k[X_1, \ldots, X_n]$ and let $>$ stand for a monomial term order on R. A weight function $\lambda : \mathbb{Z}^n \to \mathbb{Z}$ induces a partial order $>_\lambda$ on the set of monomials of R, to wit, $\mathbf{X}^\alpha >_\lambda \mathbf{X}^\beta \Leftrightarrow \lambda(\alpha) > \lambda(\beta)$ (in the ordinary order of $\mathbb{Z}$). Let λ be one such function. Given $g = \sum c_\alpha \mathbf{X}^\alpha \in R$, let $\mu(g) = \max_{c_\alpha \neq 0}\{\lambda(\alpha)\}$; set $\operatorname{in}_{>_\lambda}(g) = \sum_{\lambda(\alpha)=\mu(g)} c_\alpha \mathbf{X}^\alpha$ and $\tilde{g} = t^{\mu(g)} g(t^{-\lambda(e_1)}X_1, \ldots, t^{-\lambda(e_n)}X_n)$, where t is an indeterminate over R and $\{e_1, \ldots, e_n\}$ is the canonical basis of $\mathbb{Z}^n$. Further, for an ideal $I \subset R$, we set

$$\operatorname{in}_{>_\lambda}(I) := (\{\operatorname{in}_{>_\lambda}(g) \mid g \in I\}) \quad \text{and} \quad \tilde{I} := (\{\tilde{g} \mid g \in I\}) \subset R[t].$$

The basic result of Bayer can be summarized as follows.

Proposition 11.1.9. *Given a term order $>$ on R and a finite collection $\{I_1, \ldots, I_s\}$ of ideals, there exists a weight function $\lambda : \mathbb{Z}^n \to \mathbb{Z}$ such that, if $>_\lambda$ denotes the induced partial order on the monomials of R, then $\operatorname{in}_{>_\lambda}(I_j) = \operatorname{in}_>(I_j)$, $j = 1, \ldots, s$. Moreover, for any such λ, one has:*

(i) *$R[t]/\tilde{I}_j$ is a free $k[t]$-module, for $j = 1, \ldots, s$.*
(ii) *$(R[t]/\tilde{I}_j) \otimes_{k[t]} k[t, t^{-1}] \simeq (R/I_j) \otimes_k k[t, t^{-1}]$, for $j = 1, \ldots, s$.*
(iii) *$(R[t]/\tilde{I}_j) \otimes_{k[t]} (k[t]/(t)) \simeq R/\operatorname{in}_>(I_j)\ (= R/\operatorname{in}_{>_\lambda}(I_j))$, for $j = 1, \ldots, s$.*
(iv) *If $\{g_{j,1}, \ldots, g_{j,r_j}\}$ is a Gröbner basis of I_j with respect to $>$ then $\{\tilde{g}_{j,1}, \ldots, \tilde{g}_{j,r_j}\}$ is a Gröbner basis of $\tilde{I}_j$, for $j = 1, \ldots, s$, with respect to any monomial order on $R[t]$ that induces $>$ on R and has $\mathbf{X} \gg t$.*

Additional comments on the items of the proposition are as follows:

- In (i) a free $k[t]$-basis of $R[t]/\tilde{I}_j$ is the set of monomials of R not belonging to $\operatorname{in}_>(I_j)$, while t is a nonzero divisor on $R[t]/\tilde{I}_j$.

- The isomorphism in (iii) is induced by mapping $t \mapsto 0$.
- The isomorphism in (ii) is induced by the automorphism of $R[t,t^{-1}]$ that maps $X_i \mapsto t^{\lambda(e_i)}X_i$, $t \mapsto t$.
- By the isomorphism in (ii), $\tilde{g} \mapsto t^{\mu(g)}g$, hence $\tilde{I}_jR[t,t^{-1}]$ maps isomorphically onto $I_jR[t,t^{-1}]$, for $j = 1,\dots,s$.
- The weights $\lambda(e_i)$ may be chosen to be positive ([57, Proposition 15.16] and [142, Corollary 1.12]). Thus, when setting $\deg t = 1$ and $\deg X_i = \lambda(e_i)$, the rings $R[t]/\tilde{I}_j$ become $\mathbb{N}$-graded algebras over the field k.

From now on, we set $\mathrm{in}(I_j) = \mathrm{in}_>(I_j) = \mathrm{in}_{>_\lambda}(I_j)$, both the term order and the partial order being self-understood.

The main result of this part is the following "exchange" device.

Theorem 11.1.10. *Let k be a field and let $I,J \subset R = k[\mathbf{X}]$ be homogeneous ideals. Then $\mathfrak{J}(I,J)^{\sim} = \mathfrak{J}(\tilde{I},\tilde{J})$.*

Proof. We deform the three ideals $I,J,\mathfrak{J}(I,J) \subset R$ simultaneously, using Proposition 11.1.9, thus obtaining the ideals $\tilde{I},\tilde{J},\mathfrak{J}(I,J)^{\sim} \subset R[t]$, respectively. Now, these ideals are homogeneous in the standard grading of $R[t] = k[t][\mathbf{X}]$ that has $R[t]_0 = k[t]$ (cf. the definitions before Proposition 11.1.9). In particular, the join ideal $\mathfrak{J}(\tilde{I},\tilde{J}) \subset R[t]$ is well-defined and is also homogeneous in the above grading. Further, the extensions of all these "tilde" ideals to the ring $R[t,t^{-1}] = k[t,t^{-1}][\mathbf{X}]$ are still homogeneous in its standard grading that has $R[t,t^{-1}]_0 = k[t,t^{-1}]$. Finally, it is clear that the extensions of the original ideals $I,J,\mathfrak{J}(I,J) \subset R$ to either $R[t]$ or $R[t,t^{-1}]$ are homogeneous.

Using the structural isomorphism of Proposition 11.1.9(ii), at various levels, one obtains:

$$\mathfrak{J}(I,J)^{\sim}R[t,t^{-1}] \simeq \mathfrak{J}(I,J)R[t,t^{-1}] \tag{162}$$
$$\simeq \mathfrak{J}(IR[t,t^{-1}],JR[t,t^{-1}]) \tag{163}$$
$$\simeq \mathfrak{J}(\tilde{I}R[t,t^{-1}],\tilde{J}R[t,t^{-1}]) \tag{164}$$
$$\simeq \mathfrak{J}(\tilde{I},\tilde{J})R[t,t^{-1}], \tag{165}$$

where the isomorphisms in (162) and (164) go in opposite directions, the one in (163) results from flatness and (165) is just localization at $\{t^s, s \geq 0\}$. Therefore, the final composite is the identity map on $R[t,t^{-1}]$. In other words, $\mathfrak{J}(I,J)^{\sim}$ and $\mathfrak{J}(\tilde{I},\tilde{J})$ are equal after localizing at $\{t^s, s \geq 0\}$. Thus, it suffices to show that t is a nonzero divisor on both $\mathfrak{J}(I,J)^{\sim}$ and $\mathfrak{J}(\tilde{I},\tilde{J})$. For the first, it follows from Proposition 11.1.9(i) (cf. also the comments after that proposition). By the same token, t is a nonzero divisor on both $\tilde{I},\tilde{J}$. Therefore, t is a nonzero divisor on $(\tilde{I}\otimes_{k[t]}1, 1\otimes_{k[t]}\tilde{J})$; hence, also modulo the contraction of this ideal—the nonembedded join ideal $\mathcal{D}(\tilde{I},\tilde{J})$. It follows that t is a nonzero divisor modulo $\mathcal{D}(\tilde{I},\tilde{J}) \simeq \mathfrak{J}(\tilde{I},\tilde{J})$, as required. □

Theorem 11.1.11. *With the above assumptions,* $\mathrm{in}(\mathfrak{J}(I,J)) \subset \mathfrak{J}(\mathrm{in}(I),\mathrm{in}(J))$.

Proof. One has a commutative diagram of homomorphisms of rings:

$$\begin{array}{ccc}
(R[t]/\tilde{I}) \otimes_{k[t]} (R[t]/\tilde{J}) & \overset{(\mathrm{mod}\ t)}{\twoheadrightarrow} & (R/\mathrm{in}(I)) \otimes_k (R/\mathrm{in}(J)) \\
\cup & & \cup \\
R[t]/\mathfrak{J}(\tilde{I},\tilde{J}) & \overset{t\mapsto 0}{\twoheadrightarrow} & R/\mathfrak{J}(\mathrm{in}(I),\mathrm{in}(J)) \\
\| & & \varphi\uparrow \\
R[t]/\mathfrak{J}(I,J)^{\sim} & \overset{(\mathrm{mod}\ t)}{\twoheadrightarrow} & R/\mathrm{in}(\mathfrak{J}(I,J))
\end{array}$$

where the top vertical inclusions follow from the identification of the homogeneous coordinate ring of the embedded join with the join algebra and the lower leftmost vertical map comes from Theorem 11.1.10. This diagram then forces the existence of the rightmost lower vertical map φ and the latter is induced by the identity on R. Therefore, $\mathrm{in}(\mathfrak{J}(I,J)) \subset \mathfrak{J}(\mathrm{in}(I),\mathrm{in}(J))$, as claimed. □

The inclusion in Theorem 11.1.11 is an equality if and only if the middle horizontal map in the preceding diagram is also reduction modulo (t). The following example shows that equality fails to hold in general.

Example 11.1.12 (Subsection 10.2.3). Let $I \subset k[X_1,\dots,X_6]$ be the ideal generated by the following quadrics:

$$X_2X_4 + X_3X_6, \quad X_3X_5 + X_1X_6, \quad X_1X_2 - X_2X_5 + X_3X_5 - X_5X_6,$$
$$X_2X_3 + X_2X_6 - X_3X_6 + X_6^2, \quad X_3^2 + X_3X_4 + X_3X_6 - X_4X_6,$$
$$X_1X_3 + X_1X_4 + X_4X_5 + X_1X_6.$$

Then I has the following features:
- The generators form a Gröbner basis in the lexicographic order with $X_1 > X_2 > X_4 > X_5 > X_3 > X_6$
- In the above term order, $\mathrm{in}(I) = (X_1X_2, X_1X_4, X_1X_6, X_2X_3, X_2X_4, X_3X_4)$ is the edge ideal of a graph with exactly two odd cycles, each of order 3.

By Example 10.2.14, $\mathbb{D}(I,I)$ is of linear type, hence $\mathfrak{J}(I,I) = (0)$, while $\mathfrak{J}(\mathrm{in}(I),\mathrm{in}(I)) = (X_1X_2X_4, X_2X_3X_4)$. The vanishing of $\mathfrak{J}(I,I)$ reads to the effect that its secant variety is the whole of $\mathbb{P}^5$, whereas $\mathrm{Proj}(R/\mathrm{in}(I))$ is an arithmetically Cohen–Macaulay cone whose secant variety is a non-equidimensional cone of dimension 4. One can show that R/I is a Cohen–Macaulay normal domain, locally regular in the punctured spectrum. Therefore, $\mathrm{Proj}(R/I)$ is embedded as a nondegenerate smooth arithmetically normal and arithmetically Cohen–Macaulay surface in $\mathbb{P}^5$.

11.1.4 The embedded join of a union of linear spaces

In this part, we deal with the case where the ideals are generated by monomials. Keeping the notation of the previous subsection, we have a k-algebra surjection $\mathbb{J}(I,J) \simeq R/\mathfrak{J}(I,J) \twoheadrightarrow R/I \cap J$ and we seek to determine the embedded join ideal naturally as a subideal of $I \cap J$.

Lemma 11.1.13. *If $I, J \subset k[\mathbf{X}]$ are monomial ideals then so is $\mathfrak{J}(I,J)$. Moreover, if I and J are generated by square-free monomials then so is $\mathfrak{J}(I,J)$.*

Proof. It suffices to show the assertions for the ideal $\mathcal{D}(I,J)$. Identifying $k[\mathbf{X}\otimes 1 - 1\otimes \mathbf{X}]$ with $k[\mathbf{X}-\mathbf{Y}] := k[X_i - Y_i | 1 \leq i \leq n]$, and $I \otimes 1$ and $1 \otimes J$ with $I(\mathbf{X})$ and $J(\mathbf{Y})$, respectively, let an arbitrary $F \in k[\mathbf{X}-\mathbf{Y}]$ be written as a k-linear combination of monomials in the differences $X_i - Y_i$. Expanding to get a combination of monomials in $\mathbf{X}, \mathbf{Y}$, one clearly sees that no two such monomials cancel against each other. On the other hand, if $F \in (I(\mathbf{X}), J(\mathbf{Y}))$, since the latter is generated by monomials then each individual monomial of F (in $\mathbf{X}, \mathbf{Y}$) is a multiple of some monomial of $I(\mathbf{X})$ or of $J(\mathbf{Y})$. We can therefore reconstitute any original monomial of F (in the differences $X_i - Y_i$) as a k-linear combination of monomials from either $I(\mathbf{X})$ or $J(\mathbf{Y})$. Further, it is clear in this argument that if $I(\mathbf{X})$ and $J(\mathbf{Y})$ are generated by square-free monomials then $\mathcal{D}(I,I)$ is generated by square-free monomials in the differences $X_i - Y_i$. □

Proposition 11.1.14. *Let $I, J \subset R$ be ideals generated by monomials.*

(i) *If char $k = 0$ then $\mathfrak{J}(I,J)$ is generated by the set of monomials F in $\mathbf{X}$ such that any monomial factorization $F = GH$ (allowing $G = 1$ or $H = 1$) implies that either $G \in I$ or $H \in J$.*

(ii) *If char $k = p > 0$, then $\mathfrak{J}(I,J)$ is generated by the set of monomials F in $\mathbf{X}$ satisfying the following property: if $F = (X_1^{p^{e_1}})^{\alpha_1} \cdots (X_n^{p^{e_n}})^{\alpha_n}$ is the unique expression of F, with p not dividing α_i for every i, then given any factorization $Z_1^{\alpha_1} \cdots Z_n^{\alpha_n} = G(\mathbf{Z})H(\mathbf{Z})$ (with $\mathbf{Z} = \{Z_1, \dots, Z_n\}$ new variables), either $G(X_1^{p^{e_1}}, \dots, X_n^{p^{e_n}}) \in I$ or $H(X_1^{p^{e_1}}, \dots, X_n^{p^{e_n}}) \in J$.*

(iii) *(Arbitrary characteristic) If I and J are generated by square-free monomials in $\mathbf{X}$ then $\mathfrak{J}(I,J)$ is generated by the set of square-free monomials F in $\mathbf{X}$ such that any proper factorization $F = GH$ implies that $G \in I$ or $H \in J$.*

Proof. (i) By the above lemma, it suffices to prove that a monomial F in the differences $X_i - Y_i$ is contained in $\mathcal{D}(I,J)$ if and only if $F(\mathbf{X}-\mathbf{Y}) = G(\mathbf{X}-\mathbf{Y})H(\mathbf{X}-\mathbf{Y})$ implies $G(\mathbf{X}-\mathbf{Y}) \in I(\mathbf{X}-\mathbf{Y})$ or $H(\mathbf{X}-\mathbf{Y}) \in J(\mathbf{X}-\mathbf{Y})$. So let $F = F(\mathbf{X}-\mathbf{Y})$ be such a monomial. The monomials in $\mathbf{X}, \mathbf{Y}$ that appear in F with nonzero coefficients are exactly the products $G(\mathbf{X})H(\mathbf{Y})$, where G, H are monomials, with $F(\mathbf{X}-\mathbf{Y}) = G(\mathbf{X}-\mathbf{Y})H(\mathbf{X}-\mathbf{Y})$. Thus, $F \in \mathcal{D}(I,J)$ if and only if for every such G, H, one has $G(\mathbf{X}) \in I$ or $H(\mathbf{Y}) \in J(\mathbf{Y})$, which means that $G(\mathbf{X}-\mathbf{Y}) \in I(\mathbf{X}-\mathbf{Y})$ or $H(\mathbf{X}-\mathbf{Y}) \in J(\mathbf{X}-\mathbf{Y})$.

(ii) One proceeds as in the proof of part (i), noticing that now the monomials in $\mathbf{X}, \mathbf{Y}$ that appear in $F = ((X_1 - Y_1)^{p^{e_1}})^{\alpha_1} \cdots ((X_n - Y_n)^{p^{e_n}})^{\alpha_n}$ with nontrivial coefficients are exactly the monomials $G(X_1^{p^{e_1}}, \dots, X_n^{p^{e_n}})H(Y_1^{p^{e_1}}, \dots, Y_n^{p^{e_n}})$ with $Z_1^{\alpha_1} \cdots Z_n^{\alpha_n} = G(\mathbf{Z})H(\mathbf{Z})$.

(iii) By the above, $\mathfrak{J}(I,J)$ is generated by squarefree monomials. Further, intersecting the monomial generating sets obtained in (i) or (ii), respectively, with the set of all squarefree monomials yields exactly the set stated in this part. □

In the square-free case, one has another way of looking at the join ideal.

Proposition 11.1.15. *Let $I = P_1 \cap \cdots \cap P_r$ and $J = Q_1 \cap \cdots \cap Q_s$ be the prime decompositions of ideals $I, J \subset R = k[\mathbf{X}]$ generated by square-free monomials (hence the P_k's and the Q_l's are generated by variables of R). Then $\mathfrak{J}(I,J) = \bigcap_{k,l}[P_k, Q_l]$, where $[P_k, Q_l]$ denotes the ideal generated by the variables common to P_k an Q_l.*

Proof. By Proposition 11.1.8, the argument boils down to the join $\mathfrak{J}(P,Q)$, where P, Q are (prime) ideals generated by variables. Now $R/P \otimes_k R/Q$ is a domain, hence $\mathcal{D}(P,Q)$ is a prime ideal. Thus, $\mathfrak{J}(P,Q)$ is prime and then Proposition 11.1.14(iii) implies that $\mathfrak{J}(P,Q) = [P,Q]$. □

The results of Proposition 11.1.14 lead to a formula for the dimension of the join algebra $R/\mathfrak{J}(I,J)$ of (not necessarily radical) monomial ideals I, J, in terms of the so-called excess intersections $e(P_k, Q_l) := \operatorname{ht} P_k + \operatorname{ht} Q_l - \operatorname{ht}(P_k, Q_l)$ of pairs of respective associated primes.

Remark 11.1.16. Quite generally, for two ideals I, J, one defines their *excess intersection* to be $e(I,J) := \operatorname{ht} I + \operatorname{ht} J - \operatorname{ht}(I,J)$. The fact that this integer is nonnegative when R is regular is a nontrivial success of the early work on intersection theory. For the case where I and J are prime ideals, it was first proved in [172, Chapitre V, B-6, Théorème 3]. The general case follows from this by a simple device left to the reader (see [179, Section 2.7.5.1, p. 79])

In particular, one obtains a characterization as to when this dimension has the expected (i. e., maximal) value.

Proposition 11.1.17. *Let k be a perfect field and $I, J \subset R = k[X_1, \dots, X_n]$ be ideals generated by monomials and let $P_1, \dots, P_r$ and $Q_1, \dots, Q_s$ be the minimal primes of I and J, respectively. Then $\dim R/\mathfrak{J}(I,J) = \max_{k,l}\{n - e(P_k, Q_l)\}$. In particular, one has*

$$\dim R/\mathfrak{J}(I,J) = \begin{cases} n & \Leftrightarrow \operatorname{ht}(P_k, Q_l) = \operatorname{ht} P_k + \operatorname{ht} Q_l \quad \text{for some } k,l \\ \dim R/I + \dim R/J & \Leftrightarrow \operatorname{ht}(P_k, Q_l) = n \quad \text{for some } P_k \text{ and } Q_l \text{ of minimal height} \end{cases}$$

Proof. By Proposition (ii), we may replace I and J by $\sqrt{I}$ and $\sqrt{J}$, respectively. Thus, we assume that I, J are generated by square-free monomials. In this case, the formula follows from Proposition 11.1.15 since the excess intersection is $e(P_k, Q_l) = \operatorname{ht} P_k + \operatorname{ht} Q_l - \operatorname{ht}(P_k + Q_l) = \operatorname{ht}[P_k, Q_l]$. □

Here is an application to Stanley–Reisner ideals.

Corollary 11.1.18. *Let Δ, Δ' be simplicial complexes on the same set of vertices V and let I, J stand for the respective face ideals. Then $\mathbb{J}(I,J)$ is the Stanley–Reisner ring of the simplicial complex whose facets are the maximal elements of the set $\{F \cup F' \mid F, F' \text{ facets of } \Delta, \Delta'\}$. In particular, if $\Delta = \Delta'$ then $\mathfrak{J}(I,I) = (0)$ if and only if there exist facets $F, F' \in \Delta$ such that $F \cup F' = V$.*

Proof. Recall that the minimal primes of a Stanley–Reisner ring are exactly the ideals generated by the variables corresponding to the complement of a facet of the corresponding simplicial complex. The assertion is thus a consequence of Proposition 11.1.15. □

The following result shows that well structured classes of simplicial complexes are closed under the formation of embedded joins.

Proposition 11.1.19. *For given integers $2 \le s \le n$, let $I_{n,s} \subset R = k[X_1, \dots, X_n]$ denote the ideal generated by the s-products of different variables amongst $X_1, \dots, X_n$. Then:*

(i) *$\mathfrak{J}(I_{n,s}, I_{n,s}) = I_{n,2s-1}$; in particular, this ideal is nonzero if and only if $s \le (n+1)/2$.*

(ii) *$\mathbb{J}(I_{n,s}, I_{n,s})$ is a Cohen–Macaulay ring.*

Proof. (i) This is a consequence of Proposition (iii) and Corollary 11.1.20 since, by these results, $\mathbb{J}(I_{n,s}, I_{n,s})$ is generated by squarefree monomials and $I_{n,2s-1} \subset \mathbb{J}(I_{n,s}, I_{n,s}) \subset (\mathbf{X})^{2s-1}$.

(ii) Since the face ring of an n-simplex is (trivially) Cohen–Macaulay, then the face ring of any of its skeletons is Cohen–Macaulay (cf., e. g., [20, Exercise 5.1.23]). We choose to give a direct algebraic proof of this fact that emphasizes the underlying determinantal structure of the face ring. To wit, we may assume that k is infinite and that $t := 2s - 1 \le n$. Take any $t \times n$ matrix ψ with entries in k whose $t \times t$ minors are all nonzero (such matrices exist since k is infinite). Next, let φ denote the $t \times n$ matrix with linear entries in $R = k[\mathbf{X}]$ obtained from ψ by multiplying the ith column with X_i. Then $I_{n,t} = I_t(\varphi)$ (see also [211, Proof of Lemma 4.3]). Since $\operatorname{ht} I_{n,t} \ge n - t + 1$, the ring $R/I_{n,t}$ is Cohen–Macaulay. □

11.1.5 Initial degree of a join

In this part, one looks for lower bounds of the initial degree of a join in terms of the initial degrees of its constituents.

Corollary 11.1.20. *Let $I, J \subset k[\mathbf{X}] = k[X_1, \dots, X_n]$ be ideals generated by square-free monomials of degrees at least r and s, respectively. Then $\mathfrak{J}(I,J)$ is generated by square-free monomials of degrees at least $r+s-1$. In particular, if $r+s > n+1$ then $\mathfrak{J}(I,J) = (0)$.*

Proof. The assertion is an immediate consequence of Proposition 11.1.14(iii). □

Corollary 11.1.21. *Let $I, J \subset k[\mathbf{X}] = k[X_1, \dots, X_n]$ be homogeneous ideals. If $\sqrt{\mathrm{in}(I)} \subset (\mathbf{X})^r$ and $\sqrt{\mathrm{in}(J)} \subset (\mathbf{X})^s$ and $r+s > n+1$ then $\mathfrak{J}(I,J) = (0)$.*

Proof. One applies Corollary 11.1.20 taking in account that the dimension of the join algebra does not change by passing to the radicals of I and J. □

In the case of a single ideal, one can say a little more.

Proposition 11.1.22. *Let $I \subset k[\mathbf{X}] = k[X_1, \dots, X_n]$ be a homogeneous ideal of height g. If $\sqrt{\mathrm{in}(I)} \subset (\mathbf{X})^{g+1}$ then $\mathfrak{J}(I,I) = (0)$.*

Proof. As in the proof of the previous corollary, one may assume that I is generated by squarefree monomials, in which case $\mathfrak{J}(I,I)$ has the same property by Proposition 11.1.14(iii). Thus, if $\mathfrak{J}(I,I) \neq 0$ then certainly $X_1 \cdots X_n \in \mathfrak{J}(I,I)$. It follows that for any square-free monomial $G = X_{i_1} \cdots X_{i_{n-g}}$ of degree $n-g$ there exists a monomial H of degree g such that $GH = X_1 \cdots X_n \in \mathfrak{J}(I,I)$. As $H \notin I$ by hypothesis, Proposition 11.1.14(iii) implies that $G \in I$. Therefore, I contains all square-free monomials of degree $n-g$. But the latter generate an ideal of height $n-(n-g)+1 = \mathrm{ht}\, I + 1$, which is absurd. □

A question of relevance refers to how "deep" inside the ideals I, J sits the join ideal $\mathfrak{J}(I,J)$. In the homogeneous case, this can be translated in terms of the degrees of the generators of $\mathfrak{J}(I,J)$.

Corollary 11.1.23. *Let char $k = 0$ and let $I, J \subset R = k[\mathbf{X}]$ be homogeneous ideals. If $I \subset (\mathbf{X})^r$ and $J \subset (\mathbf{X})^s$, then $\mathfrak{J}(I,J) \subset (\mathbf{X})^{r+s-1}$.*

Proof. First notice that, quite generally, for an ideal $K \subset R$ and a given term order, $K \subset (\mathbf{X})^t$ if and only if $\mathrm{in}(K) \subset (\mathbf{X})^t$. Thus, we may use Theorem 11.1.11 to reduce to the case where I, J are generated by monomials. But, in this case, the assertion follows from Proposition 11.1.14(i). □

As a special case, the embedded join of two nondegenerate subschemes of projective space cannot lie on a quadric hypersurface.

11.2 Secants

The notion of a secant variety has been lightly touched upon in previous chapters, more particularly in Chapter 10.

The formal definition in the language of this chapter is that the *secant variety* (or, simply, the *secant*) to a subvariety $V \subset \mathbb{P}^n$ is the join of V with itself. It is clear that the secant is the projectively embedded version of the variety of joining lines as explained in Subsection 11.1.2. Set theoretically, the secant is the closure of the union over all secant lines to V in $\mathbb{P}^n$, where the latter are lines joining two distinct points of V.

We will use the notation $\mathrm{Sec}(V)$ already introduced in Chapter 10. Clearly, as a special case of a join property, one has $V \subset \mathrm{Sec}(V)$.

The associated *secant ideal* is the join ideal $\mathfrak{J}(I, I)$, where I is the homogeneous defining ideal of V in $\mathbb{P}^n$. It is denoted $\sigma(I)$. Likewise, $\sigma(I) \subset I$.

As a special case of Corollary 11.1.23, one has:

$$\operatorname{indeg}(\sigma(I)) \geq 2 \operatorname{indeg}(I) - 1. \tag{166}$$

Thus, $\sigma(I) = I$ if and only if V is a linear subspace.

11.2.1 Secants of determinantal and Pfaffian schemes

In this part, we discuss the secants of some structured classes of ideals, mostly generated by determinants.

A first case in the agenda is the ideal generated by the maximal minors of a catalecticant matrix—the case of a generic matrix will be a special case here (cf. [188, Theorem 4.12] where slightly different methods have been developed). We note that the notion of a catalecticant matrix in this book may be different from elsewhere.

Definition 11.2.1. Let $r \geq 1$ be an integer. An *r-catalecticant matrix* of size $m \times n$ ($2 \leq m \leq n$) is a matrix of the form

$$\mathbf{X} = \begin{pmatrix} X_1 & X_2 & \cdots & X_n \\ X_{1+r} & X_{2+r} & \cdots & X_{n+r} \\ X_{1+2r} & X_{2+2r} & \cdots & X_{n+2r} \\ \vdots & \vdots & & \vdots \\ X_{1+(m-1)r} & X_{2+(m-1)r} & \cdots & X_{n+(m-1)r} \end{pmatrix}$$

Note that the case $r = 1$ gives a Hankel matrix.

Proposition 11.2.2. *Let $\mathbf{X}$ be an r-catalecticant matrix of size $m \times n$ ($2 \leq m \leq n$) and let $I = I_m(\mathbf{X}) \subset k[X_1, \ldots, X_{n+(m-1)r}]$. Suppose that $n - m - 1 \leq r$.*

(i) *If $m \geq 3$, then $\sigma(I) = (0)$.*

(ii) *If $m = 2$, and $n-3 < r$ then $\sigma(I) = (0)$; if $m = 2$ and $n-3 = r$, then $\sigma(I)$ is the principal ideal generated by the determinant of the following $(n-3)$-catalecticant matrix of size 3×3:*

$$\begin{pmatrix} X_1 & X_2 & X_3 \\ X_{n-2} & X_{n-1} & X_n \\ X_{2n-5} & X_{2n-4} & X_{2n-3} \end{pmatrix}.$$

In particular, $\ell(\mathbb{D})$ has always the expected value.

Proof. In the range $n-m-1 \le r$, if $>$ is the lexicographic term order on $X_1 > \cdots > X_{n+(m-1)r}$ then $\mathrm{in}(I)$ is generated by the monomials along the diagonals of the maximal minors of $\mathbf{X}$ (cf. [135, Proposition III.3.8]). It follows, in particular, that the ideals $(X_1,\dots,X_{n-m+1})$ and $(X_{m+(m-1)r},\dots,X_{n+(m-1)r})$ are prime ideals containing $\mathrm{in}(I)$. Moreover, they are minimal over $\mathrm{in}(I)$. The condition $n-m-1 \le r$ implies $n-m+1 \le r+2$. Clearly, for $m \ge 3$, one has $r+2 < (m-1)r+m$. Therefore, for $m \ge 3$, we have the strict inequality $n-m+1 < m+(m-1)r$. But the latter means that these two prime ideals have no common variables. By Proposition 11.1.15, $\sigma(\mathrm{in}(I)) = (0)$. Then, by Theorem 11.1.11, $\sigma(I) = (0)$. This proves (i).

To prove (ii), note that for $m = 2$, our hypothesis boils down to $n-3 \le r$. Now, for $n-3 < r$ the above argument still goes through and, consequently, $\sigma(I) = (0)$ in this case. So, assume that $n-3 = r$. Here, the given matrix has the following shape:

$$\begin{pmatrix} X_1 & X_2 & \cdots & X_{n-2} & X_{n-1} & X_n \\ X_{n-2} & X_{n-1} & \cdots & X_{2n-5} & X_{2n-4} & X_{2n-3} \end{pmatrix}.$$

Since $\mathrm{in}(I)$ is generated by the products X_iX_j along the diagonals of the 2×2 minors, one sees that the corresponding graph has only one odd cycle, namely, the triangle with vertices corresponding to the variables X_1, X_{n-1}, X_{2n-3}. By Proposition 11.2.12, $\sigma(\mathrm{in}(I)) = (X_1X_{n-1}X_{2n-3})$. Since $\mathrm{in}(\sigma(I)) \subset (X_1X_{n-1}X_{2n-3})$ by Theorem 11.1.11, then $\mathrm{ht}\,\sigma(I) = \mathrm{ht}\,\mathrm{in}(\sigma(I)) \le 1$. One knows that I remains a prime ideal after going to the algebraic closure of k. Therefore, $\sigma(I)$ is a prime ideal by Proposition 11.1.8(ii); hence, it must be principal. On the other hand, the 2×2 minors of the 3×3 matrix in the statement belong to I; hence, the determinant of the matrix is an element of $\sigma(I)$, which is seen by substituting $X_i - Y_i$ for X_i and expanding the determinant (cf. also [188, the proof of Proposition 4.11]). Since the determinant is an irreducible polynomial, it must generate the join ideal of I. □

Remark 11.2.3. For $m = 2$, we can rewrite the r-catalecticant matrix $\mathbf{X}$ in the following way:

$$\left(\begin{array}{ccc|ccc|c|ccc} x_1 & x_{1+r} & \cdots & x_2 & x_{2+r} & \cdots & \cdots & x_r & x_{2r} & \cdots \\ x_{1+r} & x_{1+2r} & \cdots & x_{2+r} & x_{2+2r} & \cdots & \cdots & x_{2r} & x_{3r} & \cdots \end{array}\right),$$

where, for $n = qr + s$ with $s < r$, the first s blocks have $q+1$ columns and the last $r-s$ have q columns. In particular, according to [87], the ideal $I_2(\mathbf{X})$ defines a rational normal scroll of dimension r in $\mathbb{P}^{n+r-1}$. Note that these ideals $I_2(\mathbf{X})$ are special cases of defining ideal of rational normal scrolls. Using different methods, the higher secant ideals of the defining ideal of any rational scroll were calculated in [33, Proposition 2.2 and Lemma 3.1].

Our next application is to generic (resp., symmetric generic) determinantal ideals and to Pfaffian ideals of arbitrary size. A number of these results appears to be essentially known by geometric arguments (cf., e. g., [87, p. 145]), however our purpose is to give a direct algebraic proof which emphasizes the method of descending to the initial ideal.

Proposition 11.2.4. *Let $1 \leq s \leq m \leq n$, let $\mathbf{X}$ be a generic matrix of size $m \times n$, and let $I = I_s(\mathbf{X}) \subset R = k[\mathbf{X}]$. Then:*

(i) *$\sigma(I) = I_{2s-1}(\mathbf{X})$; in particular, $\ell(\mathbb{D}(I,I))$ has the expected value $\Leftrightarrow s > (m+1)/2$ or $s = 1 \Leftrightarrow \sigma(I) = (0)$ or $I = (\mathbf{X})$.*

(ii) *The secant algebra $\mathbb{J}(I,I)$ is a Cohen–Macaulay ring.*

(iii) *The values of the reduction number of the diagonal ideal are*

$$r(\mathbb{D}) = \begin{cases} 2(m-2s+2)(s-1) & \text{if } s \leq (m+1)/2 \\ 0 & \text{otherwise} \end{cases}$$

Proof. (i) Consider the so-called diagonal term order $>$ on $\mathbf{X}$ (i. e., the lexicographic term order induced by the decreasing ordering of the variables along successive rows (top to bottom)). It is well known that a Gröbner basis of I for this term order is given by the $s \times s$ minors of the matrix (cf., e. g., [201, Theorem 1]). Clearly, $\mathrm{in}(I)$ is generated by the square-free monomials obtained by reading the main diagonals of the $s \times s$ minors. In particular, the ideals $(X_{ij} \mid 1 \leq i \leq m-s+1,\ 1 \leq j \leq n-s+1)$ and $(X_{ij} \mid s \leq i \leq m,\ s \leq j \leq n)$ are among the minimal primes of $\mathrm{in}(I)$. By Proposition 11.1.15, $\sigma(\mathrm{in}(I))$ is contained in the ideal $(X_{ij} \mid s \leq i \leq m-s+1,\ s \leq j \leq n-s+1)$, and by Theorem 11.1.11, $\mathrm{in}(\sigma(I)) \subset \sigma(\mathrm{in}(I))$. Therefore, $\mathrm{ht}\,\sigma(I) \leq \max\{0, (m-2s+2)(n-2s+2)\} = \mathrm{ht}\, I_{2s-1}(\mathbf{X})$.

On the other hand, by [188, the proof of Proposition 4.11], the $(2s-1)\times(2s-1)$ minors yield elements of the ideal $\sigma(I)$. Since the minors generate the prime ideal $I_{2s-1}(\mathbf{X})$, the equality $\sigma(I) = I_{2s-1}(\mathbf{X})$ follows suit. Now, the expected value of $\ell(\mathbb{D})$ is

$$\min\{2 \dim k[\mathbf{X}]/I_s(\mathbf{X}), \dim k[\mathbf{X}]\} = \min\{2(m+n)(s-1) - 2(s-1)^2, mn\}.$$

If $s > (m+1)/2$ we have shown that $\sigma(I) = (0)$; hence, $\ell(\mathbb{D}) = \dim \mathbb{J}(I,I) = mn$ has the expected value. If $s \leq (m+1)/2$, then $I_{2s-1}(\mathbf{X}) \neq (0)$ and we have seen that $\sigma(I) = I_{2s-1}(\mathbf{X})$. Therefore, $mn > \ell(\mathbb{D}) = \dim k[\mathbf{X}]/I_{2s-1}(\mathbf{X}) = 2(m+n)(s-1) - 4(s-1)^2 = 2 \dim k[\mathbf{X}]/I_s(\mathbf{X}) - 2(s-1)^2$. It follows that $\ell(\mathbb{D})$ is strictly less than the expected value provided $s \geq 2$.

(ii) This follows immediately from (i).

(iii) We may assume that $s \leq (m+1)/2$ and that k is infinite. By (i), $\mathbb{J}(I,I) = k[\mathbf{X}]/I_{2s-1}(\mathbf{X})$. Factoring out a maximal regular sequence on $k[\mathbf{X}]/I_{2s-1}(\mathbf{X})$ consisting of linear forms we obtain an Artinian local ring whose socle is concentrated in degree $2(m-2s+2)(s-1)$ (cf. [21, Corollary 1.5] and [215, Proposition 5.1.3]). □

Remark 11.2.5. Part (i) of Proposition 11.2.4 implies that the corresponding secant variety is deficient with deficiency $2(s-1)^2$ if $s \leq (m+1)/2$. In particular, one obtains that the secant variety to the Segre embedding $\mathbb{P}^{m-1} \times \mathbb{P}^{n-1} \hookrightarrow \mathbb{P}^{mn-1}$ is deficient (with deficiency 2) for all values $m, n \geq 3$. Similar remarks can be made concerning the remaining applications of this section (e. g., secants to Veronese embeddings and to Pfaffian schemes).

Proposition 11.2.6. *Let* $\mathbf{X}$ *be a generic* $m \times m$ *symmetric matrix, and let* $I = I_s(\mathbf{X}) \subset R = k[\mathbf{X}]$, *with* $1 \leq s \leq m$. *Then:*
(i) $\sigma(I) = I_{2s-1}(\mathbf{X})$; *in particular,* $\ell(\mathbb{D}(I,I))$ *has the expected value* $\Leftrightarrow s > (m+1)/2$ *or* $s = 1 \Leftrightarrow \sigma(I) = (0)$ *or* $I = (\mathbf{X})$.
(ii) *The secant algebra* $\mathbb{J}(I,I)$ *is a Cohen–Macaulay ring.*
(iii) *The values of the reduction number of the diagonal ideal are*

$$r(\mathbb{D}) = \begin{cases} (m-2s+3)(s-1) & \text{if } s \leq (m+1)/2 \text{ and } m \text{ is odd} \\ (m-2s+2)(s-1) & \text{if } s \leq (m+1)/2 \text{ and } m \text{ is even} \\ 0 & \text{otherwise} \end{cases}$$

Proof. The proof is literally the same as the previous one, by replacing references [201] and [21] by [40, Theorem 2.8] and [41, Theorem 2.4], respectively. □

Proposition 11.2.7. *Let* $\mathbf{X}$ *be a generic* $m \times m$ *alternating matrix, and let* $I = \mathrm{Pf}_{2s}(\mathbf{X}) \subset R = k[\mathbf{X}]$, *with* $1 \leq 2s \leq m$, *be the ideal generated by the* $2s \times 2s$ *Pfaffians of* $\mathbf{X}$. *Then:*
(i) $\sigma(I) = \mathrm{Pf}_{4s-2}(\mathbf{X})$; *in particular,* $\ell(\mathbb{D}(I,I))$ *has the expected value* $\Leftrightarrow s > (m+2)/4$ *or* $s = 1 \Leftrightarrow \sigma(I) = (0)$ *or* $I = (\mathbf{X})$.
(ii) *The secant algebra* $\mathbb{J}(I,I)$ *is a Cohen–Macaulay ring.*
(iii) *The values of the reduction number of the diagonal ideal are*

$$r(\mathbb{D}) = \begin{cases} 2(m-4s+3)(s-1) & \text{if } s \leq (m+2)/4 \\ 0 & \text{otherwise} \end{cases}$$

Proof. (i) We may assume at the outset that k is a perfect field. With the lexicographic term order induced by ordering the variables $X_{1,m} > X_{1,m-1} > \cdots > X_{1,2} > X_{2,m} > X_{2,m-1} > \cdots > X_{2,3} > \cdots > X_{m-1,m}$, the $2s \times 2s$ Pfaffians of $\mathbf{X}$ form a Gröbner basis of I (cf. [100, Theorem 5.1]). Then, $\mathrm{in}(I)$ is generated by squarefree monomials, and the ideals $(X_{i,j} \mid s \leq i < j \leq m-s+1)$ and $(X_{i,j} \mid 1 \leq i < j-2s+2 \leq m-2s+2)$ are among its minimal primes. As before, one deduces that

$$\mathrm{ht}\,\sigma(I) \leq \mathrm{ht}\,\sigma(\mathrm{in}(I)) \leq \max\left\{0, \frac{(m-4s+4)(m-4s+3)}{2}\right\} = \mathrm{ht}\,\mathrm{Pf}_{4s-2}(\mathbf{X}).$$

On the other hand, notice that $I_{2s-1}(\mathbf{X}) \subset I$. It follows by [188, the proof of Proposition 4.11] that $I_{4s-2}(\mathbf{X}) \subset \sigma(I)$. Therefore, $\mathrm{Pf}_{4s-2}(\mathbf{X}) \subset \sqrt{I_{4s-2}(\mathbf{X})} \subset \sqrt{\sigma(I)} = \sigma(I)$, the last equality following from Proposition 11.1.8: (i) Therefore, we have the required equality. The argument on the expected value is similar.

(ii) This follows from (i) via [124, Theorem 15].

(iii) It is a consequence of (i) by using [21, Corollary 1.7]. □

11.2.2 Secants of varieties cut by quadrics

In this part, we will mainly deal with homogeneous ideals generated by quadrics in a polynomial ring $R = k[\mathbf{X}]$, with k a field. As one will see, they are special in a variety of ways.

Given a module E over $R = k[\mathbf{X}] = k[X_1, \dots, X_n]$ with a free presentation

$$R^r \xrightarrow{\varphi} R^m \longrightarrow E \longrightarrow 0, \tag{167}$$

where the entries of φ are linear forms, we write $k[\mathbf{T}] = k[T_1, \dots, T_m]$ and let ψ be the matrix of linear forms in $k[\mathbf{T}]$ such that $\mathbf{T} \cdot \varphi = \mathbf{X} \cdot \psi$, and consider the cokernel F of ψ

$$k[\mathbf{T}]^r \xrightarrow{\psi} k[\mathbf{T}]^n \longrightarrow F \longrightarrow 0.$$

This setting has been first introduced in [187], where F has been dubbed the *Jacobian dual* of E, and briefly discussed in Chapter 3 in greater generality.

We will apply this notion to the following setting: $I = (\mathbf{f}) \subset R = k[\mathbf{X}]$, where $\mathbf{f} = \{f_1, \dots, f_r\}$ are forms of degree 2; $E = \operatorname{coker}\Theta$, where Θ denotes the transposed Jacobian matrix of $\mathbf{f}$.

We first notice the following self-duality result.

Lemma 11.2.8. *Upon identification $k[\mathbf{X}] = k[\mathbf{T}]$ via $X_i \leftrightarrow T_i$, the R-module* $\operatorname{coker}\Theta$ *is isomorphic to its own Jacobian dual.*

Proof. Let $f(\mathbf{X}) \in k[\mathbf{X}]$ be any of the given quadratic polynomials. We need to show that $\sum_i (\partial f(\mathbf{X})/\partial X_i)\, T_i = \sum_i (\partial f(\mathbf{T})/\partial T_i)\, X_i$. Note for this that the polynomials on each side are bilinear in $\mathbf{X}$ and $\mathbf{T}$. The coefficient of the monomial $X_j T_i$ on the left-hand side is $\partial^2 f(\mathbf{X})/\partial X_i \partial X_j = \partial^2 f(\mathbf{X})/\partial X_j \partial X_i = \partial^2 f(\mathbf{T})/\partial T_j \partial T_i$, which is the coefficient of the monomial $T_i X_j$ on the right-hand side. □

As an application, we have the following.

Theorem 11.2.9 (char $k \neq 2$). *Let* $\operatorname{depth} R/I > 0$. *Then* $(0) :_R \operatorname{coker}\Theta \subset \sigma(I)$. *In particular, the $n \times n$ minors of Θ are contained in $\sigma(I)$.*

Proof. We may assume that k is infinite. Let F denote the Jacobian dual of $\operatorname{coker}\Theta$ and let $A = R/I$. We write $x_i \in A$ for the residue class of X_i. Let $\mathbf{Y} = \{U_1, \dots, U_n\}$ be new indeterminates over $k[\mathbf{T}]$ and let $k[\mathbf{T}, \mathbf{Y}]$ be graded by having $\mathbf{T}$ in degree 0 and $\mathbf{Y}$

in degree 1. In turn, $(A \otimes_k A)[\mathbf{T}]$ will be graded by having $A \otimes_k A$ in degree 0 and $\mathbf{T}$ in degree 1. Define a k-algebra homomorphism $\theta : k[\mathbf{T}, \mathbf{Y}] \to (A \otimes_k A)[\mathbf{T}]$ that "exchanges" degrees by setting $\theta(T_i) = T_i$, $\theta(Y_i) = 1/2(x_i \otimes 1) + 1/2(1 \otimes x_i)$.

Claim 1. θ induces a k-algebra homomorphism from the symmetric algebra $\mathcal{S}_{k[\mathbf{T}]}(F)$ to the symmetric algebra $\mathcal{S}_{A\otimes_k A}(\mathbb{D})$, where $\mathbb{D} \subset A \otimes_k A$ is the diagonal ideal.

Indeed, from one side, by Lemma 11.2.8 there is a presentation

$$\mathcal{S}_{k[\mathbf{T}]}(F) \simeq k[\mathbf{T}, \mathbf{Y}]/I_1(\mathbf{Y} \cdot \Theta(\mathbf{T})),$$

where $\Theta(\mathbf{T})$ is the transposed Jacobian matrix of $\mathbf{f}$ read in the $\mathbf{T}$'s. On the other hand, the presentation ideal of $\mathcal{S}_{A\otimes_k A}(\mathbb{D})$ on the polynomial ring $(A \otimes_k A)[\mathbf{T}]$ contains the set $\mathcal{T}$ of Taylor relations of the generators of $\mathbb{D}$ (cf. Proposition 10.2.7, by noting that in the present case one only needs char $k \neq 2$ as the polynomials have degree 2). But

$$\begin{aligned}\theta((\mathbf{Y}) \cdot \Theta(\mathbf{T})) &= (1/2(x_1 \otimes 1) + 1/2(1 \otimes x_1), \dots, 1/2(x_n \otimes 1) + 1/2(1 \otimes x_n)) \cdot \Theta(\mathbf{T}) \\ &= (\mathbf{T}) \cdot \Theta(1/2(x_1 \otimes 1) + 1/2(1 \otimes x_1), \dots, 1/2(x_n \otimes 1) + 1/2(1 \otimes x_n)),\end{aligned}$$

where the second equality follows from the definition of the Jacobian dual. On the other hand, by the well-known polarization process, it follows that $\mathcal{T}$ are the entries of the row vector $(\mathbf{T}) \cdot \Theta(1/2(x_1 \otimes 1) + 1/2(1 \otimes x_1), \dots, 1/2(x_n \otimes 1) + 1/2(1 \otimes x_n))$. This proves the claim.

Next, we compose the obtained map $\mathcal{S}_{k[\mathbf{T}]}(F) \to \mathcal{S}_{A\otimes_k A}(\mathbb{D})$ with the canonical surjection $\mathcal{S}_{A\otimes_k A}(\mathbb{D}) \twoheadrightarrow \mathcal{R}_{A\otimes_k A}(\mathbb{D})$ onto the Rees algebra of $\mathbb{D}$.

Claim 2. Under this composition, the annihilator $(0) : F$ maps to (0).

To see this, it suffices to show that the ideal generated by the elements $\{1/2(x_1 \otimes 1) + 1/2(1 \otimes x_1), \dots, 1/2(x_n \otimes 1) + 1/2(1 \otimes x_n)\}$ has positive grade in the Rees algebra. Let then $\mathfrak{m} = (\mathbf{x})A$ and $K = A \otimes_k \mathfrak{m}$. Since k is infinite and depth $A > 0$, one may, without loss of generality, assume that x_1 is A-regular, hence $x_1 \otimes 1$ is $(A \otimes_k \mathrm{gr}_{\mathfrak{m}}(A))$-regular. But $A \otimes_k \mathrm{gr}_{\mathfrak{m}}(A) \simeq \mathrm{gr}_K(A \otimes_k A)$ by flatness and $x_1 \otimes 1$ is the leading form in $\mathrm{gr}_K(A \otimes_k A)$ of the element $x_1 \otimes 1 + 1 \otimes x_1$. Since K is graded, $\cap_s K^s = (0)$. Therefore, $x_1 \otimes 1 + 1 \otimes x_1$ is $(A \otimes_k A)$-regular and the result follows since char $k \neq 2$.

Thus, the composite map induces a homomorphism $\mathcal{S}_{k[\mathbf{T}]/(0:F)}(F) \to \mathcal{R}_{A\otimes_k A}(\mathbb{D})$ of k-algebras. Composing further with the canonical surjection $\mathcal{R}_{A\otimes_k A}(\mathbb{D}) \twoheadrightarrow \mathbb{T}_{A/k}$ onto the associated graded ring of $\mathbb{D}$, one obtains a map that sends Y_i to the residue class of the element $1/2(x_i \otimes 1) + 1/2(1 \otimes x_i)$ modulo $\mathbb{D}$ which is but the element $x_i \otimes 1 = x_i \in A \subset \mathbb{T}_{A/k}$. This implies that the irrelevant ideal $(\mathcal{S}_{k[\mathbf{T}]/(0:F)}(F))_+$ maps to the maximal ideal $(\mathbf{x}) \subset A$. Thus, we finally get a k-algebra homomorphism

$$k[\mathbf{T}]/(0 : F) = \mathcal{S}_{k[\mathbf{T}]/(0:F)}(F)/(\mathcal{S}_{k[\mathbf{T}]/(0:F)}(F)_+) \to \mathbb{T}_{A/k}/(\mathbf{x})\mathbb{T}_{A/k}, \tag{168}$$

which is induced by the identity map $k[\mathbf{T}] \to k[\mathbf{T}]$. Therefore, $(0 : F)$ is contained in the $(\mathbf{T})$-presentation ideal of $\mathbb{T}_{A/k}/(\mathbf{x})\mathbb{T}_{A/k}$. Moreover, the latter is isomorphic to the fiber cone of $\mathbb{D}$, hence isomorphic to the subalgebra $k[(x_1 \otimes 1) - (1 \otimes x_1), \ldots, (x_n \otimes 1) - (1 \otimes x_n)] \subset A \otimes A$ since $f_1, \ldots, f_r$ have the same degree. Finally, applying the transfer isomorphism $X_i \otimes 1 - 1 \otimes X_i \mapsto X_i$ to this subalgebra and using Lemma 11.2.8, one gets the required result that $(0) :_R \mathfrak{J}(\Theta) \subset \sigma(I)$. □

Corollary 11.2.10. *Let $I \subset R = k[X_1, \ldots, X_n]$ be an absolutely prime ideal generated by quadrics and let Θ be the transposed Jacobian matrix of the generators of I. Assume that $\ell(\mathbb{D}) = n - 1$ and that $I_n(\Theta) \neq 0$. Then the reduction number $r(\mathbb{D}) \leq n - 1 = \operatorname{edim} R/I - 1$.*

Proof. Since $\dim R/\sigma(I) = \ell(\mathbb{D}) = n-1$ and $\sigma(I)$ is prime by Proposition 11.1.8, it follows that this ideal is principal, say, $\sigma(I) = (G)$. Using the last assertion of Theorem 11.2.9 and the present hypothesis on the 0th Fitting ideal, at least one $n \times n$ determinant of Θ yields a nonzero polynomial relation of degree n of $\mathbb{D}$; hence, must be a multiple of G. This implies that $\deg G \leq n$, in which case it readily follows that $r(\mathbb{D}) \leq n - 1$. □

There is not a great deal of many a priori conditions that guarantee the nonvanishing of the 0th Fitting ideal $I_n(\Theta)$. One of these is originally related to the so-called row condition introduced in [2], namely, one says that a matrix φ with entries in a ring R satisfies the *row condition* if, after a finite number of elementary row operations on φ, the entries of some row generate the ideal $I_1(\varphi)$.

Lemma 11.2.11 (k infinite and char $k \neq 2$). *Let $I \subset R = k[\mathbf{X}] = k[X_1, \ldots, X_n]$ be an ideal generated by quadrics. Assume that $I_1(\Theta) = (\mathbf{X})$ (which, if k is perfect, means that $\operatorname{edim}(R/I)_P < \dim R_P$ for every prime ideal $P \subsetneq (\mathbf{X})$). Then Θ satisfies the row condition if and only if $I_n(\Theta) \neq (0)$.*

Proof. To say that Θ satisfies the row condition means that, say, its first row contains n linearly independent linear forms. This in turn means that, after finitely many k-linear column operations, one has $\partial f_i/\partial X_1 = X_i$, $1 \leq i \leq n$. The latter condition is equivalent to saying that $f_1 = (1/2)X_1^2 + g_1$, $f_i = X_1 X_i + g_i$ for $2 \leq i \leq n$, where $g_1, \ldots, g_n$ are quadrics in the variables $X_2, \ldots, X_n$ alone. This finally translates into $\partial f_j/\partial X_i \equiv \delta_{ji} X_1$ (mod $(X_2, \ldots, X_n)$), $1 \leq i, j \leq n$ or, equivalently, into

$$\Theta \equiv \left(\begin{array}{ccc|c} X_1 & & 0 & \\ & \ddots & & * \\ 0 & & X_1 & \end{array} \right) \quad (\text{mod } (X_2, \ldots, X_n)).$$

But Θ has the above form after k-linear column operations if and only if $I_n(\Theta) \not\equiv 0$ (mod $(X_2, \ldots, X_n)$). Since k is infinite, the latter holds after a k-linear change of variables, i. e., after k-linear row operations on Θ, if and only if $I_n(\Theta) \neq (0)$. □

11.2.3 Secant algebras of monomial ideals

In this subsection, we deal with monomial ideals in $R = k[\mathbf{X}]$, expanding further on some special behavior of this class of ideals. For the main terminology, we refer to [219].

An ideal $I \subset k[\mathbf{X}]$ generated by square-free monomials of degree 2 is naturally associated with a simple graph G in which the vertices correspond to the variables $\mathbf{X}$ and the edges to the given generators of I. We then speak of I as of the *edge ideal* of G and denote it by $I(G)$ to keep track of the corresponding graph. A *cycle* of G is a set of edges of G forming a polygon. The algebraic counterpart is given by a subset $\{X_{i_1},\dots,X_{i_r}\}$ of the variables with the property that $\{X_{i_1}X_{i_2}, X_{i_2}X_{i_3},\dots,X_{i_{r-1}}X_{i_r}, X_{i_r}X_{i_1}\}$ correspond to edges of G. The cycle is *even* (resp., *odd*) if r is even (resp., odd). The cycle is said to be *chordless* (in the graph) if no other pair involving these variables correspond to an edge of G (in the combinatorial language this means that the induced subgraph corresponding to the given vertices is the cycle itself).

Proposition 11.2.12. *Let $I = I(G) \subset R$ be the edge ideal of a simple graph G. Then the secant ideal $\sigma(I) \subset R$ is minimally generated by the monomials $X_{i_1}\cdots X_{i_r}$ such that $\{X_{i_1},\dots,X_{i_r}\}$ correspond to the vertices of a chordless odd cycle of G.*

Proof. By Proposition 11.1.14(iii), $\mathfrak{J}(I,I)$ is generated by squarefree monomials $X_{i_1}\cdots X_{i_r}$ such that any bipartition of the set $\{X_{i_1},\dots,X_{i_r}\}$ contains a pair X_{i_j}, X_{i_k} corresponding to an edge of G. This means that the induced subgraph with vertices $\{X_{i_1},\dots,X_{i_r}\}$ is not bipartite, hence contains an *odd* cycle of G. Thus, the embedded secant ideal is generated by the monomials whose support correspond to odd cycles. Finally, if $C \subset G$ is an odd cycle having a chord which is an edge of G then one of the two subcycles determined by this chord, say C', is odd. Clearly, the monomial whose support is C is a multiple of the one whose support is C'. This procedure shows that the monomials whose support is a chordless odd cycle form a set of minimal generators of the embedded secant ideal. □

This result was extended for arbitrary secant ideals $\mathfrak{J}_t(I)$ in [203]. From graph theory, recall that a graph G is said to be *t-colorable* if it can be colored by t colors such that no two adjacent vertices have the same color. To any subset V of the vertices of G, let G_V denote the full subgraph with these vertices and set $m_V = \prod_{i\in V} X_i$.

Theorem 11.2.13 ([203, Theorem 3.2]). *Given a simple graph G, with edge ideal $I = I(G)$, the secant ideal $\mathfrak{J}_t(I)$ is generated by the square-free monomials m_V such that G_V is not $(t-1)$-colorable. In addition, a set of minimal generators of $\mathfrak{J}_t(I)$ is given by those monomials m_V such that G_V is not $(t-1)$ colorable but G_U is $(t-1)$-colorable for every proper subset $U \subset V$.*

The proof is outside the scope of this book.

11.2.4 Relation to starlike linearity

The next result is perhaps more comprehensive.

Proposition 11.2.14. *Let G be a simple connected graph and let $I \subset R = k[\mathbf{X}]$ denote the corresponding ideal of edges. Then G is bipartite if and only if $R/I(G)$ is set theoretically starlike linear.*

Proof. If $A = R/I(G)$ is set theoretic starlike linear then, as we have already remarked several times, the analytic spread $\ell(\mathbb{D})$ is maximum, hence $\mathfrak{J}(I,I) = \{0\}$. By Proposition 11.2.12, G has no odd cycles, hence is bipartite.

Conversely, if G is bipartite, we will prove, by induction on $\#\mathbf{X}$ (the number of vertices of G), that $\mathbb{D}$ is locally everywhere generated by analytically independent elements. For this, we identify $V(\mathbb{D}) \subset \operatorname{Spec}(A \otimes_k A)$ with $V(I) \subset \operatorname{Spec}(R)$. So, let $P \subsetneq (\mathbf{X})$ be a prime ideal coming from a prime ideal of $A \otimes_k A$ containing $\mathbb{D}$. Then one can invert a variable X_i in the localization at P. One gets

$$I_P \simeq I(G \setminus X_i)(k[\mathbf{X} \setminus \{X_{i_1}, \dots, X_{i_r}\}][X_i, X_i^{-1}])_Q,$$

where Q is a prime ideal of $k[\mathbf{X} \setminus \{X_{i_1}, \dots, X_{i_r}\}][X_i, X_i^{-1}]$, $G \setminus X_i$ denotes the (proper) subgraph of G obtained by removing the vertex corresponding to X_i as well as all the edges adjacent to this vertex, and $\{X_{i_1}, \dots, X_{i_r}\}$ correspond to its adjacent vertices. Thus, by induction on the number of vertices of G, $\mathbb{D}_P$ is generated by analytically independent elements. Locally at $P = (\mathbf{X})$ itself the analytic independence follows again from the above result. □

A stronger result than Proposition 11.2.14 has been conjectured in [175, Conjecture 5.12]. We extend this conjecture to a question related to Section 10.2.

Conjecture 11.2.15. *Let $I \subset R = k[\mathbf{X}]$ be an ideal generated by forms of degree 2 that form a Gröbner basis of I with respect to some term order of the monomials in R. If R/I is set theoretic starlike linear, then it is starlike linear.*

The conjecture does not hold beyond degree 2 (see Exercise 10.3.11).

The following makes a good case for the conjecture.

Proposition 11.2.16. *Let $G = (V,E)$ (V=vertices, E=edges) be a bipartite graph satisfying the following condition: there is a labeling of the vertices V, say, $V = \{v_1, \dots, v_n\}$, such that if $\{v_i, v_j\} \in E$ then also $\{v_i, v_k\} \in E$ for all $k \geq j$. If $I \subset R = k[\mathbf{X}] = k[X_1, \dots, X_n]$ is the edge ideal of G, then R/I is starlike linear.*

Proof. By Proposition 11.2.14, R/I is indeed set theoretically starlike linear. By [188, Proposition 4.1], the kernel of the canonical map $S(\mathbb{D}) \to \mathcal{R}(\mathbb{D})$ from the symmetric algebra onto the Rees algebra of $\mathbb{D}$ is nilpotent. Therefore, it suffices to show that $S(\mathbb{D})$ is reduced.

For this, we consider a suitable term order of the monomials of the polynomial ring $k[\mathbf{X}, \mathbf{Y}, \mathbf{T}]$, where $\mathbf{Y}$ and $\mathbf{T}$ are mutually independent duplicates of $\mathbf{X}$. Namely, we take the lexicographic order with respect to the ordering $X_1 > \cdots X_n > Y_1 > \cdots > Y_n > T_1 > \cdots > T_n$ of the variables. Letting x_ν denote the residue class of X_ν in R/I, we map $k[\mathbf{X}, \mathbf{Y}, \mathbf{T}]$ surjectively onto $S(\mathbb{D})$ by mapping

$$X_\nu \mapsto x_\nu \otimes 1, Y_\nu \mapsto 1 \otimes x_\nu, T_\nu \mapsto x_\nu \otimes 1 - 1 \otimes x_\nu.$$

Let $\mathcal{J}$ denote the kernel of this map. Since G is bipartite satisfying the given assumption, we have indices $1 \le i < l \le m < j < k \le n$ such that $X_iX_j \in I$ (hence $X_iX_k \in I$) and $X_lX_j \in I$. Now, we claim that the union of the following sets of polynomials constitute a Gröbner basis of $\mathcal{J}$ in the term order under consideration.

(1) X_iX_j

(2) Y_iY_j

(3) $X_iT_j + Y_jT_i$

(4) $X_jT_i + Y_iT_j$

(5) $(Y_jT_k - Y_kT_j)T_i$

(6) $(Y_iT_l - Y_lT_i)T_j$

(7) $(X_\nu - Y_\nu)T_\mu - (X_\mu - Y_\mu)T_\nu$, with $1 \le \nu < \mu \le n$

To see this, note that the union of the sets in (1), (2), (3), (4), (7) generate $\mathcal{J}$ by definition of the symmetric algebra and the form of the syzygies of the generators of I. Next, the polynomials in (5) and (6) are S-polynomials of pairs of elements taken from (3) and (4), respectively. Applying the Buchberger's criterion, along with a case by case inspection, shows that the union of the above sets of polynomials form a Gröbner basis of $\mathcal{J}$—here, one will need the hypothesis that $X_iX_j \in I$ implies that $X_iX_k \in I$ whenever $k \ge j$.

To conclude, since the leading terms of this Gröbner basis are square-free, then $\mathcal{J}$ is radical ideal. □

Corollary 11.2.17. *Let $I \subset R$ be a homogeneous ideal whose initial ideal, in some term order of R, is the edge ideal of a bipartite graph having the property stated in Proposition 11.2.16. Then R/I is starlike linear.*

Proof. It follows from Proposition 11.2.16 via Proposition 11.3.8 in the next section. □

Recall the notion of an r-catalecticant from Definition 11.2.1.

Proposition 11.2.18. *Let $\mathbf{X}$ be an r-catalecticant matrix of size $2 \times n$ and let $I = I_2(\mathbf{X}) \subset k[\mathbf{X}] = k[X_1, \ldots, X_{n+r}]$. If $r \ge n - 2$, then $k[\mathbf{X}]/I$ is starlike linear.*

Proof. It is known that the 2-minors are a Gröbner basis in the lexicographic order. It follows that the initial ideal in(I) of I is generated by the products of the entries along the main diagonals of the minors. Since $r \ge n - 2$, these square-free monomials of degree 2 define the edge ideal of a bipartite graph with partition

$$\{1,2,\ldots,n-1\} \cup \{r+2,r+2,\ldots,n+r-1,n+r\}$$

satisfying the assumption of Proposition 11.2.16. The result now follows from Corollary 11.2.17. □

Remark 11.2.19. Conjecture 11.2.15 does not reduce to the case of monomials as is shown in Example 11.1.12.

11.2.5 Higher secants

Special cases of the join construction are the higher secant varieties.

Let $I \subset R = k[X_1,\ldots,X_n]$ be a homogeneous ideal and t a positive integer. Define ideals $\mathfrak{J}_t(I) \subset R$ inductively by setting $\mathfrak{J}_1(I) = \mathfrak{J}(I,I)$ and $\mathfrak{J}_t(I) = \mathfrak{J}(I,\mathfrak{J}_{t-1}(I))$ if $t > 1$.

Definition 11.2.20. The ideal $\mathfrak{J}_t(I)$ is the *secant ideal of order t*, while the subscheme $\mathrm{Sec}_t(X) = V(\mathfrak{J}_t(I)) \subset \mathbb{P}^{n-1}$ is called the *variety of t-secants* of $X = V(I)$.

As a set, $\mathrm{Sec}_t(X)$ is the closure of the union of all t-dimensional linear subspaces spanned by $t+1$ points of X (at least if k is algebraically closed). One could set $\mathfrak{J}_0(I) := I$ for the sake of completion.

By recursion, Theorem 11.1.11 implies the following.

Proposition 11.2.21. *Let $I \subset R = k[\mathbf{X}]$ be a homogeneous ideal and $>$ a term order on R. Then, for every $t \geq 1$,*

$$\mathrm{in}_>(\mathfrak{J}_t(I)) \subset \mathfrak{J}_t(\mathrm{in}_>(I)) \tag{169}$$

Using this result, we can estimate numerical invariants of the secant ideals of I through the secant ideals of the monomial ideal $\mathrm{in}_>(I)$. For example, since the Hilbert function is preserved by initial ideals then

$$\dim k[\mathbf{X}]/\mathfrak{J}_t(\mathrm{in}_>(I)) \leq \dim k[\mathbf{X}]/\mathfrak{J}_t(I) \tag{170}$$

and, if this inequality is an equality, then

$$\deg k[\mathbf{X}]/\mathfrak{J}_t(\mathrm{in}_>(I)) \leq \deg k[\mathbf{X}]/\mathfrak{J}_t(I). \tag{171}$$

Likewise, Theorem 11.1.23 implies the following.

Corollary 11.2.22. *Let char $k = 0$ and let $I \subset k[\mathbf{X}]$ be a homogeneous ideal. If $I \subset (\mathbf{X})^r$, then $\mathfrak{J}_t(I) \subset (\mathbf{X})^{(t+1)(r-1)+1}$*

Thus the variety of t-secants of a nondegenerate subscheme of projective space cannot be contained in a hypersurface of degree $t+1$ (at least if char $k=0$) – a result originally observed by M. Catalano-Johnson [34] (a proof given by M. Catalisano appears in [71, Lecture 7]). This result has been proved in [204] in all characteristics.

Clearly, equality in (169) implies equalities both in (170) and (171). Of course, such equalities are expected to be dependable on the choice of the term order. The following cases were affirmatively established:

- Any diagonal term order for the ideal generated by the 2×2 minors of a generic $m\times n$ matrix [203, Theorem 4.9].
- The usual diagonal term order for the ideal generated by the 2×2 minors of a generic symmetric $m\times m$ matrix [203, Example 4.12].
- The usual diagonal term order for the ideal generated by the 4×4-pfaffians of a generic $m\times m$ skew-symmetric matrix [203, Example 4.13].
- The degree lexicographic monomial order on $k[x_1,\dots,x_n]$ induced by the order of the indeterminates $x_1>\cdots>x_n$ for the ideal generated by the 2×2 minors of a $2\times(n-c)$ (for a fixed $1\le c\le n-2$) extended Hankel matrix [145, Theorem 3.8].

For further results along this line, see [204].

Proposition 11.2.23. *Let k be a field of characteristic zero, let $J\subset I$ be monomial ideals such that* $\operatorname{indeg}(I)\ge 2$. *Then* $\mathfrak{J}(I,J)\subset I^{(2)}$.

Proof. By the well-known Nagata–Zariski theorem on symbolic powers, it suffices to prove that

$$\mathfrak{J}(I,J)\subset\left\{f\in I\mid\frac{\partial f}{\partial X_i}\in I,\ 1\le i\le n\right\}.$$

Thus, let $\mathbf{X}^\alpha\in\mathfrak{J}(I,J)$. We know that $\mathfrak{J}(I,J)\subset I\cap J=J\subset I$. To show that $\frac{\partial f}{\partial X_i}\in I$, we may assume that $X_i|\mathbf{X}^\alpha$. In this case, $\frac{\partial f}{\partial X_i}=\alpha_i\mathbf{X}^\alpha/X_i$, so it suffices to show that $\mathbf{X}^\alpha/X_i\in I$. But notice the easy factorization $\mathbf{X}^\alpha=X_i\cdot\mathbf{X}^\alpha/X_i$. Since $X_i\notin I$ by assumption (hence neither $X_i\in J$), then Proposition 11.1.14 implies that $\mathbf{X}^\alpha/X_i\in I$. □

11.3 Excerpts on the diagonal ideal

This section is based on unpublished joint work of the senior author with Bernd Ulrich. The authors thank Ulrich for agreeing to these insertions. The collected material below gives but a glimpse of the properties of the diagonal ideal of two ideals. As one will see, a fuller coverage of these is rather delicate.

11.3.1 Syzygies and dimension

Recall that R is a k-algebra essentially of finite type and I, J are ideals of R. One sets $A := R/I$, $B := R/J$, $C := R/I \cap J$. Assume that C and either A or B are flat over k.

The following result captures the nature of the syzygies of $\mathbb{D}(I, J)$, pretty much as in the case of one single algebra.

Proposition 11.3.1. *Let k is a ring containing the rationals. Then the $(A \otimes_k B)$-module of first syzygies of $\mathbb{D}(I, J)$ is generated by the Koszul relations on (the residues of) the elements $\mathbf{X} \otimes 1 - 1 \otimes \mathbf{X} = \{X_i \otimes 1 - 1 \otimes X_i \mid 1 \leq i \leq n\}$ and the syzygies coming from the Taylor expansion of $f(\mathbf{X} \otimes 1)$ at $1 \otimes \mathbf{X}$, for all $f \in I \cap J$.*

Proof. Let us argue that one may replace $A \otimes_k B$ by $C \otimes_k C$. This follows from the following.

Claim. $\mathbb{D}(I, J) \simeq \mathbb{D}(I \cap J, I \cap J) \otimes_{C \otimes_k C} (A \otimes_k B)$.

To prove this assertion, consider the free C-resolutions $F_.$ and $G_.$ of A and B, respectively. Then $F_. \otimes_k G_.$ is a free resolution of $A \otimes_k B$ over $C \otimes_k C$ since $\mathrm{Tor}_i^k(A, B) = (0)$ for every $i > 0$ by flatness. Using these resolutions to compute Tor, one sees that $\mathrm{Tor}_1^{C \otimes_k C}(C, A \otimes_k B) \simeq \mathrm{Tor}_1^C(A, B) = (0)$. This shows that the exact sequence

$$(C \otimes_k C)^m \to (C \otimes_k C)^n \to C \otimes_k C \to C \to 0$$

of $(C \otimes_k C)$-modules remains exact after tensoring with the $(C \otimes_k C)$-module $A \otimes_k B$. Since the ideal $\mathbb{D}(I \cap J, I \cap J)$ is the kernel of the multiplication homomorphism $C \otimes_k C \to C$, the assertion of the claim ensues.

To continue the proof of the proposition, note the natural isomorphism

$$\mathbb{D}((0), I \cap J)/(I \cap J) \otimes_k C \simeq \mathbb{D}(I \cap J, I \cap J)$$

and that $\mathbb{D}((0), I \cap J)$ is generated by the image of the $R \otimes_k R$-regular sequence $\mathbf{X} \otimes 1 - 1 \otimes \mathbf{X}$, whereas for $f \in I \cap J$ the Taylor expansion of $f(\mathbf{X} \otimes 1)$ at $1 \otimes \mathbf{X}$ rewrites $f(\mathbf{X} \otimes 1)$ as a linear combination of the elements of $\mathbf{X} \otimes 1 - 1 \otimes \mathbf{X}$. This proves the contention. □

As a matter of notation, given a prime ideal $p \in \operatorname{Spec} R/(I, J)$, we denote by the same symbol its natural inverse image in any of the rings $A \otimes_k B$, A and R.

Lemma 11.3.2. *With the above notation, assume in addition that k is a field and A and B are equidimensional. For any prime ideal $p \in \operatorname{Spec} R/(I, J)$, one has*

$$\dim(A \otimes_k B)_p = \dim A_p + B_p + \mathrm{tr.deg.}_k R_p/pR_p \tag{172}$$

Proof. The argument is the same as in the proof of (157) with the required changes. □

Proposition 11.3.3. *Let k be a perfect field and let $A = R/I$, $B = R/J$, $C = R/I \cap J$ be as above. Assume that A and B are equidimensional. The following are equivalent:*

(a) $\mathbb{D}(I,J)$ *satisfies the property* F_1.
(b) $\operatorname{edim} C_p \le \dim A_p + \dim B_p$, *for every* $p \in \operatorname{Spec} R/(I,J)$.

Proof. Using (172), (i) means that

$$\mu(\mathbb{D}(I,J)_p) \le \dim(A \otimes_k B)_p = \dim A_p + \dim B_p + \operatorname{trdeg}_k R/p$$

for every $p \in \operatorname{Spec} R/(I,J) = \operatorname{Spec}\,(A \otimes_k B)/\mathbb{D}(I,J)$. On the other hand, by the claim in the proof of Proposition 11.3.1 and by the definition of the Kähler differentials in terms of the single diagonal ideal, one has

$$\mu(\mathbb{D}(I,J)_p) = \mu(\mathbb{D}(I \cap J, I \cap J)_p) = \mu(\Omega_k(C_p)).$$

Since $\mu(\Omega_k(C_p)) = \operatorname{edim} C_p + \operatorname{trdeg}_k R/p$ for k perfect [73], one is clearly through. □

11.3.2 The linear type property

Recall that $\mathbb{D}(I,J)$ is the kernel of the multiplication map $A \otimes_k B \to R/(I,J)$. The role of the linear type property for the diagonal ideal $\mathbb{D}(I,J)$ is that it implies that the latter has maximal analytic spread, and hence, the join ideal of I, J is a polynomial ring.

11.3.2.1 Via the excess intersection

Assume that k is a field. If A and B are equidimensional, then one has

$$\begin{aligned}\operatorname{ht} \mathbb{D}(I,J) &= \dim A \otimes B - \operatorname{ht}(I,J) = 2\dim R - (\operatorname{ht} I + \operatorname{ht} J) - (\dim R - \operatorname{ht}(I,J))\\ &= \dim R + \operatorname{ht}(I,J) - \operatorname{ht} I - \operatorname{ht} J.\end{aligned}$$

Thus, $\dim R - \operatorname{ht} \mathbb{D}(I,J) = \operatorname{ht} I + \operatorname{ht} J - \operatorname{ht}(I,J)$. One recognizes the right-hand side of this equality as the excess intersection of I and J (see Remark 11.1.16).

Throughout the rest of this part, we assume for simplicity that $R = k[X_1, \dots, X_n]$ ($n \ge 2$), a polynomial ring over the field k. We next examine the cases of low value of the excess intersection. The basic argument will invoke the criteria for linear type property as stated in [96] in terms of the property F_1 and the so-called sliding depth property.

Proposition 11.3.4. *If* $A = R/I$ *and* $B = R/J$ *are Cohen–Macaulay and* $e(I,J) = 0$ *then* $\mathbb{D}(I,J)$ *is generated by a regular sequence.*

Proof. As seen above, $e(I,J) = 0$ implies that $\operatorname{ht} \mathbb{D}(I,J) = \dim R$, while $\mu(\mathbb{D}(I,J)) = \operatorname{edim} R$. Therefore, $\mathbb{D}(I,J)$ is an ideal of the principal class in the Cohen–Macaulay ring $A \otimes_k B$; hence, it is generated by a regular sequence. □

Proposition 11.3.5. *Let k be a perfect field. Assume that:*

(i) $e(I,J) = 1$.
(ii) $A = R/I$ *and* $B = R/J$ *are Cohen–Macaulay.*
(iii) $\operatorname{edim}(R/I \cap J)_p \leq \operatorname{ht} p - 1$ *for every minimal prime* p *of* $R/(I,J)$ *of least height.*

Then:
(1) $\mathbb{D}(I,J)$ *is of linear type.*
(2) *If, in addition,* $\operatorname{depth} R/(I,J) \geq \dim R/(I,J) - 1$ *then the associated graded ring* $\operatorname{gr}_{\mathbb{D}(I,J)}(A \otimes_k B)$ *and the Rees algebra* $\mathcal{R}_{A\otimes_k B}(\mathbb{D}(I,J))$ *are both Cohen–Macaulay.*

Proof. (1) Since excess intersection one implies that the ideal $\mathbb{D}(I,J)$ is an almost complete intersection, by Proposition 2.3.2 it is of linear type if (and only if) condition (F_1) holds for this ideal (see [96, Theorem 5.1 and Theorem 5.4]). By Corollary 11.3.3, this will take place if and only if $\operatorname{edim}(R/I \cap J)_p \leq \dim A_p + \dim B_p$ for every prime $p \in \operatorname{Spec}(R/(I,J))$. Again, since $\mathbb{D}(I,J)$ is an almost complete intersection, it suffices to consider such inequalities in the case where $\operatorname{ht} p = \operatorname{ht}(I,J) = \operatorname{ht} I + \operatorname{ht} J - 1$. But here

$$\begin{aligned}\dim A_p + \dim B_p &= \operatorname{ht} p - \operatorname{ht} I + \operatorname{ht} p - \operatorname{ht} J = 2\operatorname{ht} p - \operatorname{ht} I - \operatorname{ht} J \\ &= 2(\operatorname{ht} I + \operatorname{ht} J - 1) - \operatorname{ht} I - \operatorname{ht} J = \operatorname{ht} I + \operatorname{ht} J - 2 \\ &= \operatorname{ht} p - 1,\end{aligned}$$

as required.

(2) With the assumption that $\operatorname{depth} R/(I,J) \geq \dim R/(I,J) - 1$, $\mathbb{D}(I,J)$ satisfies the sliding-depth condition [96, Theorem 5.4]. In such a case, one knows that $\operatorname{gr}_{\mathbb{D}(I,J)}(A \otimes_k B)$ is Cohen–Macaulay. In addition, since $n \geq 2$ then $\operatorname{ht} \mathbb{D}(I,J) > 0$, and thus $\mathcal{S}_{A\otimes_k B}(\mathbb{D}(I,J)) = \mathcal{R}_{A\otimes_k B}(\mathbb{D}(I,J))$ is Cohen–Macaulay as well [96, Theorem 9.1]. □

Proposition 11.3.6. *Let k be a perfect field. Assume that:*
(i) $e(I,J) = 2$.
(ii) $A = R/I$ *and* $B = R/J$ *are Gorenstein.*
(iii) $R/(I,J)$ *is Cohen–Macaulay.*
(iv) $\operatorname{edim}(R/I \cap J)_p \leq \operatorname{ht} p - i$ *for every prime* $p \supset (I,J)$ *such that* $\operatorname{ht} p = \operatorname{ht} I + \operatorname{ht} J - i$, *with* $i = 1, 2$.

Then
(1) $\mathbb{D}(I,J)$ *is of linear type.*
(2) *The associated graded ring* $\operatorname{gr}_{\mathbb{D}(I,J)}(A \otimes_k B)$ *is Gorenstein.*
(3) $(n \geq 3)$ *The Rees algebra* $\mathcal{R}_{A\otimes_k B}(\mathbb{D}(I,J))$ *is Cohen–Macaulay.*

Proof. The argument is similar to the one in the previous proposition by drawing upon [96, Proposition 5.5 and Theorem 9.1]. Therefore, the assertions on $\mathbb{D}(I,J)$ being of linear type and on $\mathcal{S}_{A\otimes_k B}(\mathbb{D}(I,J))$ or $\operatorname{gr}_{\mathbb{D}(I,J)}(A \otimes_k B)$ being Cohen–Macaulay or Gorenstein, respectively, follow from that result. □

The assumptions in the last proposition are in general essential (see Exercise 11.4.10).

From the previous example, it seems natural to look for conditions in order that the diagonal ideal $\mathbb{D}(I,J)$ satisfy the sliding-depth property if at least one of the rings R/I and R/J is not Gorenstein. In the special case of nested ideals, one can give a reasonably precise picture.

Proposition 11.3.7. *Let k be a perfect field. Let $I \subset J \subset R$ be ideals of height 2 such that $A = R/I$ is Cohen–Macaulay and $B = R/J$ is a complete intersection. Then:*

(1) *$\mathbb{D}(I,J)$ satisfies the sliding-depth property if and only if $R/(I:J)$ is Cohen–Macaulay.*

(2) *Suppose that the equivalent conditions of item (1) hold. Assume moreover that A is regular locally at the minimal primes of B and that $\operatorname{edim} A_p \leq 2$ for every prime $p \supset J$ of height three. Then*

   (a) *$\mathbb{D}(I,J)$ is of linear type.*

   (b) *The associated graded ring $\operatorname{gr}_{\mathbb{D}(I,J)}(A \otimes_k B)$ is Cohen–Macaulay.*

   (c) *($n \geq 3$) The Rees algebra $\mathcal{R}_{A\otimes_k B}(\mathbb{D}(I,J))$ is Cohen–Macaulay.*

Proof. (1) Let $\dim R = n$. From a resolution $0 \to R^{m-1} \to R^m \to R \to R/I = A \to 0$, one gets an exact sequence

$$0 \to \operatorname{Tor}_2^R(A,B) \to B^{m-1} \to B^m \to \operatorname{Tor}_1^R(A,B) \to 0. \tag{173}$$

Since the Koszul homology modules of $\mathbb{D}(I,J)$ are identified with the Tor modules ([172, V-5] or [188, Theorem 2.5]), the local sliding-depth property translates into the bound $\operatorname{depth} \operatorname{Tor}_1^R(A,B) \geq n-3$ and the Cohen–Macaulayness of $\operatorname{Tor}_2^R(A,B)$. Moreover, from (173), $\operatorname{Tor}_2^R(A,B)$ is a second syzygy of $\operatorname{Tor}_1^R(A,B)$ over the Cohen–Macaulay $(n-2)$-dimensional ring B. Therefore, we are left with the condition $\operatorname{depth} \operatorname{Tor}_1^R(A,B) \geq n-3$.

In the present case, $\operatorname{Tor}_1^R(A,B) = I \cap J/IJ = I/IJ$. Since $\operatorname{depth} I = n-1$, the condition reads $\operatorname{depth} IJ \geq n-2$. We are thus reduced to the following equivalence:

$$\operatorname{depth} IJ \geq n-2 \Leftrightarrow R/(I:J) \text{ is Cohen–Macaulay.}$$

For that purpose we set $J = (f,g)$, where $\{f,g\}$ is a regular sequence, and look into the exact sequence of R-modules:

$$0 \to fI \cap gI \to fI \oplus gI \to (fI,gI) = IJ \to 0. \tag{174}$$

Since the middle term of (174) has depth $n-1$, it is enough to show that $fI \cap gI \simeq I :_R J$. For this we argue that $fg(I :_R J) = fI \cap gI$. The inclusion $fg(I :_R J) \subset fI \cap gI$ is obvious. Conversely, let $af = bg \in fI \cap gI$, where $a,b \in I$. Since $\{f,g\}$ is a regular sequence, $a = cg$, $b = cf$, for some $c \in R$. Thus, $c\,(f,g) = (a,b) \subset I$, hence $c \in (I :_R J)$. It follows that $af = fgc \in fg(I :_R J)$.

(2) This follows, as in the previous results, by the techniques of sliding depth as described in [96]. The details are left to the reader. □

11.3.2.2 Via deformation

We now apply the deformation tool of Section 11.1.3 to the diagonal ideal. Set $A = R/I$ and $B = R/J$, where $I, J \subset R = k[\mathbf{X}]$ are homogenous ideals, and $A' = R/\operatorname{in}(I)$, $B' = R/\operatorname{in}(J)$. The goal is to compare the various constructions stemming from the diagonal ideal $\mathbb{D}(I,J)$ to the corresponding ones for the ideal $\mathbb{D}(\operatorname{in}(I), \operatorname{in}(J))$. Here, $\mathbb{D}(I,J)$ and $\mathbb{D}(\operatorname{in}(I), \operatorname{in}(J))$ are ideals of $A \otimes_k B$ and $A' \otimes_k B'$, respectively. We also use the "tilde" version of the diagonal ideal

$$\mathbb{D}(\tilde{I}, \tilde{J}) = (\overline{\mathbf{X} \otimes_{k[t]} 1 - 1 \otimes_{k[t]} \mathbf{X}}) \subset R[t]/\tilde{I} \otimes_{k[t]} R[t]/\tilde{J},$$

with $\bar{}$ denoting residue class in $R[t]/\tilde{I} \otimes_{k[t]} R[t]/\tilde{J}$ as discussed in Subsection 11.1.9.

For simplicity, set $\mathbb{D} = \mathbb{D}(I,J)$, $\mathbb{D}' = \mathbb{D}(\operatorname{in}(I), \operatorname{in}(J))$, $\tilde{\mathbb{D}} = \mathbb{D}(\tilde{I}, \tilde{J})$.

Proposition 11.3.8. *With the above notation:*
(i) $\mathbb{D} \otimes_k k[t, t^{-1}] \cong \tilde{\mathbb{D}} \otimes_{k[t]} k[t, t^{-1}] (= \tilde{\mathbb{D}}_t)$. *In particular, if $\tilde{\mathbb{D}}$ is of linear type (resp., syzygetic), then $\mathbb{D}$ is of linear type (resp., syzygetic).*
(ii) *Let $I = J$ and assume that the given term order $>$ on R is graded. If $\mathbb{D}'$ is of linear type (resp. syzygetic), then $\mathbb{D}$ is of linear type (resp., syzygetic).*

Proof. (i) One has $R[t]/\tilde{I} \otimes_{k[t]} R[t]/\tilde{J} \otimes_{k[t]} k[t, t^{-1}] \simeq (A \otimes_k B) \otimes_k k[t, t^{-1}]$, hence the required isomorphism follows. Since, moreover, $k[t, t^{-1}]$ is k-free, $\mathbb{D}$ is of linear type if and only if $\mathbb{D} \otimes_k k[t, t^{-1}]$ is of linear type. Therefore, the second statement is clear.

(ii) By part (i), it suffices to show that $\tilde{\mathbb{D}}$ is of linear type (resp., syzygetic). First, since t is a nonzero divisor on $(R[t]/\tilde{I} \otimes_{k[t]} R[t]/\tilde{I})/\tilde{\mathbb{D}} \simeq R[t]/\tilde{I}$, Proposition 11.1.9(iii) shows that $\tilde{\mathbb{D}} \otimes_{k[t]} (k[t]/(t)) \simeq \mathbb{D}'$. This says that the diagonal ideal $\tilde{\mathbb{D}}$ specializes under the given assumptions. Second, by Proposition 11.1.9 (i), t is a nonzero divisor on $R[t]/\tilde{I} \otimes_{k[t]} R[t]/\tilde{I}$, and hence is a nonzero divisor on $\tilde{\mathbb{D}}^j$ for every j. Now the natural map from the symmetric power $S_j(\tilde{\mathbb{D}})$ onto $\tilde{\mathbb{D}}^j$ induces an exact sequence

$$0 \to U_j \to S_j(\tilde{\mathbb{D}}) \to \tilde{\mathbb{D}}^j \to 0.$$

Tensoring this sequence with $k[t]/(t)$ over $k[t]$ one sees that $U_j \otimes_{k[t]} (k[t]/(t))$ embeds into $U_j' = \ker(S_j(\tilde{\mathbb{D}}') \to (\tilde{\mathbb{D}}')^j)$. Finally, recall that t is a homogeneous element in the positively graded k-algebra $R[t]/\tilde{I} \otimes_{k[t]} R[t]/\tilde{I}$. Thus, according to the graded version of Nakayama's Lemma, $U_j' = (0)$ implies $U_j = (0)$. □

11.4 Exercises

Exercise 11.4.1. Give a complete argument (possibly, with computer assistance) on the details of Example 10.2.14 (resp., Example 11.1.12).

Exercise 11.4.2. Let $I = (x_0x_1, x_1x_2, x_0x_2, x_0x_3) \subset k[x_0, x_1, x_2, x_3]$. Determine the secant $\sigma(I)$ in two ways: by Proposition 11.1.14(iii) and by definition (with computer assistance). Explain the outcome geometrically inside $\mathbb{P}^3$.

Exercise 11.4.3. Let $I = (x_0x_1, x_1x_2, x_0x_2, x_0x_3, x_1x_4, x_2x_5) \subset k[x_0, x_1, x_2, x_3, x_4, x_5]$. Determine the secant $\sigma(I)$ in two ways as in the previous exercise and explain the outcome in $\mathbb{P}^5$. Is there some combinatorial thread between the two examples?

Exercise 11.4.4 (J. Harris). Let

$$I = (x_2^2 + x_3x_4, x_0 - x_2, x_1 - x_2) \quad \text{and} \quad J = (x_2^2 + x_0x_1, x_3, x_4)$$

be ideals in $R = k[x_0, x_1, x_2, x_3, x_4]$.

(1) Compute the join ideal of the initial ideals of I and J.
(*Hint*: apply one of the results on joins of monomial ideals in Subsection 11.1.4.)

(2) Prove that equality holds in Theorem 11.1.11.
(*Hint*: Show that, in the proof of the theorem, the middle homomorphism is reduction modulo t.)

(3) Deduce that the join ideal of I and J is generated by a form of degree 4.

(4) What sort of modifications of I and J could imply a form of degree ≤ 3?
(*Hint*: Try the excess intersection.)

Exercise 11.4.5. Consider the generic Hankel matrix

$$\begin{pmatrix} X_1 & X_2 & X_3 & X_4 & X_5 \\ X_2 & X_3 & X_4 & X_5 & X_6 \end{pmatrix}$$

and let $I \subset R = k[X_1, \ldots, X_6]$ denote its ideal of 2-minors.

(1) Prove that the 2-minors are a Gröbner basis in the lexicographic order.

(2) Is the initial ideal of I in the lexicographic order the edge ideal of a bipartite graph?

(3) With computer assistance, use Theorem 10.2.5 to show that R/I does not have the expected star dimension.
(*Hint*: condition (i') is the easiest.)

Exercise 11.4.6. Consider the 2-catalecticant

$$\begin{pmatrix} X_1 & X_2 & X_3 & X_4 & X_5 \\ X_3 & X_4 & X_5 & X_6 & X_7 \end{pmatrix}$$

and let $I \subset R = k[X_1, \ldots, X_7]$ denote its ideal of 2-minors.

(1) Prove that R/I is locally regular on the punctured spectrum.

(2) Deduce that R/I has the expected star dimension.
(*Hint*: Use condition (i') of Theorem 10.2.5 at the maximal ideal.)

(3) By Proposition 11.2.18, in a rather roundabout way, one knows that R/I is even starlike linear. Can we get this directly in the present case?

Exercise 11.4.7. Let $\mathbf{X}$ be an r-catalecticant matrix of size $m \times n$ $(2 \le m \le n)$ and let $I = I_m(\mathbf{X}) \subset k[X_1, \ldots, X_{n+(m-1)r}]$. Suppose that $n - m - 1 \le r$ as in Proposition 11.2.2. Prove:

(1) $\operatorname{edim} k[\mathbf{X}]/I \le 2 \dim k[\mathbf{X}]/I$ whenever $m \ge 3$ or $m = 2$, $r \ge n - 2$.
(*Hint*: Bring in the condition that the ring $k[\mathbf{X}]/I$ have the expected star dimension [188].)

(2) The inequality in (1) does not localize in general.
(*Hint*: take $n = 6$, $m = 3$, $r = 1$.)

(3) If $m = 2$ and $r \ge n - 2$, then $k[\mathbf{X}]/I$ is starlike linear.

Exercise 11.4.8. Let V, W be subvarieties of $\mathbb{P}^n$. Consider the variety $\mathcal{J}(V, W) \subset \mathbb{G}(1, n)$ of joining lines of V and W as introduced in Definition 11.1.4.

(1) Formulate an effective way to compute the defining equations of the image of the composite rational map $\mathbb{P}^n \times \mathbb{P}^n \dashrightarrow \mathbb{P}^{\binom{n+1}{2}-1}$ obtained by means of the Plücker embedding of $\mathbb{G}(1, n)$.
(*Hint*: think of the starting map $\mathbb{P}^n \times \mathbb{P}^n \dashrightarrow \mathbb{G}(1, n)$ as $(v, w) \mapsto v \wedge w$, where v, w are the respective defining vectors of points $p \in V$ and $q \in W$; algebraically, convert the wedge to 2-minors.)

(2) Apply the method to deduce that if $V = W$ is the rational normal curve in $\mathbb{P}^3$ then the variety of joining lines in $\mathbb{P}^5$ is the Veronese embedding of $\mathbb{P}^2$.

(3) The twisting of the rational normal cubic in $\mathbb{P}^3$ intuitively forces its secant variety to be the whole of $\mathbb{P}^3$, but show this from the above by the incidence method explained in the appropriate section of the book.

Exercise 11.4.9. Let $R = k[\mathbf{X}]$ be a polynomial ring of dimension 11. Let $I \subset R$ be the edge ideal of a nonagon with a two edge path issuing from one of its vertices. Let Θ denote the transpose Jacobian matrix of the generators of I. Prove:

(1) By a suitable numbering of the vertices of the graph, $(\det \Theta) = (X_1 \cdots X_{11}) \subsetneq (0) :_R \mathfrak{J}(\Theta) = (X_1 \cdots X_{10}) \subsetneq \sigma(I) = (X_1 \cdots X_9)$.

(2) Give the analytic spread and the reduction number of the diagonal ideal of R/I.

Exercise 11.4.10. Let $I = (X_1X_3, X_2X_3, X_2X_4) = (X_1, X_2) \cap (X_2, X_3) \cap (X_3, X_4)$ and $J = (X_2, X_3)$ be ideals in $k[X_1, X_2, X_3, X_4]$. Prove the following facts as related to Proposition 11.3.6:

(1) $e(I, J) = 2$.

(2) $\mathbb{D}(I, J)$ is not of linear type.
(*Hint*: Consider the polynomial relation $(X_2 - Y_2)(X_3 - Y_3) \equiv 0 \pmod{(I, Y_2, Y_3)}$.)

(3) The sliding depth property does not hold.

(4) Both the associated graded ring and the Rees algebra of $\mathbb{D}(I, J)$ are Cohen–Macaulay.

12 Aluffi algebras

Paolo Aluffi [5] introduced a graded algebra with the purpose of defining the *characteristic cycle* of a hypersurface that would stand for the characteristic classes à la Schwartz–MacPherson the same way the conormal cycle stands for the characteristic classes à la Chern–Mather. The purpose of this chapter is to provide an algebraic analogue of Aluffi's in the case of ideals and introduce an analogue for modules, hoping this will serve as a template for coherent sheaves of modules.

12.1 Relation to Chern classes

In a nutshell, Aluffi's definition goes as follows.

Let $\mathcal{X}$ denote a scheme of finite type over a field and let $\mathcal{E}$ stand for a locally free $\mathcal{O}_{\mathcal{X}}$-module of rank $e+1$. Set $\mathbb{P}(\mathcal{E})$ for the corresponding $\mathbb{P}$-bundle of rank e over $\mathcal{X}$ and $\mathbb{P}(\mathcal{E}) \xrightarrow{\pi} \mathcal{X}$ for the associated structural map.

Let

$$\mathcal{A}_r\, \mathbb{P}(\mathcal{E}) \overset{\pi^*}{\underset{\simeq}{\longleftarrow}} \mathcal{A}_{r-e}\, \mathcal{X},$$

be the isomorphism of Chow groups in degree r given by the Segre pullback operator. Thus, any cycle C in the left hand side has a unique expression of the form

$$C = \sum_{j=0}^{e} c_1(\mathcal{O}_{\mathbb{P}(\mathcal{E})}(1))^j \cap \pi^*(C_{r-e+j}).$$

Aluffi calls the sum $C_{r-e} + C_{r-e+1} + \cdots + C_r$ the *shadow* of C. Among his main results in [5], he has the following.

Theorem 12.1.1. *Let $\mathcal{X} \subset \mathcal{M}$, where $\mathcal{X}$ is a hypersurface and $\mathcal{M}$ is a smooth variety. Let $\mathcal{Y} \subset \mathcal{X}$ stand for the singular scheme of $\mathcal{X}$. Then:*
(a) $(-1)^{\dim \mathcal{X}} \check{c}_{\mathrm{Ma}}(\mathcal{X})$ = shadow of the conormal cycle $[\mathrm{Bl}_{\mathcal{Y}}(\mathcal{X})] \subset \mathbb{P}(T^*\mathcal{M}|_{\mathcal{X}})$.
(b) $(-1)^{\dim \mathcal{X}} \check{c}_{\mathrm{SM}}(\mathcal{X})$ = shadow of the charact. cycle $[\mathrm{qS}_{\mathcal{Y}}(\mathcal{X})] \subset \mathbb{P}(T^*\mathcal{M}|_{\mathcal{X}})$.

Here, "Ma" stands for "Mather," while "SM" stands for "Schwartz–Macpherson."

The mysterious algebra $\mathrm{qS}_{\mathcal{Y}}(\mathcal{X})$ is what Aluffi called *quasi-symmetric*.

The definition of $\mathrm{qS}_{\mathcal{Y}}(\mathcal{X})$ involves taking the limit of a directed family of embedded algebras. This makes the commutative algebra transcription a bit cumbersome but, fortunately, one can actually reduce the problem to one such embedded algebra.

https://doi.org/10.1515/9783110640694-012

12.2 The case of ideals: embedded version

In this section, we only deal with an embedded version of Aluffi's quasi-symmetric algebra. The original definition in terms of a certain inverse limit of such embedded algebras is actually dealt with in the case of modules later on—clearly, containing the ideal case in particular.

12.2.1 Preliminaries

The material of this part is lifted from [149]. Let A be a ring and $\mathfrak{a}$ an ideal of A. Let R be a ring surjecting onto A and let I denote the inverse image of $\mathfrak{a}$ in R. Note that the symmetric algebra $\mathcal{S}_R(I)$ maps surjectively both to the Rees algebra $\mathcal{R}_R(I)$ and (by functoriality of the symmetric algebra) to $\mathcal{S}_A(\mathfrak{a})$.

Definition 12.2.1. The *R-embedded Aluffi algebra* of $\mathfrak{a}$ is the A-algebra

$$\mathcal{A}_{R\twoheadrightarrow A}(\mathfrak{a}) := \mathcal{S}_A(\mathfrak{a}) \otimes_{\mathcal{S}_R(I)} \mathcal{R}_R(I).$$

We develop a few general preliminaries about the R-embedded Aluffi algebra. The first is a useful presentation that has already been observed in [5, Theorem 2.9] in the context of schemes.

Lemma 12.2.2. *In the above setup, write $I/J := \mathfrak{a} \subset R/J := A$, where $I \subset R$ is the inverse image of $\mathfrak{a}$ in R. There are natural A-algebra isomorphisms*

$$\mathcal{A}_{R\twoheadrightarrow A}(\mathfrak{a}) \simeq \frac{\mathcal{R}_R(I)}{(J,\tilde{J})\,\mathcal{R}_R(I)} \simeq \bigoplus_{t\geq 0} I^t/JI^{t-1},$$

where J is in degree 0 and $\tilde{J}$ is in degree 1. In particular, there is a surjective A-algebra homomorphism $\mathcal{A}_{R\twoheadrightarrow A}(\mathfrak{a}) \twoheadrightarrow \mathcal{R}_A(\mathfrak{a})$.

Proof. By the universal property of the symmetric algebra, one sees that

$$\mathcal{S}_{R/J}(I/J) \simeq \frac{\mathcal{S}_R(I)}{(J,\tilde{J})\,\mathcal{S}_R(I)}.$$

Tensoring with $\mathcal{R}_R(I)$ over $\mathcal{S}_R(I)$ gives the first isomorphism. The second one is now immediate from the definition of $\tilde{J}$. □

From the definition and Lemma 12.2.2, the Aluffi algebra is squeezed as

$$\mathcal{S}_{R/J}(I/J) \twoheadrightarrow \mathcal{A}_{R\twoheadrightarrow R/J}(I/J) \twoheadrightarrow \mathcal{R}_{R/J}(I/J) \tag{175}$$

and is moreover a residue ring of the ambient Rees algebra $\mathcal{R}_R(I)$.

If no confusion arises, for a fixed ambient R we will simply refer to this algebra as the Aluffi algebra of $\mathfrak{a} = I/J$. Unless stated otherwise, we will assume that $J \subsetneq I \subsetneq R$.

Note that if the ideal I is of linear type—i. e., the natural surjection $\mathcal{S}_R(I) \twoheadrightarrow \mathcal{R}_R(I)$ is injective—then trivially $\mathcal{S}_A(I/J) \simeq \mathcal{A}_{R\twoheadrightarrow A}(I/J)$. The following example shows that in general there is no converse to this statement even when R is a hypersurface domain.

Example 12.2.3. Let $R = k[x,y,z] = k[X,Y,Z]/(XY - Z^2)$, with $J = (x,z)$ (the ideal of a ruling in the affine cone) and $I = (x,y,z)$. Then $R/J \simeq k[Y]$ and $I/J \simeq (Y)$. Therefore, I/J is of linear type, hence $\mathcal{S}_{R/J}(I/J) \simeq \mathcal{A}_{R\twoheadrightarrow R/J}(I/J) \simeq \mathcal{R}_{R/J}(I/J)$.

This example is a member of the following large class: take $(R, \mathfrak{m})$ to be a nonregular local ring—or a nonregular standard graded algebra over a field and its irrelevant ideal—with $J \subset I = \mathfrak{m}$ such that R/J is regular. Then I/J is generated by a regular sequence on R/J, hence is of linear type, while $\mathfrak{m}$ is never of linear type.

It would be interesting to find such examples with $(R, \mathfrak{m})$ a regular local ring and $J \subset \mathfrak{m}I$. In the special case where J is a hypersurface, no such examples exist as we now prove.

Proposition 12.2.4. *Let R be a Noetherian ring and let I denote an ideal. If $a \in I$ is a regular element, then I is of linear type if and only if the natural surjection*

$$\mathcal{S}_{R/(a)}(I/(a)) \twoheadrightarrow \mathcal{A}_{R\twoheadrightarrow R/(a)}(I/(a))$$

is an isomorphism.

Proof. The trivial implication has already been mentioned above. For the reverse direction, consider an R-algebra presentation $S := R[\mathbf{T}] \twoheadrightarrow \mathcal{R}_R(I)$ based on a set of generators $\mathbf{b} = \{b_1, \ldots, b_n\}$ of I. Write $\mathcal{J} = \bigoplus_{i\geq 1} \mathcal{J}_i$ for the kernel of this map, where $\mathcal{J}_i$ stands for the homogenous part of degree i of $\mathcal{J}$ in the standard grading of S. Note that $\mathcal{J}_1 S \subset \mathcal{J}$ defines likewise the symmetric algebra of I on S, so we need to show that for any $r \geq 0$, $\mathcal{J}_r \subset \mathcal{J}_1 S$.

We induct on r, the result being trivial if $r = 1$. Thus, let $r \geq 2$. By Lemma 12.2.2 one has

$$\mathcal{J} \subset (\mathcal{J}_1, \tilde{a}, a)S.$$

Let $F = F(\mathbf{T}) \in \mathcal{J}_r$. Then $F = L + \tilde{a}G + aH$ where $L \in \mathcal{J}_1 S_{r-1}$, $G \in S_{r-1}$ and $H \in S_r$. Note that, if $a = \sum_{j=1}^n c_j b_j$ then $\tilde{a} = \sum_{j=1}^n c_j T_j$, hence $\tilde{a}(\mathbf{b}) = a$, i. e., evaluating $\tilde{a}$ on the generators of I gives back a. Therefore,

$$\begin{aligned} 0 = F(\mathbf{b}) &= L(\mathbf{b}) + \tilde{a}(\mathbf{b})G(\mathbf{b}) + aH(\mathbf{b}) \\ &= a \cdot (G + H)(\mathbf{b}), \end{aligned}$$

since $L \in \mathcal{J}$. As a is a regular element, $G + H \in \mathcal{J}$, hence, by homogeneity, $G \in \mathcal{J}_{r-1}$ and $H \in \mathcal{J}_r$.

By the inductive hypothesis, $G \in \mathcal{J}_1 S_{r-2}$, hence $\tilde{a}G \in \mathcal{J}_1 S_{r-1}$. Therefore, $F \in (\mathcal{J}_1 S_{r-1})S + a\,\mathcal{J}_r S$, thus showing the equality of ideals $\mathcal{J}_r S = (\mathcal{J}_1 S_{r-1})S + a\,\mathcal{J}_r S$. By the graded version of Nakayama's lemma, this implies that $\mathcal{J}_r S = (\mathcal{J}_1 S_{r-1})S$, as was to be shown. □

Corollary 12.2.5. *Let R be a Noetherian ring and let $\{a_1, \dots, a_m\} \subset R$ be a regular sequence. If $I \subset R$ is an ideal containing $\{a_1, \dots, a_m\}$ such that $I/(a_1, \dots, a_m)$ is of linear type on $R/(a_1, \dots, a_m)$ then I is of linear type on R.*

Proof. Induct on m. For $m = 1$, it readily follows from (175) and Proposition 12.2.4.

Next assume that $m \geq 2$ and write $J = (a_1, \dots, a_m)$. Set $\bar{R} := R/(a_1, \dots, a_{m-1})$ and, likewise,

$$\bar{J} := J/(a_1, \dots, a_{m-1}) = (\overline{a_m}) \subset \bar{I} := I/(a_1, \dots, a_{m-1}).$$

Clearly, $\bar{I}/(\overline{a_m}) \simeq I/J$ in $\bar{R}/(\overline{a_m}) \simeq R/J$. Therefore, the assumption that I/J is of linear type on R/J implies that $\bar{I}/(\overline{a_m})$ is of linear type on $\bar{R}/(\overline{a_m})$, where $\overline{a_m}$ is a regular element in $\bar{R}$. By the first part, $\bar{I}$ is of linear type on $\bar{R}$. Then, by the inductive hypothesis, I is of linear type on R. □

12.2.2 Dimension

Proposition 12.2.6. *Let $J \subsetneq I \subsetneq R$ be ideals of the Noetherian ring R.*
(a) *If J has a regular element then* $\dim \mathcal{A}_{R \twoheadrightarrow R/J}(I/J) \leq \dim R$.
(b) *If I/J has a regular element then*

$$\min\{\dim R + 1,\ \dim \mathcal{S}_{R/J}(I/J)\} \geq \dim \mathcal{A}_{R \twoheadrightarrow R/J}(I/J) \geq \dim R/J + 1.$$

Proof. (a) Since $\mathcal{R}_R(I)$ is R-torsion-free, one has $\operatorname{ht} J\mathcal{R}_R(I) \geq 1$. Therefore,

$$\dim \mathcal{A}_{R \twoheadrightarrow R/J}(I/J) \leq \dim \mathcal{R}_R(I)/J\mathcal{R}_R(I) \leq \dim R + 1 - 1 = \dim R.$$

(b) This follows immediately from (175) by the well-known dimension formula for the Rees algebra of an ideal containing a regular element. □

Note that (a) is all one can assert in such generality because if, e. g., a power of the ideal I is contained in J, then $\dim \mathcal{A}_{R \twoheadrightarrow R/J}(I/J) = \dim R/J$.

Perhaps more interesting is the following.

Theorem 12.2.7. *Let R be a catenary, equidimensional and equicodimensional Noetherian ring and let $I \subsetneq R$ be an ideal containing a regular element a. Then $\mathcal{A}_{R \twoheadrightarrow R/(a)}(I/(a))$ is equidimensional and* $\dim \mathcal{A}_{R \twoheadrightarrow R/(a)}(I/(a)) = \dim R$.

Proof. Under the assumptions on R, $I/(a)$ and a, one can apply Proposition 12.2.6(a), and the right-hand inequality of (b) to conclude that $\dim \mathcal{A}_{R\twoheadrightarrow R/(a)}(I/(a)) = \dim R - 1 + 1 = \dim R$.

To prove the equidimensionality part, we will show that $\mathcal{A}_{R\twoheadrightarrow R/(a)}(I/(a))$ is equidimensional locally at every prime ideal $\mathcal{P} \subset \mathcal{R}_R(I)$ in its support. Localizing first at $\mathcal{P}\cap R$ in the base ring one can assume that $(R, \mathfrak{m})$ is local, with $\mathcal{P} \cap R = \mathfrak{m}$ and $I \subset \mathfrak{m}$. Now, $\mathcal{M} = (\mathfrak{m}, Iu) \subset \mathcal{R}_R(I) \subset R[Iu]$ is not a minimal prime of $\mathcal{A}_{R\twoheadrightarrow R/(a)}(I/(a))$. This is because the Aluffi algebra is graded, with grading induced from $\mathcal{R}_R(I)$; hence, $\mathcal{M}$ would actually be its unique associated prime, which is impossible as $\dim R \geq 1$. Thus, for the purpose of showing equidimensionality, we may assume that $\mathcal{P}$ is a homogeneous ideal properly contained in $\mathcal{M}$.

Let $I = (b_1, \dots, b_n)$. Note that in the present situation, one has by Lemma 12.2.2:

$$\mathcal{A}_{R\twoheadrightarrow R/(a)}(I/(a)) \simeq \mathcal{R}_R(I)/(a, \tilde{a}).$$

Write $\mathcal{R}_R(I) = R[b_1u, \dots, b_nu] \subset R[u]$, so that $\tilde{a} = \sum_{j=1}^n c_jb_ju$, for suitable $c_j \in R$.

Suppose first that $(Iu) \not\subset \mathcal{P}$. Say, $b_1u \notin \mathcal{P}$. Localizing at $\mathcal{P}$ yields

$$\begin{aligned}\mathcal{A}_{R\twoheadrightarrow R/(a)}(I/(a))_{\mathcal{P}} &\simeq R[Iu]_{\mathcal{P}}/(a, \tilde{a})_{\mathcal{P}} \simeq R\left[\frac{I}{b_1}, b_1u, (b_1u)^{-1}\right]_{\mathcal{P}'} \Big/ \left(a, c_1 + \sum_{j=?}^{m} c_j\frac{b_j}{b_1}\right)_{\mathcal{P}'} \\ &= R\left[\frac{I}{b_1}, b_1u, (b_1u)^{-1}\right]_{\mathcal{P}'} \Big/ \left(a, \frac{a}{b_1}\right)_{\mathcal{P}'} \\ &= R\left[\frac{I}{b_1}, b_1u, (b_1u)^{-1}\right]_{\mathcal{P}'} \Big/ \left(\frac{a}{b_1}\right)_{\mathcal{P}'},\end{aligned}$$

where $\mathcal{P}'$ denotes the corresponding image of $\mathcal{P}$. The rightmost ring above is a factor ring of a catenary, equidimensional and equicodimensional ring by a principal ideal generated by a regular element, hence it is equidimensional and so is $\mathcal{A}_{R\twoheadrightarrow R/(a)}(I/(a))_{\mathcal{P}}$.

Suppose now that $(Iu) \subset \mathcal{P}$. Then $\mathfrak{m} \not\subset \mathcal{P}$ since $\mathcal{P} \subsetneq \mathcal{M}$, hence $p := \mathcal{P} \cap R \subsetneq \mathfrak{m}$. Note that p is a prime containing a.

If $I \not\subset p$, then $\mathcal{A}_{R\twoheadrightarrow R/(a)}(I/(a))_{\mathcal{P}}$ is a localization of the ring

$$R_p[I_pu]/(a, \tilde{a}) = R_p[u]/(a, au) = R_p[u]/(a)$$

and we conclude as above. If $I \subset p$, then $\mathcal{A}_{R\twoheadrightarrow R/(a)}(I/(a))_{\mathcal{P}}$ is a localization of the Aluffi algebra $\mathcal{A}_{R_p\twoheadrightarrow R_p/(a)}(I/(a))$ and we conclude by induction on $\dim R$. □

A geometric version of Theorem 12.2.7 case is stated in [5, Corollary 2.18]. We will have more to say about the equidimensionality of the Aluffi algebra in upcoming subsections.

We now assume that $(R, \mathfrak{m})$ is a Noetherian local ring and its maximal ideal or a standard graded algebra over a field and its maximal irrelevant ideal. Throughout $R/\mathfrak{m}$ is an infinite field.

Let $J \subset I \subset \mathfrak{m}$. We confront ourselves with two quite opposite situations, namely, when $J \subset \mathfrak{m}I$ and when J contains minimal generators of I. Note that if J is a reduction of I then $J \subset \mathfrak{m}I$ would entail $I^t = JI^{t-1} \subset \mathfrak{m}I^t$, for $t \gg 0$; hence $I^t = \{0\}$, i. e., I would be nilpotent.

Drawing upon a terminology of geometry, let us agree to say that the pair $J \subset I$ of ideals is *nondegenerate* if $J \subset \mathfrak{m}I$. If on the other extreme, $J \subset I$ is generated by a subset of minimal generators of I—as is the case when J is a reduction of I—we may call the pair $J \subset I$ *totally degenerate*. The latter case can usually be disposed of by a standard argument (see Proposition 12.2.10).

Recall that I has *maximal analytic spread* if $\ell(I) = \dim R$. Note that this forces $\mu(I) \geq \dim R$.

Proposition 12.2.8. *Let $(R, \mathfrak{m})$ be as above with $R/\mathfrak{m}$ infinite. Suppose that $J \subset I$ is a nondegenerate pair and that J has a regular element. Then:*

(i) $\ell(I) \leq \dim \mathcal{A}_{R\twoheadrightarrow R/J}(I/J) \leq \dim R$; *in particular, if I has maximal analytic spread then* $\dim \mathcal{A}_{R\twoheadrightarrow R/J}(I/J) = \dim R$.

(ii) *If I has maximal analytic spread and, moreover, $\mu(I) = \dim R$, then $\mathfrak{m}\mathcal{R}_R(I)$ is a minimal prime of $\mathcal{A}_{R\twoheadrightarrow R/J}(I/J)$ of maximal dimension.*

(iii) *If $J \subset I^2$, then* $\dim \mathcal{A}_{R\twoheadrightarrow R/J}(I/J) = \dim R$.

Proof. (i) $J \subset \mathfrak{m}I$ implies $JI^{t-1} \subset \mathfrak{m}I^t$ for every $t \geq 0$. This yields a surjective homomorphism $\mathcal{A}_{R\twoheadrightarrow R/J}(I/J) \twoheadrightarrow \mathcal{R}_R(I)/\mathfrak{m}\mathcal{R}_R(I)$, from which follows the leftmost inequality.

The other inequality stems from Proposition 12.2.6(a).

(ii) The assumption $\ell(I) = \mu(I) = \dim R$ implies that I is generated by analytically independent elements and the latter entails that $\mathcal{R}_R(I)/\mathfrak{m}\mathcal{R}_R(I)$ is a polynomial ring over $R/\mathfrak{m}$. In particular, $\mathfrak{m}\mathcal{R}_R(I)$ is a prime ideal of $\mathcal{R}_R(I)$. Since $JI^{t-1} \subset \mathfrak{m}I^t \subset \mathfrak{m}$ then $(J, \tilde{J}) \subset \mathfrak{m}\mathcal{R}_R(I)$ as ideals of $\mathcal{R}_R(I)$. Therefore, $\mathcal{A}_{R\twoheadrightarrow R/J}(I/J)/\mathfrak{m}\,\mathcal{A}_{R\twoheadrightarrow R/J}(I/J) \simeq \mathcal{R}_R(I)/\mathfrak{m}\mathcal{R}_R(I)$, hence $\mathfrak{m}\mathcal{A}_{R\twoheadrightarrow R/J}(I/J)$ is a prime ideal with

$$\dim \mathcal{A}_{R\twoheadrightarrow R/J}(I/J)/\mathfrak{m}\,\mathcal{A}_{R\twoheadrightarrow R/J}(I/J) = \dim R = \dim \mathcal{A}_{R\twoheadrightarrow R/J}(I/J)$$

by the first part.

(iii) Write $\mathrm{gr}_I(R)$ for the associated graded ring of I. Since $J \subset I^2$, one has $JI^{t-1} \subset I^{t+1}$ for every $t \geq 0$. This yields a surjective homomorphism $\mathcal{A}_{R\twoheadrightarrow R/J}(I/J) \twoheadrightarrow \mathrm{gr}_I(R)$, showing that $\dim \mathcal{A}_{R\twoheadrightarrow R/J}(I/J) \geq \dim R$. The reverse inequality follows from Proposition 12.2.6(a). □

We wrap up with a comment on the last result. Namely, we actually have

$$\dim \mathcal{A}_{R\twoheadrightarrow R/J}(I/J) \geq \max\{\ell(I), \dim R/J + 1\}$$

provided I has a regular element modulo J. The interesting case is when $\ell(I) \geq \dim R/J + 1$. If, say, R is catenary and equidimensional, it would entail

$$\dim \mathcal{A}_{R\twoheadrightarrow R/J}(I/J) \geq \frac{\dim R + 1}{2}.$$

We now look more closely at the internal structure of the Aluffi algebra and relate some of the elements of this structure to well-known notions in ideal theory.

12.2.3 Torsion and minimal primes

By Lemma 12.2.2, one has

$$\mathcal{A}_{R\twoheadrightarrow R/J}(I/J) \simeq \mathcal{R}_R(I)/(J,\tilde{J})\mathcal{R}_R(I) = \bigoplus_{t\geq 0} I^t/JI^{t-1}.$$

Since $\mathcal{R}_{R/J}(I/J) = \bigoplus_{t\geq 0}(I^t, J)/J \simeq \bigoplus_{t\geq 0} I^t/J \cap I^t$, the kernel of the natural surjection $\mathcal{A}_{R\twoheadrightarrow R/J}(I/J) \twoheadrightarrow \mathcal{R}_{R/J}(I/J)$ is the homogeneous ideal

$$\mathbb{W}_{J\subset I} := \bigoplus_{t\geq 2} \frac{J \cap I^t}{JI^{t-1}}, \tag{176}$$

dubbed the *module of Valabrega–Valla* (see [212], also [215, 5.1]).

One retrieves a result of Valla [213, Theorem 2.8].

Corollary 12.2.9. *Let $J \subset I \subsetneq R$ be ideals of the local ring R. If I/J is of linear type over R/J (e. g., if I is generated by a regular sequence modulo J), then $J \cap I^t = JI^{t-1}$ for every positive integer t.*

Proof. This follows immediately from the structural "squeezing" (175). □

Note that the assumption in [96, Proposition 3.10] to the effect that I be of linear type over R does not intervene in the above statement.

Here is a useful explicit situation, where we write $I = (J, \mathfrak{a})$, with no particular care for minimal generation.

Proposition 12.2.10. *Let $I = (J, \mathfrak{a})$. If $J \cap \mathfrak{a}^t \subset J\mathfrak{a}^{t-1}$, for every $t \geq 0$, then $\mathcal{A}_{R\twoheadrightarrow R/J}(I/J) \twoheadrightarrow \mathcal{R}_{R/J}(I/J)$ is an isomorphism.*

Proof. Indeed, one has

$$\begin{aligned} J \cap I^t &= J \cap (J, \mathfrak{a})^t = J \cap (J(J, \mathfrak{a})^{t-1}, \mathfrak{a}^t) \\ &= J(J, \mathfrak{a})^{t-1} + J \cap \mathfrak{a}^t \subset JI^{t-1} + J\mathfrak{a}^{t-1} \subset JI^{t-1}. \end{aligned}$$

□

Remark 12.2.11. In the notation of the previous proposition, one of the main results of [113] is that if $\mathfrak{a}$ is generated by a d-sequence modulo J then the assumption of the proposition is fulfilled. Therefore, under the hypothesis of [*loc. cit.*], the surjection $\mathcal{A}_{R\twoheadrightarrow R/J}(I/J) \twoheadrightarrow \mathcal{R}_{R/J}(I/J)$ is an isomorphism. This result, however, is a special case of Corollary 12.2.9 if one uses that an ideal generated by a d-sequence is of linear type. Of course, the proof of this fact requires some nontrivial work on itself and is previous to the later results, such as [96].

When $J = (a)$ is a principal ideal, one has a result somewhat subsumed in the spirit of [213].

Proposition 12.2.12. *Let $\mathfrak{a}$ be an ideal in the ring R and let $a \in R$ be an element such that $\mathfrak{a}^t : a = \mathfrak{a}^t$ for every integer $t \geq 0$. Then the inclusion $(a) \subset (a, \mathfrak{a})$ induces an isomorphism $\mathcal{A}_{R\twoheadrightarrow R/(a)}((a, \mathfrak{a})/(a)) \simeq \mathcal{R}_{R/(a)}((a, \mathfrak{a})/(a))$.*

Proof. The assumption means that $(a) \cap \mathfrak{a}^t = a\mathfrak{a}^t$ for every $t \geq 0$, hence $(a) \cap (a, \mathfrak{a})^t = a(a, \mathfrak{a})^{t-1} + (a) \cap \mathfrak{a}^t = a(a, \mathfrak{a})^{t-1}$ for $t > 0$. □

The Valabrega–Valla module gives the torsion in as many cases as the ones in which the Rees algebra is the symmetric algebra modulo torsion.

Proposition 12.2.13. *Let $J \subset I \subsetneq R$ be ideals of the Noetherian ring R. If I/J has a regular element, then the R/J-torsion of the embedded Aluffi algebra of I/J is the kernel of the natural surjection $\mathcal{A}_{R\twoheadrightarrow R/J}(I/J) \twoheadrightarrow \mathcal{R}_{R/J}(I/J)$.*

Proof. Consider the following general elementary observation: given a ring A and A-modules,

$$N \twoheadrightarrow M \twoheadrightarrow K$$

such that the A-torsion of N is the kernel of the composite $N \twoheadrightarrow K$, then the A-torsion of M is the kernel of $M \twoheadrightarrow K$. We apply this to the situation in (175), by recalling that if $\mathfrak{a} \subset A$ has a regular element in the ring A then the A-torsion of the symmetric algebra $\mathcal{S}_A(\mathfrak{a})$ is the kernel of the natural surjection $\mathcal{S}_A(\mathfrak{a}) \twoheadrightarrow \mathcal{R}_A(\mathfrak{a})$. □

Recall that, given a ring S, an ideal $\mathfrak{b} \subset S$ and an S-module E, one denotes by $\mathrm{H}^0_{\mathfrak{b}}(E)$ the zeroth local cohomology of E with respect to $\mathfrak{b}$. Recall that $\mathrm{H}^0_{\mathfrak{b}}(E) \simeq E : \mathfrak{b}^\infty$.

Corollary 12.2.14. *Let $J \subset I \subsetneq R$ be ideals of the Noetherian ring R. If I/J has a regular element, then $\mathbb{V}_{J\subset I} = \mathrm{H}^0_{I/J}(\mathcal{A})$, where $\mathcal{A}$ denotes the Aluffi algebra and $\mathbb{V}_{J\subset I}$ is the Valabrega–Valla module as introduced above.*

Proof. By Proposition 12.2.13, $\mathbb{V}_{J\subset I}$ is the R/J-torsion of $\mathcal{A}$. On the other hand, localizing at primes of the base R/J not containing I/J makes the surjection $\mathcal{S}_{R/J}(I/J) \twoheadrightarrow \mathcal{R}_{R/J}(I/J)$ an isomorphism, hence also the surjection $\mathcal{A} \twoheadrightarrow \mathcal{R}_{R/J}(I/J)$. Therefore, $\mathcal{A}$ is torsion-free locally at those primes. Since I/J has regular elements, the result follows readily from Lemma 8.1.2. □

Remark 12.2.15. The last result says, in particular, that there exists an integer $k \geq 0$ such that $I^k(J \cap I^t) \subset JI^{t-1}$ for every $t \geq 1$. Later on we will relate such an exponent to the so-called Artin–Rees number.

By the same principle, one can get a hold of the minimal primes of the Aluffi algebra. Quite generally, to any ideal $\mathfrak{a} \subset R$ we associate its extended–contracted ideal

$$\tilde{\mathfrak{a}} := \mathfrak{a}R[u] \cap R[Iu] = \sum_{t \geq 0} (\mathfrak{a} \cap I^t)u^t$$

in the Rees algebra $\mathcal{R}_R(I) \simeq R[Iu] \subset R[u]$ (u a variable over R).

Proposition 12.2.16. *Let $J \subset I \subsetneq R$ be ideals of the Noetherian ring R. Any minimal prime $\wp$ of $\mathcal{A}_{R \twoheadrightarrow R/J}(I/J)$ on $\mathcal{R}_R(I)$ is either of the form $\wp = \tilde{p}$ for some minimal prime of R/J on R, or else has the form $(q, \wp_+)$ where $q := \wp \cap R$ contains a minimal prime of R/I on R and $\wp_+ = \wp \cap (Iu)$.*

Proof. By Corollary 12.2.14—rather by its proof—a power of I annihilates the kernel of $\mathcal{A}_{R \twoheadrightarrow R/J}(I/J) \twoheadrightarrow \mathcal{R}_{R/J}(I/J)$ lifted to $\mathcal{R}_R(I)$ – call it $\mathcal{K}$. If $\wp \subset \mathcal{R}_R(I)$ is a minimal prime of $\mathcal{A}_{R \twoheadrightarrow R/J}(I/J)$ it follows that $\wp$ contains either $\mathcal{K}$ or I. In the first case, it contains a minimal prime of $\mathcal{R}_R(I)/\mathcal{K} \simeq \mathcal{R}_{R/J}(I/J)$, hence must be a minimal prime of the latter on $\mathcal{R}_R(I)$. But it is well known that the above extending-contracting operation induces a bijection between the minimal primes of R/J on R and the minimal primes of $\mathcal{R}_{R/J}(I/J)$ on $\mathcal{R}_R(I)$.

In the case $I \subset \wp$, since $\wp$ is homogeneous in the natural $\mathbb{N}$-grading of $\mathcal{R}_R(I)$, then it is clear that $\wp = (q, \wp_+)$, where $\wp_+ = \wp \cap \mathcal{R}_R(I)_+$; obviously, q contains a minimal prime of R/I on R. □

Remark 12.2.17. Note that if $\wp \subset \mathcal{R}_R(I)$ is a minimal prime of $\mathcal{A}_{R \twoheadrightarrow R/J}(I/J)$ containing I then $\wp_+$ behaves erratically: it can actually be zero in certain cases (see, e. g., Proposition 12.2.8(ii)). On the other hand, its contraction $q \subset R$ may turn out to be an embedded associated prime of R/I and not a minimal one (see Section 12.2.6).

Corollary 12.2.18. *Let $J \subset I \subsetneq R$ be ideals of a Noetherian ring R such that I/J has a regular element. If $\mathcal{A}_{R \twoheadrightarrow R/J}(I/J)$ is equidimensional then*

$$\dim \mathcal{A}_{R \twoheadrightarrow R/J}(I/J) = \dim R/J + 1.$$

Proof. This readily follows from Proposition 12.2.16 and the known value of the dimension of $\mathcal{R}_{R/J}(I/J)$ under the present assumption on I/J. □

Equidimensionality of $\mathcal{A}_{R \twoheadrightarrow R/J}(I/J)$ if J has codimension ≥ 2 may be quite rare. The next result shows that, at least in the local or graded case, pure-dimensionality is really infrequent under this assumption.

Proposition 12.2.19. *Let $(R, \mathfrak{m})$ be a Noetherian local ring and its maximal ideal $\mathfrak{m}$ or a standard graded algebra over a field and its maximal irrelevant ideal $\mathfrak{m}$. Assume that $R/\mathfrak{m}$ infinite. Let $J \subset I \subset \mathfrak{m}$, with J having a regular element and I having a regular element modulo J. Suppose that:*
(i) *$J \subset I \subset \mathfrak{m}$ is a nondegenerate pair and $\ell(I) = \mu(I) = \dim R$; or else*
(ii) *$J \subset I^2$.*

If $\mathcal{A}_{R\twoheadrightarrow R/J}(I/J)$ is equidimensional, then J has height one. If $\mathcal{A}_{R\twoheadrightarrow R/J}(I/J)$ is pure-dimensional then J is an ideal of pure height one (hence, principal if R is regular).

Proof. By Proposition 12.2.16, $\mathcal{R}_{R/J}(I/J)$ is equidimensional. In particular, one has

$$\dim \mathcal{A}_{R\twoheadrightarrow R/J}(I/J) = \dim \mathcal{R}_{R/J}(I/J) = \dim R/J + 1 \leq \dim R - \operatorname{ht} J + 1.$$

On the other hand, by Proposition 12.2.8(ii) or (iii), $\dim \mathcal{A}_{R\twoheadrightarrow R/J}(I/J) = \dim R$. Therefore, $\operatorname{ht} J \leq 1$ (hence, $\operatorname{ht} J = 1$ since J has a regular element by hypothesis).

Now, assume that $\mathcal{A}_{R\twoheadrightarrow R/J}(I/J)$ is pure-dimensional. Let $p \in \operatorname{Ass}_R(R/J)$ have height ≥ 2. By Proposition 12.2.16 and the pure dimensionality of $\mathcal{A}_{R\twoheadrightarrow R/J}(I/J)$, the prime $\tilde{p}$ satisfies

$$\begin{aligned}\dim \mathcal{A}_{R\twoheadrightarrow R/J}(I/J) &= \dim \mathcal{A}_{R\twoheadrightarrow R/J}(I/J)/\tilde{p} = \dim \mathcal{R}_R(I)/\tilde{p} = \dim \mathcal{R}_{R/p}((p, I)/p)\\ &= \dim R/p + 1 \leq \dim R - \operatorname{ht} p + 1 \leq \dim R - 2 + 1 = \dim R - 1.\end{aligned}$$

Again Proposition 12.2.8(ii) or (iii) gives a contradiction. □

More difficult is to get hold of nontrivial embedded primes of $\mathcal{A}_{R\twoheadrightarrow A}(I/J)$. In the case where J is the ideal of a homogeneous hypersurface in projective space, there are often embedded primes containing the irrelevant ideal.

We wrap up with the following.

Question 12.2.20. Suppose as above that $J = (a) \subset \mathfrak{m}I$, $I/(a)$ has a regular element and I has maximal analytic spread. To what extent can we assert that, conversely, $\mathcal{A}_{R\twoheadrightarrow A}(I/J)$ is pure-dimensional?

This seems to be the case in a variety of situations such as the one considered in Section 12.2.6.

12.2.4 Relation to the Artin–Rees number

A close associate to $\mathbb{W}_{J\subset I}$ is the well-known ideal

$$\ker(\operatorname{gr}_I(R) \twoheadrightarrow \operatorname{gr}_{I/J}(R/J)) = \bigoplus_{t\geq 0} (I^{t+1}, J \cap I^t)/I^{t+1},$$

generated by the I-initial forms of elements of J. Recall that the I-initial form of an element $a \in R$ is the residue class a^* of a in $I^{\nu(a)}/I^{\nu(a)+1}$ where $\nu(a)$ is the Ith order of a (i. e., $a \in I^{\nu(a)} \setminus I^{\nu(a)+1}$, setting $\nu(a) = \infty$ and $a^* = 0$ if $a \in \bigcap_{t\geq 0} I^t$).

A set of elements of J is called an *I-standard base* of J if their initial forms in $\mathrm{gr}_I(R)$ generate the above ideal. If R is a Noetherian local ring, then an I-standard base of J is a generating set of J (see [101, Lemma 6]). We will shorten $\nu(a_i)$ to ν_i if a_i is sufficiently clear from the context.

The following basic result will be used throughout.

Theorem 12.2.21 ([212]). *Let $J = (a_1, \dots, a_m)$ be an ideal of ring R. Then $\{a_1, \dots, a_m\}$ is an I-standard base of J if and only if*

$$J \cap I^t = \sum_{i=1}^{m} a_i I^{t-\nu_i}$$

for every positive integer t.

It implies the following.

Proposition 12.2.22. *Let R be a Noetherian local ring and $J \subset I$ be ideals of R. Let $\{a_1, \dots, a_m\}$ be an I-standard base of J such that $1 \leq \nu_1 \leq \nu_2 \leq \dots \leq \nu_m$.*

(a) *The surjection $\mathcal{A}_{R\twoheadrightarrow R/J}(I/J) \twoheadrightarrow \mathcal{R}_{R/J}(I/J)$ is an isomorphism if and only if $\nu_m = 1$ (i. e., $\nu_1 = \dots = \nu_m = 1$).*

(b) *More generally,*

$$\mathfrak{R} \cap I^{\nu_m - 1}\mathcal{R}_R(I) \subset \mathfrak{A} \subset \mathfrak{R} \cap I^{\nu_1 - 1}\mathcal{R}_R(I),$$

where $\mathfrak{R} = \bigoplus_{t\geq 0} J \cap I^t$ and $\mathfrak{A} = \bigoplus_{t\geq 0} JI^{t-1}$.

(c) *If $\nu_1 > 1$ and J is not contained in any minimal prime of R, then $\dim \mathcal{A}_{R\twoheadrightarrow R/J}(I/J) = \dim R$.*

Proof. (a) One direction is obvious from Theorem 12.2.21. For the reverse implication, assume that $J \cap I^t = JI^{t-1}$ for every positive integer t. By the above remark, one may assume that $a_1, \dots, a_m$ is a minimal set of generators of J. If for some i, $a_i \in I^2$, then by hypothesis $a_i \in JI$, which clearly contradicts the minimality of $a_1, \dots, a_m$.

(b) By definition, we want to show the two inclusions

$$J \cap I^{t+\nu_m-1} \subset JI^{t-1} \subset J \cap I^{t+\nu_1-1}$$

as subideals of $J \cap I^t$, for every $t \geq 1$.

This is however a straightforward consequence of Theorem 12.2.21 as one has thereof

$$J \cap I^{\nu_m+t-1} = \sum_{i=1}^{m} a_i I^{\nu_m+t-1-\nu_i} \subset \sum_{i=1}^{m} a_i I^{t-1} = JI^{t-1},$$

and similarly,

$$J \cap I^{\nu_1+t-1} = \sum_{i=1}^{m} a_i I^{\nu_1+t-1-\nu_i} \supset \sum_{i=1}^{m} a_i I^{t-1} = JI^{t-1}.$$

(c) Quite generally, for a positive integer r one has

$$\begin{aligned}\dim \mathcal{R}_R(I)/\mathfrak{N} \cap I^r\mathcal{R}_R(I) &\geq \max\{\dim \mathcal{R}_R(I)/\mathfrak{N}, \dim \mathcal{R}_R(I)/I^r\mathcal{R}_R(I)\} \\ &= \max\{\dim \mathcal{R}_{R/J}(I/J), \dim \mathrm{gr}_I(R)\} = \dim R.\end{aligned}$$

On the other hand, since J is assumed to be of positive height, by Proposition 12.2.6(a) one has $\dim \mathcal{A}_{R\twoheadrightarrow R/J}(I/J) \leq \dim R$. The result follows now at once. □

We close with yet another condition for the surjection $\mathcal{A}_{R\twoheadrightarrow R/J}(I/J) \twoheadrightarrow \mathcal{R}_{R/J}(I/J)$ to be an isomorphism.

For this recall that, pretty generally, given ideals $J, I \subset R$ the *Artin–Rees number* of J relative to I is the integer

$$\min\{k \geq 0 \mid J \cap I^t = (J \cap I^k)I^{t-k}\ \forall\, t \geq k\}.$$

We observe that if $J \subset I \subsetneq R$, where R is Noetherian and J has regular elements then the Artin–Rees number of J relative to I is ≥ 1.

Proposition 12.2.23. *Let $J \subset I \subsetneq R$ be ideals of a Noetherian ring R and let $k \geq 1$ be an upper bound for the Artin–Rees number of J relative to I. Then I^{k-1} annihilates the kernel of the surjection $\mathcal{A}_{R\twoheadrightarrow R/J}(I/J) \twoheadrightarrow \mathcal{R}_{R/J}(I/J)$. Moreover, the latter is an isomorphism if and only if the Artin–Rees number of J relative to I is* 1.

Proof. One has $(J \cap I^t)I^{k-1} = (J \cap I^k)I^{t-k}I^{k-1} = (J \cap I^k)I^{t-1} \subset JI^{t-1}$ for $t \geq k$. On the other hand, for $t \leq k-1$, one has $I^{k-1} \subset I^{t-1}$, hence $(J \cap I^t)I^{k-1} \subset (J \cap I^t)I^{t-1} \subset JI^{t-1}$.

The second assertion is clear. □

More generally, we have the following.

Lemma 12.2.24. *Let $J \subset I \subset R$ be ideals of a ring R. Assume that ℓ is an upper bound for the Artin–Rees number of J relative to I such that $J \cap I^t = JI^{t-1}$ for every $t \leq \ell$. Then $\mathcal{A}_{R\twoheadrightarrow R/J}(I/J) \twoheadrightarrow \mathcal{R}_{R/J}(I/J)$ is an isomorphism.*

Proof. By assumption, $J \cap I^t = I^{t-\ell}(J \cap I^\ell)$ for $t \geq \ell$. Now use the assumed equality $J \cap I^\ell = JI^{\ell-1}$ to get $J \cap I^t = JI^{t-1}$ for $t \geq \ell$. □

Given a ring R and an ideal $\mathfrak{a} = (a_1, \dots, a_n) \subset R$, recall that the relation type of $\mathfrak{a}$ is the largest degree of a minimal generator of a presentation ideal of the Rees algebra of $\mathfrak{a}$ (Subsection 2.3.4).

Corollary 12.2.25. *Let R be a Notherian ring and let $J \subset I$ be ideals in R such that:*

(i) *I/J has relation type at most ℓ as an ideal of R/J.*
(ii) *$J \cap I^t = JI^{t-1}$ for every $t \leq \ell$.*

Then $\mathcal{A}_{R\twoheadrightarrow R/J}(I/J) \twoheadrightarrow \mathcal{R}_{R/J}(I/J)$ is an isomorphism.

Proof. By Lemma 12.2.24, it suffices to show that ℓ is an upper bound for the Artin–Rees number of J relative to I. Thus, let I be generated by elements $a_1, \dots, a_n$ and let $a \in J \cap I^t$, with $t \geq \ell$. Then there exists a homogeneous $F \in R[\mathbf{T}] = R[T_1, \dots, T_n]$, of degree t, such that $F(a_1, \dots, a_n) = a$. Since $a \in J$, reducing modulo J shows that F is a polynomial relation on $a_1, \dots, a_n$ modulo J. Then, by assumption there are polynomials $G_i, H_i \in R[\mathbf{T}]$ of degrees $\ell, t-\ell$, respectively, such that $G_i(a_1, \dots, a_n) \equiv 0 \pmod{JR[\mathbf{T}]}$ for $i = 1, \dots, n$ and $F \equiv \sum_i G_iH_i \pmod{JR[\mathbf{T}]}$. Write $F = \sum G_iH_i + L$ for some homogeneous polynomial $L \in JR[\mathbf{T}]$ of degree t. Since

$$L(a_1, \dots, a_n) \in JI^t \subset I^{t-\ell}(J \cap I^\ell),$$

then $G_i(a_1, \dots, a_n) \subset J \cap I^\ell$ and $H_i(a_1, \dots, a_n) \in I^{t-\ell}$ for $t \geq \ell$. This shows that the element $a = F(a_1, \dots, a_n) \in I^{t-\ell}(J \cap I^\ell)$, i. e., $J \cap I^t = (J \cap I^\ell)I^{t-\ell}$ for $t \geq \ell$. □

Using the notion of standard base we can add a tiny bit on the problem of describing the associated primes of the Aluffi algebra.

Proposition 12.2.26. *Let $J \subset I \subsetneq R$ be ideals of a Noetherian ring R such that I/J has a regular element. If $p \in \operatorname{Ass}_R(R/J)$, then $\tilde{p} \in \operatorname{Ass}_{\mathcal{R}_R(I)}(\mathcal{A}_{R\twoheadrightarrow R/J}(I/J))$.*

Proof. Let $p \in \operatorname{Ass}_R(R/J)$. Since I/J contains a regular element, one has $I \not\subset p$. Let $a_1, \dots, a_m$ be an I-standard base of p, so that $p = (a_1, \dots, a_m)$ and

$$\tilde{p} = \bigoplus_{t\geq 0} p \cap I^t = \bigoplus_{t>0}\left(\sum_{i=1}^m a_iI^{t-v_i}\right),$$

where $v_i = v_I(a_i)$. Write $v = \max_i v_i$ for $i = 1, \dots, m$ and take $b \in I^{v-1} \setminus p$ (note that if $v = 0$, which is a possibility, one means any $b \notin p$).

Write $\mathfrak{R} = \bigoplus_{t\geq 0} J \cap I^t$ and $\mathfrak{A} = \bigoplus_{t\geq 0} JI^{t-1}$.

Say, $p = J : a$. We claim that $\tilde{p} = \mathfrak{A} : ab$ which will prove that $\tilde{p}$ is an associated prime of $\mathcal{A}_{R\twoheadrightarrow R/J}(I/J)$ on $\mathcal{R}_R(I)$.

For this, let $c_t \in J \cap I^t \subset p \cap I^t = \sum_{i=1}^m a_iI^{t-v_i}$, with $t \geq 0$. Then

$$bc_t \in \sum_{i=1}^m a_i\, bI^{t-v_i} \subset \sum_{i=1}^m a_iI^{t-v_i+v-1} \subset \sum_{i=1}^m a_iI^{t-1} = pI^{t-1} = (J : a)I^{t-1} \subset JI^{t-1} : a,$$

hence, $b\tilde{p} \subset \mathfrak{A} : a$, i. e., $\tilde{p} \subset \mathfrak{A} : ab$.

For the inverse inclusion, since $\mathfrak{A} \subset \mathfrak{R}$, it follows that $\mathfrak{A} : ab \subset \mathfrak{R} : ab = (\mathfrak{R} : a) : b$. Note that $\widetilde{J : a} = (J : a)R[t] \cap \mathcal{R}_R(I) = (JR[t] : a) \cap \mathcal{R}_R(I) = (JR[t] \cap \mathcal{R}_R(I)) : a = \mathfrak{R} : a$.

Therefore $\mathfrak{A} : ba \subset \tilde{p} : b = \tilde{p}$. Thus, $\tilde{p} = \mathfrak{A} : ab$ as was to be shown. □

12.2.5 Selected examples

Let us agree to call a pair of ideals $J \subset I \subset R$ an *$\mathcal{A}$-torsionfree pair* if the map $\mathcal{A}_{R\twoheadrightarrow R/J}(I/J) \twoheadrightarrow \mathcal{R}_{R/J}(I/J)$ is injective.

The examples we have in mind in this part are of the two sorts mentioned previously, namely, of totally degenerate or nondegenerate pairs. The first kind will be based on Proposition 12.2.10. For these, we let $R = k[\mathbf{x}]$ be an $\mathbb{N}$-graded polynomial ring over a field k, $J \subset R$ is a homogeneous ideal and $I \subset R$ is the Jacobian ideal of J, by which we always mean the ideal $(J, \mathcal{I}_r(\Theta))$ where $r = \mathrm{ht}(J)$ and Θ stands for the Jacobian matrix of a set of generators of J. One knows that this maybe a slack ideal, but it is well-defined modulo J.

First, we state a general format that implies an $\mathcal{A}$-torsion-free pair $J \subset I$.

Example 12.2.27. Let $J \subset R = k[\mathbf{x}]$ be an ideal generated by forms of the same degree $d \geq 1$. If $I = (J, \mathfrak{m}^r)$, where $\mathfrak{m} = (\mathbf{x})$ and $r \geq d$, then the pair $J \subset I$ is $\mathcal{A}$-torsion-free.

To see this, one uses Proposition 12.2.10. Namely, it suffices to show that $J \cap \mathfrak{m}^{rt} \subset J\mathfrak{m}^{r(t-1)}$ for every $t \geq 1$. Let $a_1, \ldots, a_m$ be generators of J of degree d and let F be a form in the a_i's such that $F \in \mathfrak{m}^{rt}$. Then $F = \sum_{i=1}^m G_i a_i$ where $G_i = \sum a_{\underline{\alpha}} \mathbf{x}^\alpha \in R_{rt-d+\delta}$ for $\delta \geq 0$, since $R_{rt-d+\delta} = R_{r-d+\delta} R_{rt-r}$, so we can rewrite G_i as

$$G_i = \sum_{\substack{|\alpha|=r-d+\delta \\ |\beta|=rt-r}} a_{\alpha,\beta}\, \mathbf{x}^{\alpha+\beta}, \quad \text{hence } F = \sum_{|\alpha|=r-d+\delta} \mathbf{x}^\alpha \Big(\sum_{\substack{i=1 \\ |\beta|=rt-r}}^{s} (\mathbf{x}^\beta) a_i \Big)$$

Therefore, $F \in J\mathfrak{m}^{rt-r} = J\mathfrak{m}^{r(t-1)}$, as required.

Instances of this situation seem to be any of the following ideals J with respect to the respective Jacobian ideal I.

(a) The defining ideal of the rational normal curve.
(b) The defining ideal of the Segre embedding of $\mathbb{P}^r \times \mathbb{P}^s$, with $r > 1$ or $s > 1$.
(c) The defining ideal of the 2-Veronese embedding of a projective space.
(d) The ideal generated by 4×4 Pfaffians.

Let I denote the Jacobian ideal of J on R. Set $\mathfrak{m} = (\mathbf{x})$ and write $\mathrm{ht} J = r \geq 2$. We would need to prove that $I_r(\Theta) = \mathfrak{m}^r$, where Θ is the Jacobian matrix of the natural generators of J. A calculation of a good deal of cases gives evidence to this equality—actually, it may be the expression of a more general fact disguised under an inductive procedure.

Note that the case of the Segre embedding of $\mathbb{P}^1 \times \mathbb{P}^1$ is exceptional, essentially because it is a hypersurface. Here, the ideal of 1-minors is $\mathbf{x}$ which is of linear type, but clearly the defining ideal of the relative blowup contains the equation of the dual to the (self-dual) quadric surface.

Example 12.2.28. Consider the monomial $x_1 \cdots x_n \in R = k[x_1, \dots, x_n]$ ($n \geq 3$) and let $J \subset R$ be the ideal generated by its partial derivatives $a_i =: x_1 x_2 \cdots \widehat{x_i} \cdots x_n$, for $i = 1, \dots, n$. If I is the Jacobian ideal of J, the pair $J \subset I$ is $\mathcal{A}$-torsion-free.

Proof. Its well known and easy that J is a codimension 2 perfect ideal with Hilbert–Burch matrix

$$\varphi = \begin{pmatrix} x_1 & 0 & 0 & \cdots & 0 \\ 0 & 0 & 0 & \cdots & x_2 \\ \vdots & \vdots & \vdots & \cdots & \vdots \\ 0 & 0 & x_{n-2} & \cdots & 0 \\ 0 & x_{n-1} & 0 & \cdots & 0 \\ -x_n & -x_n & -x_n & \cdots & -x_n \end{pmatrix}.$$

Setting $\Delta_{i,j} := \frac{\partial a_j}{\partial x_i} = x_1 x_2 \cdots \widehat{x_i} \cdots \widehat{x_j} \cdots x_n$, and inspecting the Hessian matrix Θ of $x_1 \cdots x_n$ – a symmetric matrix—one finds three basic types of 2×2 minors, namely

- Principal minors:

$$\det \begin{pmatrix} 0 & \Delta_{i,j} \\ \Delta_{i,j} & 0 \end{pmatrix} = \Delta_{i,j}^2,$$

  one for each pair $i < j$.
- vanishing minors:

$$\det \begin{pmatrix} \Delta_{i,j} & \Delta_{i,j'} \\ \Delta_{i',j} & \Delta_{i',j'} \end{pmatrix} = 0,$$

  one for each choice of row indices $1 \leq i, i' \leq n$ and column indices $1 \leq j, j' \leq n$.
- semidiagonal minors of typical form

$$\Lambda_j := \det \begin{pmatrix} \Delta_{i,j} & 0 \\ * & \Delta_{i',i} \end{pmatrix}.$$

Since clearly, $\Lambda_j \in J$, we get that the Jacobian ideal I of J is generated by J and the squares of the second partial derivatives of $x_1 \cdots x_n$, i. e., $I = (J, \Delta_{i,j}^2)$ for $1 \leq i < j \leq n$.

As a side curiosity, we note that $I = (J, \mathcal{I}_2(\Theta)) = (J, \mathcal{I}_{n-2}(\varphi)^2)$, hence $\sqrt{I} = \mathcal{I}_{n-2}(\varphi)$, so in particular I/J has codimension one. This example will therefore yield a case of a height one ideal in R/J which is $\mathcal{A}$-torsion-free, but clearly not of linear type because its number of generators on R/J is too large.

Setting $\Delta = (\Delta_{i,j}^2 \mid 1 \leq i < j \leq n)$, the usual algorithmic procedure to find generators of the intersection of monomial ideals yields for any $t \geq 2$,

$$J \cap \Delta^t = ((x_i, x_j)\Delta_{i,j}^{2t}, (\mathfrak{F}))$$

where $\mathfrak{F}$ is the set of all monomials in Δ^t excluding the monomials $\Delta_{i,j}^{2t}$ for $1 \le i < j \le n$. Another calculation shows that $(x_i, x_j)\Delta_{i,j}^{2t} \in J\Delta^{t-1}$, for $1 \le i < j \le n$, and that $\mathfrak{F} \subset J^2\Delta^{t-2}$. This proves that $J \cap \Delta^t \subset JI^{t-1}$. □

Question 12.2.29 (*k* **algebraically closed).** Let $J \subset R = k[x_1, \dots, x_n]$ denote the homogeneous defining ideal of an arrangement of linear subspaces of dimension $n-3$ of $\mathbb{A}^n$. If I is the Jacobian ideal of J, when is $J \subset I$ an $\mathcal{A}$-torsion-free pair?

Plausibly, a similar question can be posed about the Jacobian ideal of a hyperplane arrangement.

Example 12.2.30. Let $J \subset R = k[x_1, \dots, x_n]$ denote the ideal of the coordinate points in projective space $\mathbb{P}^{n-1}$, i. e., $J = (x_i x_j \mid 1 \le i < j \le n)$. If I is the Jacobian ideal of J, the pair $J \subset I$ is $\mathcal{A}$-torsion-free.

Proof. Since J contains all square-free monomials of degree 2, it is rather transparent that the Jacobian ideal I of J is generated by J and pure powers of the variables. Moreover, a closer inspection shows that, more precisely,

$$I = (J, x_1^{n-1}, \dots, x_n^{n-1}).$$

Setting $\Delta := (x_1^{n-1}, \dots, x_n^{n-1})$, a procedure based on finding generators of the intersection of monomial ideals yields for any $t \ge 2$,

$$J \cap \Delta^t = (x_i^{t(n-1)}(x_1, \dots, \widehat{x_i}, \dots, x_n), (\mathfrak{F})),$$

where $(\mathfrak{F})$ is the set of all monomials in Δ^t excluding the monomials $x_i^{t(n-1)}$ for $i = 1, \dots, n$. A calculation shows that $x_i^{t(n-1)}(x_1, \dots, \widehat{x_i}, \dots, x_n) \in J\Delta^{n-1}$ for $i = 1, \dots, n$ and that $(\mathfrak{F}) \subset J^2\Delta^{t-2}$. This proves that $J \cap \Delta^t \subset JI^{t-1}$. □

Example 12.2.31. Let $J \subset R = k[x, y, z]$ denote the homogeneous defining ideal of the four points $(1:0:0), (0:1:0), (0:0:1)$ and $(1:1:1)$ in the projective plane $\mathbb{P}_k^2$ and let I denote its Jacobian ideal. An easy calculation gives

$$J = (x^2 - xz, y^2 - yz) \quad \text{and} \quad I = (x^2 - xz, y^2 - yz, x(2y - z), y(2x - z), (2x - z)(2y - z)).$$

In terms of the internal grading of the algebra, using the description in Proposition 12.2.13, the torsion is generated by the appropriate residues of $\{xz^2(x - z), yz^2(y - z)\} \subset J \cap I^2$.

The last two examples motivate the question as to when $J \subset I$ is an $\mathcal{A}$-torsion-free pair in the case where J is the ideal of reduced points in $\mathbb{P}^n$ and I its Jacobian ideal. Some progress has been made in [150].

12.2.6 The Aluffi algebra of a gradient ideal

In this section, we will deal with the case where J is generated by the equation of a reduced hypersurface.

Thus, let $J = (f) \subset R$ be a principal ideal, where $R = k[\mathbf{x}] = k[x_1, \dots, x_n]$. We will focus on the Jacobian ideal $J_f = (f, \partial f/\partial x_1, \dots, \partial f/\partial x_n)$. We are particularly motivated by the problem as to when J_f is an ideal of linear type. In general, if f is not Eulerian, the local number of generators of J_f maybe an early obstruction to this property. We will consider the case where f is homogeneous in the standard grading of the polynomial ring and its degree is not a multiple of the characteristic; hence, $f \in J_f$. In this context, the ideal J_f will often be of linear type. For further details, we refer to Chapter 5.

We call $\mathcal{A}_{R\twoheadrightarrow R/(f)}(J_f/(f))$, the *gradient Aluffi algebra* of f. By Proposition 12.2.6(d), it is equidimensional of dimension $\dim R = n$. This particular case was of some interest in [5].

The following result collects parts of the main backstage for the Aluffi gradient algebra.

Theorem 12.2.32. *Let k denote an infinite field, let $f \in R = k[\mathbf{x}] = k[x_1, \dots, x_n]$ be a reduced homogeneous polynomial whose degree is not a multiple of the characteristic of k and let $J_f \subset R$ denote the corresponding gradient ideal. Assume that:*

(i) *The singular locus of $V(f) \subset \mathbb{P}^{n-1}$ consists of a nonempty finite set of points (equivalently, $\dim R/J_f = 1$).*
(ii) *The partial derivatives of f are algebraically independent over k.*

Then:

(a) *The minimal primes of the gradient Aluffi algebra, as read on $\mathcal{R}_R(J_f)$, are*
   - *The minimal prime ideals of $\mathcal{R}_{R/(f)}(J_f/(f))$, all of the form $\sum_{t\geq 0}(p) \cap I^t$ for a prime factor p of f.*
   - *The extended ideal $(\mathbf{x})\mathcal{R}_R(J_f)$.*
   - *Prime ideals whose lifting to $R[\mathbf{T}] = R[T_1, \dots, T_n]$ from a presentation $R[\mathbf{T}]/\mathcal{A} \simeq \mathcal{R}_R(I)$ have the form $(P, \mathfrak{f})$, where $P \subset R$ is a minimal prime of R/J_f and $\mathfrak{f}$ is an irreducible homogeneous polynomial in $k[\mathbf{T}]$.*

(b) *The gradient Aluffi algebra has nontrivial torsion.*
(c) *J_f is an ideal of linear type (resp., weakly of linear type, i. e., the canonical map $\mathcal{S}_R(J_f) \twoheadrightarrow \mathcal{R}_R(J_f)$ is injective in all high degrees) if and only if the natural surjection*

$$\mathcal{S}_{R/(f)}(J_f/(f)) \twoheadrightarrow \mathcal{A}_{R\twoheadrightarrow R/(f)}(J_f/(f))$$

is an isomorphism (respectively, an isomorphism in all high degrees).
(d) *The symmetric algebra $\mathcal{S}_{R/(f)}(J_f/(f))$ is Cohen–Macaulay; in particular, if J_f is of linear type then the gradient Aluffi algebra is Cohen–Macaulay.*

Proof. (a) We apply Proposition 12.2.16, from which the first set of minimal primes is clear.

To see that $(\mathbf{x})\mathcal{R}_R(J_f)$ is a minimal prime thereof, one proceeds as follows. There is a presentation of the gradient Aluffi algebra

$$\mathcal{A}_{R\twoheadrightarrow R/(f)}(J_f/(f)) \simeq R[\mathbf{T}]/\left(\mathcal{J}_f, f, \sum_{i=1}^{n} x_i T_i\right), \tag{177}$$

where $\mathcal{J}_f$ denotes the defining ideal of the Rees algebra $\mathcal{R}_R(J_f)$ on $R[\mathbf{T}]$.

Since the partials are homogeneous of the same degree algebraic independence over k is tantamount to analytic independence (i. e., the relations of the generators of J_f have coefficients in the ideal $(\mathbf{x})$). Therefore, the result follows from Proposition 12.2.8(ii).

Let $\mathcal{P}$ be a minimal prime of $\mathcal{A}_{R\twoheadrightarrow R/(f)}(J_f/(f))$ whose contraction P to R contains J_f and is properly contained in $(\mathbf{x})$. By Proposition 12.2.16, $\mathcal{P} = (P, \mathcal{P}_+)$. By assumption (i), it follows that P is a minimal prime of R/J_f; hence, has height $n-1$. By Theorem 12.2.7, $\mathcal{A}_{R\twoheadrightarrow R/(f)}(J_f/(f))$ is equidimensional. Therefore, the lifting of $\mathcal{P}$ to $R[\mathbf{T}]$ has height n. Since the lifting of any minimal generator of $(\mathcal{P}_+)$ is irreducible in $k[\mathbf{T}]$ it follows that the lifting of $\mathcal{P}$ to $R[\mathbf{T}]$ has the required form.

(b) The defining equation of the dual curve to f belongs to the presentation ideal of $\mathcal{R}_{R/(f)}(J_f/(f))$ on $R[\mathbf{T}]$ and not to $(\mathbf{x})R[\mathbf{T}]$; hence, by (a) it does not belong to the defining ideal of $\mathcal{A}_{R\twoheadrightarrow R/(f)}(J_f/(f))$ on $R[\mathbf{T}]$.

(c) This is a straightforward application of Proposition 12.2.4. The argument for the weak version of the property of being of linear type is exactly the same as in [loc. cit.].

(d) We apply the criterion of [96, Theorem 10.1]. Namely, we have to verify the following conditions:

(A) $\mu(J_f/(f)_{P/f}) \leq \mathrm{ht}(P/(f)) + 1 = \mathrm{ht}\, P$, for every prime ideal $P \supset J_f$ of R.

(B) $\mathrm{depth}(H_i)_{P/(f)} \geq \mathrm{ht}(P/(f)) - \mu(J_f/(f)_{P/(f)}) + i = \mathrm{ht}\, P - \mu(J_f/(f)_{P/(f)}) + i - 1$, for every prime ideal $P \supset J_f$ of R and every i such that $0 \leq i \leq \mu(J_f/(f)_{P/(f)}) - \mathrm{ht}(J_f/(f)_{P/(f)})$, and where H_i denotes the ith Koszul homology module of the partial derivatives on $R/(f)$.

Note that the primes containing J_f are $\mathfrak{m} = (x_1, \ldots, x_n)$ and its minimal primes, the latter all of height $n-1$.

(A) Since J_f itself is generated by n elements, it suffices to check the minimal primes. Thus, let $P \subset R$ be such a prime. Say, without lost of generality, that $x_n \notin P$. Because of the Euler relation, $\partial f/\partial x_n \in J_f$ locally at P and module (f). Therefore, locally at P and module (f), J_f is generated by $n-1 = \mathrm{ht}(P)$ elements.

(B) If P is a minimal prime of J_f, we saw in (A) that $\mu(J_f/(f)_{P/(f)}) = n-1$. Since $\mathrm{ht}\, P = n-1$, the condition is trivially verified as $i = 0, 1$.

Thus, let $P = \mathfrak{m}$. Again, an easy inspection of the numbers tells us that only the case where $i = 2$ needs an argument and, in this case, one has to prove that $\operatorname{depth}(H_2)_{\mathfrak{m}/(f)} \geq 1$. Localize R at $\mathfrak{m}$ and update the notation, so $R := R_{\mathfrak{m}} \supset J_f := J_{f\,\mathfrak{m}} \supset (f) = (f)_{\mathfrak{m}}$ and $H_2 := (H_2)_{P/(f)}$.

Now, f is a nonzero element in J_f and J_f has grade $n-1$ in R. Thus, there is a regular sequence in J_f of length $n-1$ starting with f. Write $L \subset J_f$ for the ideal generated by this regular sequence.

The following isomorphism is well know (see, e. g., [21, Theorem 1.6.16]):

$$H_2 \simeq \operatorname{Hom}_{R/(f)}\left(\frac{R/(f)}{J_f/(f)}, \frac{R/(f)}{L/(f)}\right) \simeq \operatorname{Hom}_{R/(f)}(R/J_f, R/L).$$

Therefore, $\operatorname{Ass}_{R/(f)}(H_2) = \operatorname{Supp}_{R/(f)}(R/J_f) \cap \operatorname{Ass}_{R/(f)}(R/L) \subset \operatorname{Ass}_{R/(f)}(R/L)$. But L is generated by a regular sequence of length $n-2$ modulo f by construction, while $\dim R/(f) = n-1$. Then $\mathfrak{m}/(f) \notin \operatorname{Ass}_{R/(f)}(R/(J_f))$, hence $\mathfrak{m}/(f) \notin \operatorname{Ass}_{R/(f)}(H_2)$. □

Remark 12.2.33. For $n \leq 4$, if the partial derivatives are k-linearly independent then the result of Gordan–Hesse–Noether implies that they are algebraically independent over k (see [39, Proposition 2.7] for a proof of the case $n = 3$ based on an observation of Zak). Thus, the assumption in this range is just linear independence.

As to (b), it is valid with no restriction when f is reduced since the defining equations of the dual variety to the hypersurface $V(f)$ belong to the presentation ideal $\mathcal{A}_f$ and, moreover, the ideal generated by these contains properly the defining ideal of the polar map of $V(f)$ (see [39, Remark 2.4]).

Example 12.2.34. Here is a simple illustration. Let $f = x^2y^2 + x^2z^2 + y^2z^2$, the equation of a plane quartic with 3 ordinary nodes. The minimal primes of the corresponding gradient Aluffi algebra, lifted to $k[x, y, z, T, U, V]$, are $(x, y, z)k[x, y, z, T, U, V]$, (x, y, V), (x, z, U), (y, z, T) and its lifted torsion. Since J_f is of linear type (Section 5.1.3.2), these are of course the minimal primes of the symmetric algebra $\mathcal{S}_{R/(f)}(J_f/(f))$.

12.3 The case of modules

This part is entirely based on the results of [162]. Establishing a bonafide analogue of the quasi-symmetric algebra of Aluffi for modules requires additional settings. For its precise formulation, the definition hinges on quite a bit of appropriate ingredients from module theory. Albeit a certain instability in choosing these ground participants, the definition in the embedded case turns out to be an analog of the original one for ideals. To make sure this analogue is a workable notion, one has to provide at least two things: First, that the embedded version admits a simple expression in terms of the Rees algebra of an embedded pull-back of the module; second and more difficult, that the embedded algebra lives in a family that allows to take inverse limits. This will in turn provide the analog of Aluffi's quasi-symmetric algebra for a module.

12.3.1 The embedded version

Let A be a Noetherian ring and let E be a finitely generated A-module with torsion submodule denoted $\tau_A(E)$. The following notion of rank avoids the restriction that E have a generic rank.

Definition 12.3.1. The *embedding rank* of E is the least integer $s \geq 1$ for which there is an embedding $E/\tau_A(E) \hookrightarrow F$, where F is free of rank s. If $E/\tau_A(E)$ is not torsionless, then no such embedding exists, in which case one may say that E has no embedding rank (or its embedding rank is undefined).

As observed, E admits an embedding rank if and only if the torsion-free module $E/\tau_A(E)$ is torsionless. It is also clear that the embedding rank is just the rank of E if it happens to have one. As an example, a nonzero ideal $I \subset A$ without a rank (i. e., contained in some associated prime P of A) still has embedding rank one.

Now, let $R \twoheadrightarrow A$ be a surjective ring homomorphism and let $J = \ker(R \twoheadrightarrow A)$. If $\mathbf{F}$ is an R-free module of rank n and F is an A-free module of rank n, an R-module surjection $\mathbf{F} \twoheadrightarrow F$ is said to be *compatible* with the structural ring surjection $R \twoheadrightarrow A$ if $\ker(\mathbf{F} \twoheadrightarrow F) = J\mathbf{F}$.

Let E be a finitely generated A-module such that $E/\tau_A(E)$ is torsionless (but may not have a rank). Fix an embedding $E/\tau_A(E) \hookrightarrow F$, where F is A-free of rank n.

Definition 12.3.2. A *J-deformation* of (the given embedding of) E to R is a finitely generated R-module $\mathbf{E}$ such that:

(i) There is an embedding $\mathbf{E}/\tau_R(\mathbf{E}) \hookrightarrow \mathbf{F}$ where $\mathbf{F}$ has rank n and, moreover, $J \subset (\mathbf{E}/\tau_R(\mathbf{E})) : \mathbf{F}$.
(ii) For any surjective R-map $\mathbf{F} \twoheadrightarrow F$ compatible with the ring map $R \twoheadrightarrow A$, the restriction to $\mathbf{E}/\tau_R(\mathbf{E})$ maps onto $E/\tau_A(E) \subset F$.

We will say that two J-deformations $\mathbf{E}$ and $\mathbf{E}_1$ of the given embedding of E to R are *congruent* if there is an R-isomorphism $\mathbf{F} \simeq \mathbf{F}_1$ of the respective ambient free R-modules which maps $\mathbf{E}/\tau_R(\mathbf{E})$ onto $\mathbf{E}_1/\tau_R(\mathbf{E}_1)$ and which commutes with the respective structural R-module surjections $\mathbf{F} \twoheadrightarrow F$ and $\mathbf{F}_1 \twoheadrightarrow F$ (i. e., which maps $J\mathbf{F}$ to $J\mathbf{F}_1$).

Note that the conditions imply in particular that, for a given J-deformation $\mathbf{E}$ to R, there is an A-module isomorphism $(\mathbf{E}/\tau_R(\mathbf{E}))/J\mathbf{F} \simeq E/\tau_A(E)$ and that if $\mathbf{E}_1$ is another J-deformation to R, then $\mathbf{E}_1/\tau_R(\mathbf{E}_1) \simeq \mathbf{E}/\tau_R(\mathbf{E})$.

As to the existence, fix a free R-module $\mathbf{F}$ of rank n and fix an R-map surjection $\mathbf{F} \twoheadrightarrow F$ compatible with $R \twoheadrightarrow A$. Consider the inverse image of $E/\tau(E) \subset F$ by $\mathbf{F} \twoheadrightarrow F$. This is clearly a J-deformation of E to R. Therefore, any finitely generated module E with a fixed embedding $E/\tau(E) \hookrightarrow F$ admits a unique J-deformation $\mathbf{E}$ to R up to R-torsion and congruence.

Note the induced R-algebra surjection $\mathcal{S}_R(\mathbf{E}/\tau(\mathbf{E})) \twoheadrightarrow \mathcal{S}_A(E/\tau_A(E))$ of symmetric algebras.

We can now state the analogue of Aluffi's notion.

Definition 12.3.3. Let E be a finitely generated A-module such that $E/\tau_A(E)$ is torsionless. Fix an embedding $E/\tau_A(E) \hookrightarrow F$, where F is A-free of rank n. Let $R \twoheadrightarrow A$ be a ring surjection with kernel $J \subset R$. Let $\mathbf{E}$ denote a torsion-free J-deformation of E to R (which, by the preceding, is uniquely defined up to congruence). The (*R-embedded*) *Aluffi algebra* of E is

$$\mathcal{A}_{R\twoheadrightarrow A}(E) := \mathcal{S}_A(E/\tau_A(E)) \otimes_{\mathcal{S}_R(\mathbf{E})} \mathcal{R}_R(\mathbf{E}).$$

Remark 12.3.4. This treatment does not assume a priori that E is torsion-free, merely that $E/\tau_A(E)$ has a positive embedding rank r and a fixed embedding into a free A-module of rank r is given. Of course, if E is actually a nonzero ideal $\mathfrak{a} \subset A$ (i. e., an A-module with embedding rank one with a fixed embedding) then the present notion retrieves Aluffi's concept.

Lemma 12.3.5. *There is an A-algebra isomorphism*

$$\mathcal{A}_{R\twoheadrightarrow A}(E) \simeq \frac{\mathcal{R}_R(\mathbf{E})}{(J, J\mathbf{F})\,\mathcal{R}_R(\mathbf{E})}, \tag{178}$$

where J is in degree 0 and $J\mathbf{F}$ is in degree 1. In particular, there is a surjective A-algebra homomorphism $\mathcal{A}_{R\twoheadrightarrow A}(E) \twoheadrightarrow \mathcal{R}_A(E)$.

Proof. By the very definition of $\mathcal{A}_{R\twoheadrightarrow A}(E)$, it is equivalent to show that $\mathcal{S}_A(E/\tau_A(E)) \simeq \mathcal{S}_R(\mathbf{E})/(J, J\mathbf{F})\mathcal{S}_R(\mathbf{E})$ (recall that $\tau_R(\mathbf{E}) = 0$ by construction). But this follows from the ordinary universal property of the symmetric algebra.

To see the surjection, one considers any algebra in sight in terms of a presentation over R. For this, we lift a set of generators of E over A to vectors in $\mathbf{E}$. These lifted vectors together with the set of natural generators of the submodule $J\mathbf{F}$ induced from a set of generators of J will be a set of generators of $\mathbf{E}$. We accordingly take a set of presentation variables $\mathbf{T} = \{\mathbf{T} \setminus \mathbf{T}'\} \cup \mathbf{T}'$, where the variables $\mathbf{T}'$ correspond to the generators coming from $J\mathbf{F}$. Let then $\mathcal{R}_R(\mathbf{E}) \simeq R[\mathbf{T}]/\mathcal{J}$ and $\mathcal{R}_A(E) \simeq R[\mathbf{T} \setminus \mathbf{T}']/(J, \tilde{\mathcal{J}})$, where $\tilde{\mathcal{J}}$ denotes the lifting to $R[\mathbf{T} \setminus \mathbf{T}']$ of a set of generators of the presentation ideal of $\mathcal{R}_A(E)$ over $A[\mathbf{T} \setminus \mathbf{T}']$. By the above isomorphism, it is clear that $\mathcal{A}_{R\twoheadrightarrow A}(E) \simeq R[\mathbf{T}]/(J, \mathbf{T}', \mathcal{J}) \simeq R[\mathbf{T} \setminus \mathbf{T}']/(J, \overline{\mathcal{J}})$, where $\overline{\mathcal{J}}$ is generated by the generators of $\mathcal{J}$ by setting $\mathbf{T}' = 0$. Therefore, one has an inclusion of ideals $(J, \overline{\mathcal{J}}) \subset (J, \tilde{\mathcal{J}})$, as required. □

12.3.2 Taking inverse limit

As in [5], one considers the inverse system of ring surjections $R \to R'$ commuting with given fixed surjections $R \twoheadrightarrow A$ and $R' \twoheadrightarrow A$. Let E be a finitely generated A-module such that $E/\tau_A(E)$ is torsionless, with a fixed embedding $E/\tau_A(E) \hookrightarrow F$, where F is A-free of rank n. Consider the (congruence classes of) torsionfree J-deformations $\mathbf{E}$, one for each $J = \ker(R \twoheadrightarrow A)$. Clearly, any surjection $R \to R'$ in the inverse system induces a natural map $\mathbf{F} \to \mathbf{F}'$ that maps $\mathbf{E}$ to $\mathbf{E}'$ and is compatible with the structural surjections $\mathbf{F} \twoheadrightarrow F$ and $\mathbf{F}' \twoheadrightarrow F$. The resulting set is again an inverse system inducing an inverse system of A-algebra surjections $\mathcal{A}_{R\twoheadrightarrow A}(E) \to \mathcal{A}_{R'\twoheadrightarrow A}(E)$. The resulting inverse limit

$$\mathcal{A}_A(E) := \lim_{\substack{\longleftarrow \\ R\twoheadrightarrow A}} \mathcal{A}_{R\twoheadrightarrow A}(E)$$

is the (absolute) *Aluffi algebra* of a finitely generated A-module E with a fixed embedding into a free A-module.

Still as in [5], we next restrict our inverse system to the category of algebras of finite type over a ground field k. The main bulk of the theory is given by the following theorem.

Theorem 12.3.6. *Let A be an algebra of finite type over a perfect field k and let E be a finitely generated A-module with a fixed embedding into a free A-module. Let $R \twoheadrightarrow A$ be a k-algebra surjection, where R is a smooth algebra of finite type over k. Then the structural map of A-algebras $\mathcal{A}_A(E) \twoheadrightarrow \mathcal{A}_{R\twoheadrightarrow A}(E)$ is an isomorphism.*

Proof. Although the proof follows the one of [5, Theorem 2.9], by retro-tracing the argument to a situation of rings and ideals, the details of the argument are more delicate. We first slightly rephrase the ring setup in a somewhat more precise form. Thus, let R and R' be smooth algebras of finite type over k endowed with fixed surjective k-algebra homomorphisms $R \twoheadrightarrow A$ and $R' \twoheadrightarrow A$. We assume that a surjective k-algebra homomorphism $\pi : R \twoheadrightarrow R'$ is given such that the triangle

$$\begin{array}{ccc} R & \stackrel{\pi}{\twoheadrightarrow} & R' \\ & \searrow \quad \swarrow & \\ & A & \end{array}$$

is commutative. According to the remarks prior to the statement there is a surjective A-algebra homomorphism $\mathcal{A}_{R\twoheadrightarrow A}(E) \stackrel{\tilde{\pi}}{\twoheadrightarrow} \mathcal{A}_{R'\twoheadrightarrow A}(E)$ and as in the proof of [5, Theorem 2.9] it suffices to show that this map is an isomorphism.

To do this, one uses the auxiliary commutative diagram of k-algebra homomorphisms

$$\begin{array}{ccc} & R \otimes_k R' & \\ {}^{\iota}\nearrow & & \searrow^{\mu} \\ R & \stackrel{\pi}{\twoheadrightarrow} & R' \\ \searrow & & \swarrow \\ & A & \end{array}$$

where ι is the one-sided injective map $a \mapsto a\otimes 1$ and μ is the multiplication map $a\otimes a' \mapsto \pi(a)a'$, while the lower maps are the structural maps. Note first that μ is split by the other one-sided injection $R' \simeq 1\otimes_k R' \subset R\otimes_k R'$. Also, one has $R\otimes_k R'/(\mathbb{D}, 1\otimes J') \simeq A$, where $J' = \ker(R' \twoheadrightarrow A)$ and $\mathbb{D}$ is the kernel of μ.

One now claims that the diagram induces a diagram of A-algebra homomorphisms

$$\begin{array}{ccccc} & & \mathcal{A}_{R\otimes_k R'\twoheadrightarrow A}(E) & & \\ & \tilde{\iota}\nearrow & & \searrow\tilde{\mu} & \\ \mathcal{A}_{R\twoheadrightarrow A}(E) & & \stackrel{\tilde{\pi}}{\twoheadrightarrow} & & \mathcal{A}_{R'\twoheadrightarrow A}(E) \end{array}$$

where $R\otimes_k R' \twoheadrightarrow A$ is the composite map $R\otimes_k R' \stackrel{\mu}{\to} R' \twoheadrightarrow A$ and $\tilde{\mu}$ is a split A-algebra homomorphism.

We first deal with this latter map. By definition, given an embedding $E/\tau_A(E) \subset F$, we choose a map $\mathbf{F}_{R'} \twoheadrightarrow F$ compatible with the structural surjection $R' \twoheadrightarrow A$ and take the inverse image $\mathbf{E}_{R'} \subset \mathbf{F}_{R'}$, so that $\mathbf{E}_{R'}/J'\mathbf{E}_{R'} \simeq E/\tau_A(E)$. Next, we apply to $\mathbf{F}_{R'}$ base change by the one-sided inclusion $R' \hookrightarrow R\otimes_k R'$ and set $\mathbf{F}_{R\otimes_k R'} = R\otimes_k \mathbf{F}_{R'}$. Thus, we get a commutative diagram where the horizontal A-module maps are split maps

$$\begin{array}{ccccc} \mathbf{E}_{R'} & \hookrightarrow & R\otimes_k \mathbf{E}_{R'} & \twoheadrightarrow & \mathbf{E}_{R'} \\ \cap & & \cap & & \cap \\ \mathbf{F}_{R'} & \hookrightarrow & R\otimes_k \mathbf{F}_{R'} & \twoheadrightarrow & R\otimes_k \mathbf{F}_{R'} \end{array}$$

It follows that $R\otimes_k \mathbf{E}_{R'}$ is a $(\mathbb{D}, 1\otimes J')$-deformation of $E/\tau_A(E)$. Thus, set $\mathbf{E}_{R\otimes_k R'} := R\otimes_k \mathbf{E}_{R'}$. Next, since $R\otimes_k \mathcal{S}_{R'}(\mathbf{E}_{R'}) \simeq \mathcal{S}_{R\otimes_k R'}(\mathbf{E}_{R\otimes_k R'})$ and $R\otimes_k \mathcal{R}_{R'}(\mathbf{E}_{R'}) \simeq \mathcal{R}_{R\otimes_k R'}(\mathbf{E}_{R\otimes_k R'})$, this diagram induces a commutative diagram of symmetric and Rees algebras,

$$\begin{array}{ccccccc} \mathcal{R}_{R'}(\mathbf{E}_{R'}) & \hookrightarrow & \mathcal{R}_{R\otimes_k R'}(\mathbf{E}_{R\otimes_k R'}) & \stackrel{\rho}{\twoheadrightarrow} & \mathcal{R}_{R'}(\mathbf{E}_{R'}) & & \\ \uparrow & & \uparrow & & \uparrow & & \\ \mathcal{S}_{R'}(\mathbf{E}_{R'}) & \hookrightarrow & \mathcal{S}_{R\otimes_k R'}(\mathbf{E}_{R\otimes_k R'}) & \twoheadrightarrow & \mathcal{S}_{R'}(\mathbf{E}_{R'}) & \twoheadrightarrow & \mathcal{S}_A(E/\tau_A(E)) \end{array}$$

where the only nontrivial map is ρ. The existence of this map is guaranteed by Proposition 2.3.18 because $R\otimes_k \mathbf{E}_{R'}$ is locally of linear type at the associated primes of $\mathbb{D} = \ker(R\otimes_k R' \twoheadrightarrow R')$ since R' is locally regular (being smooth over a perfect field).

One then gets a split injection of Aluffi algebras:

$$\mathcal{S}_A(E/\tau_A(E)) \otimes_{\mathcal{S}_{R'}(\mathbf{E}_{R'})} (\mathcal{R}_{R'}(\mathbf{E}_{R'})) \hookrightarrow \mathcal{S}_A(E/\tau_A(E)) \otimes_{\mathcal{S}_{R\otimes_k R'}(\mathbf{E}_{R\otimes_k R'})} (\mathcal{R}_{R\otimes_k R'}(\mathbf{E}_{R\otimes_k R'}))$$

whose left inverse is precisely the required map $\tilde{\mu}$.

On the other hand, one has

$$\begin{aligned}&\mathcal{S}_A(E/\tau_A(E))\otimes_{\mathcal{S}_{R\otimes_k R'}(\mathbf{E}_{R\otimes_k R'})}(\mathcal{R}_{R\otimes_k R'}(\mathbf{E}_{R\otimes_k R'}))\\ &\simeq \mathcal{S}_A(E/\tau_A(E))\otimes_{R\otimes_k \mathcal{S}_{R'}(\mathbf{E}_{R'})}(R\otimes_k \mathcal{R}_{R'}(\mathbf{E}_{R'}))\\ &\simeq (\mathcal{S}_A(E/\tau_A(E))\otimes_k R)\otimes_{\mathcal{S}_{R'}(\mathbf{E}_{R'})}(\mathcal{R}_{R'}(\mathbf{E}_{R'}))\\ &\simeq (\mathcal{S}_A(E/\tau_A(E))\otimes_k A)\otimes_{\mathcal{S}_{R'}(\mathbf{E}_{R'})}(\mathcal{R}_{R'}(\mathbf{E}_{R'}))\\ &\simeq \mathcal{S}_A(E/\tau_A(E))\otimes_{\mathcal{S}_{R'}(\mathbf{E}_{R'})}(\mathcal{R}_{R'}(\mathbf{E}_{R'}))\end{aligned}$$

It follows that $\tilde{\mu}$ is an isomorphism.

It remains to explain the map $\tilde{\imath}$ and argue that it is injective. This will conclude the proof.

As before, we construct everything *ab initio*. This time around, we choose a map $\mathbf{F}_R \twoheadrightarrow F$ compatible with the structural surjection $R \twoheadrightarrow A = R/J$ and take the inverse image $\mathbf{E}_R \subset \mathbf{F}_R$, so that $\mathbf{E}_R/J\mathbf{E}_R \simeq E/\tau_A(E)$. Next, we apply to $\mathbf{F}_R$ base change by the one-sided inclusion $R \hookrightarrow R\otimes_k R'$ and set $\mathbf{F}_{R\otimes_k R'} = \mathbf{F}_R \otimes_k R'$. Note this induces a surjection $\mathbf{F}_{R\otimes_k R'} \twoheadrightarrow \mathbf{F}_R\otimes_R R'/J' = \mathbf{F}_R\otimes_R A = F$ compatible with the structural map $R\otimes_k R' \twoheadrightarrow A$.

We note that though $R \hookrightarrow R\otimes_k R'$ has no left inverse, it is an injective flat map because k is a field and, for any R-module M one has $M\otimes_R (R\otimes_k R') \simeq M\otimes_k R'$ as $R\otimes_k R'$-modules. We claim that $\mathbf{E}_R\otimes_k R'$ is a $(\mathbb{D}, 1\otimes J')$-deformation of $E/\tau_A(E)$. Since $J\mathbf{E}_R \subset \mathbf{F}_R$, it follows that $(\mathbb{D}, 1\otimes J')\mathbf{E}_R\otimes_k R' \subset \mathbf{F}_R\otimes_k R'$. This shows that condition (i) of the definition of deformation is satisfied.

To see the second condition of this definition, let $\psi : \mathbf{F}_R\otimes_k R' \twoheadrightarrow F$ be a map compatible with $R\otimes_k R' \twoheadrightarrow A = R\otimes_k R'/(\mathbb{D}, 1\otimes J')$. Since the image of $\mathbf{E}_R$ in $\mathbf{E}_R\otimes_k R'$ generates the latter as $R\otimes_k R'$-module and the restriction of $\mathbf{F}_R \twoheadrightarrow F$ to $\mathbf{E}_R$ maps onto $E/\tau_A(E)$—by condition (ii) applied to the J-deformation $\mathbf{E}_R$—it is also clear that the restriction of $\psi : \mathbf{F}_R\otimes_k R' \twoheadrightarrow F$ to $\mathbf{E}_R\otimes_k R'$ maps onto $E/\tau_A(E)$ as well, so condition (ii) also holds here.

Having got so far, the conclusion is easy. Indeed, first we have

$$\mathcal{S}_{R\otimes_k R'}(\mathbf{E}_R\otimes_k R') \simeq \mathcal{S}_R(\mathbf{E}_R)\otimes_R (R\otimes_k R') \simeq \mathcal{S}_R(\mathbf{E}_R)\otimes_k R'$$

and similar isomorphisms for the corresponding Rees algebra. Therefore, one has

$$\begin{aligned}&\mathcal{S}_A(E/\tau_A(E))\otimes_{\mathcal{S}_{R\otimes_k R'}(\mathbf{E}_{R\otimes_k R'})}(\mathcal{R}_{R\otimes_k R'}(\mathbf{E}_{R\otimes_k R'}))\\ &\simeq \mathcal{S}_A(E/\tau_A(E))\otimes_{\mathcal{S}_R(\mathbf{E}_R)\otimes_k R'}(\mathcal{R}_R(\mathbf{E}_R)\otimes_k R')\\ &\simeq (\mathcal{S}_A(E/\tau_A(E))\otimes_{\mathcal{S}_R(\mathbf{E}_R)}\mathcal{R}_R(\mathbf{E}_R))\otimes_k R'\end{aligned}$$

Since the map

$$\mathcal{S}_A(E/\tau_A(E))\otimes_{\mathcal{S}_R(\mathbf{E}_R)}\mathcal{R}_R(\mathbf{E}_R) \longrightarrow (\mathcal{S}_A(E/\tau_A(E))\otimes_{\mathcal{S}_R(\mathbf{E}_R)}\mathcal{R}_R(\mathbf{E}_R))\otimes_k R'$$

is injective, we are through. □

We have seen that, as long as the ambient R is a finitely generated smooth algebra over a perfect field, it is possible to study the Aluffi algebra as a single member of its defining inverse limit. Throughout the rest of the section, we assume that, in fact, $R = k[x_1, \dots, x_n]$ is a polynomial ring over a field k of characteristic zero and let $J \subset R$ be an ideal. We will be concerned with two basic modules: the module of k-derivations (vector fields) of R/J and the normal module of J in R. An advantage at the outset is that both modules are torsion free R/J-modules by definition, being R/J-duals of the well-known module of Kähler differentials and the conormal module of J in R, respectively. The complete understanding of the Aluffi algebra of any of these modules in general is a task far from immediate grasp. We will study this algebra for particular cases of the ideal J.

12.3.3 The module of derivations of a polynomial form

Set $E := \mathrm{Der}_k(R/J)$, the module of k-derivations of the k-algebra R/J. Note that $E \subset F := \sum_{i=1}^{n}(R/J)\,\partial/\partial x_i$, so in particular it is a torsion-free R/J-module. Then E has a deformation to the *module of logarithmic derivations*

$$\mathbf{E} = \mathrm{Der}_J(R) := \{\delta \in \mathrm{Der}_k(R) \mid \delta(J) \subset J\}.$$

This module is a global polynomial analogue of Saito's module $\mathrm{Der}_S(\log V(J))$ in case $S = \mathbb{A}^n$ and $V(J)$ is a hypersurface [169, (1.4) Definition]. Note that the module of logarithmic derivations sits naturally as an R-submodule of $\mathbf{F} := \sum_{i=1}^{n} R\,\partial/\partial x_i = \mathrm{Der}_k(R)$.

In this environment the presentation (178) of the Aluffi algebra is in general still out of reach due to the mysterious structure of the Rees algebra of the module of logarithmic derivations. Assuming J reduced, one has the following dimension bounds:

$$2n - 2\,\mathrm{ht}(J) \leq \dim \mathcal{A}_{R/J}(\mathrm{Der}_k(R/J)) \leq \dim R + rk\,(\mathrm{Der}_J(R)) - 1 = 2n - 1 \qquad (179)$$

which come out of the structural nature of the embedded Aluffi algebra.

We will focus on the case where J is a principal ideal generated by a reduced polynomial f. In this situation, by (179) the dimension of the corresponding Aluffi algebra gets trapped between $2n-2$ and $2n-1$. The two main ingredients to ascribe a presentation of the Aluffi algebra is a presentation of the Rees algebra $\mathcal{R}_R(\mathbf{E})$ of the deformation $\mathbf{E}$ and a set of generators of $f\mathbf{F}$ viewed as a subideal.

Now, $\mathcal{R}_R(\mathbf{E})$ can be reached in two ways since $\mathbf{E} \subset \mathbf{F}$. The first is as the algebra obtained from the symmetric algebra of $\mathbf{E}$ by killing R-torsion. The second, based on the prolegomena of Subsection 2.3.2.4, is as follows: trading notation $\partial/\partial x_i = t_i$ for visual convenience, the symmetric algebra of $\mathbf{F}$ is identified with the polynomial ring $R[t_1, \dots, t_n]$ and the Rees algebra of the module of logarithmic derivations of f gets identified with the R-subalgebra generated by the 1-forms which are the entries of the matrix product $\mathbf{t} \cdot \mathcal{D}$, where $\mathbf{t} = (t_1 \cdots t_n)$ and $\mathcal{D}$ denotes the matrix whose columns generate the module of logarithmic derivations, i. e., we get

$$\mathcal{R}(\mathrm{Der}_f(R)) \simeq R[\mathbf{t} \cdot \mathcal{D}] \subset R[\mathbf{t}].$$

In the second approach to further identify $f\mathbf{F}$ as contained in the graded component of degree one $R[\mathbf{t} \cdot \mathcal{D}]_1$, one needs to know how f sits, in the original module notation, as an element in the ith row of $\mathcal{D}$, for some $i = 1, \dots, n$.

Henceforth, we focus on the homogeneous case. Let $R = k[\mathbf{x}] = k[x_1, \dots, x_n]$ denote a standard graded polynomial ring over a field k. Let $f \in R = k[\mathbf{x}] = k[x_1, \dots, x_n]$ stand for a reduced form of degree $d \geq 2$ and let $\mathbf{E} := \mathrm{Der}_f(R) \subset \mathbf{F} := \mathrm{Der}_k(R) = \oplus_i R\,\partial/\partial x_i \simeq R^n$ and $E := \mathrm{Der}_k(R/(f)) \subset F := \oplus_i (R/(f))\,\partial/\partial x_i \simeq (R/(f))^n$ denote the R-module of logarithmic derivations and the module of derivations of $R/(f)$, respectively. Note that both modules are graded submodules of the respective graded-free modules in the grading inherited from the standard grading of R. Since both are generated in varying degrees, the respective associated R-algebras (symmetric, Aluffi, Rees, special fiber) are quasi-homogeneous (i. e., weighted homogeneous). In particular, some care has to be exercised when using their Hilbert series in order to deduce ideal theoretic properties.

First, our pet example.

12.3.3.1 Homogeneous free divisors

We refer to Section 5.2 for this notion.

Proposition 12.3.7. *Let $f \in R$ be a homogeneous-free divisor. Then*

(a) $\mathcal{A}_{R/(f)}(E) \simeq \mathcal{S}_{R/(f)}(E)$.

(b) *Let $\mathcal{A}_{R/(f)}(E) \simeq R[U_1, \dots, U_n]/\mathfrak{A}$ denote a presentation of the Aluffi algebra of E as a quasi-homogeneous R-algebra, where $U_1, \dots, U_n$ are weighted indeterminates over R with $R[U_1, \dots, U_n] \simeq \mathcal{R}_R(\mathbf{E})$ by mapping U_i to a corresponding basis element of $\mathbf{E}$ in an ordered basis. Then $\mathfrak{A}$ is a codimension 2 perfect ideal whose presentation matrix is the $(n+1) \times n$ matrix:*

$$\left(\begin{array}{ccc} U_1 & \dots & U_n \\ \hline & \mathcal{E} & \end{array} \right),$$

where $\mathcal{E}$ denotes the matrix whose columns are an ordered basis of $\mathbf{E}$ (up to signs).

(c) *The presentation matrix of $E = \mathrm{Der}_k(R/(f))$ as an R-module of projective dimension 1 is the $n \times n$ matrix of cofactors of $\mathcal{E}$.*

Proof. (a) This is trivial since $\mathbf{E}$ is free.

(c) This matrix is the content matrix of $f\mathbf{F}$ in $\mathbf{E}$. Since $f = \det \mathcal{E}$ (up to a sign perhaps), the assertion is clear.

(b) This is the core. As control steps, note that dimensions are right. Thus, with $\mathfrak{A}$ of codimension 2, the virtual dimension is $\dim \mathcal{A}_{R/(f)}(E) = 2n-2$. This is the true dimension if (F_0) holds over $R/(f)$. Now the rank of E is $\dim R/(f) = n-1$; by tensoring in (c) the presentation matrix of E over $R/(f)$ is still the matrix of cofactors modulo (f) and its rank over $R/(f)$ is 1. On the other hand, the cofactors generate an ideal of codimension 2 over R, hence of codimension 1 over $R/(f)$. Thus, (F_0) holds trivially.

The proof is also easy. Indeed, by the structural form in Lemma 12.3.5, $\mathfrak{A}$ is generated by f and the entries of the various product matrices $(U_1 \dots U_n) \cdot C_i$, where C_i stands for the content matrix of $(0 \dots 0 f 0 \dots 0)^t$ (f in the ith position) in $\mathcal{E}$. By (c) we know that the entries of C_i are suitable cofactors. □

We now deal with the main case of this subsection.

12.3.3.2 Smooth projective hypersurfaces

The following preliminary proposition substantially extends [99, Corollary 2.3] and Goto–Hayasaka–Kurano–Nakamura [76, Theorem 4.1].

Theorem 12.3.8. *Let $\mathbf{f} = \{f_1, \dots, f_n\}$ be a regular sequence of forms of same degree $d \geq 1$ in $R = k[\mathbf{x}] = k[x_1, \dots, x_n]$, $(n \geq 2)$. Consider the cokernel*

$$\bigwedge^3 R^n \xrightarrow{\kappa} \bigwedge^2 R^n \longrightarrow \mathbf{Z} \to 0, \tag{180}$$

where $\kappa = \kappa(\mathbf{f})$ denotes the second map in the Koszul complex of $\mathbf{f}$. Then the Rees algebra $\mathcal{R}_R(\mathbf{Z})$ of the R-module $\mathbf{Z}$ is a Gorenstein domain of codimension $\binom{n-1}{2}$ on the polynomial ring $R[y_{ij} \mid 1 \leq i < j \leq n]$.

Proof. Let $\mathbf{Y} := (y_{ij})_{1 \leq i<j \leq n}$ denote a generic $n \times n$ skew-symmetric matrix over R. Consider the augmented skew-symmetric matrix

$$\mathcal{P} = \begin{pmatrix} \mathbf{Y} & {}^t\mathbf{f} \\ -\mathbf{f} & 0 \end{pmatrix}, \tag{181}$$

where we have written $\mathbf{f} = (f_1 \dots f_n)$.

We will show that the defining ideal of $\mathcal{R}_R(\mathbf{Z})$ on the polynomial ring $R[y_{ij} \mid 1 \leq i < j \leq n]$ is the ideal $\mathfrak{P}$ generated by the 4-Pfaffians of (181) and that this ideal has the required properties in the statement.

In the completely generic case—i. e., if the augmented column has indeterminate entries mutually independent from the y_{ij}'s—it is fairly classical that the ideal of the 4-Pfaffians is Gorenstein, of codimension $\binom{n-1}{2}$, generated by $\binom{n+1}{4}$ quadrics [123]. In the present case, we have the same number of generators, $\binom{n}{4}$ of the 4-Pfaffians are quadrics, while the remaining $\binom{n}{3}$ ones are the 4-Pfaffians involving last column and row, hence are of degree $d+1$. Explicitly, these equations have the following respective shapes:

$$y_{ik}y_{jl} - y_{jk}y_{il} - y_{ij}y_{kl}, \quad 1 \le i < j < k < l \le n \tag{182}$$

and

$$f_i y_{jk} - f_j y_{ik} + f_k y_{ij}, \quad 1 \le i < j < k \le n. \tag{183}$$

We now prove that the defining ideal of $\mathcal{R}_R(\mathbf{Z})$ is the ideal $\mathfrak{P}$ generated by the 4-Pfaffians of the matrix (181). We may assume that $d \ge 2$ as otherwise one has n independent linear forms, so by a change of variables this is the case of Remark 12.3.10.

Claim 1. The ideal $\mathfrak{P}$ is Gorenstein of codimension $\binom{n-1}{2}$.

Proof. Let $\mathbf{X} = \{X_1, \dots, X_n\}$ be algebraically independent over $R[\mathbf{y}]$. Consider the fully generic version of (181):

$$\begin{pmatrix} \mathbf{Y} & {}^t\mathbf{X} \\ -\mathbf{X} & 0 \end{pmatrix}. \tag{184}$$

Clearly, $\mathbf{X} - \mathbf{f} = \{X_i - f_i, 1 \le i \le n\}$ is a regular sequence on $R[\mathbf{X}] = k[\mathbf{x}, \mathbf{X}]$.

In addition, $\mathbf{X} - \mathbf{f}$ is also a regular sequence on $R[\mathbf{Y}, \mathbf{X}]/\widetilde{\mathfrak{P}}$, where $\widetilde{\mathfrak{P}}$ is the ideal of 4-Pfaffians of (184). To see this, one can use a Gröbner basis sort of argument as follows. Fix the graded lex order in the standard graded polynomial ring $k[\mathbf{x}, \mathbf{Y}, \mathbf{X}]$ with the following ordering of variables: $\mathbf{x} > \mathbf{Y} > \mathbf{X}$. Recall that we are assuming that $\mathbf{f}$ is generated in degree $d \ge 2$ in $k[\mathbf{x}]$; then there is an inclusion of initial ideals $\operatorname{in}(\mathbf{f}) \subset \operatorname{in}(\mathbf{X} - \mathbf{f})$.

Indeed, first it is clear that the initial term of any f_i is the initial term of the corresponding $X_i - f_i$. Next, consider the S-polynomial $g_{i,j}$ of a pair f_i, f_j, for different indices i, j. Let the respective leading terms be denoted $\mathfrak{n}_i$ and $\mathfrak{n}_j$. We may as usual assume that $\gcd(\mathfrak{n}_i, \mathfrak{n}_j) \neq 1$; say, $e := \deg(\gcd(\mathfrak{n}_i, \mathfrak{n}_j))$. Consider in parallel the S-polynomial $h_{i,j}$ of the pair $X_i - f_i, X_j - f_j$. Then $h_{i,j} = g_{i,j} + T$, where T is certain combination of X_i, X_j with coefficients in $k[\mathbf{x}]$. Note that any nonzero term of $g_{i,j}$ has degree $d + d - e = 2d - e$. At the other end, any nonzero term of T has degree $d - (e - 1)$. Since $2d - e > d - (e - 1)$ for $d \ge 2$ and the order is graded then the leading term of $g_{i,j}$ and of $h_{i,j}$ are the same.

Next, consider the S-polynomial of $g_{i,j}$ and some f_l (possibly, $l = i$ or $l = j$). By the same token, the nonzero terms of this polynomial have degree higher that the degrees of any nonzero term in the part involving any variable form $k[\mathbf{X}]$. This argument goes on till one eventually reaches a Gröbner basis of $\mathbf{f}$, thus proving our intermediate contention.

Now we apply this as follows:

$$(\operatorname{in}(\widetilde{\mathfrak{P}}), \operatorname{in}(\mathbf{f})) \subset (\operatorname{in}(\widetilde{\mathfrak{P}}), \operatorname{in}(\mathbf{X} - \mathbf{f})) \subset \operatorname{in}(\widetilde{\mathfrak{P}}, \mathbf{X} - \mathbf{f})$$

Since the leftmost ideal is a sum of two ideals in separate variables, its codimension is the sum of the respective codimensions, i. e., equals $\binom{n-1}{2} + n$. Therefore, the codimension of the rightmost ideal is bounded below by this number. This proves that $\mathbf{X} - \mathbf{f}$ is also a regular sequence on $R[\mathbf{Y}, \mathbf{X}]/\widetilde{\mathfrak{P}}$. Therefore, specializing by $\mathbf{X} - \mathbf{f}$ yields that the specialized ideal $\mathfrak{P}$ is Gorenstein of same codimension as that of $\widetilde{\mathfrak{P}}$. □

Remark 12.3.9. The above seems to work under more general assumptions. We suspect that it suffices to assume that the ideal $(\mathbf{f})$ has codimension at least $n-2$. However, the above proof breaks down at the end.

Claim 2. The ideal $\mathfrak{P}$ is prime.

Proof. Since $\mathfrak{P}$ is homogeneous it suffices to prove that it is a normal ideal. Since it is Gorenstein (hence, Cohen–Macaulay), it certainly satisfies Serre's property (S_2). Therefore, it remains to prove that it satisfies property (R_1).

Set $N := \binom{n-1}{2}$, the codimension of $\mathfrak{P}$. Let Θ denote the Jacobian matrix of the generators of $\mathfrak{P}$ with respect to all variables in sight. The goal is to show that

$$\operatorname{cod}(I_N(\Theta) + I) \geq N + 2, \tag{185}$$

where $I_r(M)$ denotes the ideal generated by the r-minors of a matrix M. A close inspection of the generators of $\mathfrak{P}$ readily shows that

$$\Theta = \left(\begin{array}{c|c} \mathbf{0} & A \\ \hline B & C \end{array}\right)$$

for certain submatrices A, B and C, where the latter is of order $\binom{n}{3} \times \binom{n}{2}$ and each one of its entries is either some f_i $(1 \leq i \leq n)$ or else zero.

In addition, one has the following property: $(f_1^N, \ldots, f_n^N) \subset I_N(C)$. To see this, one needs to go into further detail about C by noting that C is the Jacobian matrix of the forms

$$h_{ijk} := f_i y_{jk} - f_j y_{ik} + f_k y_{ik} \quad (1 \leq i < j < k \leq n),$$

with respect to the $\mathbf{y}$ variables. Let us observe that, given $1 \leq \ell \leq n$, there are exactly N among these forms h_{ijk} such that $\ell \in \{i, j, k\}$—let $\mathbf{h}_\ell$ denote this set. Looking this way, one sees that the Jacobian matrix of $\mathbf{h}_\ell$ with respect to the variables y_{ij} such that $\ell \notin \{i, j\}$ is a diagonal submatrix of C with $\pm f_\ell$ throughout the main diagonal. Clearly, then $f_\ell^N \in I_N(C)$, showing the contention.

Applying this inclusion, one has

$$\begin{aligned} \operatorname{cod}(I_N(\Theta) + I) &\geq \operatorname{cod}(\operatorname{Pf}_4(\mathbf{y}), f_1^N, \ldots, f_n^N) \\ &= \operatorname{cod}(\operatorname{Pf}_4(\mathbf{y})) + \operatorname{cod}(f_1^N, \ldots, f_n^N) \\ &= \binom{n-2}{2} + n \end{aligned}$$

$$= N + 2,$$

as claimed in (185). This proves Claim 2. □

To conclude the proof of the theorem, we first note the codimension of the Rees algebra:

$$\begin{aligned}\operatorname{cod}(\mathcal{R}_R(\mathbf{Z})) &= \dim R[\mathbf{y}] - \dim \mathcal{R}_R(\mathbf{Z})\\ &= n + \binom{n}{2} - (\dim R + rk\,(\mathbf{Z}))\\ &= \binom{n}{2} - (n-1) = \binom{n-1}{2}\end{aligned}$$

Therefore, in order to close the argument it suffices as a consequence of Claim 1 and Claim 2 to show that $\mathfrak{P}$ is contained in the presentation ideal of $\mathcal{R}_R(\mathbf{Z})$. For this, we use the following device in order to replace the generators of $\mathbf{Z}$ by their corresponding "polynomializations." Namely, as an R-subalgebra of R^n, the Rees algebra $\mathcal{R}_R(\mathbf{Z})$ is also isomorphic to the R-subalgebra of the polynomial ring $R[\mathbf{t}] = R[t_1, \dots, t_n]$ generated by the biforms

$$\{f_i t_j - f_j t_i, 1 \le i < j \le n\}$$

coming from the Koszul relations of $\mathbf{f}$. An immediate inspection yields that the two sets (182) and (183) of equations vanish for $y_{ij} \mapsto f_i t_j - f_j t_i$, with $1 \le i < j \le n$.

This completes the proof. □

Remark 12.3.10. The proof that the defining ideal of $\mathcal{R}_R(\mathbf{Z})$ is the ideal $\mathfrak{P}$ generated by the 4-Pfaffians of the matrix (181) has been proved in [99, Corollary 2.3] and [76, Theorem 4.1] in the case where $\mathbf{f} = \{x_1, \dots, x_n\}$, assuming $\operatorname{char}(k) \neq 2$. In this situation, the argument has been based on detecting a Sagbi basis (in the case of [99]) or drawing upon generalized Grassmann algebras (in the case of [76]). Examining our line of argument above, is becomes apparent that in this case all one needs is that the ideal of 4-Pfaffians in the generic case be prime, a result due to Kleppe [123].

Theorem 12.3.11. *Let $f \in R = k[\mathbf{x}] = k[x_1, \dots, x_n]$ be a form of degree $d \ge 2$ defining a smooth projective hypersurface, where* $\operatorname{char}(k)$ *does not divide d. Keep the remaining notation as above, with $\mathbf{E} = \operatorname{Der}_f(R)$. Then:*

(i) *The Rees algebra $\mathcal{R}_R(\mathbf{E})$ is a polynomial ring in one variable T over the Rees algebra $\mathcal{R}_R(\mathbf{Z})$ of the cokernel*

$$\bigwedge^3 R^n \overset{\kappa(\partial f)}{\longrightarrow} \bigwedge^2 R^n \longrightarrow \mathbf{Z} \to 0, \tag{186}$$

where $\kappa(\partial f)$ denotes the second map in the Koszul complex of the partial derivatives of f.

(ii) *The Rees algebra $\mathcal{R}_R(\mathbf{E})$ is a Gorenstein domain of codimension $\binom{n-1}{2}$ on the polynomial ring $R[y_{ij} \mid 1 \le i < j \le n][T]$.*

(iii) *The Aluffi algebra $\mathcal{A}_{R/(f)}(E)$ is a Gorenstein ring of codimension $\binom{n-1}{2}+2$ on the polynomial ring $R[y_{ij} \mid 1 \le i < j \le n][T]$; in particular, its dimension coincides with the dimension of the Rees algebra $\mathcal{R}_{R/(f)}(E)$ of the module of derivations over $R/(f)$.*

Proof. (i) The assumption on the characteristic of k implies, via Euler's formula, that f belongs to the ideal of the partial derivatives of f, in which case the module of logarithmic derivations $\mathbf{E}$ is a direct sum of the first module of syzygies $\mathbf{Z}$ of the gradient ideal of f and the cyclic module $R\epsilon$ generated by the Euler derivation ϵ [173, Proposition 3.2]. Since taking the symmetric algebra is functorial, one has

$$\mathcal{S}_R(\mathbf{E}) \simeq \mathcal{S}_R(\mathbf{Z}) \otimes_R \mathcal{S}_R(R\epsilon) \simeq \mathcal{S}_R(\mathbf{Z}) \otimes_R R[T] \simeq \mathcal{S}_R(\mathbf{Z})[T],$$

where T is an indeterminate over R. On the other hand, $\mathcal{R}_R(\mathbf{E}) \simeq \mathcal{S}_R(\mathbf{E})/\tau_R(\mathcal{S}_R(\mathbf{E}))$, where $\tau_R(\mathcal{S}_R(\mathbf{E}))$ is the R-torsion of $\mathcal{S}_R(\mathbf{E})$. Because taking torsion commutes with a free extension, we get

$$\mathcal{R}_R(\mathbf{E}) \simeq \mathcal{S}_R(\mathbf{Z})[T]/\tau_R(\mathcal{S}_R(\mathbf{Z})[T]) \simeq (\mathcal{S}_R(\mathbf{Z})/\tau_R(\mathcal{S}_R(\mathbf{Z})))[T] \simeq \mathcal{R}_R(\mathbf{Z})[T].$$

To conclude, since $V(f)$ is smooth, the partial derivatives of f generate an $(x_1, \dots, x_n)$-primary ideal, hence form a regular sequence. This implies that the module $\mathbf{Z}$ of their syzygies coincides with the cokernel of the second Koszul map

$$\bigwedge^3 R^n \overset{\kappa(\partial f)}{\longrightarrow} \bigwedge^2 R^n.$$

(ii) By (i), this an immediate consequence of Theorem 12.3.8.

(iii) Set $\mathbf{Y} := \{y_{ij} \mid 1 \le i < j \le n\}$. By (i) and the proof of Theorem 12.3.8 one has

$$\mathcal{R}_R(\mathbf{E}) \simeq R[\mathbf{Y}, T]/(\mathfrak{P}),$$

where $\mathfrak{P} = \mathrm{Pff}_4(\mathcal{P})$ and $\mathcal{P}$ is as in (181).

Therefore, according to the presentation (178) in Lemma 12.3.5, one has

$$\mathcal{A}_{R/(f)}(E) \simeq R[\mathbf{Y}, T]/(\mathfrak{P}, f, f\mathbf{F}), \tag{187}$$

where $f\mathbf{F} \subset \mathbf{E} = \mathbf{Z} + R\epsilon$, with ϵ denoting the Euler derivation.

While the Rees algebra of $\mathbf{E}$ has a standard graded structure, the Aluffi algebra $\mathcal{A}_{R/(f)}(E)$ has a nonstandard graded structure over $R[\mathbf{Y}, T]$ if $\deg(f) \geq 3$, with T of weight 1 (for the Euler derivation ϵ) and each y_{ij} of weight $\deg(f) - 1$.

With these weights, the defining equations coming from the inclusion

$$f\mathbf{F} \subset \mathbf{E} = \mathbf{Z} + R\epsilon$$

become homogeneous. These equations are written in compact form as

$$I_1((\mathbf{Y}, T) \cdot \mathcal{C}), \tag{188}$$

with $\mathcal{C}$ the transpose of the concatenation $[\kappa \mid^t \partial f]$, where κ is the usual matrix of the first Koszul map of the variables $x_1, \dots, x_n$ and

$$\partial f := \left(\frac{\partial f}{\partial x_1} \cdots \frac{\partial f}{\partial x_n} \right).$$

Collecting (187) and (188) yields

$$\mathcal{A}_{R/(f)}(E) \simeq R[\mathbf{Y}, T]/(\mathfrak{P}, f, I_1((\mathbf{Y}, T) \cdot \mathcal{C})), \tag{189}$$

Claim. $(f, I_1((\mathbf{Y}, T) \cdot \mathcal{C})) = (I_1((\mathbf{x}, T) \cdot \mathcal{P}))$

This is an expression of the so-called Jacobian duality [187, Section 1] plus the easy observation that f is essentially the product of $(\mathbf{x}, T)$ by the rightmost column of $\mathcal{P}$.

As a consequence, one has the presentation $\mathcal{A}_{R/(f)}(E) \simeq R[\mathbf{Y}, T]/(\mathfrak{P}, I_1((\mathbf{x}, T) \cdot \mathcal{P}))$ where the presentation ideal is generated by the defining ideal of the symmetric algebra of the cokernel of $\mathcal{P}$ and the 4-Pfaffians of $\mathcal{P}$.

We now consider the generic analogue of this ideal, namely, replace $\mathcal{P}$ by the generic skew-symmetric matrix $\mathcal{G}$ of the same size – thus, the partial derivatives of f are to be replaced by new variables $y_{1,n+1}, \dots, y_{n,n+1}$.

In this regard, the following result has been proved by A. Kustin.

Theorem 12.3.12 ([129, Theorem 4.8]). *The ideal $(\mathfrak{G}, I_1((\mathbf{x}, T) \cdot \mathcal{G}))$ is Gorenstein of codimension $\binom{n-1}{2} + 2$ on the polynomial ring $R[y_{ij} \mid 1 \leq i < j \leq n+1][T]$, where $\mathfrak{G} = \mathrm{Pff}_4(\mathcal{G})$.*

We will specialize from the above generic result. For this, it suffices to prove that the sequence $\{y_{1,n+1} - \partial f/\partial x_1, \dots, y_{n,n+1} - \partial f/\partial x_n\}$ is regular on the polynomial ring $R[y_{ij} \mid 1 \leq i < j \leq n+1][T]$ and also modulo the ideal $(\mathfrak{G}, I_1((\mathbf{x}, T) \cdot \mathcal{G}))$.

The first assertion is clear. To show the second assertion, it is enough to argue that the ideal $(\mathfrak{P}, I_1((\mathbf{x}, T) \cdot \mathcal{P}))$ has codimension $\binom{n-1}{2} + 2$ on $R[y_{ij} \mid 1 \leq i < j \leq n+1][T]$. To see this, proceed as follows.

First, by (179) one has

$$2n - 2 \leq \dim \mathcal{A}_{R/(f)}(E) \leq 2n - 1.$$

Therefore, it suffices to show that $\operatorname{cod}(\mathfrak{P}, I_1((\mathbf{x}, T) \cdot \mathcal{P})) \geq \binom{n-1}{2} + 2$. At this point, we follow a similar argument as in the proof of [129, Lemma 4.1], namely, let $P \subset R[y_{ij} \mid 1 \leq i < j \leq n+1][T]$ be a prime ideal containing $(\mathfrak{P}, I_1((\mathbf{x}, T) \cdot \mathcal{P}))$.

If $T \in P$ then $Q := (\mathfrak{P}, T)$ is a prime of codimension $\binom{n-1}{2} + 1$ contained in P. But, since $f \in P \setminus Q$ then $\binom{n-1}{2} + 2 \leq \operatorname{ht} P$.

If $T \notin P$, let $\mathcal{P}'$ and $\mathcal{P}''$ denote the matrices obtained from $\mathcal{P}$ by removing, respectively, its last column and its last row. Set $\mathfrak{P}'' := \operatorname{Pff}_4(\mathcal{P}'')$. Observe the following facts: $\mathfrak{P}''$ is a prime ideal of codimension $\binom{n-2}{2}$ contained in P and the entries of $(\mathbf{x}, T) \cdot \mathcal{P}'$ form a regular sequence on $R_T/\mathfrak{P}''R_T$ contained in PR_T.

It follows that

$$\binom{n-1}{2} + 2 = \binom{n-2}{2} + n \leq \operatorname{cod} PR_T = \operatorname{cod} P,$$

as was to be shown. □

12.3.4 The normal module of a perfect ideal of codimension 2

12.3.4.1 Preliminaries

Let $J \subset R$ be an ideal in a Noetherian ring R with a given free presentation

$$R^m \xrightarrow{\varphi} R^n \longrightarrow J \to 0.$$

This induces a free presentation of the conormal module J/J^2 over R/J, which gives the normal module as a syzygy module

$$0 \to (J/J^2)^* \longrightarrow (R/J)^{n*} \xrightarrow{\bar{\varphi}^t} (R/J)^{m*},$$

where $*$ stands for (R/J)-dual. We take $E := (J/J^2)^*$ as an (R/J)-submodule of $(R/J)^{n*}$ by the above injective map, set $F := (R/J)^{n*} \simeq (R/J)^n$ and J-deform E to its inverse image $\mathbf{E}$ in $\mathbf{F} := R^n$ by the natural surjection $R^n \twoheadrightarrow (R/J)^n$.

Some of the arguments used below hinge on the following idea, originally introduced in [187] (see Section 3.2). Let $R := k[\mathbf{x}] = k[x_1, \ldots, x_d]$ stand for a polynomial ring over a field k and let M denote a linearly presented R-module generated by m elements, with presentation matrix Φ. Then the presentation ideal $I_1([\mathbf{y}] \cdot \Phi)$ of the symmetric algebra $\mathcal{S}_R(M)$ can be written in the form $I_1([\mathbf{x}] \cdot B(\Phi))$, where $[\mathbf{y}] = [y_1 \cdots y_m]$ are presentation variables and $B(\Phi)$ denotes the transposed Jacobian matrix of the generators of $I_1([\mathbf{y}] \cdot \Phi)$ with respect to the $\mathbf{x}$-variables. Note that the entries of this matrix are linear forms in the polynomial ring $k[\mathbf{y}]$.

By Cramer, the d-sized minors of $B(\Phi)$ belong to the defining ideal of the Rees algebra of M. Typically, d is too large a degree in order that these minors be minimal generators. We will say that M (or its Rees algebra) is of the *expected* fiber type provided the presentation ideal of the Rees algebra $\mathcal{R}_R(M)$ is $(I_1([\mathbf{y}]\cdot\Phi), I_d(B(\Phi)))$.

We will also need the following result.

Lemma 12.3.13 ([205]). *Let k be a perfect field and let $\Phi = (Y_{i,j})$ be the generic $n \times m$ $(n \geq m)$ matrix over k. If $m \neq n$, then the ideal $I_m(\Phi) \subset S := k[Y_{i,j}]$ of maximal minors of Φ is rigid. In particular, if $m = n-1$, then the normal module $(I/I^2)^*$ of $I := I_{n-1}(\Phi)$ is generated by the column vectors of the Jacobian matrix of the maximal minors of Φ as a submodule of the free module of $(S/I)^*$.*

One says that an ideal of a ring essentially of finite type over a field is *licci* if it belongs to the linkage class of a complete intersection. Perfect ideals of codimension two form a large class of licci ideals.

Lemma 12.3.14 (Communicated by B. Ulrich). *Let $J \subset R$ be a licci ideal. Then the normal module of J specializes. More explicitly, if $a \in R$ is a regular element in R and also on R/J then*

$$(J/J^2)^* \otimes_R R/(a) \simeq ((J,a)/(J^2,a))^*.$$

Proof. Let $\omega := \omega_{R/J}$ denote the canonical module of R/J. Since a licci ideal is strongly nonobstructed, $J/J^2 \otimes R\omega$ is a maximal Cohen–Macaulay module. Now, a is then also a regular element on $\omega := \omega_{R/J}$ as the latter is a maximal Cohen–Macaulay module. On the other hand, since $J/J^2 \otimes_R \omega$ is a maximal Cohen–Macaulay module one has $\mathrm{Ext}^1(J/J^2 \otimes_R \omega, \omega) = 0$. Therefore, the exact sequence $0 \to \omega \to \omega \to \omega/a\omega \to 0$ induces an exact sequence:

$$0 \to \mathrm{Hom}(J/J^2 \otimes_R \omega, \omega) \xrightarrow{.a} \mathrm{Hom}(J/J^2 \otimes_R \omega, \omega) \to \mathrm{Hom}\left(J/J^2 \otimes_R \omega, \frac{\omega}{a\omega}\right) \to 0.$$

On the other hand, one has

$$\begin{aligned} \mathrm{Hom}(J/J^2 \otimes_R \omega, \omega) &\simeq \mathrm{Hom}(J/J^2, R) \\ \mathrm{Hom}(J/J^2 \otimes_R \omega, \omega/a\omega) &\simeq \mathrm{Hom}((J,a)/(J^2,a), R/(a)). \end{aligned}$$

Therefore, one gets

$$\begin{aligned} ((J,a)/(J^2,a))^* = \mathrm{Hom}((J,a)/(J^2,a), R/(a)) &\simeq \mathrm{Hom}(J/J^2,R)/a\,\mathrm{Hom}(J/J^2,R) \\ &= (J/J^2)^* \otimes_R R/(a), \end{aligned}$$

as was to be shown. ☐

12.3.4.2 Main theorem

Since the normal module of a complete intersection is free, its Aluffi algebra is of no interest. The next simplest case is that of a codimension 2 perfect ideal and as it turns, not devoid of toil.

Theorem 12.3.15. *Fix integers $d \geq 1$ and $n \geq 3$. Let $R := k[\mathbf{x}] = k[x_1, \ldots, x_d]$ be a standard graded polynomial ring over a field of characteristic zero. Let $J \subset R$ denote a perfect ideal of codimension two, generated by the maximal minors of an $n \times (n-1)$ matrix φ whose entries are linear forms in R (some possibly zero). Then a J-deformation $\mathbf{E} \subset \mathbf{F} = R^n$ of the normal module $(J/J^2)^*$ satisfies the following properties:*

(i) *The R-dual $\mathbf{E}^*$ is a free module.*

(ii) *Both $(J/J^2)^*$ and $\mathbf{E}$ specialize from the generic version $\Phi = (Y_{i,j})$ of φ.*

(iii) *$\mathbf{E}$ is a linearly presented module of projective dimension 1 and minimally generated by $n(n-1)$ elements of degree $n-2$.*

(iv) *If $\mathbf{E}$ satisfies the condition G_d, then the Rees algebra of $\mathbf{E}$ is Cohen–Macaulay and of the expected fiber type.*

(v) *$\mathbf{E}$ is a module of linear type if and only if it satisfies the condition G_d and the inequality $n(n-2) \leq d-1$ holds.*

(vi) *If the dimension of the symmetric algebra $\mathcal{S}_{R/J}((J/J^2)^*)$ is $n(n-1)$, then the dimension of its Aluffi algebra $\mathcal{A}_{R/J}((J/J^2)^*)$ equals the analytic spread $\ell(\mathbf{E})$.*

Proof. (i) We argue that $\mathbf{E}$ is an *ideal module* in the sense of [189, Section 5]. For this, since J has grade 2, it suffices to show that $J \subset \mathrm{Ann}(\mathbf{F}/\mathbf{E})$ [189, Proposition 5.1(c)]. But this is obvious since $J\mathbf{F} \subset \mathbf{E}$ by the construction of a deformation.

(ii) The degree of the generators of J is $n-1$; hence, the syzygies of the transpose φ^t modulo J have initial degree bounded below by $n-2$. That the generating syzygies have exactly this degree is to be seen.

Write $\varphi = (l_{i,j})$, where $l_{i,j} \in R$ is a linear form. Harmlessly assume that $I_1(\varphi) = (x_1, \ldots, x_n)$. The map $Y_{i,j} \mapsto l_{i,j}$ is a specialization since the sequence $\{Y_{i,j} - l_{i,j}\}$, $1 \leq i \leq j \leq n-1$, is regular both on $S := R[Y_{i,j}]$ and modulo $I_{n-1}(\Phi)S$.

Set $I := I_{n-1}(\Phi)S$. By Lemma 12.3.13 and flatness, the module $(I/I^2)^*$ coincides with the image in $(S/I)^n$ of the module generated by the column vectors of the Jacobian matrix $\mathbf{\Theta}$ of I with respect to the generic variables $Y_{i,j}$. In particular, it is minimally generated by $n(n-1)$ elements.

Letting $\widetilde{\mathbf{\Theta}}$ stand for the matrix over R specialized from $\mathbf{\Theta}$, then Lemma 12.3.14 implies that the normal module $(J/J^2)^*$ over R/J is the submodule image of $\widetilde{\mathbf{\Theta}}$ read modulo $J \subset R$. As such, therefore, it is also generated by $n(n-1)$ elements.

By definition, to get a J-deformation $\mathbf{E} \subset \mathbf{F} = R^n$ of the normal module one takes the R-submodule generated by the image of $\widetilde{\mathbf{\Theta}}$ and $J\mathbf{F}$. Note however that the product of the transpose φ^t by any column vector of $\widetilde{\mathbf{\Theta}}$ not only belongs to J, it is actually a k-linear combination of its set of generators. This shows that $J\mathbf{F}$ is already contained in the R-submodule generated by the image of $\widetilde{\mathbf{\Theta}}$, hence $\mathbf{E}$ is generated by the latter.

(iii) Reset $S := k[Y_{i,j}]$ and let $\mathbf{\Theta}$ as in (ii) denote the Jacobian matrix of $I_{n-1}(\Phi)$. Let $\mathcal{E} \subset S^n$ denote the submodule generated by the column vectors of $\mathbf{\Theta}$.

It will be convenient to rewrite $\mathbf{\Theta}$ by changing basis of the free ambient S^n and generators of $\mathcal{E}$. Namely, one can write $\mathbf{\Theta}$ as the concatenation of $n-1$ matrices

$$\mathbf{\Theta} = (\,\mathbf{\Theta}_1 \,|\ldots|\, \mathbf{\Theta}_{n-1}\,)$$

where

$$\mathbf{\Theta}_j = \begin{pmatrix} 0 & \delta^{(2)}_{1,j} & \cdots & \delta^{(n)}_{1,j} \\ \delta^{(1)}_{2,j} & 0 & \cdots & \delta^{(n)}_{2,j} \\ \vdots & \vdots & \ddots & \vdots \\ \delta^{(1)}_{n,j} & \delta^{(2)}_{n,j} & \cdots & 0 \end{pmatrix}.$$

Here for each $1 \le i \le n$, the symbol $\delta^{(i)}_{k,j}$ denotes the signed cofactor of the (k,j)-entry in the $(n-1)\times(n-1)$ matrix $\Phi_{(i)}$ obtained form Φ by omitting its ith row.

We now describe a set of natural syzygies of $\mathbf{\Theta}$. Namely, let C_j denote the jth column of Φ and let $\Phi^{(j)}$ stand for the $n \times (n-2)$-submatrix of Φ by omitting C_j. The following relations are immediately seen by Cramer's rule:

$$\mathbf{\Theta}_j \cdot C_j = (D_1 \ldots D_n)^t \quad \text{and} \quad \mathbf{\Theta}_j \cdot \Phi^{(j)} = \mathbf{0} \tag{190}$$

where D_i stands for the signed $(n-1)$-minor of φ omitting the ith row.

The matrix of syzygies corresponding to these relations can be written as the concatenation of two matrices, the first an $n(n-1)\times(n-1)(n-2)$ block diagonal matrix and the second an $n(n-1)\times(n-2)$ matrix in semiblock diagonal form:

$$\mathbf{\Psi} = \left(\begin{array}{cccccc|cccc} \Phi^{(1)} & & & & & & -C_1 & & & \\ & \Phi^{(2)} & & & & & C_2 & -C_2 & & \\ & & \Phi^{(3)} & & & & & C_3 & & \\ & & & \ddots & & & & & \ddots & \\ & & & & \Phi^{(n-2)} & & & & & -C_{n-2} \\ & & & & & \Phi^{(n-1)} & & & & C_{n-1} \end{array}\right),$$

where the empty slots are zero entries.

Therefore, one has the following complex of free modules:

$$0 \to S^{n(n-2)} \xrightarrow{\mathbf{\Psi}} S^{n(n-1)} \xrightarrow{\mathbf{\Theta}} S^n.$$

To prove exactness, we draw upon the well-known Buchsbaum–Eisenbud acyclicity criterion. Thus, first compute the ranks of $\mathbf{\Theta}$ and of $\mathbf{\Psi}$. As for $\mathbf{\Theta}$ it is well known that the maximal minors of φ are algebraically independent over k; hence, the rank of is n.

As for $\mathbf{\Psi}$, from its expression above it admits the following $n(n-2) \times n(n-2)$ submatrix in semidiagonal block shape:

$$
\mathbf{\Psi}' = \left(\begin{array}{cc|cc|c|cc|c}
\Phi_i^{(1)} & C_{1,i} & & & & & & \\
\hline
 & C_{2,i} & \Phi_i^{(2)} & C_{2,i} & & & & \\
\hline
 & & & & \ddots & & & \\
\hline
 & & & & C_{n-2,i} & \Phi_i^{(n-2)} & C_{n-2,i} & \\
\hline
 & & & & & & * & \psi
\end{array}\right).
$$

Here, $*$ denotes a certain $(n-2)\times 1$ submatrix of C_{n-1} and, for $1 \leq i \leq n$ and $1 \leq j \leq n-2$, $\Phi_i^{(j)}$ and $C_{j,i}$ stand, respectively, for the submatrices obtained from $\Phi^{(j)}$ and C_j omitting their respective ith rows—note that thus $(\Phi_i^{(j)}|C_{j,i})$ is a square matrix of order $n-1$. In addition, ψ is an arbitrary $(n-2) \times (n-2)$ submatrix of $\Phi^{(n-1)}$ with nonvanishing determinant—such matrices exist since $J = I_{n-1}(\varphi) \subset I_{n-2}(\varphi^{(n-1)})$.

Then

$$
\det(\mathbf{\Psi}') = \det(\Phi_i^{(1)}|C_{1,i}) \cdots \det(\Phi_i^{(n-2)}|C_{n-2,i}) \det(\psi) = \pm(D_i)^{n-2}\det(\psi),
$$

where D_i denotes as earlier the $(n-1)$-minor of $\mathbf{\Phi}$ striking out the ith row.

It remains to check the codimensions of the relevant Fitting ideals. Since $\mathbf{\Theta}$ has rank n, clearly $I_n(\mathbf{\Theta})$ has codimension at least 1. As for $I_{n(n-2)}(\mathbf{\Psi})$, since ψ is an arbitrary $(n-2) \times (n-2)$ submatrix of $\Phi^{(n-1)}$ with nonvanishing determinant, we observe that the above determinant actually encapsulates the following ideal inclusion:

$$
(D_1{}^{n-2}, \dots, D_n{}^{n-2}) \cdot I_{n-2}(\Phi^{(n-1)}) \subset I_{n(n-2)}(\mathbf{\Psi}). \tag{191}
$$

Since $I_{n(n-2)}(\mathbf{\Psi})$ contains the product of two ideals each of codimension at least 2 then its codimension is at least 2, as required.

Claim. The cokernel of $\mathbf{\Theta}$ is a perfect S-module.

Indeed, its annihilator has same codimension as the Fitting $I_n(\mathbf{\Theta})$. But in the generic case, the latter has same codimension as the ideal $J = (D_1, \dots, D_n) = I_{n-1}(\Phi) \subset S$. To see this, one has on one hand that $\mathbf{\Theta}$ has rank 2 modulo J, which says that $I_n(\mathbf{\Theta}) \subset J$ (recall that $n \geq 3$). On the other hand, in the opposite direction, one has $(D_1{}^{n-2}, \dots, D_n{}^{n-2}) \subset I_n(\mathbf{\Theta})$; hence, the two ideals have the same codimension (in fact, share the same radical).

This latter inclusion clearly holds after specialization, hence $I_n(\widetilde{\mathbf{\Theta}})$ has codimension at least 2, where $\widetilde{\mathbf{\Theta}}$ denotes the specialized Jacobian matrix. By generic perfection (see [20, Theorem 3.5]), we get a free resolution

$$
0 \to R^{n(n-2)} \xrightarrow{\widetilde{\mathbf{\Psi}}} R^{n(n-1)} \xrightarrow{\widetilde{\mathbf{\Theta}}} R^n,
$$

where $\widetilde{\mathbf{\Psi}}$ denotes the specialization of $\mathbf{\Psi}$.

To conclude, note that the image of $\widetilde{\Theta}$ is the deformation $\mathbf{E}$ by the previous item. Therefore, this proves that $\mathbf{E}$ is a linearly presented R-module of projective dimension 1. Moreover, since all entries of the matrices in the above free resolution are homogeneous of positive degree, the assertion about the minimal number of generators of $\mathbf{E}$ also follows.

(iv) Using (iii), this reads off [189, Proposition 4.11].

(v) We first translate the condition G_d in terms of Fitting ideals in the present context. Recall the linear presentation matrix $\widetilde{\Psi}$ of $\mathbf{E}$ found in the proof of (iii).

Transcribing the equivalence explained in [189, Section 3, p. 615] into our context one sees that $\mathbf{E}$ satisfies G_d if and only if $I_t(\widetilde{\Psi}) \geq n(n-2) - t + 2$ for $n(n-2) - (d-2) \leq t \leq n(n-2)$. But since we are assuming the inequality $n(n-2) \leq d-1$, then the leftmost inequality of the above interval for t implies that the interval becomes $1 \leq t \leq n(n-2)$. In other words, $\mathbf{E}$ actually satisfies F_1. But a module of projective dimension one satisfying F_1 is of linear type [193, Theorem 3.4].

Conversely, if $\mathbf{E}$ is of linear type it satisfies G_d for even more reason and besides, the inequality $n(n-2) \leq d-1$ is equivalent to the following one:

$$n(n-1) = \mu(\mathbf{E}) \leq \dim R + rk(E) - 1 = d + n - 1,$$

the latter being satisfied by the F_1 condition on the nose, as read in terms of number of generators of $\mathbf{E}$.

(vi) Recall that

$$\begin{aligned}\mathcal{A}_{R/J}((J/J^2)^*) &\simeq \mathcal{R}_R(\mathbf{E})/(J, J\mathbf{F}) \\ &\simeq R[\mathbf{y}]/(J, I_1(\mathbf{y}\cdot\widetilde{\Psi}), \mathfrak{A}_{\geq 2}, I_1(\mathbf{y}\cdot\mathfrak{C})) \\ &\simeq \frac{\mathcal{S}_{R/J}((J/J^2)^*)}{\widetilde{\mathfrak{A}_{\geq 2}}}\end{aligned}$$

where $\mathbf{y} = \{y_{i,j} \mid 1 \leq i \leq n, 1 \leq j \leq n-1\}$, $\mathfrak{A}_{\geq 2}$ denotes the Rees equations of $\mathbf{E}$ of $\mathbf{y}$-degree at least 2, whereas the tilde over it has the obvious meaning, and finally $\mathfrak{C}$ denotes the content matrix of the inclusion $J\mathbf{F} \subset \mathbf{E}$ (a matrix with entries in R).

By assumption, $\dim \mathcal{S}_{R/J}((J/J^2)^*) = n(n-1)$.

Reading this value backwards in the above string of isomorphisms, the defining ideal $(J, I_1(\mathbf{y}\cdot\widetilde{\Psi}), I_1(\mathbf{y}\cdot\mathfrak{C}))$ of $\mathcal{S}_{R/J}((J/J^2)^*)$ has codimension $d+n(n-1)-n(n-1) = d$ on the ambient polynomial ring $R[\mathbf{y}]$. Since it is wholly contained in the extension $(\mathbf{x})R[\mathbf{y}]$ of the maximal ideal $(\mathbf{x})$ of R, it must be primary to $(\mathbf{x})R[\mathbf{y}]$. If now $P \subset R[\mathbf{y}]$ is any prime ideal containing the whole Aluffi defining ideal $(J, I_1(\mathbf{y}\cdot\widetilde{\Psi}), \mathfrak{A}_{\geq 2}, I_1(\mathbf{y}\cdot\mathfrak{C}))$ it must contain the ideal $(\mathbf{x}, \mathfrak{J})$, with $\mathfrak{J}$ standing for the ideal generated by the bigraded piece of $\mathfrak{A}_{\geq 2}$ of bidegrees $(0, \geq 2)$. But $\mathfrak{J}$ is the defining ideal of the k-subalgebra $k[\mathbf{E}] \subset \mathcal{R}_R(\mathbf{E})$ over $k[\mathbf{y}]$; i. e., the special fiber of $\mathcal{R}_R(\mathbf{E})$. In particular, $(\mathbf{x}, \mathfrak{J})$ is a prime ideal and must be the radical of the Aluffi defining ideal.

It follows that $\dim \mathcal{A}_{R/J}((J/J^2)^*) = \dim R[\mathbf{y}]/(\mathbf{x},\mathfrak{J}) = \dim k[\mathbf{y}]/\mathfrak{J}$, as was to be shown as the analytic spread is the dimension of the special fiber. □

Remark 12.3.16. A natural question arises as to whether one can treat similarly the case of the ideal J of maximal minors of an $n \times m$ $(m < n)$ matrix having codimension $n-m+1$, given that the latter is also rigid. A technical difficulty arises at the outset due to the fact that a deformation $\mathbf{E} \subset \mathbf{F}$ of the corresponding normal module will require quite a bit of minimal generators coming from the submodule $J\mathbf{F}$; hence, the numbers will be harder to control.

12.4 Exercises

Exercise 12.4.1. Let $I \subset R = k[x_0,x_1,x_2,x_3]$ denote the reduced ideal of the five points consisting of the four coordinate points and $(1:1:1:1)$. Let $J = (I, I_3(\Theta)) \subset R = k[x_0,x_1,x_2,x_3]$ be the Jacobian ideal of I, where Θ is the Jacobian matrix of I.

(1) Prove that $I = (x_ix_j - x_2x_3,\ 0 \le i < j \le 3,\ (i,j) \ne (2,3))$.
(*Hint*: Compute the intersection of the corresponding prime ideals.)

(2) Prove that I is a Gorenstein ideal.
(*Hint*: Since R/I is Cohen–Macaulay, it suffices to prove that the socle is one-generated.)

(3) Prove that $I_3(\Theta) = \mathfrak{m}^3$.
(*Hint*: Compare Θ with the well-known Jacobian matrix of the square-free Veronese ideal $(x_ix_j, 0 \le i < j \le 3)$.)

(4) Show that the Aluffi algebra of the pair $I \subset J$ is torsion-free.
(*Hint*: Argue that it follows from having the equality $J = (I, \mathfrak{m}^3)$.)

Exercise 12.4.2. Consider the following six points in $\mathbb{P}^3_k$, written as columns of a matrix:

$$\begin{pmatrix} 1 & 0 & 0 & 0 & 1 & -1 \\ 0 & 1 & 0 & 0 & 1 & 2 \\ 0 & 0 & 1 & 0 & 1 & 3 \\ 0 & 0 & 0 & 1 & 1 & 1 \end{pmatrix}$$

and let $I \subset R = k[x_0,x_1,x_2,x_3]$ denote the reduced ideal of these points.

(1) Show that these points are in linear general position.

(2) Show that $I = (x_0x_1 - x_3(ax_1-(a-1)x_2), x_0x_2 - x_3(bx_1-(b-1)x_2), x_0x_3 - x_3(dx_1 - (d-1)x_2), x_1x_2 - x_3(cx_1-(c-1)x_2))$, for certain $a,b,c,d \in k$.

(3) Prove that the Jacobian ideal of I is $J := (I, x_0^3, x_1^3, x_2^3, x_3^3, x_1x_3^2, x_2x_3^2)$.

(4) Prove that $\mathfrak{m}^3 \subset J$, where $\mathfrak{m} = (x_0,x_1,x_2,x_3)$, but $\mathfrak{m}^3 \ne I_3(\Theta)$.

(5) Show that the Aluffi algebra of the pair $I \subset J$ is torsion-free.

Exercise 12.4.3. Let $I \subset R = k[x,y,z]$ denote the reduced ideal of the four points in linear general position: $(0:0:1), (0:1:0), (1:0:0), (1:1:1)$.

(1) Prove that $I = (xz - yz, xy - yz)$.
(2) Let Θ stand for the Jacobian matrix of the above generators of I and set $J := (I, I_2(\Theta))$. Show that the embedded Aluffi algebra of the pair $I \subset J$ is not torsion-free.
(*Hint*: Argue that $I \cap J^2 \not\subset IJ$ by counting number of generators.)
(3) Show that J is a Gorenstein ideal of finite colength.
(4) Let φ be a 5×5 skew-symmetric matrix over R whose Pfaffians are the generators of J and let $\mathbf{T} = \{T_1, \ldots, T_5\}$ denote new variables. Let further ψ denote the Jacobian matrix of the entries of $(\mathbf{T}\cdot\varphi)$ with respect to x, y, z. Prove that the maximal minors of ψ are polynomial relations of the generators of J.

Exercise 12.4.4. In the notation of the previous exercise and its items, prove:

(1) The ideal $I_3(\psi) \subset k[\mathbf{T}]$ of maximal minors is the defining ideal of the special fiber of J. What is its minimal number of generators?
(*Hint*: Show it is prime of codimension 2.)
(2) Prove that J is of the fiber type, i. e.,

$$\mathcal{R}_R(J) \simeq R[\mathbf{T}]/(I_1(\mathbf{T}\cdot\varphi), I_3(\psi)).$$

(*Hint*: Show that the ideal $(I_1(\mathbf{T}\cdot\varphi), I_3(\psi))$ is prime of codimension 4.)
(3) Prove that the rational map $\mathbb{P}^2 \dashrightarrow \mathbb{P}^4$ defined by the generators of J is birational onto the image.
(4) Prove that depth $\mathcal{R}_R(J) = 1$.
(*Hint*: Let $\mathfrak{m} = (x,y,z)$. Show that $J^n = \mathfrak{m}^{2n}$, for $n \geq 2$ and deduce an exact sequence $0 \to \mathcal{R}_R(J) \to \mathcal{R}_R(\mathfrak{m}^2) \to C \to 0$, with C of finite colength.)

Exercise 12.4.5. Let $f \in R = k[\mathbf{x}] = k[x_1, \ldots, x_n]$ be a form of degree $d \geq 2$ defining a smooth projective hypersurface, where char(k) does not divide d. With the notation of Section 12.3.3, show that the defining ideal of the Rees algebra $\mathcal{R}_{R/(f)}(E)$ is the saturation by $(\mathbf{x})$ of the defining ideal of the Aluffi algebra $\mathcal{A}_{R/(f)}(E)$.

(*Hint*: Argue that $E = \mathrm{Der}_k(R/(f))$ is locally free on the punctured spectrum, thus implying that $\mathcal{A}_{R/(f)}(E) \simeq \mathcal{R}_{R/(f)}(E)$ locally on the punctured spectrum of $R/(f)$.)

Exercise 12.4.6. Let f be the equation of a cuspidal cubic in $\mathbb{P}^2$. With the notation of Section 12.3.3, argue (or compute) that:

(1) The Rees algebra $\mathcal{R}_R(\mathbf{E})$ is a hypersurface.
(2) $\mathcal{A}_{R/(f)}(E)$ is Cohen–Macaulay of codimension 3. Is it Gorenstein?
(3) $\mathcal{A}_{R/(f)}(E)$ coincides with the symmetric algebra $\mathcal{S}_{R/(f)}(E)$.
(*Hint*: $\mathbf{E}$ is of linear type because it has homological dimension 1 with presentation $0 \to R \xrightarrow{\varphi} R^4 \to \mathbf{E} \to 0$ having $I_1(\varphi)$ of codimension ≥ 2.)
(4) $\mathcal{R}_{R/(f)}(E)$ is almost Cohen–Macaulay, but not Cohen–Macaulay.

Exercise 12.4.7. Let f be the equation of a nodal cubic in $\mathbb{P}^2$. With the notation of the previous exercise, argue (or compute) that:

(1) $\mathcal{R}_R(\mathbf{E})$ is a complete intersection of codimension 2.

(2) $\mathcal{A}_{R/(f)}(E)$ is almost Cohen–Macaulay, but not Cohen–Macaulay.

Exercise 12.4.8 (The reciprocal involution of $\mathbb{P}^2$). Let $J = (xy, xz, yz) \subset R = k[x, y, z]$. With the notation of Section 12.3.4, argue (or compute) that:

(1) $k[\mathbf{E}]$ defined by the determinant of the Jacobian dual matrix.
(*Hint*: $\mathbf{E}$ satisfies the condition G_3, hence, by Theorem 7.3.17(iv), the presentation ideal of its Rees algebra has the expected fiber type.)

(2) $\mathbf{E}$ is not of linear type.

(3) Neither the Aluffi algebra nor the Rees algebra of the normal module is a Cohen–Macaulay algebra.

Exercise 12.4.9 (The reciprocal involution of $\mathbb{P}^3$). Let $J = (xyz, xyw, xzw, yzw) \subset R = k[x, y, z, w]$. With the notation of Section 12.3.4, argue (or compute) that:

(1) $\mathbf{E}$ is not of linear type.

(2) The minimal defining equations of $k[\mathbf{E}]$ are quadrics and cubics. How many?

(3) The Rees algebra of $\mathbf{E}$ is of fiber type.

(4) The rank of the Jacobian dual matrix $B(\Phi)$ is $d = 4$ but $I_d(B(\Phi))$ is no longer minimal generators.

(5) The unmixed part of the Aluffi algebra $\mathcal{A}$ has the form stated in the proof of Theorem 7.3.17(vi); in particular, $\dim \mathcal{A} = \ell(\mathbf{E}) = 7$.

(6) Neither $\mathcal{A}$ nor the Rees algebra of the normal module is a Cohen–Macaulay algebra.

Exercise 12.4.10 (The defining equations of the twisted cubic). Let $J = (xz - y^2, xw - yz, yw - z^2) \subset R - k[x, y, z, w]$. With the notation of Section 12.3.4, argue (or compute) that:

(1) $\mathbf{E}$ is of linear type.
(*Hint*: $\mathbf{E}$ satisfies the condition G_4 and $n(n-2) = 3 = d - 1$.)

(2) The analytic spread of $\mathbf{E}$ is maximal.

(3) The Aluffi algebra $\mathcal{A}$ is not Cohen–Macaulay, but the Rees algebra of the normal module is a Cohen–Macaulay domain.

Bibliography

[1] T. Abe, A. Dimca and G. Sticlaru, Addition–deletion results for the minimal degree of logarithmic derivations of hyperplane arrangements and maximal Tjurina line arrangements, J. Algebr. Comb., **54** (2021), 739–766.

[2] I. M. Aberbach and C. Huneke, An improved Briançon-Skoda theorem with applications to the Cohen-Macaulayness of Rees algebras, Math. Ann., **297** (1993), 343–369.

[3] I. Aberbach, C. Huneke and N. V. Trung, Reduction numbers, Briançon-Skoda theorems and the depth of Rees rings, Compositio Math., **97** (1995), 403–434.

[4] M. Alberich-Carramiñana, *Geometry of the Plane Cremona Maps*, Lecture Notes in Mathematics, **1769**, Springer, Berlin-Heidelberg, 2002.

[5] P. Aluffi, Shadows of blow-up algebras, Tohoku Math. J., **56** (2004), 593–619.

[6] J. Andrade and A. Simis, A complex that resolves the ideal of minors having $n-1$ columns in common, Proc. Amer. Math. Soc., **81** (1981), 217–219.

[7] M. Artin and M. Nagata, Residual intersections in Cohen–Macaulay rings, J. Math. Kyoto Univ., **12** (1972), 307–323.

[8] K. Ansaldi, K.-N. Lin and Y.-H. Shen, Generalized Newton complementary duals of monomial ideals, J. of Algebra and Its Applications, **20**, 2150021 (2021).

[9] L. Avramov and J. Herzog, The Koszul algebra of a codimension 2 embedding, Math. Z., **175** (1980), 249–280.

[10] D. Bayer, The division algorithm and the Hilbert scheme, Thesis, Harvard University, Cambridge, Mass., USA, 1982.

[11] V. Benedetti, Manifolds of low dimension with trivial canonical bundle in Grassmannians, Math. Z., **290** (2018), 251–287.

[12] T. C. Benítez and C. D'Andrea, Minimal generators of the defining ideal of the Rees Algebra associated to monoid parametrizations, Computer Aided Geometric Design, **27** (2010), 461–473.

[13] I. Bermejo and P. Gimenez, Saturation and Castelnuovo–Mumford regularity, J. Algebra, **303** (2006), 592–617.

[14] M. A. Binyamin, H. Hasan, H. Mahmood and F. U. Rehman, Characterization and Newton complementary dual of quasi f-ideals, arXiv:2101.01584 [math.AC].

[15] J. Blanc and A. Calabri, On degenerations of plane Cremona transformations, Math. Z., **282** (2016), 223–245.

[16] M. Brodmann, A few remarks on blowing-up and connectedness, J. Reine Angew. Math., **370** (1986), 52–60.

[17] M. Brodmann and R. Sharp, *Local Cohomology. An Algebraic Introduction with Geometric Applications*, 2nd Ed., Cambridge University Press, 2012.

[18] P. Brumatti, P. Gimenez and A. Simis, The Gauss algebra associated to a rational map $\mathbb{P}^d \to \mathbb{P}^n$, J. Algebra, **207** (1998), 557–571.

[19] W. Bruns and J. Gubeladze, *Polytopes, Rings, and K-Theory*, Springer Monographs in Mathematics, 2009.

[20] W. Bruns and J. Herzog, *Cohen–Macaulay Rings*, Cambridge Studies in Adv. Math., **39**, Cambridge University Press, 1993.

[21] W. Bruns and J. Herzog, On the computation of a-invariants, Manuscripta Math., **77** (1992), 201–213.

[22] R.-O. Buchweitz and A. Conca, New free divisors from old, J. Commut. Algebra, **5** (2013), 17–47.

[23] S. Budd and A. Van Tuyl, Newton complementary duals of f-ideals, Canad. Math. Bull., **62** (2019), 231–241.

https://doi.org/10.1515/9783110640694-013

[24] R. Burity, A. Simis and S. O. Tohăneanu, On a conjecture of Vasconcelos via Sylvester forms, J. Symb. Computation, **77** (2016), 39–62.

[25] R. Burity, A. Simis and Ş. O. Tohăneanu, On the Jacobian ideal of an almost generic hyperplane arrangement (earlier version: arXiv:2101.02735v1 [math.AC], 7 Jan 2021).

[26] L. Busé and M. Chardin, Implicitizing rational hypersurfaces using approximation complexes, J. Symbolic Computation, **40** (2005), 1150–1168.

[27] L. Busé, Elimination theory in codimension one and applications. [Research Report] RR-5918, INRIA. 2006 (Updated 2013), pp. 47.

[28] L. Busé, M. Chardin and A. Simis, Elimination and nonlinear equations of Rees algebras, J. Algebra, **324** (2010), 1314–1333.

[29] L. Busé and J.-P. Jouanolou, On the closed image of a rational map and the implicitization problem, J. Algebra, **265** (2003), 312–357.

[30] F. J. Calderón-Moreno and L. Narváez-Macarro, The module $\mathcal{D}f^s$ for locally quasi-homogeneous free divisors, Compositio Math., **134** (2002), 59–74.

[31] E. Casas-Alvero, Infinitely near imposed singularities and singularities of polar curves, Math. Ann., **287** (1990), 429–454.

[32] E. Casas-Alvero, *Singularities of Plane Curves*, London Math. Soc. Lecture Notes Series, **276**, Cambridge University Press, 2000.

[33] M. Catalano-Johnson, The possible dimensions of the higher secant varieties, Amer. J. Math., **118** (1996), 355–361.

[34] M. Catalano-Johnson, The homogeneous ideals of higher secant varieties, J. Pure Appl. Algebra, **158** (2001), 123–129.

[35] M. Chardin, Powers of ideals: Betti numbers, cohomology and regularity, in *Commutative Algebra*, pp. 317–333, Springer, New York, 2013.

[36] M. Chardin, S. H. Hassanzadeh and A. Simis, Degree of rational maps versus syzygies, J. Algebra, **573** (2021), 641–662.

[37] M. Chardin, Y. Cid Ruiz and A. Simis, Generic freeness of local cohomology and graded specialization, Trans. Amer. Math. Soc., https://doi.org/10.1090/tran/8316.

[38] Y. Cid-Ruiz and A. Simis, Degree of rational maps and specialization, Int. Math. Res. Notices (2020). https://doi.org/10.1093/imrn/rnaa183.

[39] C. Ciliberto, F. Russo and A. Simis, Homaloidal hypersurfaces and hypersurfaces with vanishing Hessian, Adv. Math., **218** (2008), 1759–1805.

[40] A. Conca, Gröbner bases of ideals of minors of a symmetric matrix, J. Algebra, **166** (1994), 406–421.

[41] A. Conca, Symmetric ladders, Nagoya Math. J., **136** (1994), 35–56.

[42] A. Conca, Divisor class group and canonical class of determinantal rings defined by ideals of minors of a symmetric matrix, Arch. Math., **63** (1994), 216–224.

[43] T. Cortadellas and S. Zarzuela, Burch's inequality and the depth of the blow up rings of an ideal, J. Pure Appl. Algebra, **157** (2001), 183–204.

[44] B. Costa, T. Dias, R. Gondim and R. Machado, Counting square free monomial Cremona maps, arXiv:2110.04874v1 [math.AG].

[45] B. Costa, Z. Ramos and A. Simis, A theorem about Cremona maps and symbolic Rees algebras, Internat. J. Algebra Comp., **24** (2014), 1191–1212.

[46] B. Costa and A. Simis, Cremona maps defined by monomials, J. Pure Appl. Algebra, **216** (2012), 212–225.

[47] B. Costa and A. Simis, New constructions of Cremona maps, Math. Res. Lett., **20** (2013), 629–645.

[48] R. C. Cowsik and M. V. Nori, Fibers of blowing up, J. Indian Math. Soc., **40** (1976), 217–222.

[49] E. D. Davis, A. V. Geramita and P. Maroscia, Perfect homogeneous ideals: Dubreil's theorems revisited, Bull. Sc. Math., 2e série, **108** (1984), 143–185.

[50] A. Dimca and G. Sticlaru, Saturation of Jacobian ideals: Some applications to nearly free curves, line arrangements and rational cuspidal plane curves, J. Pure App. Algebra, **223** (2019), 5055–5066.
[51] I. V. Dolgachev, in *Lectures on Cremona transformations*, Ann Arbor-Rome, 2010/2011.
[52] I. V. Dolgachev, Polar Cremona transformations, Mich. Math. J., **48** (2000), 191–202.
[53] A. Doria, H. Hassanzadeh and A. Simis, A characteristic free criterion of birationality, Adv. Math., **230** (2012), 390–413.
[54] A. Doria and A. Simis, The Newton complementary dual revisited, J. Algebra and its Applications, **17** (2018), 1850004, 1–16 pp.
[55] P. Du Val, Reducible exceptional curves, Amer. J. Math., **58** (1936), 285–289.
[56] P. Du Val, The unloading problem for plane curves, Amer. J. Math, **62** (1940), 307–311.
[57] D. Eisenbud, *Commutative Algebra (with a view toward Algebraic Geometry)*, Graduate Texts in Mathematics, **150**, Springer, Berlin-Heidelberg-New York, 1995.
[58] D. Eisenbud and S. Goto, Linear free resolutions and minimal multiplicity, J. Algebra, **88** (1984), 89–133.
[59] D. Eisenbud, C. Huneke and B. Ulrich, What is the Rees algebra of a module? Proc. Amer. Math. Soc., **131** (2002), 701–708.
[60] D. Eisenbud and B. Ulrich, Notes on regularity stabilization, Proc. Amer. Math. Soc., **140** (2012), 1221–1232.
[61] C. Escobar, J. Martínez-Bernal and R. H. Villarreal, Relative volumes and minors in monomial subrings, Linear Algebra Appl., **374** (2003), 275–290.
[62] S. Fitchett, B. Harbourne and S. Holay, Resolutions of fat point ideals involving eight general points of $\mathbb{P}^2$, J. Algebra, **244** (2001), 684–705.
[63] H. Flenner, L. J. van Gastel and W. Vogel, Joins and intersections, Math. Ann., **291** (1991), 691–704.
[64] W. Fulton, *Intersection Theory, Ergebnisse, 3. Folge*, **2**, Springer, Berlin-Heidelberg-New York-Tokyo, 1984.
[65] W. Fulton and J. Hansen, A connectedness theorem for projective varieties, with applications to intersections and singularities of maps, Ann. Math., **100** (1979), 159–166.
[66] W. Fulton and R. Lazarsfeld, Connectivity and its applications in algebraic geometry, in *Proc. Conf. Algebraic geometry*, Illinois at Chicago Circle, 1980, Lecture Notes in Mathematics, **862**, pp. 26–92, Springer, Berlin-Heidelberg-New York, 1981.
[67] F. Gaeta, Sul calcolo effetivo della forma associata $F(W^{gl}_{\alpha\,|\,\beta\ n})$ all'intersezione di due cicli effetivi pure U^g_α, V^l_β di S_n, in funzione delle $F(U^g_\alpha)$, $F(V^l_\beta)$ relative ai cicli secanti, Atti Accad. Naz. Linc., **24** (1958), 269–276.
[68] M. Garrousian, A. Simis and Ş. Tohăneanu, A blowup algebra for hyperplane arrangements, Algebra & Number Theory, **12** (2018), 1401–1429.
[69] L. van Gastel, Excess Intersections, Ph. D. Thesis, University of Utrecht, 1989.
[70] L. van Gastel, Excess intersections and a correspondence principle, Invent. Math., **103** (1991), 197–221.
[71] A. V. Geramita, Waring's problem for forms: inverse systems of fat points, secant varieties and Gorenstein algebras, Queen's Papers in Pure and Applied Mathematics, vol. **150** (1996), The Curves Seminar, vol. X.
[72] A. M. H. Gerards and A. Sebö, Total dual integrality implies local strong unimodularity, Math. Program., **38** (1987), 69–73.
[73] R. Godement, Localités simples II, Séminaire Cartan–Chevalley, **8** (1955/56).
[74] G. G. Gibson, *Elementary Geometry of Algebraic Curves: an Undergraduate Introduction*, Cambridge University Press, Cambridge, 1998.
[75] G. Gonzalez-Sprinberg and I. Pan, On the monomial birational maps of the projective space, An. Acad. Brasil. Ciênc., **75** (2003), no. 2, 129–134.

[76] S. Goto, F. Hayasaka, K. Kurano and Y. Nakamura, Rees algebras of the second syzygy module of the residue field of a regular local ring, Contemporary Math., **390** (2005), 97–108.

[77] A. V. Geramita, A. Gimigliano and Y. Pitteloud, Graded Betti numbers of some embedded rational n-folds, Math. Ann., **301** (1995), 363–380.

[78] S. Goto and Y. Shimoda, On the Rees algebras of Cohen-Macaulay local rings, in *Lecture Notes in Pure and Applied Mathematics*, **68**, pp. 201–231, Marcel Dekker, New York, 1979.

[79] D. R. Grayson and M. E. Stillman, *Macaulay2*, a software system for research in algebraic geometry. Available at http://www.math.uiuc.edu/Macaulay2/.

[80] P. Griffiths and J. Harris, Algebraic geometry and local differential geometry, Ann. scient. Éc. Norm. Sup., **12** (1979), 355–452.

[81] A. Grothendieck, *Cohomologie Locale des Faisceaux Cohérents et Théorèmes de Lefschetz Locaux et Globaux (SGA 2)*, Advanced Studies in Pure Mathematics, North-Holland, Amsterdam, 1968.

[82] F. Harary, *Graph Theory*, Addison-Wesley, Reading, MA, 1972.

[83] B. Harbourne, Free resolutions of fat point ideals on $\mathbb{P}^2$, J. Pure Appl. Alg., **125** (1998), 213–234.

[84] B. Harbourne, The ideal generation problem for fat points, J. Pure and Applied Alg., **145** (2000), 165–182.

[85] B. Harbourne, *Problems and Progress: A survey on fat points in P2*, Queen's papers in Pure and Applied Mathematics, The Curves Seminar at Queen's, **123**, 2002.

[86] B. Harbourne, S. Holay and S. Fitchett, Resolutions of ideals of quasiuniform fat point subschemes of $\mathbb{P}^2$, Trans. Amer. Math. Soc., **355** (2003), no. 2, 593–608.

[87] J. Harris, *Algebraic Geometry: A First Course*, Graduate Texts in Mathematics, **133**, Springer, Berlin-Heidelberg-New York, 1992.

[88] R. Hartshorne, *Algebraic Geometry*, Graduate Texts in Mathematics, **52**, Springer, 1975.

[89] S. H. Hassanzadeh and A. Simis, Plane Cremona maps: saturation, regularity and fat ideals, J. Algebra, **371** (2012), 620–652.

[90] S. H. Hassanzadeh and A. Simis, Implicitization of de Jonquières parametrizations, J. Commut. Algebra, **6** (2014), 149–172.

[91] S. H. Hassanzadeh and A. Simis, Bounds on degrees of birational maps with arithmetically Cohen–Macaulay graphs, J. Algebra, **478** (2017), 220–236.

[92] M. Herrmann, E. Hyry and T. Korb, On a-invariant formulas, J. Algebra, **227** (2000), 254–267.

[93] J. Herzog, R. Jafari and A. N. Nejad, On the Gauss algebra of toric algebras, J. Algebraic Combinatorics, **51** (2020), 1–17.

[94] J. Herzog, A. Simis and W. V. Vasconcelos, Approximation complexes of blowing-up rings, J. Algebra, **74** (1982), 466–493.

[95] J. Herzog, A. Simis and W. V. Vasconcelos, Approximation complexes of blowing-up rings. II, J. Algebra, **82** (1983), 53–83.

[96] J. Herzog, A. Simis and W. V. Vasconcelos, Koszul homology and blowing-up rings, in S. Greco and G. Valla (Eds.) *Commutative Algebra*, Lecture Notes in Pure and Applied Math., **84**, pp. 79–169, Marcel-Dekker, New York, 1983.

[97] J. Herzog, A. Simis and W. V. Vasconcelos, On the arithmetic and homology of algebras of linear type, Trans. Amer. Math. Soc, **283** (1984), 661–683.

[98] J. Herzog and H. Srinivasan, Bounds for multiplicities, Trans. Amer. Math. Soc., **350** (1998), 2879–2902.

[99] J. Herzog, Z. Tang and S. Zarzuela, Symmetric and Rees algebras of Koszul cycles and their Gröbner bases, Manuscripta Math., **112** (2003), 489–509.

[100] J. Herzog and N. V. Trung, Gröbner bases and multiplicity of determinantal and Pfaffian ideals, Adv. Math., **96** (1992), 1–37.

[101] H. Hironaka, Resolution of singularities of an algebraic variety over a field of characteristic zero, Ann. of Math., **79** (1964), 109–326.

[102] M. Hochster, Rings of invariants of tori, Cohen-Macaulay rings generated by monomials, and polytopes, Ann. Math., **96** (1972), 318–337.

[103] M. Hochster and C. Huneke, Tight closure, invariant theory, and the Brianǫn–Skoda theorem, J. Amer. Math. Soc., **3** (1990), 31–116.

[104] M. Hochster and C. Huneke, Indecomposable canonical modules and connectedness, in W. Heinzer, C. Huneke and J. Sally (Eds.) *Commutative Algebra: Syzygies, Multiplicities and Birational Algebra*, Contemp. Math., **159**, pp. 197–208, 1994.

[105] M. Hochster and C. Huneke, *Tight Closure in Equal Characteristic Zero: With an Introduction to the Characteristic P Theory*, Springer Monographs in Mathematics, Sprnger–Verlag, NY, 2009.

[106] M. Hochster and L. J. Ratliff Jr., Five theorems on Macaulay rings, Pacific J. Math, **44** (1973), 147–172.

[107] J. Hong, A. Simis and W. V. Vasconcelos, The homology of two-dimensional elimination, J. Symbolic Computation, **43** (2008), 275–292.

[108] J. Hong, A. Simis and W. Vasconcelos, The equations of almost complete intersections, Bull. Braz. Math. Soc. (New Series), **43** (2012), 171–199.

[109] J. Hong, A. Simis and W. Vasconcelos, Extremal Rees algebras, J. Commut. Algebra, **5** (2013), 231–267.

[110] S. Huckaba, A d-dimensional extension of a lemma of Huneke's and formulas for the Hilbert coefficients, Proc. Amer. Math. Soc., **124** (1996), 1393–1401.

[111] H. P. Hudson, *Cremona Transformations in Plane and Space*, Cambridge University Press, 1927.

[112] J. Huh, Milnor numbers of projective hypersurfaces with isolated singularities, Duke Math. J., **163** (2014), 1525–1548.

[113] C. Huneke, On the symmetric and Rees algebra of an ideal generated by a d-sequence, J. Algebra, **62** (1980), 268–275.

[114] C. Huneke and M. E. Rossi, The dimension and components of symmetric algebras, J. Algebra, **98** (1986), 200–210.

[115] C. Huneke, Hilbert functions and symbolic powers, Michigan Math. J., **34** (1987), 293–318.

[116] C. Huneke, Determinantal ideals of linear type, Arch. Math., **47** (1986), 324–329.

[117] C. Huneke and B. Ulrich, Residual intersections, J. Reine Angew. Math., **390** (1988), 1–20.

[118] C. Huneke and I. Swanson, *Integral Closure of Ideals, Rings, and Modules*, London Math. Soc., Lecture Note Series, **336**, Cambridge University Press, 2006.

[119] K. Johnson, Immersion and embedding of projective varieties, Acta Math., **140** (1978), 49–74.

[120] J. P. Jouanolou, Ideaux résultants, Adv. Math., **37** (1980), 212–238.

[121] I. Kaplansky, *Commutative Rings*, Revised Ed., The University of Chicago Press, 1974.

[122] G. Kennedy, Flatness of tangent cones of a family of hypersurfaces, J. Algebra, **128** (1990), 240–256.

[123] H. Kleppe, Deformations of schemes defined by vanishing of Pfaffians, J. Algebra, **53** (1978), 84–92.

[124] H. Kleppe and D. Laksov, The algebraic structure and deformation of Pfaffian schemes, J. Algebra, **64** (1980), 167–189.

[125] V. Kodiyalam, Integrally closed modules over two-dimensional regular local rings, Trans. Amer. Math. Soc., **347** (1995), 3551–3573.

[126] A. B. Korchagin, On birational monomial transformations of plane, Int. J. Math. Math. Sci., **32** (2004), 1671–1677.

[127] E. Kunz, *Kähler Differentials*, Adv. Lectures in Math., Vieweg, Wiesbaden, 1986.

[128] E. Kunz and R. Waldi, Regular differential forms, Contemp. Math., **79** (1988).

[129] A. Kustin, An alternating matrix and a vector, with application to Aluffi algebras, J. Algebra, **472** (2017), 115–145.

[130] A. Kustin, C. Polini and B. Ulrich, Integral extensions and the a-invariant, J. Pure Applied Algebra, **216** (2012), 2800–2806.

[131] A. Kustin, C. Polini and B. Ulrich, Blowups and fibers of morphisms, Nagoya Math. J., **224** (2016), 168–201.

[132] S. Lichtenbaum and M. Schlessinger, The cotangent complex of a morphism, Trans. Amer. Math. Soc., **128** (1967), 41–70.

[133] J. Lipman, Dualizing sheaves, differentials and residues on algebraic varieties, Astérisque, **117** (1984).

[134] D. Lira, Z. Ramos and A. Simis, Equigenerated Gorenstein ideals of codimension three, arXiv:2005.12953v2 [math.AC], 5 Jul 2020.

[135] P. F. Machado, Algebras generated by minors of catalecticant matrices, PhD Thesis (Portuguese), UNICAMP/UFBa, January 1997.

[136] H. Markwig, Free resolutions of fat points on a conic in $\mathbb{P}^2$, Rend. Sem. Mat. Univ. Pol. Torino, **63** (2005), 63–80.

[137] H. Matsumura, *Commutative Algebra*, Benjamin/Cummings, Reading, Massachusetts, 1980.

[138] H. Matsumura, *Commutative Ring Theory*, Cambridge University Press, 1986.

[139] S. McAdam, *Asymptotic Prime Divisors*, Lecture Notes in Math., **1023**, Springer, Berlin, 1983.

[140] A. Micali, Sur les algèbres universelles, Ann. L'Institut Fourier, **14** (1964), 33–87.

[141] R. Miranda, Linear systems of plane curves, Notices of the Amer. Math. Soc., **46** (1999), 192–202.

[142] F. Mora and L. Robbiano, The Gröbner fan of an ideal, J. Symb. Comp., **6** (1988), 183–208.

[143] M. Morales and A. Simis, The symbolic powers of monomial curves in P3 lying on a quadric, Comm. in Algebra, **20** (1992), 1109–1121.

[144] M. Nagata, On rational surfaces II, Memoirs of the College of Sciences, University of Kyoto, Series A, **33** (1960), 271–293.

[145] L. D. Nam, The determinantal ideals of extended Hankel matrices, J. Pure Appl. Algebra, **215** (2011), 1502–1515.

[146] R. Nanduri, A family of irreducible free divisors in $\mathbb{P}^2$, J. of Algebra and Its Applications, **14** (2015), 1550105, 11 pp.

[147] E. Negri and T. Hibi, Gorenstein algebras of Veronese type, J. Algebra, **193** (1997), 629–639.

[148] A. N. Nejad, The Aluffi algebra of an ideal, PhD Thesis, Universidade Federal de Pernambuco, Brazil, 2010.

[149] A. Nejad and A. Simis, The Aluffi algebra, J. of Singularities, **3** (2011), 20–47.

[150] A. Nejad, A. Simis and R. Zaare-Nahandi, The Aluffi algebra of the Jacobian of points in projective space: Torsion-freeness, J. Algebra, **467** (2016), 268–283.

[151] A. Ooishi, Castelnuovo's regularity of graded rings and modules, Hiroshima Math. J., **12** (1982), 627–644.

[152] A. Ooishi, Genera and arithmetic genera of commutative rings, Hiroshima Math. J., **17** (1987), 47–66.

[153] P. Orlik and H. Terao, Commutative algebras for arrangements, Nagoya Math. J., **134** (1994), 65–73.

[154] I. Pan, Les transformations de Cremona stellaires, Proc. Amer. Math. Soc., **129** (2001), 1257–1262.

[155] I. Pan and F. Russo, Cremona transformations and special double structures, Manuscripta Math., **177** (2005), 491–510.

[156] I. Pan and A. Simis, Cremona maps of de Jonquières type, Canadian J. Math., **67** (2015), 923–941.

[157] C. Peskine and L. Szpiro, Liaison des variétés algébriques, Invent. Math., **26** (1974), 271–302.

[158] H. C. Pinkham, Deformations of algebraic varieties with $\mathbb{G}_m$ action, Astérisque, **20** (1974).

[159] Z. Ramos and A. Simis, Symbolic powers of perfect ideals of codimension 2 and birational maps, J. Algebra, **413** (2014), 153–197.

[160] Z. Ramos and A. Simis, Homaloidal nets and ideals of fat points I, LMS Journal of Computation and Mathematics, **19** (2016), 54–77.

[161] Z. Ramos and A. Simis, Homaloidal nets and ideals of fat points II, Intern. J. Algebra and computation, **27** (2017), 677–715.

[162] Z. Ramos and A. Simis, An analogue of the Aluffi algebra for modules, J. Algebra, **67** (2016), 155–182.

[163] Z. Ramos and A. Simis, Equigenerated ideals of analytic deviation one, arXiv:2009.07355v1 [math.AC], 15 Sep 2020.

[164] L. Rose and H. Terao, Free resolution of the module of logarithmic forms of a generic arrangement, J. Algebra, **136** (1991), 376–400.

[165] M. E. Rossi, D. T. Trung and N. V. Trung, Castelnuovo–Mumford regularity and Ratliff–Rush closure, J. Algebra, **504** (2018), 568–586.

[166] F. Russo and A. Simis, Birational maps and Jacobian matrices, Compos. Math., **126** (2001), 335–358.

[167] F. Russo, *On the Geometry of Some Special Varieties*, Lecture Notes of the UMI, Springer, 2016.

[168] L. Reid, L. G. Roberts and M. Roitman, On complete intersections and their Hilbert functions, Canad. Math. Bull., **34** (1991), 525–535.

[169] K. Saito, Theory of logarithmic differential forms and logarithmic vector fields, J. Fac. Sci. Univ. Tokyo Sect. 1A Math., **27** (1980), 265–291.

[170] H. Schenck, A rank two vector bundle associated to a three arrangement, and its Chern polynomial, Adv. Math., **149** (2000), 214–229.

[171] A. Schrijver, *Theory of Linear and Integer Programming*, John Wiley & Sons, New York, 1986.

[172] J.-P. Serre, *Algèbre Locale–Multiplicités*, Lecture Notes in Mathematics, **11**, Springer, Berlin–Heidelberg–New York, 1965.

[173] A. Simis, Differential idealizers and algebraic free divisors, in A. Corso, P. Gimenez, M. V. Pinto and S. Zarzuela (Eds.) *Commutative Algebra: Geometric, Homological, Combinatorial and Computational Aspects*, Lecture Notes in Pure and Applied Mathematics, **244**, pp. 211–226, Chapman & Hall/CRC, 2005.

[174] A. Simis, Cremona transformations and some related algebras, J. Algebra, **280** (2004), 162–179.

[175] A. Simis, Algebraic aspects of tangent cones, in D. Avritzer and M. Spira (Eds.) *Matemática Contemporânea*, Proceedings of the XII Escola de Álgebra, Diamantina, Brazil, 1994.

[176] A. Simis, *Remarkable Graded Algebras in Commutative Algebra*, Aport. Mat. Comun., **15**, Soc. Mat. Mex., 1995.

[177] A. Simis, Directional Graded Algebras (Portuguese), National Competition for Full Professorship, Universidade Federal de Pernambuco, November, 1996.

[178] A. Simis, Remarkable graded algebras in algebraic geometry, in *XII ELAM*, IMCA, Lima, Peru, June 1999.

[179] A. Simis, *Commutative Algebra*, De Gruyter Textbook, 2020.

[180] J. Hong, A. Simis and W. Vasconcelos, On the homology of two-dimensional elimination, J. of Symb. Computation, **43** (2008), 275–292.

[181] A. Simis, K. Smith and B. Ulrich, An algebraic proof of Zak's inequality for the dimension of the Gauss image, Math. Z., **241** (2002), 871–881.

[182] A. Simis and S. Tohăneanu, The homology of homogeneous divisors, Israel J. Math., **200** (2014), 449–487.

[183] A. Simis and Ş. O. Tohăneanu, The ubiquity of Sylvester forms in almost complete intersections, Collect. Math., **66** (2015), 1–31.

[184] A. Simis, N.-V. Trung and G. Valla, The diagonal subalgebra of a blowup algebra, J. Pure Appl. Algebra, **125** (1998), 305–328.

[185] A. Simis and B. Ulrich, On the ideal of an embedded join, J. Algebra, **226** (2000), 1–14.

[186] A. Simis and B. Ulrich, The Fitting Ideal Problem, Bull. London Math. Soc., **41** (2009), 79–88.

[187] A. Simis, B. Ulrich and W. V. Vasconcelos, Jacobian dual fibrations, Amer. J. Math., **115** (1993), 47–75.

[188] A. Simis, B. Ulrich and W. V. Vasconcelos, Tangent star cones, J. Reine Angew. Math., **483** (1997), 23–59.

[189] A. Simis, B. Ulrich and W. Vasconcelos, Rees algebras of modules, Proc. London Math. Soc., **87** (2003), 610–646.

[190] A. Simis, B. Ulrich and W. Vasconcelos, Codimension, multiplicities and integral extensions, Math. Proc. Camb. Phil. Soc., **130** (2001), 237–257.

[191] A. Simis, B. Ulrich and W. Vasconcelos, Tangent algebras, Trans. Amer. Math. Soc., **364** (2012), 571–594.

[192] A. Simis and W. V. Vasconcelos, The syzygies of the conormal module, American J. Math., **103** (1981), 203–224.

[193] A. Simis and W. V. Vasconcelos, On the dimension and integrality of symmetric algebras, Math. Z., **177** (1981), 341–358.

[194] A. Simis and W. V. Vasconcelos, Krull dimension and integrality of symmetric algebras, Manuscripta Math., **61** (1988), 63–78.

[195] A. Simis and R. H. Villarreal, Constraints for the normality of monomial subrings and birationality, Proc. Amer. Math. Soc., **131** (2003), 2043–2048.

[196] A. Simis and R. Villarreal, Linear syzygies and birational combinatorics, Results Math., **48** (2005), 326–343.

[197] A. Simis and R. Villarreal, Combinatorics of monomial Cremona maps, Math. Comp., **81** (2012), 1857–1867.

[198] K. Smith, Tight closure in graded rings, J. Math. Kyoto Univ., **37** (1997), 35–53.

[199] V. Srinivas, On the embedding dimension of an affine variety, Math. Ann., **289** (1991), 125–132.

[200] R. Stanley, Hilbert functions of graded algebras, Adv. Math., **28** (1978), 57–83.

[201] B. Sturmfels, Gröbner bases and Stanley decompositions of determinantal rings, Math. Z., **205** (1990), 137–144.

[202] B. Sturmfels, *Gröbner Bases and Convex Polytopes*, University Lecture Series, **8**, American Mathematical Society, Providence, RI, 1996.

[203] B. Sturmfels and S. Sullivant, Combinatorial secant varieties, Q. J. Pure Appl. Math., **2** (2006), 285–309. Special Issue: In Honor of Robert MacPherson.

[204] S. Sullivant, Combinatorial symbolic powers, J. Algebra, **319** (2008), 115–142.

[205] T. Svanes, Some criteria for rigidity of Noetherian rings, Math. Z., **144** (1975), 135–145.

[206] H. Terao, Arrangements of hyperplanes and their freeness I, II, J. Fac. Sci. Univ. Tokyo Sect. Math., **27** (1980), 293–320.

[207] H. Terao, The exponents of a free hypersurface, Singularities Part 2, Proc. Symp. Pure Math., **40** (1983), 561–566.

[208] S. Tohăneanu, On freeness of divisors on $\mathbb{P}^2$, Comm. in Algebra, **41** (2013), 2916–2932.

[209] N. V. Trung and Sh. Ikeda, When is the Rees algebra Cohen–Macaulay? Comm. Algebra, **17** (1989), 2893–2922.

[210] N. V. Trung, The Castelnuovo regularity of the Rees algebra and the associated graded ring, Trans. Amer. Math. Soc., **350** (1998), 2813–2832.

[211] B. Ulrich, Ideals having the expected reduction number, Amer. J. Math., **118** (1996), 17–38.

[212] P. Valabrega and G. Valla, Form rings and regular sequences, Nagoya Math. J., **72** (1978), 91–101.

[213] G. Valla, On the symmetric and Rees algebras of an ideal, Manuscripta Math., **30** (1980), 239–255.

[214] B. L. van der Waerden, On varieties on multiple-projective spaces, Indagaciones Math., **51** (1978), 303–312.

[215] W. Vasconcelos, *Arithmetic of Blowup Algebras*, London Mathematical Society, Lecture Notes Series, **195**, Cambridge University Press, 1994.

[216] W. Vasconcelos, *Computational Methods in Commutative Algebra and Algebraic Geometry*, Springer, Heidelberg, 1998.

[217] W. Vasconcelos, *Integral Closure. Rees Algebras, Multiplicities, Algorithms*, Springer Monographs in Mathematics, Springer, Berlin, 2005.

[218] R. H. Villarreal, Rees algebras of edge ideals, Comm. Algebra, **23** (1995), 3513–3524.

[219] R. H. Villarreal, *Monomial Algebras*, 2nd Ed., Monographs and Research Notes in Mathematics, A Chapman & Hall Book, 2015.

[220] C. T. C. Wall, Geometry of quartic curves, Math. Proc. Camb. Phil. Soc., **117** (1995), 415–423.

[221] H. Whitney, Differential and combinatorial topology, in *Symposium in Honour of Marston Morse*, Princeton, 1965.

[222] H. Whitney, *Complex Analytic Varieties*, Addison–Wesley, 1972.

[223] S. Yuzvinsky, A free resolution of the module of derivations for generic arrangements, J. Algebra, **136** (1991), 432–438.

[224] F. L. Zak, *Tangents and Secants of Algebraic Varieties*, Translation of Mathematical Monographs, **127**, American Mathematical Society, Providence, Rhode Island, 1993.

[225] O. Zarisky and P. Samuel, *Commutative Algebra, Volume II*, Van Nostrand, 1958. Reprinted: Graduate Texts in Mathematics, vol. 29, Springer.

Index

https://doi.org/10.1515/9783110640694-014

De Gruyter Expositions in Mathematics

Volume 61/1
Mariusz Urbański, Mario Roy, Sara Munday
Non-Invertible Dynamical Systems. Volume 1: Ergodic Theory – Finite and Infinite, Thermodynamic Formalism, Symbolic Dynamics and Distance Expanding Maps, 2021
ISBN 978-3-11-070264-4, e-ISBN 978-3-11-070268-2

Volume 68
Yuri A. Bahturin
Identical Relations in Lie Algebras, 2021
ISBN 978-3-11-056557-7, e-ISBN 978-3-11-056665-9

Volume 34
Marek Jarnicki, Peter Pflug
Extension of Holomorphic Functions, 2nd Edition, 2020
ISBN 978-3-11-062766-4, e-ISBN 978-3-11-063027-5

Volume 67
Alcides Lins Neto, Bruno Scárdua
Complex Algebraic Foliations, 2020
ISBN 978-3-11-060107-7, e-ISBN 978-3-11-060205-0

Volume 66
Anke Kalauch, Onno van Gaans
Pre-Riesz Spaces, 2018
ISBN 978-3-11-047539-5, e-ISBN 978-3-11-047629-3

Volume 65
Yakov Berkovich, Zvonimir Janko
Groups of Prime Power Order. Volume 6, 2018
ISBN 978-3-11-053097-1, e-ISBN 978-3-11-053314-9

Volume 64
Yakov Berkovich, Lev S. Kazarin, Emmanuel M. Zhmud'
Characters of Finite Groups. Volume 2, 2019
ISBN 978-3-11-022408-5, e-ISBN 978-3-11-022409-2

Volume 63
Yakov Berkovich, Lev S. Kazarin, Emmanuel M. Zhmud'
Characters of Finite Groups. Volume 1, 2017
ISBN 978-3-11-022406-1, e-ISBN 978-3-11-022407-8

Printed and bound by CPI Group (UK) Ltd, Croydon, CR0 4YY
13/09/2026
14961093-0005